THIRD EDITION

DIGITAL AVIONICS HANDBOOK

THIRD EDITION

DIGITAL AVIONICS HANDBOOK

EDITED BY

Cary R. Spitzer
In Memoriam

Uma Ferrell
Ferrell and Associates Consulting, Inc.

Thomas Ferrell
Ferrell and Associates Consulting, Inc.

CRC Press
Taylor & Francis Group
Boca Raton London New York

CRC Press is an imprint of the
Taylor & Francis Group, an **informa** business

MATLAB®, Simulink®, and Stateflow® are trademarks of The MathWorks, Inc. and are used with permission. The MathWorks does not warrant the accuracy of the text or exercises in this book. This book's use or discussion of MATLAB®, Simulink®, and Stateflow® software or related products does not constitute endorsement or sponsorship by The MathWorks of a particular pedagogical approach or particular use of the MATLAB®, Simulink®, and Stateflow® software.

RTCA documents referenced in this book may be purchased from RTCA, Inc.:

1150 18th Street NW, Suite 910, Washington, DC 20036
(202) 833-9339
www.rtca.org

SAE documents referenced in this book may be purchased from SAE International:

400 Commonwealth Drive, Warrendale, PA 15096
(724) 776-4841
www.sae.org

ARINC documents referenced in this book may be purchased from SAE-ITC, ARINC Industry Activities:

16701 Melford Blvd., Suite 120,Bowie, MD 20715
(240) 334-2578
standards@sae-itc.org

Acknowledgments: ARINC standards are affiliated with the SAE. Architecture Analysis and Design Language (AADL) was standardized by SAE. SCADE is a copyrighted product of Esterel Technologies.

CRC Press
Taylor & Francis Group
6000 Broken Sound Parkway NW, Suite 300
Boca Raton, FL 33487-2742

© 2015 by Taylor & Francis Group, LLC
CRC Press is an imprint of Taylor & Francis Group, an Informa business

No claim to original U.S. Government works

Printed on acid-free paper
Version Date: 20140822

International Standard Book Number-13: 978-1-4398-6861-4 (Hardback)

Visit the Taylor & Francis Web site at
http://www.taylorandfrancis.com

and the CRC Press Web site at
http://www.crcpress.com

This book is dedicated in loving memory to Cary R. Spitzer.

Contents

SECTION I Evolution of Avionics: Safety and Certification

SECTION II   Avionics Functions: Supporting Technology and Case Studies

SECTION III Avionics Development: Tools, Techniques, and Methods

SECTION IV   Conclusion

Preface

Avionics is the cornerstone of modern aircraft control and operation. Almost every facet of the aircraft is tied to one or more avionics functions. As a result, the complexity and cost of avionics have never been greater.

Many technologies that emerged in the last decade will be utilized in the next decade. The global positioning system has enabled satellite-based precise navigation and landing, and communication satellites are now capable of supporting aviation services. Thus, the aviation world is upgrading to satellite-based communications, navigation, and surveillance for air traffic management. Both the aircraft operator and the air traffic services provider are realizing significant benefits.

Familiar technologies in this book include data buses, one type of which has been in use for over 20 years; head-mounted displays; and fly-by-wire flight controls. New bus and display concepts, however, are emerging, for example, a retinal scanning display, which may soon replace these veteran devices.

Other emerging technologies include speech interaction with the aircraft and synthetic vision. Speech interaction may soon enter commercial service on business aircraft as another way to perform some noncritical functions. Synthetic vision offers enormous potential for both military and civil aircraft for operations under reduced visibility conditions or in cases where it is difficult to install sufficient windows in an aircraft.

This book offers a comprehensive view of avionics, from the technology and elements of a system to examples of modern systems flying on the latest military and civil aircraft. The chapters have been written with the reader in mind by working practitioners in the field. This book was prepared for the working engineer and others seeking information on various aspects of avionics. The topic of avionics cannot be fully explored in a single handbook. Readers are encouraged to make use of the References and Further Reading sections to explore topics in more detail.

The book has been divided into four sections, each containing a number of related chapters.

Section I deals with the evolution of air traffic control to air traffic management and the reason for avionics support in communication, navigation, and surveillance, as well as the supporting safety and certification infrastructure.

Section II deals with specific technology and use of different avionics, many of which are discussed in the previous section in a broad sense. Topics include an overview of functions accomplished by avionics, case studies of avionics architectures, and specific instances of avionics.

Section III is all about tools, techniques, and methods used to implement avionics functions. Topics include information on data buses, avionics architectures, modeling of avionics, and some additional details on navigation algorithms.

Section IV provides a long-range look ahead. While the industry is busy implementing NextGen, it is worth considering what a FarGen might look like.

Much of the information in this book deals with equipment and systems subject to regulatory oversight and certification. Certification and regulatory framework are constantly evolving as technology and the operating environment evolve (and yes, as lessons are learned from accidents and incidents). Opinions expressed in these chapters are but one interpretation of standards used for compliance. The material presented here has been collected by the authors from many sources that are believed to be reliable. However, no expressed or implied warranty can be made as to the accuracy or completeness of the material. No responsibility or liability is assumed by the authors or their employers. While utmost care is taken to be aligned with the certification authority, the final acceptance of that interpretation within a specific context is up to the certification authority. Many of the standards discussed in this book are coupled with issue papers or certification review items to augment their application on specific projects. For these reasons, it remains the responsibility of the entity seeking approval of their avionics for flight to demonstrate compliance to all necessary regulatory requirements.

Avionics concepts are developed and regulated world-wide using numerous industry documents from organizations such as RTCA, European Organization for Civil Aviation Equipment (EUROCAE), Society of Automotive Engineers (SAE), etc. These documents are developed in committees comprised of industry, academia, and regulators. There are also numerous regulatory documents that affect how avionics are developed, used, and maintained. Naturally, a number of these documents have been referenced, elaborated, and made use of to describe concepts within many chapters in this handbook. The chapters in this handbook are not meant to substitute these industry documents or regulatory documents that promulgate the rules. Furthermore, all of these documents evolve from time to time. It must be noted that applicable regulations take precedence over materials in this handbook. Applicable industry documents must be directly referenced for application on specific projects since it is quite conceivable that a specific version of a specific industry document may be imposed in the regulation applicable to that project.

Acknowledgments

Heartfelt thanks to Tom and Uma Ferrell—your dedicated work made this book a reality. Because of his illness, Cary struggled and was unable to complete this project. He would be grateful and pleased. Thank you, Nora, for keeping me informed of the progress of this book.

Laura Spitzer

We would like to thank all of the contributing authors of this handbook. Without their dedication, this book would not have been possible. We also would like to think Taylor & Francis for seeing this project to conclusion. Finally, the resulting volume would not have been possible without the diligence and hard work of Nora Konopka and Deepa Kalaichelvan. Thank you very much.

Tom and Uma Ferrell

Editors

Cary R. Spitzer (deceased) graduated from Virginia Tech and George Washington University. After service in the Air Force, he joined NASA Langley Research Center. During the last half of his tenure at NASA, he focused on avionics. He was the NASA manager of a joint NASA/Honeywell program that made the first satellite-guided automatic landing of a passenger transport aircraft in November 1990. In recognition of this accomplishment, he was nominated jointly by ARINC, ALPA, AOPA, ATA, NBAA, and RTCA for the 1991 Collier Trophy "for his pioneering work in proving the concept of GPS-aided precision approaches." Mr. Spitzer led a project to define the experimental and operational requirements for a transport aircraft suitable for conducting flight experiments and to acquire such an aircraft. Today, that aircraft is the NASA Langley B-757 ARIES flight research platform. Mr. Spitzer was the NASA representative to the Airlines Electronic Engineering Committee. In 1988, he received the Airlines Avionics Institute Chairman's Special Volare Award. He was only the second federal government employee so honored in over 30 years. He was active in the RTCA, including serving as chairman of the Airport Surface Operations Subgroup of Task Force 1 on Global Navigation Satellite System Transition and Implementation Strategy and as technical program chairman of the 1992 Technical Symposium. He was also a member of the Technical Management Committee. In 1993, Mr. Spitzer founded AvioniCon, an international avionics consulting firm that specializes in strategic planning, business development, technology analysis, and in-house training.

Mr. Spitzer was a fellow of the Institute of Electrical and Electronics Engineers (IEEE) and an associate fellow of the American Institute of Aeronautics and Astronautics (AIAA). He received the AIAA 1994 Digital Avionics Award and the IEEE Centennial Medal and Millennium Medal. He was a past president of the IEEE Aerospace and Electronic Systems Society. Beginning in 1979, Mr. Spitzer played a major role in the highly successful Digital Avionics Systems Conferences, including serving as general chairman. Mr. Spitzer presented one-week courses on digital avionics systems and on satellite-based communication, navigation, and surveillance for air traffic management at the UCLA Extension Division.

He also lectured for the International Air Transport Association. He was the author of *Digital Avionics Systems*, the first book in the field, published by McGraw-Hill, and editor-in-chief of *The Avionics Handbook*, published by CRC Press.

He held two patents, published more than 40 papers, and was the author of several books on avionics. Having written a tutorial on avionics, he was sought after as a consultant by UCLA and lectured all over the world. He was proud of the fact that his book, *Digital Avionics Systems*, was translated into Chinese and widely distributed in the Orient.

Uma Ferrell and Thomas Ferrell are cofounders of Ferrell and Associates Consulting, Inc. (FAA), a certification and software safety consultancy serving the aerospace industry. They have over 50 years of combined experience in helping industry in accomplishing certification objectives via training, certification project support, and related research. They are frequently sought out to help with new and novel technologies and methods where traditional certification approaches break down. In recent years, their consultancy has evolved to include project management support, especially for projects in need of recovery plans, to ensure successful engine and airframe certification. Mr. and Mrs. Ferrell have collaborated together in numerous industry efforts as well as jointly authored papers and tutorials for a number of industry conferences, most notably the Digital Avionics Systems Conference (DASC) and the International System Safety Conference (ISSC). They authored the FAA's *Service History Handbook* (DOT/FAA/AR-01/116), an alternative method for DO-178B compliance. Their other research efforts have included work on the use of wireless systems onboard aircraft (accomplished through AVSI) and a comprehensive review of existing airworthiness standards for use in commercial space conducted for the FAA. Mr. and Mrs. Ferrell have authored and presented numerous courses covering various aviation engineering topics, including DO-178C, DO-178B, DO-254, DO-200A, DO-201A, FAA aircraft certification, system and software safety, engineering ethics, technology transfer, and software verification for aviation systems.

Uma Ferrell holds a master's degree in electrical engineering from Johns Hopkins University. She also holds an MSc in solid-state physics; a BSc in physics, chemistry, and mathematics; and a BSc (honors) in physics from Bangalore University in India. Mrs. Ferrell has held senior technical positions at Reliable Software Technologies (RST), The MITRE Corporation, General Sciences Corporation (GSC), and Computer Sciences Corporation (CSC). She has also contributed to engineering of large-scale, mission-critical scientific information systems for National Aeronautics and Space Administration (NASA), National Oceanic and Atmospheric Administration (NOAA), and the Federal Aviation Administration (FAA).

Mrs. Ferrell has participated in a number of industry standards efforts, including work for both the RTCA and the IEEE. She served as the U.S. cochair of the Commercial Off the Shelf Software (COTS) subgroup of RTCA SC-190 for CNS/ATM-related ground systems. She has conducted numerous engineering studies for the requirements working group of RTCA SC-147 (traffic alert and collision avoidance system (TCAS) Minimum Operational Performance Specification). She represented The MITRE Corporation at the RTCA Certification Task Force to provide policy recommendations to the FAA. She was a part of RTCA SC-200 to establish design guidance and certification considerations for integrated modular avionics in RTCA DO-297. She has also contributed to RTCA SC-205 for the formulation of DO-178C, DO-248C, and DO-278A for CNS/ATM systems.

Mrs. Ferrell is a software designated engineering representative (DER) (parts 23 and 25) Level A for both electrical and mechanical systems. She also performs similar work as an authorized representative (AR) for numerous FAA ODAs. She currently serves on the ASQ *Software Quality Professional* editorial board, providing jury review of software-related publications.

Mr. Thomas Ferrell holds a bachelor's degree in electrical engineering from Northern Illinois University, a master's degree in information technology management from Rensselaer Polytechnic Institute, and, for something a little different, a master's degree in history from George Mason University. Mr. Ferrell has held senior technical positions at Science Application International Corporation (SAIC), Iridium LLC, and the Boeing Commercial Airplane Group.

Mr. Ferrell has served in various leadership capacities for a number of RTCA committees, including SC-190, where he led the effort to create DO-278, and SC-205, where he was responsible for DO-178C document integration. He has served as a private sector advisor to the FAA at ICAO. He is frequently sought out as a guest speaker on a variety of topics associated with system and software safety, as well as aircraft certification.

Mr. Ferrell is a software and airborne electronic hardware designated engineering representative (parts 23, 25, and 33) for electrical and mechanical systems, as well as engine controls. He is also an authorized representative (AR) for numerous FAA ODAs.

Contributors

Kathy H. Abbott
Federal Aviation
    Administration
Washington, DC
and
Langley Research Centre
National Aeronautics and Space
    Administration
Hampton, Virginia

Randall E. Bailey
Langley Research Center
National Aeronautics and Space
    Administration
Hampton, Virginia

Mark G. Ballin
Langley Research Center
National Aeronautics and Space
    Administration
Hampton, Virginia

Gregg F. Bartley
Federal Aviation
    Administration
Washington, DC

Douglas W. Beeks
Beeks Engineering and Design
Beaverton, Oregon

Ivan Cibrario Bertolotti
Institute of Electronics,
    Computer and
    Telecommunication
    Engineering
National Research Council
    of Italy
Turin, Italy

Barry C. Breen
Honeywell
Monroe, Washington

Jean-Louis Camus
Esterel Technologies
Toulouse, France

Diganta Das
Department of Mechanical
    Engineering
Center for Advanced Life
    Cycle Engineering
University of Maryland
College Park, Maryland

Julien Delange
Carnegie Mellon Software
    Engineering Institute
Pittsburgh, Pennsylvania

Chris de Long
Honeywell Aerospace
Albuquerque, New Mexico

Ben Di Vito
Langley Research Center
National Aeronautics and Space
    Administration
Hampton, Virginia

Tom Erkkinen
The MathWorks, Inc.
Natick, Massachusetts

James Farrell
VIGIL, Inc.
Severna Park, Maryland

Thomas K. Ferrell
Ferrell and Associates
    Consulting, Inc.
Charlottesville, Virginia

Uma D. Ferrell
Ferrell and Associates
    Consulting, Inc.
Charlottesville, Virginia

John M. Foley
Garmin AT, Inc.
Salem, Oregon

Randall Fulton
FAA Consultant DER
SoftwAir Assurance, Inc.
Redwood City, California

Christopher J. Hegarty
The MITRE Corporation
Bedford, Massachusetts

Ann Heinke
Overlook Consulting, Inc.
Bridgeton, New Jersey

Steve Henely
Rockwell Collins
Cedar Rapids, Iowa

Richard Hess
Honeywell Aerospace
Phoenix, Arizona

Ellis F. Hitt
StratSystems, Inc.
Westerville, Ohio

Peter J. Howells
Rockwell Collins Head-Up
  Guidance Systems
Portland, Oregon

Tingting Hu
Institute of Electronics,
  Computer and
  Telecommunication
  Engineering
National Research Council of Italy
and
Department of Control and
  Computer Engineering
Polytechnic University of Turin
Turin, Italy

Mirko Jakovljevic
TTTech Computertechnik AG
Vienna, Austria

Marge Jones
Safety Analytical Technologies,
  Inc.
Huntsville, Alabama

Sai K. Kalyanaraman
Rockwell Collins
Cedar Rapids, Iowa

Myron Kayton
Kayton Engineering Company
Santa Monica, California

Lynda J. Kramer
Langley Research Center
National Aeronautics and Space
  Administration
Hampton, Virginia

Bruce Lewis
U.S. Army Aviation and Missile
  Research Development and
  Engineering Center
Madison, Alabama

Thomas M. Lippert
Microvision Inc.
Bothel, Washington

Joseph Lyvers
Xcelsi Group
Dayton, Ohio

G. Frank McCormick
Certification Services, Inc.
Eastsound, Washington

James Melzer
Rockwell Collins Optronics
Carlsbad, California

Scott Montgomery
Universal Avionics Systems
  Corporation
Tucson, Arizona

Michael J. Morgan
Honeywell
Olathe, Kansas

Pieter Mosterman
The MathWorks, Inc.
Natick, Massachusetts

Roy T. Oishi
ARINC
Annapolis, Maryland

Michael G. Pecht
Electronics and Production
  Systems Center
Center for Advanced Life Cycle
  Engineering
University of Maryland
College Park, Maryland

Peter Potocki de Montalk
Airbus Industrie
Blagnac, France

Bill Potter
The MathWorks, Inc.
Natick, Massachusetts

Paul J. Prisaznuk
ARINC
Annapolis, Maryland

Leanna Rierson
Digital Safety Consulting
Wichita, Kansas

Philip A. Scandura, Jr.
Honeywell International
Phoenix, Arizona

Andrew Shupe
Dayton, Ohio

Cary R. Spitzer
AvioniCon
Williamsburg, Virginia

Jack Strauss
Xcelsi Group
Dayton, Ohio

Donald L. Sweeney
D.L.S. Electronic Systems, Inc.
and
D.L.S. Conformity Assessment,
  Inc.
Wheeling, Illinois

Michael Traskos
Lectromec
Chantilly, Virginia

Pascal Traverse
Airbus SAS
Toulouse, France

Maarten Uijt de Haag
Department of Electrical
  Engineering
Ohio University
Athens, Ohio

P.V. Varde
Center for Advanced Life Cycle
  Engineering
University of Maryland
Washington, DC

and

Bhabha Atomic Research
  Centre
Mumbai, India

Terry Venema
Xcelsi Group
Dayton, Ohio

Nikhil Vichare
Dell Computers
Austin, Texas

David G. Vutetakis
Concorde Battery
    Corporation
West Covina, California

Randy Walter
GE Aviation Systems
Grand Rapids, Michigan

Christopher B. Watkins
Gulfstream Aerospace
    Corporation
Savannah, Georgia

Joel M. Wichgers
Rockwell Collins
Cedar Rapids, Iowa

Robert B. Wood
Rockwell Collins Head-Up
    Guidance Systems
Portland, Oregon

Steven D. Young
Langley Research Center
National Aeronautics and Space
    Administration
Hampton, Virginia

Ping Zhao
Apple Inc.
Cupertino, California

I

Evolution of Avionics: Safety and Certification

The first section deals with the evolution of air traffic control (ATC) to air traffic management (ATM) and the reason for avionics support in communication, navigation, and surveillance and the supporting safety and certification infrastructure. The chapter on NextGen/Single European Sky ATM Research (SESAR) gives a broad description of the context in which aircraft operates; while pilots retain tactical control, ATM retains the strategic control in order to promote safety and efficiency. The different chapters in this section deal mostly with the avionics portion of the functions even though none of these functions are complete without considering the ground portion also. While the chapter on communication is a broad brush on the historic narration as well as various developments in technology, the chapter on navigation is more focused on algorithms. The surveillance part of the communications, navigation, and surveillance (CNS) is dealt with in the chapters on global positioning system (GPS), automatic dependent surveillance-broadcast (ADS-B), and TCAS II. The next group of chapters deals with the basic equipment that allows the aircraft to be managed within the airspace. Safety-specific avionics are discussed in the next group of chapters. Finally, the set of certification-specific guidance that supports design and development assurance are discussed.

1

Evolving Avionics: Meeting the Challenge of NextGen and SESAR

Uma D. Ferrell
Ferrell and Associates Consulting, Inc.

Thomas K. Ferrell
Ferrell and Associates Consulting, Inc.

1.1 Avionics: A Historical Perspective

1.1.1 Term

The term avionics was coined from "aviation" and "electronics" in the 1970s. If we broaden the definition to include instruments and mechanical systems, many systems such as radios, altimeters, radar, fuel gauges, and navigation instruments were used in the cockpit before the advent of electronics driven by the military.

1.1.2 Technology, Safety, and Regulations

Ever since the advent of powered flight in 1903, aviation has continuously progressed with a variety of advances in all fields, science and engineering of flight as well as how to control the skies for safety from collisions with other aircraft, terrain avoidance, weather avoidance, and safe landing and takeoff.

Industry leaders pushed to have common safety standards imposed by a government agency—the resulting Air Commerce Act in 1926 was the beginning of the federal certification authority in the United States. ATC centers were established in Newark, Cleveland, and Chicago to help pilots with en route directions. Traffic was controlled using blackboards and "shrimp boats" (paper boats signifying the airplane position) that were manually progressed depending upon the information from dispatchers, radio operators, and airports via telephone. In the history of the Federal Aviation Administration (FAA), two separate branches were created via Civil Aeronautics Act of 1938—one with the responsibility for Air Traffic Control (ATC) and another with the responsibility for safety rulemaking, accident investigation, and economic regulation of airlines. The early ATC based on visual signals and light beacons evolved to radar signals following World War II technical developments. The advent of jets and higher density of traffic provided more challenges for safety. Avoidance of midair collisions, terrain, and weather came to sharp focus with each high-profile accident. A single department of transportation to cover all modes of transportation including air transportation was established in 1986 via congressional authorization for a cabinet department with a separate authority for accident investigation transferred to the National Transportation Safety Board. This act took the agency to become the Federal Aviation Administration that we have today. A number of rules and regulations have evolved with both the evolution of technology and the recognition of the need for new technology instigated by accidents, incidents, increase in traffic, and increase in operating costs.

1.1.3 Top Three Technologies and Air Traffic Control

The three essential technologies available to pilots in air and air traffic controllers on the ground that help tactical and strategic control of aircraft with coordinated functions are communications, navigation, and surveillance. The challenges of these systems include global compatibility and interoperability as well as affording the service to both military and civil aircraft.

The three aspects of avionics, namely, Communications, Navigation and Surveillance (CNS), have a parallel history in that advancements in one area necessitate advances in other areas. Wartime advances in navigation and radar detection required that communications be made more sophisticated. These advances were brought into civil aviation; the same radar technology that was used in military aviation was adapted to be used to control civil aviation. Increase in the air traffic necessitated more dependency on distributed CNS equipment both on ground and air to orchestrate traffic efficiently while avoiding conflicts. To address the high price of oil during the 1970s, more and more digital systems were introduced to more precisely control flight. As more and more instruments were introduced into the cockpit, some aids such as autopilots and warning systems had to be introduced to address pilot workload.

Navigation in general is determining own position and velocity so that position and velocity at a future time can be calculated with and without changes in velocity. Navigation of aircraft from a pilot point of view is to know where the aircraft is with respect to a planned track so that the aircraft motion can be controlled. Navigation consists of four functions, namely, planning, tracking, recording, and controlling the aircraft motion. Air navigation has evolved from early ship navigation. In the early days of air travel, avionics equipment for navigation was as rudimentary as following known landmarks such as rail tracks or rivers combined sometimes with sophisticated celestial navigation techniques until there was a need to fly in conditions where visibility was poor because of night flying or bad weather conditions. The basic idea of accurately locating own ship was increasingly important as the skies became crowded and they had to be strategically organized to fly specific routes via a centralized ground control. The first blind flight and landing based on navigation using gyroscopes and radio navigation aids was at the end of the 1920s. Today, aircraft could be flown under visual flight rules (VFR) or instrument flight rules (IFR). Accurate navigation technology has been used to maximize the airspace capacity while balancing safety. Navigation systems must satisfy four important categories of performance requirements, namely, accuracy, integrity, availability, and continuity of service.

The third aspect of surveillance is determination of the position and velocity of the aircraft as perceived from the outside of the aircraft (e.g., from the ground ATC). Most of the improvements in the CNS were a direct result of technologies developed during World War II. Radio beacons and directional beams came to being in the 1930s. The first ATC tower was established in 1935 at the Newark Liberty International Airport in NJ in 1935. There are two types of radar onboard the aircraft: primary surveillance radar and secondary surveillance radar. The primary surveillance radar is passive in that the ground detects the radar energy scattered from the surface of the aircraft to measure the distance and heading of the aircraft relative to the source of radar on the ground. The secondary surveillance radar is active in that the radar from the ground initiates a transponder on the aircraft that transmits a reply to the ground signal. This type of transmission can also be picked up by the surrounding aircraft to aid in collision avoidance. A unique address is given to each aircraft so that its transponder transmissions can be unambiguously identified. This addressing system is followed throughout the world so that the aircraft transponder is useful in all airspace. Transponder signals, whether from primary surveillance radar or secondary surveillance radar, will be displayed on the ATC console. Radar signals have been extended to aid in detecting weather and terrain. Navigation and surveillance ideas have been combined using technology such as Automatic Dependent Surveillance-Broadcast (ADS-B) to accomplish avionics for collision avoidance for manned as well as unmanned systems, terrain avoidance, terminal avian hazard detection, and assistance in using parallel runways. These are only some examples of the ever-growing suite of avionics.

ATC has evolved into ATM, which is using processes, procedures, and resources to assure that aircraft are guided in the sky and on the ground. Aircraft safety is a responsibility shared and divided between the ground and the air. Since these responsibilities are so interconnected, the concepts of Required Communications Performance (RCP), Required Navigation Performance (RNP), and Required Surveillance Performance (RSP) have been introduced to support efficient separation between aircraft while addressing possible constraints of the equipment and related procedures. Thus, many of the avionics that are described in this handbook ultimately support the strategic management of aircraft in giving proper tools to the pilot.

1.2 Free Flight to NextGen/SESAR

The earliest ATC guiding multiple aircraft was in the early 1930s when the controllers tracked the position of planes using maps, blackboards, and "shrimp boats" with only telephone connections to airline dispatchers, radio operators, and other airport air traffic controllers. The same basic system continued as improvements were made to communications, radar sensor data, electronic displays, etc. Many discrete automation systems were introduced in different air traffic facilities as these were helping the ever-evolving complexity and decision making of the air traffic controllers. These systems introduced complexity in configuration management of existing functionality and any new introduction of functionality for interoperability. National Airspace System (NAS) plan was introduced by the FAA to systematically manage projected growth for air travel over the next 20 years. This plan proved to be too ambitious; a new Free Flight program was introduced to take advantage of newly developed GPS technology. This concept took NAS from a centralized command-and-control system between pilots and air traffic controllers to a distributed system that allows pilots, whenever practical, to choose their own route and file a flight plan that follows the most efficient and economical route. This concept made way to a distributed ATM, which combined distributed decision making for traffic separation and self-optimization. These traits demand that the aircraft have specific capabilities measured as required performance in CNS. The resulting concept is known as performance-based navigation, which is one of the primary concepts of NextGen and SESAR. In other words, NextGen and SESAR are programs that implement technology that will allow Free Flight concepts to be used safely and securely in an environmentally responsible manner; the concepts include transformation of the air transportation system by changing technologies, infrastructure, and procedures. The resulting demands on avionics suite to be used by aircraft are RCP, RSP, and RNP.

RCP defines the required performance of each element of the communications network, including the human element. Each element must perform with certain specifications in order to maintain defined aircraft separation standards. RCP has specific time requirement in seconds for messages sent and then received. The complete end-to-end communication limit is either 240 or 400 s based on aircraft equipment and the ability to have an alternate means of establishing communication if the primary fails.

RSP defines system technical performance requirements independent of technology and architecture to be met by an air traffic service (ATS) surveillance system in order to support a particular ATS service or function. Similar to RCP, RSP also has specific performance requirements in seconds for messages sent and received. The completed end-to-end communication cycle is either 180 or 400 s based on aircraft equipment and whether there is an alternate means of establishing communication if the primary fails.

RNP is "a statement of the navigation performance accuracy necessary for operation within a defined airspace." There are two values that express this qualification—a distance in nautical miles called

"RNP type" and a probability measure that is 95% for 1 X RNP and 99.999% within 2 X RNP. For example, an airplane is qualified to operate in an RNP 4 airway; the aircraft must have a demonstrated capability of its navigation system to result in the airplane being within 4 nmi of the indicated position on the navigation system at least 95% of the flight duration within a given flight segment or phase. It is also necessary for the navigation system to issue a warning to the pilot when this condition cannot be met. Within the "containment limit" of twice the RNP, in this example, 8 nmi of the indicated position, the indicated position on the navigation display panel which includes the total system error is expected to not exceed the designated RNP number at least 99.999% of the flight duration within the same segment or phase.

Required total system performance (RTSP) is a combination of RNP, RSP, and RCP as well ATC surveillance capability, which defines a benchmark for separation minima and collision safety risk. The required performances are operationally derived without any dependency on any specific techniques, technologies, and/or architecture.

Evolution in ATM impacts airborne avionics equipment, flight planning, airspace planning, ground infrastructure, procedures used for managing flight traffic, certification, and related activities and stakeholders. This evolution can be viewed by many of these perspectives; indeed there have been many industry activities that have contributed to this evolution.

1.3 NextGen

GPS technology has given pilots the tools needed for planning point-to-point flights rather than planning via way points. This satellite-based system will allow for shorter routes, savings in time and fuel, reduction of delays, and increase in en route capacity by giving pilots the tools to fly closer together on direct routes. Better communication between pilots and ground controllers will accommodate an efficient use of airports. Data fusion is being planned to collect global weather observations into a common weather picture

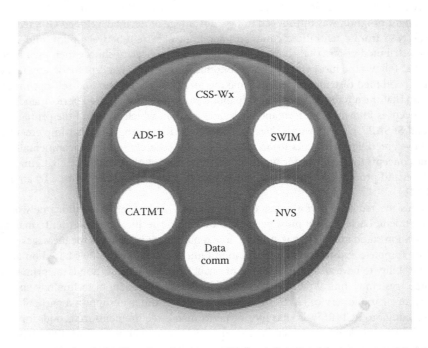

FIGURE 1.1 Core set of tools for NextGen. (Courtesy of Federal Aviation Administration, Washington, DC, Available online at: https://www.faa.gov/about/office_org/field_offices/fsdo/orl/local_more/media/fy13summit/ NEXTGEN_MCO_Safety_Summit.pdf, accessed on April 27, 2014.)

to enable better tactical and strategic traffic management decisions. These various goals are being accomplished by a core set of tools shown in Figure 1.1: NAS Voice System (NVS) accommodates the key voice communication component of NextGen, System-Wide Information Management (SWIM) allows sharing of information, Common Support Services Wx (CSS-Wx) is used for disseminating weather information via SWIM, Collaborative Air Traffic Management Technologies (CATMT) provides enhancements to existing traffic flow management systems and works with SWIM, ERAM processes flight radar data and provides real-time aeronautical information, and ADS-B increases situational awareness. While some of the following goals are still experimental, some have already been realized in the field as published/updated by the FAA:

1. Use of GPS and ADS-B to create a single real-time display of air traffic that has the same information disseminated to both pilots and air traffic controllers. The use of satellite-based precision approach procedures can be done even in low visibility, without needing ground-based landing systems; many small airports may have only one or no instrument landing system.
2. Creation and provision of a common weather picture across the national airspace. Many different tools provide icing and turbulence information at different altitudes to pilots.
3. Greater use of data communications via data link for routine messages between pilots and controllers.
4. Use of a single voice communication system for air/ground and ground/ground communications.
5. Integration of unmanned aircraft system into the airspace

NextGen utilities and concepts are being put into place in a number of airports with decidedly positive results.

SWIM is the key tool that is used in common with international systems to make NextGen compatible with SESAR. SWIM is used for collecting and sharing system-wide information of aircraft and ground facilities. SWIM services handle five data domains, namely, flight data, aeronautical data, weather data, surveillance data, and capacity and demand data. SWIM was adopted by ICAO in 2005. This service-oriented architecture uses existing networks, off-the-shelf hardware, and SWIM-compatible software tools to provide sharing of near real-time information by airline operations, air traffic managers and controllers, and the military (Figure 1.2).

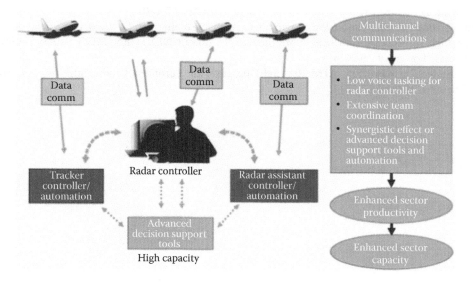

FIGURE 1.2 **(See color insert.)** Tomorrow: Evolved sector with data comm. (Courtesy of Federal Aviation Administration, Washington, DC, Available online at: https://www.faa.gov/about/office_org/field_offices/fsdo/orl/local_more/media/fy13summit/NEXTGEN_MCO_Safety_Summit.pdf, accessed on April 27, 2014.)

1.4 SESAR

Since Europe is made up of different countries, there is no single management of air navigation services. ATC over Europe is extremely busy and dense. These factors make for a very complex system. The initiative of Single European Sky was introduced to provide a uniform level of safety and efficiency for all of Europe by reorganizing and restructuring European airspace without being constrained to national borders. SESAR is a joint undertaking among all of the stakeholders to define, develop, and deploy a high-performance ATC infrastructure. Converting from airspace-based management to trajectory-based operations and employing an integrated data communication system are the important features of SESAR, just as NextGen. A set of key performance areas has been defined to focus and measure the progress. These performance areas are interdependent but have to be balanced in making trade-off decisions. The performance areas focus on accommodating increase in traffic in a safe and efficient manner with environmentally friendly methods. SESAR trials have proven success in select airports.

1.5 Summary

For both concepts, NEXTGEN and SESAR, the operations are based on shared net-centric information to aid collaborative decision making by all stakeholders and trajectory-based operations for efficient use of resources in both air and ground. There are differences between the two systems in the way the data are compiled and distributed; while NextGen is more centralized, SESAR is distributed. NextGen is mainly government controlled to ensure interoperability of components, while SESAR is a single multi-stakeholder consortium. Both NextGen and SESAR represent enormous challenges since the change is one of a paradigm shift and adaptation to new technology. European and American authorities have an agreement on interoperability between their respective ATM infrastructures, thus allowing for uniformity in required avionics capabilities as well.

Further Reading

Federal Aviation Administration, NextGen implementation plan, June 2013, Washington, DC. http://www.faa.gov/nextgen/implementation/. Accessed on April 27, 2014.

Federal Aviation Administration, Washington, DC. Available online at: https://www.faa.gov/about/office_org/field_offices/fsdo/orl/local_more/media/fy13summit/NEXTGEN_MCO_Safety_Summit.pdf. Accessed on April 27, 2014.

SESAR (Single European Sky ATM Research) program by European Union. http://www.sesarju.eu/. Accessed on April 27, 2014.

2
Communications

Roy T. Oishi
ARINC

Ann Heinke
Overlook Consulting, Inc.

2.1 Air–Ground Communications

2.1.1 History

Airplanes became a tool of war in World War I; they became a tool of commerce in the following decade, the Roaring Twenties. In Europe and America, aircraft were used for entertainment and later as a means to carry mail from city to city. As the number of airspace users grew, so did the need for communications between companies and their pilots, as well as between pilots and the nascent air traffic control (ATC) (namely, airports). Early attempts at air–ground communication used visual means: lights, flags, and even bonfires. But more was needed. Early radios used Morse code to communicate, but that was not practical in an open, bouncing cockpit. With practical voice radios, air–ground communications became a necessary element of the fledgling air transport industry, and it remains so to this day.

In the 1970s, the airlines introduced data communications to aircraft in flight for company communications using the Aircraft Communications Addressing and Recording System (ACARS). Various trials by air navigation service providers (ANSPs) have proven the effectiveness of data communications for ATC. Some oceanic and remote areas have implemented ATC operations via data link, and trials in domestic airspace have shown the value of this medium.

In the 1990s, the International Civil Aviation Organization (ICAO) developed the *Manual for the Aeronautical Telecommunication Network (ATN)*, which specified an air–ground data network based on the International Organization for Standardization (ISO) Open Systems Interconnection (OSI) model. Subsequent work extended this network to support the Internet protocol suite (IPS).

2.1.2 Introduction of Radios

Early on, the use of radio waves was recognized as an important resource. By its third decade, national and international bodies were allocating the radio spectrum. In the United States, the Federal Radio Commission (FRC), predecessor to the Federal Communications Commission (FCC), was licensing radio frequencies to operators. In 1929, the FRC directed the aircraft operating companies to band together to make consolidated frequency allocation requests. Aeronautical Radio Inc. (ARINC) was formed in that year specifically for that purpose. The department of ARINC which was responsible for aeronautical radio frequency allocation in the United States is now an independent company known as Aviation Spectrum Resources, Inc. (ASRI). To resolve any issues in the aviation spectrum associated with the use of voice and data at the national and international levels—FAA and FCC work in cooperation and negotiate at the ICAO level.

As would be expected, communications needs were greatest around airports. Since each aircraft operating company had its own frequency allocation and its own radio operators, as the industry grew, the need for more frequencies grew. This recurring problem, known as spectrum depletion, has been solved in various ways. In the early days, it was solved by banding together to use common frequencies and radio operators. ARINC was a natural choice to implement these common radio stations. At one point, ARINC had 55 such communication centers across the United States.

Another method of solving spectrum depletion was the introduction of new technology. The steady improvements in radio technology have opened higher and higher frequency bands to practical communications use, subject to physical limitations. The increased sophistication of radio circuitry allowed different ways of modulating (i.e., impressing information onto) the radio signals. Initially, these improvements allowed better, clearer, and more efficient and effective voice communications. Later technology permitted the evolution of data communications.

The combination of higher usable frequencies and improved modulation techniques has served to extend the useful life of air–ground voice communications for nearly a century. In the 1930s and 1940s, air–ground voice moved from the high-frequency (HF) to the very-high-frequency (VHF) bands. Amplitude modulation (AM) was used and continues to be used as the basic means of ATC communications in domestic airspace. Long-range voice communications relies on the properties of the HF band, as will be discussed later.

2.1.3 Introduction of Data Communications

As the airlines came to rely increasingly on information provided to and received from the aircraft in flight, voice gave way to data. Changeover from voice to data has a number of reasons such as different pilots and different air traffic controllers with various accents (even though canned words were used), clarity in transmissions, distortion, and overlap between different conversations. Data transmission is cleaner and less intrusive. Automating responses where they can be automated also improves efficiency for both the air traffic controllers and the flight crew. In 1978, the airlines initiated air–ground VHF data link, which served two purposes: (1) to allow the messages to originate automatically on the aircraft, reducing crew workload, and (2) to allow the messages to be relayed to airline computer systems without any ground radio operators. The data link was initially called the ARINC Communications Addressing and Reporting System (ACARS), but "ARINC" was soon changed to "Aircraft" in recognition of the nonproprietary nature of the new medium.

Air–ground data link has become the mainstay of airline operations. In the early days, ACARS messages consisted primarily of four automated downlinks per flight segment: the so-called OOOI (pronounced oo-ee) or Out, Off, On, and In messages. These messages allowed the airlines to better track their flights and provided automated timekeeping on the crew. The original 50 aircraft participating in ACARS have grown to almost 10,000. The number of messages now tops 20 million in a month, and the types of messages cover every imaginable facet of airline operations: flight operations, administrative information such as crew scheduling, passenger information such as connecting gates, maintenance information such as engine performance and failure reports, airport and airline coordination such as deicing and refueling, and so on. Also, now many data link transactions are two way, including uplinks as well as downlinks. In fact, some applications are interactive with requests and responses initiated by the ground or by the flight crew. Although these interactive transactions via data link lack the immediacy of voice conversations, they are asynchronous in that both requestor and responder need not be "on the line" at the same time. For non-time-critical applications, this is a significant advantage.

2.1.4 Introduction of ATC Data Link

While VHF voice remains the primary means of ATC communications in domestic airspace, ACARS was first approved for ATC in the South Pacific flight information regions (FIRs) in 1995. Initially, Boeing 747-400 aircraft flying between the United States west coast and Australia and New Zealand pioneered ATC data link by using controller–pilot data link communications (CPDLC) and automatic dependent surveillance (ADS). Boeing offered this combination of features in the Future Air Navigation System (FANS)-1 avionics package. FANS—originally an acronym coined by the ICAO—was a term covering communication and navigation using satellites; however, the term has taken on a life of its own. Airbus released a similar capability package and named it "FANS-A." So, FANS-1/A, as it is now known to acknowledge both packages of the same ADS and CPDLC applications, is being supported by air traffic service providers all over the world. The original air traffic service providers of the South Pacific have been joined by those of the North Atlantic, North Pacific, Indian Ocean, far east Russia, China, South Africa, and other regions.

Prior to the use of ACARS for airborne ATC communications, two applications were implemented on the ground between the aircraft and ATC: Pre-Departure Clearance (PDC) and Digital Automatic Terminal Information Service (D-ATIS). The receipt and acknowledgment of these messages by the flight crew is mandatory prior to takeoff and landing. Receiving these messages via ACARS has several significant advantages. For the flight crew, the message need not be transcribed for later reference, and it can be requested and received without the effort of finding the proper voice channel and requesting the PDC or listening for the beginning of the recorded ATIS. The tower controller need not line up all of the PDC slips, call each aircraft, read the clearance, and verify the readback. Plus the reduction in congestion on the clearance delivery channel is a significant advantage for situations such as peak departure times at a busy airport.

All of these ATC applications use the ACARS air–ground link, which was neither designed for nor initially approved as an ATC communication medium. For that purpose, ICAO developed standards and recommended practices (SARPs) for the Aeronautical Telecommunication Network (ATN), which was designed for both air–ground and ground–ground communication. In the latter role, it is intended to replace the Aeronautical Fixed Telecommunication Network (AFTN), which has served the industry well as a teletype-based, message-switching network for many years. However, technology has overtaken AFTN, and the more modern packet-switched technology of the ATN is seen as more appropriate. The ATN SARPs' development began in the early 1990s. In the past 15 years, the Internet, which is based on a different packet-switched technology, that is, Transmission Control Protocol/Internet Protocol (TCP/IP), has had unprecedented success. ICAO has created another set of ATN SARPs to accommodate the IPS, known as the ATN/IP SARPs. The original ATN, based on OSI protocols, is now known as the ATN/OSI.

As ACARS became essential to airline operations, the limitations of the initial VHF link became intolerable, first because of coverage limitations and then because of speed. The former was solved in two different ways. Long-range data link was implemented first using Inmarsat satellites; this was the basis for initial FANS implementations in the South Pacific. The oceanic coverage, direct pilot-to-controller communication, and improved performance provided by satellites and data link were an improvement over HF voice services. All of the advantages of data link over voice communications were highlighted in the initial FANS trials and operational use. The advantages included (1) direct controller–pilot communications; (2) consistent and rapid delivery of messages; (3) standardized message texts, which were understood by all no matter what their native language is; (4) automated delivery of position reports; and (5) integration of message content with flight management systems (FMSs). High-frequency data link (HFDL) provided another long-range ACARS subnetwork that covered the north polar regions, which are not reached by Inmarsat signals. VHF Digital Link (VDL) Mode 2 provided a higher-speed subnetwork in continental airspace. More recently, Iridium provides a satellite subnetwork able to pass ACARS and ATC messages globally, including over both polar regions. These points will be elaborated in subsequent sections.

2.2 Voice Communications

2.2.1 VHF Voice

The modern VHF transceiver provides air–ground communications for all aircraft in controlled airspace. For transport aircraft (e.g., commercial airliners), the VHF transceiver is a minimum equipment list (MEL) item, meaning the aircraft cannot take off without the requisite number of operational units, in this case two. The reason for the dual-redundancy requirement is that the VHF transceiver is the primary means of communication with ATC.

The aeronautical VHF communication band covers the frequency range 118–136.975 MHz. VHF signals are limited to line of sight between the ground station and the aircraft, usually taken as a radius of approximately 120 nmi around the ground station. Aeronautical VHF voice operations are primarily limited by the radio horizon—the lowest unobstructed path angle between the aircraft and the ground station. Other factors include the altitude of the aircraft and the power of the transmission. Practically, aeronautical communications on a given voice frequency are limited to the extent of the ATC sector, as each new sector controller will be assigned a different channel. Reuse of a given channel is appropriate at about twice the usable radius.

The aeronautical VHF band is a protected spectrum, which means that any transmission not related to the safety and regularity of flight is prohibited. The wavelength of these signals is about 2 m or 90 in., which drives antenna size. The VHF band is divided into 760 channels spaced 25 kHz apart from 118.000 to 137.000 MHz. There is a 12.5 kHz guard band on each end of the allocated band. For 8.33 kHz operation, the VHF transceiver must be capable of tuning to one of 2280 channels spaced 8.33 kHz apart in the same frequency band. This capability was developed for European airspace when the number of ATC sectors (and, therefore, the number of radio channels assigned) grew beyond the ability to assign usable 25 kHz channels. The universally recognized emergency frequency is 121.500 MHz, which is monitored by all ATC facilities.

ICAO Annex 10 to the Convention on International Civil Aviation "International Standards and Recommended Practices, Aeronautical Telecommunications" promulgates the SARPs for voice and data communications in support of international air traffic services. In the case of voice ATC, the ICAO SARPs are generally followed for domestic ATC services as well.

The VHF voice audio is impressed upon the radio-frequency (RF) signal at the carrier frequency by using double-sideband (DSB) AM. This modulation method impresses the audio signal, typically 1–2 kHz, on the RF by varying the amplitude of the RF in proportion to the amplitude of the audio signal. In the frequency domain, the signal can be seen as a peak at the carrier frequency flanked by equal peaks above and below the sidebands. Upon reception, this signal is reconverted to audio and distributed to the headsets and the cockpit voice recorder. Figure 2.1 shows an AM signal both in the time domain, where the audio signal can be seen as riding on the RF carrier, and in the frequency domain, where the peaks representing the carrier and the sidebands can be easily seen.

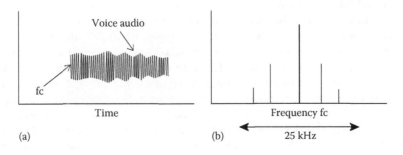

FIGURE 2.1   DSB AM voice signals: (a) time domain view and (b) frequency domain view.

Older VHF radios were tuned to a frequency (channel) from a remote radio control panel that housed a set of dials used to select each digit. The remote control panel was connected to the radio by 19 wires. Five lines, two of which were grounded for any selection, represented each decade digit of the frequency. This method originated from a scheme whereby the digit selection grounded a power connection to a motor at the radio. When the motor had driven a similar dial switch to its corresponding position, the ground was removed and the motor stopped. Presumably, the motor turned a tuning device (typically a variable capacitor). Later, non-motor-driven tuning methods kept the two-out-of-five grounded wire scheme until a digital data bus replaced it.

A modern VHF radio (e.g., one specified by ARINC characteristic 750) is connected to the radio control panel by the two wires of an ARINC 429 bus, which carry command words that perform all of the frequency select functions of the former 20 wires and others as well.

Frequency tuning is not the only element of the modern airborne VHF radio that has evolved. Whereas the motor-driven radio performed the DSB AM function using vacuum tubes or later transistors, the modern radio develops the same output signal in a radically different manner. Today's radio replaces the analog RF and modulator circuits with a high-speed microprocessor called a digital signal processor (DSP). The DSP works in conjunction with a high-speed analog-to-digital (abbreviated A to D or A/D) converter (ADC) that takes the voice audio input and converts it into a series of binary words, each representing the amplitude of the signal. A series of such samples, taken at a high enough rates, can faithfully represent the original analog waveform. This is the same method that records music onto a CD-ROM or MP3 file. A digital-to-analog (D/A) converter (DAC) performs the reverse function.

The DSP takes the digital representation of the audio input and algorithmically combines that information with the RF carrier signal and produces the DSB AM signal, which is sent to the power amplifiers. That was an easy sentence to write, but one that takes many lines of code to implement within the DSP. This method of combining information content, in this case voice audio, with the RF carrier at the selected frequency has tremendous flexibility. Within the constraints imposed by the power and speed of the DSP, the sample rates and sample size of the DAC/ADC, and other considerations, this architecture allows a great deal of flexibility. As will be seen, this type of radio is capable of producing not only DSB AM signals but other voice and data signals as well.

The terms "digital" and "analog" must be used with care. It is true to say that the modern avionics VHF radio is a "digital radio." It is also true to say that it handles voice signals digitally. However, to imply that we have "digital voice" is misleading. The signals propagated from the ground radio to the aircraft are the same DSB AM signals as the motor-tuned, vacuum-tube radio sent and received. Later, we will briefly discuss methods of sending voice over the RF in a digital manner.

2.2.2 HF Voice

HF voice is used for ATC communications in oceanic and remote airspace at various frequencies between 2.850 and 23.350 MHz with wavelengths between 100 and 10 m, respectively. The nature of the propagation of HF signals is such that it provides reliable communications at ranges of thousands of miles. This is possible because HF signals can be reflected by the bottom of the ionosphere at heights of about 70 miles. This effectively permits over-the-horizon or sky wave reception, not unlike the service performed by a satellite. At these frequencies, RF signals propagate with both a ground wave and a sky wave. The ground wave can give useful communication over the horizon, and the sky wave is usable well beyond that. Multiple hops are possible but are not reliable enough for aeronautical voice communication. Other characteristics of HF signals detract from its effectiveness for voice communications. For example, HF signals fade in a diurnal cycle and are susceptible to interference from solar activity. For example, approximately every 11 years, sunspot activity peaks causing significant effects on the ionosphere and on HF propagation.

The long wavelengths, comparable in length or longer than the aircraft, provide challenges as far as antenna placement is concerned. In the propeller-driven age, long wires from the tail forward were used. Later, long probes were mounted on the wingtips or tail. Now the HF antenna is typically installed in the

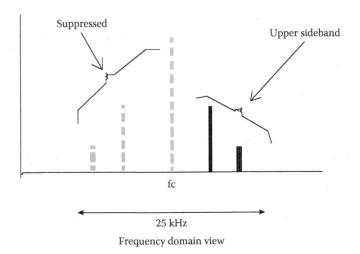

FIGURE 2.2   SSB voice signal.

forward edge of the vertical stabilizer, which is commonly made of composite materials. Ground station antennas can cover a football field.

HF voice is modulated onto the RF carrier using single-sideband (SSB) suppressed carrier modulation. Figure 2.2 shows the frequency domain view of an SSB signal. Note that only the upper sideband is used for aeronautical voice communications. Reliable HF communications requires the aircraft to transmit at 200 W peak envelope power (PEP) and the ground stations to transmit at as much as 5 kW PEP. SSB has the advantage over DSB AM of increased power in the intelligence-carrying signal as opposed to the carrier.

Flying in remote or oceanic airspace would require long hours of listening to the static-filled HF channel if it were not for selective calling (SELCAL), a technique that allows the flight crew to turn down the volume on the HF radio until the ground signals them using preselected tones. Special receive circuits are needed as these tones are not sent as SSB signals. When the preselected tones are recognized, the flight crew is alerted to come up on the HF channel.

2.2.3 Voice Developments

The proliferation of satellite telephone service on long-range aircraft has raised the question, "Why not just dial up ATC and talk to the controller?" Flight crews would like this method, but the reluctance is with the ANSPs. The management of frequencies in protected spectrum is, by now, a well-established procedure; the management of telephone numbers at each control position is not. There are other considerations, such as the use of nonprotected spectrum and the consequence of probable loss of protected spectrum, that inhibit a wholesale acceptance of telephone calls for ATC. Finally, the successful implementation of data link to the controller workstation environment has replaced the use of voice to the controller at such facilities for suitably equipped aircraft. However, carrying a voice radio remains a safety requirement.

Work is underway to create a third-party satellite voice service, similar to the HF voice service, to allow ATC satellite calls to/from the aircraft and the ground voice operator in remote and oceanic airspace. Such calls would terminate at the operator, where they would be converted to data to be sent to/from the controller.

Europe has demonstrated that the aeronautical VHF voice band can be expanded by the use of 8.33 kHz voice. The ICAO SARPs for VDL Mode 3 defines a true digital voice signal in space in which the audio signal is converted to a digital representation, is transmitted digitally across the air–ground VDL Mode 3 subnetwork, and is then reconverted to analog audio signals when it reaches its destination.

While VDL Mode 3 greatly expands the number of voice channels possible, the costs of replacing all VF radios, both airborne and ground, reduced support for this technique. This issue, along with other technical issues, caused this solution to be removed from further consideration.

The long-term possibility that broadband network connectivity to the aircraft may provide acceptable quality voice communication deserves some consideration for the far term. Meanwhile, DSB AM voice will remain the primary method of ATC voice communications for the foreseeable future.

2.3 Data Communications

2.3.1 ACARS Overview

Today, ACARS provides worldwide data link coverage. Five distinct air–ground subnetworks are available for suitably equipped aircraft: original VHF, Inmarsat satcom, HFDL, VDL Mode 2, and Iridium satellite. In order to understand the function of the avionics for ACARS, it is necessary to see the larger network picture. Figure 2.3 shows an overview of the ACARS network showing the aircraft, the four air–ground subnetworks, the central message processor, and the ground message delivery network.

The ACARS message-passing network is an implementation of a star topology with the central message processor as the hub. The ground message network carries messages to and from the hub, and the air–ground subnetworks all radiate from the hub. There are a number of ACARS network service providers, and their implementations differ in some details, but all have the same star topology.

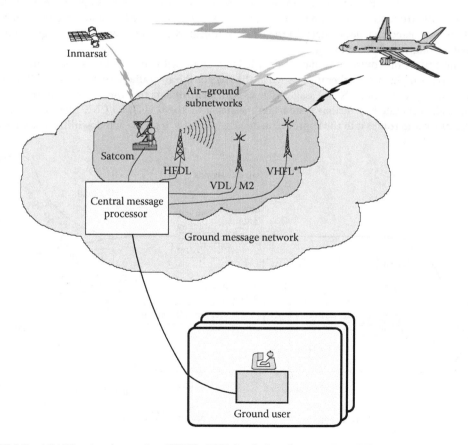

FIGURE 2.3   ACARS network overview. *VHFL, VHF data link; either ACARS or VDLM2.

Two data link service providers provide worldwide ACARS coverage, with several others providing regional coverage. Any given ACARS message can be carried over any of the air–ground subnetworks, a choice configured by the aircraft operator. It should be noted that ACARS is a character-oriented network, which means that only valid ASCII characters are recognized and that certain control characters are used to frame a valid message.

2.3.2 ACARS Avionics

The ACARS avionics architecture is centered on the management unit (MU), communications management unit (CMU), or communications management function (CMF), which acts as an onboard router. All air–ground radios connect to the MU or CMU/CMF to send and receive messages. The CMU/CMF is connected to all of the various radios that communicate to the ground. Figure 2.4 illustrates the avionics architecture.

2.3.3 ACARS Management Unit

The MU or CMU/CMF acts as the ACARS router onboard the aircraft. All messages to or from the aircraft, over any of the air–ground subnetworks, pass through the MU or CMU/CMF. Although the MU or CMU/CMF handles all ACARS message blocks, it does not perform a message-switching function because it does not recombine multiple message blocks into a "message" prior to passing it along. It passes each message block in accordance with its "label" identifier, and it is up to the receiving end system (ES) to recombine message blocks into a complete message. The original OOOI messages were formatted and sent to the MU from an avionics unit that sensed various sensors placed around the aircraft and determined the associated changes of state. In the modern transport aircraft, many other avionics units send and receive routine ACARS messages.

The multifunction control and data unit (MCDU), along with the printer, is the primary ACARS interface to the flight crew. Other units, such as the FMS or the air traffic services unit (ATSU), will also interact with the crew for FANS messages. The vast majority of data link messages today are downlinks automatically generated by various systems on the airplane. The MU/CMU/CMF identifies each uplink message block and routes it to the appropriate device. Similarly, it takes each downlink, adds associated

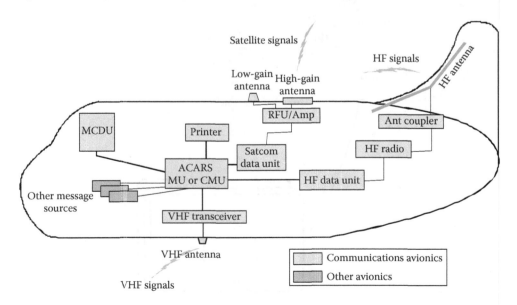

FIGURE 2.4   ACARS avionics architecture.

aircraft information such as the tail number, and sends it to one of the air–ground subnetworks. The latest avionics for each of the four subnetworks accepts an ACARS block as a data message over a data bus, typically ARINC 429. The subnetwork avionics will then transform the message block into the signals needed to communicate with the ground radio. Each subnetwork has its own protocols for link layer and physical layer exchange of a data block.

2.3.4 VHF Subnetwork

The original VHF subnetwork that was pioneered in 1978 uses the same 25 kHz VHF channels used by ATC and aeronautical operational communication (AOC) voice; the signal in space is sometimes called plain old ACARS (POA) for reasons that will become clearer when we discuss VDL Mode 2. The VHF subnetwork uses a form of frequency shift keying (FSK) called minimum shift keying (MSK) wherein the carrier is modulated with either a 1200 or 2400 Hz tone. Each signaling interval represents one bit of information, so the 2400 baud (i.e., rate of change of the signal) equals the bit rate of 2400 bps. After initial synchronization, the receiver then can determine whether a given bit is a one or a zero.

VHF ACARS uses the carrier-sensed multiple access (CSMA) protocol to reduce the effects of two transmitters sending a data block at the same or overlapping times. CSMA is nothing more than the automated version of voice radio protocols wherein the speaker first listens to the channel before initiating a call. Once a transmitter has begun sending a block, no other transmitter will "step on" that transmission. The VHF ACARS subnetwork is an example of a connectionless link layer protocol in that the aircraft does not "log in" to each ground station along its route of flight. The aircraft does initiate a contact with the central message processor, and it does transmit administrative message as it changes subnetworks. A more complete description of the POA signal and an ACARS message block as it is transmitted over a VHF channel can be found in ARINC 618, Appendix B.

In congested airspace, such as the northeastern United States or Europe, multiple VHF ACARS channels are needed to carry the message traffic load. For example, in the Chicago area, 10 channels are needed and a sophisticated frequency management scheme has been put in place, which automatically changes the frequency used by individual aircraft to balance the loads.

Initial ACARS MUs worked with VHF radios that were little modified from their voice-only cousins. The ACARS modulation signal was created as two-tone audio by the MU (e.g., ARINC 724 MU) and sent to the radio (e.g., ARINC 716 VHF radio), where it modulated the RF, just as voice did from a microphone. Later evolutions of the ACARS interface between the CMU (e.g., ARINC 758 CMU) and the latest radio (e.g., ARINC 750 VDR [VHF data radio]) sent ACARS message blocks between the CMU and the radio over a serial data bus (i.e., ARINC 429 Digital Information Transfer System [DITS]), and the radio modulated the RF directly from the data.

2.3.5 Satcom

The first satellite ACARS subnetwork uses the Inmarsat constellations. In the I-3 constellation, four satellites in geosynchronous orbit provide global beam and spot beam coverage of the majority of the globe (up to about 82° latitude) with spot beam coverage over the continents. In the I-4 constellation, three satellites in geosynchronous orbit provide global beam and spot beam of the major landmasses and northern oceans. The Inmarsat constellation provides telephone circuits as well as data link, so it uses a complex set of protocols over several different types of channels using different signals in space. In the aeroclassic services, a packet channel is used to send and receive ACARS or cabin packet data messages. The packet channel is established when the avionics satellite data unit (SDU) logs on to a satellite ground earth station (GES). Each frame is acknowledged between the SDU and GES at the data link layer. Any ACARS data link message block generated by the C/MU for transfer over the satcom subnetwork is sent to the SDU for transfer over this channel to the GES, where it is then forwarded to the ACARS central message processor. The message forwarding function requires advance coordination for appropriate routing and billing to take place. In the SwiftBroadband data service, which is a 432 kbps packet data service over the I-4 constellation, the ACARS or cabin packet data messages will be sent on available IP bandwidth as connectionless datagrams. The Inmarsat satellite access nodes (SANs) route the message on the ground to appropriate gateway services.

The Inmarsat aeroclassic services operate in the L-band, around 1 GHz on frequencies reserved for aeronautical mobile satellite (route) services, or AMS(R)S, which are protected for safety and regularity of flight. Satcom avionics have been purpose built, meaning that they did not evolve from the previous use of L-band radios for voice as VHF ACARS and (as we shall see) HFDL radios evolved from voice radios. In the Aero classic services, the RF unit (RFU), along with high-gain and low-noise amplifiers and the diplexer, sends and receives signals over the various L-band channels defined for Inmarsat services.

In 1995, the use of ACARS messages over satcom was certified for use in the south Pacific for long-range ATC communications with the FAA (Oakland Center), Fiji, New Zealand (Auckland Center), and Australia (Brisbane Center). The message set used was called the FANS-1 message set and mirrored HF voice messaging in oceanic airspace. Boeing 747-400 aircraft were the first to implement FANS-1, but long-range Airbus aircraft soon followed with the FANS-A implementation. Since that time, FANS-1/A has been implemented by many CAAs around the world where the message set supports local ATC procedures.

2.3.6 HFDL

The HFDL ACARS subnetwork uses channels in the HF voice band. The HFDL radio can be a slightly modified HF voice radio connected to the HF data unit (HFDU). Alternatively, an HF data radio (HFDR) can contain both voice radio and data link functions. In either case, the HF communication system must be capable of independent voice or data operation.

HFDL uses phase-shift modulation (PSK) and time-division multiple access (TDMA). A 32 s frame is divided into 13 slots, each of which can communicate with a different aircraft at a different data rate. Four data rates (1800, 1200, 600, and 300 bps) use three different PSK methods (8PSK, 4PSK, and 2PSK). The slowest data rate is affected by doubling the power of the forward error-correcting code. All of these techniques (i.e., multiple data rates, forward error correction, TDMA) are used to maximize the long-range properties of HF signals while mitigating the fade and noise inherent in the medium. Twelve HFDL ground stations provide worldwide coverage, including good coverage over the North Pole but excluding the south polar region. More details on HFDL may be found in ARINC 753: HF Data Link System.

The need for a large antenna, plus the fact that even a quarter-wavelength antenna is problematic, necessitates an antenna coupler that matches the impedance of the feed line to the antenna. The RFU, whether it is a separate unit or incorporated in the HFDR, combines the audio signal representing the data modulation with the carrier frequency, suppresses the carrier and lower sidebands with appropriate filtering, and amplifies the resultant signal.

2.3.7 VDL Mode 2

VDL Mode 2 operates in the same VHF band as POA. Four channels have been reserved worldwide for VDL Mode 2 services. Currently, the only operating frequency is 136.975 MHz. VDL Mode 2 uses differential 8-level phase-shift keying (D8PSK) at a signaling rate of 10.5 kbaud to modulate the carrier. Since each phase change represents one of eight discernible phase shifts, three bits of information are conveyed by each baud or signal change; therefore, the data rate is 31.5 kbps. With about 10 times the capacity of a POA channel, VDL Mode 2 has the potential to significantly reduce channel congestion for ACARS. CSMA is used for media access, but a connection-oriented link layer protocol called the aviation VHF link control (AVLC) is established between the VDR and the ground station. ACARS over AVLC (AOA) is the term used to distinguish ACARS message blocks from other data packets that can also be passed over AVLC. By using AOA, an aircraft equipped with VDL Mode 2 may take advantage of a higher-speed VHF link without any changes to the AOC messages passed to or from the aircraft.

It should be noted that VDL Mode 2 has been implemented in accordance with the ICAO SARPs as a subnetwork of the ATN. The ARINC 750 radio is capable of supporting 25 and 8.33 kHz voice, POA, and AOA. It may only be used for one of these functions at any given time.

2.3.8 Iridium

The Iridium system is capable of connecting telephone calls and data messages to and from aircraft in flight anywhere on earth. ACARS uses the short burst data (SBD) capability of the Iridium system to carry ACARS blocks between the MU or CMU and the central processor of the airline-selected ACARS service provider.

The Iridium constellation consists of 66 satellites in low earth orbits (LEO) at about 485 miles altitude, in six polar orbital planes. LEO satellites travel rapidly across the sky relative to a ground or airborne subscriber. The connection from the aircraft for telephone calls and the point-to-point protocol (PPP) connection for data are maintained by cross-linking between satellites and then downlinking to the Iridium gateway in Arizona. LEO satellites require less transmit power from the avionics than geosynchronous satellite data links.

2.3.9 ATN

2.3.9.1 ATN History and Overview

In the 1980s, the ICAO Air Navigation Commission (ANC) recognized the need to assure commonality among future data links used for air traffic communications. In 1989, the ANC tasked the secondary surveillance radar (SSR) improvement and collision avoidance panel (SICASP) to develop material to assure that commonality. By 1991, the automatic dependent surveillance panel (ADSP) had produced the *Manual of Data Link Applications*, defining message sets for use by ANSPs. In 1997, the ANC approved SARPs for the ATN as the framework for all future ATC data communications.

2.3.9.2 ATN Architecture

The ATN architecture is based on the OSI model for data communications that was published by the ISO. This architecture, as shown in the following figure, identifies seven layers that provide flexibility in implementation while maintaining an orderly flow of message traffic to and from the ES. Other basic characteristics of the ATN include bit-oriented messaging and packet-switched routing.

The ATN is based on multiple air–ground subnetworks, to facilitate communication to a wide variety of aircraft in widely varying airspace, and multiple ground–ground networks to allow for independent domains for air navigation and other service providers.

The structure of the ATN includes ESs, which originate and receive ATN messages with each having a seven-layer ISO stack, and intermediate systems (IS) also called routers, which assure that message packets get to the proper destination ES within the domain. If a message is directed to an ES outside the domain, it is directed to a boundary intermediate system (BIS) for transmission to the proper domain.

The aforementioned architecture applies to all ground and airborne ESs. For aircraft in flight, the ATN connection is maintained by one or more of the ATN subnetworks. For ground ESs, normal telecommunications infrastructure may be used.

2.3.9.3 ATN Subnetworks

At the data link layer (layer 2) and the physical layer (layer 1), the ATN includes SARPs for the following air–ground data links:

- VDL
- Geosynchronous satellite (satcom)
- HF data link (HFDL)
- Iridium satellite

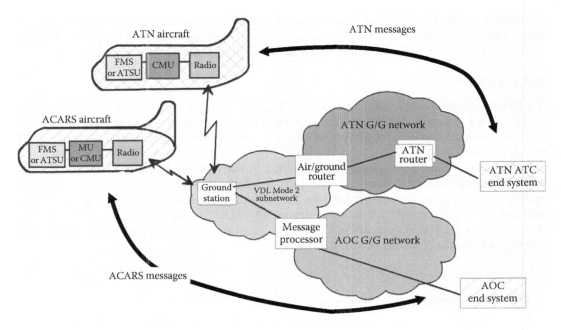

FIGURE 2.5   VDL Mode 2 Subnetwork supports both ACARS and ATN.

Each of these subnetworks is implemented with a unique RF modulation and protocol. VDL operates line of sight and therefore requires multiple ground stations to assure continuous coverage. The other three subnetworks may be used in remote and oceanic airspace, but each has its unique advantages and disadvantages.

2.3.9.4 VDL Subnetwork

As of this writing, the VDL Mode 2 is in operation and is the only ATN air–ground subnetwork being used for ATN message traffic. In Europe, VDL Mode 2 is being used for operational ATC data link messages, while in the United States, ATC data link trials are underway providing departure clearances.

Figure 2.5 shows how the VDL Mode 2 subnetwork has been designed to carry both ACARS messages and ATN messages. VDL Mode 2 is a bit-oriented data link layer protocol, which, in the case of AOA, happens to be carrying ACARS message blocks. ACARS message blocks are directed to the message processor for forwarding over the AOC ground–ground network. ATN packets are directed to an air/ground router that forwards them to an ATN router for delivery via the ATN ground–ground network.

2.3.10 Data Communications Developments

The implementation of broadband Internet connections in the aircraft while in flight has the potential to provide versatile, fast, and cheap connectivity between the aircraft and the ground. Since the earliest voice radio links, through all of the ACARS air–ground subnetworks, air–ground communications has been so specialized that the equipment has been specially designed and built at great cost. If broadband Internet (meaning TCP/IP) connectivity can be made reliable and secure, there is no reason this medium could not be made usable for air–ground data link communication. The definition of the IPS for the ATN has the potential to add near-universal connectivity for ATC communications.

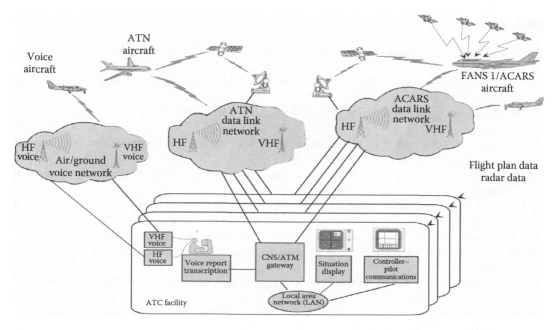

FIGURE 2.6 National ATC facility supporting multiple voice and data networks.

The trend in the telecommunications industry is toward high-speed, high-capacity, general-purpose connectivity. For example, fiber optic links installed to carry cable TV are being used, without significant change, as Internet connections or telephone lines. Sophisticated high-capacity RF modulation techniques are permitting the broadcast of digital signals for high-definition TV and radio. Mobile telephone technology carries digital voice and data messages over the same network. The Internet itself carries far more than the text and graphics information it was originally designed to carry.

Figure 2.6 shows a notional ATC facility of the future, which is able to use voice, ATN data link, and FANS-1/A data links to communicate with suitably equipped aircraft traversing its airspace. The transfer of the majority of routine communications to data link, often with automatic exchanges between the ground and the aircraft, will reduce workload for aircrews and controllers. This will increase the number of aircraft participating in air traffic management (ATM) that will allow benefits for all involved: airlines, aircrews, controllers, and airspace managers.

2.4 Summary

The airlines will continue to increase their dependence on air–ground data link to send and receive information necessary to efficiently operate their fleets. ATC will increase its dependence on air–ground communications, even as the number of voice transactions is reduced. Looking 10–20 years ahead, data link will increasingly be used for ATC communications. If the concept of ATM is to become the rule instead of the exception, the ground automation systems and the FMSs will no doubt be in regular contact, exchanging projected trajectory, weather, traffic, and other information. Voice intervention will be minimal and likely still be over DSB AM in the VHF band.

The modern transport aircraft is becoming a flying network node that will inevitably be connected to the ground for seamless data communications. It's only a matter of time and ingenuity. When that happens, presuming there is sufficient bandwidth, availability, and reliability for each use, many applications will migrate to that link.

References

American Radio Relay League, *The Radio Amateur's Handbook*, 36th ed., The Rumford Press, Concord, NH, 1959.

ARINC Characteristic 566A-9, Mark 3 VHF Communications Transceiver, Aeronautical Radio, Inc., Annapolis, MD, January 30, 1998.

ARINC Characteristic 719-5, Airborne HF/SSB System, Aeronautical Radio, Inc., Annapolis, MD, July 6, 1984.

ARINC Characteristic 724-9, Aircraft Communications Addressing and Reporting System, Aeronautical Radio, Inc., Annapolis, MD, October 9, 1998.

ARINC Characteristic 724B-5, Aircraft Communications Addressing and Reporting System, Aeronautical Radio, Inc., Annapolis, MD, February 21, 2003.

ARINC Characteristic 741P2-7, Aviation Satellite Communication System Part 2 System Design and Equipment Functional Description, Aeronautical Radio, Inc., Annapolis, MD, December 24, 2003.

ARINC Characteristic 750-4, VHF Data Radio, Aeronautical Radio, Inc., Annapolis, MD, August 11, 2004.

ARINC Characteristic 753-3, HF Data Link System, Aeronautical Radio, Inc., Annapolis, MD, February 16, 2001.

ARINC Characteristic 758-2, Communications Management Unit Mark 2, Aeronautical Radio, Inc., Annapolis, MD, July 8, 2005.

ARINC Specification 410-1, Mark 2 Standard Frequency Selection System, Aeronautical Radio, Inc., Annapolis, MD, October 1, 1965.

ARINC Specification 618-5, Mark 2 Standard Frequency Selection System Air/Ground Character-Oriented Protocol Specification, Aeronautical Radio, Inc., Annapolis, MD, August 31, 2000.

ARINC Specification 619-2, ACARS Protocols for Avionic End Systems, Aeronautical Radio, Inc., Annapolis, MD, March 11, 2005.

ARINC Specification 620-4, Data Link Ground System Standard and Interface Specification, Aeronautical Radio, Inc., Annapolis, MD, November 24, 1999.

ARINC Specification 720-1, Digital Frequency/Function Selection for Airborne Electronic Equipment, Aeronautical Radio, Inc., Annapolis, MD, July 1, 1980.

Institute of Electrical and Electronics Engineers and Electronic Industries Association (IEEE and IEA), Report on Radio Spectrum Utilization, Joint Technical Advisory Committee, Institute of Electrical and Electronics Engineers, New York, 1964.

The ARINC Story, The ARINC Companies, Annapolis, MD, 1987.

3

Navigation

Myron Kayton
*Kayton Engineering
Company*

3.1 Introduction

"Navigation" is the determination of the position and velocity of a moving vehicle, on land, at sea, in the air, or in space. The three components of position and the three components of velocity make up a six-component *state vector* whose time variation fully describes the translational motion of the vehicle. With the advent of the global positioning system (GPS), surveyors use the same sensors as navigators but achieve higher accuracy as a result of longer periods of observation and more complex postprocessing.

In the usual navigation system, the state vector is derived on board, displayed to the crew, recorded on board, or transmitted to the ground. Navigation information is usually sent to other onboard subsystems such as waypoint steering, communication control, display, weapon-control, and electronic warfare (emission detection and jamming) computers. Some navigation systems, called *position-location systems*, measure a vehicle's state vector using sensors on the ground or in another vehicle. The external sensors usually track passive radar returns or a transponder. Position-location systems usually supply information to a dispatch or control center.

The term *guidance* has two meanings, both of which are different than *navigation*. In the *first*, steering is toward a destination of known position from the vehicle's present position. The steering equations are derived from a plane triangle for nearby destinations or from a spherical triangle for distant destinations. In the *second* definition, steering is toward a destination without calculating the state vector explicitly. A guided vehicle *homes* on radio, infrared, or visual emissions. Guidance toward a *moving* target is usually of interest to military tactical missiles in which a steering algorithm assures impact within the maneuver and fuel constraints of the interceptor. Guidance toward a *fixed* target involves beam riding, as in the instrument landing system (ILS), Section 3.5.

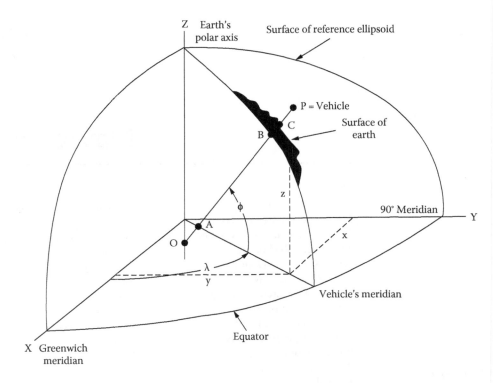

FIGURE 3.1 Latitude–longitude–altitude and XYZ coordinate frames. *Note:* ϕ, geodetic latitude; OP is normal to the ellipsoid at B; λ, geodetic longitude; h = BP = altitude above reference ellipsoid = altitude above mean sea level.

The term *flight control* refers to the deliberate rotation of an aircraft in three dimensions, which causes its velocity vector to change direction.

3.2 Coordinate Frames

Navigation is with respect to a coordinate frame of the designer's choice. For navigation over hundreds of kilometers (e.g., helicopters), various map grids exist whose coordinates can be calculated from latitude and longitude. NATO helicopters and land vehicles use a Universal Transverse Mercator grid. Long-range aircraft navigate relative to an Earth-bound coordinate frame, the most common of which are (Figure 3.1) as follows:

1. Latitude–longitude–altitude, measured with respect to a reference ellipsoid. The most useful reference ellipsoid is described in WGS-84 (1991). Longitude becomes indeterminate in polar regions, rendering these coordinates unsuitable there.
2. Earth-centered rectangular (xyz). These coordinates are valid worldwide; hence, GPS calculates in them and often converts to latitude–longitude–altitude for readout.

3.3 Categories of Navigation

Navigation systems can be categorized as absolute, dead reckoning, or mapping. *Absolute navigation systems* measure the state vector without regard to the path travelled by the vehicle in the past. These are of two kinds: radio systems (Section 3.5) and celestial systems (Section 3.6). Radio systems consist of a network of transmitters (sometimes transponders) on the ground or in satellites. An aircraft detects the transmissions and computes its position relative to the known positions of the stations in the navigation coordinate frame. The aircraft's velocity is measured from the Doppler shift of the transmissions or from a sequence of position measurements.

The second absolute system, celestial (Section 3.6), measures the elevation and azimuth of celestial bodies relative to the local level and true north. Electronic star sensors are used in special-purpose high-altitude aircraft and in spacecraft. Manual celestial navigation was practiced at sea for millennia (Bowditch, 1995).

Dead-reckoning navigation systems derive their state vectors from a continuous series of measurements beginning at a known initial position. There are two kinds: those that measure heading and either speed or acceleration (Section 3.4) and those that measure emissions from continuous-wave radio stations whose signals create ambiguous "lanes" (Section 3.5). Dead-reckoning systems must be *updated* as errors accumulate and if electric power is lost. The only radio dead-reckoning system, Omega (Section 3.5), was decommissioned in 1997.

Lastly, *mapping navigation systems* observe and recognize images of the ground, profiles of altitude, sequences of turns, or external features (Section 3.7). They compare their observations to a stored database. They are mostly used in cruise missiles.

3.4 Dead Reckoning

The simplest dead-reckoning systems measure aircraft heading and speed, resolve speed into the navigation coordinates, and then integrate to obtain position (Figure 3.2). The oldest heading sensor is the magnetic compass: a magnetized needle, an electrically excited toroidal coil (called a *flux gate*), or an electronic magnetometer, as shown in Figure 3.3. Magnetometers are Hall effect or magnetoresistors that measure the change in electrical resistance of a magnetic substance as the magnetic field changes. Three orthogonal magnetoresistors are usually mounted on a circuit board (Figure 3.3) that measures the direction of the Earth's field to an accuracy of about 2° at a steady aircraft speed. Hall-effect

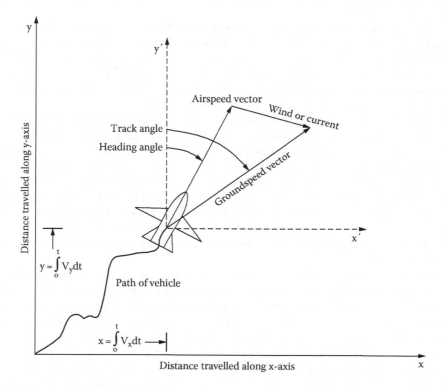

FIGURE 3.2 Geometry of dead reckoning.

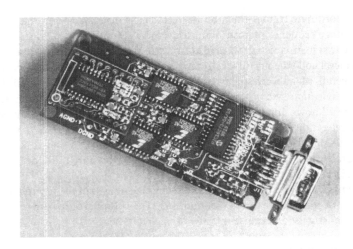

FIGURE 3.3 Circuit board from 3-axis digital magnetometer. A single-axis sensor chip and a 2-axis sensor chip are mounted orthogonally at the end opposite the connector. The sensor chips are magnetoresistive bridges with analog outputs that are digitized on the board. (Photo courtesy of Honeywell, Morristown, NJ.)

magnetometers are less common in aircraft. The horizontal component of the magnetic field points toward *magnetic north*. The angle from true to magnetic north is called *magnetic variation* and is stored in the navigation computers as a function of position over the region of anticipated travel (Quinn, 1996). Magnetic deviations caused by iron and motors in the vehicle can exceed 30° and must be compensated in the navigation computer.

A more complex heading sensor is the *gyrocompass*, consisting of a spinning wheel whose axle is constrained to the horizontal plane by a pendulous weight. The aircraft version (more properly called a *directional gyroscope*) holds any preset heading relative to the Earth, drifts at more than 50°/h, and must be reset periodically. Attitude and heading reference systems (AHRS) use inexpensive gyroscopes (some built on silicon chips as vibrating beams) and silicon accelerometers that are coupled to magnetic compasses to reduce maneuver-induced errors and long-term drift.

The usual speed sensor on an aircraft or helicopter is a *pitot tube* that measures the dynamic pressure of the air stream from which airspeed is derived in an *air-data computer*. To compute ground speed, the velocity of the wind must be vectorially added to that of the aircraft (Figure 3.2). Hence, unpredicted wind will introduce an error into the dead-reckoning computation. Most pitot tubes are insensitive to the component of airspeed normal to their axis, called *drift*. Pitot tubes must be heated to prevent ice from blocking the orifices.

A Doppler radar measures the frequency shift in radar returns from the ground or from water below the aircraft, from which ground speed is inferred. Multibeam Doppler radars can measure all three components of the aircraft's velocity relative to the reflecting surface below. Doppler radars are widely used on military helicopters because of the difficulty of locating pitot tubes on a vehicle that can fly sideways or backward.

The most accurate dead-reckoning system is an *inertial navigator* (Figure 3.4) in which precise accelerometers measure the vehicle's acceleration, while precision gyroscopes measure the orientation of the accelerometers. An onboard computer resolves the accelerations into navigation coordinates and integrates the components of acceleration once to obtain velocity and again to obtain position. When spinning-wheel gyroscopes were used, they and the accelerometers were mounted in servoed gimbals that angularly isolated them from rotations of the aircraft. Since about 1990, only super-precision star trackers and naval navigators use gimbals. All modern aircraft inertial navigators fasten the gyroscopes and accelerometers directly to the aircraft ("strap down"), thereby exposing these instruments to the angular rates and angular accelerations of the aircraft.

FIGURE 3.4 GPS-inertial navigator. The inertial instruments are mounted at the rear with two laser gyroscopes and electrical connectors visible. The input–output board is next from the rear; it excites and reads the inertial sensors. The computer board is next closest to the observer and includes MIL-STD-1553 and RS-422 external interfaces. The power supply is in front. Between them is the single-board shielded GPS receiver. Round connectors on the front are for signals and electric power. A battery is in the case behind the handle. Weight 10 kg, power consumption 40 W. This navigation set is used in the F-22 and many other military aircraft and helicopters. (Photo courtesy of Northrop-Grumman Corporation, Falls Church, VA.)

Attitude is computed by a *quaternion* algorithm (Kayton and Fried, 1997, pp. 352–356) that integrates measured angular increments in three dimensions. A slower algorithm calculates the navigation coordinates. Inertial systems measure aircraft orientation within 0.1° for steering and pointing. Most accelerometers consist of a gram-sized proof mass that is mounted on a flexure pivot. The least expensive ones measure the deflection of the proof mass; the most precise accelerometers restore the proof mass to null electrically, thus not relying on the structural properties of the flexure. In the newest accelerometers, the proof masses are etched into silicon chips.

The oldest gyroscopes contained metal wheels rotating in ball bearings or gas bearings that measured angular velocity or angular increments relative to *inertial space*. More recently, gyroscopes contained rotating, vibrating rings whose frequency of oscillation measured the instrument's angular rates. The most precise gyroscopes are evacuated cavities or optical fibers in which counterrotating laser beams are compared in phase to measure the sensor's angular velocity relative to *inertial space* about an axis normal to the plane of the beams. Vibrating hemispheres and rotating, vibrating bars are the basis of some navigation-quality gyroscopes (drift rates less than 0.1°/h).

Fault-tolerant configurations of cleverly oriented redundant gyroscopes and accelerometers (typically four to six of each) detect and correct sensor failures. Inertial navigators are used in long-range airliners, in business jets, in most military fixed-wing aircraft, in space boosters and entry vehicles, and in manned spacecraft.

3.5 Radio Navigation

Scores of radio navigation aids have been invented and many of them have been widely deployed, as summarized in Table 3.1. The most precise is the GPS, a network of 24 satellites and 4 or more on-orbit spares, 17 ground monitor stations for monitoring, a master control station, and its backup station.

TABLE 3.1   Worldwide Radio Navigation Aids for Aviation

System	Frequency Hz	Band	Number of Stations	Number of Aeronautical Users
Loran-C/Chaika	100 kHz	LF	50	10,000
Beacon[a]	200–1600 kHz	MF	4000	130,000
Instrument landing System (ILS)[a]	108 – 112 MHz	VHF	1500	200,000
	329 – 335 MHz	UHF		
VOR[a] VORTAC	108–118 MHz	VHF	1500	200,000
SARSAT/COSPAS	243,406 MHz	UHF	5 satellites	200,000
JTIDS	960–1213 MHz	L	None	500
DME[a]	962–1213 MHz	L	1500	120,000
Tacan[a]	962–1213 MHz	L	120	15,000
Secondary surveillance Radar (SSR)[a]	1030, 1090 MHz	L	800	250,000
GFS-GLONASS	1227, 1575 MHz	L	24 + 24 satellites	200,000
Radar altimeter	4200 MHz	C	None	40,000
MLS[a]	5031–5091 MHz	C	5	200,000
		X		
Airborne Doppler radar	13–16 GHz	Ku	None	40,000
SPN-41 carrier-landing monitor	15 GHz	Ku	25	1600
SPN-42/46 carrier-landing radar	33 GHz	Ka	25	1600

[a] Standardized by International Civil Aviation Organization.

An aircraft derives its 3D position and velocity from one-way ranging signals received from four or more satellites. Each satellite transmits a civil frequency (L1 = 1575.42 MHz), a military frequency (L2 = 1227.60 MHz) with a superimposed L2C civil modulation, and on the newest satellites, a second civil frequency (L5 = 1176.45 MHz). Each spacecraft carries multiple precise atomic clocks (one part in 10E13), whereas the aircraft carry temperature-controlled quartz clocks whose accuracy is 10E8. The largest source of error is delay as the radio signals pass through the ionosphere. The delay is frequency dependent. Hence, by receiving any two of L1, L2, and L5, an aircraft corrects the ionospheric delay to first order. By receiving all three, it can correct the ionospheric delay to second order. Misra and Enge (2001) and Parkinson and Spilker (1996) describe the GPS in detail.

GPS offers better than 20 m ranging errors on a moving aircraft to civil users and 5 m ranging errors to military users. Single-frequency receivers were available in 2013 for less than $100. GPS receivers are built into cell phones at 100 m accuracy without the ability to track at aircraft speeds.

GPS provides continuous worldwide navigation for the first time in history. It has displaced dead reckoning on many aircraft and reduced the cost of most navigation systems. Figure 3.5 is an artist's drawing of a GPS Block 2F spacecraft, first launched in 2010. During the 1990s, Russia deployed a satellite navigation system, incompatible with GPS, called GLONASS (Urlichich, 2011, May 2012). In 2013, more than 20 operating GLONASS were in orbit. In 2013, the European Union had launched test versions of its own navigation satellite system, called Galileo, which will offer free and paid services (Hein et al., 2003; Anonymous, 2009). The United States plans a major upgrade of GPS by 2015 to reduce vulnerability to jamming (Enge, 2004). The International Global Positioning Service (IGS) issues post-processing (six hour delay and longer) corrections to satellite positions to centimeter accuracy (Novatel, 2014).

Differential GPS (DGPS) employs ground stations at known locations that receive GPS signals and transmit measured ranging errors on a radio link to nearby aircraft via geosynchronous satellites or via Mode-S transponders (1090 MHz). DGPS improves accuracy (centimeters for fixed observers close to a DGPS station) and, more importantly for aeronautical use, detects faults in GPS satellites immediately. Status messages from the satellites can be hours out of date. In 2003, the United States created a nationwide

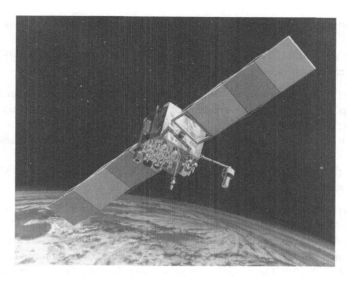

FIGURE 3.5 GPS, Block 2F. (Photo courtesy of Rockwell, Milwaukee, WI.)

aeronautical DGPS consisting of about 50 stations and monitoring sites. This *Wide-Area Augmentation System* (WAAS) (Walter et al., 2012) will eventually replace VORTAC on less-used airways and category I ILS (providing navigation signals down to 200 ft aboveground in the United States). The European WAAS is called EGNOS based on position fixes from Galileo satellites. In 2006, the United States began experimenting with a dense network of DGPS sites at airports called the *Local-Area Augmentation System* (LAAS). The intent is to replace several ILS landing aids at each airport with a small number of less expensive DGPS stations. The experiments show that in order to achieve category II and III landing (*all-weather*), inertial aiding will be needed. Error detection for WAAS and LAAS occurs in about 1 s, hence making GPS safe for aircraft operating near the ground.

Loran is used by general-aviation aircraft for en route navigation and for nonprecision approaches to airports (in which the cloud bottoms are more than 400 ft above the runway). It is principally a maritime navaid but may become a backup to GPS for aeronautical users.

Loran's 100 kHz signals are usable within 1000 nmi (a nautical mile is 1852 m exactly) of a station. Stations were once groups into *chains*, consisting of 3 or 4 synchronized stations (Kayton and Fried, 1997, Chapter 4.5). Loran stations have been upgraded to Enhanced Loran (eLoran) in which atomic clocks are placed at all stations; hence, an aircraft can use any stations in reception range (Narins, 2002); the need for chains no longer exists. Russia has a compatible system called *Chaika that still relies on chains*. When receiving eLoran or Chaika, the airborne receiver measures the difference in time of arrival of pulses emitted by any two ground stations, thus locating the vehicle on one branch of a hyperbola whose foci are at the stations. Two or more station pairs give a 2D position fix at the intersection of the hyperbolas whose best accuracy is 0.25 nmi, limited by propagation uncertainties over the terrain between the transmitting station and the aircraft. The measurement of 100 μs time differences is made with a low-quality clock (one part in 10E4) in the aircraft. American Loran stations were upgraded in the late 1990s and decommissioned in 2010 as a cost-saving measure. Thousand-foot tall antennas are being demolished, that would have sent eLoran signals. eLoran is operational in South Korea, England, Netherlands, and Saudi Arabia for maritime traffic in densely travelled straits. They are usually differential eLoran networks whose corrections are broadcast on 300 kHz marine becons. eLoran is a candidate to be a coarse monitor of GPS and as a stand-alone navigation aid whenever GPS is out of service. GPS monitor functions might alternatively be provided by European or Russian navigation satellites, by private communication–navigation (comm–nav) satellites, or by multilateration to DME stations.

Satellite-based monitors are more accurate than Loran but are subject to the same outages as GPS: solar flares and jammers, for example.

The most widely used aircraft radio aid at the start of the third millennium is VORTAC (Kayton and Fried, 1997, Chapter 4.4), whose stations offer three services:

1. Analog bearing measurements at 108–118 MHz (very high frequency omni-range, VOR). The vehicle compares the phases of a rotating cardioid pattern and an omnidirectional sinusoid emitted by the ground station.
2. Pulse distance measuring equipment (called DME) at 960–1215 MHz, by measuring the time delay for an aircraft to interrogate a VORTAC station and receive a reply.
3. TACAN bearing information, conveyed in the amplitude modulation of the DME replies from the VORTAC stations.

On short flights over oceans, the inertially derived state vector drifts 1–2 nmi/h. When an aircraft approaches shore, it acquires a VORTAC station and continues to its destination using VORTAC for navigation and perhaps ILS for landing. On long over-ocean flights (e.g., trans-Pacific), GPS is usually used with one or more inertial navigators to protect against failures in GPS satellites or in airborne receivers. Long-range transport aircraft typically carry three inertial navigators; some have not been upgraded to GPS.

Over land in developed areas, air-traffic control is based on position determination by radars located en route and in the terminal area. In the United States and Europe, most radars are expected to be phased out in favor of automatic dependent surveillance broadcasts (ADS-B-out [see Chapter 23]). Aircraft broadcast their GPS or inertial or multilateration fix on 1090 MHz at random intervals ("squittering" at 1 s intervals) to a network of 800 ground stations in the United States that send the data to air-traffic control centers from which it is sent on 1090 and 978 MHz to other aircraft that are equipped with ADS-in receiving equipment. The ADS ground stations are supposed to be cheaper to maintain than are the radars. Over ocean and in undeveloped areas (such as Arctic and Africa), low-altitude IRIDIUM satellites will receive ADS messages and forward them to air-traffic control facilities. This will allow worldwide coverage of aircraft positions by ADS. The same transponders are used for traffic alert and collision avoidance (TCAS) transmissions (Kayton and Fried, 1997, Chapter 14). Lo and Enge (2012a) simulated congestion at 1090 MHz caused by radar replies, ADS, TCAS, and DME.

Throughout the western world, civil aircraft use VOR/DME, whereas military aircraft use Tacan/DME for en route navigation. In the 1990s, China and the successor states to the Soviet Union began to replace their direction-finding stations with ICAO-standard navigation aids (VOR, DME, and ILS) at their international airports and along the corridors that lead to them from their borders. DGPS sites (WAAS) will eventually replace most VORTACs; 50 DGPS sites replace a thousand VORTACs thus saving an immense sum of money for maintenance. Nevertheless, VORTAC stations are likely to be retained worldwide on important routes through 2025. WAAS and LAAS allow direct routing and continuous descent and curved approaches, thus saving fuel by eliminating repeated altitude holds on descent.

Specially equipped aircraft are used for the routine calibration of radio navigation aids, speed and velocity sensors, heading sensors, and new algorithms. Airborne test beds and hardware–software integration laboratories are routinely used to develop algorithms and sensor–software interfaces. Test ranges sometimes employ highly accurate multilateration to local ground beacons (Anonymous, 2013) against which to measure the performance of operational navigation systems.

Operational navigation aids are becoming so accurate that finding standards against which to measure them is difficult. The Locata reference (Anonymous, 2013) claims better than 10 cm accuracy airborne.

Omega was a worldwide radio aid that consisted of eight ground stations each of which emitted continuous sine waves at 10–13 KHz (Kayton and Fried, 1997, pp. 155–171). Most vehicles measured the range differences between two stations by observing the phase differences between the received sinusoids. Ambiguous lanes were created when phase differences exceeded 360°. Errors were about 2 nmi due to radio propagation irregularities in the spherical waveguide above the Earth. Omega was used by submarines, over-ocean general-aviation aircraft, and a few international air carriers. It was decommissioned in 1997.

Landing guidance throughout the world (in the 2000s, even in China, India, and the former Soviet Union) is with the ILS (Kayton and Fried, 1997, pp. 608–620). Transmitters adjacent to each runway create a horizontal guidance signal near 110 MHz and a vertical guidance signal near 330 MHz. Both signals are modulated such that the nulls intersect along a line in space that leads an aircraft from a distance of about 15 nmi to within 50 ft above the runway. ILS gives no information about where the aircraft is located along the beam except at two or three vertical *marker beacons*. Most ILS installations are certified to the International Civil Aviation Organization's *category I*, where the pilot must abort the landing if the runway is not visible at an altitude of 200 ft while descending. In the United States, about 100 ILSs are certified to *category II*, which allows the aircraft to descend to 100 ft above the runway before aborting for lack of visibility. *Category III* allows an aircraft to land at still lower weather ceilings. A few ILSs are certified to category III, mostly in Western Europe, which has the worst flying weather in the developed world. Category III ILSs detect their own failures and switch to a redundant channel within 1 s to protect aircraft that are flaring out (within 50 ft above the runway) and can no longer execute a missed approach. Once above the runway, the aircraft's bottom-mounted radio altimeter measures altitude and either the autopilot or the pilot guides the flare maneuver. Landing aids are described by Kayton and Fried (1997). GPS is allowed as the sole navigation aid for non-precision approaches. Aided GPS (WAAS or EGNOS) is allowed for precision approaches to 200 ft altitude.

The U.S. Navy aircraft find their aircraft carriers at sea with TACAN and use a microwave scanning system at 15.6 GHz to land; NASA's space shuttle used the Navy system to land at its spaceports but an inertially aided DGPS eventually replaced it. Another microwave landing system (MLS) at 5 GHz was supposed to replace the ILS in civil operations, especially for categories II and III. However, experiments during the 1990s showed that DGPS with a coarse inertial supplement could achieve an in-flight accuracy of better than 3 m as a landing aid and could detect satellite errors within a second. At most, MLS is deployed at fewer than five airports worldwide. Hence, it is likely that LAAS will replace or supplement ILS, which has been guaranteed to remain in service at least until the year 2020. NATO uses portable MLS and may use portable LAAS for flights into tactical airstrips.

Position-location systems monitor the state vectors of aircraft and usually display the data in a control room or dispatch center. Examples are civil air-traffic control systems and naval combat control centers. Some aircraft derive their state vector from the ranging modulations (e.g., DME); others merely report an independently derived position (e.g., GPS reports on ADS-B-out). *Secondary surveillance radars* interrogate aircraft on 1030 MHz and receive coded replies from aircraft on 1090 MHz so they can be identified by human air-traffic controllers and by collision-avoidance algorithms.

Some commercial communication satellites will offer digital-ranging services worldwide for a fee. The intermittent nature of these commercial fixes would require that vehicles dead-reckon between fixes, perhaps using solid-state inertial instruments. Thus, if taxpayers ever insist on collecting fees for navigation services (as does Galileo), private comm–nav networks may supplement the government-funded GPS and air-traffic communication networks in the mid-twenty-first century. In 2014, IRIDIUM and INMARSAT offer position reporting over oceans via its satellites.

INMARSAT, being in geosynchronous orbit, requires a tracking antenna on the aircraft.

Military comm–nav systems measure the position of air, land, and naval vehicles on battlefields and report to headquarters; examples are the American Joint Tactical Information Distribution System (JTIDS) (for aircraft) and the Position Location Reporting System (PLRS) (for helicopters and land vehicles). Their terminals were said to cost hundreds of thousands of dollars each in 2014.

A worldwide network of approximately 40 SARSAT-COSPAS stations monitors signals from emergency location transmitters (on aircraft, ships, and land users) on 243 and 406 MHz, two of the three international distress frequencies, relayed via low-orbit satellite-based transponders. Software at the listening stations calculates the position of the emergency location transmitters within 5–15 km at 406 MHz, and within 15–30 km at 243 MHz, based on the Doppler-shift history observed by the satellites so that rescue vehicles can be dispatched. Some 406 MHz emergency location transmitters contain GPS sets that transmit their position to satellites. SARSAT-COSPAS has saved more than 33,000 lives worldwide since 1982 from arctic bush pilots to tropical fishermen (NASA/NOAA website).

3.6 Celestial Navigation

Human navigators use sextants to measure the elevation angle of celestial bodies above the visible horizon. The peak elevation angle occurs at local noon or midnight:

$$\text{Elev angle (degrees)} = 90 - \text{latitude} + \text{declination}$$

Thus, at local noon or midnight, latitude can be calculated by simple arithmetic from a table of declination (the angle of the sun or star above the Earth's equatorial plane). When time began to be broadcast to vehicles in the 1930s, off-meridian observations of the elevation angles of two or more celestial bodies became possible at any known time of night (cloud cover permitting). These fixes were hand calculated using logarithms, then plotted on charts by a navigator. In the 1930s, handheld bubble-level sextants were built that measured the elevation of celestial bodies from an aircraft without the need to see the horizon. The human navigator observed sun and stars through an *astrodome* on top of the aircraft. The accuracy of airborne celestial fixes was 5–30 miles in the air, limited by the uncertainty in the horizon and the inability to make precise angular measurements on a pitching, rolling vehicle. Kayton (1990) reviews the history of celestial navigation at sea and in the air.

The first automatic star trackers were built in the late 1950s (Kayton and Fried, 1997, Chapter 12.4). They measured the azimuth and elevation angles of stars relative to a gyroscopically stabilized platform. Approximate position measurements by dead reckoning allowed the telescope to point within a fraction of a degree of the desired star. Thus, a narrow field of view was possible, permitting the telescope and photodetector to track stars in the daytime through a window on top of the aircraft. An onboard computer stored the right ascension and declination of 20–100 stars and computed the vehicle's position. Automatic star trackers are used in long-range military aircraft and on space shuttles, physically mounted on the stable element of a gimballed inertial navigator. Clever design of the optics and of stellar-inertial signal-processing filters achieves accuracies better than 500 ft (Kayton and Fried, 1997). Future lower-cost systems may mount the star tracker directly to the vehicle.

Automatic star trackers mounted on gimbaled inertial navigators are said to cost more than a million dollars in 2014.

3.7 Map-Matching Navigation

On manned aircraft, mapping radars and optical sensors present an image of the terrain to the crew whereas on unmanned aircraft, navigation must be autonomous. Automatic map matchers have been built since the 1960s that correlate the observed image to stored images of patches of distinctive terrain, choosing the closest match to update the dead-reckoned state vector. Since 1980, aircraft and cruise missiles have measured the vertical profile of distinctive patches of terrain below the aircraft and match it to a stored profile. *Updating* with the matched profile, perhaps hourly, reduces the long-term drift of the inertial navigator. The profile of the terrain is measured by subtracting the readings of a baro-inertial altimeter (calibrated for altitude above sea level) and a radio altimeter (measuring terrain clearance). An onboard computer calculates the cross-correlation function between the measured profile and each of many stored profiles on possible parallel paths of the vehicle. The onboard inertial navigator usually contains a digital filter that corrects the drift of the azimuth gyroscope as a sequence of fixes is obtained. Hence, the direction of flight through the stored map is known, saving the considerable computation time that would be needed to correlate for an unknown azimuth of the flight path.

The most complex mapping systems observe their surroundings by digitized video (often stereo) and create their own map of the navigated space. In the 2000s, optical mappers were developed, which allow landings at fields that are not equipped with electronic aids. Systems that use sensor data (usually millimeter wave or infrared) to create an image to assist pilots to land at such fields are called "enhanced vision" (Read, 2002). Systems that use stored terrain to create the image are called "synthetic vision." These images are often projected onto a head-up display. The synthetic vision image is sharper and more detailed than the enhanced image but it will not show mobile equipment (taxying aircraft or construction cranes) on the runway. The trend is to superimpose both images onto the head-up display if budgets permit.

3.8 Navigation Software

Navigation software is sometimes embedded in a central processor partitioned with other avionic-system software, sometimes confined to one or more navigation computers. The navigation software contains algorithms and data that process the measurements made by each sensor (e.g., GPS, inertial, or air data). It contains calibration constants, initialization sequences, self-test algorithms, reasonability tests, and alternative algorithms for periods when sensors have failed or are not receiving information. In the simplest systems, a state vector is calculated independently from each sensor while the navigation software calculates the best estimate of position and velocity. Prior to 1970, the best estimate was calculated from a least-squares algorithm with constant weighting functions or from a frequency-domain filter with constant coefficients. Since the 1970s, a multisensor algorithm calculates the best estimate of position and velocity. The most widely used multisensor filter is the *Kalman filter* that calculates the best estimate from mathematical models of the dynamics of each sensor (Kayton and Fried, 1997, Chapter 3).

Digital maps are carried on ever-more aircraft so position can be visually displayed to the crew and terrain warnings issued. Many civil aircraft superimpose their navigated position, weather, and ADS-B-in traffic reports on a stored map of the terrain. In 2014, general-aviation aircraft are the principal users of superimposed displays. Military aircraft superimpose their navigated position on a stored map of terrain and cultural features to aid in the penetration of and escape from enemy territory. Algorithms for waypoint steering and for control of the vehicle's attitude are contained in the software of the *flight management* and *flight-control* subsystems.

Attempts to improve the reliability of avionic software led to the Radio Technical Commission for Aeronautics (RTCA) standard DO-178C (see Chapter 13). It defines five levels of software reliability. Level A failures result in a catastrophic failure of the aircraft. Level B failures cause a loss of system function, for example, the ability to complete a mission. These levels differ in the amount of documentation and testing required. Navigation software is usually rated for level B, while flight-control and air-data software are usually level A.

3.9 Navigation Hardware

Navigation hardware is installed in the avionics bays and instrument panel. Hardware and wiring are built to the same standards as other avionics equipment. Navigation is dependent on clock accuracy; hence, aircraft often carry quartz crystal clocks in a thermally controlled or thermally monitored oven. Numerous antennas are mounted on top and bottom of the fuselage, on the fin, and on wings. Unbroken point-to-point wiring for power, signals, and coaxial radio is the most reliable but new aircraft are often built in prewired sections and assembled, requiring wiring connectors. The comparative reliability is yet to be established. Civil aviation hardware is usually qualified to RTCA standard DO-160 (Chapter 11).

3.10 Design Trade-Offs

The navigation-system designer conducts trade-offs for each aircraft to determine which navigation systems to use and how to interface them. Trade-offs must be consistent with the current edition of the U.S. Federal Radionavigation Plan. Trade-offs consider the following attributes:

1. *Cost*, including the construction and maintenance of transmitter stations and the purchase of onboard electronics and software. Users are concerned only with the costs of onboard hardware and software.

FIGURE 3.6 Navigation displays in the U.S. Air Force C-5 transport showing flat-panel displays in front of each pilot; vertical situation display outboard and horizontal situation display inboard. Waypoints are entered on the horizontally mounted control-display unit just visible aft of the throttles. In the center of the instrument panel are status and engine displays and backup analog instruments. (Photo courtesy of Honeywell, Morristown, NJ.)

2. *Accuracy* of position and velocity, which is specified as a circular error probable (CEP) (in meters or nautical miles). The maximum allowable CEP is often based on the calculated risk of collision on a typical mission. Over land, modern navigation systems and their pilots are often asked to achieve a required navigation performance (RNP) of 0.3 nmi for which the aircraft and crew are rated. Over-ocean RNPs of 20 nmi are common.
3. *Autonomy*, the extent to which the vehicle determines its own position and velocity without external aids. Autonomy is important to unmanned air vehicles popularly called "drones."
4. *Time delay* in calculating position and velocity, caused by computational and sensor delays. Time delays are sometimes called "latency."
5. *Geographic coverage*: Radio systems operating below 100 KHz can be received beyond line of sight on Earth; those operating above 100 MHz are confined to line of sight. GPS is usable globally because many satellites give users line-of-sight coverage.
6. *Automation*: The vehicle's operator (onboard crew or ground controller) receives a direct reading of position, velocity, and equipment status, usually without human intervention. The navigator's crew station disappeared in the 1970s because electronic equipment automatically selects stations, calculates waypoint steering, and accommodates failures (Figure 3.6). New aircraft have multiple screens ("glass panels") that show navigation data, engine status, airborne radar images, and other data. They can be switched for redundancy.

Bibliography

Articles

Anonymous, Galileo slips, EGNOS operates. *GPS World*, November 2009, 12–14.
Anonymous, Iridium to put ADS-B receivers on next constellation. *Avionics Magazine*, August 2012, 8.
Anonymous, Locata tests lead to air force contract. *GPS World*, January 2013, 29–30.
Billingsley, T.B. et al., Collision avoidance for general aviation. *IEEE Aerospace Magazine*, July 2012, 4–12.
Bowditch, N., The American practical navigator. U.S. Government Printing Office. Re-issued approximately every five years. Marine focus but good discussion of celestial navigation, 1995, 873pp.

Braff, R., Description of the FAA's Local Area Augmentation System (LAAS). *Navigation*, Winter 1997–1998, 411–423.

Craig, D., USAF's new reference system. *Inside GPS*, May 2012, 37–48.

Enge, P., Global positioning system. *Scientific American*, May 2004, 91–97.

Hein, G.W. et al., Galileo frequency and signal design. *GPS World*, January 2003, 30–45.

Kayton, M., *Navigation: Land, Sea, Air, and Space*. IEEE Press, New York, 1990, 461pp.

Kayton, M. and W.R. Fried, *Avionics Navigation Systems*, 2nd edn. John Wiley, New York, 1997, 650pp.

Lo, S.C. and P. Enge, Assessing the capability of DME to support future air traffic capacity. *Navigation*, Winter 2012a, 249–260.

Lo, S.C. and P. Enge, Capacity study of multilateration-based navigation FR alternate position, navigation, and timing services for aviation. *Navigation*, Winter 2012b, 263–279.

Lo, S.C. et al., Loran data modulation: A primer. *IEEE Aerospace and Electronic Systems Magazine*, September 2007, 31–50.

May, M., GLONASS in perspective. *ION Newsletter*, Fall 2012, 4–19.

Minzner, R.A., The U.S. Standard Atmosphere 1976. NOAA report 76-1562, NASA SP-390. 1976 or latest edition, 1976, 227pp.

Misra, P. and P. Enge, *Global Positioning System*. Ganga-Jamuna Press, Lincoln, MA, 2001, 390pp.

Narins, M., Status of loran-C modernization testing. *Ion Newsletter*, Winter 2001–2002, 4–5.

Novatel, Multi-GNSS monitoring, *INSIDE GNSS*, January 2014, 30–34.

Parkinson, B.W. and J.J. Spilker, Eds. *Global Positioning System, Theory and Applications*. American Institute of Aeronautics and Astronautics, Reston, VA, 1996, 1300pp., 2 volumes.

Quinn, J., 1995 Revision of joint US/UK geomagnetic field models. *Journal of Geomagnetism and Geo-Electricity*, Fall 1996.

Read, B., Seeing clearly. *Aerospace International*, May 2002, 30–33.

Taylor, J. et al., GPS control segment upgrade details. *GPS World*, June 2008, 27–33.

Urlichich, V.S. et al., GLONASS modernization. *GPS World*, November 2011, 34–39.

U.S. Air Force, *Navstar-GPS Space Segment/Navigation User Interfaces*. IRN-200c-004. ARINC Research, Annapolis, MD, 2000, 160pp.

U.S. Government, Federal radionavigation plan. Departments of Defense and Transportation, issued biennially, 200pp.

U.S. Government, Federal radionavigation systems. Departments of Defense and Transportation, issued biennially, 200pp.

U.S. Government, WGS-84 World Geodetic System. U.S. Defense Mapping Agency, Washington, DC, 1991.

Van Graas, F. et al., Ohio University/FAA flight test demonstration of local area augmentation system (LAAS). *Navigation*, Summer 1998, 129–135.

Walter, T. et al., Evolving WAAS to service L1/L5 users. *Navigation*, Winter 2012, 317–325.

Zhao, Y., *Vehicle Location and Navigation Systems*. Artech House, Norwood, MA, 1997, 345pp.

Journals

IEEE Transactions on Aerospace and Electronic Systems; bimonthly through 1991, now quarterly.

Proceedings of the IEEE Position Location and Navigation Symposium (PLANS), biennially.

Navigation, Journal of the U.S. Institute of Navigation, Quarterly.

Journal of Navigation, Royal Institute of Navigation (UK), Quarterly.

AIAA Journal of Guidance, Control, and Dynamics, bimonthly.

GPS World, monthly.

Inside GPS, monthly.

Commercial aeronautical standards produced by International Civil Aviation Organization (ICAO, Montreal), Aeronautical Radio, Inc. (ARINC, Annapolis, MD), Radio Technical Committee for Aeronautics (RTCA, Inc., Washington, DC), and European Commission for Aviation Electronics (EUROCAE, Paris).

Websites

www.faa.gov (landing category I, II, III requirements).
www.navcen.uscg.gov (almanac and constellation status).
www.gps.losangeles.af.mil.
www.inmarsat.org.
www.sarsat.noaa.gov.
www.arinc.com.
www.loran.org.
www.gpsworld.com/the-almanac.
www.garmin.com/adsb.
http://microwavelandingsystem.com.

4

Global Positioning System

Christopher J. Hegarty
The MITRE Corporation

John M. Foley
Garmin AT, Inc.

Sai K. Kalyanaraman
Rockwell Collins

4.1 Introduction

The global positioning system (GPS) [1–4] is a satellite navigation system operated by the United States. The system consists of a constellation of nominally 24 satellites in medium Earth orbit, as well as a worldwide ground network to monitor and control the satellites. The GPS program began in the early 1970s and the system was declared fully operational in 1995. Internationally, the GPS constellation is considered to be just one component within the global collection of navigation satellites that is referred to as the global navigation satellite system (GNSS).

GPS has found widespread use on both civilian and military aircraft. This chapter provides an overview of the GPS system, aviation augmentations, GPS avionics, and aviation applications. The chapter concludes with a discussion of future trends.

4.2 GPS System

The following sections provide short descriptions of the GPS constellation, signals, principle of operation, services, and performance. The interested reader is referred to [2–4] for more comprehensive treatments.

4.2.1 Constellation

The GPS constellation nominally consists of 24 satellites in circular orbits with an orbital radius of 26,559 km [5]. Redundant atomic clocks, rubidium, and/or cesium are key components of each satellite so that signals that are precisely synchronized to a common timescale can be broadcast. The satellite orbits are inclined 55° with respect to the equatorial plane. Four satellites are contained in each of six orbital planes, which are equally spaced with respect to their orientation around the Earth's spin axis.

In recent years, the constellation has been overpopulated with up to 31 operational satellites. The first 3 satellites beyond 24 are placed into *expandable slots* of the baseline 24-satellite constellation (see [5]). Surplus satellites beyond 27 are typically launched into locations adjacent to satellites that are expected to require replacement the soonest.

The GPS satellites were procured over time in blocks, with increasing capabilities from each block to the next. Presently, there are 31 operational GPS satellites including 7 Block IIA, 12 Block IIR, 7 modernized Block IIR (IIR-M), and 5 Block IIF vehicles. The Block IIA satellites were built by Rockwell International and launched between 1990 and 1997. The Block IIR and IIR-M satellites were built by Lockheed Martin and launched between 1998 and 2009. The Block IIFs are, at the present time, still being built by Boeing. Twelve IIFs are planned, and thus far, five have been launched. The first launch was in August 2010. The GPS satellites to follow the Block IIFs were originally referred to as Block III but are now referred to as GPS III. A contract to build the first GPS III satellites was awarded to Lockheed Martin in May 2008. The first GPS III satellite is anticipated to be launched in 2015.

4.2.2 Navigation Signals

Code division multiple access (CDMA) is utilized for all the GPS navigation signals, that is, all of the satellites broadcast their signals upon the same carrier frequencies. Up through the Block IIR vehicles, the GPS satellites broadcast what are referred to now as the *legacy* navigation signals upon two carrier frequencies: Link 1 (L1) at 1575.42 MHz and Link 2 (L2) at 1227.6 MHz. The legacy navigation signals include two direct sequence spread spectrum (DSSS) signals with rectangular symbols that are broadcast in phase quadrature on L1 [6]. The coarse/acquisition (C/A) code signal has a 1.023 MHz chipping rate and the precision (P) code signal has a 10.23 MHz chipping rate. The C/A code is generated using length-1023 Gold codes [7], which repeat every millisecond. The P code is 1 week long when unencrypted but is normally encrypted to deter spoofing, and when it is, it is referred to as the Y code. An identical P(Y) code signal is also broadcast on the L2 carrier. Both the C/A and P(Y) code signals are further modulated by the same 50 bps data. This 50 bps data stream includes information required for navigation including the ephemeris, clock corrections, and health information for the broadcasting satellite, as well as almanac data for the entire constellation.

The Block IIR-M satellites introduced two new navigation signals—a new military signal on L1 and L2 referred to as the M code [8] and a new civil signal on L2 referred to as L2C [6,9]. Both of these new signal types have advanced designs that include a dataless signal component and forward error correction of the navigation data to enable robust tracking and data demodulation by user equipment. L2C uses DSSS modulation with rectangular symbols and a 1.023 MHz chipping rate. The M code uses DSSS modulation with a 5.115 MHz chipping rate and a spread spectrum symbol that is two cycles of a 10.23 MHz square wave. This DSSS modulation variant is referred to as binary offset carrier (BOC) [10] and may alternatively be viewed and in practice may be generated as the product of (1) an ordinary DSSS signal using rectangular symbols and (2) a square wave subcarrier.

The Block IIF satellites add an additional civil navigation signal on a new carrier frequency. The new carrier frequency, at 1176.45 MHz, and signal are both referred to as Link 5 (L5) [11,12]. L5 is generated with DSSS modulation using rectangular symbols and a 10.23 MHz chipping rate. As with the other modernized GPS signals (e.g., L2C and M code), L5 includes a dataless signal component and forward error correction of the navigation data for robust tracking and data demodulation.

The GPS III spacecraft will broadcast an additional L1 civil signal (L1C), which will employ a signal that is created using BOC modulations with a 1.023 MHz chipping rate and time-multiplexed mixture of symbols that are derived from 1.023 and 6.138 MHz square wave subcarriers [13,14].

Figure 4.1 illustrates the overall evolution of the GPS navigation signals. The vertical axis of the plot represents the power spectrum of each signal and the horizontal axis is not contiguous. The bandwidth spanned by the C/A code and L2C signals, as measured between the first spectrum null on either side of the carrier frequency, is 2.046 MHz. The null-to-null bandwidth for the P(Y) code and L5 signals is 20.46 MHz.

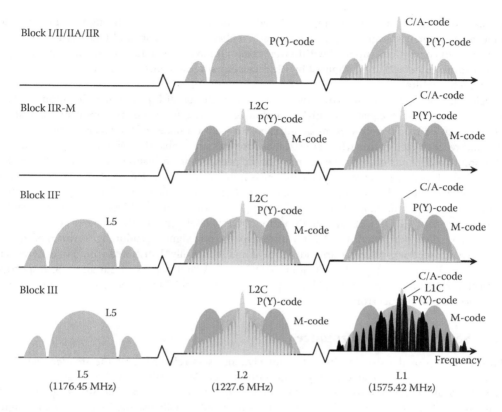

FIGURE 4.1 Evolution of the GPS signals.

4.2.3 Principle of Operation

GPS receivers determine their position and precise time using passive range measurements to visible satellites. These range measurements are obtained through a measurement of the transit time of the broadcast navigation signals and are referred to as *pseudoranges* since all simultaneous measurements may have a large common bias due to a low-cost receiver clock. The pseudorange measurements available at one instant in time with N visible satellites may be modeled as

$$\rho_1 = R_1 + c \cdot b + \varepsilon_1$$
$$\rho_2 = R_2 + c \cdot b + \varepsilon_2$$
$$\vdots$$
$$\rho_N = R_N + c \cdot b + \varepsilon_N$$

(4.1)

where for the ith visible satellite, ρ_i is the pseudorange in meters, R_i is the true range between the satellite and receiver antennas in meters, c is the speed of light in meters per second, b is the receiver clock error in seconds, and ε_i (meters) is the contribution from all other error sources.

The true range between the receiver antenna at position coordinates (x, y, z) and the ith satellite antenna at coordinates (x_i, y_i, z_i) is given by

$$R_i = \sqrt{\left(x - x_i\right)^2 + \left(y - y_i\right)^2 + \left(z - z_i\right)^2}$$

(4.2)

The measurement errors are generally reduced to the extent possible through various methods including equipment design, models (e.g., for atmospheric delay errors and relativistic effects), and application of corrections supplied within the broadcast navigation data. The satellite antenna position coordinates in (4.2) are computed by the receiver using elements of the navigation data broadcast by each satellite.

Neglecting the measurement errors and with the substitution of Equation 4.2, Equation 4.1 can be viewed as a system of N equations with four unknowns (x, y, z, b). Although nonlinear, the system of equations can be readily solved in most situations to yield estimates of the receiver antenna coordinates and receiver clock error when at least four satellites are visible (i.e., $N \geq 4$) [2–4]. The accuracy of the solution depends on both the accuracy of the pseudorange measurements, that is, the statistics of the residual measurement errors after the measurements are corrected by all available means, and the geometry of the visible satellites in the sky. The latter accuracy factor is often quantified by various metrics referred to as *dilutions of precision* (DOPs) [2–4]. Commonly used DOPs include vertical DOP (VDOP), horizontal DOP (HDOP), and position DOP (PDOP) that relate, respectively, the ratios of one-sigma vertical, horizontal, and 3-D position errors to one-sigma pseudorange errors. For instance, if a receiver's pseudorange errors are independent and identically distributed with a one-sigma value of 5.0 m, and assuming typical midlatitude HDOP and VDOP values of 1.0 and 1.5, respectively, the receiver's position estimates would be expected to exhibit a 5.0 m horizontal one-sigma error and 7.5 m vertical one-sigma error.

4.2.4 Services and Performance

GPS presently provides two services—one for civilian users referred to as the *standard positioning service* (SPS) [5] and one available only to authorized users (primarily the U.S. military and the militaries of U.S. allies) referred to as the *precise positioning service* (PPS) [15]. The SPS is based upon use of only the L1 C/A code signal, whereas PPS users may additionally use the L1/L2 P(Y) code signals. The United States has pledged to make the GPS SPS available for civil aviation use on a continuous worldwide basis, free of direct user fees, with a minimum of 6 years advance notice to be provided in the event that this service will be terminated. This commitment was initially made by the administrator of the Federal Aviation Administration (FAA) in 1994 [16]. The commitment to provide GPS SPS service was reiterated in 2007 [17], with an additional commitment made at that time to provide GPS–satellite-based augmentation system (SBAS) services in North America, free of direct user charges, through the FAA's Wide Area Augmentation System (WAAS) (see Section 4.3 for a description of SBASs, including WAAS).

At one time, the accuracy of the SPS was intentionally degraded using a technique referred to as selective availability (SA), which was observed to be implemented as a pseudorandom dithering of the satellite clock that could be removed only by PPS receivers with knowledge of the generation algorithm and cryptographic keys [2]. On May 1, 2000, the intentional degradation of SPS performance due to SA was ceased [18] and in September 2007, the United States announced that the capability to implement SA will be removed from future GPS satellite procurements [19].

The specified accuracy for GPS SPS pseudoranges is better than or equal to 7.8 m, 95% [5]. This specification is for the signal in space (SIS) only (i.e., it does not include errors due to the atmosphere, multipath, or user equipment) and is based upon a global average. With typical GPS DOP values, the specified SPS pseudorange accuracy would be expected to yield 95% horizontal and vertical positioning accuracies on the order of 8 and 12 m, respectively. Actual performance is typically significantly better than this. For instance, the observed 95% horizontal and vertical positioning accuracies for 28 GPS SPS receivers distributed throughout North America from July 1 to September 30, 2012, were 2.8 and 4.3 m, respectively [20]. Further, the data reported in [20] include all real-world errors, whereas the accuracy specification in the SPS performance standard [5] only includes SIS errors.

4.3 Aviation Augmentations

While GPS provides a robust, global positioning and timing capability, it does not at present meet all of the requirements for navigation and other airborne applications (see Section 4.5) without augmentation. In particular, although GPS is normally extremely accurate, there have been rare instances when extremely large range errors have occurred without any indication to the receiver that the associated satellite is unhealthy. The SPS performance standard [5] includes a listing of potential failure modes and a discussion of their characteristics.

The International Civil Aviation Organization (ICAO) defines three augmentations to GPS that alleviate the problems discussed earlier. These are referred to as aircraft-based augmentation systems (ABAS), SBAS, and ground-based augmentation systems (GBAS). These augmentations are described, in turn, in the following sections.

4.3.1 Aircraft-Based

ICAO defines an ABAS as "an augmentation system that augments and/or integrates the information obtained from the other GNSS elements with information available onboard the aircraft." ABAS includes methods to provide integrity monitoring through either the exploitation of redundant GNSS measurements referred to as *receiver autonomous integrity monitoring* (RAIM) [21] or through the use of onboard sensors (e.g., barometric altimeters, inertial navigation systems, other navigation systems). As noted in Section 4.2.3, only four visible satellites are required for positioning with GPS, so redundant measurements are available anytime five or more satellites are visible. With exactly five visible satellites and good geometry, receivers can detect the presence of one abnormally biased pseudorange. This function is referred to as either RAIM or *fault detection* (FD). With at least six visible satellites and good geometry, receivers can not only detect the presence of one bad pseudorange but can also determine which satellite the error corresponds to and exclude it (see Figure 4.2). This enhanced capability is often referred to as *fault detection and exclusion* (FDE).

ABAS also includes the use of other onboard sensors to enhance continuity, availability, or accuracy over that provided by the other elements of GNSS [22].

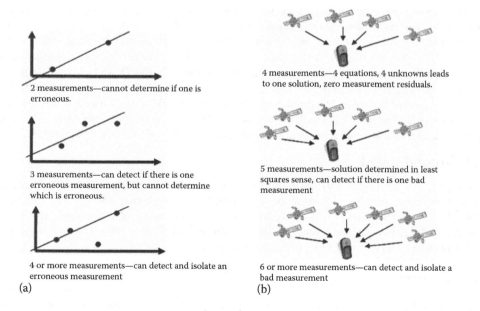

2 measurements—cannot determine if one is erroneous.

3 measurements—can detect if there is one erroneous measurement, but cannot determine which is erroneous.

4 or more measurements—can detect and isolate an erroneous measurement

(a)

4 measurements—4 equations, 4 unknowns leads to one solution, zero measurement residuals.

5 measurements—solution determined in least squares sense, can detect if there is one bad measurement

6 or more measurements—can detect and isolate a bad measurement

(b)

FIGURE 4.2 Illustration of RAIM concept through an analogy of (a) a 2-D problem involving noisy measurements of a linear relationship and (b) the 4-D problem of solving user position and clock error in GNSS.

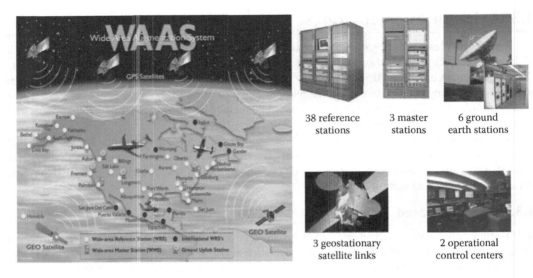

38 reference   3 master   6 ground
stations   stations   earth stations

3 geostationary   2 operational
satellite links   control centers

FIGURE 4.3 **(See color insert.)** WAAS. (Courtesy of the Federal Aviation Administration, Washington, DC.)

4.3.2 Satellite-Based

SBAS [23] are being developed around the world by the global civil aviation community to augment GPS so that it may be used for aircraft navigation in instrument conditions for en route through precision approach. An SBAS consists of a widespread ground network of monitoring stations that collect GNSS satellite measurements (today primarily GPS-only) and geostationary (GEO) satellites to broadcast differential corrections, integrity parameters, and ionospheric data to end users. Current-generation SBAS GEOs broadcast directly on the GPS L1 carrier frequency of 1575.42 MHz. The SBAS signal [24] is open (i.e., unencrypted), resembles the GPS C/A code signal, and may be used for ranging. As compared with GPS C/A code, a higher data rate of 250 bps is employed with rate ½ forward error correction encoding to enable all the requisite system data to be provided to the user. User equipment process GPS C/A code and SBAS signals on L1 only.

Several SBAS systems are either already operational or in the final development stage. These include the WAAS [25–27] in the United States (see Figures 4.3 and 4.4), the European Geostationary Navigation Overlay Service (EGNOS) in Europe [28], the Multifunctional Transport Satellite (MTSAT)-Based Augmentation System (MSAS) in Japan [29], the GPS and GEO Augmented Navigation (GAGAN) system in India [30], and the Russian System of Differential Correction and Monitoring (SDCM) [31]. China has recently announced its intentions to develop an SBAS as well. It is envisioned that SBAS services will eventually migrate toward supporting dual-frequency user equipment with signals at both 1575.42 and 1176.45 MHz [32].

4.3.3 Ground-Based

A GBAS provides differential corrections and integrity data for the GPS or other GNSS signals using redundant reference stations situated at an airport and a very high frequency (VHF) data broadcast (VDB). The GBAS architecture is illustrated in Figure 4.5. GBAS is intended to provide area navigation in the terminal area and support category I–III precision approach operations. A detailed description of the GBAS concept and various implementations may be found in [33].

4.4 Avionics

This section addresses GPS avionics. The first section provides an overview of prevalent international and domestic standards. Subsequent sections describe, in order, typical equipment found on general aviation (GA), air transport, and military aircraft.

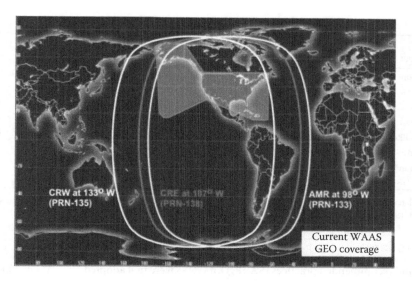

FIGURE 4.4 (See color insert.) WAAS GEO coverage. (Courtesy of the Federal Aviation Administration, Washington, DC.)

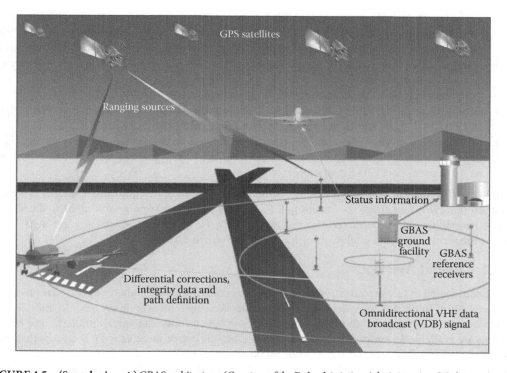

FIGURE 4.5 (See color insert.) GBAS architecture. (Courtesy of the Federal Aviation Administration, Washington, DC.)

TABLE 4.1 FAA TSOs for GNSS Navigation Equipment

Equipment	TSO	Invoked RTCA Document	Date First Published	Status
Stand-alone GPS	TSO-C129	DO-208	1992	Cancelled
Stand-alone GPS	TSO-C196	DO-316	2009	Active
Antennas	TSO-C144[a]	DO-228	1998	Active
Antennas	TSO-C190[a]	DO-301	2007	Active
GPS/satellite-based augmentation system (SBAS)	TSO-C145	DO-229	1998	Active
GPS/SBAS	TSO-C146	DO-229	1998	Active
GPS/ground-based augmentation system (GBAS)	TSO-C161	DO-253	2003	Active
GBAS VDB	TSO-C162	DO-253, DO-246	2003	Active

[a] The latest revision of TSO-C144 (-C144a) is intended only for new passive antenna models. The latest revision of TSO-C190 should be used for new active antenna models.

4.4.1 Standards

Standards and recommended practices (SARPs) for GNSS were first adopted by ICAO in 2001. The GNSS SARPs have subsequently been amended numerous times and are contained within Annex 10 to the International Convention on Civil Aviation [34]. Current SARPs only include standards for two core constellations (GPS and the Russian GLONASS system) and also for the aviation augmentations introduced earlier in Section 4.3.

Relevant FAA technical standard orders (TSOs) are listed in Table 4.1 [35–42]. As with many FAA TSOs, those listed are typically short documents but invoke by reference much longer documents produced by RTCA, Inc. [24,43–48]. These are cross-referenced in the table. The FAA standards address four types of equipment:

- Antennas—TSO-C144 originally provided requirements for passive and active airborne GNSS antennas, but the latest revision (-C144a) may only be used for new passive antenna models. TSO-C190 is an updated standard for active antennas. Both of these standards are for L1-only equipment.
- Stand-alone GPS receivers—TSO-C129 was the first FAA TSO for stand-alone GPS equipment and has been replaced by TSO-C196. Both of these standards are for GPS L1 C/A-code-only equipment that meets applicable integrity requirements using ABAS/RAIM.
- GPS/SBAS receivers—TSO-C145 and TSO-C146 provide standards for airborne receivers relying on GPS L1 C/A code signals augmented by SBASs such as WAAS in the United States.
- GPS/GBAS receivers—TSO-C161 provides standards for airborne GPS receivers relying on L1 C/A code signals augmented by GBAS. TSO-C162 provides standards for the VDB link that is required to support GBAS operation.

As noted earlier, current-generation airborne equipment standards only require processing of L1 signals. Aviation standards organizations, including RTCA and ICAO, are currently working on updated standards that will include airborne receiver processing of the GPS L5 signal, as well as L1/L5 signals that will be broadcast by foreign satellite navigation systems such as Europe's Galileo.

4.4.2 General Aviation

GPS avionics used in GA aircraft fall into two main categories: integrated flight decks and panel-mounted GPS navigators. Integrated flight decks have become standard equipment on many new GA aircraft. Such systems are generally forward-fit avionics installed by the aircraft manufacturer, although some retrofit options are available. Integrated flight decks typically employ two or more large LCD screens that serve as a primary flight display (PFD) and a multifunction display (MFD). The GPS receivers themselves may be remote mounted and combined with VHF omnidirectional range (VOR)/instrument landing system (ILS) and VHF communication receivers. Figure 4.6 shows one example of such a system.

FIGURE 4.6 GA integrated flight deck (Garmin G1000).

GPS is used in conjunction with attitude and heading reference systems (AHRS) and air data computers to drive the PFD and MFD. State-of-the-art systems provide synthetic vision displays that combine GPS position with attitude and terrain data to provide realistic 3-D depiction of ground and water features, airports, obstacles, and traffic. Integrated flight decks employ dual GPS receivers and antennas for redundancy, with one unit feeding the PFD on the pilot side, and the other feeding the second PFD on the copilot side.

The other main category of GA GPS receiver is the panel-mounted GPS receiver. A typical configuration will combine GPS, VOR/ILS, and VHF communications functionality into a single unit. Panel-mounted GPS receivers are available as both forward-fit and retrofit equipment and are by far the most common type of receiver found in the GA fleet. Well over 100,000 panel-mounted receivers have been sold to date. More than 30,000 of these units include SBAS functionality. Figures 4.7 and 4.8 depict two examples of common panel-mounted GPS navigators.

Both GA integrated flight decks and modern panel-mounted GPS receivers include moving map displays that depict the aircraft position relative to the flight path, landmarks, and ground-based navigation aids. Many units are able to depict detailed terrain data, with some providing a TSO-compliant

FIGURE 4.7 Panel-mounted GPS (Garmin GTN 750).

FIGURE 4.8   Panel-mounted GPS (Garmin GNS 530W).

terrain awareness and warning system (TAWS) function. When coupled to appropriate external sensors, these displays can also overlay real-time weather and traffic data onto the moving map. The inclusion of airport surface charts can improve situational awareness during ground operations and help prevent inadvertent runway incursions.

GA GPS receivers are typically certified for IFR en route, terminal, and nonprecision approach capability. SBAS-capable receivers certified under TSO-C146 can also support vertically guided approaches to lateral/vertical navigation (LNAV/VNAV) and localizer precision with vertical guidance (LPV) minimums. Even though GA GPS is certified for IFR use, VFR operations also benefit from the enhanced situational awareness GPS provides. The ability to fly direct to any given waypoint can greatly simplify flight planning. In the event of an in-flight emergency, GPS allows the pilot to quickly identify the nearest airport and provides guidance at the push of a button.

Due to the wide variety of missions GA aircraft undertake, there is a great deal of variation in GA installations and the GPS receiver may interface with many other types of equipment in the cockpit. Table 4.2 lists some of the external equipment interfaces found in a common GA panel mount GPS receiver.

GPS installations in GA aircraft almost universally use active antennas that are powered by DC voltage passed through the antenna coax cable. These antennas may use an ARINC 743A [49] form factor

TABLE 4.2   Panel Mount GPS Receiver External Interfaces

Type of Equipment	Interface
Air data computer	ARINC 429, RS-232
Altimeter	Parallel (Gray code), RS-232
Attitude and heading reference system (AHRS)	ARINC 429, RS-232
Autopilot	Analog, ARINC 429
Course deviation indicator (CDI)/vertical deviation indicator (VDI)	Analog
Distance measuring equipment (DME)	Parallel, serial
Electronic flight instrument system (EFIS)	ARINC 429
Fuel management system	RS-232
Horizontal situation indicator (HSI)	Analog, ARINC 429
Mode-S transponder	ARINC 429, RS-232
Satellite weather receiver	RS-232
Traffic advisory system	ARINC 429

or similar sized tear drop configuration. Many GA aircraft are relatively small in size, so antenna placement can be a challenge. A common solution is the use of combination antennas that include not only a GPS antenna but also VHF communications and/or satellite data antennas.

Portable GPS receivers are commonly used in GA aircraft, sometimes sharing the cockpit with more traditional installed GPS equipment. Portable receivers are stand-alone units that have built-in antennas and batteries, making them a valuable source of accurate navigation in the event of an aircraft power failure. They are often specifically designed for aviation use and have many of the features of a panel-mounted GPS, including moving map displays, satellite weather, and ADS-B traffic, at a significantly lower price point. However, portable GPS receivers are not certified for installation in an aircraft and cannot be used as a navigation source in instrument meteorological conditions (IMC).

4.4.3 Air Transport

Air transport aircraft typically carry redundant multimode receivers (MMRs) as the onboard GNSS sensor (see Figure 4.9). As of 2012, over 28,000 MMRs have been purchased for use in the worldwide air transport fleet. These receivers are referred to as multimode, because they also provide other navigation sensor functionality. Two major form factors in use include the digital MMR [50] and the analog MMR [51]. The digital MMR provides GNSS, ILS, and optional microwave landing system (MLS) receiver capabilities within a single unit. A typical analog MMR additionally provides VOR and marker beacon (MB) functionality.

Although some MMRs include hardware to process GLONASS signals, this capability is largely a growth path and current-generation MMRs rely primarily if not exclusively on the GPS L1 C/A code signals for their GNSS functionality. Older MMRs with GPS capability were primarily based on unaugmented GPS. These were certified under TSO-C129, while the newer MMRs include SBAS and/or GBAS capabilities and have a combination of TSO-C145/-C146/-C161/-C196 functionality as applicable. Until recently, the majority of air transport aircraft operators had shown increased interest in GBAS than in SBAS due to the greater perceived operational benefits for the former system versus the latter. However, with the mature status of WAAS (and growth of other SBAS systems around the world), the availability of more published WAAS LPV approaches and the FAA's ADS-B OUT mandate; air transport operators are displaying additional interest in GBAS and SBAS-capable MMRs. MMRs that support GBAS augmentations also implement VDB receiver functionality, which receives and processes the differential eight-phase-shift-keyed (PSK)

FIGURE 4.9 GLU 925 digital MMR. (Courtesy of Rockwell Collins, Cedar Rapids, IA.)

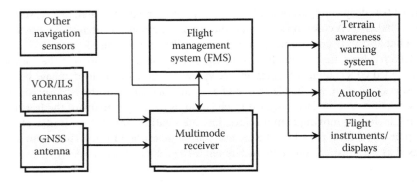

FIGURE 4.10   Typical integration of MMR within air transport aircraft navigation system.

signal carrying the GBAS corrections. All fielded MMRs, at a minimum, use RAIM for integrity monitoring. The MMRs are capable of accepting external aiding inputs (such as barometric altimeter inputs) in order to provide augmented availability. They are also capable of accepting inputs from inertial reference sensors and provide position and velocity outputs in aided navigation modes.

A typical integration of MMRs within an air transport aircraft's navigation system is shown in Figure 4.10. Redundant GNSS and VOR/ILS antennas supply the requisite inputs to the redundant MMRs. The GNSS antennas are top mounted on the aircraft for good visibility of the satellites, typically near the centerline of the fuselage, fore of the wings to avoid blockage and multipath from the wings and tail structure. A common form factor for airborne GPS antennas is specified in ARINC 743A [49]. This form factor calls for a conformal antenna that is 4.7 in. × 2.9 in. × 0.75 in., with the height dimension (0.75 in.) only accounting for the portion of the unit protruding above the fuselage. Transport aircraft typically carry aeronautical mobile satellite service (AMSS) communication equipment that transmits in the frequency band close to GPS (1626.5–1660.48 MHz). As a result, these GPS antennas are located at a minimum distance from the Satcom antenna in order to provide sufficient isolation.

Additional inputs to the MMRs may be supplied from the flight management system (FMS) or other navigation sensors for initialization purposes, as well as from control units (not shown) for, for example, mode selection and channel tuning. The ARINC standard (743A, 755, 756) outputs of the MMRs are provided to the FMS and also to flight displays, autopilot, and terrain awareness warning system (TAWS). MMRs that are SBAS and/or GBAS capable can provide GPS-based approach guidance to the FMS/flight director (FD). They use the SBAS/GBAS augmented GPS position information and runway database parameters to compute lateral and vertical deviations that are transmitted on the ARINC compliant redundant deviation bus outputs. These deviations may be angular or rectilinear deviations as required by the FMS/FD. Table 4.3 shows some of the labels that are provided by the MMR on the deviation output bus:

TABLE 4.3   MMR Output Labels

Octal Label	Message Description	Maximum Message Transmit Interval (ms)
116	Horizontal GLS deviations—rectilinear (BNR)	50
117	Vertical GLS deviation—rectilinear (BNR)	50
126	Computed vertical alert limit	200
127	FAS vertical alert limit	200
166	Computed lateral alert limit	200
167	FAS lateral alert limit	200
173	Localizer deviation	50
174	Glide slope deviation	50
377	Equipment ID	1000

As seen in Table 4.3, the MMR will provide downstream avionics equipment with lateral and vertical deviations (116 and 117/173 and 174) based on the runway reference parameters and augmented GNSS position in GLS (GNSS Landing System) mode. When the MMR is tuned to the ILS/MLS mode (controlled by the tuning head at the aircraft level), it will provide labels 173 and 174 based on the corresponding mode. In some aircraft configurations, the MMR will also compute FLS (FMS Landing System) deviations that are based on position and reference path inputs provided to the MMR by the FMS. This allows the crew to see deviations with respect to a virtual "beam" and perform necessary corrections similar to flying an ILS approach. However, this mode is typically used for nonprecision approaches. In principle, immaterial of the approach used to compute the deviation information (SBAS GLS/GBAS GLS/ILS/MLS/FLS), the data in labels 173/174 would be used to provide course deviation indication at the flight deck thereby reducing crew workload during critical phases of flight. In addition to the deviation outputs, the MMRs also provide the FMS/FD with the computed lateral and vertical integrity limits and relevant final approach segment (FAS) alert limits for decision making at the aircraft level.

The FMS may implement integrity monitoring or performance enhancement of the GNSS input through cross-checking with other redundant onboard GNSS sensors or blending with other available navigation sensor inputs. As a function of the target level of safety to be met at the sensor level, additional cross checks may be performed within the MMR by comparing position and deviation outputs across multiple redundant solutions. Today's MMRs are capable of performing ILS cat I/II/III and GBAS cat I approaches, while the next generation of MMRs will be designed to support GBAS cat II/III capabilities and have the ability to process signals at multiple frequencies from different GNSS constellations. Future MMRs will also be capable of computing an integrated GNSS–inertial solution in support of worldwide aircraft operations that require tighter RNP thresholds.

Another variety of GNSS solution present in air transport and business and regional aircraft is a stand-alone GNSS sensor such as the GPS4000(S) shown in Figure 4.11. This is a standard 2 MCU configuration and provides ARINC 743A compliant redundant navigation data outputs. These outputs are provided to the FMS that may fuse the GNSS sensor output along with other navigation sensor information (barometric altimeter/inertial sensor) and/or perform cross checks across redundant GNSS sensors. The GNSS

FIGURE 4.11   Stand-alone GPS sensor—GPS4000(S). (Courtesy of Rockwell Collins, Cedar Rapids, IA.)

TABLE 4.4   Stand-Alone GPS Sensor Output Data

Octal Label	Parameter	Nominal Transmit Interval (s)	Max. Transmit Delay (ms)
060	SV Measurement status	1.0	200
061	Pseudorange	1.0	200
062	Pseudorange fine	1.0	200
063	Range rate	1.0	200
064	Delta range	1.0	200
065	SV position X	1.0	200
066	SV position X fine	1.0	200
070	SV position Y	1.0	200
071	SV position Y fine	1.0	200
072	SV position Z	1.0	200
073	SV position Z fine	1.0	200
074	UTC measurement time	1.0	200
076	GPS altitude (MSL)	1.0/0.2	50
101	HDOP	1.0/0.2	N/A
102	VDOP	1.0/0.2	N/A
103	Track angle—true (GPS)	1.0/0.2	50
110	Present position latitude (GPS)	1.0/0.2	50
111	Present position longitude (GPS)	1.0/0.2	50
112	Ground speed (GPS)	1.0/0.2	50
120	Present position latitude, fine (GPS)	1.0/0.2	50
121	Present position longitude, fine (GPS)	1.0/0.2	50
124	Digital time mark (GPS)	1.0	25
125	UTC (GPS)	1.0/0.2	50
130	Horizontal integrity limit autonomous	1.0/0.2	50
133	Vertical integrity limit autonomous	1.0/0.2	50

position information provided by this sensor can also be used by the FMS in conjunction with its runway database information in order to provide approach guidance information at the cockpit level.

A subset of labels seen on the navigation data bus is shown in Table 4.4.

In essence, the navigation information comprises of the phase center location of GPS antenna mounted atop the aircraft, time as derived from GPS, the protection level of computed position, user velocity, and the raw satellite measurements from all the GPS satellites that are tracked by the receiver. It is to be noted that in addition to the data provided on the deviation bus, the GNSS-capable MMRs transmit the aforementioned labels on the ARINC 743A bus. In order to be able to support aircraft operations based on augmented GNSS functionality (e.g., LPV with SBAS augmentation), the update rate of the augmented solutions has to be higher and is typically required to be no less than 5 Hz.

4.4.4 Military

GPS signals have an inherent level of interference resistance by virtue of their CDMA signal structure. As mentioned earlier (Section 4.2.2), the GPS satellites transmit signals in both L1 and L2 frequency bands in addition to some satellites that transmit in the L5 band. The United States and Allied military's GPS receivers process the encrypted P(Y) code signals at both L1 and L2 frequencies using hardware architectures that support PPS Security Module (PPS-SM) or the more modern SA Anti-Spoofing Module (SAASM). These receivers are designed with specific mission capabilities and meet performance requirements in the presence of jammer and/or spoofing threats.

Airborne military GPS solutions can be broadly separated into weapons and aircraft platforms. The top row of Figure 4.12 shows two views of a receiver that is capable of processing GPS signals in both L1 and L2 frequency bands in the presence of interfering signals that are 90 dB more powerful than the desired GPS signal. Receivers of this type are designed to be small (size of a hockey puck) and fit onto projectiles and spinning platforms that would experience significant accelerations (up to 20,000G's) over short time intervals. In order to perform under these conditions, these keyed receivers employ digital nulling capabilities (with up to five antenna inputs in this case) along with the ability to support ultratight GPS–inertial coupling that enables the receiver design to reduce the carrier tracking loop's equivalent noise bandwidth. These receivers communicate with the rest of the munitions platform via high-speed serial interfaces and provide a 1 pulse per second (PPS) signal for the purposes of synchronization. The middle row of Figure 4.12 shows a sectioned view of this gun-hardened GPS receiver.

When size, weight, and power (SWAP) constraints are relaxed, additional capabilities can be realized in a precision-guided missile or munitions platform. The bottom row in Figure 4.12 shows an integrated

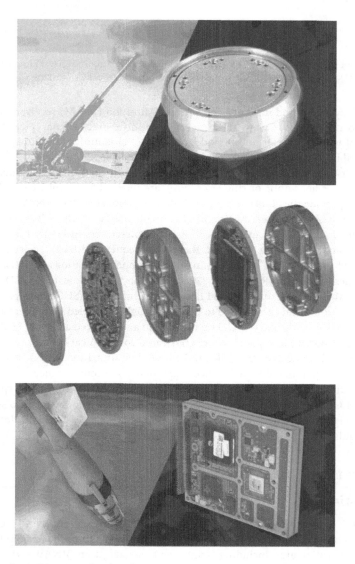

FIGURE 4.12 NavStorm™ +, gun-hardened GPS receiver (top row), integrated GPS AJ system (IGAS—bottom row). (Courtesy of Rockwell Collins, Cedar Rapids, IA.)

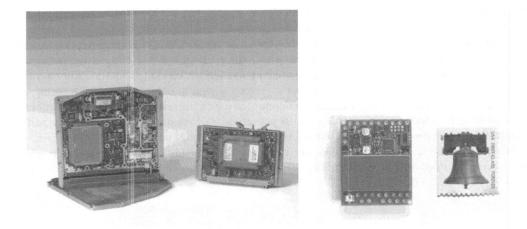

FIGURE 4.13   From left to right—GEM (GPS embedded module), ASR (airborne SAASM receiver), and MicroGRAM (shown along with U.S. postage stamp for size comparison). (Courtesy of Rockwell Collins, Cedar Rapids, IA.)

GPS antijam (AJ) system found in Joint Direct Attack Munition (JDAM) receivers that are capable of steering beams to all satellites in view. These receivers perform direct Y (encrypted P) code acquisition in the presence of jammer threats and provide a high level of jamming immunity while tracking the desired GPS signal in challenging signal operating environments.

Military aircraft are also equipped with SAASM GPS receivers that, when integrated with high-grade inertial sensors (either fiber-optic or ring laser gyro), are referred to as embedded GPS–inertial (EGI) units. These units provide the ability to support both navigational warfare (NAVWAR) and global air traffic management (GATM) requirements and can also be interfaced with advanced antenna electronics for applications requiring AJ capabilities. The first two receivers in Figure 4.13 (from the left) are used on both fixed and rotary wing platforms. These dual-frequency (L1/L2) receivers have the ability to operate on both PPS and SPS (L1 C/A) signals in order to support operations in civil airspace by providing FDE capabilities in line with RTCA/DO-229 MOPS. They communicate via ICD-GPS-155 compatible dual-port RAM (DPRAM) and/or serial host control interfaces (SHCI) and provide pseudorange and carrier phase outputs that can be used to compute a tightly coupled GPS–inertial navigation solution. In addition, they provide one PPS I/O capability and an enhanced accuracy HAVEQUICK output (a frequency-hopped system) to support secure UHF radio communications.

Although GPS receivers such as the ASR can be hosted on unmanned airborne platforms, the next generation of lightweight (<3 lb) intelligence, surveillance, reconnaissance, and targeting (ISRT) unmanned micro air vehicles demand small, lighter GPS solutions that are jam and spoof resistant. This is met by SAASM GPS receivers such as the MicroGRAM (Figure 4.13), which is a small (stamp size), optimized lightweight (0.25 oz), low power (<0.5 W) SAASM-capable GPS unit designed to perform direct Y code acquisition.

4.5 Applications

4.5.1 Navigation

The primary application of GNSS for civil aviation is as a navigation sensor in IMC for all phases of flight: departure, en route, nonprecision approach, and precision approach. GNSS offers many benefits over traditional navigation aids, including facilitating area navigation (RNAV)—the ability to fly arbitrary routes rather than being constrained by the location of ground navigation facilities. Other benefits include the provision of improved navigation services in areas that are not presently covered by ground navigation aids, and the possibility to alleviate some of the expense of maintaining expansive networks of ground navigation facilities.

Performance requirements for navigation sensors generally fall within four categories: accuracy, integrity, continuity, and availability. Accuracy is the degree of conformance between true aircraft position and that position estimate provided by the navigation sensor. Since navigation sensor errors are probabilistic, accuracy requirements are typically specified as horizontal and vertical position error levels that are achieved with high probability (e.g., 95%).

Integrity is the ability of navigation system to provide timely warnings when the system cannot be safely used for navigation. Integrity requirements for safety-critical navigation applications are commonly specified using three parameters: (1) an alert limit, (2) time to alert, and (3) integrity level or probability of hazardously misleading information. An alert limit is the maximum allowable navigation system position error before safety would be unacceptably compromised if the user is not promptly notified. The time to alert is the maximum allowable period from the onset of an out-of-tolerance condition until an alert is provided. The integrity level or probability of hazardously misleading information is the maximum acceptable probability of occurrence of an out-of-tolerance condition without a timely alert.

Continuity is the capability of a navigation system to perform its function without unscheduled interruptions during the intended operation. Availability is the fraction of time during which a system is usable to perform an intended operation.

Table 4.5 summarizes ICAO's GNSS SIS performance requirements. Note that although the GPS SPS can meet the accuracy requirements for many phases of flight, it cannot achieve the integrity requirements for any phase of flight without augmentation (ABAS, SBAS, or GBAS). For instance, the integrity requirements within the GPS SPS performance standard are based upon the possible occurrence of up to three major service failures per year, each of 6 h in duration, where a major service failure is defined as the presence of a large (over 30 m) range error for measurements to a satellite without the user being able to detect this from the satellite's broadcast navigation data [5].

Most of the operations listed in Table 4.5 have been defined for decades and may be performed using traditional ground navigation aids where available. En route through nonprecision approach applications only require horizontal position estimates from the navigation sensor, because these operations rely on vertical position estimates from an onboard barometric altimeter. The approach with vertical guidance (APV) operations are defined to tailor operations to the capabilities provided by GNSS.

TABLE 4.5 ICAO GNSS Signal-in-Space Performance Requirements

Operation	Horizontal/ Vertical Accuracy (95%)	Integrity Level	Horizontal/ Vertical Alert Limit	Time to Alert	Continuity	Availability
En route (oceanic, remote)	7.4 km	$1-1 \times 10^{-7}$/h	7.4 km N/A	5 min	$1-1 \times 10^{-4}$/h to $1-1 \times 10^{-8}$/h	0.99–0.99999
En route	3.7 km N/A	$1-1 \times 10^{-7}$/h	3.7 km N/A	5 min	$1-1 \times 10^{-4}$/h to $1-1 \times 10^{-8}$/h	0.99–0.99999
Terminal	0.74 km N/A	$1-1 \times 10^{-7}$/h	1.85 km N/A	15 s	$1-1 \times 10^{-4}$/h to $1-1 \times 10^{-8}$/h	0.999–0.99999
Nonprecision approach	220 m N/A	$1-1 \times 10^{-7}$/h	556 m N/A	10 s	$1-1 \times 10^{-4}$/h to $1-1 \times 10^{-8}$/h	0.99–0.99999
Approach with vertical guidance (APV)-I	16 m 20 m	$1-2 \times 10^{-7}$/ approach	40 m 50 m	10 s	$1-8 \times 10^{-6}$ per 15 s	0.99–0.99999
Approach with vertical guidance (APV)-II	16 m 8 m	$1-2 \times 10^{-7}$/ approach	40 m 20 m	6 s	$1-8 \times 10^{-6}$ per 15 s	0.99–0.99999
Category I	16 m 4–6 m	$1-2 \times 10^{-7}$/ approach	40 m 10–35 m	6 s	$1-8 \times 10^{-6}$ per 15 s	0.99–0.99999

Source: Anonymous, Annex 10 to the Convention of International Civil Aviation, Vol. I (Radio Navigation Aids), Amendment 87, International Civil Aviation Organization, Montreal, Quebec, Canada, November 2012.

Currently available user equipment can achieve the integrity requirements for en route through nonprecision approach using RAIM, SBAS, or GBAS, whereas SBAS or GBAS is required for vertically guided operations except for aircraft with sophisticated ABAS capabilities (e.g., barometric vertical navigation [baro-VNAV]). Researchers are currently exploring whether RAIM techniques can be applied to meet precision approach (e.g., APV or category I) requirements (see, e.g., [52]).

Many nations have approved GNSS operations in IMC. Figure 4.14 depicts nations that have approved aviation operations using GPS, as compiled by the U.S. FAA from data available circa 2005. An important concept in air navigation at the present time is required navigation performance (RNP), which is defined to be RNAV operations with navigation containment and monitoring [53]. Many RNP procedures have been developed or are planned worldwide with GNSS as one enabling technology.

A number of nations, including the United States, are planning to decommission significant numbers of ground-based navigation aids in the future as the GNSS infrastructure advances. A key concern with the increasing reliance on satellite navigation is the vulnerability of GNSS signals to unintentional or intentional radio-frequency interference. Prudent means to address this concern include the retention of a subset of existing ground-based navigation aids and the development of operational procedures to mitigate the impact of an event in which GNSS service is lost over a large geographic area.

4.5.2 Automatic Dependent Surveillance

Automatic dependent surveillance (ADS) is a concept whereby aircraft continually transmit position, intent, and other data to air traffic service facilities or to other aircraft. ADS provides a number of benefits over radar-based surveillance. These benefits include the provision to air traffic controllers of the location of aircraft in areas where radar coverage is infeasible or impractical, for example, oceanic and remote airspace. ADS systems can also supply aircraft intent information, that is, the planned trajectory of the aircraft, which is not available from radars. Modern ADS implementations also allow pilots to view the locations of nearby aircraft on cockpit displays, enhancing situational awareness.

There are two main types of ADS systems. The first type, referred to as ADS-addressed (ADS-A) or ADS-contract (ADS-C), involves transmitting an aircraft's location to a single air traffic services recipient over a point-to-point data link. The second type is ADS-broadcast (ADS-B), in which an aircraft continually broadcasts its position over a data link to air traffic services and other nearby aircraft. GNSS is the most commonly used onboard position sensor for the various ADS services that are currently implemented worldwide.

The use of ADS for civil aviation was first studied in depth upon the establishment in 1983 of the ICAO Special Committee on Future Air Navigation Systems (FANS) [55]. At that time, GPS and GLONASS were just two of several navigation inputs for ADS that were considered. Other candidates included OMEGA, inertial navigation systems, VOR, distance measuring equipment (DME), and Loran-C. ADS-C implementations using GPS as the primary navigation input were tested in the early 1990s and implemented in some regions of the world shortly after the certification in 1995 of Boeing's FANS-1 navigation system. Airbus later developed an ADS-C and GPS-capable avionics package, FANS-A, which was first certified in 2000 on the Airbus A340/A330 family of aircraft. Implementations involving FANS-1 and FANS-A equipped aircraft combined with compatible ground systems are collectively referred to as FANS-1/A. FANS-1/A ADS-C implementations follow standards developed by RTCA [55] and ARINC [56] that are based upon a dedicated data link connection between each equipped aircraft and air traffic service provider and are still in operation.

The second form of ADS, ADS-B, is currently being implemented in many areas of the world. ADS-B equipped aircraft continually broadcast their position, intent, and other information over a data link to nearby air traffic facilities and other suitably equipped aircraft. The broadcasting function is known as ADS-B OUT, whereas the function enabling an aircraft to listen to ADS-B broadcasts from other aircraft and air traffic facilities is referred to as ADS-B IN. Chapter 33 provides a comprehensive overview of ADS-B.

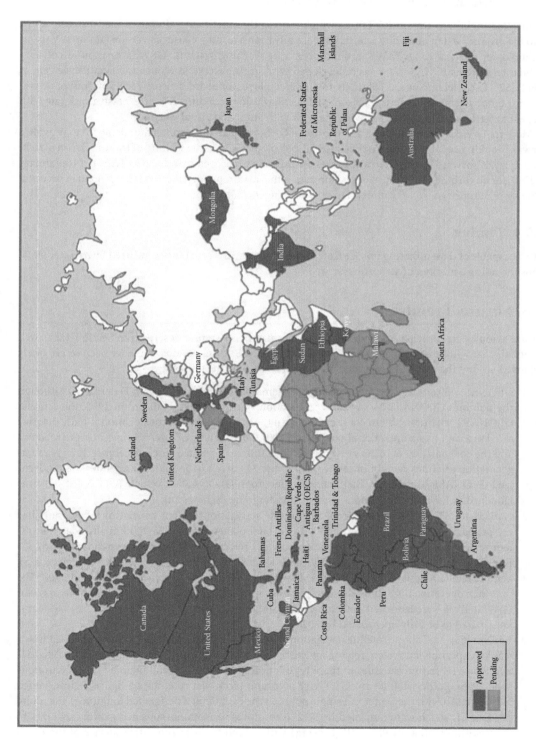

FIGURE 4.14   (See color insert.) Nations that have approved GPS for aircraft navigation in IMC. (Courtesy of the Federal Aviation Administration, Washington, DC.)

4.5.3 Terrain Awareness Warning Systems

Controlled flight into terrain (CFIT), in which a perfectly functioning aircraft is inadvertently flown into the ground, water, or an obstacle, has historically been a leading cause of aviation fatalities. Various technologies referred to as TAWS, ground proximity warning systems (GPWS), or ground collision avoidance systems (GCAS) have been developed with great success to reduce the occurrence rate of CFIT [57]. Early technological solutions to CFIT employed onboard sensors including radio altimeters, air data systems, and inertial sensors to detect hazardous conditions (e.g., excessive sink rate with respect to terrain clearance) and provide the crew with aural and visual warnings.

Modern TAWS implementations add the use of GNSS with an onboard terrain database to provide a forward-looking capability and in some cases also depict terrain in the vicinity of the aircraft on a cockpit display. A forward-looking terrain awareness capability has been mandated by ICAO, since January 1, 2007, for all turbine- or piston-engined aircraft with a maximum certificated takeoff mass in excess of 5700 kg or authorized to carry more than nine passengers [58].

4.5.4 Timing

GPS is capable of disseminating precise time and frequency. This capability is utilized by many applications including a number of airborne systems.

4.6 Future Trends

Future avionics are anticipated to utilize the modernized GPS signals described in Section 4.2 as well as signals that will be broadcast by a number of other GNSS constellations that have been or are being deployed around the world. These other constellations include the following:

1. GLONASS [59,60]—a satellite navigation system operated by the Russian Federation. Although the nominal GLONASS 24-satellite constellation was first fully populated in 1995, the original GLONASS satellites exhibited a short life span (1–3 years), and the constellation rapidly deteriorated to as few as six operational satellites in 2001. Fortunately, GLONASS is now on the rebound and the constellation of MEO satellites has been fully repopulated since December 2011. No first-generation satellites remain in service. All of the 24 currently operational satellites are modernized GLONASS-M vehicles. The first next-generation GLONASS-K vehicle (GLONASS-K1) was launched in February 2011 and is currently undergoing flight testing.

2. Galileo [61,62]—a planned European satellite navigation system consisting of 30 satellites (27 primary plus 3 active spares) in 3 MEO orbital planes at about 23,000 km altitude. The system will provide four distinct navigation services referred to as the (1) Open Service (OS), (2) Integrity Monitoring Service (formerly the Safety of Life service), (3) Commercial Service (CS), and (4) Public Regulated Service (PRS). Two test satellites were launched in 2005 and 2008. The first four operational satellites are referred to as in-orbit validation (IOV) vehicles. The first two IOV satellites were launched in October 2011 and the second pair in October 2012. The system is anticipated to be fully operational by 2020.

3. BeiDou [63]—a Chinese satellite navigation system that is being deployed in three phases. Phase I was an experimental system relying on active (two-way) ranging to GEO orbit satellites. Phase II and Phase III rely on satellites in three types of orbit: MEO, GEO, and inclined geosynchronous orbit (IGSO). The satellites in IGSO have a ground track that resembles a figure-eight-oriented North–South with the center at a single point on the equator at a designated longitude. The intent of Phase II, which was completed in 2012, is to provide an autonomous navigation capability to users within China. In Phase III, coverage will expand to cover the entire surface of the Earth

using a combination of 27 MEO, 3 IGSO, and 5 GEO satellites. To date, four MEOs, five IGSOs, and five GEOs have been launched. Phase III is planned to be completed by 2020.

4. QZSS [64,65]—a satellite navigation system that is being developed by the government of Japan. QZSS is not intended to provide a stand-alone navigation capability but rather to improve the performance of GPS in Japan, particularly in urban environments where buildings obscure visibility of much of the sky. QZSS was originally planned to be a three IGSO-satellite constellation with coverage optimized for Japan. Recently, Japan has announced its intent to expand QZSS to seven satellites, but the details of the new design are not yet available. One satellite has been launched thus far in September 2010, and the full constellation is expected to be deployed within a decade.

5. Indian Regional Navigation Satellite System (IRNSS) [30]—a satellite navigation system planned by India. The overall system will consist of seven satellites. Three of the satellites will be placed in GEO with longitudes of 34°E, 83°E, and 132°E. The four remaining satellites will be placed in IGSO with a pair of figure-eight ground tracks centered on the equator at 55°E and 111°E in longitude. The intended service volume is bounded in longitude between 40°E and 140°E and in latitude between 40°S and 40°N. Current plans call for the first satellite to be launched in 2012 and the entire constellation to be operational around 2020.

One additional future trend, as mentioned earlier, may be the use of advanced RAIM (ARAIM) techniques to enable a precision approach capability for equipment capable of tracking the signals broadcast from multiple GNSS constellations. Recent results in this area may be found in [66].

References

1. Parkinson, B. and S. Gilbert, NAVSTAR: Global positioning system—Ten years later, *Proceedings of the IEEE*, New York, October 1983, pp. 1117–1186.
2. Parkinson, B. and J. J. Spilker, Jr. (eds.), *Global Positioning System: Theory and Applications*, Vol. I, American Institute of Aeronautics and Astronautics, Washington, DC, 1996.
3. Kaplan, E. and C. Hegarty (eds.), *Understanding GPS: Principles and Applications*, 2nd edn., Artech House, Norwood, MA, 2006.
4. Misra, P. and P. Enge, *Global Positioning System: Signals, Measurements, and Performance*, 2nd edn., Ganga-Jamuna Press, Lincoln, MA, 2006.
5. Anonymous, *Global Positioning System Standard Positioning Service Performance Standard*, 3rd edn., U.S. Department of Defense, Washington, DC, September 2008.
6. Anonymous, Navstar GPS space segment/user navigation user interfaces, IS-GPS-200F, U.S. Air Force, GPS Directorate, Los Angeles Air Force Base, El Segundo, CA, September 21, 2011.
7. Gold, R., Optimal binary sequences for spread spectrum multiplexing, *IEEE Transactions on Information Theory*, IT-13, 619–621, October 1967.
8. Barker, B., J. Betz, J. Clark, J. Correia, J. Gillis, S. Lazar, K. Rehborn, and J. Straton, Overview of the GPS M code signal, *Proceedings of the Institute of Navigation National Technical Meeting*, Anaheim, CA, January 2000, pp. 542–549.
9. Fontana, R. D., W. Cheung, and T. Stansell, The new L2 civil signal, *GPS World*, 12(9), September 2001, pp. 28–34.
10. Betz, J. W., Binary offset carrier modulations for radionavigation, *NAVIGATION: Journal of the Institute of Navigation*, 48(4), 227–246, Winter 2001–2002.
11. Van Dierendonck, A. J. and C. Hegarty, The new L5 civil GPS signal, *GPS World*, 11(9), June 2000, pp. 64–72.
12. Anonymous, Navstar GPS space segment/user navigation user segment L5 interfaces, IS-GPS-705B, U.S. Air Force, GPS Directorate, Los Angeles Air Force Base, El Segundo, CA, September 21, 2011.

13. Betz, J., M. Blanco, C. Cahn, P. Dafesh, C. Hegarty, K. Hudnut, V. Kasemsri et al., Description of the L1C signal, *Proceedings of the Institute of Navigation ION GNSS 2006*, Fort Worth, TX, September 2007, pp. 2080–2091.

14. Anonymous, Navstar GPS space segment/user navigation user segment L1C interfaces, IS-GPS-800B, U.S. Air Force, GPS Directorate, Los Angeles Air Force Base, El Segundo, CA, September 21, 2011.

15. Anonymous, *Global Positioning System Standard Positioning Precise Positioning Service Performance Standard*, 1st edn., U.S. Department of Defense, Washington, DC, February 2007.

16. Hinson, D. R. (Administrator of the Federal Aviation Administration), Letter to Dr. Assad Kotaite (President of the Council, International Civil Aviation Organization), Federal Aviation Administration, Washington, DC, October 14, 1994.

17. Blakey, M. C. (Administrator of the Federal Aviation Administration), Letter to Dr. Roberto Kobeh (President of the Council, International Civil Aviation Organization), Federal Aviation Administration, Washington, DC, September 10, 2007.

18. Clinton, W. J., Statement by the President regarding the United States' decision to stop degrading global positioning system accuracy, United States White House, Office of the Press Secretary, Washington, DC, May 1, 2000.

19. Perino, D., Statement by the Press Secretary, United States White House, Office of the Press Secretary, Washington, DC, September 18, 2007.

20. Anonymous, Global Positioning System (GPS) Standard Positioning Service (SPS) performance analysis report, Report #79, Federal Aviation Administration, William J. Hughes Technical Center, Atlantic City, NJ, October 2012.

21. Anonymous, *Global Positioning System: Papers Published in NAVIGATION*, Vol. V (RAIM), The Institute of Navigation, Fairfax, VA, 1998.

22. Murphy, T., M. Harris, and M. Braasch, Availability of GPS/INS integration methods, *Proceedings of the Institute of Navigation's ION GPS 2001*, Salt Lake City, UT, September 2001, pp. 600–609.

23. Walter, T. and M. B. El-Arini (eds.), *Global Positioning System: Papers Published in NAVIGATION*, Vol. VI (Satellite-Based Augmentation Systems), The Institute of Navigation, Fairfax, VA, 1999.

24. RTCA, Minimum operational performance standards for global positioning system/wide area augmentation system airborne equipment, RTCA DO-229D, RTCA, Inc., Washington, DC, December 13, 2006.

25. Enge, P., T. Walter, S. Pullen, C. Kee, Y. Chao, and Y. Tsai, Wide area augmentation of the global positioning system, *Proceedings of the IEEE*, 84(8), 1063–1088, August 1996.

26. Walter, T. and P. Enge, The wide-area augmentation system, in *EGNOS: The European Geostationary Overlay System*, Javier Ventura-Traveset (ESA) and Didier Flament (Alcatel Alenia Space) (eds.), European Space Agency Publication SP-1303, Noordwijk, the Netherlands, December 2006, pp. 395–412.

27. Lawrence, D., D. Bunce, N. Mathur, and C. E. Sigler, Wide area augmentation system (WAAS)— Program status, *Proceedings of the Institute of Navigation ION GNSS 2007*, Fort Worth, TX, September 2007, pp. 892–899.

28. ESA, *EGNOS: The European Geostationary Overlay System*, European Space Agency Publication SP-1303, Noordwijk, the Netherlands, December 2006.

29. Manabe, H., MSAS programme overview, in *EGNOS: The European Geostationary Overlay System*, Javier Ventura-Traveset (ESA) and Didier Flament (Alcatel Alenia Space) (eds.), European Space Agency Publication SP-1303, Noordwijk, the Netherlands, December 2006, pp. 417–422.

30. Ganeshan, A., On Indian satellite based navigation system and implementation status, *Seventh Meeting of the United Nations International Committee on Global Navigation Satellite Systems (ICG)*, Tokyo, Japan, September 2011. http://www.oosa.unvienna.org/pdf/icg/2012/icg-7/3-2.pdf. Accessed on April 3, 2014.

31. Stupak, G., SDCM status and plans, *Seventh Meeting of the United Nations International Committee on Global Navigation Satellite Systems*, Beijing, China, November 4–9, 2012. http://www.oosa. unvienna.org/pdf/icg/2012/icg-7/3-2.pdf. Accessed on April 3, 2014.

32. Van Dierendonck, A. J., C. Hegarty, and R. Niles, Next generation satellite based augmentation system signal specification, *Proceedings of the Institute of Navigation National Technical Meeting*, San Diego, CA, January 2005, pp. 371–384.

33. Murphy, T. and T. Imrich, Implementation and operational use of ground-based augmentation systems (GBAS)—A component of the future air traffic management system, *Proceedings of the IEEE*, 96(12), December 2008, pp. 1936–1957.

34. Anonymous, Annex 10 to the Convention of International Civil Aviation, Vol. I (Radio Navigation Aids), Amendment 87, International Civil Aviation Organization, Montreal, Quebec, Canada, November 2012.

35. FAA, Airborne supplemental navigation equipment using the global positioning system (GPS), TSO-C129a, Federal Aviation Administration, Washington, DC, February 20, 1996.

36. FAA, Airborne supplemental navigation sensors for global positioning system equipment using aircraft-based augmentation, TSO-C196a, Federal Aviation Administration, Washington, DC, February 15, 2012.

37. FAA, Passive airborne global navigation satellite system (GNSS) antenna, TSO-C144a, Federal Aviation Administration, Washington, DC, March 30, 2007.

38. FAA, Active airborne global navigation satellite system (GNSS) antenna, TSO-C190, Federal Aviation Administration, Washington, DC, March 30, 2007.

39. FAA, Airborne navigation sensors using the global positioning system augmented by the satellite based augmentation system, TSO-C145c, Federal Aviation Administration, Washington, DC, May 2, 2008.

40. FAA, Stand-alone airborne navigation equipment using the global positioning system augmented by the satellite based augmentation system, TSO-C146c, Federal Aviation Administration, Washington, DC, May 9, 2008.

41. FAA, Ground based augmentation system positioning and navigation equipment, TSO-C161a, Federal Aviation Administration, Washington, DC, December 17, 2009.

42. FAA, Ground based augmentation system very high frequency data broadcast equipment, TSO-C162a, Federal Aviation Administration, Washington, DC, December 17, 2009.

43. RTCA, Minimum operational performance standards for airborne supplemental navigation equipment using global positioning system (GPS), DO-208 including Change 1, RTCA, Inc., Washington, DC, September 21, 1993.

44. RTCA, Minimum operational performance standards for global positioning system/aircraft based augmentation system airborne equipment, DO-316, RTCA, Inc., Washington, DC, April 14, 2009.

45. RTCA, Minimum operational performance standards for global navigation satellite system (GNSS) airborne antenna equipment, DO-228 with Change 1, RTCA, Inc., Washington, DC, January 11, 2000.

46. RTCA, Minimum operational performance standards global navigation satellite system (GNSS) airborne antenna equipment for the L1 frequency band, DO-301, RTCA, Inc., Washington, DC, December 13, 2006.

47. RTCA, Minimum operational performance standards for GPS local area augmentation system, DO-253, RTCA, Inc., Washington, DC, December 16, 2008.

48. RTCA, GNSS-based precision approach local area augmentation system (LAAS) signal-in-space interface control document, DO-246, RTCA, Inc., Washington, DC, December 16, 2008.

49. Anonymous, Global navigation satellite system (GNSS) sensor, ARINC Characteristic 743A-5, ARINC, Inc., Annapolis, MD, May 2009.

50. Anonymous, Multi-mode receiver (MMR)—Digital, ARINC Characteristic 755-3, ARINC, Inc., Annapolis, MD, February 2005.

51. Anonymous, GNSS navigation and landing unit (GNLU), ARINC Characteristic 756-3, ARINC, Inc., Annapolis, MD, February 2004.

52. Walter, T., J. Blanch, B. Pervan, and P. Enge, The GPS Evolutionary Architecture Study (GEAS), *Proceedings of the IEEE*, 96(12), December 2008, pp. 1918–1935.

53. Anonymous, *Performance Based Navigation Manual*, 3rd edn., Doc. 9613-AN/937, International Civil Aviation Organization, Montreal, Quebec, Canada, 2008.

54. Massoglia, P. L., M. T. Pozesky, and G. T. Germana, The use of satellite technology for oceanic air traffic control, *Proceedings of the IEEE*, 77(11), 1695–1708, November 1989.

55. Special Committee 170, Minimum operational performance standards for airborne automatic dependent surveillance (ADS) equipment, RTCA DO-212, RTCA, Inc., Washington, DC, October 26, 1992.

56. Airlines Electronic Engineering Committee, Automatic dependent surveillance (ADS), ARINC Characteristic 745-2, ARINC Incorporated, Annapolis, MD, June 1993.

57. Breen, B. C., Controlled flight into terrain and the enhanced ground proximity warning system, *Proceedings of the 16th AIAA/IEEE Digital Avionics Systems Conference*, Irvine, CA, October 1997, pp. 3.1-1–3.1-7.

58. Anonymous, Annex 6 to the Convention of International Civil Aviation, Part I (International Commercial Air Transport—Aeroplanes), 8th edn., Amendment 29, International Civil Aviation Organization, Montreal, Quebec, Canada, November 24, 2005.

59. Revnivykh, S. G., GLONASS status and modernization, *Seventh Meeting of the United Nations International Committee on Global Navigation Satellite Systems (ICG)*, Beijing, China, November 4–9, 2012.

60. Anonymous, Global navigation satellite system GLONASS interface control document, version 5.1, Coordination Scientific Information Center, Russian Federation Ministry of Defence, Moscow, Russia, 2008.

61. Anonymous, Galileo open service signal in space interface control document (OS SIS ICD), Issue 1.1, European Space Agency/European GNSS Supervisory Authority, Prague, Czech Republic, September 2010.

62. Hayes, D., Status of Galileo and EGNOS, *Seventh Meeting of the United Nations International Committee on Global Navigation Satellite Systems (ICG)*, Beijing, China, November 4–9, 2012. http://www.oosa.unvienna.org/pdf/icg/2012/icg-7/4.pdf. Accessed on April 3, 2014.

63. Huang, Q., Development of BeiDou navigation satellite system, *Seventh Meeting of the United Nations International Committee on Global Navigation Satellite Systems (ICG)*, Beijing, China, November 4–9, 2012.

64. Nomura, E., Quasi-Zenith satellite system, *Seventh Meeting of the United Nations International Committee on Global Navigation Satellite Systems (ICG)*, Beijing, China, November 4–9, 2012.

65. Anonymous, Quasi Zenith satellite system navigation service: Interface specification for QZSS (IS-QZSS), Version 1.4, Japan Aerospace Exploration Agency, Chofu, Tokyo, February 28, 2012.

66. Anonymous, EU-US cooperation on satellite navigation: Working group C, ARAIM Technical Subgroup Interim Report, Issue 1.0, December 19, 2012, available at http://www.gps.gov/policy/cooperation/europe/2013/working-group-c/ARAIM-report-1.0.pdf. Accessed on April 2, 2014.

5

Fault-Tolerant Avionics

Ellis F. Hitt
StratSystems, Inc.

5.1 Introduction

Fault-tolerant designs are required to ensure safe operation of digital avionics systems performing flight-critical functions. This chapter discusses the motivation for fault-tolerant designs and the many different design practices evolving to implement a fault-tolerant system. The designer needs to make sure the fault-tolerance requirements are fully defined to select the design concept to be implemented from the alternatives available. The requirements for a fault-tolerant system include performance, dependability, and a methodology to assure that the design, when implemented, meets all requirements. The requirements must be documented in a specification of the intended behavior of a system, specifying the tolerances imposed on the various outputs from the system (Anderson and Lee, 1981). The development of the design proceeds in parallel with the development of the methods of assurance to validate that the design meets all requirements, including the fault tolerance. The chapter concludes with references to further reading in this developing field.

A fault-tolerant system provides continuous, safe operation in the presence of faults. A fault-tolerant avionics system is a critical element of flight-critical architectures, which include the fault-tolerant computing system (hardware, software, and timing), sensors and their interfaces, actuators, aircrew, components, and data communication among the distributed components. The fault-tolerant avionics system ensures integrity of output data used to control the flight of the aircraft, whether operated by the pilot or by autopilot. A fault-tolerant system must detect errors caused by faults, isolate the fault, assess the damage caused by the fault, and gracefully recover from the error. The fault-tolerant system must indicate the correct action to be taken by the aircrew if their action is required to recover from the error. It is generally not economical to design and build a system that is capable of tolerating all possible faults in the universe. The faults the system is to be designed to tolerate must be defined based on an analysis of requirements including the probability of each fault occurring and the impact of not tolerating the fault.

A user of a system may observe an error in its operation that is the result of a fault being triggered by an event. Stated another way, a fault is the cause of an error, and an error is the cause of a failure. A mistake made in designing or constructing a system can introduce a fault into the design of the system, either because of an inappropriate selection of components (hardware, software algorithms, or incorrect definition of aircrew interface and interactions with the system) or because of inappropriate (or missing) interactions between components. On the other hand, if the design of a system is considered to be correct, then an erroneous transition can occur only because of a failure of one of the components of the system. Design faults require more powerful fault-tolerance techniques than those needed to cope with component faults. Design faults are unpredictable; their manifestation is unexpected, and they generate unanticipated errors. In contrast, component faults can often be predicted, their manifestation is expected, and they produce errors that can be anticipated (Anderson and Lee, 1981).

In a non-fault-tolerant system, diagnosis is required to determine the cause of the fault that was observed as an error. Faults in avionics systems are of many types; they generally can be classified as hardware, software, and timing related or aircrew interaction. Faults can be introduced into a system during any phase of its life cycle, including requirements definition, design, production, or operation.

In the 1960s, designers strived to achieve highly reliable safe systems by avoiding faults or masking faults. The Apollo guidance and control system employed proven, highly reliable components and triple modular redundancy (TMR) with voting to select the correct output. Improvements in hardware reliability, and our greater knowledge of faults and events that trigger them, have led to improved design methods for affordable fault-tolerant systems.

In any fault-tolerant system, the range of potential fault conditions that must be accommodated is quite large; enumerating all such possibilities is a vital, yet formidable, task in validating the system's airworthiness or its readiness for deployment. The resultant need to handle each such fault condition prompts attention to the various assurance-oriented activities that contribute to certification system airworthiness.

5.1.1 Motivation

Safety is of primary importance to the economic success of the aviation system. The designer of avionics systems must assure that the system provides the required levels of safety to passengers, aircrew, and maintenance personnel. Fault-tolerant systems are essential given the trend toward increasingly complex digital systems.

Many factors necessitate fault tolerance in systems that perform functions that must be sustained without significant interruption. In avionics systems, such functions are often critical to continued safe flight or to the satisfactory conduct of a mission, hence the terms flight critical and mission critical. The first compelling reality is that physical components are nonideal; that is, they are inescapably disposed to physical deterioration or failure. Clearly, then, components inherently possess a finite useful life, which varies with individual instances of the same component type. At some stage, then, any physical component will exhibit an abrupt failure or excessive deterioration such that a fault may be detected at some level of system operation. Due to low demand for high-reliability, military-qualified integrated circuits (ICs), the avionics industry and their customers have elected to use commercial off-the-shelf (COTS) ICs in their designs for new avionics used in new aircraft and retrofit into existing aircraft to upgrade their functional capability. These COTS ICs have increasing clock frequencies, decreasing process geometries, and decreasing power supply voltages. Studies (Constantinescu, 2002) of these trends conclude that the dependability of COTS ICs is decreasing and the expected life of ICs is also decreasing (Driscoll et al., 2004) with the anticipated working life in the range of 5–10 years.

The second contributing factor to physical faults is the less-than-ideal environment in which an avionics system operates. Local vibrations, humidity, temperature extremes and cycling, electrical power transients, electromagnetic interference, and so on tend to induce stress on the physical component, which may cause abrupt failure or gradual deterioration. The result may be a transient or a permanent variation in output, depending on the nature and severity of the stress. The degree of induced deterioration encountered may profoundly influence the useful life of a component. Fortunately, design measures can be taken to reduce susceptibility to the various environmental effects. Accordingly, a rather comprehensive development approach is needed for system dependability, albeit fault tolerance is the most visible aspect because it drives the organization and logic of the system architecture (Avizienis et al., 1987).

The major factor necessitating fault tolerance is design faults. Tolerance of design faults in hardware and software and the overall data flow is required to achieve the integrity needed for flight-critical systems. Relying on the hardware chip to produce correct output when there is no physical failure is risky, as demonstrated by the design error discovered in the floating point unit of a high-performance microprocessor in wide use. Because of the difficulty in eliminating all design faults, dissimilar redundancy is used to produce outputs that should be identical even though computed by dissimilar computers. Use of dissimilar redundancy is one approach to tolerating common-mode failures (CMFs). A CMF occurs when copies of a redundant system suffer faults nearly simultaneously, generally due to a single cause (Lala and Harper, 1994).

5.1.2 Definitional Framework

A digital avionics system is a "hard real-time" system producing time-critical outputs that are used to control the flight of an aircraft. These critical outputs must be dependable—both reliable and safe. Reliability has many definitions and is often expressed as the probability of not failing; another definition is the probability of producing a "correct" output (Vaidya and Pradhan, 1993). Safety has been defined as the probability that the system output is correct or the error in the output is detectable (Vaidya and Pradhan, 1993). Correctness is the requirement that the output of all channels agrees bit for bit under no-fault conditions (Lala, 1994). Another design approach, approximate consensus, considers a system to be correct if the outputs agree within some threshold. Both approaches are in use.

Hardware component faults are often classified by extent, value, and duration (Avizienis, 1976). Extent applies to whether the errors generated by the fault are localized or distributed; value indicates whether the fault generates fixed or varying erroneous values; duration refers to whether the fault is transient or permanent. Several studies have shown that permanent faults cause only a small fraction of all detected errors, as compared with transient faults (Sosnowski, 1994). A recurring transient fault is often referred to as intermittent (Anderson and Lee, 1981). Figure 5.1 depicts these classifications in the tree of faults.

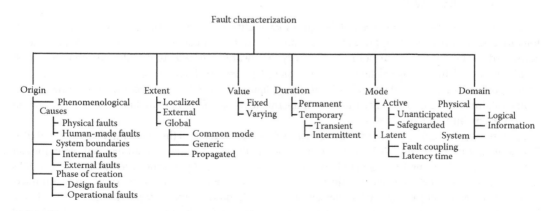

FIGURE 5.1   Fault classification.

Origin faults may result from a physical failure within a hardware component of a system or may result from human-made faults. System boundary internal faults are those parts of the system's state that, when invoked by the computation activity, will produce an error, while external faults result from system interference caused by its physical environment or from system interaction with its human environment. Origin faults classified by the time phase of creation include design faults resulting from imperfections that arise during the development of the system (from requirements specification to implementation), subsequent modifications, the establishment of procedures for operating or maintaining the system, or operational faults that appear during the system operation (Lala and Harper, 1994).

A fault is in the active mode if it yields an erroneous state, either in hardware or in software, that is, a state that differs from normal expectations under extant circumstances. Alternatively, a fault is described as latent when it is not yielding an erroneous state. Measures for error detection that can be used in a fault-tolerant system fall into the following broad classification (discussed in Section 5.4) (Anderson and Lee, 1981):

1. Replications checks
2. Timing checks
3. Reversal checks
4. Coding checks
5. Reasonableness checks
6. Structural checks
7. Diagnostic checks

During the development process, it is constructive to maintain a perspective regarding the fault attributes of domain and value in Figure 5.1. Basically, domain refers to the universe of layering of fault abstractions that permit design issues to be addressed with a minimum of distraction. Value simply refers to whether the erroneous state remains fixed or whether it indeterminately fluctuates. While proficient designers tend to select the proper abstractions to facilitate particular development activities, the following associated fault domains should be explicitly noted:

- PHYSICAL FAILURES—elemental physical failures of hardware components (underlying short, open, ground faults)
- LOGICAL FAULTS—manifested logical faults per device behavior (stuck-at-one, stuck-at-zero, inverted)
- ERROR STATES—exhibited error states in interpreted results (incorrect value, sign change, parity error)
- SYSTEM FAILURE—resultant system failure that provides unacceptable service (system crash, deadlock, hardover)

These fault domains constitute levels of design responsibilities and commitments as well as loci of fault-tolerance actions per se. Thus, fault treatment and, in part, fault containment are most appropriately addressed in the physical fault domain. Similarly, hardware fault detection and assessment are most readily managed in the logical fault domain, where the fixed or fluctuating nature of the erroneous value(s) refines the associated fault identification mechanism(s). Lastly, error recovery and perhaps some fault containment are necessarily addressed in the informational fault domain and service continuation in the system fault domain.

For safety-critical applications, physical hardware faults no longer pose the major threat to dependability. The dominant threat is now CMFs, which result from faults that affect more than one fault-containment region (FCR) at the same time, generally due to a common cause. Approaches used in tolerating CMFs include fault avoidance, fault removal through test and evaluation or via fault insertion, and fault tolerance implemented using exception handlers, program checkpointing, and restart (Lala and Harper, 1994). Table 5.1 presents a classification of common-mode faults; the X indicates the possible combinations of faults that must be considered that are not intentional faults.

TABLE 5.1 Classifications of Common-Mode Faults

Phenomenological Cause		System Boundary		Phase of Creation		Duration		Common-Mode Fault Label
Physical	Human Made	Internal	External	Design	Operational	Permanent	Temporary	
X			X		X		X	Transient (external) CMF
X			X		X	X		Permanent (external) CMF
	X	X		X			X	Intermittent (design) CMF
	X	X		X		X		Permanent (design) CMF
	X		X		X		X	Interaction CMF

Physical, internal, and operational faults can be tolerated by using hardware redundancy. All other faults can affect multiple FCRs simultaneously. Four sources of CMFs need to be considered:

1. Transient (external) faults, which are the result of temporary interference to the system from its physical environment such as lightning, high-intensity radio frequencies (HIRFs), and heat
2. Permanent (external) faults, which are the result of permanent system interference caused by its operational environment such as heat, sand, salt water, dust, vibration, and shock
3. Intermittent (design) faults, which are introduced due to imperfections in the requirements specifications, detailed design, implementation of design, and other phases leading up to the operation of the system
4. Permanent (design) faults, which are introduced during the same phases as intermittent faults but manifest themselves permanently (Lala and Harper, 1994)

An *elemental physical failure*, which is an event resulting in component malfunction, produces a physical fault. These definitions are reflected in Figure 5.2, a state transition diagram that portrays four fault status conditions and associated events in the absence of fault tolerance. Here, for example, a latent fault condition transitions to an active fault condition due to a potentiating event. Such an event might be a functional mode change that caused the fault region to be exercised in a revealing way. Following the incidence of a sufficiently severe active fault from which spontaneous recovery is not forthcoming, a *system failure* event occurs wherein expected functionality can no longer be sustained. If the effects of a particular active fault are not too debilitating, a system may continue to function with some degradation in services. Fault tolerance can, of course, forestall both the onset of system failure and the expectation of degraded services.

The spontaneous recovery event in Figure 5.2 indicates that faults can sometimes be transient in nature when a fault vanishes without purposeful intervention. This phenomenon can occur after an external disturbance subsides or an intermittent physical aberration ceases. A somewhat similar occurrence is provided through the incidental fault remission event in Figure 5.2. Here, the fault does not vanish but rather spontaneously reverts from an active to a latent mode due to the cessation or removal of fault excitation circumstances. Table 5.2 complements Figure 5.2 by clarifying these fault categories. Although these transient fault modes are thought to account for a large proportion of faults occurring in deployed systems, such faults may nonetheless persist long enough to appear as permanent faults to the system. In many cases then, explicit features must be incorporated into a system to ensure the timely recovery from faults that may induce improper or unsafe system operation.

Three classes of faults are of particular concern because their effects tend to be global regarding extent, where "global" implies impact on redundant components present in fault-tolerant systems.

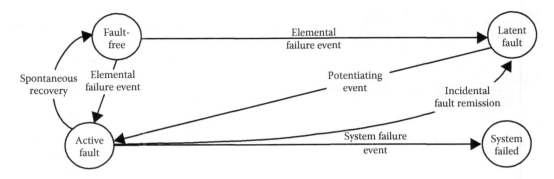

FIGURE 5.2   Hardware states (no corrective action).

TABLE 5.2   Delineation of Fault Conditions

Recovered Mode	Latent Mode	Active Mode
Spontaneous recovery following disruption	—	Erroneous state induced by transient disturbance
or		or
marginal physical fault recovery	—	passing manifestation of marginal fault
—	Hard physical fault remission	Passing manifestation of hard fault
	or	or
	hard physical fault latency	persistent manifestation of hard fault

A common-mode fault is one in which the occurrence of a single physical fault at a one particular point in a system can cause coincident debilitation of all similar redundant components. This phenomenon is possible where there is a lack of protected redundancy and the consequence would be a massive failure event. A generic fault is a development fault that is replicated across similar redundant components such that its activation yields a massive failure event like that due to a common-mode fault. A propagated fault is one wherein the effects of a single fault spread out to yield a compound erroneous state. Such fault symptoms are possible when there is a lack of fault-containment features. During system development, particular care must be exercised to safeguard against these classes of global faults, for they can defeat fault-tolerance provisions in a single event. Considerable design assessment is therefore needed, along with redundancy provisions, to ensure system dependability.

Byzantine faults are of great concern with the move away from ICs built to avionics standards to the use of COTS ICs in avionics systems. A Byzantine fault presents different symptoms to different observers. A Byzantine failure is the loss of a system service due to a Byzantine fault (Driscoll et al., 2004). If a system uses a voting mechanism to reach consensus, Byzantine faults can cause Byzantine failures. Safety-critical functions for avionics are required to have a failure probability of $<10^{-9}$ per flight hour. Research has identified Byzantine faults due to a digital signal stuck at 1/2, a CMOS open, and slightly-off-specification (SOS) transmission timing as just a few faults that designers need to consider to achieve the required system dependability.

5.1.3 Dependability

Dependability is an encompassing property that enables and justifies reliance upon the services of a system. Hence, dependability is a broad, qualitative term that embodies the aggregate nonfunctional attributes, or "-ilities," sought in an ideal system, especially one whose continued safe performance is critical. Thus, attributes like safety, reliability, availability, and maintainability, which are quantified using

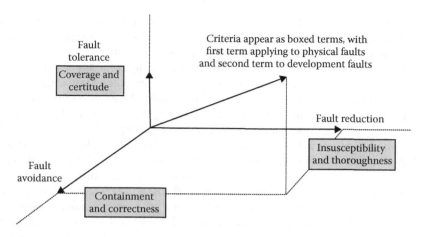

FIGURE 5.3 Dependability.

TABLE 5.3 Ensuring Dependability

	Physical Faults	Development Faults
Fault avoidance	Minimize by analysis	Prevent by rigor development errors
Fault reduction	Selectively reduce the incidence of faults	Remove by verification
Fault tolerance	Ensure by redundancy	Testing

Note: Both physical and developmental fault handling may be present, but any deficiency revealed is a developmental defect.

conditional probability formulas, can be conveniently grouped as elements of dependability. As a practical matter, however, dependability usually demands the incorporation of fault tolerance into a system to realize quantitative reliability or availability levels. Fault tolerance, moreover, may be needed to achieve maintainability requirements, as in the case of online maintenance provisions.

For the sake of completeness, it should be noted, as depicted in Figure 5.3, that the attainment of dependability relies on fault avoidance and fault reduction, as well as on fault tolerance. This figure is complemented by Table 5.3, which emphasizes the development activities, such as analyzing fault possibilities during design to minimize the number and extent of potential fault cases. This activity is related to the criteria of containment in Figure 5.3 in that the analysis should ensure that both the number and propagation of fault cases are contained. The overall notion here is to minimize the number of fault possibilities, to reduce the prospects of their occurrences, and to ensure the safe handling of those that do happen in deployed systems.

5.1.4 Fault-Tolerance Options

System reliability requirements derive from the function criticality level and maximum exposure time. A flight-critical function is one whose loss might result in the loss of the aircraft itself and possibly the persons on board as well. In the latter case, the system is termed safety critical. Here, a distinction can be made between a civil transport, where flight critical implies safety critical, and a combat aircraft. The latter admits to the possibility of the crew ejecting from an unflyable aircraft, so its system reliability requirements may be lower. A mission-critical function is one whose loss would result in the compromising or aborting of an associated mission. For avionics systems, a higher cost usually is associated with the loss of an aircraft than with the abort of a mission (an antimissile mission to repel a nuclear weapon could be an exception). Thus, a full-time flight-critical system would normally pose much more demanding reliability requirements than a flight-phase mission-critical system. Next, the system reliability requirements coupled with the marginal reliabilities of system components influence the design of the fault-tolerant system architecture and determine the level of redundancy to ensure system survivability. To ensure meeting system reliability requirements then, the evolution of a fault-tolerant design must be based on interplay between design configuration commitments and substantiating analyses.

For civil transport aircraft, the level of redundancy for a flight-phase critical function like automatic all-weather landing is typically single fail operational, meaning that the system should remain operable after any potential single elemental failure. Alternatively, a full-time critical function like fly-by-wire (FBW) primary flight controls is typically double fail operational. It should be noted that a function that is not flight critical itself can have failure modes that threaten safety of flight. In the case of active controls to alleviate structural loads due to gusts or maneuvering, the function would not be critical where the purpose is merely to reduce structural fatigue effects. If the associated flight control surfaces have the authority during a hardover or runaway fault case to cause structural damage, then such a failure mode is safety critical. In such cases, the failure modes must be designed to be fail passive, which precludes any active mode failure effect like a hardover control surface. A failure mode that exhibits active behavior can still be failsafe, however, if the rate and severity of the active effects are well within the flight crews' capability to manage safely.

Complexity of avionics communication, navigation, and surveillance functions increases as the avionics system on a single aircraft functions as part of a net-centric operational system involving a distribution of functions across other aircraft- and ground- and space-based systems. Cooperative exchange of correct time- and position-referenced information necessitates that the system have the capability to tolerate faults that arise internal to the avionics system on an aircraft as well as faults that occur external to the aircraft. These faults may originate in any node (aircraft-, ground-, or space-based) in the network and include network operation faults as well as faulty information. Net-centric operations options for ensuring a dependable fault-tolerant system include fault avoidance by intentionally disabling vulnerable network elements when a threat is detected or predicted and fault elimination by reconfiguring the system to replace faulty components with known good components (Knight et al., 2002). Fault-tolerant network survivability involves a control loop structure in which the network state is sensed and analyzed and required changes effected through reconfiguration of the network (Hill and Knight, 2003). The level of integrity required for net-centric operations information flow between multiple aircraft operating in highly congested airspace creates the need for designers to not only include fault tolerance but also ensure that network survivability and resilience is embedded in the avionics system design.

5.1.5 Flight Systems Evolution

Beginning in the 1970s, NASA's F-8 digital fly-by-wire (DFBW) flight research program investigated the replacement of the mechanical primary flight control systems with electronic computers and electrical signal paths. The goal was to explore the implementation technology and establish the practicality of replacing mechanical linkages to the control surfaces with electrical links, thereby yielding significant weight and maintenance benefits. The F-8 DFBW architecture relied on bit-wise exact consensus of the outputs of redundant computers for fault detection and isolation (Lala and Harper, 1994).

The Boeing 747, Lockheed L-1011, and Douglas DC-10 used various implementations to provide the autoland functions, which required a probability of failure of $<10^{-9}$ during the landing. The 747 used triply redundant analog computers, the L-1011 used digital computers in a dual–dual architecture, and the DC-10 used two identical channels, each consisting of dual-redundant fail-disconnect analog computers for each axis. Since that time, the Airbus A-320 uses a full-time DFBW flight control system. It uses software design diversity to protect against CMFs. The Boeing 777 flight control computer architecture uses a three-by-three matrix of nine processors of three different types. Multiversion software is also used.

The Airbus A-380 and Boeing 787 avionics architectures use the Avionics Full-Duplex Switched Ethernet (A664-P7)/ARINC 664 for data communication between the integrated modular avionics cabinets. The A-380 and Boeing 787 avionics vendors use the ARINC Specification 653, "Avionics Application Software Standard Interface" (see Chapter 36). The 787 uses a Common Core System (CCS) that is similar to the Airplane Information Management System (AIMS) of the 777 but hosts more functions and provides greater communication bandwidth. The CCS includes the common computing resource (CCR) cabinet, the A664-P7, and remote data concentrators.

The move away from the federated avionics architectures to the integrated architectures of the A-380, and the 787 may also move dedicated functions such as the flight data acquisition unit and flight data recorder to a distributed implementation. Integrated Vehicle Health Management (IVHM) functions detect, identify, log, and isolate and contain faults. Some IVHM designs include recovery and repair to a nominal system state (Scandura and Garcia-Galan, 2004). The implementation of a dependable fault-tolerant design, IVHM, and digital flight data recorder functions requires an integrated design approach to ensure that the errors observed do not result in the IVHM and fault-tolerant system detecting different faults for the same observed error. Various errors and faults are displayed to the aircrew, and the aircrew is trained to implement various actions based on the displayed information.

5.1.6 Design Approach

The design of a dependable fault-tolerant avionics system must be based on proven systems engineering processes and tools. The designers must identify all of the functions and the information and data flow between processes that implement these functions. Functions involving fault detection, identification, isolation, and recovery must use tools that accurately document the allocation of the processes to hardware, software, and the human (aircrew and maintenance personnel). The design, development, integration, and testing must trace to these allocations.

It is virtually impossible to design a complex avionics system that will tolerate all possible faults. Faults can include both permanent and transient faults and hardware and software faults, and they can occur singularly or concurrently. Timing faults directly trace to the requirement of real-time response within a few milliseconds and may be attributable to both hardware and software contributions to data latency as well as incorrect data. Incorrect data faults will be of increasing importance as the demands for more efficient use of air space reduce spacing between aircraft en route and in the terminal areas. Required navigation performance and 4-D (aircraft position as a function of time) control of flight increase the demand for accurate transmission of aircraft state information when interrogated by air traffic control or another aircraft. Other data faults of concern are use of databases containing invalid data, such as terrain and man-made object locations and flight plan information used by the flight management system.

The implementation of fault tolerance entails increased system overhead, complexity, and validation challenges. The overhead, which is essentially increased system resources and associated management activities, lies in added hardware, communications, and computational demands. The expanded complexity derives from more intricate connectivity and dependencies among both hardware and software elements, the greater number of system states that may be assumed, and appreciably more involved logic necessary to manage the system. Validation challenges are posed by the need to identify, assess, and confirm the capability to sustain system functionality under a broad range of potential fault cases. Hence, the design approach taken for fault-tolerant avionics must attain a balance between the costs incurred in implementing fault tolerance and the degree of dependability realized.

Design approach encompasses system concepts, development methodology, and fault-tolerance elements. The system concepts derive largely from the basic fault-tolerance options introduced in Section 5.1.4, with emphasis on judicious combinations of features that are adapted to given application attributes. The development methodology reduces to mutually supportive assurance-driven methods that propagate consistency, enforce accountability, and exact high levels of confidence in system dependability. Fault-tolerance design elements tend to unify the design concepts and methods by providing an orderly pattern of system organization and evolution. The following are fault-tolerance elements, which, in general, should appear in some form in any fault-tolerant system:

- Error detection—recognition of the incidence of a fault
- Damage assessment—diagnosis of the locus of a fault
- Fault containment—restriction of the scope of effects of a fault
- Error recovery—restoration to a restartable error-free state
- Service continuation—sustained delivery of system services
- Fault treatment—repair of fault

A fundamental design parameter that spans these elements and constrains their mechanization is the granularity of fault handling. Basically, the detection, isolation, and recovery from a fault should occur at the same level of modularity to achieve a balanced and coherent design. In general, it is not beneficial or justified to discriminate or contain a fault at a level lower than that of the associated fault-handling boundary. There may be exceptions, however, especially in the case of fault detection, where a finer degree of granularity may be employed to take advantage of built-in test features or to reduce fault latency.

Depending on the basis for its instigation, fault containment may involve the inhibition of damage propagation of a physical fault and the suppression of an erroneous computation. Physical fault containment has to be designed into the hardware, and software error-state containment has to be designed into the applications software. In most cases, an erroneous software state must be corrected because of applications program discrepancies introduced during the delay in detecting a fault. This error recovery may entail resetting certain data object values and backtracking a control flow path in an operable processor. At this point, the readiness of the underlying architecture, including the coordination of operable components, must be ensured by the infrastructure. Typically, this activity relies heavily on system management software for fault tolerance. Service continuation, then, begins with the establishment of a suitable applications state for program restart. In an avionics system, this sequence of fault-tolerance activities must take place rather quickly because of real-time functional demands. Accordingly, an absolute time budget must be defined with tolerances for worst-case performance for responsive service continuation (Shin and Parmeswaran, 1994).

5.2 System-Level Fault Tolerance

5.2.1 General Mechanization

As discussed in Section 5.1.2, system failure is the loss of system services or expected functionality. In the absence of fault tolerance, a system may fail after just a single crucial fault. This kind of system, which in effect is zero fail operational, would be permissible for noncritical functions. Figure 5.2, moreover, characterizes this kind of system in that continued service depends on spontaneous remission of an active fault or a fault whose consequences are not serious enough to yield a system failure.

Where the likelihood of continued service must be high, redundancy can be incorporated to ensure system operability in the presence of any permanent faults. Such fault-tolerant systems incorporate an additional fault status state, namely, that of recovery, as shown in Figure 5.4. Here, system failure occurs only after the exhaustion of spares or an unhandled severe fault. The aforementioned level of redundancy can render it extremely unlikely that the spares will be exhausted as a result of hardware faults alone. An unhandled fault could occur only as a consequence of a design error, like the commission of a generic error wherein the presence of a fault would not even be detected.

This section assumes a system-level perspective and examines fault-tolerant system architectures and examples from this perspective. Still, these examples embody and illuminate general system-level principles of fault tolerance. Particular prominence is directed toward flight control systems, for they have motivated and pioneered much of the fault-tolerance technology. In the past, such systems have been functionally dedicated, thereby providing a considerable safeguard against malfunction due to extraneous causes.

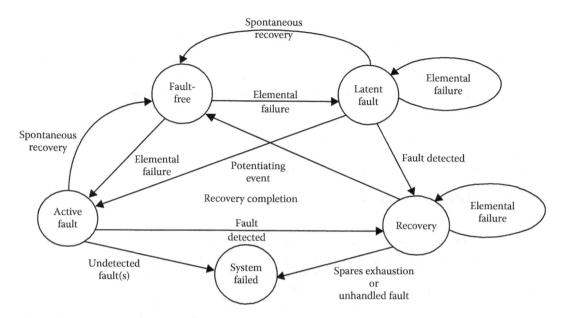

FIGURE 5.4 Hardware states (with corrective action).

With the increasing prevalence of integrated avionics, however, the degree of function separation is irretrievably reduced. This is actually not altogether detrimental; more avionics functions than ever are critical, and a number of benefits accrue from integrated processing. Furthermore, the system developer can exercise prerogatives that afford safeguards against extraneous faults.

5.2.2 Redundancy Options

Fault tolerance is usually based on some form of redundancy to extend system reliability through the invocation of alternative resources. The redundancy may be in hardware, software, time, or combinations thereof. There are three basic types of redundancy in hardware and software: static, dynamic, and hybrid. Static redundancy masks faults by taking a majority of the results from replicated tasks. Dynamic redundancy takes a two-step procedure for detection of and recovery from faults. Hybrid redundancy is a combination of static and dynamic redundancy (Shin and Hagbae, 1994).

In general, much of this redundancy resides in additional hardware components. The addition of components reduces the mean time between maintenance actions, because there are more electronics that can, and at some point will, fail. Since multiple, distinct faults can occur in fault-tolerant systems, there are many additional failure modes that have to be evaluated in establishing the airworthiness of the total system. Weight, power consumption, and cooling are examples of other penalties for component redundancy. Other forms of redundancy also present system management overhead demands, like computational capacity to perform software-implemented fault-tolerance tasks. Like all design undertakings, the realization of fault tolerance presents trade-offs and the necessity for design optimization. Ultimately, a balanced, minimal, and validatable design must be sought that demonstrably provides the safeguards and fault survival margins appropriate to the subject application.

A broad range of redundancy implementation options exist to mechanize desired levels and types of fault tolerance. Figure 5.5 presents a taxonomy of redundancy options that may be invoked in appropriate combinations to yield an encompassing fault-tolerance architecture. This taxonomy indicates the broad range of redundancy possibilities that may be invoked in system design. Although most of these options are exemplified or described in later paragraphs, it may be noted briefly that redundancy

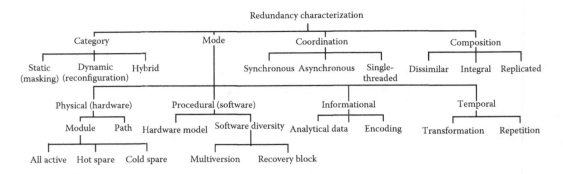

FIGURE 5.5 Redundancy classifications.

is characterized by its category, mode, coordination, and composition aspects. Architectural commitments have to be made in each aspect. Thus, a classical system might, for example, employ fault masking using replicated hardware modules that operate synchronously. At a lower level, the associated data buses might use redundancy encoding for error detection and correction. To safeguard against a generic design error, a backup capability might be added using a dissimilar redundancy scheme.

Before considering system architectures per se, however, key elements of Figure 5.5 need to be described and exemplified. Certain of these redundancy implementation options is basic to formulating a fault-tolerant architecture, so their essential trade-offs need to be defined and explored. In particular, the elements of masking versus reconfiguration, active versus standby spares, and replicated versus dissimilar redundancy are reviewed here.

No unifying theory has been developed that can treat CMFs the same way the Byzantine resilience (BR) treats random hardware or physical operational faults. Three techniques—fault avoidance, fault removal, and fault tolerance—are the tools available to design a system tolerant of CMFs. The most cost-effective phase of the total design and development process for reducing the likelihood of CMFs is the earliest part of the program. Table 5.4 presents fault avoidance techniques and tools that are being used (Lala and Harper, 1994).

Common-mode fault removal techniques and tools include design reviews, simulation, testing, fault injection, and a rigorous quality control program. Common-mode fault tolerance requires error detection and recovery. It is necessary to corroborate the error information across redundant channels to ascertain which recovery mechanism (i.e., physical fault recovery or CMF recovery) to use. Recovery from CMF in real time requires that the state of the system be restored to a previously known correct point from which the computational activity can resume (Lala and Harper, 1994).

TABLE 5.4 Fault Avoidance Techniques and Tools

Technique
Use of mature and formally verified components
Conformance to standards
Formal methods
Design automation
Integrated formal methods and VHDL design methodology
Simplifying abstractions
Performance CMF avoidance
Software and hardware engineering practice
Design diversity

5.2.3 Architectural Categories

As indicated in Figure 5.5, the three categories of fault-tolerant architectures are masking, reconfiguration, and hybrid.

5.2.3.1 Fault Masking

The masking approach is classical per von Neumann's TMR concept, which has been generalized for arbitrary levels of redundancy. The TMR concept centers on a voter that, within a spare exhaustion constraint, precludes a faulted signal from proceeding along a signal path. The approach is passive in that no reconfiguration is required to prevent the propagation of an erroneous state or to isolate a fault.

Modular avionics systems consisting of multiple identical modules and a voter require a trade-off of reliability and safety. A "module" is not constrained to be a hardware module; a module represents an entity capable of producing an output. When safety is considered along with reliability, the module design affects both safety and reliability. It is usually expected that reliability and safety should improve with added redundancy. If a module has built-in error detection capability, it is possible to increase both reliability and safety with the addition of one module providing input to a voter. If no error detection capability exists at the module level, at least two additional modules are required to improve both reliability and safety. An error control arbitration strategy is the function implemented by the voter to decide what the correct output is, and when the errors in the module outputs are excessive so that the correct output cannot be determined, the voter may put out an unsafe signal. Reliability and safety of an n-module safe modular redundant (nSMR) architecture depend on the individual module reliability and on the particular arbitration strategy used. No single arbitration strategy is optimal for improving both reliability and safety. Reliability is defined as the probability the voter's data output is correct and the voter does not assert the unsafe signal. As system reliability and safety are interrelated, increasing system reliability may result in a decrease in system safety and vice versa (Vaidya and Pradhan, 1993).

Voters that use bit-for-bit comparison have been employed when faults consist of arbitrary behavior on the part of failed components, even to the extreme of displaying seemingly intelligent malicious behavior (Lala and Harper, 1994). Such faults have been called Byzantine faults. Requirements levied on an architecture tolerant of Byzantine faults (referred to as BR) comprise a lower bound on the number of fault containment regions (FCRs), their connectivity, their synchrony, and the utilization of certain simple information exchange protocols. No *a priori* assumptions about component behavior are required when using bit-for-bit comparison. The CMF is the dominant contributor to failure of correctly designed BR system architecture.

Fault effects must be masked until recovery measures can be taken. A redundant system must be managed to continue correct operation in the presence of a fault. One approach is to partition the redundant elements into individual FCRs. An FCR is a collection of components that operates correctly regardless of any arbitrary logical or electrical fault outside the region. A fault-containment boundary requires the hardware components be provided with independent power and clock sources. Interfaces between FCRs must be electrically isolated. Tolerance to such physical damage as a weapon hit necessitates a physical separation of FCRs such as different avionics bays. In flight control systems, a channel may be a natural FCR. Fault effects manifested as erroneous data can propagate across FCR boundaries; therefore, the system also must provide error containment by using voters at various points in the processing, including voting on redundant inputs, voting the result of control law computations, and voting at the input to the actuator. Masking faults and errors provides correct operation of the system with the need for immediate damage assessment, fault isolation, and system reconfiguration (Lala and Harper, 1994).

5.2.3.2 Reconfiguration

Hardware interlocks provide the first level of defense prior to reconfiguration or the use of the remaining nonfaulty channels. In a triplex or higher redundancy system, the majority of channels can disable the output of a failed channel. Prior to taking this action, the system will determine whether the failure is permanent or transient. Once the system determines a fault is permanent or persistent, the next step is to ascertain what functions are required for the remainder of the mission and whether the system needs to invoke damage assessment, fault isolation, and reconfiguration of the remaining system assets. The designer of a system required for long-duration missions may implement a design with reconfiguration capability.

5.2.3.3 Hybrid Fault Tolerance

Hybrid fault tolerance uses hybrid redundancy, which is a combination of static and dynamic redundancy—masking, detection, and recovery that may involve reconfiguration. A system using hybrid redundancy will have N-active redundant modules, as well as spare (S) modules. A disagreement detector detects if the output of any of the active modules is different from the voter output. If a module output disagrees with the voter, the switching circuit replaces the failed module with a spare. A hybrid (N, S) system cannot have more than (N − 1)/2 failed modules at a time in the core, or the system will incorrectly switch out the good module when two out of three have failed.

Hybrid fault tolerance employs a combination of masking and reconfiguration, as noted in Section 5.2.3. The intent is to draw on strengths of both approaches to achieve superior fault tolerance. Masking precludes an erroneous state from affecting system operation and thus obviates the need for error recovery. Reconfiguration removes faulted inputs to the voter so that multiple faults cannot defeat the voter. Masking and reconfiguration actions are typically implemented in a voter–comparator mechanism, which is discussed in Section 5.3.1.

Figure 5.6 depicts a hybrid TMR arrangement with a standby spare channel to yield double-fail-operational capability. Upon the first active channel failure, it is switched out of the voter-input configuration, and the standby channel is switched in. Upon a second channel failure, the discrepant input to the voter is switched out. Only two inputs remain then, so a succeeding (third) channel failure can be detected but not properly identified by the voter per se. When a voter selects the lower of two remaining signals and precludes a hardover output, a persistent miscomparison results in a fail-passive loss of system function. An alternative double-fail-operational configuration would forego the standby channel switching and simply employ a quadruplex voter. This architecture is actually rather prevalent in dedicated flight-critical systems like FBW flight control systems. This architecture still employs reconfiguration to remove faulty inputs to the voter.

The fault-tolerance design elements described in Section 5.1.6 are reflected in the fault-tolerant architecture in Figure 5.6 by way of annotations. For example, error detection is provided by the comparators;

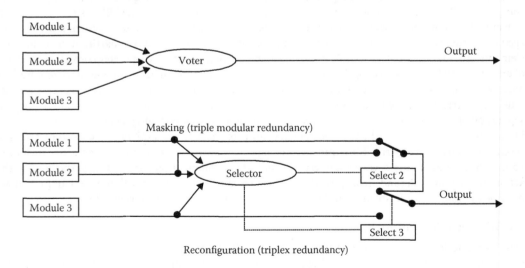

FIGURE 5.6 Masking versus reconfiguration. *Note:* 3-Channel redundancy usually provides single fail-operational fail-passive capability.

damage assessment is then accomplished by the reconfiguration logic using the various comparator states. Fault containment and service continuation are both realized through the voter, which also obviates the need for error recovery. Lastly, fault treatment is accomplished by the faulty path switching prompted by the reconfiguration logic. Thus, this simple example illustrates at a high level how the various aspects of fault tolerance can be incorporated into an integrated design.

5.2.4 Integrated Mission Avionics

In military applications, redundant installations in some form will be made on opposite sides of the aircraft to avoid loss of functionality from battle damage to a single installation. Vulnerability to physical damage also exists in the integrated rack installations being used on commercial aircraft. Designers must take these vulnerabilities into account in the design of a fault-tolerant system.

5.2.5 System Self-Tests

Avionics system reliability analyses are conditional on assumptions of system readiness at dispatch. For lower-criticality systems, certain reductions in redundancy may sometimes be tolerable at dispatch. For full-time flight-critical systems, however, a fully operable system with all redundancy intact is generally assumed in a system reliability prediction (Tomek et al., 1994). This assumption places appreciable demands on the coverage and confidence values of system preflight self-testing. Such a test is typically an end-to-end test that exercises all elements in a phased manner that would not be possible during flight. The fault-tolerance provisions demand particular emphasis. For example, such testing deliberately seeks to force seldom-used comparator trips to ensure the absence of latent faults like passive hardware failures. Analysis of associated testing schemes and their scope of coverage is necessarily an ongoing design analysis task during development. These schemes must also include appropriate logic interlocks to ensure safe execution of the preflight test, for example, a weight-on-wheels interlock to preclude testing, except on the ground. Fortunately, the programming of system self-tests can be accomplished in a relatively complete and high-fidelity manner.

Because of the discrete-time nature of digital systems, all capacity is not used for application functions. Hence, periodic self-tests are possible for digital components like processors during flight. Also, the processors can periodically examine the health status of other system components. Such tests provide a self-monitoring that can reveal the presence of a discrepancy before error states are introduced or exceed detection thresholds. The lead time afforded by self-tests can be substantial because steady flight may not stimulate comparator trips due to low-amplitude signals. The longer a fault remains latent, the greater the possibility that a second fault can occur; therefore, periodic self-tests can significantly enhance system reliability and safety by reducing exposure to coincident multiple fault manifestations.

Self-monitoring may be employed at still lower levels, but there is a trade-off as to the granularity of fault detection. This trade-off keys on fault detection and recovery response and on the level of fault containment selected. In general, fault containment delineates the granularity of fault detection unless recovery response times dictate faster fault detection that is best achieved at lower levels.

5.3 Hardware-Implemented Fault Tolerance: Fault-Tolerant Hardware Design Principles

5.3.1 Voter Comparators

Voter comparators are very widely used in fault-tolerant avionics systems, and they are generally vital to the integrity and safety of the associated systems. Because of the crucial role of voter comparators, special care must be exercised in their development. These dynamic system elements, which can be implemented in software as well as hardware, are not as simple as they might seem. In particular, device integrity and threshold parameter settings can be problematic.

Certain basic elements and concerns apply over the range of voter–comparator variants. A conceptual view of a triplex voter–comparator is depicted in Figure 5.7. The voter here is taken to be a middle signal selector, which means that the intermediate level of three inputs is selected as the output. The voter section precedes the comparators because the output of the voter is an input to each comparator. Basically, the voter output is considered the standard of correctness, and any input signal that persists in varying too much from the standard is judged to be erroneous.

In Figure 5.7, the respective inputs to each of the signal paths are an amplitude-modulated pulse train, as are normal in digital processing. Each iteration of the voter is a separate selection, so each voter output is apt to derive from any input path. This is seen in Figure 5.8, where the output pulse train components are numbered per the input path selected at each point in time. At each increment of time, the voter output is applied to each of the comparators, and the difference with each input signal is fed to a corresponding amplitude threshold detector. The amplitude threshold is set so that accumulated tolerances are not apt to

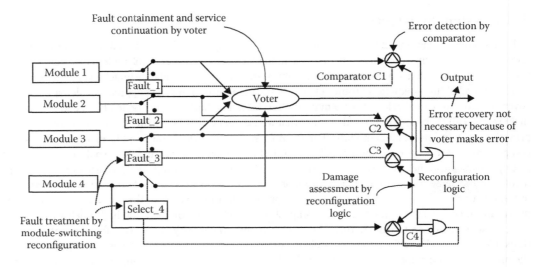

FIGURE 5.7   Hybrid TMR arrangement.

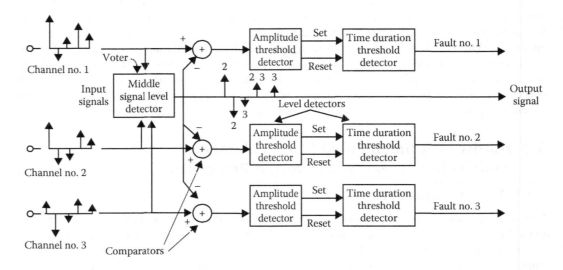

FIGURE 5.8   Triplex voter–comparator.

trip the detector. As shown here, the amplitude detector issues a set output when an excessive difference is first observed. When the difference falls back within the threshold, a reset output is issued.

Because transient effects may produce short-term amplitude detector trips, a timing threshold is applied to the output of each amplitude detector. Typically, a given number of consecutive out-of-tolerance amplitude threshold trips are necessary to declare a faulty signal. Hence, a time duration threshold detector begins a count whenever a set signal is received and, in the absence of further inputs, increments the count for each sample interval thereafter. If a given cycle count is exceeded, an erroneous state is declared and a fault logic signal is set for the affected channels. Otherwise, the count is returned to zero when a reset signal is received.

The setting of both the timing and amplitude thresholds is of crucial importance because of the trade-off between nuisance fault logic trips and slow response to actual faults. Nuisance trips erode user confidence in a system; their unwarranted trips can potentially cause resource depletion. On the other hand, a belated fault response may permit an unsafe condition or catastrophic event to occur. The allowable time to recover from a given type of fault, which is application dependent, is the key to setting the thresholds properly. The degree of channel synchronization and data skewing also affects the threshold settings, because they must accommodate any looseness. The trade-off can become quite problematic when fast fault recovery is required.

Because the integrity and functionality of the system is at stake, the detailed design of a voter–comparator must be subject to careful assessment at all stages of development. In the case of a hardware-implemented device, its fault detection aspects must be thoroughly examined. The main concern is passive failures in circuitry that is not normally used. Built-in test, self-monitoring, or fail-hard symptoms are customary approaches to device integrity. In the case of software-implemented voter comparators, their dependability can be reinforced through formal proof methods and in-service verified code.

5.3.2 Watchdog Timers

Watchdog timers can be used to catch both hardware and software wandering into undesirable states (Lala and Harper, 1994). Timing checks are a form of assertion checking. This kind of check is useful because many software and hardware errors are manifested in excessive time taken for some operation. In synchronous data flow architectures, data are to arrive at a specific time. Data transmission errors of this type can be detected using a timer.

5.4 Software-Implemented Fault Tolerance: State Consistency

Software performs a critical role in digital systems. The term "software-implemented fault tolerance" is used in this chapter in the broader sense, indicating the role software plays in the implementation of fault tolerance, and not as a reference to the SRI International project performed for NASA in the late 1970s and referred to as SIFT.

5.4.1 Error Detection

Software plays a major role in error detection. Error detection at the system level should be based on the specification of system behavior. The outputs of the system should be checked independent of the system to assure that the outputs conform to the specification. Since they are implemented in software, the checks require access to the information to be checked and therefore may have the potential of corrupting that information. Hence, the independence between a system and its check cannot be absolute. The provision of ideal checks for error detection is rarely practical, and most systems employ checks for acceptability (Anderson and Lee, 1981).

Deciding where to employ system error detection is not a straightforward matter. Early checks should not be regarded as a substitute for last-moment checks. An early check will of necessity be based on knowledge of the internal workings of the system and hence will lack independence from the system. An early check could detect an error at the earliest possible stages and minimize the spread of damage; a last-moment check ensures that none of the output of the system remains unchecked. Therefore, both last-moment and early checks should be provided in a system (Anderson and Lee, 1981).

In order to detect software faults, it is necessary that the redundant versions of the software be independent of each other, that is, of diverse design (Avizienis and Kelly, 1982) (see Section 5.5).

5.4.1.1 Replication Checks

If design faults are expected, replication must be provided using versions of the system with different designs. Replication checks compare the two sets of results produced as outputs of the replicated modules. The replication check raises an error flag and initiates the start of other processes to determine which component or channel is in error (Anderson and Lee, 1981).

5.4.1.2 Timing Checks

Timing checks are used to reveal the presence of faults in a system, but not their absence (Anderson and Lee, 1981). In synchronous hard real-time systems, messages containing data are transmitted over data buses at a specific schedule. Failure to receive a message at the scheduled time is an error, which could be caused by faults in a sensor or data bus, for example. In this case, if the data were critical, a method of tolerating the fault may be to use a forward state extrapolation.

5.4.1.3 Reversal Check: Analytical Redundancy

A reversal check takes the outputs from a system and calculates what the inputs should have been to produce that output. The calculated inputs can then be compared with the actual inputs to check whether there is an error. Systems providing mathematical functions often lend themselves to reversal checks (Anderson and Lee, 1981).

Analytic redundancy using either of two general error detection methods—multiple model (MM) or generalized likelihood ratio (GLR)—is a form of reversal check. Both methods make use of a model of the system represented by Kalman filters. The MM attempts to calculate a measure of how well each of the Kalman filters is tracking by looking at the prediction errors. Real systems possess nonlinearity and the model assumes a linear system. The issue is whether the tracking error from the extended Kalman filter corresponds to the linearized model "closest to" the true, nonlinear system and is markedly smaller than the errors from the filters based on "more distant" models. Actuator and sensor failures can be modeled in different ways using this methodology (Willsky, 1980).

The GLR uses a formulation similar to that for MM, but different enough that the structure of the solution is quite different. The starting point for GLR is a model describing normal operation of the observed signals or of the system from which they come. Since GLR is directly aimed at detecting abrupt changes, its applications are restricted to problems involving such changes, such as failure detection. GLR, in contrast to MM, requires a single Kalman filter. Any detectable failure will exhibit a systematic deviation between what is observed and what is predicted to be observed. If the effect of the parametric failure is "close enough" to that of the additive one, the system will work.

Underlying both the GLR and MM methods is the issue of using system redundancy to generate comparison signals that can be used for the detection of failures. The fundamental idea involved in finding comparison signals is to use system redundancy—known relationships among measured variables to generate signals that are small under normal operation and display predictable patterns when particular anomalies develop. All failure detection is based on analytical relationships between sensed variables, including voting methods, which assume that sensors measure precisely the same variable. Using analytical relationships, we can reduce hardware redundancy and maintain the same level of fail operability. In addition, analytical redundancy allows extracting more information from the data, permitting detection of subtle changes in system component characteristics. On the other hand, the use of this information can cause problems if there are large uncertainties in the parameters specifying the analytical relationships (Willsky, 1980).

The second part of a failure detection algorithm is the decision rule, which uses the available comparison signals to make decisions on the interruption of normal operation by the declaration of failures. One advantage of these methods is that the decision rule—maximized and compared to a threshold—is simple, while the main disadvantage is that the rule does not explicitly reflect the desired trade-offs. The Bayesian sequential decision approach, in which an algorithm is used for the calculation of the approximate Bayes decision, has exactly the opposite properties—that is, it allows for a direct incorporation of performance trade-offs—but it is extremely complex. The Bayes sequential decision problem is to choose a stopping rule and terminal decision rule to minimize the total expected cost and the expected cost that is accrued before stopping (Willsky, 1980).

5.4.1.4 Coding Checks

Coding checks are based on redundancy in the representation of an object in use in a system. Within an object, redundant data are maintained in some fixed relationship with the (nonredundant) data representing the value of the object. Parity checks are a well-known example of a coding check, as are error detection and correction codes such as the Hamming, cyclic redundancy check, and arithmetic codes (Anderson and Lee, 1981).

5.4.1.5 Reasonableness Checks

These checks are based on knowledge of the minimum and maximum values of input data, as well as the limits on rate of change of input data. These checks are based on knowledge of the physical operation of sensors and employ models of this operation.

5.4.1.6 Structural Checks

Two forms of checks can be applied to the data structures in a computing system. Checks on the semantic integrity of the data will be concerned with the consistency of the information contained in a data structure. Checks on the structural integrity will be concerned with whether the structure itself is consistent. For example, external data from subsystems are transmitted from digital data buses such as MIL-STD-1553, ARINC 429, or ARINC 629. The contents of a message (number of words in the message, contents of each word) from a subsystem are stored and the incoming data checked for consistency.

5.4.1.7 Diagnostic Checks

Diagnostic checks create inputs to the hardware elements of a system, which should produce a known output. These checks are rarely used as the primary error detection measure. They are normally run at startup and may be initiated by an operator as part of a built-in test. They may also run continuously in a background mode when the processor might be idle. Diagnostic checks are also run to isolate certain faults.

5.4.2 Damage Confinement and Assessment

When an error has been discovered, it is necessary to determine the extent of the damage done by the fault before error recovery can be accomplished. Assessing the extent of the damage is usually related to the structure of the system. Assuming timely detection of errors, the assessment of damage is usually determined to be limited to the current computation or process. The state is assumed consistent on entry. An error detection test is performed before exiting the current computation. Any errors detected are assumed to be caused by faults in the current computation.

5.4.3 Error Recovery

After the extent of the damage has been determined, it is important to restore the system to a consistent state. There are two primary approaches—backward and forward error recovery. In backward error recovery, the system is returned to a previous consistent state. The current computation can then be retried with existing components (retry) and with alternate components (reconfigure), or it can be ignored (skip frame). The use of backward recovery implies the ability to save and restore the state. Backward error recovery is independent of damage assessment. Forward error recovery attempts to continue the current computation by restoring the system to a consistent state, compensating for the inconsistencies found in the current state. Forward error recovery implies detailed knowledge of the extent of the damage done and a strategy for repairing the inconsistencies. Forward error recovery is more difficult to implement than backward error recovery (Hitt et al., 1984).

5.4.4 Fault Treatment

Once the system has recovered from an error, it may be desirable to isolate and correct the component that caused the error. Fault treatment is not always necessary because of the transient nature of some faults or because the detection and recovery procedures are sufficient to cope with other recurring errors. For permanent faults, fault treatment becomes important because the masking of permanent faults reduces the ability of the system to deal with subsequent faults. Some fault-tolerant software techniques attempt to isolate faults to the current computation by timely error detection. Having isolated the fault, fault treatment can be done by reconfiguring the computation to use alternate forms of the computation to allow for contin-ued service. (This can be done in series, as in recovery blocks, or in parallel, as in N-version programming.) The assumption is that the damage due to faults is properly encapsulated to the current computation and that error detection itself is faultless (i.e., it detects all errors and causes none of its own) (Hitt et al., 1984).

5.4.5 Distributed Fault Tolerance

Multiprocessing architectures consisting of computing resources interconnected by external data buses should be designed as a distributed fault-tolerant system (Best et al., 1988). The computing resources may be installed in an enclosure using a parallel backplane bus to implement multiprocessing within the enclosure. Each enclosure can be considered a virtual node in the overall network. A network operating system, coupled with the data buses and their protocol, completes the fault-tolerant distributed system. The architecture can be asynchronous, loosely synchronous, or tightly synchronous. Maintaining con-sistency of data across redundant channels of asynchronous systems is difficult (Papadopoulos, 1985).

5.5 Software Fault Tolerance

Software faults, considered design faults, may be created during the requirements development, specifica-tion creation, software architecture design, code creation, and code integration. While many faults may be found and removed during system integration and testing, it is virtually impossible to eliminate all possible software design faults. Consequently, software fault tolerance is used. Table 5.5 lists the major fault-tolerant software techniques in use today. The two main methods that have been used to provide software fault tolerance are N-version software and recovery blocks (Hudak, 1993).

5.5.1 Multiversion Software

Multiversion software is any fault-tolerant software technique in which two or more alternate versions are implemented and executed and the results are compared using some form of a decision algorithm. The goal is to develop these alternate versions such that software faults that may exist in one version are not contained in the other version or versions and the decision algorithm determines the correct value from among the alternate versions. Whatever means are used to produce the alternate versions, the common goal is to have distinct versions of software such that the probability of faults occurring simultaneously is small and that faults are distinguishable when the results of executing the multiversions are compared with each other.

The comparison function executes as a decision algorithm once it has received results from each ver-sion. The decision algorithm selects an answer or signals that it cannot determine an answer. This decision

TABLE 5.5 Categorization of Fault-Tolerant Software Techniques

Multiversion software
    N-version program
    Cranfield algorithm for fault-tolerance (CRAFT) food taster
    Distinct and dissimilar software
Recovery blocks
    Deadline mechanism
    Dissimilar backup software
Exception handlers
    Hardened kernel
    Robust data structures and audit routines
    Run-time assertions[a]
Hybrid multiversion software and recovery block techniques
    Tandem
    Consensus-recovery blocks

Source: Hitt, E. et al., *Study of Fault-Tolerant Software Technology*, NASA CR 172385, Langley Research Center, Hampton, VA, 1984.

[a] Not a complete fault-tolerant software technique as it only detects errors.

algorithm (Gu et al., 1994), and the development of the alternate versions constitute the primary error detection method. Damage assessment assumes the damage is limited to the encapsulation of the individual software versions. Faulted software components are masked so that faults are confined within the module in which they occur. Fault recovery of the faulted component may or may not be attempted.

N-versions of a program are independently created by N-software engineering teams working from a (usually) common specification. Each version executes independently of the other versions. Each version must have access to an identical set of input values, and the outputs are compared by an executive that selects the result used. The choice of an exact or inexact voting check algorithm is influenced by the criticality of the function and the timing associated with the voting.

5.5.2 Recovery Blocks

The second major technique shown in Table 5.5 is the recovery block and its subcategories—deadline mechanism and dissimilar backup software. The recovery block technique recognizes the probability that residual faults may exist in software. Rather than develop independent redundant modules, this technique relies on the use of a software module, which executes an acceptance test on the output of a primary module. The acceptance test raises an exception if the state of the system is not acceptable. The next step is to assess the damage and recover from the fault. Given that a design fault in the primary module could have caused arbitrary damage to the system state and that the exact time at which errors were generated cannot be identified, the most suitable prior state for restoration is the state that existed just before the primary module was entered (Anderson and Lee, 1981).

5.5.3 Trade-Offs

Coverage of a large number of faults has an associated overhead in terms of redundancy and the processing associated with the error detection. The designer may use modeling and simulation to determine the amount of redundancy required to implement the fault tolerance versus the probability and impact of the different types of faults. If a fault has minimal or no impact on safety or mission completion, investing in redundancy to handle that fault may not be effective, even if the probability of the fault occurring is significant.

5.6 Summary

Fault-tolerant systems must be used whenever a failure can result in loss of life or loss of a high-value asset. Physical failures of hardware decreased when analog avionics were replaced with the first- and second-generation digital avionics. The third-generation digital avionics use of highly integrated multifunction COTS ICs and circuit cards implementing multiple functions has more complex failure mechanisms and may, at the card level, have higher failure rates than the second-generation digital avionics. Physical and design faults are virtually impossible to completely eliminate, so increased reliance on fault tolerance must occur to meet safety mandates.

5.6.1 Design Analyses

In applying fault tolerance to a complex system, there is a danger that the new mechanisms may introduce additional sources of failure due to design and implementation errors. It is important, therefore, that the new mechanisms be introduced in a way that preserves the integrity of a design with minimum added complexity. The designer must use modeling and simulation tools to assure that the design accomplishes the needed fault tolerance.

Certain design principles have been developed to simplify the process of making design decisions. Encapsulation and hierarchy offer ways to achieve simplicity and generality in the realization of particular fault-tolerance functions. Encapsulation provides the following:

- Organization of data and programs as uniform objects, with rigorous control of object interaction
- Organization of sets of alternate program versions into fault-tolerant program modules (e.g., recovery blocks and N-version program sets)
- Organization of consistent sets of recovery points for multiple processes
- Organization of communications among distributed processes as atomic (indivisible) actions
- Organization of operating system functions into recoverable modules

The following are examples of the hierarchy principle used to enhance reliability of fault-tolerance functions:

- Organization of all software, both application and system type, into layers with unidirectional dependencies among layers
- Integration of service functions and fault-tolerance functions at each level
- Use of nested recovery blocks to provide hierarchical recovery capability
- Organization of operating system functions so that only a minimal set at the lowest level (a "kernel") needs to be exempted from fault tolerance
- Integration of global and local data and control in distributed processors

That portion of the operating system kernel that provides the basic mechanisms the rest of the system uses to achieve fault tolerance should be "trusted." This kernel should be of limited complexity so that all possible paths can be tested to assure correct operation under all logic and data conditions. This kernel need not be fault tolerant if the foregoing can be assured.

5.6.2 Safety

Safety is defined in terms of hazards and risks. A hazard is a condition or set of conditions that can produce an accident under the right circumstances. The level of risk associated with the hazard depends on the probability that the hazard will occur, the probability of an accident taking place if the hazard does occur, and the potential consequence of the accident (Williams, 1992).

MIL-STD-882 is the primary guidance for Department of Defense weapons systems' safety. Safety is designed into the system via engineering as well as through management control of risks, and safety requirements are balanced with other requirements for mission success. There have been instances of use of civil standards to accomplish a level of confidence in military equipment. This must be done with judgment of balancing safety with performance requirements necessary for that equipment. More flexibility in consideration of system safety should be acceptable for military aviation to account for differences in operational environment, use of cutting-edge technology, use of weapons, use of self-defense, and focus on mission.

For civil aviation, safety is imposed a legal recognition of "certification" by regulators. There are three aspects that are required throughout the world for avionics:

1. The system performs its intended function.
2. The system as designed does not represent or contain a single point of failure.
3. The system has been shown, to the extent possible, to be free of unintended function.

In civil aviation, avionics system development is accomplished using various design assurance standards or equivalent alternate methods. SAE Aerospace Recommended Practice (ARP) 4754 is a system development methodology used to systematically find and mitigate hazards and improve the architecture and design in the system development process. This document also introduces the notion of "gradation" of development effort matching with the consequences of faults in specific systems. These notions of gradation, fault avoidance, and fault tolerance are further continued in RTCA DO-178C for airborne software development and RTCA DO-254 for airborne electronic hardware development. Specific uses of avionics such as unmanned aerial systems that interface with ground control for safe flight have specific hazards and specific fault-tolerance issues. Similarly, specific architectures such as integrated modular avionics have drawn attention to common cause and CMFs.

5.6.3 Validation

Validation is the process by which systems are shown through operation to satisfy the specifications. The validation process begins when the specification is complete. The difficulty of developing a precise specification that will never change has been recognized. This reality has resulted in an iterative development and validation process. Validation requires developing test cases and executing these test cases on the hardware and software comprising the system to be delivered. The tests must cover 100% of the faults the system is designed to tolerate and a very high percentage of possible design faults, whether hardware, software, or the interaction of the hardware and software during execution of all possible data and logical paths. Once the system has been validated, it can be put in operation. In order to minimize the need to validate a complete operational flight program (OFP) every time it is modified, development methods attempt to decompose a large system into modules that are independently specified, implemented, and validated. Only those modules and their interfaces that are modified must be revalidated using this approach (see Chapter 28).

Rapid prototyping, simulation, and animation are all techniques that help validate the system. Formal methods are being used to develop and validate digital avionics systems. There are arguments both in favor of and against the use of formal methods (Rushby, 1993; Williams, 1992).

5.6.4 Conclusion

For safety-critical systems, fault tolerance must be used to tolerate design faults that are predominately software and timing related. It is not enough to eliminate almost all faults introduced in the later stages of a life cycle; assurance is needed that they have been eliminated or are extremely improbable. Safety requirements for commercial aviation dictate that a failure causing loss of life must be extremely improbable—on the order of 10^{-9} per flight hour. The designer of safety-critical, fault-tolerant systems should keep current with new development in this field since both design and validation methods continue to advance in capability.

Further Reading

A good introduction to fault-tolerant design is presented in *Fault Tolerance, Principles and Practices*, by Tom Anderson and P. A. Lee. Hardware-specific design techniques are described by P. K. Lala (1985) in *Fault Tolerant and Fault Testable Hardware Design*. Other excellent journals are the IEEE publications *Computer, IEEE Micro, IEEE Software, IEEE Transactions on Computers*, and *IEEE Transactions on Software Engineering*.

References

Anderson, T. and Lee, P.A., *Fault Tolerance, Principles and Practices*, Prentice Hall, London, U.K., 1981.

Avizienis, A., Fault-tolerant systems, *IEEE Trans. Comput.*, C-25(12):1304–1312, 1976.

Avizienis, A. and Kelly, J., *Fault-Tolerant Multi-Version Software: Experimental Results of a Design Diversity Approach*, UCLA Computer Science Department, Los Angeles, CA, 1982.

Avizienis, A., Kopetz, H., and Laprie, J.C., Eds., *The Evolution of Fault-Tolerant Computing*, Springer-Verlag, New York, 1987.

Best, D.W., McGahee, K.L., and Shultz, R.K.A., *Fault Tolerant Avionics Multiprocessing System Architecture Supporting Concurrent Execution of Ada Tasks*, Collins Government Avionics Division, AIAA 88-3908-CP, 1988.

Constantinescu, C., Impact of deep submicron technology on dependability of VLSI circuits, in *Proc. Dependable Systems and Networks*, 2002.

Driscoll, K. et al., Byzantine fault tolerance, from theory to reality, *Int. Conf. Computer Safety, Reliability and Security*, 2004.

Gu, D., Rosenkrantz, D.J., and Ravi, S.S., Construction of check sets for algorithm-based fault tolerance, *IEEE Trans. Comput.*, 43(6):641–650, 1994.

Hill, J. and Knight, J.C., *Selective Notification: Combining Forms of Decoupled Addressing for Internet-Scale Command and Alert Dissemination*, Computer Science Department, University of Virginia, Charlottesville, VA, 2003.

Hitt, E., Webb, J., Goldberg, J., Levitt, K., Slivinski, T., Broglio, C., and Wild, C., *Study of Fault-Tolerant Software Technology*, NASA CR 172385, Langley Research Center, Hampton, VA, 1984.

Hudak, J., Suh, B.H., Siewiorek, D., and Segall, Z., Evaluation and comparison of fault tolerant software techniques, *IEEE Trans. Reliability*, 1993.

Knight, J. et al., The Willow Architecture: Comprehensive Survivability for Large-Scale Distributed Applications, Intrusion Tolerance Workshop, DSN-2002, *International Conference on Dependable Systems and Networks*, Washington, DC, June 2002.

Lala, J.H. and Harper, R.E., Architectural principles for safety-critical real-time applications, *Proc. IEEE*, 82(1):25–40, 1994.

Lala, P.K., *Fault Tolerant and Fault Testable Hardware Design*, Prentice Hall, London, U.K., 1985.

Papadopoulos, G.M., Redundancy Management of Synchronous and Asynchronous Systems, Fault Tolerant Hardware/Software Architecture for Flight Critical Functions, AGARD-LS-143, 1985.

Rushby, J., *Formal Methods and Digital Systems Validation for Airborne Systems*, NASA CR 4551, Langley Research Center, VA, 1993.

Scandura, P. and Garcia-Galan, C., A unified system to provide crew alerting electronic checklists and maintenance using IVHM, in *Proceedings of the 23rd Digital Avionics Systems Conference*, 2004.

Shin, K.G. and Hagbae, K., A time redundancy approach to TMR failures using fault-state likelihoods, *IEEE Trans. Comput.*, 43(10):1151–1162, 1994.

Shin, K.G. and Parmeswaran, R., Real-time computing: A new discipline of computer science and engineering, *Proc. IEEE*, 82(1):6–24, 1994.

Sosnowski, J., Transient fault tolerance in digital systems, *IEEE Micro*, 14(1):24–35, 1994.

Tomek, L., Mainkar, V., Geist, R.M., and Trivedi, K.S., Reliability modeling of life-critical, real-time systems, *Proc. IEEE*, 82(1):108–121, 1994.

Vaidya, N.H. and Pradhan, D.K., Fault-tolerant design strategies for high reliability and safety, *IEEE Trans. Comput.*, 42(10):1195–1206, 1993.

Williams, L.G., *Formal Methods in the Development of Safety Critical Software Systems*, UCRL-ID-109416, Lawrence Livermore National Laboratory, Livermore, CA, 1992.

Willsky, A.S., Failure Detection in Dynamic Systems, Fault Tolerance Design and Redundancy Management Techniques, AGARD-LS-109, 1980.

6

Electromagnetic Environment

Richard Hess
*Honeywell Aerospace**

6.1 Introduction

The advent of digital electronic technology in electrical/electronic systems has enabled unprecedented expansion of aircraft system functionality and evolution of aircraft function automation. As a result, systems incorporating such technology are used more and more to implement aircraft functions, including Level A systems that affect the safe operation of the aircraft; however, such capability is not free. The electromagnetic environment (EME) is a form of energy that is the same type (electrical) used by electrical/electronic equipment to process and transfer information. As such, this environment represents a fundamental threat to the proper operation of systems that depend on such equipment. It is a common mode threat that can defeat fault-tolerant strategies reliant upon redundant electrical/electronic systems.

Electrical/electronic systems, characterized as Level A, provide functions that can affect the safe operation of an aircraft and depend upon information (i.e., guidance, control, etc.) processed by electronic equipment. Thus, the EME threat to such systems may translate to a threat to the airplane itself. The computers associated with modern aircraft guidance and control systems are susceptible to upset from lightning and sources that radiate radio frequencies (RF) at frequencies predominantly between 1 and several GHz and produce aircraft internal field strengths of 5–200 V/m or greater. Internal field strengths >200 V/m are usually periodic pulses with pulsewidths <10 ms. Internal lightning-induced voltages and currents can range from approximately 50 V and 20 A to over 3000 V and 5000 A.

Electrical/electronic system susceptibility to such an environment has been the suspected cause of nuisance disconnects, hardovers, and upsets. Generally, this form of system upset occurs at significantly lower levels of electromagnetic (EM) field strength than that which could cause component failure, leaves no trace, and is usually nonrepeatable.

---

* This is a reprinted chapter with minor editorial changes and changes due to technology or regulatory evolution by technical editors since the original author could not be reached. The author's affiliation is from the original contribution; current affiliation of the author is unknown.

6.2 EME Energy Susceptibility

It is clear that the sources of EM threats to the electrical/electronic system, whether digital or analog, are numerous. Although both respond to the same threats, there are factors that can make the threat response to a momentary transient (especially intense transients like those that can be produced by lightning) far more serious in digital processing systems than in analog systems. For example, the information bandwidth, and therefore the upper noise response cutoff frequency, in analog devices is limited to, at most, a few MHz. In digital systems, it is often in excess of 100 MHz and continues to increase. This bandwidth difference, which is at least 10 times more severe in digital systems, allows substantially more EM energy of many types (modulated, pulse, etc.) to be coupled into the digital system. Moreover, the bandwidths of analog circuits associated with autopilot and flight management systems are on the order of 50 Hz for servo loops and much less for other control loops (<1 Hz for outer loops). Thus, if the disturbance is short relative to significant system time constants, even though an analog circuit device possessing a large gain and a broad bandwidth may be momentarily upset by an electromagnetic transient, the circuit will recover to the proper state. It should be recognized that, to operate at high speeds, proper circuit card layout control and application of high-density devices is a must. When appropriate design tools (signal integrity, etc.) are applied, effective antenna loop areas of circuit card tracks become extremely small, and the interfaces to a circuit card track (transmission line) are matched. Older (1970s–1980s) technology with wirewrap backplanes and processors built with discrete logic devices spread over several circuit cards were orders of magnitude more susceptible. Unlike analog circuits, digital circuits and corresponding computational units, once upset, may not recover to the proper state and may require external intervention to resume normal operation. It should be recognized that (for a variety of reasons) large-gain bandwidth devices are and have been used in the digital computing platforms of aircraft systems. A typical discrete transistor can be upset with 10^{-5} J (2000 V at 100 µA for 50 µs). The energy to upset a "typical" integrated circuit can range from 10^{-9} J (20 V at 1 µA for 50 µs), to $<10^{-10}$ J (5 V at <0.4 µA for 50 µs). As time goes on and processor semiconductor junction feature sizes get smaller and smaller, this problem becomes worse.

It should be noted that, in addition to upset, lightning-induced transients appearing at equipment interfaces can, because of the energy they possess, produce hard faults (i.e., damage circuit components) in interface circuits of either analog or digital equipment. For instance, mechanical, electromechanical, and electrohydraulic elements associated with conventional (not electronic primary flight controls with associate "smart" actuators) servo loops and control surface movement are either inherently immune or vastly more robust when exposed to EME energy effects than the electronic components in an electrical/electronic system.

Immunity of electronic components to damage is a consideration that occurs as part of the circuit design process. The circuit characteristic (immunity to damage) is influenced by a variety of factors:

1. Circuit impedances (resistance, inductance, capacitance), which may be distributed as well as lumped
2. The impedances around system component interconnecting loops along with the characteristic (surge) impedance of wiring interfacing with circuit components
3. Properties of the materials used in the construction of a component (e.g., thick-film/thin-film resistors)
4. Threat level (open circuit voltage/short circuit current), resulting in a corresponding stress on insulation, integrated circuit leads, PC board trace spacing, etc.
5. Semiconductor device nonlinearities (e.g., forward biased junctions, channel impedance, junction/gate breakdown)

Immunity to upset for analog processors is achieved through circuit design measures, and for digital processors it is achieved through architectural as well as circuit design measures.

6.2.1 Soft Faults

Digital circuit upset, also known in the avionics digital computer and information processing community as a "soft fault," is a condition known to occur even in relatively benign operating environments. Soft faults occur despite the substantial design measures (timing margins, transmission line interconnects, ground and power planes, clock enablers of digital circuits) to achieve a relatively high degree of integrity in digital processor operation.

In a normal operating environment, the occurrence of soft faults within digital processing systems is relatively infrequent and random. Such occasional upset events should be treated as probabilistic in nature and can be the result of the following:

- Coincidence of EME energy with clocked logic clock edges, etc.
- Occasional violation of a device's operational margin (resulting margin from the design, processing, and manufacturing elements of the production cycle)

From this perspective, the projected effect of a substantial increase in the severity of the electromagnetic environment will be an increased probability of a soft fault occurrence. That is, in reality, a soft fault may or may not occur at any particular point in time; but, on the average, soft faults will occur more frequently with the new environmental level.

Once developed, software is "burned into nonvolatile" memory (becomes "firmware"); the result will be a special-purpose real-time digital-electronic-technology data processing machine with the inherent potential for "soft faults." Because it is a hardware characteristic, this potential exists even when a substantial amount of attention is devoted to developing an "error-free" operating system and application programs (software) for the general purpose digital machine (computing platform, digital engine, etc.).

6.2.2 MTBUR/MTBF

In the past, service experience with digital systems installed on aircraft has indicated that the confirmed failure rates equal or exceed predicted values that were significantly better than previous generation analog equipment. However, the unscheduled removal rate remains about the same. In general, the disparity in Mean Time Between Unscheduled Removal (MTBUR) and the Mean Time Between Failure (MTBF) continues to be significant. The impact of this disparity on airline direct operating costs is illustrated in Figure 6.1.

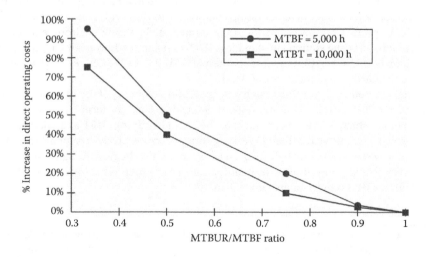

FIGURE 6.1   MTBUR/MTBF ratio impact of operating costs.

To the extent that soft faults contribute to the MTBUR/MTBF disparity, any reduction in soft fault occurrence and propagation could translate into reduction of this disparity.

6.3 Civil Airworthiness Authority Concerns

The following groups have identified the lightning and High-Intensity Radiated Field (HIRF) elements of the EME as a safety issue for aircraft functions provided by electrical/electronic systems:

- Federal Aviation Administration (FAA)
- Joint Aviation Authorities (JAA): Non-European Union*
- European Aviation Safety Agency (EASA): European Union

The following factors, identified by the FAA and JAA, have led to this concern about lightning and HIRF effects:

- Increased reliance on electrical and electronic systems to perform functions that may be necessary for the continued safe flight and landing of the aircraft.
- Reduction of the operating power level of electronic devices that may be used in electrical and electronic systems, which may cause circuits to be more reactive to induced lightning and RF voltages and currents leading to malfunction or failure.
- Increased percentage of composite materials in aircraft construction. Because of their decreased conductivity, composite materials may result in less inherent shielding by the aircraft structure.
- Since current flowing in the lightning channel will be forced (during lightning attachment) into and through the aircraft structure without attenuation, decreased conductivity for aircraft structure materials can be particularly troubling for lightning.

The direct effects of lightning (dielectric puncture, blasting, melting, fuel ignition, etc.) have been recognized as flight hazards for decades, and in 1972 the Society of Automotive Engineers (SAE) formed the AE4 Special Task F (which later became AE4L and is now AE2) to address this issue. In the early 1980s, the FAA began developing policy relative to the effects of lightning on electrical/electronic systems (indirect effects), and AE4L supported the FAA and JAA by providing the technical basis for international standards (rules and regulations) and guidance material that, for aircraft type certification, would provide an acceptable means for demonstrating compliance to those rules and regulations. AE4L also supported RTCA Special Committee 135 (SC-135) to integrate lightning environment conditions and test procedures into airborne equipment standards (DO-160) and the European Organization for Civil Aviation Equipment (EUROCAE) standards counterpart (ED-14). In 1987, EUROCAE formed Working Group 31 to be the European counterpart of AE4L.

In 1986, the FAA and JAA identified High-Energy Radio Frequency (HERF) electromagnetic fields as an issue for aircraft electrical/electronic systems. Some time later, the term HERF was changed to its present designation, which is HIRF. Subsequent to the FAA identifying HIRF as a safety issue, SAE and EUROCAE formed Committee AE4R and Working Group 33, respectively, to support the FAA and JAA in much the same way as AE4L and lightning. In addition, unlike the case for lightning, RTCA SC-135 formed a HIRF working group (the corresponding European group was already part of EUROCAE/WG33) to integrate HIRF requirements into DO-160/ED-14.

---

* At the time of publication of this handbook, EASA has subsumed the regulatory functions of JAA. Currently, there are activities within the regulatory and industry bodies concerning electromagnetic environment because of new electrical systems (lithium batteries, electrical wiring interconnection system, high intensity radiated field and lightning, power supply systems for portable electronic devices, wireless local area network, in flight entertainment, global system for mobile communication, etc.).

Prior to the existence of the rule for lightning and HIRF, special conditions were issued to applicants for aircraft-type certification (TC, STC, Amended Type Certificate [ATC]). What follows is the rationale for the special condition which still holds true as the rationale for the rules for lightning and HIRF:

These series of aircraft will have novel or unusual design features associated with the installation of new technology electrical and electronic systems, which perform critical or essential functions. The applicable airworthiness regulation does not contain adequate or appropriate safety standards for the protection of these systems from the effects of lightning and radio frequency energy. This notice contains the additional safety standards that the Administrator considers necessary to ensure that critical and essential functions the new technology electrical and electronic systems perform are maintained when the airplane is exposed to lightning and RF energy.

The FAA established the Aviation Rule-Making Advisory Committee, which in turn established the Electromagnetic Effects Harmonization Working Group (EEHWG) to develop the rule-making package for HIRF and for amendments to the lightning rules. Presently, the code of federal regulations associated with the FAA have been updated to include HIRF (Title 14, Code of Federal Regulations [14 CFR] §§ 23.1308, 25.1317, 27.1317, and 29.1317). Advisory Circulars have been released to address these topics.

Regarding portable electronic devices (PEDs) such as tablets, smartphones, e-readers, and MP3 players, in 1992, the FAA requested the RTCA to study the EME produced by PEDs. In response to an FAA request, RTCA formed Special Committee 177 (SC-177) in 1992. In 1996, SC-177 issued a report titled "Portable Electronic Devices Carried Onboard Aircraft" (DO-233). At that point, control of PEDs and their associated EM emissions were handled by integrating some of the RTCA recommendations into airline policy regarding instructions given to passengers (prohibiting personal cellular phone use, asking passengers to turn off PEDs during taxi, take-off, and landing, etc.). SC-202, reestablished in 2003, reexamined the PED issue. With the October 2004 publication of its Phase 1 report, Guidance on Allowing Transmitting Portable Electronic Devices (T-PEDs) on Aircraft (RTCA document DO-294), SC-202 has developed testing procedures to assess the risk of interference for particular PEDs onboard aircraft, and operational procedures for mitigating operational impacts (human factors). In Phase 2, the committee focused on emerging PED technologies, for example, ultra-wideband devices or pico-cells for telephone use on board aircraft. The final report—DO-294B, Guidance on Allowing Transmitting Portable Electronic Devices (T-PEDs) on Aircraft, was issued on December 13, 2006. On October 11, 2007, the committee published a second guidance document DO-307, Aircraft Design and Certification for Portable Electronic Device (PED) Tolerance. This document gives the aircraft design and certification criteria to tolerate the operation of PEDs. These aircraft design and certification recommendations, when implemented, would reduce the need for restricting the use of PEDs. The FAA also formulated PED Aviation Rulemaking Committee (ARC) that concluded in a report released on September 30, 2013, that most commercial airplanes can tolerate radio interference signals from PEDs. Both the FAA and the EASA have issued guidance for the use of PEDs in "aircraft mode" or "flight mode" in specific phases of flight. The final decision, however, is up to the aircraft owner/operator to decide how they want to incorporate this guidance into their policy. There are many operational and technical concerns related to onboard PED usage. These include passenger compliance to PED operating mode restrictions, distraction from emergency announcements due to PED usage, and disturbance to surrounding passengers.

6.3.1 EME Compliance Demonstration for Electrical and Electronic Systems

FAA, JAA, and EASA FARs and JARs require compliance demonstration either explicitly or implicitly for the following EME elements:

- Lightning
- HIRF (FAA)
- HIRF (JAA/EASA)
- Electromagnetic Compatibility (EMC)

TABLE 6.1 Nomenclature Cross Reference between AC25.1309 and SAE ARP4754[a]

Failure Condition Classification	Development Assurance Level
Catastrophic	Level A
Severe major/hazardous	Level B
Major	Level C
Minor	Level D
No effect	Level E

[a] ARP 4754 has been updated to ARP 4754A but still keeping the basic perspective of five levels.

At the aircraft level, the emphasis should be on lightning and HIRF because most of the energy and system hazards arise from these threats. Their interaction with aircraft systems is global and also the most complex, requiring more effort to understand. Intrasystem electromagnetic emissions fall under the broad discipline of EMC. PEDs are a source of EM emissions that fall outside of the categories of equipment normally included in the EMC discipline. Like lightning and HIRF, the interaction of PED emissions with aircraft electrical/electronic systems is complex and could be global.

The electrical and electronic systems that perform functions considered to be flight critical must be identified by the applicant with the concurrence of the cognizant Federal Aviation Administration Aircraft Certification Office (FAA ACO) by conducting a functional hazard assessment and, if necessary, preliminary system safety assessments (see SAE Aerospace Required Practices [SAE ARP4761]). The term critical refers to those functions whose failure would contribute to, or cause, a catastrophic failure condition (loss of aircraft). Table 6.1 provides the relationship between function failure effects and development assurance levels associated with those systems that implement functions that can affect safe aircraft operation.

Terms such as "Level A" designate particular system development assurance levels, which refer to the rigor and discipline of processes used during system development (design, implementation, verification and certification, production, etc.). It was deemed necessary to focus on the development processes for systems based upon "highly integrated" or "complex" (those whose safety cannot be shown solely by test and whose logic is difficult to comprehend without the aid of analytical tools) elements; that is, primarily digital electronic elements.

Development assurance activities are ingredients of the system development processes. As has been noted, systems and appropriate associated components are assigned "development assurance levels" based on failure condition classifications associated with aircraft-level functions implemented by systems and components. The rigor and discipline needed in performing the supporting processes will vary depending on the assigned development assurance level. There is no development process for aircraft functions. Basically, they should be regarded as intrinsic to the aircraft and are categorized by the role they play for the aircraft (control, navigation, communication, etc.). Relative to safety, they are also categorized (from FAA advisory material) by the effect of their failures, such as catastrophic, severe major/hazardous, major and minor.

EMC has been included in FAA regulations since the introduction of radio and electrical/electronic systems into aircraft. Electrical equipment, controls, and wiring must be installed so that operation of any one unit or system of units will not adversely affect the simultaneous operation of any other electrical unit or system essential to aircraft safe operation. Cables must be grouped, routed, and located so that damage to essential circuits will be minimized if there are faults in heavy current-carrying cables. Critical environmental conditions must be considered in demonstrating compliance with aircraft electrical/electronic system safety requirements with respect to radio and electronic equipment and their installations. Radio and electronic equipment, controls, and wiring must be installed so that operation of any one component or system of components will not adversely affect the simultaneous operation of any other radio or electronic unit or system of units required by aircraft functions.

Relative to safety and electrical/electronic systems, the systems, installations, and equipment whose functioning is required for safe aircraft operation must be designed to ensure that they perform their intended functions under all foreseeable operating conditions. Aircraft systems and associated components that are part of any Level A–C function, considered separately and in relation to other systems, must be designed so that the following are true:

- The occurrence of any failure condition that would prevent the continued safe flight and landing of the airplane is extremely improbable.
- The occurrence of any other failure condition that would reduce the capability of the airplane or the ability of the crew to cope with adverse operating conditions is improbable.

6.3.2 EME Energy Propagation

As has been noted in Section 6.1 and illustrated in Figure 6.2, lightning and HIRF are threats to the overall aircraft. Since they are external EME elements, of the two, lightning produces the most intense environment, particularly by direct attachment. Both lightning and HIRF interactions produce internal fields. Lightning can also produce substantial voltage drops across the aircraft structure. Such structural voltages provide another mechanism (in addition to internal fields) for energy to propagate into electrical/electronic systems. Also, the poorer the conductivity of structural materials, the greater the possibility that the following conditions exist:

- Voltage differences across the structure
- Significant lightning diffusion magnetic fields
- Propagation of external environment energy

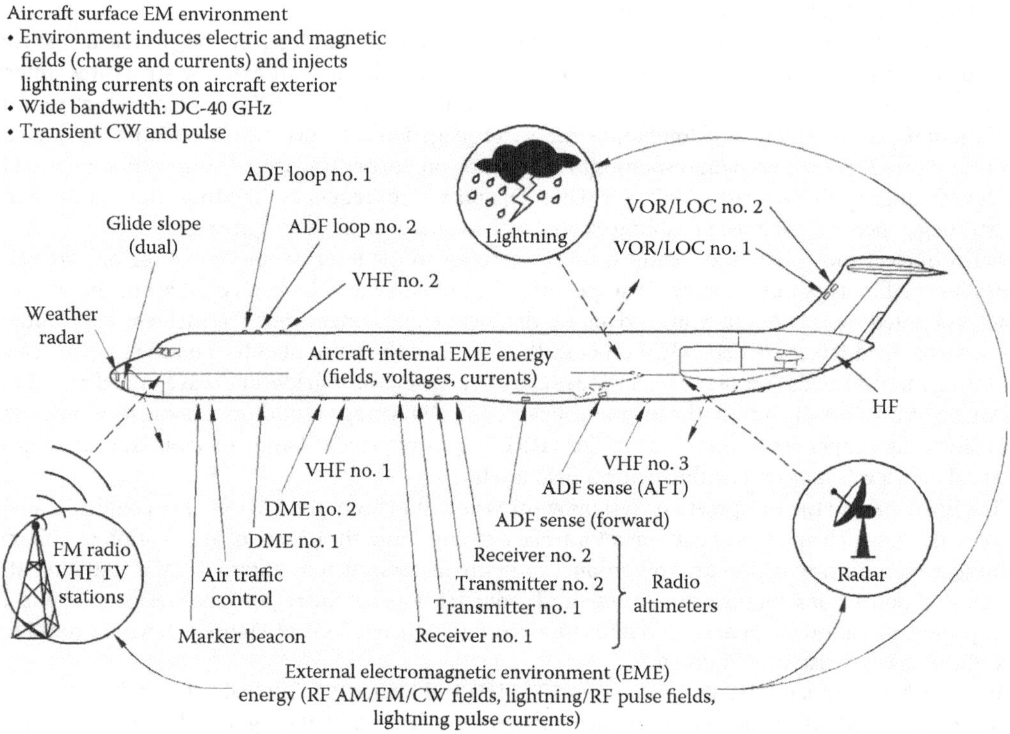

Aircraft surface EM environment
- Environment induces electric and magnetic fields (charge and currents) and injects lightning currents on aircraft exterior
- Wide bandwidth: DC-40 GHz
- Transient CW and pulse

ADF loop no. 1

Glide slope (dual)   ADF loop no. 2   VOR/LOC no. 2

Lightning   VOR/LOC no. 1

VHF no. 2

Weather radar

Aircraft internal EME energy (fields, voltages, currents)

HF

VHF no. 1   VHF no. 3
ADF sense (AFT)

DME no. 2   ADF sense (forward)

DME no. 1

FM radio
VHF TV
stations   Air traffic control   Receiver no. 2
Transmitter no. 2   Radio altimeters   Radar
Transmitter no. 1

Marker beacon   Receiver no. 1

External electromagnetic environment (EME) energy (RF AM/FM/CW fields, lightning/RF pulse fields, lightning pulse currents)

FIGURE 6.2   External EME (HIRF, lightning) interaction.

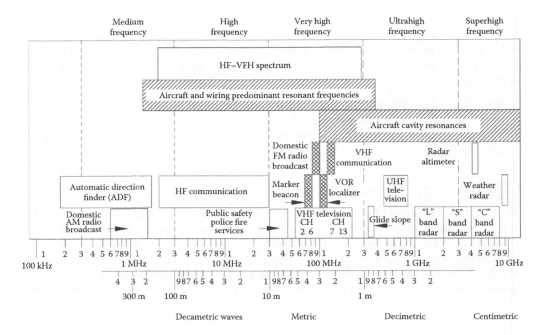

FIGURE 6.3   RF spectrum and associated installation dimensions of interest.

Figure 6.3 gives the HIRF spectrum and associated aircraft/installations features of interest. Not shown are GPS and Mode 5 frequencies (1–2 GHz).

In general, the propagation of external EME energy into the aircraft interior and electrical/electronic systems is a result of complex interactions of the EME with the aircraft exterior structures, interior structures, and system installations (see Figures 6.3 through 6.7). Figure 6.8 gives representative transfer functions, in the frequency domain, of energy propagation into electrical/electronic systems, and Figure 6.9 provides time domain responses to a lightning pulse resulting from transfer functions having the low-frequency characteristic Vo(f) = K(f)[Hi(f)] and a high frequency "moding" (resonant) characteristic (e.g., open loop voltage of cabling excited by a magnetic field; see Figure 6.8).

Paths of electromagnetic wave entry from the exterior to the interior equipment regions are sometimes referred to as points of entry. Examples of points of entry may be seams, cable entries, and windows. As noted, points of entry are driven by the local environment, not the incident environment. The internal field levels are dependent on both the details of the point of entry and the internal cavity. Resulting internal fields can vary over a wide range of intensity, wave shape, and wave impedance. Below 10 MHz within a metal aircraft, the magnetic fields due to lightning predominate because of the electric field shielding properties of metal skins. For HIRF "high-frequency" bands in some internal regions, internal field levels may exceed the incident field levels.

The EME local to the equipment or system within the installation (the EME energy coupled to installation wiring, which appears at equipment interface circuits) and the degree of attenuation or enhancement achieved for any region are the product of many factors, such as external EME characteristics, materials, bonding of structure, dimensions and geometric form of the region, and the location and size of any apertures allowing penetration into the aircraft (G0 through G5 of Figure 6.4, which could have any of the characteristics of Figure 6.8).

In HIRF high-frequency bands (frequencies 100 MHz and higher) the internal field resulting from such influences, as noted, will in most cases produce a nonuniform field within the region or location of the system or equipment. The field cannot be considered as uniform and homogeneous. The field will not necessarily allow the adoption of single-point measurement techniques for the accurate determination of the equivalent

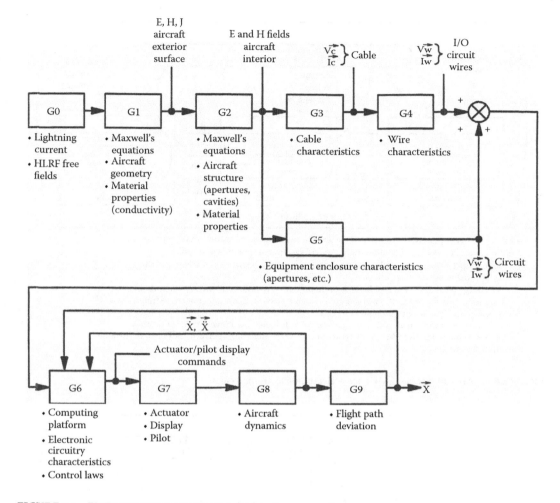

FIGURE 6.4 EME propagation process-transfer function perspective.

internal field to be used as the test level for systems. Several hot spots typically exist within any subsection of the aircraft. This is particularly true at cavity-resonant conditions. Intense local effects are experienced at all frequencies in the immediate vicinity of any apertures for a few wavelengths away from the aperture itself. For apertures that are small with respect to wavelength, measurements of the fields within the aperture would yield fields much larger than those further inside the aircraft because the fields fall off inversely proportional to radius cubed. For apertures a wavelength in size or larger, the fields may penetrate unattenuated.

The HIRF spectrum of RF energy that couples into aircraft wiring and electrical/electronic systems can be summarized into three basic ranges:

- *HIRF energy below 1 MHz*—Induced coupling at these frequencies is inefficient and thus will be of lesser concern.
- *HIRF energy between 1 and 400 MHz*—Induced coupling is of major concern since aircraft wiring acts as a highly efficient antenna at these frequencies.
- *HIRF energy above 400 MHz*—Coupling to aircraft wiring drops off at frequencies above 400 MHz. At these higher frequencies, the EM energy tends to couple through equipment apertures and seams and to the quarter wavelength of wire attached to the Line Replaceable Unit (LRU). In this frequency range, aspects of equipment enclosure construction become important.

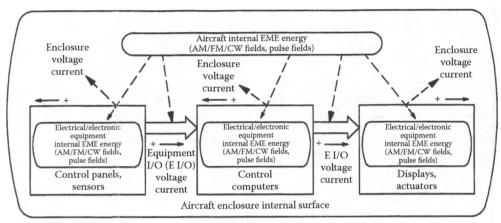

- External energy penetrates to interior via apertures, composites, seams, joints, and antennas.
- Voltages and currents induced on flight control system components and cables.
  - RF energy below 1 MHz induced coupling at these frequencies is inefficient and thus will probably be of lesser concern.
  - RF energy between 1 and 300 MHz is of major concern as aircraft wiring, when their lengths are on the order of a wavelength divided by two ($\lambda/2$) or longer at these frequencies, acts as a highly efficient antenna.
  - RF energy coupling to aircraft wiring drops off at frequencies above 300 MHz (at these higher frequencies, the EM energy tends to couple through box apertures rather than through aircraft wiring).

FIGURE 6.5 Aircraft internal EME energy electrical/electronic system.

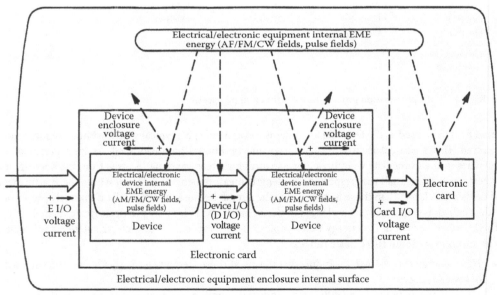

- Voltage, fields, currents, and charge on system components penetrate into equipment enclosure interiors via holes, seams, and airplane wiring (cables).
- Energy (voltage and current) picked up by wires and printed conductors on cards and carried to electronic devices.

FIGURE 6.6 Electrical/electronic equipment internal EME interaction electrical/electronic circuitry.

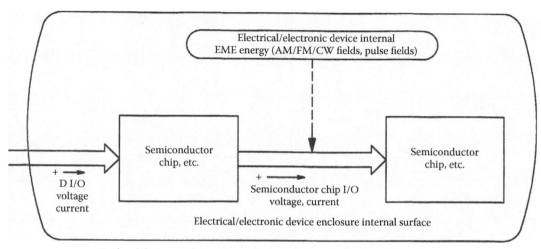

- Card and device conductors carry energy to the semiconductor chips, etc.
- Possible effects
  - Damage
  - Upset

FIGURE 6.7 Electrical/electronic device internal EME interaction electrical/electronic circuitry.

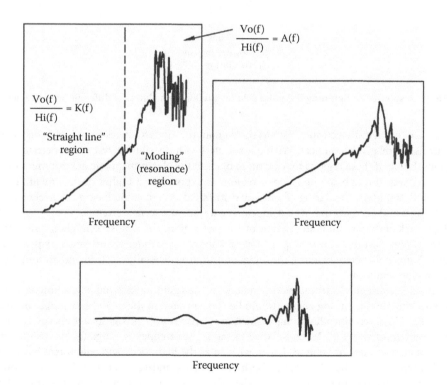

FIGURE 6.8 Frequency domain representation of EME energy attenuation/coupling transfer functions.

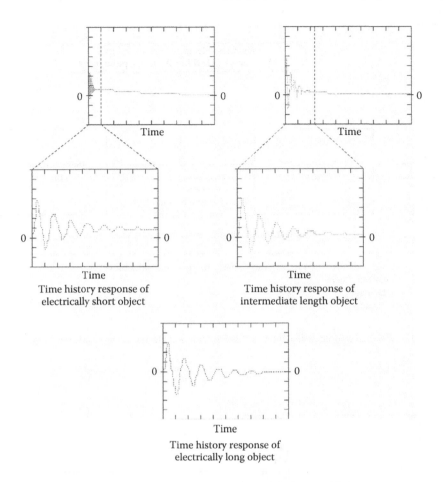

FIGURE 6.9   Responses for lightning EM pulse field interaction with objects of different "electrical lengths."

The extension of electrical/electronic systems throughout the aircraft ranges from highly distributed (e.g., flight controls) to relatively compact. Wiring associated with distributed systems penetrates several aircraft regions. Some of these regions may be more open to the electromagnetic environment than others, and wiring passing through the more open regions is exposed to a higher environment. Thus, at frequencies below 400 MHz, the wiring of a highly distributed system could have a relatively wide range of induced voltages and currents that would appear at equipment interface circuits.

The flight deck of the aircraft is an example of an open region. The windscreen "glass" presents approximately zero attenuation to an incoming field at and above the frequency for which its perimeter is one wavelength. Some enhancement above the incident field level generally exists in and around the aperture at this resonance condition.

Lightning is a transient electromagnetic event, as is the resulting internal environment. Relative to a spectral representation, lightning energy would be concentrated in the 0–50 MHz range (most energy is below 3 MHz). However, since lightning is such an intense transient, significant energy can be present up to and sometimes above 10 MHz. Relative to the higher frequency range (above 100 MHz), strong resonances of aircraft interior volumes (cavities) such as the flight deck and equipment bay could occur. At very high frequencies, the EME can be both very intense and very short in duration. From a cavity resonance issue, since the time constant of a relatively good cavity resonator is on the order of 1 ms, the pulse can be gone before significant field energy is developed within the cavity.

6.4 Architecture Options for Fault Mitigation

New system architecture measures have been evolving that could complement and augment traditional schemes to provide protection against EME energy effects. Architecture options can be applied at the overall system level or within the digital computing platform for the system. These options include the following:

- Distributed bus architecture
- Error detection and corrective (EDC) schemes
- Fiber optic data transfer
- Computation recovery

6.4.1 Electrical/Electronic System

In the past, soft faults in digital avionics were physically corrected by manual intervention, recycle power, and so on. More recently, system-level measures for the automatic correction of soft faults have begun to be developed. It is perceived that significant benefits can be gained through soft fault protection measures designed into the basic system mechanization. System-level soft fault protection methodologies provide the ability to tolerate disruption of either input/output data or internal computation. Accordingly, there are two distinct classes of disruption:

- Disruption at the system equipment interface boundary, causing corruption of data flowing to or from the affected subsystem.
- Disruption that reaches within system equipment to corrupt internal data and computation. As a worst-case scenario, it must be presumed that any logic state elements within the computational machine (registers, memory, etc.) may be affected at the time of disruption.

The short-term disruption of input/output data at an equipment boundary can be managed via a variety of existing methodologies. Data errors must be detected and the associated data suppressed until the error status is cleared. The data processing algorithm should tolerate data loss without signaling a hard fault. The length of time that can be tolerated between valid refreshes depends on the data item and the associated time constants (response) of the system and corresponding function being implemented.

The ability to tolerate disruption that reaches computation and memory elements internal to system equipment without propagation of the associated fault effect is a more difficult problem. For systems with redundant channels, this means tolerance of the disruption without loss of any of the redundant channels. Fault clearing and computation recovery must be rapid enough to be "transparent" relative to functional operation and flight deck effects. Such computational recovery requires that the disruption be detected and the state of the affected system be restored. Safety-critical systems are almost always mechanized with redundant channels. Outputs of channels are compared in real time, and an errant channel is blocked from propagating a fault effect. One means available for safety-critical systems to detect disruption is the same cross-channel monitor. If a miscompare between channels occurs, a recovery is attempted. For a hard fault, the miscompare condition will not have been remedied by the recovery attempt.

A basic approach to "rapid" computational recovery would be to transmit function state variable data from valid channels to the channel that has been determined faulted and for which a recovery is to be attempted (Figure 6.10). However, the cross-channel mechanization is ineffective against a disruption that has the potential to affect all channels.

6.4.2 Digital Computing Platform

The platform for the Airplane Information Management System (AIMS) used on Boeing 777 aircraft and Versatile Integrated Avionics (VIA) technology is an example of an architectural philosophy in the design of computing platforms. Essentially, VIA is a repackaged version of the AIMS technology.

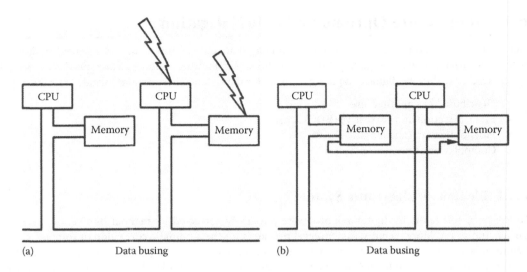

FIGURE 6.10　Redundant CPU cross-lane recovery (can accomplish some degree of "rapid" recovery). (a) Disruption and (b) transfer of state variables.

As mentioned, first-generation digital avionics have been plagued with high MTBUR (no-fault-found) rates. One primary goal of the Boeing 777 program was to greatly improve operational readiness and associated life cycle cost performance for the airlines. The AIMS functionally redundant, self-checking pairs architecture was specifically selected to attack these problems. The high integration supported by AIMS required a very comprehensive monitoring environment that is ideal for in-channel "graceful" recovery.

In AIMS, the more dramatic step of making hardware monitoring active on every CPU clock cycle was taken. All computing and input–output (I/O) management resources are lockstep compared on a processor cycle-by-cycle basis. All feasible hardware soft or hard faults are detected. In this approach, if a soft or hard fault event occurs, the processor module is immediately trapped to service handlers, and no data can be exported. In past systems, the latency between such an event and eventual detection (or washout) was the real culprit. The corrupted data would propagate through computations and eventually affect some output. To recover, drastic actions (reboots or rearms) were often necessary. In AIMS, critical functions such as displays (because the flight crew could "see" hiccups) have a "shadowing" standby computational resource. The shadow sees the same input set at the same time as the master self-checking pair. If the master detects an event, within nanoseconds the faulty unit is blocked from generating outputs. The Honeywell SAFEbus® system detects the loss of output by the master and immediately passes the shadow's correct data for display.

In the faulted processor module, the core system has two copies of processor "state data" fundamental in the self-checking pair. Unlike past systems, in which the single thread processor may be so defective it cannot record any data, at least one half of the AIMS self-checking pair should be successful. Thus, the process of diagnosing hardware errors involves comparing what each half of the pair thought was going on. Errors, down to processor address, control, or data bits, can be easily isolated. If the event was a soft fault, the core system allows a graceful recovery before the processor module is again allowed to export data. On the surface, it appears to be a more sensitive system; however, even with the comprehensive monitoring (potentially a brittle operation), from the standpoint of a self-checking (dual-lockstep) pairs processor data comparison, in these platforms the automatic recovery capabilities should provide a compensating, more robust operation. In other words, from a macro time perspective, system functions will continue to be performed even though, on a micro time basis, a soft fault occurred.

In addition to the isolation of hardware faults (hard or soft), effective temporal and physical partitioning for execution of application software programs involving a variety of software levels has been achieved by the monitoring associated with the self-checking pairs processor and a SAFEbus® communication technology approach.

Definitions

DO-160: RTCA Document 160, Environmental Conditions and Test Procedures for Airborne Equipment, produced by RTCA Special Committee 135. Harmonized with ED-14.

ED-14: EUROCAE Document 14, Counterpart to DO-160, produced by EUROCAE Working Groups 14, 31, and 33. Harmonized with DO-160.

EMC: Electromagnetic compatibility is a broad discipline dealing with EM emissions from and susceptibility to electrical/electronic systems and equipment.

EME: Electromagnetic environment, which for commercial aircraft, consists of lightning, HIRF, and the electrical/electronic system and equipment emissions (intra and inter) portion (susceptibility not included) of EMC.

EUROCAE: European Organization for Civil Aviation Equipment; for the European aerospace community, serving a role comparable to that of the RTCA and SAE.

MTBF: Mean Time Between Failures (World Airlines Technical Operations Glossary).

MTBUR: Mean Time Between Unscheduled Removals (World Airlines Technical Operations Glossary).

PED: Portable Electronic Device such as tablets, smartphones, e-readers and MP3 players.

Further Reading

14 CFR § 25.581, *Lightning Protection*, Code of Federal Regulations for Lightning Protection of Part 25 aircraft.

14 CFR § 23.1306, 25.1306, 27.1306, 29.1306, *Electrical and Electronic System Lightning Protection as they Pertain to Different Types of Aircraft*, Code of Federal Regulations.

AC 20-136B, *Aircraft Electrical and Electronic System Lightning Protection*, issued on September 07, 2011, The FAA.

AC 20-158, *The Certification of aircraft Electrical and Electronic Systems for Operation in the High-Intensity Radiated Fields (HIRF) Environment*, issued July 20, 2007, The FAA.

AC 20-174, *Development of Civil Aircraft and Systems*, issued September 30, 2011 to recognize the use of SAE ARP 4754A, The FAA.

AC-21-16G RTCA Document DO-160 versions D, E, F, and G, *Environmental Conditions and Test Procedures for Airborne Equipment*, issued June 22, 2011 to cover the use of later versions of DO-160 for lightning and HIRF.

AC 33.4-3, *Instructions for Continued Airworthiness, Aircraft Engine High Intensity Radiated Fields (HIRF) and Lightning Protection Features*, issued September 16, 2005, The FAA.

Clarke, C.A. and Larsen, W.E., FAA Report DOT/FAA/CT 86/40, Aircraft electromagnetic compatibility, June 1987.

Committee to the Federal Aviation Administration, A report from the Portable Electronic Devices Aviation Rulemaking, September 30, 2014. http://www.faa.gov/about/initiatives/ped/media/ped_arc_final_report.pdf (accessed April 28, 2014).

DO-307, *Aircraft Design and Certification for Portable Electronic Device (PED) Tolerance*, issued October 11, 2007, Prepared by SC-202.

EUROCAE ED-14D/RTCA DO-160G, *Environmental Conditions and Test Procedures for Airborne Equipment*, issue date December 6, 2007, Prepared by RTCA SC-135.

FAA AC 20-158A, *The Certification of Aircraft Electrical and Electronic Systems for Operation in the High-Intensity Radiated Fields (HIRF) Environment*, issued on May 30, 2014, The FAA.

FAA Notice 8900.240, *Expanded Use of Passenger Portable Electronic Devices (PED)*, issued October 31, 2013.

Hess, R.F., Implications associated with the operation of digital data processing in the relatively harsh EMP environments produced by lightning, *International Aerospace and Ground Conference on Lightning and Static Electricity*, Paris, France, June 1985.

Hess, R.F., Computing platform architectures for robust operation in the presence of lightning and other electromagnetic threats, *Proceedings of the Digital Avionics Systems Conference*, Irvine, CA, October 1997.

Hess, R.F., Options for aircraft function preservation in the presence of lightning, *International Conference on Lightning and Static Electricity*, Toulouse, France, June 1999.

MIL-STD-464C, *Electromagnetic Environmental Effects, Requirements for systems, Department of Defense Interface Standard*, December 1, 2010, US DoD.

Report AC25.1309-1A, System design and analysis, Advisory Circular, U.S. Department of Transportation, Washington, DC, 1988.

SAE AE4L Report: AE4L-87-3 ("Orange Book"), *Certification of Aircraft Electrical/Electronic Systems for the Indirect Effects of Lightning*, September 1996 (original publication February 1987).

SAE ARP4754, *Certification Consideration for Highly Integrated or Complex Aircraft Systems*, SAE, Warrendale, PA, issued November 1996.

SAE ARP4761, *Guidelines and Tools for Conducting the Safety Assessment Process on Civil Airborne Systems and Equipment*, SAE, Warrendale, PA, issued December 1996.

SAE ARP5412, *Aircraft Lightning Environment and Related Test Waveforms*, SAE, Warrendale, PA, 1999.

SAE ARP5413, *Certification of Aircraft Electrical/Electronic Systems for the Indirect Effects of Lightning*, SAE, Warrendale, PA, issued August 1999.

SAE Report: AE4L-97-4, *Aircraft Lightning Environment and Related Test Waveforms Standard*, July 1997.

Society of Automotive Engineers (SAE) Aerospace Required Practice 4754A *Guidelines for Development of Civil Aircraft and Systems*, dated December 21, 2010, SAE.

7

Vehicle Health Management Systems

Philip A.
Scandura, Jr.
Honeywell International

7.1 Introduction

The notion of vehicle health management (VHM) should be familiar to anyone who has ever operated or ridden in a vehicle, whether it was an automobile, truck, boat, or aircraft. Even the Wright Brothers performed VHM in their bicycle shop over 100 years ago. VHM includes the set of *activities that are performed to identify, mitigate, and resolve faults with the vehicle* [1]. These activities can be grouped into four phases, as illustrated in Figure 7.1.

The first phase, *health state determination*, is primarily concerned with determining the overall health state of the vehicle. By using diagnostic and prognostic algorithms, the vehicle and its systems are monitored to detect and isolate failures. This can be performed manually using procedures and observations, automatically using embedded hardware and software, or by some combination of these means.

The next phase, *mitigation*, involves the real-time assessment of the impact of the failures on the vehicle and its current mission. Once the impact is assessed, system redundancy management reconfigures the vehicle to maintain a safe operating condition and continue the mission, if possible. In those cases in which reconfiguration is not sufficient to continue the mission, the flight crew may modify the

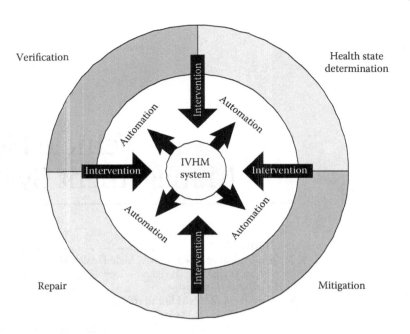

FIGURE 7.1   Health management activity model. (From Aaseng, G.B., Blueprint for an integrated vehicle health management system, *IEEE 20th Digital Avionics Systems Conference*, Daytona Beach, FL, October 2001.)

mission to compensate. In the case of an aircraft, for example, an engine failure may require the crew to modify the original route to accommodate diversion to a closer airport.

Following mission completion, the *repair* phase conducts activities to return the vehicle to its nominal operating state. This typically involves the repair or replacement of the failed components. Depending upon the situation, such as a long-duration space mission to Mars, repairs may actually take place during the mission, as well as once the mission is completed.

The final phase, referred to as *verification*, consists of activities to ensure that all repairs have been performed correctly and that the system has been returned to full operational status. Depending upon the requirements of the governing regulatory authority, such as the Federal Aviation Administration (FAA), the verification phase may require the use of independent inspectors.

7.2 Definition of Integrated Vehicle Health Management

Although the foregoing discussion paints a reasonable picture of the activities involved in operating and maintaining a vehicle in an acceptable operating condition, there is actually more to VHM than just supporting vehicle operation. Integrated vehicle health management (IVHM) spans the entire life cycle of the vehicle, including design, operation (as already discussed), and improvements. Not just focused on the vehicle itself, IVHM must also address the supporting infrastructure necessary to operate the vehicle. As illustrated in Figure 7.2, this requires taking an "enterprise-wide" approach to IVHM, addressing both the business cycle (vehicle development and continuous improvement) and the mission cycle (vehicle mission planning and execution).

7.2.1 System Engineering Discipline

It is important to understand that IVHM is not just a stand-alone subsystem added onto an existing vehicle nor should a group of sensors and related instrumentation system be considered IVHM. From a software

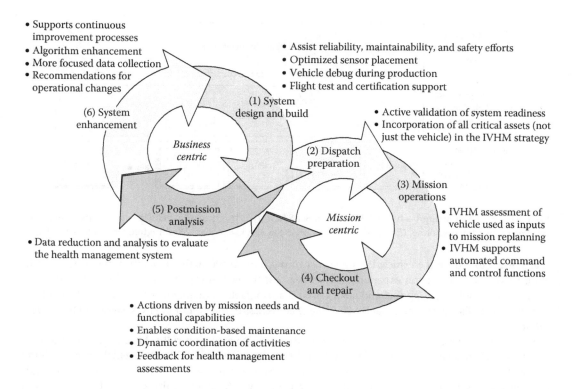

- Supports continuous
  improvement processes
- Algorithm enhancement
- More focused data collection
- Recommendations for
  operational changes

- Assist reliability, maintainability, and safety efforts
- Optimized sensor placement
- Vehicle debug during production
- Flight test and certification support

(6) System
enhancement

(1) System
design and build

- Active validation of system readiness
- Incorporation of all critical assets (not
  just the vehicle) in the IVHM strategy

*Business
centric*

(2) Dispatch
preparation

(3) Mission
operations

(5) Postmission
analysis

*Mission
centric*

- IVHM assessment of
  vehicle used as inputs
  to mission replanning
- IVHM supports
  automated command
  and control functions

- Data reduction and analysis to evaluate
  the health management system

(4) Checkout
and repair

- Actions driven by mission needs and
  functional capabilities
- Enables condition-based maintenance
- Dynamic coordination of activities
- Feedback for health management
  assessments

FIGURE 7.2 Enterprise-wide approach to IVHM. (From Scandura, P.A. Jr., Integrated vehicle health management as a system engineering discipline, *IEEE 24th Digital Avionics Systems Conference*, Washington, DC, October 2005.)

perspective, IVHM is more than just fault models, algorithms, and sensor processing software. Although it is accurate to state that IVHM may use these components to perform its intended function, a true IVHM system is more than just a collection of "pieces and parts." In actual practice, IVHM must incorporate a philosophy, methodology, and process that focuses on design and development for safety, operability, maintainability, reliability, and testability. To be most effective, IVHM must be "designed into" the vehicle and its supporting infrastructure from the beginning of the program, not "added on" along the way.* IVHM principles must permeate the culture and mind-set of the organization and be held in similar regard to safety. In summary, *IVHM must be elevated to the status of a system engineering discipline* [2].

7.2.2 Layered Approach

IVHM can be thought of as a distributed system, implemented as a series of layers, in which each layer performs a portion of the overall IVHM function, as illustrated in Figure 7.3. The first layer requires the establishment of a strong foundation of subsystem health management (SHM), provided by embedded built-in test (BIT) and fault detection, isolation, and recovery (FDIR) capabilities that monitor the components within each subsystem boundary. A subsystem can be thought of as a component or collection of components that provide a higher-level vehicle function. The primary purpose of SHM is to ensure *safe operation* of the subsystem by providing the necessary subsystem monitors and functional tests as directed by the safety analysis for that subsystem. Often the functional design of a subsystem must be enhanced by BIT to mitigate latent faults or hazardous conditions. The secondary purpose of SHM is *economic* in

---

* Although there are instances of successful add-on VHM systems, typically they are not as effective or efficient as those
  designed into the vehicle and its supporting infrastructure.

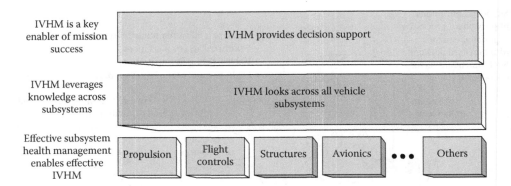

FIGURE 7.3 Layered approach to IVHM. (From Scandura, P.A. Jr., Integrated vehicle health management as a system engineering discipline, *IEEE 24th Digital Avionics Systems Conference*, Washington, DC, October 2005.)

that it helps to reduce the vehicle life-cycle cost through improved maintainability, testability, and reliability. Commercial aviation experience has shown that the largest expense incurred over the life cycle of an aircraft (not including fuel and labor costs) is attributable to maintenance activities.* Without the establishment of accurate and reliable SHM, the effectiveness of the overall IVHM system will be severely limited.

In the middle layer, IVHM looks across all subsystems to assess overall vehicle health. It is important that SHM information from all subsystems be made visible to IVHM at the vehicle level, enabling IVHM to detect and isolate faults that may occur "between" subsystems or whose effects impact multiple subsystems. In this way, IVHM is able to determine overall vehicle health and use that information to annunciate possible vehicle-wide mitigation strategies. Many modern aircraft employ some type of central maintenance system that fulfills the role of collecting faults from all subsystems, performing root cause determination, and recommending repair actions.

Finally, at the highest layer, vehicle health and operations are integrated to maximize the benefits to the overall system. As illustrated in Figure 7.4, IVHM can provide decision support capabilities to all facets of the enterprise. It is important to note that the use of IVHM data depends upon the business case for the particular market segment and application. One must determine if the available IVHM data create the necessary value for the intended user community, be it maintenance crew, flight crew, airline operator, vehicle manufacturer, or even passenger. In addition to business needs, the IVHM system design (both the physical architecture and the processes used to develop the architecture) must support the safety and criticality needs of the consumers of IVHM data. Traditional maintenance systems have been used at the end of the flight to guide the repair of the aircraft, rather than during the flight to aid in the off-nominal operation of the aircraft, and have therefore been classified as noncritical systems. On the other hand, the crew-alerting system (CAS) is used to determine the overall functional capability of the aircraft and plays a significant role in its operation; therefore, CAS has been classified as a critical system [3]. Satisfying such a broad spectrum of users emphasizes the need to determine the entire user community for IVHM data and ensure that the proper level of design assurance is applied to address their needs [2].

7.2.3 Health Management Integration

As previously discussed, IVHM must be designed into the system from the onset of the program, not added on as an afterthought. To do so, coordination and integration of IVHM efforts must occur across

---

* In the year 2001, the average age of in-service commercial aircraft was approximately 12 years old [7]. Those readers who drive older model automobiles should agree that as a vehicle ages, it costs more to maintain it in a safe and operational condition.

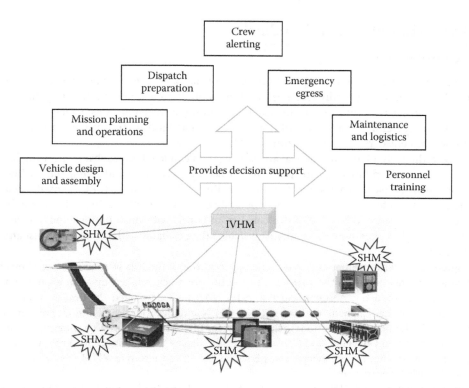

FIGURE 7.4 IVHM provides decision support. (From Scandura, P.A. Jr., Integrated vehicle health management as a system engineering discipline, *IEEE 24th Digital Avionics Systems Conference*, Washington, DC, October 2005.)

the program. The notion of *health management integration* helps to ensure that IVHM is properly designed into the system and involves the establishment of health management policies and processes that are enforced* across the design of the vehicle and supporting systems. Examples of health management policies and processes include the following:

- Fault detection and isolation philosophy
- Optimal sensor quantity and placement guidelines
- Standard designs and practices for developing BIT and FDIR
- IVHM/SHM metrics (e.g., fault coverage percentage, fault isolation accuracy percentage)
- IVHM/SHM test plans and procedures
- Fault modeling guidelines
- Interface standards between SHM and IVHM

Defining a comprehensive set of health management policies and processes requires the cooperation of the vehicle manufacturer and subsystem providers. Once established, periodic program and technical oversight helps to ensure consistent and correct application of the policies and processes across all subsystems. To provide such oversight, often a *health management integrator* is employed who is ultimately responsible for the coordination and integration activities that are crucial to the successful deployment of the IVHM system.

The ultimate goal of health management integration is to ensure an optimal IVHM balance across the system, resulting in improved system safety and reliability and reduced life-cycle costs [2].

---

* Enforcement is typically implemented contractually in the form of requirements flowed down from the vehicle manufacturer to the various subsystem vendors and associated design reviews of their designs against the requirements.

7.3 Evolution of VHM Standards

7.3.1 Commercial

Early commercial aircraft were composed primarily of mechanical and analog devices. Testing the functionality of a device typically employed nothing more than a simple push button that supplied current to the internal circuitry of the device. If continuity of the circuit was detected, a green light would illuminate, signifying a successful test. Referred to as Push-to-Test or Go/No-Go Test, these could loosely be considered as the beginning of BIT.

Starting in the early 1980s, commercial aircraft began to employ digital subsystems using hardware and software to perform functions previously performed by mechanical and analog means. These new digital subsystems, typically consisting of one or more line-replaceable units (LRUs), posed special challenges to aircraft mechanics as the ability to troubleshoot a "black box" was limited to the indications provided by the subsystem. The use of dedicated front panels with push buttons and simple display capability (e.g., lights, alphanumeric readouts) provided the mechanic with the ability to test and query the subsystem.

As digital subsystems proliferated, it became apparent that standards were necessary, as mechanics were being overwhelmed with the varied and differing approaches taken by avionics manufactures. Working with the industry, Aeronautical Radio Inc. (ARINC) developed the first aviation industry standard for health management, entitled ARINC-604 "Guidance for Design and Use of Built-In Test Equipment."* Major document sections include Goals for BITE (Built-In Test Equipment), Maintenance Concept, BITE System Concepts, and Centralized Fault Display System Concepts. From this standard, the field of VHM in commercial aviation was born, although the acronym VHM would not come into usage until nearly 20 years later.

Following the release of ARINC-604, the advent of centralized display panels shared by several LRUs provided the mechanic with a single access point to several systems, theoretically reducing the amount of training required to learn the individual systems. It was not until the emergence of centralized maintenance computers in the late 1980s and early 1990s that mechanics would truly benefit. These centralized systems gathered health and status data from several LRUs and performed fault consolidation and root cause analysis, directing the mechanic to the offending system that required repair or replacement† and pointing to the applicable maintenance procedure. Referred to as the central maintenance computer (CMC) or onboard maintenance system (OMS), these new systems were the result of further work by the aviation industry to produce updated standards, including ARINC-624, "Design Guidance for Onboard Maintenance System." Major document sections in ARINC-624 include Maintenance Concept, OMS Description, CMC Design Considerations, OMS Member System BITE, OMS Communications Protocol, Onboard Maintenance Documentation, and Airplane Condition Monitoring Function (ACMF) [4].

Several commercial aircraft use subsystems that follow ARINC-604-1. Newer aircraft include a CMC based upon ARINC-624. Table 7.1 provides a partial listing of these commercial aircraft.

7.3.2 Military

The ARINC standards discussed previously have been used by some military programs; however, many programs have transitioned to a more modern standard that traces its roots to network protocol layered architectures. The standard, known as open systems architecture for condition-based maintenance (OSA-CBM), was developed originally for use on navy ships and is now steadily spreading to military ground vehicles and aircraft. A consortium including Boeing, Caterpillar, Mimosa, Oceana Sensor,

---

* ARINC-604 was first published in 1985, soon after revised as ARINC-604-1 in 1988, as it still exists today.
† For more information regarding the emergence of the central maintenance computer, see Section 7.3.1.1.

TABLE 7.1   Commercial Aircraft Usage of ARINC-604-1 and ARINC-624 (Partial Listing)

Aircraft	Entry into Service (Approximate)	ARINC-604-1	ARINC-624	Simplified Variant[a] of ARINC-604-1 and ARINC-624
Boeing 757/767	1982/1983	X		
Airbus A320	1988	X		
Boeing 747–400	1989	X		
McDonnell Douglas MD-11	1991	X		
Boeing 777	1995		X	
Boeing 717	1999	X		
Cessna Sovereign (business jet)	2004			X
Agusta AB-139 (helicopter)	2004			X
Gulfstream 450/500/550 (business jet)	2004			X
Embraer ERJ-170/190 (regional jet)	2004/2005			X

[a] Specialized standard developed for use with Honeywell International Primus Epic® systems.

Rockwell, Penn State, and the Office of Naval Research developed OSA-CBM. The consortium's mission statement declares the following:

Condition-Based Maintenance (CBM) is becoming more widespread within US industry and military. A complete CBM system comprises a number of functional capabilities and the implementation of a CBM system requires the integration of a variety of hardware and software components. There exists a need for an Open System Architecture to facilitate the integration and interchangeability of these components from a variety of sources [5].

Shown in Figure 7.5, OSA-CBM is a layered architecture approach to VHM in which each layer is viewed as a collection of similar tasks or functions. Following a hierarchical relationship, data flow through logical transitions from sensor outputs at the lowest layer to decision support at the highest layer through various intermediate layers. The intention of such an approach is to encourage standardization of product offerings for each layer.

A sample list of military programs currently using OSA-CBM or variants includes onboard CBM for the U.S. Navy; the Joint Strike Fighter (JSF F-35) for the U.S. Navy, Air Force, and Marines; Future Combat Systems for the U.S. Army; and the Expeditionary Fighting Vehicle for the U.S. Marines.

7.4 Maintaining Federated versus Modular Avionics Systems

For many years, the traditional approach to avionics systems has been federated; that is, one or more LRUs have been dedicated to each aircraft function. For example, flight management hosted in its own flight management computer (usually dual); flight controls function hosted in a triplex configuration of flight control computers, cockpit displays with individual signal generator LRUs, and display control panels. In summary, each avionics subsystem resided in its own physical LRU with dedicated connections to control panels, sensors, and actuators.

In the early 1990s, the Boeing 777 broke with the federated tradition, using instead a modular avionics approach in which multiple functions were hosted on generic line-replaceable modules (LRMs) installed in two modular racks, designated left and right. Examples of LRMs include power supply modules, display processing modules, aircraft input/output (I/O) modules, communication modules, and database modules. These first-generation LRMs were not totally generic, however, as many contained unique hardware components restricting avionics functions to specific modules. Regardless of this limitation, the 777

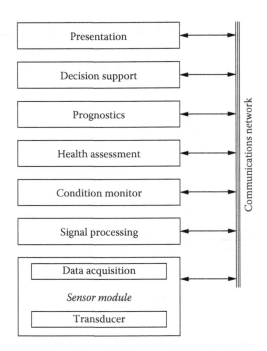

FIGURE 7.5   OSA-CBM layered approach.

approach boldly moved away from the one avionics function per LRU relationship and toward multiple avionics functions per LRM.

Subsequent generations of modular avionics systems further refined the generic nature of the LRMs, creating generic computing modules for processing both generic and custom I/O modules, network interface modules, data storage modules, and so on. In addition, the modular rack concept was expanded to allow the use of multiple racks networked together via an aircraft-wide redundant databus. The key to the success of these newer systems was their scalability and extensibility to numerous aircraft and helicopter models produced by differing manufacturers for various markets (e.g., business aviation, general aviation, regional airlines, commercial airlines, search and rescue, military) rather than targeted to specific aircraft (such as the 777).

Although modular avionics systems provided many benefits, the challenge of maintaining the system had to be addressed. Line maintenance performed by the aircraft mechanic is relatively simple in a federated environment. Because most functions are hosted in their own dedicated LRU, failures in a particular system typically result in the removal and replacement of that LRU. "Shotgun troubleshooting" was common; that is, if a problem occurs in the left LRU, it can be swapped with the right LRU. If the problem follows the LRU, it is most likely being caused by that LRU.

In a modular system, however, functions no longer enjoy dedicated LRUs. Rather than a flight management function being hosted in a dedicated computer, it now lives in a processor module, receives aircraft data from an I/O module, and uses navigation data stored on a database module; all three receive power from a power supply module. Failures of the flight management function can be attributed to faults in any of the modules or communication paths between them. It can be difficult, if not impossible, for the mechanic to determine which module is the cause of the problem without additional information from the system. The emergence of the CMC solved this problem by collecting health state data from all LRUs and LRMs, performing fault consolidation and root cause analysis to determine the faulty unit, and finally directing the line mechanic to the unit that required replacement. In this way, the CMC provides an essential role in the troubleshooting and repair of aircraft with modular avionics systems [4].

7.5 Key Technologies

To this point, numerous process-related aspects of IVHM have been discussed, including design philosophy, health management policies, the role of the health management integrator, and various industry standards. Although each is essential to the success of an IVHM system, various key technologies enable IVHM to accomplish its main objective of determining the health state of the vehicle. A brief overview of these technologies is provided in the sections that follow. For more in-depth information, consult the References and Further Readings at the end of this chapter.

7.5.1 Member System Concept

The definition of a subsystem requires the establishment of an arbitrary boundary in which components within the boundary are part of the subsystem and those outside are part of other subsystems. Establishment of boundaries is important because each defines ownership of components for SHM purposes. Without clear boundaries, "orphan" components may result that are not monitored by any particular subsystem. This should be discouraged during the vehicle design phase, as these orphan components typically cause an unmaintainable and potentially unsafe condition.

A member system is defined as a subsystem that complies with the health management policies and processes established for the vehicle. Member systems are responsible for performing SHM upon the hardware and software components within the member system boundary and reporting faults and health status to the IVHM system, especially those that impact the intended function of the member system.

Member systems are categorized in terms of their level of compliance with the established health management policies and processes. This compliance falls into one of three categories:

1. Fully compliant member system—those supporting all IVHM features and interfaces
2. Partially compliant member system—those supporting a selected subset of IVHM features and interfaces
3. Noncompliant (or nonmember) system—those that do not to interface to the IVHM system

While the latter category may seem counterintuitive given the importance of IVHM, there are business reasons (e.g., not cost-effective) or technical reasons (e.g., increase in complexity) that dictate when an IVHM system interface is not added. In these cases, it is deemed acceptable to use the subsystem as is rather than incur the cost of updating and recertifying the system. It should be noted that existing or legacy subsystems chosen for use in a new aircraft design often fall into this category.

7.5.2 Diagnostics

A simple, straightforward definition of diagnostics is the detection of faults or anomalies in vehicle subsystems. The following sections address diagnostic methods to detect faults caused by hardware failures, software errors, and those occurring at the system level.

7.5.2.1 Hardware Diagnostic Methods

While not intended to be all-inclusive, many of the classical diagnostic methods for hardware have been listed in Table 7.2. In addition, detailed examples are provided for a selected subset of I/O methods, including output wraparound and input stimulus.

7.5.2.1.1 *Output Wraparound Example*

The detection of output faults is an essential part of vehicle fault isolation. Output faults typically propagate downstream, appearing as input faults on other subsystems, which makes it very important for an LRU to detect the occurrence of an output failure, isolate the failed circuitry (if possible), and report

TABLE 7.2   Hardware Diagnostic Methods

Hardware Type	Diagnostic Method
Memory	ROM checksums, CRCs; RAM pattern tests (stuck address line, stuck data bit); parity, error detect and correct (EDAC)
CPU core	Instruction sets; register/cache tests; PCI access; interrupts; watchdog timer
I/O testing	Output wraparounds; input stimulus; A/D and D/A conversion; reference voltages
Power supply	Voltage monitors; current monitors; temperature monitors

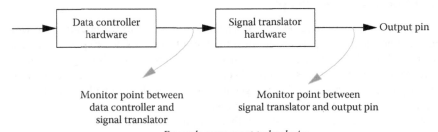

Example component technologies
Data controllers: ARINC-429 chip, discrete chip, A/D converter chip
Signal translators: Op amps, discrete drivers (ground/open or 28 V/open)
Output pins: Card-level pin, LRU pin

FIGURE 7.6   Output wraparound monitoring, general approach.

the failure to the CMC. Figure 7.6 shows the general approach to output wraparound monitoring. The output signal can be monitored at two points: the first as it leaves the data controller and the second after it undergoes signal translation and is ready to leave the card or LRU. As suggested by Figure 7.6, placing the wraparound point closer to the external output pin maximizes the amount of circuitry tested by the monitor. Depending upon the specific circuit design, one monitor point may be more feasible to implement than the other. Regardless of the point chosen, designers must guard against adding more test circuitry than functional circuitry; that is, they must balance the amount of extra hardware added for testing versus the resulting reduction in circuit reliability due to the added components.

Figure 7.7 shows a simple example of wraparound circuitry for discrete outputs, and Figure 7.8 depicts a more complex example of wraparound circuitry used for digital bus outputs (in this case, ARINC-429 transceivers). The left-hand portion of the figure illustrates output transmitter wraparounds. The use of a multiplexed test receiver (test multiplexer [Test MUX] and test receiver [Test Rx]) allows one receiver to listen to any selected transmitter, which permits the software to compare the input pattern at the test receiver with the intended output pattern of the transmitter. The remainder of Figure 7.8 illustrates the ability to stimulate the input receivers, as discussed in the next section.

7.5.2.1.2 Input Stimulus Example

Without the ability for an LRU to stimulate its own inputs, the only way to test them is to rely on the presence of external data from an upstream subsystem. If no data are present, the LRU cannot determine if its input circuitry is faulty or if the upstream subsystem is not sending data. By adding the ability to stimulate the inputs, the LRU can test them during the power-up sequence or following the detection of data loss from the external subsystem.

The design of most discrete input circuitry is simple enough that the addition of self-stimulating components violates the rule of thumb to "avoid adding more test hardware than functional hardware." Consequently, one rarely encounters self-stimulating discrete inputs. Digital bus inputs, on the other hand, are subject to various failure modes making it advisable to add self-stimulus circuitry.

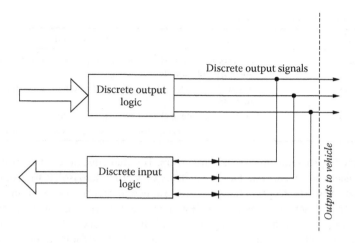

FIGURE 7.7   Simple discrete output wraparound.

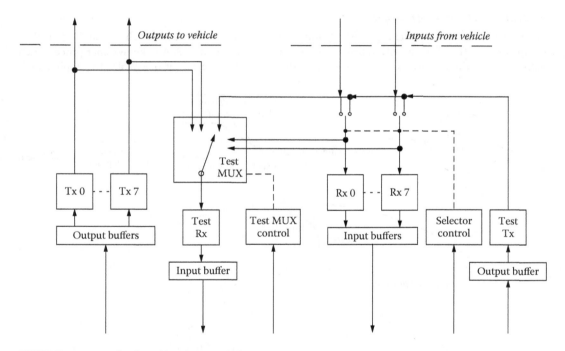

FIGURE 7.8   Complex digital bus wraparound.

As illustrated in the right-hand portion of Figure 7.8, the addition of a test transmitter (Test Tx) can route test data to any selected input receiver. If the transmitted data do not match the received data, a hardware fault has occurred. In addition, the Test Tx is used to test the Test Rx prior to testing the other transmitters and receivers.

7.5.2.2 Software Diagnostic Methods

The topic of software diagnostic methods can be somewhat controversial, since classical design assurance processes (e.g., validation and verification) are intended to prevent software design errors from being released into the field. Regardless of the process employed, experience shows that escapes do occur,

TABLE 7.3   Software Diagnostic Methods

Software Function	Diagnostic Method
Exception handlers	Fatal exceptions; processor-specific exceptions; access violation exceptions; floating point exceptions
Execution monitors	Software heartbeat; software order; ticket punching; execution traceback and state capture
Threshold monitors	Level sensing; rate of change; comparison/voting
Data validation	Reasonableness checks; parity/checksums/CRCs; source selection/reversion

resulting in in-service problems due to software errors. The challenge here is that software errors typically result in transient or intermittent faults, which are very difficult to isolate to a root cause; therefore, it is important that software exception handlers are designed to capture the maximum amount of isolation and debug data as practical, allowing the design engineers to recreate the scenario and determine the offending software code. Table 7.3 provides a sample listing of several diagnostic methods for software. What follows are detailed examples of methods this author has found to be very effective over the years, including ticket punching, execution traceback, and state capture.

7.5.2.2.1 Ticket-Punching Example

When using the ticket-punching method, software processes or threads "punch a ticket" when they complete execution during a given time slice, as illustrated in Figure 7.9. A ticket watcher process then gathers up all the required tickets and strobes the application heartbeat or enables the application's

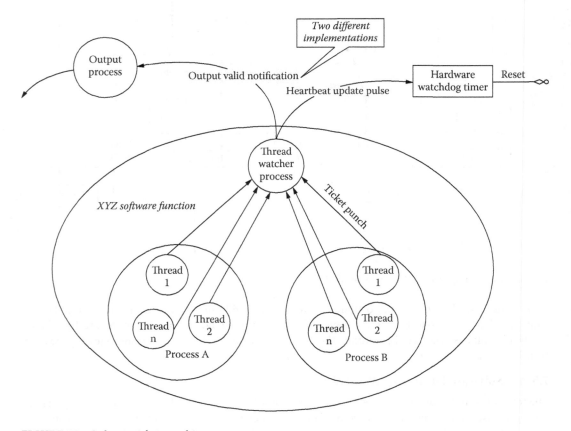

FIGURE 7.9   Software ticket punching.

output process. In the case of a heartbeat implementation, missed ticket punches result in no heartbeat updates, typically causing a system reset. In an output process implementation, missed ticket punches result in the transmission of no data (the output goes silent), stale data (data are unchanged from the previous transmission), or data that have been flagged invalid, depending upon the desired output response. The ticket-punching method protects against nonresponsive software by detecting it and responding accordingly.

7.5.2.2.2 Execution Traceback and State Capture Examples

The execution traceback method (Figure 7.10) is triggered by a software exception (e.g., a divide-by-zero violation), which causes the operating system to walk back through the execution stack, determining "who-called-who," and logging the resultant stack trace with the exception fault. Depending upon the operating system design, the stack trace may be captured in real time using a sliding window technique, or it may be constructed on demand when the exception occurs. Not all commercial operating systems provide this useful feature; however, it can be incorporated into existing homegrown operating systems in use by many LRUs.

The state capture method (Figure 7.11) involves the collection of key state data on a periodic basis. When a software exception occurs, the latest data snapshot is logged with the exception fault. Examples of state data include vehicle-specific data (altitude, heading, position, etc.), software execution data (processor registers, system timers, stack pointers, etc.), or both.

When used together, the execution traceback and state capture methods are powerful allies in the struggle to debug in-service software errors.

7.5.2.3 System-Level Diagnostic Methods

System-level diagnostic methods are intended to detect faults occurring between two or more subsystems. Table 7.4 lists many of the methods currently in use, followed by examples explaining compatibility/configuration checking and redundancy management.

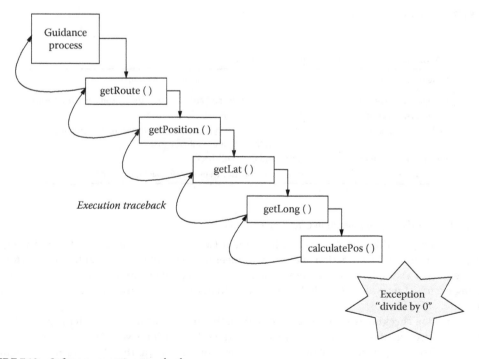

FIGURE 7.10 Software execution traceback.

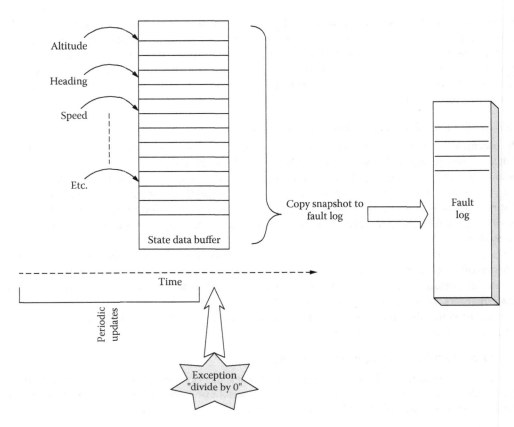

FIGURE 7.11   Software state capture.

TABLE 7.4   System-Level Diagnostic Methods

System Function	Diagnostic Method
Compatibility checks	Hardware to hardware; hardware to software; software to software
Configuration checks	Subsystem to vehicle; vehicle to mission
Redundancy management	Comparison monitors (between two subsystems); voting monitors (between three or more subsystems)
Integrity management	End-to-end protocol checks
Mode-specific tests	On-ground; safe mode; power-up; background; power save

7.5.2.3.1 Compatibility/Configuration Checking Example

The goal of compatibility checking is to determine if two or more components play nicely together, that is, to determine if they function and interface together according to the established rules for those components. For example, a 10 mm metric socket is not compatible with a 1/2 in. bolt. Compatibility checking occurs at many levels, including hardware to hardware, hardware to software, and software to software.

Configuration checking is a specialized form of compatibility checking in which the goal is to determine if the collection of systems is the correct set for the vehicle or if the vehicle is the correct choice for the mission. From the perspective of the certification authorities, such as the FAA, an aircraft is equipped

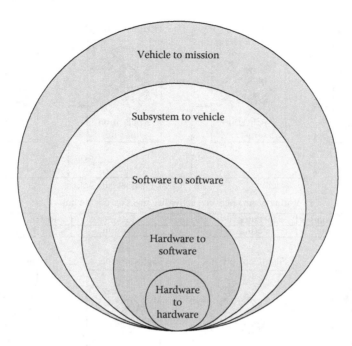

FIGURE 7.12 Compatibility/configuration checking.

with an approved set of systems and software (as documented in the type certificate). Although other versions of these systems and software may also be compatible with each other, they are not approved as part of that aircraft's configuration. In other words, a successful compatibility check does not ensure a successful configuration check.

As summarized by Figure 7.12, the various levels of compatibility and configuration checks start at the innermost layer of the onion and build upon each other, working from the inside out.

7.5.2.3.2 Redundancy Management Example

Redundancy management is often used as an architectural method of increasing safety levels. Rather than depending upon a single instance of a subsystem to provide autopilot functionality, for example, modern aircraft often use multiple instances in which one instance is in control of the aircraft, while the others monitor its performance. In the event of a subsystem failure, the commanding instance is taken offline and the control is transferred to another instance.

Redundancy management schemes can be implemented using simple comparison monitors between two instances of a subsystem, in which a miscompare event indicates a fault, although it cannot be determined which subsystem is in error. In these cases, the subsystems may retry the operation or take themselves offline and attempt recovery. More complex redundancy management schemes typically involve voting between three or more instances of a subsystem, in which case the majority rules.

Shown in Figure 7.13a through c is a notional quad-redundant scheme (typically used in human-rated spacecraft control systems) in which three instances are active, with the fourth remaining as a hot spare (Figure 7.13a). Voting information is exchanged via a cross-channel data link bus. In the event of a voting failure, the minority system is taken offline and the hot spare is made active (Figure 7.13b). If a second failure occurs, the system reverts to comparison monitoring between the remaining two instances (Figure 7.13c). This quad-redundant scheme is also referred to as "two-fault tolerant" in that it can experience two failures and still provide safe operation.

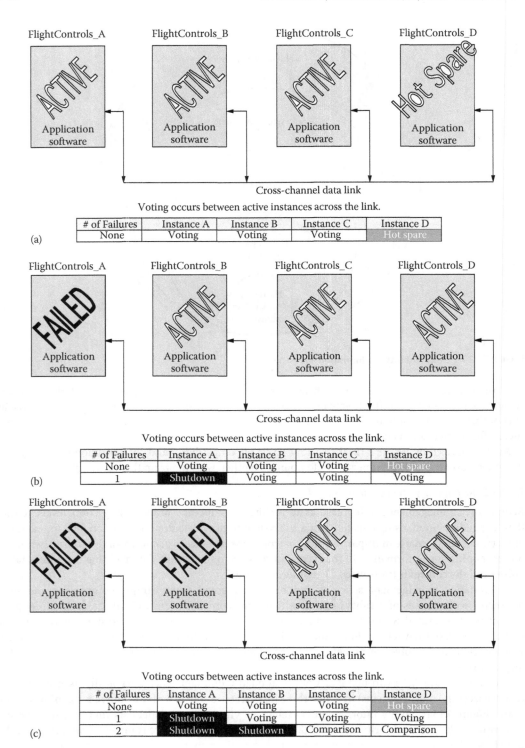

FIGURE 7.13 Redundancy management example—(a) no failures (three-way voting), (b) one failure (three-way voting), and (c) two failures (two-way comparison).

7.5.3 Prognostics

Unlike the field of diagnostics in which many of the methods discussed thus far are mature and in widespread use, the field of prognostics is still relatively young. In fact, some refer to prognostics as both "art" and "science." While there are differing definitions, this text has adopted the following definition from Engel:

> The capability to provide early detection of the precursor and/or incipient fault condition (very 'small' fault) of a component, and to have the technology and means to manage and predict the progression of this fault condition to component failure [6].

Common prognostic methods focus on the detection and diagnosis of events that mark the early stages of component failure, for example, via parameter trending of vibrations, pressures, and temperatures. This knowledge can be used to accelerate vehicle maintenance, encouraging the correction of problems while they are still small rather than waiting for them to escalate into major failures.

There are those in the prognostic community that are also interested in solving the more difficult problem of estimating the remaining life of a component or system in an effort to extend the vehicle maintenance cycle. For example, while traditional maintenance methods may have required the replacement of component X every 500 h, armed with a reliable estimate of remaining life, the interval might be extended to 600 h. It should be noted that beyond the technical challenges of estimating remaining life, there are certification challenges in proving to the authorities that it is safe to extend the maintenance interval.

Examples of prognostic methods include probability density functions, fuzzy logic, neural networks, genetic algorithms, state-based feature recognition, and Dempster–Shafer theory, all of which are beyond the scope of this text. For more information, the reader is encouraged to consult the References and Further Readings at the end of this chapter.

7.5.4 Intelligent Reasoners

The concept of "intelligent reasoning" refers to the collection of symptoms (e.g., vehicle status, diagnostic/prognostic results) from vehicle subsystems, the application of relationship-based rules or models to filter out cascaded effects, and the isolation of the incipient fault condition (i.e., the root cause) behind those symptoms. Before the advent of centralized maintenance computers (introduced previously as CMCs), the flight crew and/or aircraft mechanic manually performed these activities. Once CMCs came into existence, these activities were automated using intelligent reasoners to assist the aircraft mechanic. Functionally, these reasoners are categorized as either "rule-based" or "model-based."

7.5.4.1 Rule-Based Reasoners

Rule-based reasoners typically employed hard-coded logic in the form of "if-then-else" statements. System engineers serving as domain experts in each vehicle system would create logic rules to diagnose thousands of symptom scenarios. Software engineers would codify the logic rules, often resulting in highly complex logic. While the rule-based approach works for "small" systems, it doesn't scale up to handle the more complex systems and their interactions typically found on aircraft. In addition, updating rule-based systems has proven troublesome, as the impacts of changes are not always well understood. The seemingly simple change of a single rule could actually break many other rules making engineers very reluctant to update them.

7.5.4.2 Model-Based Reasoners

The advent of model-based reasoners attempted to address many of the limitations of the rule-based approach. While there are many types of models, most aircraft CMCs use fault propagation models, also referred to as "cause and effect" models. The basic premise is that internal LRU failures manifest themselves as output faults, the effects of which propagate to downstream systems and are either detected on their inputs or pass through undetected and cause subsequent failures within the downstream system.

By modeling the paths, these faults can take and report the presence of interface faults to the CMC, and it can filter out the cascaded faults and isolate the faulty LRU. In the event of an ambiguity that results in more than one possible LRU at fault, procedures are provided to the aircraft mechanic to assist in fault isolation.

Other model-based approaches include functional models and parametric models. Functional models attempt to model the actual operation of the vehicle and its subsystems, as opposed to modeling just their faults (as in the fault propagation approach). Parametric models use mathematical equations, created using live vehicle data collected during "nominal" operational scenarios, to predict the behavior of the vehicle in actual use. Both the functional and parametric model approaches consume input data in parallel with the vehicle, process those data via the model, and compare the model outputs against vehicle outputs. Resulting residual data between the model and vehicle outputs suggest the vehicle is performing off-nominal.

Similar to fault propagation models, the functional model approach provides the ability to determine the cause of the off-nominal performance, although the engineering resources required for creating and testing the model can be considerably approaching the amount of effort required for creating and testing the actual vehicle. Further, providing enough computing resources to execute the model in real time may not be feasible on board the vehicle. Depending upon the complexity of the vehicle and its subsystems, the use of functional modeling may be cost prohibitive.

Unlike fault propagation models and functional models, the parametric model approach does not specifically diagnose the cause of fault; rather, it identifies how the vehicle performance has deviated from nominal. For example, engine thrust is 5% below nominal for current operating mode. Because the parametric model is developed based on nominal vehicle operation, it is typically developed after the vehicle has been in-service for some period of time and has stabilized in terms of operational reliability. Parametric modeling can also be viewed as a "black box" or "gray box" testing method in that knowledge of the relationship between inputs and outputs is needed, rather than intimate details of the inner workings of the subsystems. For more in-depth information on parametric modeling, the reader is encouraged to consult the References and Further Readings list at the end of this chapter.

7.5.4.3 Fielded Intelligent Reasoners

At the time of this writing, numerous intelligent reasoners have been fielded for use in the aviation and space industry. While not intended to be all-inclusive, listed in Table 7.5 is a sample of these products and their usage.

TABLE 7.5 Sample Listing of Fielded Intelligent Reasoners

Intelligent Reasoner	Type	Known Applications	Company Information
CMC	Fault propagation modeling	Boeing 777; Primus Epic (business jets, regional jets, helicopters)	Honeywell International (www.honeywell.com)
TEAMS® toolset	Multisignal dependency modeling (advanced form of fault propagation modeling)	Consult company website	Qualtech Systems Inc. (www.teamqsi.com)
eXpress™ design toolset	Dependency modeling (similar to fault propagation modeling)	Consult company website	DSI International (www.dsiintl.com)
Livingstone	Artificial intelligence–based reasoner (mixture of functional and parametric modeling)	Deep Space 1 spacecraft; Earth Observing One (EO-1) satellite	NASA Ames Research Center (http://ti.arc.nasa.gov/)
BEAM and SHINE	Artificial intelligence–based reasoner (mixture of functional and parametric modeling)	NASA Deep Space Missions (Voyager, Galileo, Magellan, Cassini, and Extreme Ultraviolet Explorer)	NASA Jet Propulsion Laboratory (www.jpl.nasa.gov)

7.6 Examples of State-of-the-Art IVHM Systems

What follows is a brief overview of several IVHM systems, which reflect the state of the art for aviation. Additional examples can also be found in the automotive and process control industries.

7.6.1 Honeywell Primus Epic® Aircraft Diagnostic Maintenance System

The Primus Epic avionics system represents the next generation in modular avionics systems. Receiving initial FAA certification and entering into service in 2003, Honeywell's Primus Epic system was designed to be scalable and extensible to aircraft and helicopters produced by many manufacturers for various markets, rather than targeting a specific aircraft. Today, Primus Epic systems are flying on numerous business aircraft, regional aircraft, and helicopters.

Integrated within the Primus Epic system, the Aircraft Diagnostic Maintenance System (ADMS) represents an evolution of several maintenance features used in previous systems. It is comprised of the CMC, Aircraft Condition Monitoring Function, and the BIT functionality of the various member systems on the aircraft. Illustrated in Figure 7.14, the Primus Epic ADMS provides coverage of more than 200 aircraft subsystems, provided by both Honeywell and various third-party subsystem suppliers. ADMS serves as the maintenance access point to all subsystems via a point-and-click graphical user interface (GUI), allowing literally nose-to-tail coverage on most aircraft. ADMS performs root cause diagnostics to eliminate cascaded faults and provides correlation between flight deck effects and system faults. It is configurable via a separately loadable diagnostic database, provides fault information to the ground via aircraft data link, generates reports to the cockpit printer, and provides onboard data loading of aircraft subsystems.

7.6.2 Honeywell Health and Usage Monitoring for Helicopters

Health and usage monitoring systems (HUMSs) for helicopters enable the transition from traditional maintenance philosophy, which is largely based on fleet statistics and scheduled maintenance intervals, to a more efficient CBM philosophy that focuses on the usage of individual vehicles and their components. The Honeywell system monitors specific helicopter components and subsystems to form the basis for maintenance actions. Shown in Figure 7.15 is an example of the numerous sensors installed in the Bell 407 helicopter in support of HUMS. These sensors and the HUMS processing algorithms enable various types of monitoring including engine condition and performance, continuous vibration, engine exceedance, and rotor track and balance. In addition, HUMS provides ground-based tools that perform fleet data analysis. This combination of onboard and ground-based functionality allows more efficient, proactive use of maintenance assets and leads to higher fleet-readiness levels.

7.6.3 Boeing 787 Crew Information System/Maintenance System

The Boeing 787 crew information system (CIS) [4] provides a networking infrastructure, as illustrated in Figure 7.16, enabling airborne applications to interact seamlessly with ground applications. Hosted within the CIS are various applications, including the maintenance system, electronic flight bag, software distribution system, crew information services, flight deck printer, and wireless LAN support.

The lineage of the Boeing 787 maintenance system portion can be traced to the approach used in the Boeing 777, which pioneered many of the maintenance concepts standardized by ARINC-624 (discussed earlier in this chapter). The maintenance system for 787 improves on many of these concepts by adopting the flexibility and power of today's wireless communication and web-based technologies. Maintenance access is provided via Internet browser technology hosted on laptop personal computers, which are members of the secure wireless network surrounding each aircraft. Using this network, maintenance access and the ability to upload software updates from secure Internet distribution servers are provided as well as the ability to download maintenance data into the aircraft operator's maintenance and logistics network.

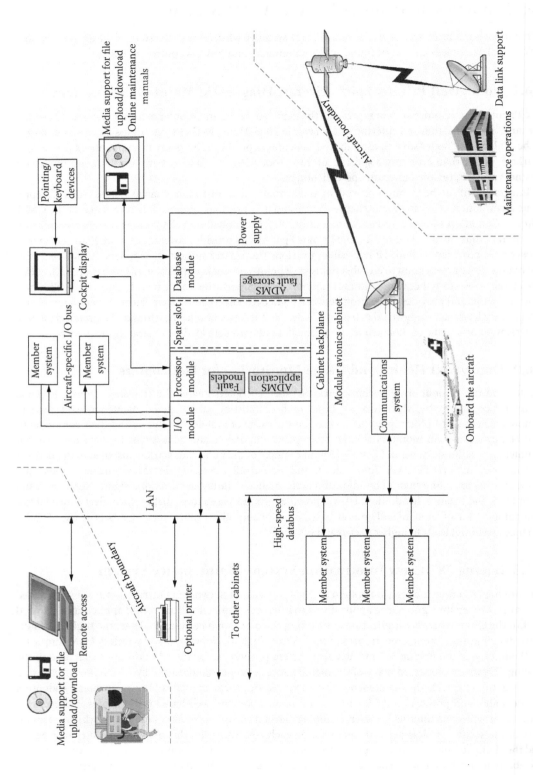

FIGURE 7.14 Honeywell Primus Epic ADMS.

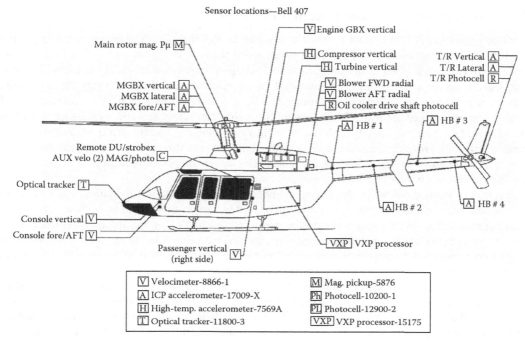

FIGURE 7.15 Honeywell HUMS for helicopters.

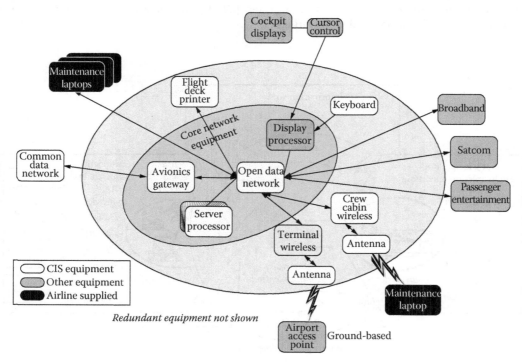

FIGURE 7.16 Boeing 787 CIS/maintenance system. (From Bird, G. et al., Use of integrated vehicle health management in the field of commercial aviation, *NASA First International Forum on Integrated System Health Engineering and Management in Aerospace*, Napa, CA, November 2005.)

7.6.4 Honeywell Sense and Respond Program

Current-generation IVHM solutions are constrained by the fact that high-fidelity access to many of the IVHM functions requires one to be in close proximity of the vehicle. Yet the majority of the expertise and resources best equipped to analyze, decide, and act upon the outputs are widely distributed in time and space from the aircraft. Consider Formula 1 racing operation in which each race car is outfitted with hundreds of sensors wirelessly streaming data back to an operational and analytical hub located in the team van, parked in the infield. These data are analyzed by computers and vehicle experts who in turn forward instructions directly to the pit crew, the race strategists, and the driver enabling real-time adjustments to improve performance and capability. Bulk data are also collected and forwarded immediately to automotive engineers back at the home office where the planning begins for vehicle design modifications and upgrades before the next race.

The innovation highlighted by this example is not a technological innovation; rather, it is actually an operational and process innovation. It is through operational innovations such as this that commercial aviation can realize the next wave of improvements enabled by IVHM and information technology. Sensing and responding to anomalies in real time, coupled with the ability to engage a global community of resources and expertise to resolve issues, enable industry to move from reactive to proactive to predictive strategies.

Honeywell's sense and respond solution, illustrated in Figure 7.17, closes the loop between vehicle, supplier, and operator. In this system, onboard diagnostic data are collected from various aircraft subsystems (including a CMC, if so equipped) and automatically transmitted to a central data repository. From there, the diagnostic data are parsed, decoded, normalized, and stored in a database that supports industry-standard queries using a web-based service-oriented architecture. Users include both internal Honeywell organizations (engineering, reliability, product improvement, and warranty services) and external customers (original equipment manufacturers [OEMs], aircraft owners, and operators) in support of their manufacturing and fleet operations [4].

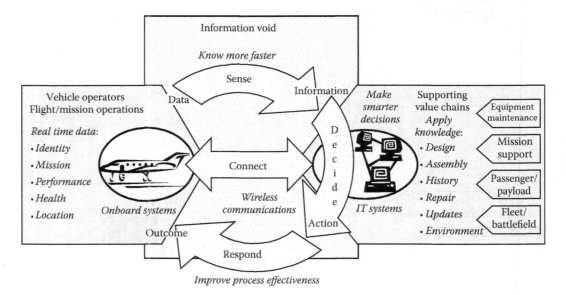

FIGURE 7.17 Honeywell *sense and respond* program. (From Bird, G. et al., Use of integrated vehicle health management in the field of commercial aviation, *NASA First International Forum on Integrated System Health Engineering and Management in Aerospace*, Napa, CA, November 2005.)

7.7 Summary

The fundamental purpose of VHM is the determination of the health state of the target vehicle. Once determined, that knowledge can be used to ensure mission success by mitigating the cause of the failure(s) via system reconfiguration or, if necessary, modifying the mission to make the best use of the remaining healthy systems and still accomplish the mission objectives. VHM is later used to repair the vehicle and verify that it has been returned to nominal operational status.

VHM determines health state based on the use of diagnostics, prognostics, and intelligent reasoners, which work in cooperation to detect faults, to determine the root cause, and to recommend the necessary mitigation actions. Following a layered approach, SHM performs health assessment at the subsystem level, providing that information to VHM that performs health assessment at the vehicle level. Depending upon the level of business enterprise integration, IVHM may be used to integrate vehicle health state into the design, production, planning, logistics, and training processes of the enterprise.

IVHM is used in various applications, as shown by the examples in this chapter, spanning the range from noncritical maintenance-only usage to safety-critical/mission-critical decision-making. The IVHM design processes and architecture used must be appropriately matched to the level of criticality intended, to ensure a safe system.

Maximizing the effectiveness of IVHM requires that it be designed into the vehicle, rather than added on at some later date. Treating IVHM as a system engineering discipline is essential to designing such a system. Keeping the proper focus on IVHM is achieved by health management integration, whose ultimate goal is to ensure an optimal IVHM balance across the system, resulting in improved system safety, improved reliability, and reduced life-cycle costs.

Definitions

ACMF:	Airplane condition monitoring function
A/D:	Analog to digital (conversion)
ADMS:	Aircraft Diagnostic and Maintenance System
AIAA:	American Institute of Aeronautics and Astronautics
AIMS:	Airplane Information Management System
ARINC:	Aeronautical Radio Inc.
BEAM:	Beacon-based exception analysis for multimissions
BIT(E):	Built-in test (equipment)
CAS:	Crew-alerting system
CBM:	Condition-based maintenance
CMC:	Central maintenance computer
CPU:	Central processing unit
CRC:	Cyclic redundancy checking
D/A:	Digital to analog (conversion)
EDAC:	Error detect and correct
FAA:	Federal Aviation Administration
FDIR:	Fault detection, isolation, and recovery
GUI:	Graphical user interface
HUMS:	Health and usage monitoring system
IEEE:	Institute of Electrical and Electronics Engineers
I/O:	Input/output
IVHM:	Integrated vehicle health management
LAN:	Local area network
LRM:	Line-replaceable module
LRU:	Line-replaceable unit

MUX:	Multiplexer
NASA:	National Aeronautics and Space Administration
OMS:	Onboard maintenance system
OSA-CBM:	Open systems architecture for condition-based maintenance
PCI:	Peripheral component interconnect
RAM:	Random-access memory
ROM:	Read-only memory
Rx:	Receiver
SHINE:	Spacecraft Health Inference Engine
SHM:	Subsystem health management
Tx:	Transmitter
VHM:	Vehicle health management

Further Reading

The author recommends the books, websites, and papers listed in the following texts for more information on various VHM-related topics.

VHM Desk References

Jennions, I.K. (ed.), *Integrated Vehicle Health Management: Perspectives on an Emerging Field*, SAE International, Warrendale, PA, 2011.

Johnson, S.B. (ed.), Gormley, T., Kessler, S., Mott, C., Patterson-Hine, A., Reichard, K., and Scandura, P. (Co-eds.), *System Health Management: With Aerospace Applications*, John Wiley & Sons, Chichester, U.K., 2011.

Aerospace Standards

ARINC Avionics Maintenance Conference (www.arinc.com/amc).

Military OSA-CBM

Discenzo, F., Nickerson, W., Mitchell, C.E., and Keller, K.J., Open Systems Architecture Enables Health Management for Next Generation System Monitoring and Maintenance, OSA-CBM Development Group.

Machinery Information Management Open Systems Alliance (MIMOSA) (www.mimosa.org).

Diagnostics and Prognostics

Aaseng, G.B., Patterson-Hine, A., and Garcia-Galan, C., A review of system health state determination methods, presented at *AIAA First Space Exploration Conference*, Orlando, FL, January 31, 2005.

"Fault Diagnostics/Prognostics for Machine Health Maintenance," Four-day short course offered by Georgia Tech (www.pe.gatech.edu).

Park, H., Mackey, R., James, M., and Zak, M., BEAM: Technology for autonomous self-analysis, presented at *IEEE Aerospace Conference*, Big Sky, MT, 2001.

"Prognostics and Health Management and Condition Based Maintenance," Two-day design course offered by Impact Technologies (www.impact-tek.com).

Scandura, P.A. Jr., Thomas, D., Luis, H., and Roger, V., Enterprise health management framework—A holistic approach for technology planning, R&D Collaboration and Transition, *IEEE International Conference on Prognostics and Health Management*, Denver, CO, October 2008.

Parametric Modeling

Ganguli, S., Deo, S., and Gorinevsky, D., Parametric fault modeling and diagnostics of a turbofan engine, *IEEE CCA*, Taipei, Taiwan, September 2004, [On-Line]. Available at www.stanford.edu/~gorin/. Accessed September, 2005.

References

1. Aaseng, G.B., Blueprint for an integrated vehicle health management system, *IEEE 20th Digital Avionics Systems Conference*, Daytona Beach, FL, October 2001.
2. Scandura, P.A. Jr., Integrated vehicle health management as a system engineering discipline, *IEEE 24th Digital Avionics Systems Conference*, Washington, DC, October 2005.
3. Scandura, P.A. Jr. and Garcia-Galan, C., A unified system to provide crew alerting, electronic checklists and maintenance using IVHM, *IEEE 23rd Digital Avionics Systems Conference*, Salt Lake City, UT, October 2004.
4. Bird, G., Christensen, M., Lutz, D., and Scandura, P.A. Jr., Use of integrated vehicle health management in the field of commercial aviation, *NASA First International Forum on Integrated System Health Engineering and Management in Aerospace*, Napa, CA, November 2005.
5. Open System Architecture for Condition Based Maintenance, www.mimosa.org
6. Engel, S.J., Gilmartin, B.J., Bongort, K., and Hess, A., Prognostics, the real issues involved with predicting life remaining, *Proceedings of the* IEEE Aerospace Conference, Vol. 6, Big Sky, MT, March 2000, pp. 457–469.
7. U.S. Department of Transportation, Transportation statistics annual report, September 2004. [On-Line]. Available at www.bts.gov/publications/transportation_statistics_annual_report/2004. Accessed September 2005.

8

Cockpit Voice Recorders and Data Recorders

Scott Montgomery
Universal Avionics
Systems Corporation

8.1 Introduction

The most common recorder used in the aviation industry is the cockpit voice recorder (CVR). Although different airlines and installation facilities will refer to these two units by different names or acronyms (such as DVR for digital voice recorder, DVDR for digital voice data recorder), CVR is the most common references and is recognizable industry-wide.

CVRs come in a variety of shapes and sizes. Specifications dictate size restrictions, that they are colored bright orange, and have a specific text shown on the outside of the unit. But, other than that, avionics manufacturers have design leeway on the recorder appearance as shown in Figure 8.1.

The primary function of the CVR is to gather and store recorded audio to assist investigators in the event of an incident or accident. These data can provide the investigators with a large amount of information to help determine what happened during an incident, what caused the incident, and what could be done in the future to prevent it from happening again.

8.2 System Architecture

The CVR system consists of the recorder unit, often referred to as the CVR, and the control unit. In many systems, the control unit will consist of a test switch, an erase switch, a headset jack, and a visual pass/fail indicator. Some control units will have a signal level indicator that will illuminate to display the audio level being received by the CVR. The CVR unit itself is comprised of most of the electronics controlling the system and the crash-protected enclosure that stores all of the data being recorded. Typically, there is also an underwater locator beacon (ULB) attached to the CVR unit.

Although the primary function of the CVR is to record audio, other inputs are recorded as well. A typical CVR has the ability to record

- Four channels of audio, one each from the pilot, copilot, third crew member or passenger address (PA) system, and an area microphone mounted in the cockpit

FIGURE 8.1 (**See color insert.**) Cockpit voice recorders.

- Coordinated universal time (UTC), often referred to as Greenwich mean time (GMT)
- Rotor speed in rotorcraft installations
- Data link communication messages (newer CVR models)

There are several other inputs to the CVR that assist in the operation of the CVR. One such input is the recorder independent power supply (RIPS). This is a relatively newer mandate for CVR systems. RIPS is a backup power supply, either internal to the CVR or provided as an external input to the CVR, that will provide power to critical CVR recording systems for an added time period of between 9 and 11 min. This new requirement stemmed from several instances where power fluctuations or a complete loss of power caused the loss of critical audio in aircraft incidents. This loss of critical data hampered the investigation of the incident or accident.

Other inputs to the CVR that assist in its operation are as follows:

- On-ground interlock discrete input
- Interface to an onboard maintenance system
- Stop recording discrete input
- On-aircraft discrete input

The weight on wheels (WOW) input will typically be used as the on-ground interlock discrete input to allow or prevent the erase feature. CVRs will often have a bulk erase function that allows the aircrew or maintenance personnel to delete or erase the audio stored in the CVR. To prevent this function from being accidentally or purposely activated in flight when potential recordings could be needed for investigation, the WOW input is often routed to the CVR to disable the erase function when not present. Furthermore, the WOW input is usually routed through one or more aircraft interlocks to assure that the aircraft is on the ground before enabling the erase function. The European Organization for Civil Aviation Electronics (EUROCAE) document ED-112 provides three specific examples of aircraft interlocks acceptable for this function:

- The parking brake and the WOW sensor
- The main cabin door position sensor (door open) and the WOW sensor
- In rotorcraft installations, the rotor brake and the WOW sensor

Each of these instances would ensure that the aircraft is on the ground and erasure of the recorded CVR information would be allowed.

An interface to an onboard maintenance system is often available. This will allow the CVR to report status to an existing maintenance display on the aircraft. Oftentimes, the maintenance display or crew-alerting system will have the ability to display a CVR fault message. This interface will allow the CVR to communicate status information to the existing system. This interface may also allow existing systems on the aircraft to command a self-test of the CVR.

Per regulations, there must be a way to stop the CVR from recording within 10 min after impact. Many CVRs have a stop CVR input discrete to comply with this regulation. In many previous CVR installations, facilities have used some type of impact (g) switch that will provide a ground input discrete to the CVR. However, some governing authorities have started discouraging the use of impact switches due to reliability issues. One example of the use of this input without the use of an impact switch is to tie the input to engine oil pressure.

The on-aircraft discrete is used to "tell" the CVR that it is installed on an aircraft. Most CVRs will not start the recording function unless this input is true. There are two reasons for this particular input. The first pertains to current CVR design regulations detailed in ED-112. These regulations state that audio cannot be downloaded from a CVR unless it is removed from the aircraft. The CVR will disable any audio downloading function while this discrete is true.

The second function of this discrete will be to allow the CVR to enter the recording phase. This would be beneficial when the CVR is removed from the aircraft for maintenance or download purposes. The CVR should not enter the recording phase and possibly overwrite the audio previously recorded if it is not on the aircraft. Therefore, if this discrete is not true, the CVR will not begin recording.

The following is a representation of a typical CVR and the external interfaces:

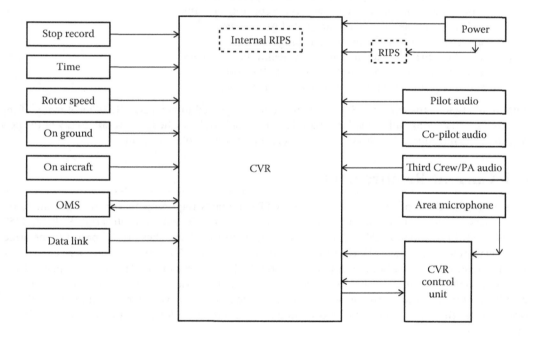

8.3 History

There have been a number of technical standard orders (TSOs) issued by the FAA regarding CVR design requirements. TSO-C84 was issued in September 1964. The referenced minimum performance standards for CVRs, issued in November 1963, detailed the design requirements of CVRs using a magnetic tape to record four channels of cockpit audio data, recording speed, distortion levels, and other details associated with analog tape recording medium. This TSO was cancelled on May 18, 1995, by Federal Register Vol. 60, No. 74 to "ensure that future…cockpit voice recorders are produced under TSO-C123a."

TSO-C123a was released in August 1996. This TSO did not have its own associated minimum performance standards. It referenced EUROCAE document ED-56A "Minimum Operational Performance requirements for Cockpit Voice Recorder Systems" Chapters 2 through 6 as the basis for CVR design (ED-56A, 1993). This specification was different than the previous CVR specification in several ways:

- It introduced specifications for solid-state recording technology.
- It had specifications for 30 min, 1 h, and 2 h duration CVRs.
- It specified the need for CVR and flight data recorder (FDR) synchronization.

All recorders up until this point had used tape to record the voice recordings. This was an unreliable way to record audio and avionics technology was progressing. This new solid-state recording technology did away with the old tape recording technology and led to the option of having a longer-duration recording stored in the CVR.

TSO-C123b is the current CVR TSO. It became effective in June 2006. This revision also did not have its own associated minimum performance standards. This revision referenced EUROCAE document ED-112 as the minimum performance standard. This new specification, released in March 2003, had many differences and improvements upon the previous ED-56A Specifications (ED-112, 2003):

- It combined the CVR and FDR specifications into one document harmonizing survivability and environmental testing requirements for both types of units.
- It prohibited the use of magnetic tape as a recording medium.
- It added requirements for recording data link messages.
- It added requirements for image recording.
- It added requirements for automatically deployable recorders.
- It added requirements for combined (CVFDR) recorders.
- It added requirements for independent power supplies.

Although C123b is the current TSO, CVRs previously approved under TSO-C123a meet many of the required specifications and can still be manufactured, sold, and installed on aircraft. Many of the new requirements introduced in TSO-C123b are not required (yet) on older aircraft.

8.4 Current Requirements

Title 14 of the Code of Federal Regulations (CFRs) dictates requirements for aircraft and rotor-craft. Parts 23 and 25 deal with airplanes, while Parts 27 and 29 deal with rotorcraft. FAR §23.1457, §25.1457, §27.1457, and §29.1457 spell out the requirements for CVRs. These are all basically the same, with very minor differences. Within the last several years, the CVR requirements were changed to add new requirements for both aircraft and rotorcraft. There weren't too many major changes. There were two major additions to these specifications. The first addition was the requirement to record data link communications. The second was a requirement to have a backup power supply (RIPS) that would power the CVR and the cockpit-mounted area microphone for 9–11 min after a loss of CVR main power.

These new additions were not required on all aircraft. Specifications in Parts 91, 121, 125, and 135 were also modified at the same time to detail which airplanes/rotorcraft had to incorporate which requirements and by when. Some of these requirements were "forward fit," or only required on new aircraft manufactured after a certain date. Other requirements were retrofit, or required to be incorporated on aircraft already in service. The following paragraphs contain a summary of CVR requirements.

8.4.1 14 CFR §91.609

1. CVR is required on an aircraft with two pilots *and* that has six or more seats.
2. Retrofit: by April 7, 2012, all aircraft manufactured before April 7, 2010, must have CVR and FDR in a separate container when *both* CVR and FDR are required.

3. New aircraft manufactured after April 7, 2010
   A. Must meet the entire CVR requirements detailed in 14 CFR §23.1457, §25.1457, §27.1457, or §29.1457 *excluding* data link recording and RIPS
4. New aircraft manufactured on or after April 6, 2012
   A. Must meet the entire CVR requirements detailed in 14 CFR §23.1457, §25.1457, §27.1457, or §29.1457 *including* data link recording and RIPS
5. Data link forward fit
   A. If data link is installed on an aircraft that requires *both* a CVR and FDR after April 6, 2012, data link info must be recorded.

8.4.2 14 CFR §121.309

1. CVR is required. Requirements are dependent upon seating configuration (10–19 *passenger seats or 20 or more passenger seats*).
2. Retrofit: by April 7, 2012, all aircraft manufactured before April 7, 2010, must
   A. Have CVR and FDR in a separate container when both are required
   B. Retain the last 2 h of recorded information
   C. Have a CVR of TSO-C123a or later
   D. Record additional audio
      a. Flight crew interphone communication
      b. Audio from navaids
      c. Passenger loudspeaker (PA) if fourth channel of CVR is available
3. Forward fit
   A. On or after April 7, 2010, must
      a. Comply with §23.1457 or §25.1457 in its *entirety* with only one exception: data link recording
         i. *This includes* the RIPS requirement
      b. Retain the last 2 h of recorded information
      c. Have a CVR of TSO-C123a or later
   B. On or after December 6, 2010, must
      a. Comply with §23.1457 or §25.1457 in its *entirety*, including the data link requirement
4. Data link forward fit
   A. If data link is installed after December 6, 2010 (regardless of aircraft manufacture date), data link info must be recorded.

8.4.3 14 CFR §125.227

1. CVR is required.
2. Retrofit: by April 7, 2012, all aircraft manufactured before April 7, 2010, must
   A. Have CVR and FDR in a separate container when both are required
   B. Retain the last 2 h of recorded information
   C. Have a CVR of TSO-C123a or later
   D. Record additional audio
      a. Flight crew interphone communication
      b. Audio from navaids
      c. Passenger loudspeaker (PA) if fourth channel of CVR is available
3. Forward fit
   A. On or after April 7, 2010, must
      a. Comply with §25.1457 in its *entirety* with only one exception: data link recording
         i. This *includes* the RIPS requirement

b. Retain the last 2 h of recorded information
c. Have a CVR of TSO-C123a or later
   B. On or after December 6, 2010, must
      a. Comply with §25.1457 in its *entirety*, including the data link requirement
4. Data link forward fit
   A. If data link is installed after December 6, 2010 (regardless of aircraft manufacture date), data link info must be recorded.

8.4.4 14 CFR §135.151

1. CVR is required on aircraft with 6 or more passenger seats *and* 2 pilots are required *or* 20 or more passenger seats. Specific requirements for each seating configuration are spelled out in detail.
2. Retrofit: by April 7, 2012, all aircraft manufactured before April 7, 2010, must
   A. Have CVR and FDR in a separate container when both are required
   B. Record additional audio
      a. Flight crew interphone communication
      b. Audio from navaids
      c. Passenger loudspeaker (PA) if fourth channel of CVR is available
3. Forward fit
   A. On or after April 7, 2010, must
      a. Comply with §23.1457, §25.1457, §27.1457, or §29.1457 in its *entirety* with only one exception: data link recording
         i. This *includes* the RIPS requirement
      b. Be operated continuously from checklist before flight to completion of final checklist at the end of flight
      c. Retain the last 2 h of recorded information
      d. Have a CVR of TSO-C123a or later
   B. On or after December 6, 2010, must
      a. Comply with §23.1457, §25.1457, §27.1457, or §29.1457 in its *entirety*, including the data link requirement
4. Data link forward fit
   A. If data link is installed after December 6, 2010 (regardless of aircraft manufacture date), data link info must be recorded.

8.5 Periodic Inspection Requirements

Along with the initial installation requirements, many areas of the world are beginning to require annual inspections of the CVR system. This is a recommendation from a large number of accident investigators. In numerous accident investigations, it was discovered that the CVR system was not fully functional. Some of the audio inputs to the CVR were not present when audio recordings were downloaded from these units. As a result, recommendations were made to periodically perform a full functional test of the entire CVR system. However, with differing governing authorities came slightly different inspection requirements.

ICAO Annex 6 Part I Attachment D provides some detailed guidelines on what should be done to check the CVR and FDR systems. It dictates an annual examination of the CVR audio by replaying a CVR recording. The testing can be done by providing test signals from each aircraft source. Or the preferred method is to examine actual in-flight recordings and verify intelligibility. India adopted the ICAO inspection requirements word for word in their Civil Aviation Requirements, Section 2—Airworthiness Series "I" Part VI Issue II, the latest revision of which was issued in November 2012.

EASA issued SIB 2009-28 in December 2009. As with ICAO, this *Safety Information Bulletin* was issued to try to reduce the number of FDR and CVR dormant failures. However, EASA inspection requirements differ slightly from those listed in ICAO Annex 6. In this bulletin, it lays out the inspection requirements for CVR and FDR systems. Per the bulletin, a CVR system must be inspected every 6 months. In this inspection, using the CVR controller monitor jack, it must be confirmed that audio from each channel being routed to the CVR for recording is present and being recorded. Along with this check, a functional test of the bulk erase inhibit logic is required.

The CVR inspection requirements are also slightly different than ICAO procedures in the Canadian Aviation Regulations (CARs) Part VI—General Operating and Flight Rules, Standard 625 Appendix C—Out of Phase Tasks and Equipment Maintenance Requirements. The CARs describe four types of periodic maintenance of the CVR system—operational check, functional check, intelligibility check, and unit overhaul—and detail when these checks are to be performed. The functional and intelligibility check is to be performed every 12 months or 3000 h, whichever comes first. The CARs are less descriptive about how the check is to be performed detailing only that "the test procedure....shall enable verification of intelligible recorded audio information from all the various input sources required by the regulations."

The FAA is in the planning stages of drafting guidance requiring periodic CVR inspections. It is unsure at the present time if these requirements will strictly copy those that are in ICAO Annex, as India did, or if they will use that as a basis of determining their own particular set of requirements. Common sense dictates that the FAA will more than likely do something similar to what EASA did in SIB 2009-28 and detail their own requirements.

8.6 Conclusion

The CVR system is a very useful tool on board many aircraft and rotorcraft flying today. The initial technology of these systems has evolved over many years. New digital technologies interface with onboard aircraft systems and have rendered the first magnetic tape CVRs obsolete. The newer recorders are smaller and lighter and can record far more information than their predecessors. However, as the recorders have evolved, so have the mandated requirements. It is now not acceptable for the CVR to just record audio. It must have backup power supplies, record information from other systems on the aircraft, and record all of this information for a longer period of time. As time goes on, CVRs will continue to evolve, record more information, and be interfaced to more aircraft systems.

References

ED-56A, Minimum operational performance specification for cockpit voice recorder system, December 1993.

ED-112, Minimum operational performance specification for crash protected airborne recorder systems, March 2003.

Electronic Code of Federal Regulations, Title 14 Aeronautics and Space, http://www.ecfr.gov/cgi-bin/text-idx?&c=ecfr&tpl=/ecfrbrowse/Title14/14tab_02.tpl.

Federal Register, Vol. 60, No. 74, April 18, 1995.

TSO-C84, http://www.airweb.faa.gov/Regulatory_and_Guidance_Library/rgTSO.nsf/MainFrame?OpenFrameSet.

TSO-C123a, http://www.airweb.faa.gov/Regulatory_and_Guidance_Library/rgTSO.nsf/MainFrame?OpenFrameSet.

TSO-C123b, http://www.airweb.faa.gov/Regulatory_and_Guidance_Library/rgTSO.nsf/MainFrame?OpenFrameSet.

9

Certification
of Civil Avionics

G. Frank McCormick
Certification Services, Inc.

9.1 Introduction

Almost all aspects of the design, production, and operation of civil aircraft are subject to extensive regulation by governments. This chapter describes the most significant regulatory involvement a developer is likely to encounter in the certification of avionics.

Certification is a critical element in the safety-conscious culture on which civil aviation is based. The purpose of avionics certification is to document a regulatory judgment that a device meets all applicable regulatory requirements and can be manufactured properly. At another level, beneath the legal and administrative machinery of regulatory approval, certification can be regarded differently. It can be thought of as an attempt to predict the future. New equipment proposed for certification has no service history. Certification tries, in effect, to provide credible predictions of future service experience for new devices—their influences on flight crews, their safety consequences, their failure rates, and their maintenance needs. Certification is not a perfect predictor, but historically it has been a good one.

In this chapter, for the most part, certification activities appropriate to the U.S. Federal Aviation Administration (FAA) are discussed. However, be aware that the practices of civil air authorities elsewhere, while generally similar to those of the FAA, often differ in detail or scope. In the original edition of *The Avionics Handbook*, this chapter discussed some representative differences between FAA practices and those of Europe's Joint Aviation Authorities, or JAA. The second edition of the *handbook* highlighted the European Union's shift of European certification responsibilities to a new organization, the European Aviation Safety Agency, or EASA. That shift is now essentially complete.

Toward the end of this chapter, we will examine some implications of EASA's activities. The FAA and EASA are both world leaders within their discipline, and regulatory differences tend to be minor. In moving their certification focus from the JAA to EASA, European regulators are simply transferring long experience and mature capabilities to a new organizational structure. Nevertheless, FAA and EASA's regulations and guidance could change in significant and unforeseeable ways, and expensive misunderstandings can result from differences among regulators. The rules and expectations of every authority, the FAA included, change over time. For current guidance, authoritative sources should always be consulted.

This chapter discusses the following topics:

- The FAA regulatory basis
- The technical standard order (TSO) system for equipment approval
- The supplemental type certificate (STC) system for aircraft modification
- Use of FAA designees in lieu of FAA personnel
- Definition of system requirements
- Safety assessments
- Environmental qualification
- Design assurance of software
- Design assurance of complex electronic hardware
- Production approvals
- The EASA

Conceptually, the certification of avionics is straightforward, indeed almost trivial: an applicant simply defines a product, determines the applicable regulatory requirements, and demonstrates that those requirements have been met. The reality is, of course, more problematic.

It is a truism that, for any proposed avionics system, a suitable market must exist. As with any commercial pursuit, adequate numbers of avionics units must be sold at margins sufficient to recover investments made in the product. Development costs must be controlled if the project is to survive. Warranty and support costs must be predicted and managed. The choices made in each of these areas will affect and be affected by certification.

This chapter is an introduction to certification of avionics. It is not a complete treatment of the subject. Some important topics are discussed only briefly, and many situations that come up in real-life certification projects are not addressed.

Good engineering should not be confused with good certification. A new avionics device might be brilliantly conceived and flawlessly designed yet ineligible for certification. Good engineering is a prerequisite to successful certification, but the two are not synonymous.

Certification has a strong legalistic element and is more craft than science. It is not unusual for projects to raise odd regulatory-approval quirks during development. Certification surprises are rarely pleasant, but surprises can be minimized or eliminated by maintaining open and honest communication with the appropriate regulators.

9.2 Regulatory Basis of the FAA

The FAA, created in 1958, acts primarily through publication and enforcement of the Federal Aviation Regulations, or FARs.* FARs are organized by sections known as Parts. The FAR Parts covering most avionics-related activity are the following:

- Part 1—Definitions and abbreviations
- Part 21—Certification procedures for products and parts
- Part 23—Airworthiness standards: normal, utility, acrobatic, and commuter category airplanes
- Part 25—Airworthiness standards: transport category airplanes
- Part 26—Continued airworthiness and safety improvements for transport category airplanes
- Part 27—Airworthiness standards: normal category rotorcraft
- Part 29—Airworthiness standards: transport category rotorcraft
- Part 33—Airworthiness standards: aircraft engines
- Part 34—Fuel venting and exhaust emission requirements for turbine engine powered airplanes
- Part 39—Airworthiness directives
- Part 43—Maintenance, preventative maintenance, rebuilding, and alteration
- Part 91—General operating and flight rules
- Part 121—Operating requirements: domestic, flag, and supplemental operations
- Part 183—Representatives of the administrator

Only a subset of these regulations will apply to any given project. A well-managed certification program identifies and complies with the complete but minimum set of applicable regulations.

9.3 FAA Approvals of Avionics Equipment

The FARs provide several different forms of approval for electronic devices installed aboard civil aircraft. Of these, most readers will be concerned primarily with approvals under the TSO system, approvals under an STC, or approvals as part of a type certificate (TC), amended type certificate (ATC), or service bulletin.†

9.4 Technical Standard Order

Approvals under the TSO system are common for materials, parts, and appliances. TSOs are regulatory instruments that recognize the broad use of certain classes of products, parts, and devices. TSOs apply to more than avionics; they can apply to any civil-aircraft article with the potential for wide use, from seat belts and fire extinguishers to tires and oxygen masks. Indeed, that is the guiding principle behind TSOs—they must be widely useful. Considerable FAA effort goes into the sponsorship and adoption of a TSO. The agency would have little interest in publishing a TSO for a device with limited application.

TSOs contain product specifications, required data submittals, marking requirements, and various instructions and limitations. Many TSOs are associated with avionics: flight-deck instruments, communications radios, ILS receivers, navigation equipment, collision avoidance systems, and flight data recorders, to name just a few.

---

* While the FAA and aviation industry continue to use the acronym FAR to mean Federal Aviation Regulation, the official U.S. government reference for these regulations is Title 14 of the Code of Federal Regulations (CFR). The acronym FAR is typically reserved for Federal Acquisition Regulation. Individuals needing to research the aviation regulations should use the form 14 CFR and the corresponding part, for example, 14 CFR 21.

† In the past, newly developed equipment was sometimes installed as part of a field approval under an FAA Form 337. This practice is now disallowed in almost all cases.

TSO-C113a, "Airborne Multipurpose Electronic Displays," is representative of avionics TSOs. Electronic display systems are used for a variety of purposes: display of attitude, airspeed, or altitude, en route navigation display, guidance during precision approach, display of engine data or aircraft status, maintenance alerts, passenger entertainment, and so on. The same physical display device could potentially be used for any or all of these functions and on many different aircraft types. Recognizing this broad applicability, the FAA published TSO-C113a so that developers could more easily adapt a generic display device to a variety of applications. TSO-C113a is typical, specifying requirements for the following data:

- Explanation of applicability
- References to related regulations, data, and publications
- Requirements for environmental testing
- Requirements for design assurance of software and complex electronic hardware
- Requirements for the marking of parts
- Operating instructions
- Equipment limitations
- Installation procedures and limitations
- Schematics and wiring diagrams
- Equipment specifications
- Parts lists
- Drawing list
- Process specifications
- Functional test specifications
- Equipment calibration procedures
- Maintenance procedures

If an avionics manufacturer applies for a TSO approval and that manufacturer's facilities, capabilities, and data comply with the terms of the TSO, the manufacturer receives a TSO authorization (TSOA) from the FAA. A TSOA represents approval of both design data and manufacturing rights. That is, the proposed device is deemed to be acceptable in its design, and the applicant has demonstrated the ability to produce identical units.

In TSO-based projects, the amount of data actually submitted to the FAA varies by system type, by the FAA's experience with particular applicants, and by FAA region. In one case, an applicant might be required to submit a great deal of certification data; in another, a one-page letter from an applicant might be adequate for issuance of a TSOA. On any new project, it is unwise to presume that all regulatory requirements are known. Consistency is a goal of the FAA, but regional differences among agency offices do exist. Early discussion with the appropriate regulators will ensure that the expectations of agency and applicant are mutually understood and agreed on.

For more information on TSOs, search for "technical standard order" on the FAA's website, <www.faa.gov>. See also FAA Order 8150.1 (Revision C, Change 1, or later), "Technical Standard Order Program." That order is particularly helpful to TSO planning in which the proposed device meets only a subset of its TSO specification, performs non-TSO functions, or performs functions covered by multiple TSOs.

Note that a TSO does *not* grant approval for installation in an aircraft. Although data approved under a TSO can be used to support an installation approval, the TSOA itself applies only to the equipment in question. Installation approvals must be pursued through other means (see next section) and are not necessarily handled by an avionics equipment manufacturer.

9.5 Supplemental Type Certificate

An STC is granted to a person or organization, usually other than the aircraft manufacturer, that wishes to modify the design of an existing aircraft. Retrofits and upgrades of avionics equipment are common motivations for seeking STC approvals from the FAA.

In an STC, the applicant is responsible for all aspects of an aircraft modification. Those aspects typically include the following:

- Formal application for an STC
- Negotiation of the certification basis of the relevant aircraft with the FAA
- Identification of any items requiring unusual regulatory treatment
- Preparation of a certification plan
- Performance of all analyses specified in the certification plan
- Coordination with the FAA throughout the project
- Physical modification of aircraft configuration
- Performance of all conformity inspections
- Performance of all compliance inspections
- Performance of all required lab, ground, and flight testing
- Preparation of flight manual supplements
- Preparation of instructions needed for continued airworthiness
- Preparation of a certification summary
- Support of all production approvals

An applicant for an STC must be "a U.S. entity," although the exact meaning of that phrase can be unclear. One common case is that of a nominally foreign firm with an office in the United States. It is acceptable to the FAA for that United Stated-based office to apply for and to hold an STC.

An applicant for an STC begins the process officially by completing and submitting FAA Form 8110-12, "Application for Type Certificate, Production Certificate, or Supplemental Type Certificate," to the appropriate FAA Aircraft Certification Office (ACO). Accompanying that application should be a description of the project and the aircraft type(s) involved, the project schedule, a list of locations where design and installation will be performed, a list of proposed designees (discussed later in this chapter), and, if desired, a request for an initial meeting with the FAA. The FAA will then assign a project number, appoint a manager for the project, schedule a meeting if one was requested, and send to the applicant an acknowledgement letter with these details. A lengthy or complex project might require the FAA regional office to coordinate with FAA national headquarters before committing its support.

The applicant must determine the certification basis of the aircraft to be modified. The "certification basis" is the sum of all applicable FAA regulations (at specified amendment levels) and any binding guidance that applies to the aircraft and project in question. Regulations tend to become more stringent over time with the intent generally being to increase safety margins or make compliance more consistent. Complying with later rules may be more time consuming and expensive than with earlier rules.

The starting point for a certification basis may be established by reference to the type certificate data sheet (TCDS) for each affected aircraft type. In most cases, the certification basis will simply be those rules currently in effect at the time of application and applicable to the STC in question. In the past, aircraft modifiers were allowed to comply with an aircraft's original certification basis. Long obsolete requirements were thus "grandfathered" into current acceptability. In recent years, the regulatory authorities have moved to limit grandfathering, instead requiring new modifications to comply with current rules.

Complex avionics systems, extensive aircraft modifications, and novel system architectures all raise the odds that something in a project will be unusual, that something will not fit neatly into the normal regulatory framework. For such activities, an applicant might wish to propose compliance based on other regulatory mechanisms: alternative means of compliance, findings of equivalent safety, exemptions, or special conditions. Any such approach is generally considered unique and thus requires close coordination with the FAA.

An STC applicant must prepare a certification plan. The plan should include the following:

- A brief description of the modification and how compliance is to be substantiated.
- A summary of the functional hazard assessment (FHA) (see Section 9.9).

- A list of proposed compliance documentation, including document numbers, titles, authors, and whether designees, if applicable, will be used to approve or recommend approval of the data (the role of designees is described in more detail later in this chapter).
- When designees are proposed, additional information should be provided including their name, designee number, appointing FAA office, classification, authorized areas, and authorized functions.
- A compliance checklist, listing the applicable regulations from the certification basis, their amendment number, subject, means of compliance, substantiating documents, and relevant designees.
- A definition of minimum dispatch configuration to meet the master minimum equipment list (MMEL).
- A project schedule, including dates for data submittals, test plan submittals, tests (with their locations), conformity inspections, installation completion, ground and flight testing, and project completion.

Some FAA ACOs require all designated engineering representatives (DERs) (see next section) participating in a project to sign an FAA Form 8110-3, "Statement of Compliance with the Federal Aviation Regulations," recommending approval of the project's certification plan.

Extensive analysis and testing are generally required to demonstrate compliance. Results of these analyses and tests must be preserved. Later in this chapter, four of the most important of these activities—safety assessments, environmental qualification, software assurance, and design assurance of complex electronic hardware—will be discussed, along with another engineering topic, development and handling of system requirements.

The FAA's involvement in an STC is a process, not an act. FAA specialists support multiple projects concurrently, and matching the schedules of applicant and agency requires planning. That planning is the applicant's responsibility. Missed deadlines and last-minute surprises on the part of an applicant can result in substantial delays to a project, as key FAA personnel are forced to reschedule their time, possibly weeks or months later than originally planned.

The STC process assumes modification of at least one prototype aircraft. It is in the aircraft modification that all the engineering analysis—aircraft performance, structural and electrical loading, weight and balance, human factors, and so on—comes together. Each component used in an aircraft modification must either be individually examined for conformance to its complete specifications or, more commonly, manufactured under an approved production system. Individual examination is known as "parts conformity inspection." A completed aircraft modification is then subject to an "installation conformity inspection." In complex installations or even complex parts, progressive conformity inspections may be required. Conformity inspections are conducted by an FAA inspector or a designee authorized by the FAA—a designated manufacturing inspection representative (DMIR) or designated airworthiness representative (DAR) (see next section).

As noted earlier, a conformity inspection ensures that a part matches its specifications. By contrast, a compliance inspection verifies through physical examination that a modification complies with applicable FARs. Typical of compliance inspections are examinations of modified wiring on an aircraft or of visibility of required placards. A compliance inspection is conducted by an FAA specialist or by an authorized DER.

For significant projects involving ground and flight testing, the FAA will issue a type inspection authorization (TIA). The TIA details all the inspections, ground tests, and flight tests necessary to complete the certification program. Prior to issuing a TIA, the FAA should have received and reviewed all of the descriptive and compliance data for the project. The FAA now requires a flight test risk assessment be included as part of the TIA procedures. The risk assessment seeks to identify and mitigate any perceived risks in flight tests that include FAA personnel, based on data supplied by the applicant.

New avionics equipment installed as part of an STC will often lead to new and different procedures for flight crews. An applicant will, in most cases, document new procedures in a supplement to an approved flight manual. In some cases, it may also be necessary to provide a supplement to an operations manual.

An applicant must also provide instructions for continued airworthiness (ICAs) of a modified airplane. For example, penetrations of the pressure vessel by, say, wiring or tubing, may require periodic inspection. Actuators associated with a new subsystem may need scheduled maintenance. ICAs are usually supplements to maintenance manuals but may also include supplements to illustrated parts catalogs, structural repair manuals, structural inspection procedures, or component maintenance manuals.

Much of this discussion has been more applicable to transport aircraft than to smaller aircraft. Regulatory requirements for the smaller (FAR Part 23) aircraft are, in some respects, less stringent than for transport aircraft. Yet, even for transports, not everything described earlier is required in every circumstance. Early discussion between applicant and regulator is the quickest way to determine what actually needs to be done.

Beginning in the 1960s, some avionics developers found it desirable to seek STCs through a type of firm called a Designated Alteration Station (DAS). A DAS could, if properly authorized by the FAA, perform all the work associated with a given aircraft modification and issue an STC. In this approach, the avionics developer might not deal with FAA personnel at all. When working with a DAS, it is important to address the issues of STC ownership rights and how production approvals will be handled.

Since the original publication of *The Avionics Handbook*, the FAA has overhauled its organizational delegations (primarily DASs, Delegation Option Authorizations, and SFAR 36 Authorizations). In mid-October 2005, the FAA published final rules (Federal Register, Volume 70, Number 197, October 13, 2005) for replacement of previous organizational delegations with a new framework known as Organization Designation Authorization (ODA). ODA consolidates previous organizational authorizations in a new Subpart D of FAR Part 183 and, in the process, addresses new flight standards functions. ODA can now be constituted to handle type certification, production certification, supplemental type certification, TSOA, major repair, major alteration, airworthiness, parts manufacturer approval (PMA), and flight standards operational authorization. For more information, see FAA Order 8100.15A (or later), "Organization Designation Authorization Procedures."

9.6 Type Certificate, Amended Type Certificate, and Service Bulletin

Approvals as part of a TC, ATC, or service bulletin are tied to the certification activities of airframers or engine manufacturers, referred to collectively as original equipment manufacturers, or OEMs. On programs involving TCs, ATCs, or service bulletins, an avionics supplier's obligations are roughly similar to those imposed by an STC project, though detailed requirements can vary greatly. Avionics suppliers participating in an aircraft- or engine-development program can and should expect to receive certification guidance from their OEM customers.

In such programs, suppliers should be cautious in their evaluations of proposals that shift extraordinary or ill-defined responsibilities for certification to suppliers. Outsourcing is a useful tool and can be implemented and managed effectively. Doing so requires a comprehensive understanding of obligations on the part of every party concerned. But comprehensive understanding is rarely available during, say, contract negotiations, when that understanding is crucial. Without it, projects can plunge ahead, fed more by optimism than by a clear-eyed view of potentially expensive obligations. Suppliers who accept certification responsibilities on behalf of an OEM—particularly those suppliers with little or no prior certification experience that is directly applicable to the systems proposed—should be prepared to do nontrivial homework in the subject and to address hard questions candidly with their customers and regulators, before agreements are signed.

9.7 FAA Designees

In the United States, any applicant may deal directly with the FAA. The FAA does not collect fees for its services from applicants. This provision of services at no charge to the user is in contrast to the practices of other civil air authorities, some of which do charge set fees for routine regulatory matters. At its discretion, however, the FAA can appoint individuals who meet certain qualifications to act on behalf of the agency.

These appointees, called designees,* receive authorizations under FAR Part 183 and serve in a variety of roles. Some are physicians who issue medical certificates to pilots. Others are examiners who issue licenses to new pilots. Still others are inspectors authorized to approve maintenance work.

Avionics developers who choose to work with and through designees are most likely to encounter FAA DERs, DMIRs, or DARs. If the avionics developer is associated with an ODA (or is having their equipment approved by an OEM that has an ODA), the compliance finding activity is likely to be performed by a unit member (UM) or authorized representative (AR) of the ODA. ARs and UMs exist for all the same areas as FAA designees.

All designees must possess authorizations from the FAA appropriate to their activities. DERs can approve engineering data. Flight test pilot DERs can conduct and approve the results of flight tests in new or modified aircraft. DMIRs and DARs can perform conformity inspections of products and installations. DARs can issue airworthiness certificates. When acting in an authorized capacity, a designee is legally a representative of the FAA; in most respects, he or she *is* the FAA for an applicant's purposes. Nevertheless, there are practical differences in conduct between the FAA and its designees.

The most obvious difference is that an applicant actually hires and pays a designee and thus has more flexibility in managing the designee's time on a project. The resulting benefits in project scheduling can more than offset the costs of the designee. Experienced designees can be sources of valuable guidance and recommendations, whereas the FAA generally restricts itself to findings of compliance—that is, the agency will simply tell an applicant whether or not submitted data comply with the regulations. If data are judged noncompliant, the FAA will not, in most cases, tell an applicant how to bring it into compliance. By contrast, a designee can assist the applicant with recovery strategies or, better yet, steer an applicant toward approaches that are both cost efficient and predictably compliant in the first place.

The FAA often encourages the use of designees by applicants. An applicant must define and propose the use of designees, by name, to the FAA ACO for each project. If the proposed designees are acceptable to the ACO, the ACO will coordinate with its manufacturing counterpart and delegate certain functions to the specified designees. Those designees then act as surrogates for the relevant FAA personnel on the project, providing oversight and ultimately approving or recommending approval of compliant data.

Although an applicant's use of designees is discretionary, the realities of FAA workload and scheduling may make the use of designees a pragmatic necessity. Whenever designees are considered for inclusion in a project, their costs and benefits should be evaluated with the same care devoted to any other engineering resource. For more information, see FAA Order 8100.8 (Revision D or later), *Designee Management Handbook*, and FAA Order 8110.37 (Revision E or later), *Designated Engineering Representatives (DER) Guidance Handbook*.

This chapter has so far dealt mainly with the definitions and practices of FAA regulation. There is, of course, a great deal of engineering work to be done in any avionics development. Five engineering topics of great interest to the FAA are the handling of system requirements, performance of a safety assessment, environmental qualification, software assurance, and design assurance of so-called "complex electronic hardware."

9.8 System Requirements

Avionics developers must document the requirements of their proposed systems, ideally in ways that are easily controlled and manipulated. Many experienced practitioners regard the skillful capture of requirements as the single most important technical activity on any project. A system specification is the basis for descriptions of normal and abnormal operation, for testing, for training and maintenance procedures, and much else. Our brief treatment of the topic here does not imply that it can be

---

* The FAA designee system is unusual in the scope of activities that can be performed by the designee. However, other regulatory authorities use similar delegation approaches. For example, EASA may delegate similar compliance finding activities to a delegated individual referred to as a certification verification engineer (CVE).

approached superficially. On the contrary, system specification is so important that a large body of literature exists for it elsewhere. Requirements definition is supported by many acceptable methods. Each company evolves its own practices in this area.

Over the years, many types of avionics systems have come to be described by de facto standardized requirements, easing the burden of both engineering and certification. New systems, though, are free to differ from tradition in arbitrary ways. Applicants should expect such differences to be scrutinized closely by regulators and customers, who may demand additional justification and substantiation of the changes.

Proper requirements are the foundation for well-designed avionics. Whatever the sources of requirements and whatever the methods used for their capture and refinement, an applicant must be able to demonstrate that a new system's requirements—performance, safety, maintenance, continued airworthiness, and so on—have been addressed comprehensively. Some projects simply tabulate requirements manually, along with one or more means of compliance for each requirement. Others implement large, sophisticated databases to control requirements and compliance information. Compliance is generally shown through analysis, test, inspection, demonstration, or some combination thereof.

9.9 Safety Assessment

Early in a project—the earlier the better—developers should consider the aircraft-level failure conditions associated with their proposed equipment. This is the first of possibly several steps in the safety assessment of a new system.

There is an explicit correlation between the severity of a system's failure conditions and the scrutiny to which that system is subjected. With a few notable exceptions,* systems that are inconsequential from a safety standpoint receive little attention. Systems whose improper operation can result in aircraft damage or loss of life receive a great deal of attention and require correspondingly greater engineering care and substantiation.

Unsurprisingly, there is an inverse relationship between the severity of a system's failure conditions and the frequency at which those failure conditions are tolerated. Minor annoyances might be tolerable every thousand or so flight hours. Catastrophic failure conditions must arise less frequently than once in every billion flight hours, thus rendering it highly improbable that such a hazard would be encountered in the entire lifetime of a fleet of that particular aircraft type. Most failure conditions fall somewhere between those two extremes. For transport aircraft, the regulations also require that no single random failure, regardless of probability, result in a catastrophic hazard, implying that any such hazard must arise from two or more independent and random failures.

Initial considerations of failure conditions should be formalized in an FHA for the proposed system. An FHA should address failure conditions only at levels associated directly with operation of the system in question. For example, an autopilot FHA would consider the failure conditions of an uncommanded hardover or oscillation of a control surface. A display-system FHA would consider the failure conditions of blank, frozen, and active-but-misleading displays during various phases of flight.

In general, if an FHA concludes that misbehavior of a system has little or no effect on continued safe flight and landing, no further work is needed for its safety assessment. Conversely, if the FHA confirms that a system can pose nontrivial risk to aircraft or occupants, then investigation and analysis must continue. Additional work, if needed, will likely involve preparation of a preliminary system safety assessment, a fault tree analysis, a failure modes and effects analysis, a common cause analysis, and a final system safety assessment.

---

* For example, failures of flight data recorders, cockpit voice recorders, and emergency locator transmitters have no effect on continued safe flight and landing. Conventional safety-assessment reasoning would dismiss these devices from failure–effect considerations. Such systems obviously perform important functions, however, and the FAA defines them as worthy of more attention than suggested by a safety assessment.

In the absence of a specific aircraft installation, assumptions must be made regarding avionics usage in order to make progress on a safety assessment. This is true in TSO approvals, for example, if design-assurance levels are not specified in the TSO or if developers contemplate failure conditions or usage different from those assumed in the TSO. There are pitfalls* in unthinking acceptance and use of generic hazard classifications and software levels (see Section 9.11), even for standard products. Technologies change quickly; regulations do not. The gap between what is technically possible and what can be approved sometimes leads to conflicting requirements, bewildering difficulties, and delays in bringing to market devices that offer improvements to safety, operating economics, or both. Once again, the solution is early agreement with the appropriate regulators concerning the requirements applicable to a new device.

The details of safety assessments are outside the scope of this chapter. For an introduction to safety-related analysis, refer to the following:

- Chapter 10 of this book
- ARP4754A—Guidelines for Development of Civil Aircraft and Systems; Society of Automotive Engineers, Inc., December 2010
- ARP4761[†]—Guidelines and Methods for Conducting the Safety Assessment Process on Civil Airborne Systems and Equipment; Society of Automotive Engineers Inc., December 1996
- NUREG-0492—Fault Tree Handbook; U.S. Nuclear Regulatory Commission, 1981
- FAA Advisory Circular 25.1309-1A—System Design Analysis, 1988
- FAA Advisory Circular 23.1309-1E[‡]—System Safety Analysis and Assessment for Part 23 Airplanes, 2011
- Safeware: System Safety and Computers—Nancy G. Leveson, Addison-Wesley Publishing Company, 1995
- Systematic Safety: Safety Assessment of Aircraft Systems—Civil Aviation Authority (UK), 1982

Aircraft operators may demand that some failures, even those associated with minor hazards, occur less frequently than tolerated by regulation. In other words, the customer's requirement may be more stringent than the FAA's. Economic issues, such as dispatch reliability and maintenance costs, are the usual motivations. In such cases, meeting the customer's specification automatically satisfies the regulatory requirement assuming that the customer specification has considered the assurance level from the aircraft point of view.

Some TSOs refer to third-party guidance material, usually in the form of equipment-performance specifications from organizations such as RTCA[§] and the Society of Automotive Engineers. TSOs, advisory circulars (ACs), and third-party specifications can explicitly call out hazard levels and software

---

* A given TSO might specify a software level (see "Software Assurance" section in this chapter), and a TSOA could certainly be granted on that basis. Note, though, that actual installation of such a device on an aircraft might require a higher software level. For example, an airspeed sensor containing Level C software could be approved under TSO-C2d. That approved sensor could not, however, be used to supply a transport aircraft with primary air data. The primary air data function on Part 25 aircraft invariably requires at least some Level A software.

† ARP4754A and ARP4761 (or their successors) are expected to be recognized by a new FAA AC, AC 25.1309-1B. At this writing, that AC has not been adopted but is considered to exist as a relatively mature draft referred to as the *arsenal version*. The FAA has accepted proposals by applicants to use arsenal on recent development programs, while EASA has incorporated similar guidance within CS-25, *Certification Specification for Large Aeroplanes*. At this writing, the FAA has not announced a date for formal release of AC 25.1309-1B.

‡ An applicant developing avionics exclusively for general aviation should pay special attention to AC 23.1309-1E (or later). That AC offers regulatory relief from many requirements that would otherwise apply. In particular, for some functions on several classes of small airplanes, the AC allows software assurance at lower levels than would be the case for transport aircraft.

§ RTCA, Inc., formerly known as the Radio Technical Corporation of America, is a nonprofit association of United States-based aeronautical organizations from both government and industry. RTCA seeks sound technical solutions to problems involving the application of electronics and telecommunications to aeronautical operations. RTCA tries to resolve such problems by mutual agreement of its member and participating organizations (cf. EUROCAE). RTCA's recommendations are typically created as part of its responsibilities under the U.S. Federal Advisory Committee Act (FACA).

assurance levels. If those specifications prescribe hazard levels and assurance levels appropriately for a given project, developers may simply adopt the prescriptions given for use in their own safety assessments. Developers must still, of course, substantiate their claims to the levels called for.

In addition to a safety assessment, analysis of equipment reliability may be required, in order to predict average times between failures of the equipment. Although this analysis is often performed by safety analysts, the focus is different. Whereas a safety assessment is concerned with the operational consequences and probabilities of system failures, or specific equipment failure modes related to a specific failure condition, a reliability analysis is concerned with the frequency of any failures of a particular component that can affect ability to dispatch aircraft (availability) or trigger a maintenance action.

9.10 Environmental Qualification

Environmental qualification is required of avionics. The standard in this area is RTCA/DO-160 (Revision G or later), "Environmental Conditions and Test Procedures for Airborne Equipment" (RTCA, 2010). DO-160G specifies testing for temperature range, humidity, crashworthiness, vibration, susceptibility to radiated and conducted radio frequencies, lightning tolerance, and other environmental factors.

It is the responsibility of applicants to identify environmental tests appropriate to their systems. Whenever choices for environmental testing are unclear, guidance from FAA personnel or DERs is in order.

To receive certification credit, environmental testing must be performed on test units whose configurations are controlled and acceptable for the tests in question. An approved test plan, conformity inspection of test articles and test setups, and formal witnessing of tests by FAA specialists or designees are generally required. In all cases, an applicant must document and retain evidence of equipment configurations, test setups, test procedures, and test results.

For further information on environmental testing, see Chapter 11.

9.11 Software Assurance

Software has become indispensable to avionics development and has a correspondingly high profile in certification. Airborne software is invariably a critical path in certification planning.

For software, regulatory compliance can be shown by conforming to the guidelines described in RTCA/DO-178 (Revision B or later), "Software Considerations in Airborne Systems and Equipment Certification." DO-178B was developed jointly by RTCA and the European Organisation for Civil Aviation Equipment (EUROCAE).* In late 2011, a long-awaited update, DO-178C, was published by RTCA, along with several companion documents: (a) DO-330, "Software Tool Qualification Considerations," (b) DO-331, "Model-Based Development and Verification Supplement to DO-178C and DO-278A," (c) DO-332, "Object-Oriented Technology and Related Techniques Supplement to DO-178C and DO-278A," and (d) DO-333, "Formal Methods Supplement to DO-178C and DO-278A." As this edition of *The Avionics Handbook* is being prepared, many existing development programs continue to use DO-178B. New programs will typically be expected to adopt DO-178C and its related guidance. A limited discussion of DO-178x appears in the following; see Chapters 12 and 13 for a more detailed treatment of software design assurance.

---

* Many of the special committees established by RTCA are joint committees with their European counterpart, EUROCAE. When successful, this results in the same set of guidelines being published in both the United States and Europe. For example, RTCA/DO-178B is equivalent to EUROCAE/ED-12B, *Considerations sur le Logiciel en Vue de la Certification des Systemes et Equipements de Bord* (EUROCAE, 1992). RTCA/DO-178C (2011) is similarly equivalent to EUROCAE/ED-12C. This type of arrangement may also be encountered for other standards-making bodies such as the Society for Automotive Engineers. For example, ARP 4754 is the equivalent to ED-79.

The FAA has accepted RTCA/DO-178C as a means of regulatory compliance, yet the use of DO-178B remains common at this writing. Hence, most of the following discussion will be in terms of DO-178x. For certification purposes, each applicant must identify the version of the software design-assurance standard appropriate to each project.

DO-178x is not a development standard for software. It is an assurance standard. DO-178x is neutral with respect to development methods. Developers are free to choose their own methods, provided the results satisfy the assurance criteria of DO-178x in the areas of planning, requirements definition, design and coding, integration, verification, configuration management, and quality assurance.

DO-178x defines five software levels, A through E, corresponding to severity classifications derived from the safety assessment discussed earlier. At one extreme, Level A software is associated with functions whose anomalous behavior could cause or contribute to a catastrophic failure condition for the aircraft. Obvious examples of Level A software include fly-by-wire primary control systems and full-authority digital engine controllers. At the other extreme, passenger entertainment software is almost all Level E, because its failure has no safety-related effects.

A sliding scale of effort exists within DO-178x: the more critical the software, the more scrutiny that must be applied to it. Level A software generates more certification data than does Level B software, Level B generates more than does Level C, and so on.

Avionics customers sometimes insist on software assurance levels higher than those indicated by a safety assessment. This is purely a contractual matter. Confusion can be avoided by separating a customer's contractual wishes from regulatory-compliance data submitted to the FAA or to DERs. Certification submittals should be based on the safety assessment rather than on the contract. If a safety assessment concludes that a given collection of software should be Level C but that software's customer wants it to be Level B, then in general, the applicant should submit to the FAA plans and substantiating data for Level C software. Any additional evidence needed to demonstrate contractual compliance to Level B should be an issue between the supplier and customer. That evidence is not required for certification and should become a regulatory matter only in unusual circumstances.*

FAA guidance itself sometimes requires that software be assured to a level higher than indicated by conventional safety assessment. This is not uncommon in equipment required for dispatch but whose failures do not threaten continued safe flight and landing.

For example, a flight data recorder must be installed and operating in most scheduled-flight aircraft. Note, however, that functional failure of a recorder during flight has no effect on the ability of a crew to carry on normally. Thus, from a safety-assessment viewpoint, a flight data recorder has no safety-related functional failure conditions (physical hazards must also be addressed). Based on that, the recorder's software would be classified as Level E, implying that the software needs no FAA scrutiny. This, of course, violates common sense. The FAA plainly has a regulatory interest in the proper operation of flight data recorders. To resolve such mismatches, the FAA requires at least Level D for any software associated with a dispatch-required function.

Digital technology predates DO-178x. Many software-based products were developed and approved before DO-178x became available. If an applicant is making minor modifications to equipment approved under an older standard, it may be possible to preserve that older standard as the governing criteria for the update. More typically, the FAA will require new or changed software to meet the guidelines of DO-178x, with unchanged software "grandfathered" in the new approval. When transport airplanes are involved in such cases, an issue paper dealing with the use of "legacy" software may be included in the certification basis of the airplane. In a few cases, the FAA may require wholesale rework of a product to meet current standards.

---

* It is usually prudent to avoid setting precedents of additional work beyond that required by regulation. Applicants are always free to do additional work, of course—developers often do, for their own reasons—and if the regulations seem inappropriate or inadequate, applicants should seek to improve the regulations. However, precedents are powerful things, for good and ill, in any regulatory regime. New precedents can have unintended and surprising consequences.

How much software data should be submitted to the FAA for certification? It is impractical to consider submitting all software data to regulators. An applicant can realistically submit—and regulators can realistically review—only a fraction of the data produced during software development. Applicants should propose and negotiate that data subset with the FAA. Whether submitted formally or not, an applicant should retain and preserve all relevant data (see DO-178B, Section 9.4, as a starting point). The FAA can examine an applicant's facilities and data at any time. It is each applicant's responsibility to ensure that all relevant data are controlled, archived, and retrievable.

9.12 Complex Electronic Hardware

For more than two decades, regulators have watched with growing concern as digital devices have become increasingly complex and ever more common in systems that perform critical aircraft functions. Today, most aircraft systems rely in whole or in part on digital logic, and design errors in such logic imply the potential for failure conditions.

Because of their discrete logic, digital devices bear a strong intellectual resemblance to software. That discrete nature makes digital devices behaviorally unlike traditional, physical parts whose form and function are inseparable and whose input/output relationships are (at least piecewise) mathematically continuous. For example, a given aircraft structural element might be characterized fully and effectively in terms of its linearized bending, shear, and torsional loads and stresses, while the thermal properties of an anti-ice heating element would inherently be governed by material choices and energy dissipation rather than by a designer's specification of imagined behavior.

Software and digital devices enjoy no such simplifications and no such constraints by the physical world. As with software, the behavior of a digital device is essentially arbitrary. The designer of digital logic may freely specify any relationship desired among inputs, outputs, and timing. The difficulty lies in getting those relationships right in all circumstances.

Complex digital hardware can rarely be "proven" correct. Instead, designers and regulators accept certain forms of design assurance as evidence that the device in question is suitable for use on civil aircraft. The standard governing hardware design assurance is RTCA/DO-254, "Design Assurance Guidance for Airborne Electronic Hardware."

DO-254 is oriented toward process. It aims to remove errors from hardware products through scrutiny of planning, requirements, design, validation, verification, configuration management, test criteria, documentation, and so on. Like DO-178x, five design-assurance levels, A through E, are recognized. These levels correspond to failure conditions of catastrophic, hazardous, major, minor, and no safety effect, respectively. The assurance level assigned to a device determines its depth of documentation, its requirements for independence, and the applicability of particular objectives within DO-254.

In recent years, the FAA has paid ever greater attention to programmable logic devices: application-specific integrated circuits, field-programmable gate arrays, and so on. These devices are referred to collectively as complex electronic hardware. The agency's increased scrutiny of complex electronic hardware is intended to ensure that acceptable processes are being followed during development of such devices. FAA AC 20-152 authorizes the use of RTCA/DO-254 as a means of showing compliance to the airworthiness requirements of complex electronic hardware and describes application of DO-254 to those technologies. For more information, see Chapter 14 of this handbook.

9.13 Manufacturing Approvals

It is not enough to obtain design approval for avionics equipment. Approval to manufacture and mark production units must be obtained as well. Parts manufactured in accordance with an approved production system do not require parts conformity inspection.

With a TSO, as explained earlier, the approvals of design and manufacturing go together and are granted simultaneously. In order to receive a TSOA, the applicant must demonstrate not just an acceptable prototype but also an ability to manufacture the article.

An STC holder must demonstrate production capabilities separately. After obtaining an STC approval, the holder may apply for PMA authority, in order to produce the parts necessary to support the STC. PMA approvals are issued by the FAA Manufacturing Inspection District Office responsible for the applicant. An STC applicant who will need subsequent PMA authority should plan and prepare for PMA from the beginning of a project. Alternatively, an STC holder may assign production rights to others, who would then hold PMA authority for the parts in question. For more information, see FAA Order 8110.42 (Revision C or later), "Parts Manufacturer Approval Procedures."

9.14 European Aviation Safety Agency

In the first edition of this handbook, a grouping of Europe's national civil air authorities—the JAA—was discussed. The purpose of that discussion was to highlight a few representative similarities and differences between the FAA and other influential regulatory authorities.

Since that time, many European civil certification responsibilities have been shifted to the EASA. Through new legislation by the European Parliament and the Council of the European Union, EASA began work officially in late September 2003. EASA is headquartered in Cologne, Germany.

EASA has responsibility for design approvals, continued airworthiness, design organization approvals, and environmental certification within the European Union. In addition, EASA will approve certain activities—production facilities, repair stations, and maintenance training organizations—outside the European Union. The agency is charged with standardization and oversight of all aviation safety certification activities, operational approvals, and personnel licensing among its member states.

The national civil aviation authorities of European Union countries will retain responsibility for many production activities and maintenance functions within their own borders and for acceptance of products into their own national registries. These national authorities will follow EASA procedures. The JAA can represent states that are members of the JAA but not of the European Union and will accept EASA approvals for recommendation to those states.

For more information, see http://www.easa.eu.int/ and FAA Order 8110.52, "Type Validation and Post-Type Validation Procedures."

9.15 Summary

Certification can be straightforward, but like any other developmental activity, it must be managed. At the beginning of a project, applicants should work with their regulators to define expectations on both sides. During development, open communication should be maintained among suppliers, customers, and regulators. In a well-run project, evidence of compliance with regulatory requirements will be produced with little incremental effort, almost as a side effect of good engineering. The cumulative result will, in the end, be a complete demonstration of compliance, soon followed by certification.

Regulatory officials, whether FAA employees or designees, work best and are most effective when they are regarded as part of an applicant's development team.

An applicant is obliged to demonstrate compliance with the applicable regulations, nothing more. However, partial information from an applicant can lead to misunderstandings and delays, and attempts to resolve technical disagreements with regulators through nontechnical means rarely have the desired effect.

In the past, regrettably, large investments have been made in systems that could not be approved by the FAA. In order to avoid such outcomes, applicants are well advised to hold early discussions with appropriate FAA personnel or designees.

Definitions

Certification: Legal recognition, through issuance of a certificate, by a civil aviation authority that a product, service, organization, or person complies with that authority's requirements.

Certification basis: The sum of all current regulations applicable to a given project at the time application is made to a civil aviation authority to begin a certification process.

Designee: An individual authorized by the FAA under FAR Part 183 to act on behalf of the agency in one or more specified areas.

Issue paper: An instrument administered by an FAA directorate to define and control a substantial understanding between an applicant and the FAA, such as formal definition of a certification basis or a finding of equivalent safety, or to provide guidance on a specific topic, such as approval methods for programmable logic devices.

PMA: Parts manufacturer approval, by which the FAA authorizes the production of parts for replacement and modification, based on approved designs.

Special condition: A modification to a certification basis, necessary if an applicant's proposed design features or circumstances are not addressed adequately by existing FAA rules; in effect, a new regulation, administered by an FAA directorate, following public notice and a public comment period of the proposed new rule.

STC: Supplemental type certificate, by which the FAA approves the design of parts and procedures developed to perform major modifications to the design of existing aircraft.

TSOA: Technical standard order authorization, the mechanism by which the FAA approves design data and manufacturing authority for products defined by a TSO.

Websites

Certification Services, Inc.: www.certification.com.

European Aviation Safety Agency (EASA): www.easa.eu.int.

European Organisation for Civil Aviation Equipment (EUROCAE): www.eurocae.org.

Federal Aviation Administration (FAA): www.faa.gov.

Joint Aviation Authorities (JAA): www.jaa.nl.

RTCA: www.rtca.org.

Society of Automotive Engineers (SAE): www.sae.org.

10

System Safety and System Development

Marge Jones
Safety Analytical Technologies, Inc.

Today's aircraft utilize integrated, electronic-based equipment to implement multiple system and subsystem functionalities. Understanding interrelationships and dependencies of these designs can be difficult and, if improperly accommodated or ignored, can lead to unintended consequences. In evaluating today's integrated and electronic-based designs from a safety perspective, consideration must be given as to how the implementation can contribute to and potentially cause aircraft or system adverse effects (i.e., cause or contribute to failure conditions). The system development and system safety processes that are applied are key elements and are used as a means of compliance to 14 CFR Parts 23, 25, 27, and 29, Subpart F Equipment, Section 2X.1309* regulation.

The commercial aircraft safety assessment process is based on a top-down concept of identifying failure conditions at the aircraft and system levels, classifying the severity of the effects of the failure conditions, and establishing top-level safety requirements that must be achieved in order for the failure condition to be considered to comply with "1309." There are two types of top-level

---

* 2X.1309 represents 14 CFR 23.1309, 14 CFR 25.1309, 14 CFR 27.1309, or 14 CFR 29.1309 and associated ACs. "1309" is used as a generic reference to any of these regulations; however, 14 CFR 27.1309 does not include severity and probability language in the regulation. AC27-1B does refer to safety and assurance processes discussed in this chapter for use when integrated or avionics-based equipments are used, particularly in critical applications. Please refer to the specific AC for the discussion of specific applicability and exclusions.

safety requirements: random failure–based requirements and systemic (also referred to as systematic) failure–based requirements. Both must be specified for each failure condition identified.

Random failures occur as a result of a physical cause typically traceable to a degradation or wear-out mechanism. Because these failures occur due to degradation, exhaustive testing could be accomplished to define and understand the specific modes of failures and the rate at which they occur. Historical, in-service data are also available for random failures that define component failure modes (open, short, parameter change, etc.) and statistical failure rate data. Quantitative probability requirements defined in the "1309" regulation and advisory circulars (ACs) are focused to addressing the occurrence of random failures.

Systemic failures are created by errors. Errors are deficiencies in requirements that can include missing or incomplete requirements, ambiguous requirements (requirements that lead to different interpretations), or incorrect requirements. Errors can occur in systems, equipment designs, software or Airborne Electronic Hardware (AEH), manufacturing processes, maintenance processes, testing procedures, etc. If a systemic failure exists, it will always occur when exposed to the same set of circumstances or conditions. Due to the nature of systemic failures, exhaustive testing of failures and traditional safety analysis techniques are not considered sufficient. It is difficult to predict when the needed set of circumstances or conditions will be present in order to reveal the systemic failure. Therefore, in order to show compliance to "1309" for systemic failures, processes are applied to the aircraft, system, equipment, and software/AEH development to provide some assurance that errors have been minimized to a required level of rigor.

Table 10.1 provides the "1309" top-level safety requirements based on severity categorization defined in AC25.1309-1B (Arsenal Draft).* The criterion shown is referred to in ACs as the "inverse relationship" between severity and probability. Similar criteria are provided in AC23.1309-1E and AC29-2C. However, in AC23.1309-1E, quantitative probability and assurance requirements are different depending on the class of the aircraft.

10.1 System Development and Assurance

The concept of assurance-based certification for systemic failures was first introduced by RTCA DO-178 for software and later by DO-254 for AEH. These documents define a development and verification process with five levels (level A through level E) of assurance rigor based on the severity of the software's/ AEH's contribution to a failure condition. The more severe the consequence of the effects, the higher level of assurance is needed so more objectives have to be satisfied (refer to Chapters 12 through 14). The challenge with using only the RTCA DO processes is that they are focused on the development and verification of software/AEH to a specific, allocated set of requirements. They do not address any assurance requirements for the processes used to define the set of requirement allocated to software/AEH. Systemic failures/errors can be introduced throughout the development process for aircraft, systems, and equipment, not just the software and AEH. In the mid-1990s, industry developed SAE ARP4754 and SAE ARP4761 to define a system development process to cover assurance requirements for aircraft, systems, and equipment, as well as establish a safety process that aids in determining the assurance requirements. These two standards complete the needed pieces to define an assurance process that jointly can be used to show compliance to "1309" for both systemic and random failures.

In 2010, SAE ARP4754 was revised to further stress the importance of process assurance as a method for complying with "1309." ARP4754A provides a complete description of an aircraft and aircraft–system development process that involves planning, requirement capture, requirement validation, and requirement verification. The concept as shown in Figure 10.1 is that the top-level concept requirements are defined (captured), design decisions are made, architecture is developed, and requirements are reviewed and analyzed for feasibility, accuracy, and completeness (i.e., validation). Derived requirements that are

---

* AC25.1309-1A is the current revision; however, Federal Register Volume 68, Number 82, Proposed Rules, allows the use of "Arsenal Draft" version of AC25.1309-1B. The severity definitions used in the *Arsenal Draft* are harmonized with EASA AMC25.1309, ARP4761/4754A, and the RTCA DO-178C and DO-254 and are commonly used for certification projects.

TABLE 10.1 Safety Criteria

Severity	Potential Effects			Probability Requirements			FDAL	Typical Safety Analyses
	Passengers or Cabin Crew	Aircraft (Safety Margin)	Flight Crew	Term	Quantitative	Qualitative		
Catastrophic	Multiple fatalities	Loss of aircraft; unable to continue safe flight and landing	Unable to respond or compensate for failure condition	Extremely improbable	On the order of 1×10^{-9} or less	Not anticipated to occur during the entire operational life of all airplanes of one type. No single failure can be assessed as extremely improbable.	A	Qualitative design and installation analyses (PRA, ZSA, CMA) and quantitative design analyses (FTA with supporting source data; FMEA and predictions)
Hazardous	Serious or fatal injury to a relatively small number of the occupants other than the flight crew	Large reduction in safety margins or functional capabilities	Physical distress or excessive workload such that the flight crew cannot be relied upon to perform their tasks accurately or completely	Extremely remote	On the order of 1×10^{-7} or less	Not anticipated to occur to each airplane during its total life but which may occur a few times when considering the total operational life of all airplanes of the type.	B	Same as catastrophic

(Continued)

TABLE 10.1 (*Continued*)　Safety Criteria

Severity	Potential Effects			Probability Requirements			FDAL	Typical Safety Analyses
	Passengers or Cabin Crew	Aircraft (Safety Margin)	Flight Crew	Term	Quantitative	Qualitative		
Major	Physical distress to passengers or cabin crew, possibly including injuries	Significant reduction in safety margins or functional capability of the airplane	Significant increase in workload or in conditions impairing crew efficiency; discomfort to the flight crew	Remote	On the order of 1×10^{-5} or less	Unlikely to occur to each airplane during its total life but which may occur several times when considering the total operational life of a number of airplanes of the type	C	Depends on complexity and redundancy of design. Range from qualitative design and installation appraisal to quantitative analysis (FTA with supporting source data; FMEA and predictions)
Minor	Some physical discomfort to passengers or cabin crew	Slight reduction in safety margins or functional capability of the airplane	Crew actions well within capabilities; slight increase in workload (such as routing flight plan change)	Probable	$>1 \times 10^{-5}$ per flight hour	Anticipated to occur one or more times during the entire operational life of each airplane.	D	Design and installation appraisal to verify adequately isolated from more critical equipment or functions
No effect	*Not defined*	No effect on safety or operational capability of the aircraft	No increase on crew workload	N/A	N/A	N/A	E	Same as minor

Sources:　Based on Advisory Circular 25.1309–1B Arsenal Draft, *System Design and Analysis*, U.S. Department of Transportation Federal Aviation Administration, 2002; SAE ARP4754A, *Guidelines for Development of Civil Aircraft and Systems*, Society of Automotive Engineers, Warrendale, PA, 2010; Advisory Circular 20-136B, *Aircraft Electrical and Electronic System Lightning Protection*, 2011.

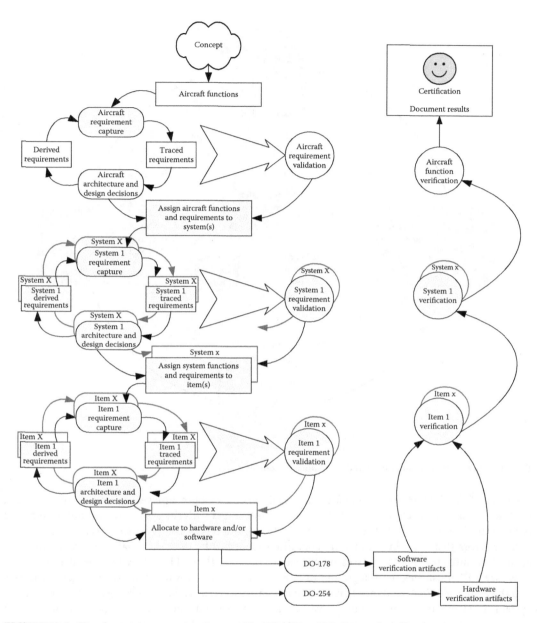

FIGURE 10.1 Development process: requirement identification, validation, and verification.

created by design decisions and are not traceable to a higher-level requirement are evaluated for impact on and compatibility with the existing requirements. All requirements are then allocated to the next lower indenture level and the capture and validation process is repeated. This capture and validation of requirements process is repeated at each indenture level until the specific architecture and implementation decisions are made as to what functionality is implemented in hardware, software, AEH, etc. Once the development process has captured, validated, and allocated the requirements top-down, the requirements that are allocated to the software and AEH development processes (DO-178, DO-278, DO-254, etc.) have been "assured" to be correct and complete. The RTCA DO verification artifacts/data are developed at the lowest indenture level and begin the bottom-up verification process. Results from

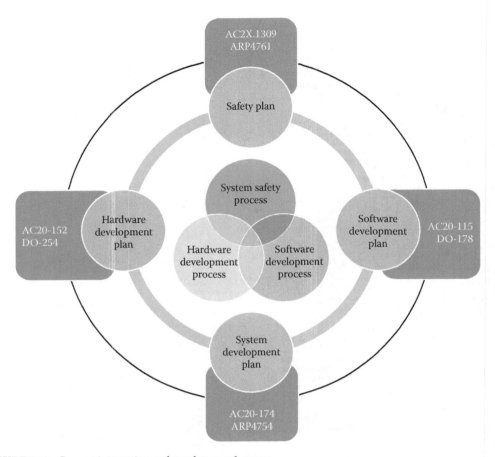

FIGURE 10.2   Process integration and regulatory references.

the RTCA DO-178/-254 verification are integrated into the equipment-level verification results (tests, analyses, reviews/inspections, etc.) that in turn get integrated with the system verification results and finally into the aircraft level. The combined ARP4754A and RTCA DO processes ensure that systemic failures/errors that contribute to failure conditions identified by the system safety process have been adequately mitigated. Both processes are needed to comply with the 14 CFR 2X.1309 regulation.

Figure 10.2 illustrates the integration of system, hardware, software, and system safety processes and the associated FAA guidance materials that identify them as an acceptable means of compliance.* System development planning must consider each of these and integrate the processes for effective development and certification. Because the system, hardware, and software development processes are actually part of the mitigation for failure conditions and compliance for 14 CFR 2X.1309, the safety requirements that establish the use of the processes must be specified early in the development process. Development assurance requirements are derived from the safety process.†

---

* ACs are identified as "an acceptable means but not the only means of compliance." However, if alternate means are planned, coordinate early with certification authority for concurrence.

† Some ACs and Technical Standard Orders (TSOs) may define an assurance level for specific functionality. However, these should always be reviewed in the safety process for compatibility to safety requirements. It may be that a design decision or architecture may require a higher assurance level. AC and TSO assurance or functional criticality requirements should be considered minimum requirements. Using any safety-derived DALs in lieu of AC or TSO levels should be coordinated early with certification authorities.

The ARP4754A development process is tailored based on the top-level safety requirements established in the safety process. An aircraft or system function that has high criticality is required to have a rigorous, structured development process, whereas a function with minor criticality has minimal development requirements. The level of development process that needs to be implemented to mitigate errors to an acceptable level from a safety perspective is referred to as "assurance level." ARP4754A establishes criteria and methodology for assigning development assurance levels (DALs) to functions (FDALs) and items (IDALs). The architecture that implements a function is key consideration in establishing the FDAL and IDAL requirements. Once the DAL is identified, ARP4754A Appendix A defines the development process requirements for each assurance level.

10.2 System Safety Concept

System safety is the process of trying to postulate all the various ways accidents/incidents can occur and ensure that the design/function, operation, and physical attributes of a "system"* mitigate the risk. As engineers, most of the training and education we receive is focused on how to make things work and not as much about how they might not work as intended or what happens when they experience failure. System safety aids the development process in trying to do just that: understand how the aircraft, systems, and equipment could perform in unintended ways and establish requirements that ensure the likelihood of these unintended functions complies with the civil aviation regulations. System safety should not be considered as a separate entity that just does all these safety documents that the civil aviation authorities require, but rather as an integral part of the overall system development. System safety is an engineering process and includes the same tasks of any engineering process: identifying the problem; developing alternative solutions to the problem; determining the most effective solutions within constraints of performance, cost, etc.; and implementing the best solution.

Figure 10.3 shows a generic model of the system safety engineering concept. In system safety engineering, different terminologies are used to describe the "problem": hazards and failure conditions are the most common. Essentially, these are terms used to describe potential "threats" or unintended consequences associated with a "system." So the first step in a safety engineering process is to identify the "threat" or unintended function. Once identified, the "threat" is assessed as to the severity of its effect on the "system." Based on the severity, a risk requirement is established. This is often referred to as "acceptable risk criteria" or in civil aviation as "inverse relationship" between severity and probability. Risk requirements may also include specific fault tolerance requirements such as no single failure for catastrophic severity.

Once the "problem" is identified, the next step in the process is to define the solution. In safety engineering, the solution comes from the risk criteria. Higher risks require more solutions or mitigations and lesser risk requires fewer mitigations. Mitigation (also referred to as "fail-safe design" or "hazard controls") can include such things as redundancy, independence, detection/indication, safety margins, and procedures. These mitigations are used to build a functional and physical architecture for the "system" that can achieve the risk requirements.

After mitigations are proposed, they are evaluated for effectiveness against meeting the risk criteria. This is part of the "determining the most effective solution." Safety analysis techniques are used to identify what within the architecture can cause the "threat" to occur and define mitigation requirements. Once the mitigations are identified and requirements allocated, the system development progresses through the design phase, and ultimately the as-designed implementation is verified to be effective.

This safety engineering process is shown as iterative. As mitigation strategies and architecture are developed, new "threats" might be created that require changes to the architecture. If mitigation strategies are

---

* "System" is used to refer to a generic reference to what is being analyzed in system safety. System, without quotations, is used when referring to aircraft system level.

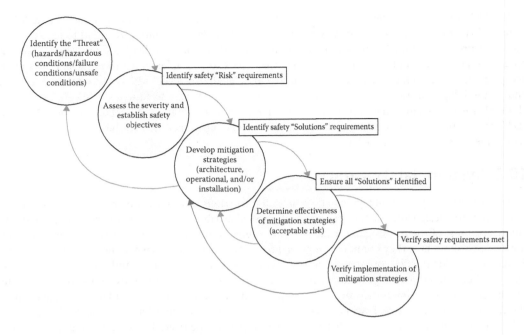

FIGURE 10.3   **(See color insert.)** Generic system safety process model.

found to be ineffective or incomplete, architecture changes may need to be made. It might also be that architecture implementation may change the severity of an existing "threat" that causes a change (more or less mitigations required). These iterations are considered a normal part of the process (reflected by the green arrows). If the process from "threat" identification to determining mitigation effectiveness is robust and occurs early in design, then it is less likely that costly design changes during final design and verification will be needed (red arrow showing iteration from verification back to design).

From Figure 10.3, it is easy to see how the safety engineering process is easily linked to the system development process: identifying safety "risk" requirements (capturing safety requirements), identifying safety "solutions" requirements and ensuring all "solutions" requirements are identified (validating safety requirements), and verifying safety requirements are met (verifying requirements). Although the safety engineering process always follows this generic model, different industries have different definitions of "risk criteria", different preferred analytical techniques that are used to identify threats and solutions, and different documentation requirements.

10.3 Civil Aviation System Safety Process

For civil aviation, the tailoring of the generic system safety or safety engineering process begins with the "1309" regulation. It, along with further definition provided by its AC 2X.1309, requires identification of failure conditions at the aircraft and systems levels including consideration of operating and environmental conditions (Figure 10.3: "Threat"), defines the "inverse relationship" between severity and probability of occurrence requirements that each failure condition has to achieve (Figure 10.3: Identify safety "risk" requirements), discusses application of a fail-safe design concept (Figure 10.3: Identify safety "solutions" requirements), and establishes the consideration that must be included in showing compliance to the "inverse relationship" criteria such as single failure modes, multiple failure modes, external failure modes (i.e., installation or physical hazards), latent failures, and crew interactions (Figure 10.3: Ensure all "solutions" identified and verify safety requirements met). Additionally, AC 2X.1309 outlines various analysis techniques commonly used to accomplish the safety process task of identifying, validating, and verifying safety requirements.

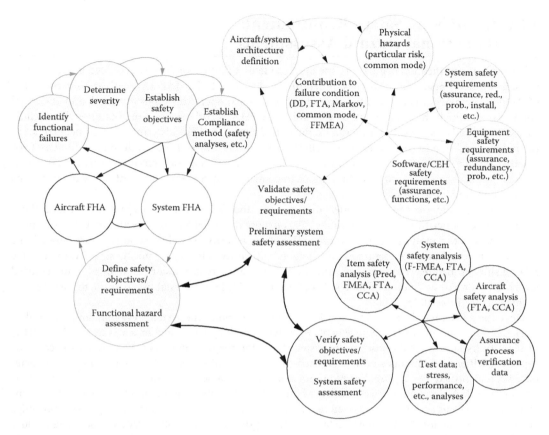

FIGURE 10.4 **(See color insert.)** Safety assessment process.

Figure 10.4 illustrates the aircraft system safety process as defined by AC2X.1309, SAE ARP4754A, and ARP4761. It should be noted that 14 CFR 25.1309(d) requires that an analysis must be performed to show compliance. A functional test by itself, even if it simulates failures of the system/equipment, is generally not sufficient. An analysis of the aircraft- and system-related functional failures is always required to establish its top-level safety probability (qualitative or quantitative) requirements and the minimum required safety analyses to show compliance.

The process illustrated in Figure 10.4 shows the activities and analyses that are accomplished as part of the requirement capture, validation, and verification activities. The system safety tasks are represented by the circles located above each of the three activities. The completion at each stage is documented by a functional hazard assessment (FHA), preliminary system safety assessment (PSSA), and system safety assessment (SSA). While the process is divided into three separate data products, the tasks and analyses of the FHA and PSSA are performed somewhat concurrently. A high-level (aircraft) FHA with primary functions is performed to identify requirements. To comply with requirements, certain architecture features are developed. Safety models of the architecture are created for each FHA failure condition to identify additional safety requirements. Architecture decisions and features may create new failure conditions that have to be added to the FHA. The process is repeated at the system level. Within system architectures, subfunctions (or secondary functions) are identified and need to be evaluated to determine their severity in order to establish the subfunction/equipment requirements (such as commands, monitors, redundant equipment) so they become part of the FHA or PSSA.

10.4 Identifying Safety Requirements (Functional Hazard Assessment)

Requirement identification begins with the FHA to identify failure conditions. As noted in Figure 10.4, the safety process begins at the highest indenture level (Aircraft FHA [AFHA]) with an understanding of the basic primary functions of the aircraft and associated operating, environmental, and abnormal condition for which the aircraft will be approved. The FHA postulates functional failures that include "loss of," "partial loss," and "malfunction" failure conditions. The FHA also considers the occurrence of the failure condition in the context of the effect if the crew is aware of the condition to compensate or whether the condition is not known to the crew so there is no crew compensation. This consideration of aware versus unaware failure conditions aids in assuring proper identification of safety requirements for monitors and indicators.

The FHA should rigorously consider all functions and failure conditions and not exclude those thought to be "noncritical" or "noncredible." The purpose of the FHA is to show all functions have been considered and the analysis identifies those with highest severity. As failure conditions are defined, the effects of the condition are identified and assigned a severity category. From the severity, the inverse relationship probability requirements, functional assurance requirements, and minimum required level of safety analysis ("depth of analysis") defined in AC2X.1309 are established as the top-level safety requirements (Table 10.1).

Once the AFHA has developed the top-level failure conditions, the system development process allocates the functions and safety requirements to systems. Each system undergoes a System FHA (SFHA) to identify its failure conditions. At the system level, interdependencies between systems need to be captured. Different techniques have been used to show that all system interactions and dependencies have been identified and captured as requirements. These include generating "exchanged functions" as part of the SFHA, developing cross matrices of the system and functions, or elevating all interdependencies to the AFHA as architecture-derived functions. Examples of interdependencies or interrelationships are exchanged functions that define interfaces *to* other systems and *from* other systems and system failure condition effects that rely on another system to compensate or reduce the overall severity of the effect.

Because the FHA is functional, it can be and should be initiated at concept development and continued to be refined throughout the architecture development process. Many development programs wait until the certification plan is complete or at least in draft form to begin the system safety process; this is too late. Reasons for the delay are often to save budget until it is determined that the concept being developed is a "viable" program. However, by this time, architectures have been defined and often agreement with equipment suppliers may be in place without consideration for the top-level safety requirements. The system safety process and the top-level safety requirements greatly influence the chosen architecture and its associated development. Defining architectures and equipment functional allocations without consideration of safety requirements fails to provide an adequate trade-off evaluation. One architecture implementation may result in function and/or item DAL requirements greatly different from other architectures. This directly corresponds to complexity and cost of the development and certification programs. Early identification of top-level safety requirements, including assurance levels, makes it less likely for costly delays and redesigns later in the development and certification processes.

10.5 Safety Requirement Validation (Preliminary System Safety Assessment)

Validation of the aircraft-level safety requirements is assuring that the set of defined safety requirements are correct and complete. Validation can be accomplished using several different methods including multidiscipline (safety, systems, crew/flight test, aerodynamics, etc.) review of functions, failure conditions, and effects; review of analyses (usually using FTAs) that allocate probability and FDAL requirements; review of requirement traces to assure all aircraft safety requirements are assigned to systems; and other engineering analysis, simulations, or testing that validate failure condition effect severity classification (as required). Results of aircraft-level validation are documented in a preliminary aircraft safety assessment (PASA).* Resulting system-level safety requirements should be captured in corresponding specifications. The final PASA may not be completed before the SFHAs are completed to ensure system interdependencies are captured but should be sufficient enough to support requirement reviews, preliminary design reviews, and completion of the certification plan.

The system-level PSSA is usually more detailed than a PASA. All SFHA failure conditions with a severity of major and higher are modeled (most typically using FTAs or dependency diagrams [DDs]) or qualitatively discussed in regard to the architecture and fail-safe design principles that support compliance to the 1309 regulation. The purpose of the PSSA is to document the capture and allocation of the safety requirements. It focuses on defining subfunctions within the systems, identifying common mode failures within the system architecture, and identifying safety requirements for equipment items. The PSSA is often considered to be a snapshot in time of the design: a draft version of an SSA. However, this is not correct. Don't be misled by the use of "preliminary" in its title. The PSSA is an architecture safety assessment with different contents and objectives from the SSA. If design changes occur to the architecture, the impact on the safety requirements must be evaluated and requirements revised accordingly. Therefore, the PSSA should be maintained current over the life of the product.†

There are many analytical techniques that can be used to aid in developing the FTA or DD models. Functional failure mode and effect analysis (FMEA), both hardware and software, aid in understanding the architecture, understanding information exchanged between different equipment items, and understanding how failures are detected. Failure rate allocations or predictions are used for determining reliability and safety requirements for equipment. The models can also be used to tailor the type of follow-on safety analyses required for subsystems or equipment. For example, if the model for loss of communications includes multiple radio equipment items with failure detection outside of the radio equipment, a simple functional FMEA with an equipment-level failure rate may be sufficient. On the other hand, if the failure condition model includes an equipment item that has internal redundancy or internal detection, then a more detailed FMEA and prediction supported with a common mode analysis to verify independence may be more appropriate. Using the early PSSA to refine the specific types of safety analyses and level of detail specified in Suppliers' statements of work and contract deliverables can reduce costs and improve the likelihood that contract data deliverables will support the system and aircraft-level analyses. Issues regarding level of detailed required versus proprietary/intellectual property issues can be identified and managed early. Limitations can be reflected in the safety analysis and allocation of safety requirements.

The PSSA validation process can be used at the equipment level. If an equipment item integrates many different functions and safety requirements or contains architecture with multiple failure mitigations (independent, redundant items), it is wise to review the requirements against the equipment architecture to ensure the allocated safety requirements to the equipment can be achieved. It also documents lower-level equipment safety requirements.

In addition to the functional safety requirements, the PASA and PSSA also document the physical installation safety requirements. As soon as aircraft systems and equipment locations are defined, "external events" can be identified and requirements defined (refer to ARP4761, common cause analysis—particular risk and zonal). Architecture requirements to limit the effects of common physical hazards are established (i.e., all failure condition mitigations are not located in the same area and affected by a single external

---

* The PASA was introduced in ARP4754A to document the allocation of aircraft safety requirements to systems and capture system interdependencies. The PASA is being added to ARP4761 currently under revision. Refer to ARP4754A for more information on contents.

† Maintenance of FHAs and PSSAs over the type certification life needs to be part of the system development and safety planning process. Databases may be used to accomplish the maintenance of safety requirements and document review of derived requirements.

event). Installation locations may create unique internal aircraft or microenvironments (loss of cooling, specific fluid exposure, HIRF/lightning, etc.). Early identification of physical "threats" to failure condition mitigations is needed to assure adequate separation or segregation* requirements are specified.

The important final step of the PASA/PSSA is to actually incorporate the safety requirements into specifications or requirements databases. It seems like an obvious task but one that is often overlooked. Usually this is because the PASA/PSSA is being done too late in the development and specifications or contracts have already been put in place. Without incorporation into specifications, then allocation of an assured (validated as correct and complete) set of requirements to software/ AEH cannot be established. Additionally, if safety requirements are not clearly defined and maintained, then future modification projects tend to continually "reinvent the wheel" in terms of safety assessments.

10.6 Derived Requirements

Requirements that are derived during the design process and are not traceable to another requirement need to be evaluated for impact on the failure conditions (FHAs) and safety requirements (PSSA). Documenting these safety impact evaluations in the PSSA provides the artifacts/data to support the DO-178/254 objectives for safety review of derived requirements. The safety impact review of lower-level requirements may be documented outside the PSSA; however, actual analysis results need to be captured. The review analysis should be a systematic, structured consideration of the requirements impact on functions, failure conditions, effects and severity classifications, fault tolerance, failure rates, failure detection, safety assumptions, and independence requirements. Safety analyst/engineer attendance at requirement reviews alone would not be sufficient. A completed safety impact review checklist documented with the minutes of the meeting would provide more sufficient data/artifacts for demonstrating the derived requirements have adequately been reviewed for impact.

10.7 Defining Development Assurance Levels

Each failure condition identified in the FHAs is assigned an FDAL based on severity. The FDAL assignment is applied to the aircraft- and system-level development processes. Functional independence ensures that the functional requirements that are implemented in the design are different. From this FDAL, individual equipment, software, and AEH can be assigned an IDAL considering the architecture related to the failure condition. Item design independence ensures that the hardware or software design, in which the functions are implemented, is different between independent items. Typically used aircraft and system architectures include dissimilar redundancy, command/monitor, and mechanical devices protecting from software/AEH errors, etc. With use of these types of architecture, an error in development of one item wouldn't affect or create the same error in the other item. Because of the independence and containment of the error effects, one or more IDALs may be at a level lower than the failure condition FDAL. ARP4754A defines the criteria, limitations, and methodology that are used to allocate system FDALs and assign IDALs.

In order to allocate DALs, an FHA and architecture definition is needed. Another key aspect that has to be understood is the architecture's susceptibility to common mode failures. Common modes are common characteristics or potential failures (random or systemic) that affect multiple items thought

---

* *Separation* is where independence is achieved through distance. The concept is that any failure in one item/equipment will not affect the independent item based on the distance between them. *Segregation* is achieving independence through a physical barrier. If independence by segregation is desired, the safety requirements should specify segregation. Otherwise, separation most likely will be used because it is often easier, cheaper, and lighter.

(or desired) to be independent. ARP4761* includes a common mode checklist that can be used to aid in identifying susceptibilities to common mode failures. When looking for common modes in architecture, it helps to adopt the philosophy of "guilty until proven innocent." The common mode exists unless it can be shown not to exist or has been minimized through assurance processes.

For Part 23 aircraft, AC23.1309-1E defines the DAL assignment requirements that are to be used for each aircraft class. These are used instead of the ARP4754A process. The development objectives to be accomplished still correlate to the ARP4754A and DO-178/DO-254 objectives for the AC-defined levels. For example, in a class I aircraft with a catastrophic failure condition, the FDAL and IDAL will be for level C. ARP4754A Appendix A requirement capture, validate, and verify recommended objectives for level C would be accomplished. DO-178 and DO-254 level C objectives would be followed for the items. For class IV aircraft, the ARP4754A criteria and method may be used if approved by the small aircraft directorate or aircraft certification (refer to paragraph 21.e(3) in AC23.1309-1E for more details and specific approval requirements).

FDALs and IDALs are identified as information required in certification plans for a project. It is premature to review DALs with certification authorities until there is sufficient safety analysis to provide rationale for the DAL assignments. To justify DALs, an FHA and PSSA will be required. Therefore, the FHA and PSSA must be completed (or at least a fairly well-developed draft) early in the project.

10.8 Lightning Certification Level

Protection against the effects of lightning for aircraft electrical and electronic systems, regardless of whether "indirect" or "direct," is addressed under 14 CFR 23.1306, 25.1316, 27.1316, and 29.1316. The terms "indirect" and "direct" are often used to classify the effects of lightning. However, the regulations do not differentiate between the effects of lightning. The focus is to protect aircraft and systems from effects of lightning transients induced in electrical and electronic system wiring and equipment and lightning damage to aircraft external equipment and sensors that are connected to electrical and electronic systems, such as radio antennas and air data probes.[†]

In preparing a lightning certification plan, results of the FHA and PSSA can be used to aid in identifying the effects of lighting and the required system lightning certification level. These levels are defined in AC20-136. AC20-136 also notes: "The specific aircraft safety assessment related to lightning effects required by 14 CFR 23.1306, 25.1316, 27.1316, and 29.1316 takes precedence over the more general safety assessment process described in AC 23.1309-1, AC 25.1309-1, AC 27-1, and AC 29-2. Lightning effects on electrical and electronic systems are generally assessed independently from other system failures that are unrelated to lightning, and do not need be considered in combination with latent or active failures unrelated to lightning." Lightning is considered in the PSSA as an "external event" as part of common cause analysis (particular risk and zonal analyses). Compliance to 14 CFR 23.1306, 25.1316, 27.1316, and 29.1316 is used as mitigation for the common cause failures.

As with the DAL rationale, the FHA and PSSA need to be initiated early to be available to support the lightning certification planning process. Chapter 11 (DO-160) provides additional discussion on HIRF/lightning.

10.9 Safety Verification (System Safety Assessment)

Figure 10.5 provides an illustration of the various analyses performed and documentation prepared in the system safety process. The horizontal lanes represent different activities based on the indenture level of the items under analysis. The safety requirements and validation discussions included the allocation

---

* ARP4761 CMA example shows what *justification* is used to minimize the common mode as an element of verification as part of the SSA. It is equally important to complete the checklist during the PSSA process and specify a requirement to mitigate the susceptibility.

† Based on AC20-136, paragraphs 5.b. and 6.c

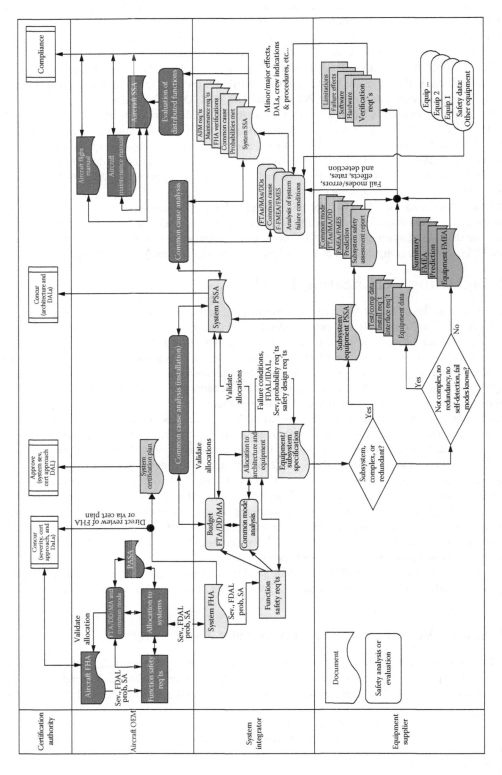

FIGURE 10.5   **(See color insert.)** Integrated safety assessment process from highest to lowest indenture.

of requirements from the aircraft down to equipment items. Verification is the bottom-up process shown as starting at the equipment level and rolling up to the system and aircraft level.

The SSA documents all activities, analyses, tests, crew procedures, maintenance actions, limitations, etc., that are used to show compliance to the safety requirements identified by the capture and validation phases. It integrates results from all the verification activities as they relate to each failure condition for the purpose of showing that the failure condition is compliant with all probability, fault tolerance, and assurance requirements defined by the "1309" regulation. ARP4761 provides further definition of the SSA and includes an example. As a minimum, the SSA should provide the following:

1. A functional and physical system description consistent with the level of detail provided in the safety analyses. This is to aid the document reviewer in understanding the analyses and the design mitigations, as well as being able to conclude that the analyses are complete and correct for the defined failure conditions.

2. A definition of the configuration analyzed including installations, equipment, and software/AEH including revision levels. The assessment covers a specific configuration or group of configurations. If the configuration or design changes, then impact on the safety assessment must be accomplished.

3. A list of all assumption and conditions (sometimes referred to as ground rules) used in the SSA and any safety analyses contained in the SSA.

4. A failure condition verification summary—a list of safety requirements from FHA and PSSA and verification results referencing the detailed analyses. The actual safety analyses may be part of the SSA or separate reports and referenced from SSA.

5. Discussion of the assurance process results with reference to documentation of compliance to the process objectives. Any open items from the item development should be reviewed for impact on overall safety but also impact on the safety analysis results and assumptions.

6. Discussion of common cause analysis (common mode, particular risk, and zonal) and how these can affect catastrophic and hazardous failure conditions.

7. Discussion of latent failures and whether it is acceptable for them to remain latent based on the probability analyses, as well as consideration of unsafe system operating conditions (i.e., 25.1309(c)).

8. A list of crew actions that were used to mitigate or reduce the severity of failure condition and reference to the airplane flight manual procedure (or requirement for procedure if Airplane Flight Manual [AFM] isn't available or released) that implements the action or the crew evaluation that determined a specific AFM entry was not required (accepted as normal crewmanship that is part of normal training programs).

9. A list of maintenance actions used in the safety analysis for hazardous and catastrophic failure conditions that are used to limit failure exposure time. These are defined as candidate certification maintenance requirements (CMRs) and will be reviewed per AC25-19A.

10.10 EWIS

14 CFR 25 Amendment 25-123 added new regulations related to electrical wiring interconnect system (EWIS). It establishes specific regulations and also adds 14 CFR 25.1709 EWIS system safety that requires an analysis to demonstrate compliance to "inverse relationship" criteria. AC25.1701-1 highlights that both the functional aspects of wiring and physical failure modes of wiring must be addressed in the safety assessment process.

The functional aspects can be captured as an integral part of the 25.1309 system safety process. Wire bundle/connector architecture can be included in PSSA as contributors to failure conditions and separation requirements established. The SSA analyses can include results from the EWIS separation and physical hazard analysis. AC25.1701-1 provides additional guidance on EWIS safety analysis. See Chapter 25 for further information on EWIS.

10.11 Use of TSOs in Safety Assessments

Many technical standard orders (TSOs) include severity classifications for functions defined by the TSOs. This allows the equipment developed to meet the TSO to follow DO-178/254 for the software and AEH. However, an aircraft- and system-level safety assessment is still required to determine if the assurance levels published in the TSO are adequate for the specific application and architecture. There may be some disparities between TSO functional severity and the level required based on architecture implementation.

Care should also be used in making assumptions in the aircraft- or system-level safety analyses regarding what aspects are covered by the TSO designation. Many TSOs do not specify a safety analysis as part of the data to be submitted to the FAA. The failure modes, rates, and means of detection can be different between different designs that both meet the same TSO functional requirements. The TSO item's safety analysis must be obtained and used as source data for the system-level analyses.

10.12 MIL-STD-882 System Safety

MIL-STD-882E defines system safety as used by the Department of Defense. It covers all aspects of safety from cradle to grave from the perspective of the developer, manufacturer or installer, operator or user, and maintainer for a wide variety of "systems." So its scope is broader than aircraft certification and airworthiness. MIL-STD-882 defines a different approach to risk criteria and software safety assessments and has several different hazard analysis techniques that are not used in the commercial process. However, as illustrated in Figure 10.6, if the tasks are related back to the generic model, you can see how the various analysis tasks of MIL-STD-882 are designed to accomplish similar objectives.

Figure 10.6 is not intended to imply that the analyses are identical in methodology but rather indicate that with some tailoring, the two processes could be blended for systems or equipment that will be used in both applications. Figure 10.7 provides additional cross-references between the two processes. If the desire is to use MIL-STD-882 as a governing system safety process while developing data that could also be used to support commercial certification, the analysis tasks and data item descriptions will need to be tailored to ensure that the FHA (task 208), system requirements hazard analysis (task 203), and system-of-systems hazard analysis (task 209) and the safety assessment report (task 301) follow the ARP4761 process. Similarly, if the goal is to use the ARP4761 process for satisfying MIL-STD-882 tasks, remember that an ARP4761 SSA doesn't discuss residual risk. It only addresses compliance to the "1309" regulations. Also, the "1309" regulations and ACs do not require system safety program plans (although ARP4754A does include an example as part of the integrated planning discussions). The ARP4761 process also does not produce analyses equivalent to the operating and support hazard analysis, health hazard analysis, environmental hazard analysis, or explosives data tasks. Some aspects of the operating and support hazard analysis objectives are accomplished as part of zonal analysis, but the zonal analysis may not accomplish the procedure's review and analysis aspects that would be addressed in the operating and support hazard analysis.

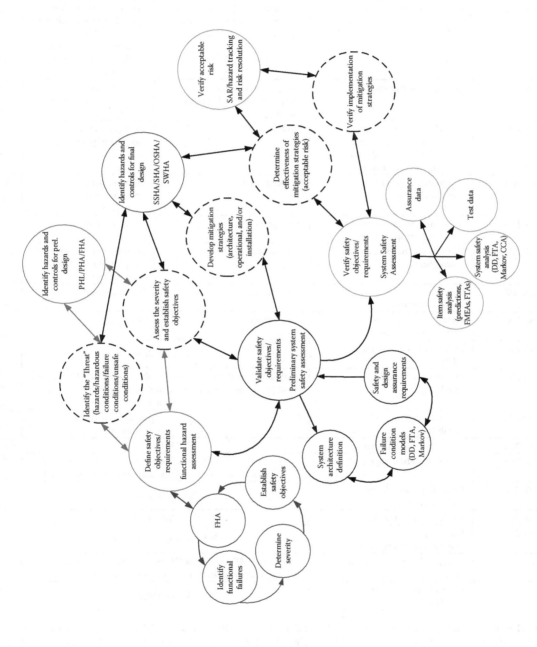

FIGURE 10.6 Relationship of commercial aviation and MIL-STD-882 safety assessment processes.

	MIL-STD-882E	ARP4761	ARP4754A
Task 101	Hazard identification and mitigation effort using the system safety methodology	Top-down requirement allocation and bottom-up verification	Integration of system development and system safety
Task 102	System safety program plan	No equivalent (accomplished in certification plans with certification authority)	System safety program plan (content is different due to system safety process differences and only addressed aircraft certification)
Task 103	Hazard management plan	Integration of system development and system safety	Integration of system development and system safety
Task 103	Integration/management of associate contractors, subcontractors, and architect and engineering firms	No equivalent	No equivalent; only as part of planning
Task 104	Support of government reviews/audits	No equivalent	Only as part of requirement and program reviews
Task 105	Integrated product team/working group support	No equivalent	No equivalent
Task 106	Hazard tracking and risk resolution	No equivalent	No direct equivalent; only as part of requirement traceability
Task 107	Hazard management progress report	No equivalent	No equivalent
Task 108	Hazardous materials management plan	No equivalent	No equivalent
Task 201	Preliminary hazard list (PHL)	No equivalent	
Task 202	Preliminary hazard analysis	Early tasks for common mode analysis, zonal safety analysis, and particular risk analysis that examine physical (intrinsic) hazards	
Task 203	Safety requirements hazard analysis	PSSA	Requirement capture and validation
Task 204	Subsystem hazard analysis	FMEA	
Task 205	System hazard analysis	System FMEA, FTA (or DD or Markov), common mode analysis, zonal safety analysis, and particular risk analysis	
Task 206	Operating and support hazard analysis	No equivalent	
Task 207	Health hazard assessment	No equivalent	
Task 208	Functional hazard analysis (system and component)	FHA (aircraft and system)	
Task 209	System-of-systems hazard analysis	System FMEA, FTA (or DD or Markov), common mode analysis, zonal safety analysis, and particular risk analysis	
Task 210	Environmental hazard analysis	No equivalent	

FIGURE 10.7 Cross-reference between MIL-STD-882E and ARP4761 and ARP4754A. (Extracted from SED-PMHDBK-ASA, U.S. Army Software Engineering Directorate, *Program Manager's Handbook for Aviation Software Airworthiness.*) *(Continued)*

	MIL-STD-882E	ARP4761	ARP4754A
Task 301	Safety assessment report	Aircraft SSA and System-level SSA Same except for residual risk, risk acceptance, hazardous materials, and maintenance personnel safety concerns	
Task 302	Hazard management assessment report	No equivalent	
Task 303	Test and evaluation participation	No equivalent (some for companies and risk analysis prior to flight with certification authority imposed by other FAA policies and ACs)	
Task 304	Safety review of engineering change proposals, change notices, deficiency reports, mishaps, and requests for deviation/waiver	No equivalent; however, required for maintaining continued airworthiness regulations	Modification to aircraft analysis tailoring guidelines
Task 401	Safety verification	Part of SSA and/or compliance reports (address compliance to applicable regulations not just safety requirements)	Requirement verification
Task 402	Explosive hazard classification data	No equivalent	
Task 403	Explosive ordnance disposal data	No equivalent	

FIGURE 10.7 (*Continued*) Cross-reference between MIL-STD-882E and ARP4761 and ARP4754A. (Extracted from SED-PMHDBK-ASA, U.S. Army Software Engineering Directorate, *Program Manager's Handbook for Aviation Software Airworthiness.*)

10.13 Concluding Summary

For effective aircraft and system development, the system safety process must be integrated with the development process. The system safety tasks and objectives are dependent upon a top-down development process. The delay of FHAs/PSSAs means the development process requirements aren't being followed and system and equipment designs will be impacted.

STCs must follow the same process as TC for both system development and system safety. Objectives of the safety and development processes are the same only that their complexity may be less depending upon the scope of the change and severity of related or impacted functions. Regardless of the functional severity classification of the modification, physical installation should always be addressed in the context of 2X.1309.

The system development and safety processes are to ensure designs and installations comply with regulations. Documented results of the process are needed to show compliance.

Further Reading

Ericson, Clifton A. II, *Hazard Analysis Techniques for System Safety*, John Wiley & Sons, Hoboken, NJ, 2005.

Fault Tree Handbook with Aerospace Applications, NASA-FTH Version 1.1, August 2002.

Fault Tree Handbook, NUREG 0492, January 1981.

Handbook of Reliability Prediction for Mechanical Equipment, Naval Surface Warfare Center, Dahlgren, VA, 2011.

Kritzinger, Duane, *Aircraft System Safety, Military and Civil Aeronautical Applications*, CRC Press, Boca Raton, FL, 2006.

O'Connor, Patrick D. T., *Practical Reliability Engineering*, 4th Edition, John Wiley & Sons Ltd., Chichester, U.K., 2007.

Rierson, Leanna, *Developing Safety-Critical Software: A Practical Guide for Aviation Software and DO-178C Compliance*, CRC Press, Boca Raton, FL, 2013.

SAE ARP5580, *Recommended Failure Modes and Effects Analysis (FMEA) Practices for Non-Automotive Applications*, Society of Automotive Engineers, Warrendale, PA, 2001.

SAE ARP5150, *Safety Assessment of Transport Airplanes in Commercial Service*, Society of Automotive Engineers, Warrendale, PA, 2003.

System Reliability Toolkit, SRKIT, Reliability Information Analysis Center (No Date).

References

Advisory Circular 20-115C, *Airborne Software Assurance*, 2013.

Advisory Circular 20-136B, *Aircraft Electrical and Electronic System Lightning Protection*, 2011.

Advisory Circular 20-152, RTCA, Inc., *Document RTCA/DO-254, Design Assurance Guidance/or Airborne Electronic Hardware*, 2005.

Advisory Circular 20-174, *Development of Civil Aircraft and Systems*, 2011.

Advisory Circular 23.1309-1E, *System Safety Analysis and Assessment for Part 23 Airplanes*, U.S. Department of Transportation Federal Aviation Administration, 2011.

Advisory Circular 25-19A, *Certification Maintenance Requirements*, U.S. Department of Transportation Federal Aviation Administration, 2011.

Advisory Circular 25.1309-1A, *System Design and Analysis*, U.S. Department of Transportation Federal Aviation Administration, 1988.

Advisory Circular 25.1309-1B Arsenal Draft, *System Design and Analysis*, U.S. Department of Transportation Federal Aviation Administration, 2002.

Advisory Circular 25.1701-1, *Certification of Electrical Wiring Interconnect Systems on Transport Category Aircraft*, U.S. Department of Transportation Federal Aviation Administration, 2002.

Advisory Circular AC 27-1B Change 3, *Certification of Normal Category Rotorcraft, Section 27.1309 Equipment, Systems, and Installations*, U.S. Department of Transportation Federal Aviation Administration, 2008.

Advisory Circular 29-2C Change 3, *Certification of Transport Category Rotorcraft, Section 29.1309 Equipment, Systems, and Installations*, U.S. Department of Transportation Federal Aviation Administration, 2008.

Code of Federal Regulations, Title 14, *Aeronautics and Space*.

MIL-STD-882E, *System Safety*, Department of Defense, 2012.

RTCA DO-178C, *Software Considerations in Airborne Systems and Equipment Certification*, RTCA, Inc., Washington, DC, 2011.

RTCA DO-254, *Design Assurance Guidance for Airborne Electronic Hardware*, RTCA, Inc., Washington, DC, 2000.

RTCA DO-278A, *Software Integrity Assurance Considerations for Communication, Navigation, Surveillance and Air Traffic Management (CNS/ATM) Systems*, RTCA, Inc., Washington, DC, 2011.

SAE ARP4754A, *Guidelines for Development of Civil Aircraft and Systems*, Society of Automotive Engineers, Warrendale, PA, 2010.

SAE ARP4761, *Guidelines and Methods for Conducting the Safety Assessment Process on Civil Airborne Systems and Equipment*, Society of Automotive Engineers, Warrendale, PA, 1996.

11

Understanding the Role of RTCA DO-160 in the Avionics Certification Process

Donald L. Sweeney

D.L.S. Electronic Systems, Inc.

D.L.S. Conformity Assessment, Inc.

11.1 Introduction

RTCA SC-135 is a standing committee updating requirements and technical concepts for aviation document (RTCA DO-160) as new environmental issues are noted in use or when new technology is introduced that deserves a new method of testing. DO-160 standards are coordinated by RTCA SC-135 and the European Organisation for Civil Aviation Equipment (EUROCAE) Working Group (WG) 14.

Please note that there are many versions of DO-160 that are regularly used. It is imperative that the specific version required for a particular application (defined by the certification basis of the aircraft) be used. In 2013, testing is rarely done to version D, while E, F, and G are regularly used. G is the most recent version and will most likely be used for new systems. There are, in some cases, significant changes from one version to another; therefore, testing to the latest version might not meet the intended requirements for the product. You can find the latest version of DO-160 that is currently in effect, listed at www.dlsemc.com/DO-160testing along with alerts of anticipated changes and possible errata to the standard and this write-up.

11.1.1 Why Use DO-160?

RTCA DO-160 environmental testing is required for various components used in an avionics system to determine their ability to perform as specified by the product's specification defined by the purchaser, manufacturer, or governmental agency, such as the federal aviation administration (FAA). In keeping with the aviation regulations that have two parts—one to prove intended use and another to assure, to the extent that is practicable, that there are no unintended functions—environmental testing checks performance of a device in the following ways:

1. The device must not interfere with other equipment near it, such as radio-frequency emission energy (Section 21.2).
2. The device must operate as intended when acted upon by environmental effects from nearby equipment, such as radio-frequency susceptibility (radiated and conducted), Section 20; shock and vibration, Sections 7 and 8; sand and dust, Section 12; and even the shock and the effects of being involved in a plane crash, Section 7.

DO-160 allows one to develop requirements that match with the actual environment in which the device will be operated. The entire goal of these requirements is to make the aircraft as safe as is reasonably possible.

11.1.2 Reason for Testing

Each required test has a specific purpose. DO-160 defines a series of minimum standard environmental test conditions (categories) and applicable test procedures for airborne equipment. The purpose of these tests is to provide a laboratory the means of determining the performance characteristics of airborne equipment in environmental conditions that may be encountered.

The standard environmental test conditions and test procedures contained herein may be used in conjunction with applicable equipment performance standards as a minimum specification under environmental conditions, which can ensure a sufficient degree of confidence in performance during operations.

11.1.3 Test Procedure

In preparation for performing the tests, a test procedure should be generated and it should be agreed upon by all parties involved, from the manufacturer/designer to the FAA or the agency granting final approval of the airworthiness of the components or final assembly.

If one waits until the final submittal before having the tests approved, there will very likely be discussions that could require that some or all of the testing be performed over. Any retesting will add cost and, even worse, delay the overall project.

11.1.4 Guidelines

There are user guidelines at the end of various sections within DO-160. They provide background information for the associated test procedures and requirements described in the main body of that section. This information includes rationale for requirements, guidance in applying the requirements, commentary, possible troubleshooting techniques, and lessons learned from previous laboratory experience. The user's guidelines follow a parallel format to the main body of that section for easy cross-reference. Paragraph numbers correspond to each paragraph in the section and in the user's guidelines.

11.1.5 Sequencing

Often it is very important to consider the sequence of the test. Sequencing may require one test to be performed before another in order not to interfere or bias the results of the next test with residual effects from a previous test. For example, fungus test must be before salt spray, sand, and dust. It is advisable to save destructive tests for last especially in cases where only a limited amount of equipment is available for testing. If testing is allowed on multiple units, then the order of the test might not be important. Generally, environmental tests start with radiated emissions, then conducted emissions. Problems found here when mitigated will make other tests more likely.

11.1.6 Shared Information

There are parts of the DO-160 standard that are stated in one section yet could relate to many or all sections. For example, many sections comment that information contained in this section is pertinent to all test procedures described in the other sections of this document. However, this information may not be stated in the other sections. For example, Appendix A identifies environmental tests that may need to be performed in other sections. Appendix C provides the names of RTCA special committee members who lead specific subject matter. When problems are found or improvements are recognized in these subjects, specific individuals listed in this appendix should be notified so that appropriate actions are taken and information is shared with the community.

11.2 Table of Contents from DO-160G

The following table of contents is taken from the latest standard (Rev G).

Section 1, Purpose and Applicability
Section 2, Definitions of Terms—General
Section 3, Conditions of Tests
Section 4, Temperature and Altitude
Section 5, Temperature Variation
Section 6, Humidity
Section 7, Operational Shocks and Crash Safety
Section 8, Vibration
Section 9, Explosion Proofness
Section 10, Water Proofness
Section 11, Fluids Susceptibility
Section 12, Sand and Dust
Section 13, Fungus Resistance
Section 14, Salt Spray
Section 15, Magnetic Effect
Section 16, Power Input
Section 17, Voltage Spike
Section 18, Audio Frequency Conducted Susceptibility—Power Inputs
Section 19, Induced Signal Susceptibility
Section 20, Radio Frequency Susceptibility (Radiated and Conducted)
Section 21, Emission of Radio Frequency Energy
Section 22, Lightning Induced Transient Susceptibility
Section 23, Lightning Direct Effects
Section 24, Icing
Section 25, Electrostatic Discharge
Section 26, Fire, Flammability
Appendix A, Environmental Test Identification
Appendix B, Membership
Appendix C, Change Coordinators

Note: A detailed table of contents appears in most sections.

The following is a review of each section of DO-160 and is intended as an introduction and overview of the standard, not a substitute for having a copy of the official standard.

Section 1: Purpose and Applicability

DO-160 defines a series of minimum standard environmental test conditions (categories) and applicable test procedures for airborne equipment. The purpose of these tests is to provide a laboratory means of determining the performance characteristics of airborne equipment in environmental conditions representative of those that may be encountered in airborne operation of the equipment.

The standard environmental test conditions and test procedures contained in DO-160 may be used in conjunction with applicable equipment performance standards as a minimum specification under environmental conditions, which can ensure a sufficient degree of confidence in performance during operations.

Note:

In each of the test procedures contained in DO-160, the following phrase will be seen several times: "DETERMINE COMPLIANCE WITH APPLICABLE EQUIPMENT PERFORMANCE STANDARDS."

The "applicable equipment performance standards" referred to are either of the following:

a. EUROCAE minimum operational performance specifications (MOPS), formerly requirements (MOPR)
b. RTCA minimum performance standards (MPSs) and/or RTCA minimum operational performance standards (MOPS)
c. The manufacturer's equipment specification(s), where applicable

Some of the environmental conditions and test procedures contained in DO-160 are not necessarily applicable to all airborne equipment. The selection of the appropriate and/or additional environmental conditions and test procedures is the responsibility of the writers (authors) of the performance standards for the specific airborne equipment.

There are several additional environmental conditions (categories) that specific airborne equipment may need to be evaluated against that have not been included in DO-160. These include, but are not limited to, hail, acceleration, and acoustic vibration. The procedures for testing airborne equipment for special environmental conditions that are uniquely related to that specific type of airborne equipment should be the responsibility of the writer (author) of the performance standard for that specific equipment. The International System of Units is usually used throughout the document as the primary value. In certain instances, however, when the primary values were derived in the U.S. units, the U.S. units are used as the primary values. Subject to the provisions of Subsection 3.2, it is permissible to use more than one test article. It is the responsibility of those who wish to apply the test conditions and procedures contained in DO-160 to determine the applicability of these test conditions and procedures to a specific equipment intended for installation on, or within, a specific or general class or type of aircraft.

The applicant will need to read each section of DO-160 and base decisions on who is buying the device and what their specification calls out as applicable environmental limits within which the aircraft will operate. Test procedures should be written and approved before beginning the testing. Questions that need to be answered before the testing starts, in order to determine the category of testing, are as follows:

Where will it be installed on and/or in the aircraft?
Will it be internal or external to the aircraft? If internal, will it be in the cockpit, avionics bay, or main cabin?
What is the orientation of installation?

The aircraft manufacturer has the right to massage, manipulate, or add to the test requirements of DO-160 as long as these changes are justified for the specific installation. Companies have the right to require the test procedures be followed in the order they wish except where dictated by the standard. Usually, environmental testing begins with testing to support safety of flight when the equipment has to pass a set of minimum tests before the aircraft can fly test flights. Worst-case testing is required to prove safety.

MOPS prepared by RTCA, Inc., for airborne equipment contain requirements that the equipment must meet to ensure reliable operation in actual aeronautical installations. These equipment requirements must be verified in ambient and stressed environmental conditions. The MOPS typically contain recommended bench test procedures for ambient conditions and refer to RTCA Document DO-160, Environmental Conditions and Test Procedures for Airborne Equipment, for the stressed environmental testing. The test categories defined in DO-160 are intended to encompass the full spectrum of environmental conditions that airborne equipment may experience from benign to very hostile.

The environmental conditions and test procedures defined in DO-160 are intended to determine only the performance of the airborne equipment under the stated environmental conditions and are not intended to be used as a measure of service life of the airborne equipment subjected to these tests.

Any regulatory application of DO-160 is the sole responsibility of appropriate governmental (regulatory) agencies.

Historical Note and General Guidance to Users

DO-160 (or its precursor, DO-138) has been used as a standard for environmental qualification testing since 1958. It has been referenced in MOPS for specific equipment designs and is referenced in FAA Advisory Circulars as a means of environmental qualifications for Technical Standard Order (TSO) authorization. It has been subjected to a continuous process of upgrade and revision as new needs arise within the aviation community, as improved test techniques have emerged, and as the reality of equipment operation under actual environmental conditions has become better understood.

Environmental stresses can result from natural forces or man-made effects and may be mitigated by details of the equipment installation. The categories that have been developed over time reflect a reasonably mature understanding as to the severity of the stresses, the degrees of mitigation achievable in the design of an installation, and the robustness that must be designed into equipment in order to perform under the resultant stress. In order to fully reap the advantages of the maturity of this document, it is incumbent upon the designers of the installed equipment, as well as the designers of the host installation, to consider the categories defined herein as early in their programs as feasible. The categories defined within each environmental test procedure have proved to be a practical set of boundary conditions between the requirements of real-world installations and the performance of installed equipment. Effective dialogue between the airframe and equipment designers is essential to ensure that correct categories are utilized.

DO-160 User Guides

Various sections within DO-160 now contain user's guide material. This material appears at the end of the respective section. The user guides provide background information for the associated test procedures and requirements in the main body of DO-160. This information includes rationale for requirements, guidance in applying the requirements, commentary, possible troubleshooting techniques, and lessons learned from previous laboratory experience. This information is intended to help users understand the intent behind the requirements and to help users develop detailed test procedures based on the general test procedures in this document. These user guides are provided for guidance purposes. The user guides do not contain requirements. Only the procedures in the body of each section are required to be applied to the equipment under test (EUT).

The user guides follow the same general format as the main body of the section. Main body paragraph numbers are repeated in the user guide.

Section 2: Definition of Terms, General

This section contains the definitions of general terms that are utilized throughout this document. The definition of terms specific to a particular section may be found in the appropriate section. Only those definitions that contribute to this article are further explained in the following.

Category of Tests and Declarations
For each environmental condition addressed in this document, the equipment supplier should select, from categories defined within the particular sections, a category that best represents the *most severe environment* to which the equipment is expected to be regularly exposed during its service life. An exception to this rule is the use of Category X as explained in the following. The selected categories are to be tabulated on the environmental qualification form in accordance with the guidelines presented in Appendix A, which can be found at the end of the standard. For any category listed on the environmental qualification form, it can be derived that the equipment would also be able to perform its intended function(s) when exposed to categories that can be assessed as being less severe. Use of Category X on the environmental qualification form in association with any environmental test procedure of DO-160 is reserved for the case where the equipment supplier wishes to indicate that compliance with equipment performance standards has not been demonstrated under the environmental conditions addressed by that particular procedure. If any category listed on the environmental qualification form is not sufficient for a specific aircraft certification, then additional testing may be required.

When the statement "DETERMINE COMPLIANCE WITH APPLICABLE EQUIPMENT PERFORMANCE STANDARDS" is found at the end of or during the test procedures, it should be understood that performance compliance and verification is considered to be the requirement that allows the equipment to be certified as to its ability to perform its intended function(s) during and/or after a specific test category.

Applicability of Test Results
The results of the testing using these procedures are only valid for the test configuration (test setup, external configuration, and internal configuration) used during the qualification tests. Any change from this configuration, either externally or internally (such as PCB layout, component change inside the unit, installation wiring), must be assessed to ensure that the test results are still applicable. If an assessment cannot show that the results are still applicable, retesting is required.

Section 3: Conditions of Tests
This is an introduction to how the equipment might be configured and conditions measured during the test. This section deals with connection and orientation of equipment, order of tests, multiple test articles, combining tests, measurement of air temperature in the test chamber, ambient conditions, test condition tolerances, and configuration for susceptibility tests. These are all the test setups and measurements that are expected for environmental testing.

Unless otherwise stated, the equipment's electrical and mechanical connections including any cooling provisions (see Figure 11.1) must be identical to how it is intended to be mounted on an actual aircraft.

Where not specified, power leads shall be 1.0 m ± 10 cm and interconnecting cables shall be at least 3.3 m. Any inputs or outputs to or from other equipment normally associated with the EUT shall be connected or adequately simulated. *Note*: Paragraphs 19.3 and 20.3, if applicable, will require an interconnecting cable longer than these minimums (see Figure 11.2).

Additional lengths may be required in some sections. If installation is known, use the actual cable lengths for testing. In general, the setup for the testing for the worst case should be to choose a cable length that avoids the need to retest the system multiple times. If EUT products will be sold both with and without shielded cables, the unshielded cable must be tested. It is prudent to test both cases as it is possible that the shielded cable might be the worst case. The test setup is aircraft specific in some cases, such as direct lightning.

It is acceptable to employ alternate procedures developed as combinations of the procedures provided it can be demonstrated that all applicable environmental conditions specified in the original procedures are duplicated or exceeded in the combined procedure. If alternate procedures are used, appropriate information should be provided along with the environmental qualification form (see Appendix A of DO-160).

FIGURE 11.1 Cooling provisions during testing. (Courtesy of D.L.S. Electronic Systems, Inc., Wheeling, IL.)

FIGURE 11.2 Complete wiring harness from an aircraft. (Courtesy of D.L.S. Electronic Systems, Inc., Wheeling, IL.)

All stimulus and measurement equipment used in the performance of the tests should be identified by make, model, serial number, and the calibration expiration date and/or the valid period of calibration where appropriate. When appropriate, all test equipment calibration standards should be traceable to national and/or international standards.

If the equipment to be tested consists of several separate units, these units may be tested separately, provided the functional aspects are maintained as defined in the relevant equipment specification.

During any environmental test of electrical and electronic equipment (e.g., susceptibility test) conducted per DO-160, the EUT should be configured in the most sensitive functioning mode that could be encountered during its normal operation in the field, that is, the test should be conducted to uncover vulnerabilities. When the equipment is embedded with basic and/or application software, it must be tested with software functions operating (or simulated) to exhibit the maximum sensitivity to the test environment. The description and the justification for the functional configuration, including software (if it is different from flight configuration and flight software), should be reported or referenced in the test report. Unless otherwise specified in the applicable MPS, special-purpose test firmware/software is acceptable only if the hardware and interfaces are comprehensively exercised and yield good test coverage based on validated requirements and that the configuration is controlled. The special-purpose test firmware/software, when used, should be capable of thoroughly exercising the hardware functions.

From this point on, specific tests are described in more detail for each section of DO-160.

Section 4: Temperature and Altitude

Equipment is categorized depending upon the expected flight profiles for the actual use of the EUT in order to cover the wide range of environments known to exist in the majority of aircraft types and installation locations. It should be recognized that not all possible combinations of temperatures and altitude limits are covered in these equipment categories. These profiles and categories are noted in Table 4-1 of DO-160. The actual temperature and pressure values are noted in Table 4-2 of DO-160. The user is expected to determine the category depending upon the location of use. Depending upon the category, the actual test conditions are obtained using Table 4-2. The operational period for monitoring is obtained from the unit's design and specification. Higher pressures require added safety measures to prevent injury and damage to test equipment and personnel. As in all other tests, the user is expected to be mindful of units and conversion of both input data and observed results.

Section 5: Temperature Variation

This test determines performance characteristics of the equipment during temperature variations between high and low operating temperature extremes. The test is not intended to verify the behavior of the equipment in wet or icing conditions. In conducting this test, the test chamber may incorporate the capability of controlling or altering humidity to the extent that condensation is minimized or does not occur. The user is expected to find the category of the EUT using Table 4-1 of DO-160. Different rates of change of temperature apply to different categories. The test duration depends upon the expected temperature variation or the rate of change of temperature. The EUT's operation/powered events must be considered during the test.

These tests take a considerable amount of time. Often they will be run unattended, depending on the length of test cycle; the ambient conditions and EUT operational requirements can be maintained between cycles to allow the test to be carried out when personnel are available to operate test equipment. The 2 min nonoperational/unpowered segment is a critical requirement and programming/operation of test equipment should be given careful attention.

Section 6: Humidity

This test determines the ability of the equipment to withstand either natural or induced humid atmospheres. The main adverse effects to be anticipated are corrosion and change of equipment characteristics resulting from the absorption of humidity. For example, there may be adverse effects due to exposure of humidity in mechanical (metals) properties, electrical (conductors and insulators) properties, chemical (hygroscopic elements) properties, and thermal (insulators) properties.

Most tests are unpowered during exposure, but spot-checks (powered and operating) can be performed at specified times during cycling and for limited durations. Sealing interfaces on powered EUTs with the appropriate cable/plug are recommended if available. The use of clay, silicone plugs, and butyl rubber tapes provides a weather-tight seal for orifices. This sealing is for the support equipment, not the EUT.

Section 7: Operational Shocks and Crash Safety

The operational shock test verifies that the equipment will continue to function within performance standards after exposure to shocks experienced during normal aircraft operations. These shocks may occur during taxiing, landing, or when the aircraft encounters sudden gusts in flight. This test applies to all equipment installed on fixed-wing aircraft and helicopters. For fixed-wing aircraft, a complete installation demonstration, that is, including aircraft acceleration loads (such as flight maneuvering, gust and landing) in addition to the crash safety loads, may be accomplished by using the "unknown or random" orientations for the "sustained" test procedure. The crash safety test verifies that certain equipment will not detach from its mountings or separate in a manner that presents a hazard during an emergency landing.

Determination of category and test type (number of shocks and severity) is specified by the standard and is dependent on the EUT's location of use. Operational shocks have requirements for the EUT's functionality during the applied shocks. Crash safety shocks do not require operation during applied shocks. The EUT's functionality criteria can be evaluated after testing. Shocks are applied in three orthogonal axes, in both positive and negative directions. Sustained level shocks are performed with the use of a centrifuge.

Some EUT applications require operation/power during crash safety shock levels. EUTs can be operated and powered during sustained acceleration on a centrifuge. Single accelerometer control is utilized when performing testing. Response accelerometers on the EUT are not required but can be used if called out in the test procedure. If the crash safety test is applicable, both the impulse and sustained test procedures shall be performed.

Section 8: Vibration

Vibration tests demonstrate that the equipment complies with the applicable equipment performance standards (including durability requirements) when subjected to vibration levels specified for the appropriate installation as determined by the category. Sinusoidal vibration, random vibration, and sine vibration on random categories are specified. Sine vibration can identify design characteristics over a frequency band. Resonant frequencies are defined as acceleration of 2× the input acceleration. Random vibration simulates real-world excitation per application. Sine on random tests combine characteristics for robust excitations frequently seen on helicopters. The EUT is tested in three orthogonal axes (if applicable). Response accelerometers are attached to the EUT to verify forces applied during the test. The EUT is operational/powered during the test to ensure functionality.

Proper fixturing is required to measure the EUT's response characteristics during the test. Fixturing should mimic the EUT's mounting application to simulate its in-use characteristics. Fixturing can also rigidly isolate the EUT for durability during product screening. Averaging control can eliminate off-axis cross talk that could induce transient forces into the EUT.

The vibration tests described in DO-160 apply to equipment installed on fixed-wing propeller aircraft, fixed-wing turbojet, turbofan, and propfan aircraft and helicopters.

Note: A full analysis of vibration levels related to some specific engine imbalance conditions has not been evaluated against these limits. Therefore, this test alone may not be sufficient for some applications without additional tests or analysis.

Section 9: Explosive Atmosphere

This test specifies requirements and procedures for aircraft equipment that may come into contact with flammable fluids and vapors such as those specified in DO-160. It also refers to normal and fault conditions that could occur in areas that are or may be subjected to flammable fluids and vapors during flight operations. The flammable test fluids, vapors, or gases referred to in DO-160 for this section simulate those that are normally used in conventional aircraft and that require oxygen for combustion (e.g., monofuels [containing an oxidizer] are not included). These tests also do not relate to potentially dangerous environments occurring as a result of leakage from goods carried on the aircraft as baggage or cargo.

Tests are dependent on the EUT's location of use (category). The containment chamber is designed such that the explosion is contained within the chamber so that only the EUT is affected. The fuel mixture is specified and ignition source is controlled. The explosiveness (concentration) of the mixture is to be measured. The EUT operation can, depending on the category, ignite the explosive atmosphere or it can be ignited from an external source. The EUT can, depending on the category, be required to survive the ignited explosive atmosphere and remain functional.

The EUT case can be opened or altered to allow circulation of the explosive mixture. The EUT can be operated/powered and monitored externally during testing. The chamber can be equipped with a sight glass or video monitoring system to collect visual data.

Sequencing Note: The explosive atmosphere tests should not be conducted prior to any other DO-160/EUROCAE ED-14 tests except flammability test (see DO-160 Subsection 3.2, "Order of Tests").

In order to properly interpret the procedures and requirements outlined in Section 9, a user's guide is provided in Appendix 9. The user's guide includes rationale, guidance and background information for the environment, test procedures and requirements, guidance in applying the requirements, and lessons learned from aircraft and laboratory experience. This information should help users understand the intent behind the requirements, aid in tailoring (making adjustments to the test procedure, which deviates from the standard) the requirements as necessary for particular applications, and should help user's develop detailed test procedures based on the general test procedures in Section 9. The user's guide follows a parallel format to the main body of Section 9 for easy cross-reference. Paragraph numbers corresponding to each paragraph in Section 9 are included in the user's guide. Where there is no additional information provided in the user's guide, the paragraph header is provided but the following paragraph is left blank.

It is the responsibility of the installer to make sure the test results satisfy the certification requirements of the proposed installation. This user's guide does not contain requirements; it is intended to provide background information and considerations that improve the likelihood of successful test results.

Section 10: Water Proofness
These tests determine whether the equipment can withstand the effects of liquid water being sprayed or falling on the equipment or the effects of condensation. These tests are not intended to verify performance of hermetically sealed equipment. Therefore, hermetically sealed equipment may be considered to have met all water proofness requirements without further testing. Equipment shall be considered hermetically sealed when the seal is permanent and airtight.

As in the other types of tests, these tests also depend upon the EUT's location of use (category). Procedures include

1. Condensing waterproof test—EUT operational last 10 min of exposure
2. Drip proof test—EUT nonoperational during test
3. Spray proof test—EUT operational during test
4. Continuous stream proof test—EUT nonoperational during test

Tester should perform calibration runs of flow rates/drip fallout (collected moisture) prior to testing the EUT. Testers should ensure that measurements are verified with calibrated equipment. Temperature-controlled water supply is necessary for continuous stream proof test. The specified drip tray mounted level delivers the uniform drip rate across the EUT.

Section 11: Fluids Susceptibility
These tests determine whether the materials used in the construction of the equipment can withstand the effects of aircraft fluid exposure.

Since many contaminants may have flash points within the test temperature range, care should be taken to ensure that adequate safety measures are taken to limit the possibility of fire or explosion. Some contaminants may themselves, or in combination with other contaminants, or with the test sample, be toxic. Due consideration should be given to this possibility before commencing the tests. Depending on the EUT's category, fluid susceptibility is administered via spray or immersion. The EUT is subjected to fluid testing that is to operate per specification. This test is used to determine the adverse affects that might occur if the EUT is exposed to various fluids.

Spray test is cycled between fluid applications and a drying period. Three 24 h cycles per fluid type with 8 h of spray and 16 h drying at 65°C define a cycle. During the last 10 min of the third cycle of spray, the EUT is operated. The EUT shall operate for 30 min after the third drying cycle, but only after ambient stabilization of the EUT has occurred. This procedure is repeated for each fluid tested.

Immersion testing requires an EUT to be completely immersed in a specified fluid. The EUT is immersed for 24 h. The EUT is to operate at the completion of 24 h for 10 min while completely immersed. The EUT is placed in a constant temperature of 65°C for 160 h. The EUT is to return to ambient temperature and operate for 2 h. The preceding procedure is to be repeated for each fluid tested.

DO-160 covers seven general classes of other contaminating fluids. In addition, there are 25 specific fluids that are used to test these classes. Table 11-1 of DO-160 contains the class of fluids, the specific fluids, and the temperatures required in these tests. It is required to specify fluids used, test method, and temperature of exposure in the test procedure and in the environmental qualification form.

The material specimen tests may be used in place of equipment tests. The results of these tests shall assure that the material will protect the equipment from harmful effects after being exposed to the relevant fluid in the manner defined in the equipment test procedures.

Section 12: Sand and Dust (See Figure 11.3)

This test determines the resistance of the equipment to the effects of blowing sand and dust where carried by air movement at moderate speeds. The main adverse effects to be anticipated are

1. Penetration into cracks, crevices, bearings, and joints, causing fouling and/or clogging of moving parts, relays, filters, etc.
2. Formation of electrically conductive bridges
3. Action as nucleus for the collection of water vapor, including secondary effects of possible corrosion
4. Pollution of fluids

Tests depend upon the EUT's location of use (category).

FIGURE 11.3   Sand and dust. (Courtesy of D.L.S. Electronic Systems, Inc., Wheeling, IL.)

Dust testing requires that chamber temperature and relative humidity be monitored and controlled and the EUTs are exposed for 1 h along each orthogonal axis.

For sand testing, sand concentrations vary depending on the EUT's intended use and location. The EUT should be 3 m from the sand injection point. Wind velocity should range between ~40 and 60 mph. Chamber temperature and relative humidity is monitored and controlled.

EUTs may be operated/powered during the test. EUTs should be verified visually after each axis.

Sequencing Note: Consideration must be given in determining where in the sequence of environmental tests to apply this test procedure, as dust residue from this test procedure, combined with other environmental synergistic effects, may corrode or cause mold growth on the test item and adversely influence the outcome of succeeding test procedures. Sand abrasion may also influence the results of the salt spray, fungus, or humidity test procedures.

Dust used in a suitable test chamber shall be raised and maintained at a concentration of 3.5–8.8 g/m^3 and shall be 97%–99% silicon dioxide. The dust characteristics are detailed.

If sand concentration data are available for the worst-case field environment the equipment is expected to properly function in that sand concentration level. Material likely to be used close to helicopters operating over unpaved surfaces is defined. Material sand and dust never used or exposed in the vicinity of operating aircraft, but which may be used or stored unprotected near operating surface vehicles, is defined. Material that will be subjected only to natural conditions is defined.

Note: Unless otherwise required in the relevant specification, the equipment is not required to operate during the exposure period.

At the end of this exposure period, the equipment shall be removed from the chamber and cooled to room temperature. Externally accumulated sand only on surfaces of the equipment required to verify proper operation (e.g., displays, connectors, keyboards, test ports) shall be removed by brushing, wiping, or shaking with care being taken to avoid introducing additional sand into the equipment. Under no circumstances shall sand be removed by either air blast or vacuum cleaning.

Section 13: Fungus Resistance

These tests determine the effect of fungi on equipment material under conditions favorable for their development, namely, high humidity, warm atmosphere, and presence of inorganic salts. It should be noted that secondary effects of fungi are also possible via exposure to fluids during operation and maintenance or exposure to solar actinic effects via breaking down of materials to reduce the material to be a fungus nutrient. Further, these tests must follow the sequencing rules.

Equipment that is installed in an environment where it will be exposed to severe fungus contamination is identified as Category F and shall be subjected to the fungus resistance test. This test shall *not* be conducted after salt spray or sand and dust. A heavy concentration of salt may affect the fungal growth, and sand and dust can provide nutrients, which could compromise the validity of this test. Materials can be directly and indirectly attacked by fungus. Most susceptible to attack are mineral-based materials. Natural materials (carbon based), cellulose materials (wood, paper, fiber textiles, cordage), animal- and vegetable-based adhesives, grease, oils, hydrocarbons, and leather are prone to attack. Synthetic materials, PVC formulations, polyurethanes, plastics, paints, and varnishes also can be attacked by fungus. This test can be performed on a nonfunctioning device or component. A specified fungus is grown to specifications and applied to control and test items. The control and test items are incubated per the standard. Items are evaluated for type of fungal growth and their effect on the test items.

The fungal growth on a test item is analyzed to determine the species and if the growth is on the test item material(s) or on contaminants. Fungal growth on test item materials is examined for the extent of growth. The extent of growth on susceptible components or materials must be completely described, for example, the immediate effect that the growth has on the physical characteristics of the material, the long-range effect that the growth could have on the material, and the specific material (nutrient[s]) supporting the growth. The evaluation of human factor effects (including health risks) should also be performed.

Section 14: Salt Fog

This test determines the effects on the equipment of prolonged exposure to a salt atmosphere or to salt fog experienced in normal operations. The main adverse effects to be anticipated are

1. Corrosion of metals
2. Clogging or binding of moving parts as a result of salt deposits
3. Insulation fault
4. Damage to contacts and uncoated wiring

A salt solution of 5% by weight of sodium iodine and distilled or demineralized water is atomized at 35°C. EUT is positioned inside the chamber so that there is no overlapping or test apparatuses that will shield or cause runoff onto the EUT. The EUT's topmost surface is placed to be parallel to the top of the fog-emitting nozzle. Collection receptacles (2) are placed within the chamber: one placed at the perimeter edge of the EUT nearest to the nozzle and another at the perimeter of the EUT but at the farthest point from the nozzle.

Section 15: Magnetic Effect (See Figure 11.4)

This test determines the magnetic effect of the equipment. The test is intended for finding the closest distance to magnetic sources such as compasses or compass sensors (flux gates) at which that unit can be installed so that it operates without interference.

Tests should be performed away from any sources of magnetic field so that the test is not contaminated. The mode of operation can affect magnetic emissions of the EUT. To prevent excessive setup time, the magnetic effect tests should be performed first or last. In this test, the compass not the EUT is moved to find the distance where the compass deflection occurs.

Section 16: Power Input (See Figure 11.5)

This test is intended to simulate everything a device in an airplane might see during operation, such as when the aircraft is sitting at the terminal receiving terminal power, the possible voltage sags from starting the engines when disconnected from outside power sources, the inductive kick from another system

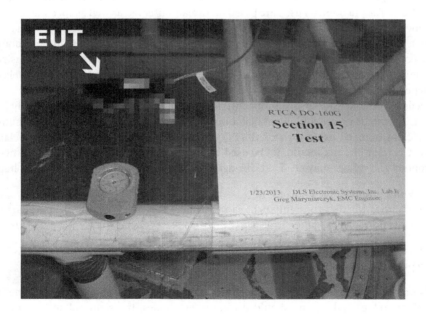

FIGURE 11.4   Magnetic effect being performed. (Courtesy of D.L.S. Electronic Systems, Inc., Wheeling, IL.)

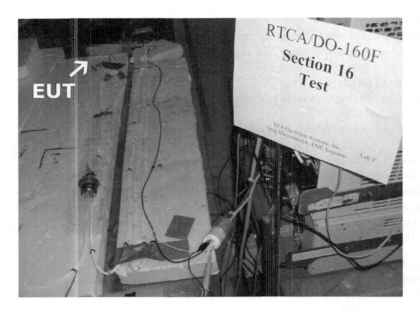

FIGURE 11.5   Power input testing. (Courtesy of D.L.S. Electronic Systems, Inc., Wheeling, IL.)

inside the aircraft, or any systems that might come on line during normal operation and after landing including emergency conditions. It covers the following electrical power supplies:

- 14 V dc, 28 V dc, and 270 V dc
- 115 V_{rms} ac and 230 V_{rms} ac either at a nominal 400 Hz frequency or over a variable frequency range that includes 400 Hz

This tests power quality issues *that might* be experienced by the EUT in an installation from the power feeding the EUT. The phenomena used are based on threats observed as well as speculated. They include normal, abnormal, and emergency operation running on auxiliary generators, batteries, and battery packs, which might include overvoltages, undervoltages, disruptions of power, and harmonic content. An oscilloscope is used to verify proper application of test waveforms. Power line transients for ac-powered equipment can be difficult to capture effectively on an oscilloscope. Various levels of severity are used depending on the category.

Care needs to be taken to ensure proper wiring to the programmable power supply, specifically reverse polarity with dc power equipment. Solution to problems might relate to sense and cutoff circuits, overvoltage clamps, metal oxide varistor (MOV), protection diodes, or something that clamps to the rail voltage. Crowbar circuits should not be used, as they would shut the power down. When there is a voltage dropout, the voltage may need to be bridged with large capacitors to supply the voltage.

Section 17: Voltage Spike (See Figure 11.6)
This test determines whether the equipment can withstand the effects of voltage spikes arriving at the equipment on its power leads, either ac or dc. The main adverse effects to be anticipated are

1. Permanent damage, component failure, and insulation breakdown
2. Susceptibility degradation or changes in equipment performance

Two levels of voltage spikes are used: Category A-600 volts and Category B double-line voltage. To protect against voltage spikes, one should use transient diodes, transient voltage suppressive (TVS) device, and MOVs. With the EUT disconnected, the transient wave shape is verified to be the one that

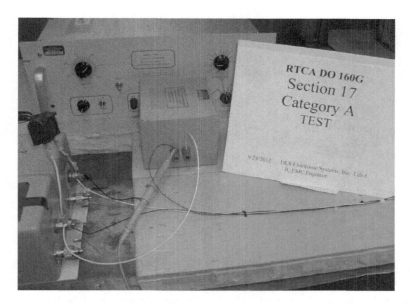

FIGURE 11.6 Voltage spike testing. (Courtesy of D.L.S. Electronic Systems, Inc., Wheeling, IL.)

is required by DO-160. If performance is measured during the application of this test, then the performance requirements contained in the applicable equipment performance requirements apply.

Section 18: Audio Frequency Conducted Susceptibility (CS) (See Figures 11.7 and 11.8)
This test is meant to simulate normally expected fluctuations and harmonics from the mains ac and dc power buses to determine whether the equipment will function properly. These frequency components are normally harmonically related to the power source fundamental frequency.

DC testing starts at 30 Hz and ac starts at the second harmonic. For each test, there is an expected dwell time of 1 min at each frequency making this a long test.

FIGURE 11.7 Audio frequency CS. (Courtesy of D.L.S. Electronic Systems, Inc., Wheeling, IL.)

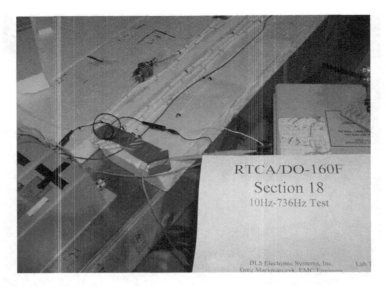

FIGURE 11.8   Audio frequency CS. (Courtesy of D.L.S. Electronic Systems, Inc., Wheeling, IL.)

Section 19: Induced Signal Susceptibility (See Figure 11.9)

This test is used to determine if the equipment can withstand various signals that may be introduced by failures in other equipment in an aircraft. The equipment interconnect circuit configuration accepts a level of induced voltages caused by the installation environment. This section relates specifically to interfering signals related to the power frequency and its harmonics, audio frequency signals, and electrical transients that are generated by other onboard equipment or systems and coupled to sensitive circuits within the EUT through its interconnecting wiring. Power leads are exempt from this test.

Section 20: Radio Frequency Susceptibility (Radiated and Conducted) (See Figure 11.10)

EUT's ability to operate within performance specifications is checked during its exposure to a specific level of radio-frequency-modulated power either by a radiated radio frequency (RF) field or by injection

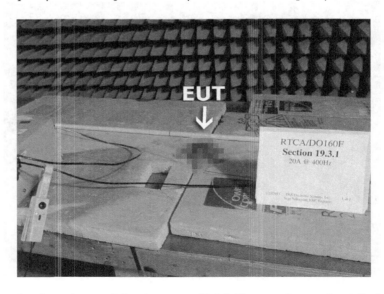

FIGURE 11.9   Induced signal susceptibility. (Courtesy of D.L.S. Electronic Systems, Inc., Wheeling, IC.)

FIGURE 11.10 Entire aircraft wire harness ready to test. (Courtesy of D.L.S. Electronic Systems, Inc., Wheeling, IL.)

probe induction onto the power lines and interface circuit wiring. In this test, electric fields are generated and exposed to circuits, cables, and systems. The fields can be picked up by the wires that create unwanted signals. These unwanted signals can cause a device, circuit, or system to malfunction.

Very often it is the modulation that is combined with the RF field that enters a circuit via the original RF. The modulation is stripped off through a diode somewhere in the device causing an analog level to shift creating a failure.

In the user guide, there is a discussion of the use of reverberation chambers (RCs) (See Figure 11.11). These chambers are an alternate means of generating high-level fields.

FIGURE 11.11 RC testing. (Courtesy of D.L.S. Electronic Systems, Inc., Wheeling, IL.)

FIGURE 11.12   Setup for emissions. (Courtesy of D.L.S. Electronic Systems, Inc., Wheeling, IL.)

Section 21: Emission of Radio Frequency Energy (See Figure 11.12)

These tests determine whether or not the EUT emits any undesired RF noise in excess of the levels specified in DO-160, which are set to protect aircraft RF sensors' operating frequencies.

Many electronic devices generate harmonic frequencies of the normal frequencies being processed. These harmonic frequencies can be hundreds of times greater than the original frequencies.

Section 22: Lightning Induced Transient Susceptibility (See Figure 11.13)

Lightning induced transient susceptibility tests are used to check the capability of EUT to withstand a selection of defined test transients that are intended to represent the induced effects of lightning.

FIGURE 11.13   Induced lightning. (Courtesy of D.L.S. Electronic Systems, Inc., Wheeling, IL.)

The waveforms and levels and the pass/fail criteria for equipment performance during the test shall be listed in the applicable equipment test procedure. Two groups of tests may be used for equipment qualification. The first is a damage tolerance test performed using pin injection. The second group evaluates the functional upset tolerance of equipment when transients are applied to interconnecting cable bundles. Cable bundle tests include single-stroke, multiple-stroke, and multiple-burst response tests. Cable bundle tests can also provide an indication of damage tolerance. *Note:* These tests may not cover all aspects of lightning induced interaction and effects on equipment. Additional tests such as tests of the equipment in a complete system may be required to achieve certification of a specific installation, depending upon the functions performed. For externally mounted equipment, direct effects tests may also be required (see Section 23).

Whenever high energy is exposed to an aircraft, there will be some energy that penetrates into the aircraft. This test is an attempt to simulate these effects and determine if the EUT malfunctions. Whenever these high energy levels flow, it will be imperative to make sure the grounding is correct and representative of the aircraft ground.

This test can be destructive; so it needs to be considered in sequencing of tests as well as availability of multiple units.

Section 23: Lightning Direct Effects (See Figure 11.14)

The tests described in this section are intended to determine the ability of externally mounted equipment to withstand the direct effects of a lightning strike externally to the main skin of the aircraft and include all such equipment that is covered only by a dielectric skin or fairing that is an integral part of the equipment. This section also includes connecting cables and associated terminal equipment furnished by the equipment manufacturer as a part of the equipment.

These are very high energy levels. Current levels can reach 200,000 amps and testing to hundreds of thousands of volts. It has the potential to crack glass, damage housing, and even melt metal. The test relates to a limited number of devices that might be mounted externally to the aircraft. The EUT is usually powered during the test. The EUT must have actual fasteners, gaskets, etc., that are used on the aircraft. This test can be destructive; the sequencing of tests and availability of multiple EUTs should be considered.

Section 24: Icing

These tests determine performance characteristics for equipment that must operate when exposed to icing conditions that would be encountered under conditions of rapid changes in temperature, altitude, and humidity.

FIGURE 11.14   Direct lightning. (Courtesy of D.L.S. Electronic Systems, Inc., Wheeling, IL.)

Moving an EUT between a (cold) temperature chamber and a warm, humid chamber to form condensation and refreezing of condensation requires two chambers relatively close together. Monitoring the surface temperature of the EUT is critical to determine at what temperature it is ready to be cycled back into the cold chamber for freezing of condensation. The selection of icing category depends on equipment location in (or on) the aircraft. Three icing test procedures are specified according to the category for which the equipment is designed to be used and installed in the aircraft and the type of icing conditions expected. These conditions must be considered by the equipment designer in evaluating these requirements, which are determined by the end application and use of the equipment. These tests generally apply to equipment mounted on external surfaces or in non-temperature-controlled areas of the aircraft where rapid changes in temperature, altitude, and humidity are generally encountered.

Section 25: Electrostatic Discharge (See Figure 11.15)

The electrostatic discharge (ESD) test measures the effects of static electricity discharges from human contact on the performance of EUT without permanent degradation of performance. Low relative humidity, temperature, and use of low-conductivity (artificial fiber) carpets, vinyl seats, and plastic structures, which may exist in all locations within an aircraft, are some of the factors that contribute to static electricity discharge. This test is applicable for all equipment and surfaces, which are accessible during normal operation and/or maintenance of the aircraft. This test is not applicable to connector pins.

This is one of the last tests to perform since it might be destructive.

Section 26: Fire, Flammability

Fire and flammability tests measure functionality and the level of fire resistance depending upon the equipment category. There are three categories—fireproof, fire resistant, and flammability. Flammability and fire tests apply to equipment installed on fixed-wing propeller-driven aircraft, fixed-wing turbojet aircraft, turbofan aircraft, propfan aircraft, and helicopters. These tests are applicable for equipment mounted in pressurized zones, in fire zones, and in nonpressurized and nonfire zones.

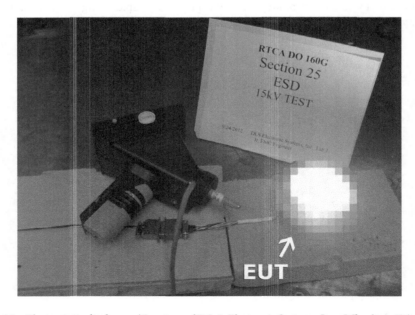

FIGURE 11.15   Electrostatic discharge. (Courtesy of D.L.S. Electronic Systems, Inc., Wheeling, IL.)

Testing is not necessary on enclosures housing electronic or nonmetallic material if the following applies:

1. The enclosures are constructed of metal with a metal finish that is nonflammable on all sides and has no vent holes.
2. The enclosures are constructed of metal (metal finish that is nonflammable) on five sides, and one side is constructed of glass polycarbonate (display) that has met the 12 s vertical test and has no vent holes.

Parts/materials that are considered small may be exempt due to their small size and amount, because they would not contribute significantly to the propagation of a fire. Examples of small parts could be knobs, handles, rollers, fasteners, clips, grommets, rub strips, and pulleys. Consideration must be given when more than one small part is located in the same proximity with the same or other small parts (one part may ignite the other part) as the combined fuel load may contribute to propagation of a flame. In this case, the aforementioned small parts exemption would not apply. Small parts exemption does not apply to wire and cable.

Appendix A: Environmental Test Identification
This information for documentation of environmental testing exists in the Appendix A of DO-160.

The documentation of environmental testing is used in postincident or accident investigation, installation certification, repair, etc. The environmental qualification form is the evidence of environmental testing that is to be included in the equipment data package submitted for Technical Standard Order (TSO) authorization and in the installation and maintenance instructions. Manufacturers should identify the method used to establish traceability to the environmental test categories to which the equipment was tested, including the applicable revision number of the test procedure used (Section of RTCA/DO-160), and association must be documented through the equipment type, model, or part number. The manufacturer may wish to qualify the equipment to more than one category for a particular environmental test. If all requirements for one category are clearly the most severe, only the most severe category need be identified. Information such as vibration tests conducted with or without shock mounts, fluid tests conducted with Jet A fuel, type of deicing fluid, and other parameters pertinent to the tests shall be included on the form. Any other details should be conveyed within a chronologically detailed test log pertaining to the coordinating test section performed.

11.3 Conclusion

RTCA DO-160 is the standard to use for assuring the airworthiness of the device in its intended environment in the avionics certification process. Applicants use this document to assist in the requirements definition, design, and construction of the device as well as in testing so that the device is put through the harshest environmental conditions that the aircraft may be exposed to and assure the desired level of safety.

12

RTCA DO-178B/ EUROCAE ED-12B

Thomas K. Ferrell
Ferrell and Associates Consulting, Inc.

Uma D. Ferrell
Ferrell and Associates Consulting, Inc.

12.1 Introduction

This chapter provides a summary of the document RTCA document (RTCA DO) 178B, Software Considerations in Airborne Systems and Equipment Certification [1], with commentary on the most common mistakes made in understanding and applying DO-178B. The joint committee of RTCA Special Committee 167 and EUROCAE* Working Group 12 prepared RTCA DO-178B[†] (also known as EUROCAE ED-12B), and it was subsequently published by RTCA and by the EUROCAE in December of 1992. DO-178B provides guidance for the production of software for airborne systems and equipment such that there is a level of confidence in the correct functioning of that software in compliance with airworthiness requirements. DO-178B represents industry consensus opinion on the best way to ensure safe software. It should also be noted that although DO-178B does not discuss specific development methodologies or management activities, there is clear evidence that by following rigorous processes, cost and schedule benefits may be realized. The verification activities specified in DO-178B are particularly effective in identifying software problems early in the development process.

---

* European Organisation for Civil Aviation Equipment.
[†] DO-178B and ED-12B are copyrighted documents of RTCA and EUROCAE, respectively. In this chapter, DO-178B shall be used to refer to both the English version and the European equivalent. This convention was adopted solely as a means for brevity.

DO-178B does not go into details of certification aspects of field-loadable software, user-modifiable software, reuse of software life-cycle data to be able to take certification credit, management of open problem reports, and supplier management to name a few. Treatment of such topics is imposed through using other certification vehicles such as the Federal Aviation Administration (FAA) Orders and European Aviation Safety Agency (EASA) Certification Memoranda. If there are particular technical concerns unique to a project, requirements to mitigate these concerns may be imposed on that specific project using agency-specific vehicles such as FAA Issue Papers or EASA Certification Review Item. For these reasons, DO-178B may not be the only compliance document to be satisfied on a software development project for an airborne system or equipment. All of the compliance documents should be considered holistically for planning purposes.

It is important to note that software is never approved as a stand-alone entity; it is always considered within the context of the system in which it is used.

12.2 Comparison with Other Software Development Standards

DO-178B is a mature document, having evolved over the last 20 years through 2 previous revisions: DO-178 and DO-178A. It is a consensus document that represents the collective wisdom of both the industry practitioners and the certification authorities. DO-178B is self-contained, meaning that no other software development standards are referenced except for those that the applicant produces to meet DO-178B objectives. Comparisons have been made between DO-178B and other software development standards such as MIL-STD-498, MIL-STD-2167A, IEEE/EIA-12207, IEC 61508, and UK Defense Standard 0-55. All of these standards deal with certain aspects of software development covered by DO-178B. None of them have been found to provide complete coverage of DO-178B objectives. In addition, these other standards lack objective criteria and links to safety analyses at the system level. However, organizations with experience applying these other standards often have an easier path to adopting DO-178B.

Advisory Circular (AC) 20-115B specifies DO-178B is an acceptable means, but not the only means, for receiving regulatory approval for software in systems or equipment being certified under a Technical Standard Order (TSO) authorization, type certificate (TC), or supplemental type certificate (STC). Most applicants use DO-178B to avoid work in showing that other means are equivalent to DO-178B. Even though DO-178B was written as a guideline, it has become the standard practice within the industry especially since there had not been clear policy on what is expected in an alternate method. AC 20-171 has been released by the FAA to correct this problem. Any proposed alternative method for software approval should be evaluated in conjunction with the certification process of the system and AC 20-171 to demonstrate equivalent level of safety assurance.

DO-178B is officially recognized as a de facto international standard by the International Organization for Standardization (ISO).

12.3 Document Overview

DO-178B consists of 12 sections, 2 annexes, and 4 appendices as shown in Figure 12.1.

Sections 2 and 10 are designed to illustrate how the processes and products discussed in DO-178B relate to, take direction from, and provide feedback to the overall certification process. Integral processes detailed in Sections 6, 7, 8, and 9 support software life-cycle processes noted in Sections 3, 4, and 5. Section 11 provides details on the life-cycle data, and Section 12 gives guidance to any additional considerations. Annex A discussed in more detail in the following provides a leveling of objectives. Annex B provides the document's glossary. The glossary deserves careful consideration since much effort and care was given to the precise definition of terms. Appendices A, B, C, and D provide additional material including a brief history of the document, the index, a list of contributors, and a process improvement form, respectively. It is important to note that with the exception of the appendices and

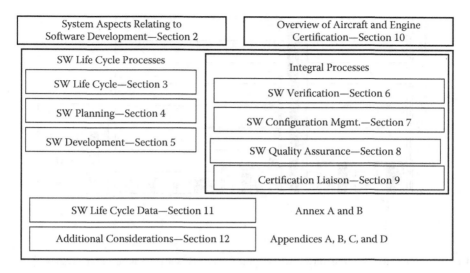

FIGURE 12.1 Document structure.

some examples embedded within the text, the main sections and the annexes are considered normative, that is, required to apply DO-178B.

The 12 sections of DO-178B describe processes and activities for the most stringent level of software.* Annex A provides a level-by-level tabulation of the objectives for lower levels of software.[†] This leveling is illustrated in Table 12.1 extracted from Annex A, Table A-4, Verification of Outputs of Software Design Process.

In addition to the leveling of objectives, Annex A tables serve as an index into the supporting text by way of reference and illustrate where independence is required in achieving the objective, which data item should include the objective evidence, and how that evidence must be controlled. More will be said on the contents of the various Annex A tables in the corresponding process section of this text. If an applicant adopts DO-178B for certification purposes, Annex A may be used as a checklist to achieve these objectives. The FAA's position as expressed in the FAA Order 8110.49 and Software Job aid is that if an applicant provides evidence to satisfy the objectives, then the software is DO-178B compliant. In this vein, the FAA's tables of checklist questions as documented in Software Job aid for performing audits of DO-178B developments are based on Annex A tables. Software Job aid clearly states that these checklist questions are *not* to be used as the definitive checklist but to be used as a starting point to build a set of questions to address during the audit that is specific to that particular project under review with the intent of accomplishing compliance to DO-178B Annex A objectives.

TABLE 12.1 Example Objective from Annex A

	Objective		Applicability by SW Level				Output		Control Category by SW Level			
	Description	Ref.	A	B	C	D	Description	Ref.	A	B	C	D
1	Low-level requirements comply with high-level requirements.	6.3.2a	●	●	○		Software verification results	11.14	②	②	②	②

Note: SW, software.

---

* Levels are described in the section "Software as Part of the System."
[†] Software that is determined to be at level E is outside the scope of DO-178B.

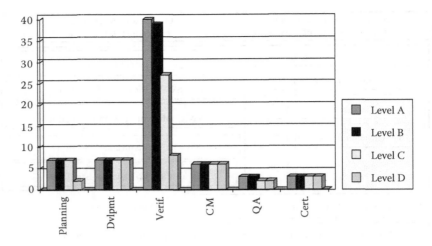

FIGURE 12.2 Objectives over the software development life-cycle software as part of the system.

Before discussing the individual sections, it is useful to look at breakout of objectives as contained in Annex A. While DO-178B contains objectives for the entire software development life cycle, there is a clear focus on verification as illustrated by Figure 12.2. Although at first glance it appears that there is only one objective difference between levels A and B, additional separation between the two is accomplished through the relaxation of independence requirements. Independence is achieved by having the verification or quality assurance performed by a person other than the one who developed the data item. Tools may also be used to achieve independence.

12.4 Software as Part of the System

Application of DO-178B fits into a larger system of established or developing industry practices for systems development and hardware. The system-level standard is SAE ARP4754, "Certification Considerations for Highly Integrated or Complex Aircraft Systems" [2]. The relationship between system characteristics, software, and hardware is illustrated in Figure 12.3.

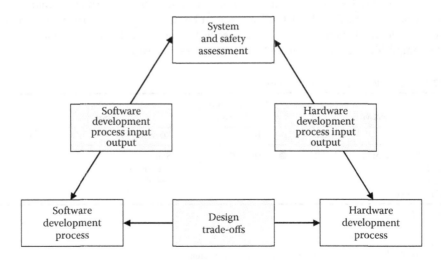

FIGURE 12.3 Relationship between systems development process and the software development process.

TABLE 12.2   Software Levels

Failure Condition	DO-178B Software Level
Catastrophic	A
Hazardous	B
Major	C
Minor	D
No effect	E

The interfaces to the systems development process were not well defined at the time DO-178B was written. This gap was filled when ARP4754 was published in 1996. DO-178B specifies the information flow between system processes and software processes. The focus of the information flow from the system process to the software process is to keep track of requirements allocated to software, particularly those requirements that contribute to the system safety. The focus of information flow from software process to system process is to ensure that changes in the software requirements including the introduction of derived requirements (those not directly traceable to a parent requirement) do not adversely affect system safety. Changes to software requirements may occur during requirement reviews and testing at various levels where there may be a discovery that the requirements as specified need to be modified. DO-178B is written with the idea of strictly monitoring the software requirements (derived or not), their changes, and their transformation to an executable image. This activity is strictly supporting the assurance that requirements were implemented correctly; there are no objectives related to whether correct requirements were implemented. This later notion lies squarely with systems engineering and system safety.

The idea of system safety, although outside the scope of DO-178B, is crucial to understanding how to apply DO-178B. The regulatory materials governing the certification of airborne systems and equipment define five levels of failure conditions. The most severe of these is catastrophic, meaning failures that result in the loss of ability to continue safe flight and landing. The least severe is no effect where the failure results in no loss of operational capabilities and no increase in crew workload. The remaining levels define various levels of loss of functionality resulting in corresponding levels of workload and potential for loss of life. These five levels map directly to the five levels of software defined in DO-178B. This mapping is shown in Table 12.2.

It is important to note that software is never certified as a stand-alone entity. A parallel exists for hardware development process and flow of information between hardware processes and system process. Design trade-offs between software processes and hardware processes are also taken into consideration at the system level. Software levels may be lowered by utilizing protective software or hardware mechanisms elsewhere in the system. Such architectural methods include partitioning, use of hardware or software monitors, and architectures with built-in redundancy.

12.5 Software Life-Cycle Processes

The definition of how data are exchanged between the software and systems development processes is part of the software life-cycle processes discussed in DO-178B. Life-cycle processes include the planning process, the software development processes (requirements, design, code, and integration), and the integral processes (verification, configuration management, software quality assurance [SQA], and certification liaison process). DO-178B defines objectives for each of these processes as well as outlining a set of activities for meeting the objectives.

DO-178B discusses the software life-cycle processes and transition criteria between life-cycle processes in a generic sense without specifying any particular life-cycle model. Transition criteria are defined as "the minimum conditions, as defined by the software planning process, to be satisfied to enter a process." Transition criteria may be thought of as the interface points between all of the processes discussed in DO-178B. Transition criteria are used to determine if a process may be entered or reentered. They are a mechanism for knowing when all of the tasks within a process are complete and may be used to allow processes to execute in parallel. Since different development models require different criteria to be satisfied for moving from one step to the next, specific transition criteria are not defined in DO-178B. However, it is possible to define a set of characteristics that all well-defined transition criteria should meet. For transition criteria to successfully assist in entry from one life-cycle process to another, they should be quantifiable, flexible, well documented, and present for every process. It is also crucial that the process owners agree upon the transition criteria between their various processes. Transition criteria assist in the analysis of extent of rework in the various life-cycle phases in case of process defects or process violations that affect the product.

12.6 Software Planning Process

DO-178B defines five types of planning data* for a software development. They are

1. Plan for software aspects of certification (PSAC)
2. Software development plan
3. Software verification plan
4. Software configuration management plan
5. SQA plan

These plans should include consideration of methods, languages, standards, and tools to be used during the development. A review of the planning process should have enough details to assure that the plans, proposed development environment, and development standards (requirements, design, and code) comply with DO-178B.

Although DO-178B does not discuss the certification liaison process until Section 9, the intent is that the certification liaison process should begin during the project's planning phase. The applicant outlines the development process and identifies the data to be used for substantiating the means of compliance for the certification basis. It is especially important that the applicant outline specific features of software or architecture that may affect the certification process.

12.6.1 Software Development Process

Software development processes include requirements, design, coding, and integration. DO-178B allows for requirements to be developed that detail the system's functionality at various levels. DO-178B refers to these levels as high- and low-level requirements. System complexity and the design methodology applied to the system's development drive the requirement's decomposition process. The key to understanding DO-178B's approach to requirement's definition can be summed up as "one person's requirements are another person's design." Exactly where and to what degree the requirements are defined is less important than that all requirements are accounted for in the resulting design and code and that traceability is maintained to facilitate verification.

Some requirements may be derived from the design, architecture, or implementation nuances of the software and hardware. It is recognized that such requirements will not have a traceability to the high-level requirements. However, these requirements must be verified and must also be considered for safety effects in the system safety assessment process.

DO-178B provides only a brief description of the design, coding, and integration processes since these tend to vary substantially between various development methodologies. The one exception of this is in

---

* The authors of DO-178B took great plans to avoid the use of the term document when referring to objective evidence that needed to be produced to satisfy DO-178B objectives. This was done to allow for alternative data representations and packaging as agreed upon between the applicant and the regulatory authority. For example, the four software plans outlining development, verification, QA, and CM may all be packaged in a single plan, just as the PSAC may be combined with the system certification plan.

the description of the outputs of each of the processes. The design process yields low-level requirements and software architecture. The coding process produces the source code, typically either in a high-order language or in an assembly code. The result of the integration effort is executable code resident on the target computer along with the various build files used to compile and link the executable. Each of these outputs is verified, assured, and configured as part of the integral processes.

12.6.1.1 Integral Processes

DO-178B defines four processes as integral, meaning that they overlay and extend throughout the software life cycle. These are the software verification process, software configuration management, SQA, and certification liaison process.

12.6.2 Software Verification

As noted earlier, verification objectives outnumber all others in DO-178B, accounting for over two-thirds of the total. DO-178B defines verification as a combination of reviews, analyses, and testing. Verification is a technical assessment of the results of both the software development processes and the software verification process. There are specific verification objectives that address the requirements, design, coding, integration, and the verification process itself. Emphasis is placed at all stages to assure that there is traceability from high-level requirements to the final "as-built" configuration.

Reviews provide qualitative assessment of a process or product. The most common types of reviews are requirements reviews, design reviews, and test procedure reviews. DO-178B does not prescribe how these reviews are conducted or what means are employed for effective reviews. Best practices in software engineering process state that for reviews to be effective and consistent, checklists should be developed and used for each type of review. Checklists provide

- Objective evidence of the review activity
- A focused review of those areas most prone to error
- A mechanism for applying "lessons learned"
- A practical traceable means for ensuring that corrective action is taken for unsatisfactory items

Review checklists can be common across projects, but they should themselves be reviewed for appropriateness and content for a particular project.

Analyses provide a repeatable evidence of correctness and are often algorithmic or procedural in nature. Common types of analyses used include timing, stack, dataflow, and control flow analyses. Race conditions and memory leakage should be checked as part of the timing and stack analysis. Data and control coupling analysis should include, at a minimum, basic checks for set/use and may extend to a full model of the system's behavior. Many types of analyses may be performed using third-party tools. If tools are used for this purpose, DO-178B rules for tool qualification must be followed.

The third means of verification, testing, is performed to demonstrate that

- The software product performs its intended function
- The software does not demonstrate any unintended actions

The key to accomplishing testing correctly to meet DO-178B objectives in a cost-effective manner is to maintain a constant focus on requirements. This requirements-based test approach represents one of the most fundamental shifts from earlier versions of the document. As test cases are designed and conducted, requirements coverage analysis is performed to assess that all requirements are tested. A structural coverage analysis is performed to determine the extent to which the requirements-based test exercised the code. In this manner, structural coverage is used as a means of assessing overall test completion. The possible reasons for lack of structural coverage are shortcomings in requirements-based test cases or procedures, inadequacies in software requirements, and compiler-generated code, dead code, or deactivated code. As part of the test generation process, tests should be written for both normal range and abnormal inputs (robustness). Tests should also be conducted using the target environment whenever possible. Certification credit can be taken for tests conducted at the system level that clearly demonstrate software properties at system-level testing.

Structural coverage and how much testing is required for compliance at the various levels are misunderstood topics. Level D software verification requires test coverage of high-level requirements only. No structural coverage is required. Low-level requirements testing is required at level C. In addition, testing of the software structure in order to show proper data and control coupling is introduced. This coverage involves coverage of dependencies of one software component on other software component via data and control. Decision coverage is required for level B, while level A code requires the modified condition–decision coverage (MCDC). For level A, structural coverage analysis may be performed on source code only to the extent that the source code can be shown to map directly to object code. The reason for this rule is that some compilers may introduce code or structure that is different from source code. MCDC coverage criteria were introduced to retain the benefits of multiple-condition coverage while containing the exponential growth in the required number of test cases required. MCDC requires that each condition must be shown to independently affect the outcome of the decision or that the outcome of a decision changes when one condition is changed at a time. Many tools are available to determine the minimum test case set needed for DO-178B compliance. There is usually more than one set of test cases that satisfy MCDC coverage [3]. There is no set policy on which set should be used for compliance. It is best to get an agreement with the certification authorities in algorithms and tools used to determine compliance criteria.

12.6.3 Software Configuration Management

Verification of the various outputs discussed in DO-178B is only credible when there is clear definition of what has been verified. This definition or configuration is the intent of the DO-178B objectives for configuration management. The six objectives in this area are unique, in that they must be met for all software levels. This includes identification of what is to be configured, how baselines and traceability are established, how problem reports are dealt with, how the software is archived and loaded, and how the development environment is controlled. Some developers may use a developmental configuration management library to work with developmental artifacts and use a release-based configuration management library to interface with other related projects or with the customer. Regardless of whether one or more configuration libraries are used, the purpose remains to establish a clear trail of products that have been transformed because of the process, as well as the evidence of the process itself. While configuration management is a fairly well-understood concept within the software engineering community (as well as the aviation industry as a whole), DO-178B does introduce some unique terminology that has proven to be problematic. The concept of control categories is often misunderstood in a way that overall development costs are increased, sometimes dramatically. DO-178B defines two control categories (CC1 and CC2) for data items produced throughout the development. The authors of DO-178B intended the two levels as a way of controlling the overhead costs of creating and maintaining the various data items. Items controlled as CC2 have less requirements to meet in the areas of problem reporting, baselining, change control, and storage. The easiest way to understand this is to provide an example. Problem reports are treated as a CC2 item. If problem reports were a CC1 item and a problem was found with one of the entries on the problem report itself, a second problem report would need to be written to correct the first one. CC1 items are usually those that directly affect the product and they are not allowed to be altered without a problem report. A second nuance of control categories is that the reader of DO-178B may define what CC1 and CC2 are within their own configuration management (CM) system as long as the DO-178B objectives are met. One example of how this might be beneficial is in defining different retention periods for the two levels of data. Given the long life of airborne systems, these costs can be quite sizeable. Other matters of consideration for archival systems are technology obsolescence of the archival medium as well as means of retrieval.

12.6.4 Software Quality Assurance

SQA objectives provide oversight of the entire DO-178B process and require independence at all levels. It is recognized that it is prudent to have an independent assessment of quality. SQA is active from the beginning of the development process. SQA assures that any deviations during the development process from plans and standards are detected, recorded, evaluated, tracked, and resolved. For levels A and B, SQA is required to assure transition criteria are adhered to throughout the development process. SQA works with the CM process to assure that proper controls are in place and applied to life-cycle data. This last task culminates in the conduct of a software conformity review. SQA is responsible for assuring that the as-delivered product matches the as-built and as-verified product. The common term used for this conformity review in commercial aviation industry is "first article inspection."

12.7 Certification Liaison Process

As stated earlier, the certification liaison process is designed to streamline the certification process by ensuring that issues are identified early in the process. While DO-178B outlines 20 distinct data items to be produced in a compliant process, 3 of these are specific to this process and must be provided to the certifying authority. They are

1. PSAC
2. Software configuration index
3. Software accomplishment summary

Other data items may be requested by the certification authority, if deemed necessary. As mentioned earlier, applicants are encouraged to start a dialogue with certification authorities as early in the process as possible to reach a common understanding of a means of achieving compliance with DO-178B. This is especially important as new technology is applied to avionics and as new personnel enter the field. Good planning up front, captured in the PSAC, should minimize surprises later in the development process, thus minimizing cost. Just as the PSAC says "what you intend to do," the accomplishment summary captures "what you did." It is used to gauge the overall completeness of the development process to ensure that all objectives of DO-178B have been satisfied. Finally, the configuration index serves as an overall accounting of the content of the final product as well as the environment needed to recreate it. Life-cycle data related to type design have regulations governing their retrieval and approval. These data items are listed in DO-178B.

Note that software developed for TSO equipment does not necessarily need a certification liaison since TSO compliance is perceived as self-certification by the certification authorities. The planning and development data are expected to be reviewed by the certification authorities for approval. However, if these data are reviewed only at the end of the project, there is a risk of rework imposed on the developers. Approval process appears to be eased by using a certification liaison to conduct the audits so that DO-178B compliance can be demonstrated and rework can be avoided.

12.7.1 Additional Considerations

During the creation of DO-178B, it was recognized that new development methods and approaches existed for developing avionics. These included incorporation of previously developed software (PDS), use of tools to accomplish one or more of the objectives required by DO-178B, and application of alternate means in meeting an objective such as formal methods. In addition, there are a small class of unique issues such as field-loadable and user-modifiable software. Section 12 collects these items together under the umbrella title of Additional Considerations. Two areas, PDS and tool qualification, are common sources of misunderstanding in applying DO-178B.

12.8 Previously Developed Software

PDS is a software that falls in one of the following categories:

- Commercial-off-the-shelf (COTS) software (e.g., shrink-wrap)
- Airborne software developed to other standards (e.g., MIL-STD-498)
- Airborne software that predates DO-178B (e.g., developed to the original DO-178)
- Airborne software previously developed at a lower software level

The use of one or more of these types of software should be planned for and discussed in the PSAC. In every case, some form of gap analysis must be performed to determine where specific objectives of DO-178B have not been met. It is the applicant's responsibility to perform this gap analysis and propose to the regulatory authority a means for closing any gaps. Alternate sources of development data, service history, additional testing, reverse engineering, and wrappers* are all ways of ensuring the use of PDS is safe in the new application. In all cases, usage of PDS requires that the safety assessment process be reentered. A special instance of PDS usage occurs when software is used in a system to be installed on an aircraft other than the one for which it was originally designed. Although the function may be the same, inter-faces with other aircraft systems may behave differently. As before, the system safety assessment process must be repeated to assure that the new installation operates and behaves as intended. If service history is employed in making the argument that a PDS component is safe for use, the relevance and sufficiency of the service history must be assessed. Two tests must be satisfied for the service history approach to work. First, the application for which history exists must be shown to be similar to the intended new use of the PDS. Second, there should be data, typically problem reports, showing how the software has preformed over the period for which credit is sought. The authors of DO-178B intended that any use of PDS be shown to meet the same objectives required of newly developed code. Prior to identifying PDS as part of a new system, it is prudent to investigate and truly understand the costs of proving that the PDS satisfies the DO-178B objectives. Sometimes, it is easier and cheaper to develop the code again!

12.9 Tool Qualification

DO-178B requires qualification of tools when the processes noted by DO-178B are eliminated, reduced, or automated by a tool without its output being verified according to DO-178B. If the output of a tool is demonstrated to be restricted to a particular part of the life cycle, the qualification can also be limited to that part of the life cycle. Only deterministic tools can be qualified. Tools are classified as development tools and verification tools. Development tools produce output that is part of airborne system and thus can introduce errors. Rules for qualifying development tools are fashioned after the rules of assurance for generating code. Once the need for tool qualification is established, a tool qualification plan must be written. The rigor of the plan is determined by the nature of the tool and the level of code upon which it is being used. A tool accomplishment summary is used to show compliance with the tool qualification plan. The tool is required to satisfy the objectives at the same level as the software it produces, unless the applicant can justify a reduction in the level to the certification authority.

Verification tools cannot introduce errors but may fail to detect or mask their presence. Qualification criterion for verification tools is the demonstration of its requirements under normal operational conditions. Compliance is established by noting tool qualification within PSAC and software accom-plishment summary.

---

* Wrappers is a generic term used to refer to hardware or software components that isolate and filter inputs to and from the PDS for the purposes of protecting the system from erroneous PDS behavior.

12.9.1 Additional Guidance

RTCA SC-190/EUROCAE WG-52 was formed in 1997 to address issues that were raised by the industry and certification authorities in the course of applying DO-178B since its release in 1992. This committee produced two outputs. The first, DO-248B/ED-94B, provides clarifications and corrections to DO-178B in the form of errata, frequently asked questions, and discussion papers. DO-248B/ED-94B is titled Final Report for Clarification of DO-178B "Software Considerations in Airborne Systems and Equipment." The second document, DO-278/ED-109, provides assurance objectives for ground- and space-based systems. The document is titled Guidelines for Communication, Navigation, Surveillance and Air Traffic Management (CNS/ATM) Systems Software Integrity Assurance. DO-278/ED-109 is based on and must be used in conjunction with DO-178B. DO-278/ED-109 introduces an additional assurance level for compatibility with existing ground-based safety regulations, as well as guidelines for the use of COTS software and adaptation data.

RTCA SC-205/EUROCAE WG-71 was formed in 2005 for the purposes of updating DO-178B. The primary reason for this update cycle has been to address industry concerns related to emerging technology including formal methods and model-based development. Specific areas of DO-178B that were improved include the use of tools, object-oriented technology, relationship between software and system, traceability, and parametric data. After the set of documents including DO-178C and the four supplements were published by RTCA, the FAA has released AC 20-115C to officially recognize these RTCA documents. At the time of this writing, EASA is in the process of making updates to a certification memorandum to accommodate the use of DO-178C and the supplements. DO-248C/ED-94C for explanations of DO-178C topics, and DO-278A/ED-109A for the use of ground and space software, also have been released.

12.10 Conclusions

DO-178B provides objectives for software life-cycle processes and activities to achieve these objectives and outlines objective evidence for demonstrating that these objectives were accomplished. The purpose of software compliance to DO-178B is to provide considerable confidence that the software is suitable for use in airborne systems. DO-178B should not be viewed as a documentation guide. Compliance data are intended to be a consequence of the process. Complexity and extent of the required compliance data depends upon the characteristics of the system/software, associated development practices, and the interpretation of DO-178B especially when it is applied to new technology and no precedence is available. Finally, it has to be emphasized that DO-178B objectives do not directly deal with safety. Safety is dealt with at the system level via the system safety assessment. DO-178B objectives help verify the correct implementation of safety-related requirements that flow from the system safety assessment. Like any standard, DO-178B has good points and bad points (and even a few errors). However, careful consideration of its contents taken together with solid engineering judgment should result in better and safer airborne software.

Further Reading

The Federal Aviation Administration Web Page: www.faa.gov. Accessed on April 28, 2014.

The RTCA Web Page: www.rtca.org. Accessed on April 28, 2014.

Spitzer, C.R., *Digital Avionics Systems Principles and Practice*, 2nd edn., McGraw-Hill, Inc., New York, 1993.

Wichmann, B.A., A review of a safety-critical software standard, National Physical Laboratory, Teddington, Middlesex, U.K., n.d.

References

1. RTCA DO-178B, *Software Considerations in Airborne Systems and Equipment Certification*, RTCA Inc., Washington, DC, 1992; ED-12B, *Software Considerations in Airborne Systems and Equipment Certification*, EUROCAE, Paris, France, 1992.
2. SAE ARP4754, *Certification Considerations for Highly-Integrated or Complex Aircraft Systems*, SAE International, Warrendale, PA, November 1996.
3. Chilenski, J.J. and Miller, P.S., Applicability of modified condition/decision coverage to software testing, *Software Engineering Journal*, 9(3), 193–200, September 1994.

13

RTCA DO-178C/ EUROCAE ED-12C and the Technical Supplements

Thomas K. Ferrell
*Ferrell and Associates
Consulting, Inc.*

Uma D. Ferrell
*Ferrell and Associates
Consulting, Inc.*

13.1 Introduction

In the preceding chapter, we covered DO-178B, the prevailing software design assurance guidelines for over two decades. At the end of 2012, an updated version of this venerable "standard" was published by the RTCA. Both the Federal Aviation Administration (FAA) and European Aviation Safety Agency (EASA) have since formally recognized this update for use of future projects, although many projects will likely be grandfathered under the B version for many years to come. There is not a lot of difference between the B and C versions. This was by design and not without controversy. The real change that comes with the C version is the companion documents created at the same time. These include

1. DO-330, Software Tool Qualification Considerations
2. DO-331, Model-Based Development and Verification Supplement to DO-178C and DO-278A
3. DO-332, Object-Oriented Technology and Related Techniques Supplement to DO-178C and DO-278A
4. DO-333, Formal Methods Supplement to DO-178C and DO-278A

All of these documents originated as technical supplements to the "core" DO-178C. The "supplement" status was dropped for the tool qualification document in recognition that it really is a stand-alone guideline that can be used with other guidelines (e.g., DO-254, ARP-4754A).

This chapter is intended to provide a solid introduction to the changes in the core DO-178 document, as well as the technical supplements. Individuals unfamiliar with DO-178 should read this chapter in conjunction with the preceding chapter as a basic knowledge of DO-178B is assumed in this chapter.

13.2 Core Document Changes

When one picks up DO-178C for the first time, one could be forgiven for thinking the committee simply slapped a new cover page on and called it good. The overall structure of DO-178B was retained, and there were no section deletions, additions, or major reordering. The committee focused their efforts primarily on making the document work better within the broader framework of design assurance and in cleaning up areas of the document that had caused interpretation problems in the past. The largest technical addition centered on documenting and verifying configuration data that have become common in modern avionics design. To help in the rapid understanding of the changes that were made, the next few sections follow the table of contents of DO-178C. The intent is to draw attention to those items that are most likely to affect how a company applies the document and how compliance is likely to be assessed by the regulatory authority.

13.2.1 Introductory Material

There are a number of subtle changes here that are worth noting. First, this document now includes language similar to that found in DO-254 that attempts to eliminate the gray area of firmware. Digital design and implementation must be declared as being either software or hardware and the appropriate guidance applied. A clear statement concerning flow down of DO-178C objectives to all suppliers for which compliance to the guidelines will need to be demonstrated is included. The greater emphasis on suppliers is a theme that carries through many of the chapters in DO-178C. Language, making it clear that Annex A, while providing the leveling for various objectives, is not intended as a stand-alone "checklist" for DO-178C compliance, has also been added. The user of the document is expected to read, understand, and apply all of the guidance contained in DO-178C. Finally, a general pointer to the existence of supplements has been added. This pointer notes that supplements may "add, delete, or otherwise modify objectives, activities, explanatory text, and software life cycle data" in DO-178C for a specific development technique.

13.2.2 The System/Software Interface

While arguably not substantially different in intent, there were more changes in specific wording throughout Section 2 of DO-178C than in any other part of the document. The purpose of all of these changes was to better align and define the interface with the system development process. This is critically important as there have been numerous accidents and incidents where issues arising in the system/software interface have been implicated as causal factors.

Many of the updates in this section are focused on identification and treatment of safety-related requirements. This term appeared numerous times in DO-178B but was never well explained. It was also unclear whether they warranted special treatment such as dedicated tagging (for traceability) or greater verification. DO-178C indicates that such requirements flow from the system safety assessment process and "may include functional, integrity, and reliability requirements, as well as design constraints." While still not explicitly requiring such requirements to be identified separate from the rest of the software requirements, discussions later in the document concerning low-level requirement creation and subsequent verification all suggest that doing so is a good practice.

Section 2 of DO-178C provides clarity on the terms error, fault, and failure with the latter being the connecting point with the higher-level regulatory framework where potential aircraft-level failure conditions are discussed. Using these terms as a basis, software levels are defined and an expanded discussion on the role of architectural mitigation is provided. This includes serial and parallel path processing, as well as the various forms of partitioning, now a common element of most advanced avionics architectures.

Perhaps the biggest addition to Section 2 is the first introduction on DO-178C of the concept of a parameter data item or PDI. The glossary in DO-178C defines a PDI as

> A set of data that, when in the form of a Parameter Data Item File, influence the behavior of the software without modifying the Executable Object Code and that is managed as a separate configuration item. Examples include databases and configuration tables.

PDIs have proliferated as a way of allowing software to be configured for different types of aircraft, for different combinations of equipment onboard a single aircraft, and for providing maximum configurability of the functions within a piece of software without having to create separate and distinct software part numbers, each requiring its own approval as part of a certification process. In short, PDIs enable a significant degree of flexibility in how software is deployed within different instances of the same avionics platform. There are two key things to keep in mind concerning PDI. First, for configuration data to be considered a true PDI, they must be separately loadable from the executable object code (EOC). If it is compiled with the EOC, it should be considered part of the EOC and verified with the EOC. Second, although separately loadable from the EOC, whether or not the PDI needs to be tested with the EOC depends on multiple criteria. More on this point in the discussion of Section 6 changes as follows. One or more PDIs will generally be captured in a PDI file (PDIF), one of two new data items introduced in DO-178C.

13.2.3 Software Planning and Development

Section 3 of DO-178C is a general introductory section on software life cycles. There are no changes to this chapter as part of the update. Section 4 discusses software planning. This section continues the increased emphasis on supplier utilization and adds requirements for planning for PDI usage. In addition, new hooks have been added in the planning process to formally require tool errata analysis and to ensure robustness is considered as part of the software development standards. Finally, a note has been added in the standards section to remind designers that standards may be needed for the detection and control of single event effects should these actions be allocated to software through the systems engineering process.

Although the actual amount of modified text is small in Section 5, the additions and modifications are significant. The first change to note is a new requirement to ensure that the rationale behind the creation of a derived requirement is documented. This has been added for both high- and low-level requirements. A new activity has been added to ensure that high-level requirements are written for any planned use of PDI. This includes the definition of the PDI's structure, the various data attributes, and the allowable values for such data. A new activity has been identified to explicitly capture low-level requirements for interfaces between software components, something that should support the conduct of data and control coupling analysis. A new subsection has been introduced that identifies specific activities to be accomplished whenever deactivated code is planned. These new guidelines should greatly reduce the variability that these authors have seen in how the regulatory authorities have treated this topic over the years. For the first time, the Coding subsection formally acknowledges the use of autocode generators, a topic discussed in more detail in Section 13.4. Finally, a new subsection (§5.5) is introduced that draws together for the first time in one place all of the guidelines for traceability throughout the development process. While not really new, the presentation of a concise description requiring "bidirectional association" between system requirements and high-level software requirements, high-level and low-level software requirements, and low-level requirements and code accompanied by a clear requirement to document this information is the second new data item, trace data, that removes any ambiguity over what needs to be documented in the way of traceability information.

13.2.4 Verification

Changes, while typically surgical in nature, a new bullet here, an added note there, significantly strengthen many aspects of the verification process. Examples include

- Greater focus on robustness and clear linkage between robustness behavior and requirements
- Requirement for justifying the adequacy and fidelity of alternative test environments
- Requirement for reviews and analysis to address "cache management, unused variables, and data corruption due to task or interrupt conflicts"
- Explicit requirement to analyze compiler warnings

In addition, numerous small modifications have been made to both the functional and structural coverage activities. The responsibility of the verifier to clearly resolve or justify any coverage gaps is now made explicit. As was done in planning, more time and attention is devoted to deactivated code with two categories now being identified. Category 1 covers things like debug code and requires that the deactivation mechanisms be proven and that the presence of such code be taken into account by the system safety assessment process. Category 2 covers code that is present to support alternative configurations (e.g., different engine types possible for a given aircraft type). In this case, complete verification for the different configurations is expected.

Finally, two new subsections are added to DO-178C. The first of these is a continuation of the trace data requirements with traceability now explicitly called out for requirements to test cases, test cases to test procedures, and test procedures to test results. The second new subsection addresses verification of PDIs. As noted previously, there are cases when PDIs may be tested independent of the EOC. The criteria to be satisfied to allow this to happen include the following:

- The EOC has already been tested to address the full range of values possible in a PDI that has been structured per its defined requirements.
- The EOC testing included the required robustness testing relative to and defined PDI.
- The EOC behavior based on PDI content is fully verifiable.
- The PDI is fully manageable separate from the PDI.

If any of these criteria are not satisfied, the expectation is that the PDI will be tested with the EOC. If separate testing is possible, then two additional objectives are identified to govern such independent verification. The first objective entails ensuring the PDI satisfies its required structure, has all the necessary attributes, has the correct value, and does not contain extraneous information. The second objective is a coverage objective intended to ensure all such PDI is, in fact, verified.

13.2.5 Configuration Management

There are very few material changes added in this section. However, at least one of them should go a long way in clearing up confusion from the original DO-178B. Table 7-1 of DO-178B has been retitled in DO-178C to read "SCM Process Activities Associated with CC1 and CC2," whereas the old title referred to objectives. Aside from the obvious fact the list in the table did not align cleanly with the objectives given in either the preceding sections or the associated Annex A table, it was never very clear what the table was trying to represent. With the shift to activities, it becomes apparent that the intent is to truly lighten the burden when working with CC2 data. For example, problem reporting as an activity is not required for CC2 data, meaning that changes may be made and simply noted in revisioning information or via notes during a check-in process with the project's configuration management system rather than needing an explicit problem report (PR) to serve as a change driver.

13.2.6 SQA, CL, and the Certification Process

The committee opted to leave these three sections almost completed untouched. For SQA, the biggest change comes in the form of a slightly reordered set of objectives. The changes here were introduced to allow correction of a long-standing contradiction concerning the oversight of standards compliance presence in DO-178B. Supplier oversight also gets some additional emphasis.

For certification liaison, only very minor wording changes were introduced, the one exception being the inclusion of PDI as part of the type design data.

13.2.7 Software Life-Cycle Data

As previously mentioned, two new data items have been added: trace data and PDI files. In addition, numerous small changes were made throughout the life-cycle data section relating to the increased emphasis on supplier oversight and the life-cycle environment. Of particular note are two items added concerning PRs. Under PR content, a clarifying phrase was added to now require "sufficient detail to facilitate the assessment of the potential safety or functional effects of the problem." Expect some debate and negotiation with the regulatory authority on what constitutes "sufficient detail" in the years ahead. In the description of the Software Status section of the software accomplishment summary (SAS), a detailed list of information required concerning any open PRs at type of certification is now included. This formalizes what had already become a standing expectation imposed through other regulatory guidance. The additions include a description of any "functional limitations, operational restrictions," as well as "potential adverse effect(s) on safety." Pulling this information together will absolutely require involvement from the systems engineering team, in addition to the software engineers typically responsible for authoring these data.

13.2.8 Additional Considerations

The major changes in this section center on tool qualification and product service history (PSH). For tools, the previous definitions of development and verification tools have been retired in favor of three different tool criteria. A criteria 1 tool is any tool that creates a portion of the flying code and therefore could insert on error in that code. This is synonymous with tools previously defined as development tools. Criteria 3 tools are any tools that could fail to detect a latent error in the inputs being processed by the tool. This category is synonymous with tools previously defined as verification tools. In between these two is a new category, criteria 2, which can be thought of as a "super" verification tool. Such tools not only could fail to detect latent errors, but their use is further exploited to eliminate other downstream activities, thus possibly increasing the likelihood that an error or group of errors would go undetected. As you might expect, qualification activities for criteria 1 tools are the most stringent, criteria 2 tools less so, and criteria 3 tools the least stringent. This is shown in the new leveling approach built around a newly defined tool qualification level (TQL). Table 12-1 from DO-178C illustrates this leveling (Table 13.1).

With the tool criteria and TQL established, readers of DO-178C are then pointed to DO-330 for the details of the qualification process to be employed for each tool type and level. DO-330 contents are discussed later in this chapter.

While the majority of the Alternate Methods subsection was left unchanged, the guidelines for PSH were rewritten and expanded. The additional focus on this topic is the recognition that many companies have software developed for other domains, often military, that could be used in new avionics applications. As before, the relevance of the prior usage history and the number and nature of problems

TABLE 13.1   TQL Determination (Table 12-1 from DO-178C)

Software Level	Criteria		
	1	2	3
A	TQL-1	TQL-4	TQL-5
B	TQL-2	TQL-4	TQL-5
C	TQL-3	TQL-5	TQL-5
D	TQL-4	TQL-5	TQL-5

encountered during this earlier operational period are of paramount concern. The completeness of this problem history and the nature of each encountered problem are all expected to be analyzed. While no explicit time durations are stipulated for specific software levels, a framework for having a meaningful dialogue with the regulatory authority is provided. This alone should help companies with useful and significant PSH find a way forward without having to completely reverse engineer the product.

13.3 DO-330: Software Tool Qualification Considerations

For users of DO-178B, performing developmental tool qualification was a challenge. The typical problems stemmed from contradictions in the tool section with normative guidance found elsewhere in the document, all of which was applicable at the software level at which the tool was to be employed. Bringing clarity to this the topic of developmental tool qualification was one of the driving reasons behind the effort to update DO-178B. Starting out first as a technical supplement, DO-330 graduated to a stand-alone document intended for use not only with DO-178C but with a variety of other guidelines including DO-278A, DO-200A, DO-254, and ARP4754A. It has been written to completely address all levels of tool qualification for all three criteria tools mentioned in DO-178C.

To create DO-330, each part of DO-178B was reviewed, interpreted, and, where necessary, rewritten in the context of authoring a tool rather than airborne software. As with DO-178C, significant emphasis is placed on properly and completely planning the tool qualification effort. In general, TQL 5 tools can be handled just as was done previously with tool qualification being planned for in the project's PSAC and the results being summarized in the project's SAS. A Tool Operational Requirements (TOR) is written and verification of those requirements is subsequently accomplished. Above this TQL, a dedicated tool qualification plan (TQP) is required along with a corresponding set of data items clearly identified by a set of tool-specific Annex A tables present in the back of DO-331.

One very important aspect of DO-331 is the specific treatment given to commercial-off-the-shelf (COTS) tools. In Section 11.3 of DO-331, distinct roles are defined for both a tool developer and a tool user. With these roles defined, two tables (11-1 and 11-2) are introduced, which identify which of the tool Annex A objectives are to be accomplished by each role. As an example of how this works, objectives 3 and 5 of Table T-1 in DO-331 are concerned with the identification of the tool development environment and tool development standards, respectively. These objectives are identified as being the responsibility of the tool developer. For a COTS tool where the user is simply procuring the tool, the tool user is relieved of the responsibility for these two objectives. The tool developer is responsible for providing evidence that they accomplished these objectives as part of the qualification package provided to the tool user in support of their certification effort. This role-based approach should greatly facilitate the qualification of third-party tools for use in avionics development.

13.4 DO-331: Model-Based Development and Verification Supplement

DO-331 defines a model as "an abstract representation of a given set of aspects of a system that is used for analysis, verification, simulation, code generation, or any combination thereof." Modern software development is often model-based. It is fully recognized that models bring many benefits to software engineering. The major concerns for model-based development are system to software allocation, validation of this allocation, use of system models that allow direct software implementation, use of appropriate development standards, use of symbols libraries, verification, and traceability. This supplement provides a way to assure model-based development and verification by interpreting the DO-178C core objectives especially with respect to these major concerns. There are many different ways of using models; some examples are given in the supplement. The application of the supplement and related objectives depends upon the way in which models are used on a particular project.

Software models need a modeling standard that should include

- Methods and tools for developing the models
- Modeling languages to be used
- Style guides and complexity restrictions
- Design constraints
- Requirements identification and traceability representation
- Identification of derived requirements
- Mechanisms for identifying nonrequirement/nonarchitecture data in the model

The modeling standard needs to consider any tools and methodology that is employed in order to implement any constraints and rules that may help avoid known problems and complexities or constructs that would make verification difficult.

DO-331 addresses the difficulty in granularity of model representation and traceability by requiring both specification model and design model. A single model cannot be classified as both a specification model (high-level requirements) and a design model (low-level requirements). Bidirectional traceability requirement dictates that high-level requirements trace to system-level requirements and that low-level requirements trace to code. It should be demonstrated that all system-level requirements allocated to software have been represented by specification and design models, and there is no code that cannot be traced back to a model. If the design models are developed as an output of the system process, MB.A-3 objectives 1 and 6 are implicitly satisfied.

Two major sections extend the core document verification contents: model coverage analysis and simulation. The purpose of model coverage analysis is to detect unintended function in the design model, as well as assessment of thoroughness of the model verification activities. Simulation may be used to provide repeatable evidence of compliance of the model to the requirements from which the model was derived. New objectives have been added to cover simulation cases and procedures, as well as simulation results (Tables MB.A-3, MB.A-4, MB.A-6, and MB.A-7).

13.5 DO-332: Object-Oriented Technology and Related Techniques Supplement

DO-332 provides guidance on the use of object-oriented techniques (OOTs) and related techniques (RTs). OOT in the context of this supplement includes

- Class and object definition/manipulation
- Typing and type safety (inclusive of Liskov substitution principle [LSP])
- Hierarchical encapsulation
- Polymorphism
- Function passing, closure, and dispatch

RTs in the context of this supplement include

- Inheritance
- Parametric polymorphism
- Type conversion
- Method-level exception handling
- Dynamic memory management

DO-332 also includes explanatory text in Annex OO.D concerning the pitfalls (vulnerabilities) of using these concepts. The technical information in this annex was derived from the earlier FAA–industry collaboration on this topic. The intent is to explicitly highlight the problems that can arise in the use of a specific OOT/RT aspect and to provide supporting information for the activities described elsewhere in the supplement. Because of these vulnerabilities, use of OOT in safety-critical development often complicates assurance as well as demonstration of assurance. Applicants using this supplement should expect to demonstrate how the vulnerabilities discussed have been addressed and mitigated. Planning on an OOT project should address the use of OOT including vulnerability analysis. For example, traceability can be complicated when inheritance is used. Inheritance creates a relationship between requirements in a subclass with those of a parent class—the effect is to multiply the numbers of trace links. Specific OO language features such as constructors, destructors, in-lines, run-time checks, and implicit type conversion all complicate source to object code mapping. Behavior and structure captured in models (e.g., UML) create a need for traceability to and within the model. Planning for meeting the traceability objectives should explicitly treat each of these items.

Requirements need to address any OOT aspects in the code (e.g., dynamic memory allocation). Verification objectives have only modest impacts with exception of two new A-7 objectives both required at levels A and B, namely, verify local-type consistency and verify that the use of dynamic memory management is robust. Verification of dynamic memory allocation is a complex topic that needs to address ambiguity, fragmentation starvation, deallocation starvation, memory exhaustion, premature deallocation, lost updates, stale references, and unbounded allocation or deallocation times. Verification activities must include assurance of sufficient memory to accommodate the maximum storage required. Successful allocation of memory for every request must also be verified. Calculation of memory should be accurate and free from leakage problems.

The key to dealing with OOT is to understand how its use complicates traditional DO-178C topics.

13.6 DO-333: Formal Methods Supplement

DO-333 defines formal methods (FMs) as mathematically based techniques for the specification, development, and verification of software aspects of digital systems.

FM provides an opportunity for

- Unambiguously describing requirements of software systems
- Enabling precise communication between engineers
- Providing verification evidence such as consistency and accuracy of a formally specified representation of software
- Providing verification evidence of the compliance of one formally specified representation
- Demonstrating system properties such as freedom from exceptions, freedom from deadlock, isolation between different levels, worst case execution timing (WCET), stack usage, and correct synchronous/asynchronous behavior

Application of FM usually involves two types of activities: modeling and analysis, where the models and analysis procedures are derived from or defined by an underlying mathematically precise foundation. Not all models meet the requirements for "formality"; the supplement has guidance for when a model/analysis is to be considered an FM. This supplement is tied to model-based supplement in that the model development may be assured using this supplement as one method of assurance.

FM can be used to analyze models for correctness either fully or complemented with other verification including testing on target. Formal analysis may be classified under three categories:

- Deductive methods—theorem proving; proofs are constructed by using rigorous mathematical methods, typically using automated systems
- Model checking to explore all of the properties of the model to find problematic properties
- Abstract interpretation is a theory for formally constructing conservative representations (i.e., enforcing soundness) of the semantics of programming languages

DO-333 not only gives the guidance where FM can be used but also where the guidance from DO-178C core document guidance must be used. Annex FM.A tables are so designed to indicate the FM objectives that must be followed within the context of the objectives of the core document. The intent of verification objectives and activities as in the core document is unchanged. But the process with which verification is carried out for FM is different and these differences are highlighted in the changed paragraphs. Formal analysis can be used to satisfy many of the verification objectives, completely in some cases and only partly in others. FM may be partially or completely used to model requirements, design, and architecture in which case a mix of FM guidance and core document can be used for verification. Target compatibility should be checked by running test cases on target even though a number of analyses can augment the results for WCET, stack analysis, and state transition analysis. The key items to focus for FM verification are requiring verification of the analysis methods and associated transformation and allowing credit for the formal proof of correctness in lieu of other forms of verification including reviews and testing. Four new objectives are added to each of the Annex A tables—A-3, A-4, and A-5. The interpretation of the core Annex table A-6 is changed; no new objectives are added to this table. A-7 is most heavily impacted. Core test objectives 1 through 8 are deleted for FMs. Core objective 9 is accomplished using FM guidance.

A-7 adds new objectives, FM1 through 9:

- FM cases and procedure correctness
- Results correctness and explanation of discrepancies
- Coverage of HLR
- Coverage of LLR
- Complete coverage of each requirement
- Completeness of requirements
- Unintended dataflow detected
- Dead/deactivated code detected
- FM that is correctly defined, justified, and appropriate

In general, FM supplement allows the user to take advantage of the unambiguous manner in which formal models can be expressed provided that all assumptions are justified.

13.7 Applying DO-178C

The FAA formally recognized DO-178C, DO-330, and the three technical supplements via Advisory Circular (AC) 20-115C in July 2013. This AC essentially requires DO-178C's use for brand new developments while providing a set of criteria whereby previously approved software, referred to as Legacy System Software (LSS), can be assessed to determine if DO-178C needs to be applied for changes. The AC also provides a small number of clarifications concerning the use of the supplements including a requirement for the applicant to address the specific supplement applicability by software component in their PSAC, as well as how compliance to multiple supplements will be shown (if applicable). This is most easily accomplished through a set of detailed compliance matrices. Finally, AC 20-115C discusses the use of DO-330 for previously qualified or "legacy" tools.

Similar recognition from EASA has been drafted and is likely to be published by the time this Handbook goes to print. Compliance finding activities for DO-178C are expected to be similar to those already in use for DO-178B.

14

RTCA DO-254/ EUROCAE ED-80

Randall Fulton
FAA Consultant DER
SoftwAir Assurance, Inc.

14.1 Introduction

This chapter provides a summary of the document RTCA DO-254, *Design Assurance Guidance for Airborne Electronic Hardware.*[1] The joint committee of Radio Technical Commission for Aeronautics (RTCA) Special Committee 180 and European Organization for Civil Aviation Equipment (EUROCAE) Working Group 46 prepared RTCA DO-254 (also known as EUROCAE ED-80), and it was subsequently published by RTCA and by EUROCAE in April 2000. DO-254 provides guidance for the production of electronic hardware items used in airborne systems and equipment such that there is a level of confidence in the correct functioning of that hardware in compliance with airworthiness requirements. The hardware items encompass line-replaceable units (LRUs), circuit card assemblies, and custom microcoded devices such as field-programmable gate arrays (FPGAs) and application-specific integrated circuits (ASIC). DO-254 represents industry consensus opinion on the best way to ensure that electronic hardware performs its intended function and meets industry regulations. DO-254 recommends typical activities performed during the hardware development life cycle. The document also embodies the industry consensus on engineering best practices for electronic hardware development. The verification activities specified in DO-254 are particularly effective in identifying hardware problems early in the development process.

In this chapter, design assurance level (DAL) and its shortened version—Level, as in Level A and Level B—are used interchangeably.

14.2 Scope of DO-254

DO-254 was created to address certification and safety concerns about aircraft systems with ever-increasing complexity of electronic hardware. Electronics complexity increases at the device level, the circuit level, and the circuit card level. As silicon feature size decreases, the number of gates available on programmable logic devices (PLDs) increases. Integrated circuits and packages continue to shrink, allowing more functionality on the same board space. Technology advances in the production of resistors, capacitors, and planar magnetics along with evolving circuit card manufacturing processes and device packaging allow circuit densities on boards to increase. The industry working group was formed in the 1990s and convened for 7 years to create the guidance.

DO-254 is written with a top-down perspective on electronic hardware, starting from the system/LRU level. The guidance, objectives, and activities discussed are meant to be inclusive of all complex electronic hardware that performs safety critical system functions. While certain aspects of ASIC and PLDs are discussed in DO-254, these aspects are considered in the context of system and equipment development. DO-254 is most comprehensible when considered from the perspective and philosophy in which it was conceived and in which it is intended.

Problems arise interpreting and applying guidance for electronic systems to singular components (ASIC or PLDs) within the system. These problems include determining the scope of the life cycle data produced in a program using DO-254, determining how to interpret and apply DO-254 Table A-1 to PLD life cycle data, sorting out which activities and aspects apply to the system level versus a PLD component level (such as acceptance test aspects, environmental tests, or thermal analysis), and understanding the intent of elemental analysis in DO-254 Appendix B. Philosophical and/or intellectual discord is possible when considering how guidance and practices applied to a lone component, or several PLDs, on a complex circuit card makes the entire circuit card or system it is used in perform its intended function and meet safety objectives.

The application of DO-254 results in approved electronic hardware. It is important to note that electronic hardware is not certified as a stand-alone entity.

14.3 Make Everything as Simple as Possible, But Not Simpler—Albert Einstein

DO-254 is written to address design assurance of complex electronic hardware and makes a distinction in classifying hardware as simple or complex. Classifying hardware as simple allows for a reduction in the design assurance activities and documentation described in DO-254. Hardware that is fully testable, through comprehensive and deterministic verification tests that address all foreseeable operating conditions, is classified as simple. The verification of simple hardware needs to demonstrate that the hardware has deterministic behavior and is free of anomalies. Hardware is classified as complex if it cannot be classified as simple.

DO-254 recommends that an agreement with the certification authority on the design assurance approach for complex hardware should occur early in the program to mitigate risk. This recommendation is prudent and equally true for simple hardware. The applicant and the certification authority need to agree on the classification as simple and the design assurance approach, commensurate with the DAL.

Chapters 4 and 5 of Federal Aviation Administration (FAA) Order 8110.105[2] include guidance for simple electronic hardware (PLDs). Order 8110.105 states that a classification of simple should be made early in the program, during the certification planning. The Plan for Hardware Aspects of Certification

(PHAC) needs to state the classification of the hardware and explain the approach for hardware classified as simple. The Order also states the applicants should perform more thorough and rigorous testing for DAL A and B hardware. Specifics on the testing, analyses, test environment, and documentation are also provided in the Order.

It turns out that in some cases, simple is not quite so simple.

14.4 View from 35,000 ft

DO-254 fits in to a top-down system perspective of development and design assurance. Performing the system engineering and system safety activities prior to undertaking design assurance for complex electronic hardware promotes a streamlined and organized effort. Keeping the system perspective in mind will guide the usage of the document, the organization of the life cycle data, and the management of the activities.

Systems aspects are considered first:

- DO-254 is applied in the context of a system approach to development and certification. The systems aspects are addressed by ARP-4754A.[3] System or function DAL is determined from aircraft level effects of a hazard.
- The functional hazard assessment is performed in accordance with ARP-4761.[4] The hazard classification (catastrophic, hazardous, major, or minor) for each system function is assigned from a system perspective. Functional failure paths and associated DAL for the function are first identified in the preliminary system safety assessment.
- System functions, and their respective DAL, are allocated to hardware. The designers select which aspects of functionality will be implemented in electronic hardware—discrete, programmable, simple, or complex.
- The requirements for the system functions are allocated to the electronic hardware and include functional, performance, and safety aspects.

Functions and their effects are identified:

- Functions are expressed by requirements.
- Each function has an associated DAL.
- A functional failure path analysis (FFPA) is used to determine which hardware functions are in catastrophic or hazardous aircraft functions. The FFPA is used to identify the circuits associated with the functions.
- Functions are optimally organized or grouped by functions. Circuits or circuit elements are designed to implement the requirements.
- Traceability is established from requirements to the circuits that implement the functions.

Design assurance approach for hardware is defined and performed:

- Determine which hardware functions and associated circuits implement DAL A and B functions.
- Verification is requirements based. Verification for DAL A or catastrophic functions is the most stringent and rigorous. Level B is the next most stringent and so on down through Level C and Level D.
- PLDs are assigned the highest DAL for the functional failure path(s) associated with the device.
- Additional design assurance approaches are required for Level A and B electronic hardware.
- Incorporate the activities, data, and transition criteria for the selected life cycle and design assurance approach in the hardware management plans. Execute the program according to these plans.
- Create hardware standards that define the criteria that a design needs to meet. Standards also define the criteria for reviewing and verifying electronic hardware. Produce hardware, design data, and verification data that meet or exceed the standards.

14.5 Electronic Hardware and System Development

The application of DO-254 is part of the established industry practices for systems development and hardware. When DO-254 was published, the system-level aspects of certification were described in SAE ARP4754, *Certification Considerations for Highly-Integrated or Complex Aircraft Systems*.[5] The current version of ARP4754 is Revision A, published in 2010.

The interfaces to the system development process are defined in ARP4754, which was originally published in 1996. DO-254 specifies the information flow between system processes and hardware processes. The focus of the information flow from the system process to the hardware process is to keep track of requirements allocated to hardware, particularly those requirements that contribute to the system safety or originate in the system safety processes. The focus of information flow from the hardware process to the system process is to ensure that changes in the hardware requirements, including the introduction of derived requirements, do not adversely affect system safety. Derived requirements are those additional requirements that result from the design process or design decisions and might not directly trace to an upper-level requirement. The system process determines the DAL of the system function and the DAL of the hardware item implementing the function. This information flow is shown in Figure 14.1.

System safety methods and analyses, although outside the scope of DO-254, are crucial to understanding how to apply DO-254. The regulatory materials governing the certification of airborne systems and equipment define five classifications of failure conditions. The most severe of these is *catastrophic*, meaning failures that result in the loss of ability to continue safe flight and landing. The least severe is *no effect*,

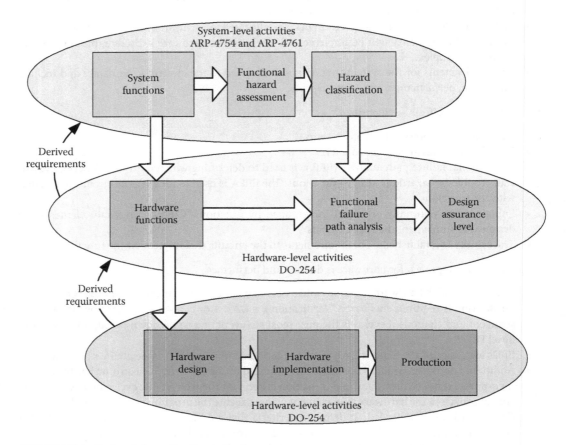

FIGURE 14.1 System information flow to hardware processes.

TABLE 14.1 Design Assurance Level and Hazard Classification

Failure Classification from Functional Hazard Assessment	Hazard Description	Hardware Design Assurance Level
Catastrophic	• Prevents continued safe flight and landing	Level A
Hazardous/Severe-Major	• Serious or fatal injuries to small number of occupants • Reduces aircraft capabilities or crew's ability to deal with adverse operating conditions • Higher crew workload • Large reduction in safety margins	Level B
Major	• Possible injuries to occupants • Reduces aircraft capabilities or crew's ability to deal with adverse operating conditions • Increase in crew workload • Significant reduction in safety margins	Level C
Minor	• Possible inconvenience to occupants • Reduces aircraft capabilities or crew's ability to deal with adverse operating conditions • Slight increase in crew workload • Slight reduction in safety margins	Level D
None	• No effect on operational capabilities • No crew workload impact	Level E

where the failure results in no loss of operational capabilities and no increase in crew workload. The other three types define various degrees of loss of functionality resulting in corresponding crew workload and potential for loss of life or injury to occupants. These five hazard classifications map directly to the five levels of hardware design assurance defined in DO-254. This mapping is shown in Table 14.1.

A parallel process exists for the software development process and flow of information between software processes and system process. Design trade-offs between software processes and hardware processes are also taken into consideration at the system level. Hardware DALs may be lowered by using protective software or hardware mechanisms elsewhere in the system. Such architectural methods include partitioning, use of hardware or software monitors, and architectures with built-in dissimilarity and/or redundancy.

14.6 Using DO-254

As stated previously, a DO-254 project starts with system-level activities. The system level determines which functions, as expressed through the associated requirements, are allocated to the electronic hardware. The preliminary system safety assessment determines the DAL for each system function. System requirements are allocated or decomposed into electronic hardware requirements. The DAL for each electronic hardware function is based on the associated system-level function and its DAL. Derived requirements are added to the electronic hardware requirements as requirements are translated between system and hardware level and to accommodate design decisions made in the system design process. The circuits, or circuit elements, to implement the electronic hardware requirements are created. Traceability is established between requirements and the circuit elements that implement the requirements. Note that for a PLD, the circuit elements would typically be the hardware design language (HDL), especially for register transfer level (RTL) design abstraction models.

Once the system-level activities and safety analyses are under way, the FFPA described in DO-254 Appendix B can be used to determine which electronic hardware performs DAL A or B functions. Section 2.3.4 of DO-254 is then used to select the design assurance strategy for Level A and B functions. Note that while varying functional failure paths may be associated with a PLD, the design assurance of the device is associated with the most severe failure. The entire PLD is then developed in accordance with the highest DAL.

DO-254 Appendix B includes guidance for several methods used for additional design assurance in Level A and B electronic hardware. Elemental analysis is frequently the selected option. For PLDs, code coverage tools used in conjunction with functional simulation can serve as elemental analysis provided that the requirements structure and design abstraction supports the approach.

DO-254 Appendix A is used to determine the life cycle data for the respective DAL. Appendix A also shows the hardware control level or degree of rigor in the configuration management and control of the hardware and associated data. Section 10 of DO-254 gives a brief description of the contents of the life cycle data.

The planning phase describes the activities, documentation, or life cycle data and transition criteria for each phase of the development. The certification liaison process should start early to get concurrence on the selected design assurance approach.

Organizing electronic hardware requirements by function allows for a logical structure in the documentation, traceability, and the verification. The verification can use test cases designed to optimally test all aspects of the requirements and thus the functionality. The tests can use the trace data to assess the effective coverage of the design associated with the requirements. The tests and additional analyses are structured and modulated depending on the DAL and associated activities.

14.7 Invocation of the Guidance

While DO-254 was published in 2000, the FAA recognition of the guidance, Advisory Circular (AC) 20-152,[6] was issued 5 years later in 2005. In the 1990s, concurrent with the creation of DO-254, airframe manufacturers and their equipment suppliers started to develop their own criteria for complex electronics, especially ASICs and custom programmable devices. Aircraft certification programs used FAA issue papers to address certification authority concerns with complex devices. Initial criteria and guidance followed the lead of DO-178B, *Software Considerations in Airborne Systems and Equipment Certification.*[7] The intent was to encourage industry to use a structured methodology for complex hardware development that emphasized requirements management, requirements-based verification, quality or process assurance, and configuration management. The lingering effect on the aerospace industry has been the impression that electronic hardware, and, in particular, devices specified with a language-based design, is like software. While software programming languages and HDLs have similar features such as constructs, keywords, and syntax, the similarities end there.

The Advisory Circular was written for ASICs and PLDs and excludes commercially available microprocessors and other types of electronic circuits. In other words, the guidance was adopted for a subset of the scope of DO-254. Advisory Circular AC 20-152 acknowledges DO-254 as an acceptable means of compliance to airworthiness regulations for DAL A, B, and C custom microcoded components in systems or equipment being certified under a Technical Standard Order (TSO) authorization, type certificate (TC), amended type certificates (ATC), amended supplemental type certifications (ASTC), supplemental type certificate (STC), or Parts Manufacturer Approval (PMA). DAL D electronic hardware may use DO-254 or applicable in-house design assurance practices.

While the FAA allows alternate means of compliance, the majority of applicants use DO-254 to avoid the work involved in showing that other means are indeed equivalent to DO-254.

14.8 Document Overview

DO-254 consists of 11 sections and 4 appendices as shown in Figure 14.2. Section 2 of DO-254 provides information on how the processes, activities, hardware, and related data discussed in DO-254 integrate with the overall system certification process. Supporting processes described in Sections 6–9 are applied to the hardware life cycle processes and outputs described in Sections 3–5. Section 10 provides details on the intended content of life cycle data and Section 11 gives guidance for additional considerations. Following Section 11 is a roster of Special Committee 180 (EUROCAE WG-46) members and administrators that created DO-254. Appendix A follows Section 11 and includes Table A-1 and guidance

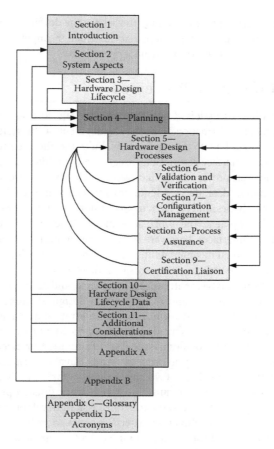

FIGURE 14.2   DO-254 document structure.

describing how life cycle data vary as a function of DAL. Included in Appendix A is a description of how independence is achieved for Levels A and B. Table A-1 also denotes the hardware control category for life cycle data. Appendix B provides considerations for the additional design assurance required for Level A and B functions. Appendix C provides the document's glossary of terms and Appendix D lists the acronyms and their definition. The glossary allows for a more uniform understanding and application of DO-254 through a precise definition of the terms in the context of hardware design assurance.

The objectives or purpose of each life cycle process defined in DO-254 are embedded within the respective section. Overall, DO-254 has 34 objectives. The objective(s) and associated life cycle data are provided in Table A-1. The following list shows where each set of objectives is defined in DO-254:

- Section 4.1   Planning Process Objectives
- Section 5.1   Requirements Capture Objectives
- Section 5.2   Conceptual Design Objectives
- Section 5.3   Detailed Design Objectives
- Section 5.4.1 Implementation Objectives
- Section 5.5.1 Production Transition Objectives
- Section 6.1   Validation Process Objectives
- Section 6.2   Verification Process Objectives
- Section 7.1   Configuration Management Objectives
- Section 8.1   Process Assurance Objectives

Sections 2–10 of DO-254 describe processes and activities for the development of Level C and D hardware. Level A and B hardware require independence, as described in Appendix A and one or more of the considerations for additional design assurance from Appendix B.

14.9 Hardware Life Cycle Processes

DO-254 starts Section 2 with a discussion of how the hardware and software life cycles fit in the overall context of system development. Information from the development of the system flows to the hardware development. The system process flow to the hardware includes the allocated requirements, DAL for the functions, failure probabilities, interface description, design and/or architecture choices or constraints, and any impact from system life cycle processes. The hardware development process flow to the system development includes implementation data, derived requirements, safety and reliability analysis data, data to support system processes or activities, and problem reports. Information from the development of the hardware flows to the software development. The information may include derived requirements or design constraints for memory mapping, interfaces, hardware/software integration, communications protocols, data to coordinate verification, functional incompatibilities, limitations or constraints, and problem reports.

For hardware development, the life cycle processes include planning, the hardware design, validation and verification, configuration management, process assurance, and certification liaison. The hardware design processes include requirements capture, conceptual design, detailed design, implementation, and production transition. Validation and verification, configuration management, process assurance, and certification liaison are collectively referred to as the supporting processes. DO-254 defines objectives for each of these processes and provides a typical set of activities for meeting the objectives.

DO-254 discusses the hardware life cycle processes and transition criteria between life cycle processes in a generic sense without requiring or favoring any particular life cycle model. Transition criteria are used to define the data and activities required to start and/or finish a process. Transition criteria promote control of the design process, can serve to reduce risk in development, and also allow for evidence to support audits. Well-defined transition criteria support coordination between the design and verification/validation activities and the configuration control of the data. The transition criteria are defined in the hardware management plans and are audited by process assurance.

14.9.1 Hardware Planning Process

DO-254 defines six types of planning data for hardware development. The hardware management plans may be separate or combined. The planning process results in the data shown in Table 14.2. The table also lists the DO-254 section where the corresponding content is described, the respective objectives satisfied by the data, and an indication as to whether the data are submitted to the certification authority.

These plans should include the hardware design methods, standards, and tools to be used during the development and verification. A review of the planning process should have enough details to assure that the plans, proposed development environment, and development standards (requirements, design, and code) comply with DO-254.

Although DO-254 does not discuss the certification liaison process until Section 9, the intent is that coordination with the certification authority begins during the planning phase. The applicant defines how the objectives of DO-254 will be satisfied and identifies the data to be used for substantiating compliance to the project certification basis. The PHAC should be submitted to the certification authority in a timely manner.

In addition, an Electronic Component Management Plan (ECMP) is used to address the procurement, substitution, and obsolescence of electronic components. While DO-254 does not specifically address the contents of an ECMP, International Electrotechnical Commission (IEC) document IEC TS 62239,[8] *Process management for avionics—Preparation of an electronic components management plan*, provides further information on the formulation of an ECMP appropriate to aerospace equipment.

TABLE 14.2   Hardware Management Plans

Plan	Content Described in DO-254 Section	Objective Satisfied	Submit to Certification Authority
PHAC	10.1.1	4.1—#1–4	Yes
Hardware Design Plan	10.1.2	4.1—#1–4	
Hardware Validation Plan	10.1.3	4.1—#1–4; 6.1.1—#1	
Hardware Verification Plan	10.1.4	4.1—#1–4; 6.2.1—#1	Yes
Hardware Configuration Management Plan	10.1.5	4.1—#1–4; 7.1—#3	
Hardware Process Assurance Plan	10.1.6	4.1—#1, 2, 4; 8.1—#1–3	

TABLE 14.3   Hardware Standards

Standard	Content Described in DO-254 Section	Objective Satisfied
Requirements Standards	10.2.1	4.1—#2
Hardware Design Standards	10.2.2	4.1—#2
Validation and Verification Standards	10.2.3	4.1—#2
Hardware Archive Standards	10.2.4	4.1—#2; 5.5.1—#1; 7.1—#1, 2

The planning process also produces the standards for the hardware development. DO-254 defines four types of standards for hardware development, which may be separate or combined. The standards produced during the planning process are listed in Table 14.3. The table also lists the DO-254 section where the corresponding content is described and the respective objectives satisfied by the data.

14.9.2 Hardware Design Processes

Hardware design processes include requirements capture, conceptual design, detailed design, implementation, and production transition. Requirements capture starts with the requirements that are allocated to the hardware from the system development process. The system designer decides which functions are put in the hardware and allocates the corresponding set of requirements. Requirements capture methods and formats are typically described in the requirements standards. Various stages of the hardware design and implementation processes may introduce additional requirements for the hardware item. The design of the upper-level hardware may impose additional, or derived, requirements on the hardware. These additional or derived requirements are fed back to the requirements capture process. Clarity on which requirements are considered derived is useful since subsequent validation processes are applied to the derived requirements. The going in assumption for DO-254 is that the hardware requirements have already been validated at the system level; hence, only the additional derived requirements need to be validated. Requirements can also change form or units from one level of abstraction in the system and hardware design processes to the next level. The change to a requirement can also be considered a derived or refined aspect.

Several examples will clarify the nature of derived requirements. Display brightness may be expressed in foot lamberts at the system level and in volts at the hardware level. The translation of the units of measurement between the two levels of requirements is considered a derived aspect of the hardware requirement and thus needs to be reviewed for correctness and completeness. A second example is the design decision for a circuit card to use a serial peripheral interface (SPI) for a digital-to-analog converter (DAC) that is controlled by a PLD such as an FPGA. The design decision at the circuit card level will necessitate additional derived requirements in the FPGA to accommodate the signals shared between the FPGA and the DAC. These FPGA-derived requirements can trace to the design or design decisions of the circuit card. Some types of derived requirements may trace back to parent or higher-tier requirements; other derived requirements are associated with the design of the parent or higher-tier design. Requirements can also decompose or break up into two or more requirements, as they propagate through lower levels of abstraction. Validation can also be applied to the decomposed requirements to ensure that the sum of the parts satisfies the whole.

The other important aspect of derived requirements is that the additional, new, or different functionalities that they introduce need to be evaluated at the system level for any system or aircraft impact.

Once the requirements are established, the first step in the design is a conceptual design to fulfill the requirements. Conceptual designs are intended to be a high(er)-level description of the design to choose and evaluate architectures, topologies, components, packages, and any implementation or production constraints. Designers often use conceptual design as a first pass, subject to revision and further refinement. Note too that the data created in this phase are subject to the less stringent configuration control objectives (hardware control category 2). Conceptual design methods and standards are typically described in the hardware design standards. Additional derived requirements identified in the conceptual design process are fed back to the requirements capture process.

The next process creates the detailed design from the requirements and the conceptual design data. The initial, or conceptual, design gets all the details added to fully express the design. DO-254 does not specify any specific design technique or methodology. The choice is up to the organization performing the design. Design methods used will vary according to the type of hardware being designed. System and assembly designs require different methods, tools, standards, and techniques than those used for the design of a circuit card or a PLD. Detailed design methods and standards are described in the hardware design standards. Additional derived requirements identified in the detailed design process are fed back to the requirements capture process.

The next process creates the hardware from the detailed design data. The implementation process uses manufacturing techniques representative of the ones that will be used during production. The hardware produced in this phase is used in the testing activities. Additional derived requirements identified in the implementation process are fed back to the requirements capture process.

The final design process gathers all the data needed for transition into production. This includes the design data necessary to manufacture the part(s) as well as the data needed to test it.

DO-254 also contains guidance on acceptance tests used during manufacturing, repair, or alteration of the hardware. The scope of DO-254 includes the acceptance test criteria, but not the actual acceptance testing. Acceptance testing is typically applied most stringently at the system level, then to some degree for circuit card and subassembly production. Acceptance tests could be used for ASICs after completion of their fabrication and packaging or for screening preprogrammed FPGAs. In most cases, requirements-based testing is applied to components such as an FPGA, while acceptance test is used for production of hardware assemblies.

The final aspect of hardware design in DO-254 is series production. No specific objectives or activities are described since the topic is considered out of the scope of DO-254. DO-254 does include consideration so that changes in the production process or of the actual hardware design do not impact safety, certification, or compliance with requirements or regulations.

14.9.3 Supporting Processes

DO-254 defines four processes as supporting, meaning that they overlay and extend throughout the hardware life cycle. These are the validation and verification process, configuration management, process assurance, and certification liaison process.

14.9.3.1 Validation and Verification

The validation and verification process evaluates the outputs of the hardware design process. The derived requirements are validated and all requirements, derived or allocated, are verified.

14.9.3.1.1 Validation

Electronic hardware development assumes that all requirements have been validated at the system level with application of ARP-4754(A). The derived requirements added during the hardware design process need to be validated. The validation shows that the derived requirements are correct and complete. Derived requirements are evaluated against the associated system requirements—the system requirements from which the hardware requirements were allocated. If the derived requirements originate from design decisions, then the derived requirements are evaluated against the associated design decision.

Derived requirements are also evaluated for any impact on safety. The derived requirements from the hardware design process were not included in the original ARP-4754(A) requirements processes. The derived requirements need to be evaluated from a systems, and aircraft, perspective to determine whether they introduce any safety issues.

DO-254 includes simulation, prototyping, modeling, and several other analyses and even testing in the validation activities. In most instances, especially in the case of PLDs, a derived requirements review will suffice. All requirements, including the derived requirements, are subsequently subjected to verification by test.

Requirements review checklists should include criteria specifically applied to requirements identified as derived.

14.9.3.1.2 Verification

Verification is the process by which it is shown that the hardware, as implemented, meets its requirements. Verification is central to compliance with airworthiness regulations for demonstration of proper system function when installed on an aircraft and performing the system's intended function under all foreseeable operating conditions. Verification methods include review, analysis, and test. The verification is applied to all hardware requirements, including the derived requirements.

A review is a qualitative assessment of the life cycle data and the hardware implementation. Reviews are performed by one or more engineering peers and use a documented set of criteria to evaluate the hardware and/or data. Reviews should be documented and clearly indicate how issues discovered in the review process were resolved. DO-254 includes guidance for requirements reviews in Section 6.3.3.1 and guidance for design reviews in Section 6.3.3.2. Reviews are performed on:

- Hardware management plans including the PHAC
- Hardware design standards
- Hardware design data including requirements, conceptual and detailed design, implementation, and production data
- Validation data including reviews of derived requirements
- Verification data including
  - Test cases, procedures, and results
  - Analysis cases, procedures, and results
- Hardware Accomplishment Summary (HAS)

An analysis is a quantitative assessment of the life cycle data and the hardware implementation. Note that analysis includes simulation, which is especially useful in PLD verification. Simulations can be used to assess whether hardware requirements are met when extremes or limits are imposed on the design and circuits. This would include worst- and best-case timing such as low-temperature/low-voltage conditions and high-temperature/high-voltage conditions. Timing analysis for PLDs can include dynamic analysis using delays determined from postlayout device models or static analysis that evaluates design tool outputs against signal constraints for setup and hold times. Extremes or limits also include component value or stress analysis to show that circuits perform as intended when derating is applied to components. Thermal analysis will evaluate whether a circuit or component will perform over the anticipated temperature range and include heat dissipation and calculations of transistor junction temperature. A reliability analysis is used to determine whether reliability requirements are met. A verification coverage analysis is used to ensure that all requirements have been verified and that the verification is comprehensive.

PLD simulations can be performed with minimum, nominal, and maximum timing. These simulations show that requirements are verified under extreme timing conditions, which would prove difficult, if not impossible, using hardware test. PLD simulation tools also support collection of code coverage metrics. The coverage metrics are useful completion criteria for verification.

Verification by test is the ultimate demonstration that the actual hardware meets its requirements and performs its intended function. Verification testing uses production or production equivalent hardware to ensure that the tests are performed on hardware representative of what will be used in the aircraft. Hardware tests conducted in alternate configurations or environments need a justification. DO-254 encourages testing at the highest integration level as possible. This test strategy allows for more coverage of possible error sources. While a PLD can be tested in a stand-alone fixture, stand-alone testing would not reveal any problems with surrounding circuits, the circuit card layout, or integration with a processor and software, when used. Also, a circuit card can be tested by itself, but ultimately, tests should be conducted with the circuit card connected to the other circuits and the actual power supply used in the production configuration. DO-254 requirements-based tests are typically conducted at room temperature with standard production parts. The previously mentioned analyses are used to show that the requirements will still be met over the allowed precision and tolerance of parts in combination with temperature, power, and clock variations. Environmental qualification testing should be performed in accordance to the applicable version of DO-160. Typically, the environmental qualification testing is covered under ARP-4754(A) and the system-level certification plan.

DO-254 makes several references to test cases and test procedures, although Table A-1 and Sections 10.4.2 and 10.4.4 do not include these distinctions. Verification can be made more efficient and effective when the requirements are structured to include specific signals or measurable events. The test cases can be constructed to fully verify the requirement(s). Test cases would include the description or purpose of the test case, the initial conditions prior to executing the test, the inputs or stimulus needed, the outputs to be monitored or measured, and the pass/fail criteria. The test procedures are then created from the test cases and include any necessary adaptations for implementing the test cases in the physical hardware test and/or simulation environments. The key to successful verification starts with the expression of the requirements and also depends on circuits and circuit designs that include test points to capture or inject test signals.

Verification results include the data or evidence from conducting all the reviews, analyses, and tests. Ideally, all simulation and test results should be captured as the raw data in a verbose or complete format. Simulation results include log files and waveform files. Care should be taken to ensure that simulation results are collected, not just a pass/fail interpretation of the results. Hardware test results can include handwritten data, photographs, screenshots from oscilloscopes or logic analyzers, data from volt/ohm-meters, or data from custom test fixtures.

Verification also includes the tracing from requirements to implementation and from the requirements to the test cases (if used), the test procedures, and test results. The requirements to implementation trace data could conceivably be created in the design process and then checked by review in the verification process.

Level A and B projects frequently use an elemental analysis advanced verification method from DO-254 Appendix B as the additional design assurance activity. The elemental analysis assesses the test coverage of the circuits, or design elements, from a bottom-up perspective. The analysis assesses which part(s) of the design is verified by the test case or test procedure that verifies the associated requirement(s). With proper requirements structure, hardware design abstraction, and traceability in a PLD, the elemental analysis could use code coverage for the elemental analysis. Code coverage would obviously not be a good choice for elemental analysis in an analog or circuit implemented with discrete components. DO-254 does not give specific criteria for the coverage achieved by elemental analysis. The criteria should be documented in the plans and standards and discussed with the certification authority early in the planning phase.

14.9.3.2 Configuration Management

Configuration management ensures that the hardware and all associated data are uniquely identified, allows replication of the hardware or data, allows changes for problem resolution or updates, provides a controlled method to modify the hardware or data, and archives the life cycle data in a retrievable format for the life of the product. Configuration management uses baselines, or sets of data, that define the revision or version level of the hardware and associated life cycle data. Baselines provide an organized way to proceed with formal verification and also support incremental changes and associated verification. DO-254 defines 11 configuration management activities: (1) configuration identification, (2) baselines, (3) baseline traceability, (4) problem reporting, (5) change control (integrity and identification), (6) change control (records, approvals, and traceability), (7) release, (8) retrieval, (9) data retention, (10) protection against unauthorized changes, and (11) media selection, refreshing, and duplication.

While most hardware engineering organizations have tools and procedures for controlling drawings, a typical DO-254 life cycle will add documentation and various types of data files (requirements and design documents, PLD design source code, test cases, test scripts, test results, simulation files). One effective way to manage the data associated with a DO-254 life cycle is a file management tool similar (or identical) to those used to manage source code and data in a software development project.

Problem reporting and change control typically use a tracking tool hosted on a database. DO-254 does not require electronic tools; a paper-based problem reporting system would suffice. If the tools are used, they allow coordination of problem reports across an organization and automation of actions. The tools also provide reports for status and tracking to manage change in a coherent manner. Some problem reporting systems also provide back-end integration with file management tools to keep control on changes made to files and data managed within the system.

DO-254 defines two categories, hardware control category one (HC1) and hardware control category two (HC2), that are associated with each life cycle data item. The hardware control category for each data item is identified in DO-254 Table A-1. Note that the hardware control category can vary as a function of the DAL. For instance, the hardware configuration management plan is HC1 for Levels A and B and HC2 for Levels C and D. Table 7-1 in DO-254 identifies which of the 11 configuration management process activities are needed for HC1 and HC2 controlled data items. The fundamental difference is that HC1 controlled data need a problem report to change the data. HC2 data do not require a problem report for a change. Process assurance, configuration management, and verification results data are HC2 controlled. HC1 is applied to some of the plans, the requirements, drawings, and verification procedures.

14.9.3.3 Process Assurance

Process assurance objectives ensure that the entire DO-254 process, as documented in the project plans, procedures, and standards, is complete. Process assurance is often synonymous with quality assurance. In DO-254, process assurance checks the design assurance activities during the development life cycle right up through the first production articles used for verification activities. Quality or process assurance differs from quality control, which extends into production.

Process assurance uses reviews, inspections, and audits to check that the processes used in the development life cycle comply with the approved plans and that the resultant life cycle data also comply with the plans. A set of checklists and review criteria are typically created to perform process assurance activities. Process assurance also ensures that hardware used in conformity reviews is built in accordance with the drawings and life cycle data. For system-level application of DO-254, the conformity can be a formal FAA process. For PLD-level application of DO-254, the conformity can be a company configuration control demonstration or audit. The conformity review in the commercial aviation industry is often referred to as a "First Article Inspection."

Another important aspect of process assurance is checking that all of the configuration management activities and objectives are met. This includes audits or reviews of problem reporting, change control, document or data release, data archives, baselines, and configuration management tools.

Process assurance can perform their duties as an exhaustive review or use a sampling method. Sample size should be large enough to cover all data types, all the tools and processes, all the various organizations, and key players. Sampling should have triggers for escalation when errors or trends are discovered.

Process assurance also tracks and evaluates any deviations from project plans and standards. The deviation can use the problem reporting system and tools, or a separate system can be implemented to track deviations.

Process assurance is usually an independently performed activity, although DO-254 does not specifically require such independence.

14.9.3.4 Certification Liaison Process

The certification liaison process is designed to ensure coordination among the stakeholders in the certification process. DO-254 lists 27 distinct data items in Table A-1 that are produced in a compliant process. Four of these data items are specified to be submitted to the certification authority. These four data items are:

- PHAC
- Hardware Verification Plan
- Top-Level Drawing
- HAS

In practice, sufficient details of the verification process may be included in the PHAC. In that case, the hardware verification plan is not submitted. Other data items may be requested by the certification authority, if deemed necessary. Applicants are encouraged to start a dialogue with certification authorities early in the process to reach a common understanding of a means of achieving compliance with DO-254 and the certification basis. Just as the PHAC states *what you intend to do*, the accomplishment summary captures *what you did*. The HAS is used to gauge the overall completeness of the development process and to ensure that all objectives of DO-254 have been satisfied. In concept, the PHAC is a project-specific instantiation of DO-254; the HAS shows compliance with the project-specific PHAC. Thus, the project has complied with DO-254 in the way it was interpreted and applied to the project.

For DO-254 applied to PLDs, a hardware configuration index can be used in lieu of a top-level drawing. The hardware configuration index provides an overall accounting of the content of the final product and status of any open problem reports and includes or references the environment needed to recreate it.

The certification liaison process also includes coordination of delegation of compliance findings to individuals, such as a designated engineering representative (DER), or to an organization, such as an organization designation authorization (ODA) program. Liaison may also include meetings, reviews, and audits conducted at the certification authorities' facility and/or on site at the applicant or supplier.

14.10 Additional Considerations

Section 11 of DO-254 addresses additional considerations or approaches that can be used to satisfy the objectives defined in previous sections of the document. Additional consideration topics include previously developed hardware (PDH), use of commercial-off-the-shelf (COTS) components, product service experience, and tool qualification. If used, these considerations should be addressed in the hardware management plans and coordinated with the certification authority in the planning phase.

14.10.1 Previously Developed Hardware

PDH is hardware that falls in one or more of the following categories:

- COTS hardware or component
- Airborne hardware developed to other standards (e.g., a military or company standard)
- Airborne hardware that predates DO-254
- Airborne hardware previously developed at a lower DAL
- Airborne hardware previously developed for a different aircraft application
- Airborne hardware previously and subsequently changed

The use of one or more of these types of hardware should be planned for and discussed in the PHAC. A gap analysis should be performed to determine what specific objectives of DO-254 have not been met. Alternate sources of life cycle data, service history, additional verification, and reverse engineering are ways of ensuring that the use of PDH is compliant in the new application.

PDH that has been modified due to parts obsolescence, error correction, functional changes, or enhancements needs to be analyzed for any effect on the system safety assessment. Changes that result in an increase in the DAL need to be addressed. A change impact analysis should be performed, and all aspects of the life cycle from planning and standards, requirements and design, verification and validation, and through release into production should be considered.

PDH that will be used on a new or different aircraft also needs to revisit the system safety assessment. This may impact the DAL or any additional activities required in the certification basis. Using PDH on a new or different aircraft could have combined considerations if the design is changed or the DAL is changed. A gap analysis should be used to determine which aspects of the verification should be repeated. The obvious minimal testing would be the hardware interfaces in the new/different aircraft if the hardware is otherwise unmodified.

PDH can be used in a new design environment or be integrated with different hardware and/or software than the original application. All new tools should be assessed in the tool qualification process (covered in subsequent paragraphs). An analysis should be performed to determine which hardware and/or software interfaces are different and the corresponding verification to perform.

Hardware can also be used in a new application and/or aircraft that requires an upgrade of the DAL (e.g., from Level C to Level A). Data from the original development may be used and may still apply. The previously developed data would need to be evaluated against relevant DO-254 objectives. The system safety assessment process should be used to determine the scope of impact. Data can be reverse engineered to fill in the areas that are deficient for the upgraded DAL. If sufficient relevant service experience exists, the service experience data can be evaluated (covered in subsequent paragraphs).

Any PDH process also needs to address configuration management aspects including traceability from the PDH baseline to the new baseline and problem reporting or change control processes that address concurrent use of the hardware in the PDH and new application.

14.10.2 Commercial-Off-the-Shelf Components Usage

Electronic hardware designs rely extensively on commercially available parts. COTS components range from resistors, capacitors, and integrated circuits up to complete power supplies and processor circuit cards. COTS components may not have been ruggedized or designed for the reliability required for avionics and safety-critical applications. FPGAs and other types of PLDs are usually purchased as a COTS component and the programming performed during the manufacture of the host circuit card assembly. COTS components need to be managed to ensure that the parts are properly selected for temperature range, packaging, power and voltage, reliability, performance, and obsolescence. The usage of COTS components in avionics applications is assured through quality procedures used by the COTS part manufacturer, testing performed by the COTS part manufacturer or the end user, proper component selection, monitoring part reliability and performance, and tracking changes or modification to COTS procured parts.

DO-254 requires management of COTS components through an electronic component management process. DO-254 does not provide complete guidance on COTS topics. Further information on the preparation of an ECMP can be found in the IEC document IEC TS 62239, *Process management for avionics—Preparation of an electronic components management plan.*

14.10.3 Product Service Experience

Service experience can be used to partially or fully satisfy the objectives of DO-254 for COTS components and PDH. The service experience requires data collection as evidence of current and/or previous usage of the hardware. The data can be collected from aerospace applications and nonaerospace applications. An example of suitable nonaerospace applications would be the use of design or verification tools in the electronics industry.

DO-254 provides a set of criteria for the acceptance of data to support service experience. Section 11.3.2 of DO-254 discusses how to assess service experience data and Section 11.3.3 describes the content of the service experience data.

The use of product service experience should be discussed in the PHAC during the planning phase of the program and coordinated with the certification authority.

14.10.4 Tool Assessment and Qualification

DO-254 requires assessment of tools used in hardware design and verification processes. Tool assessment begins with a survey to collect information on all the tools used by the design and verification team. Each tool is assessed following the flow chart provided in Figure 11-1 of DO-254.

Tools are classified as design tools and verification tools. Design tools generate the hardware design or the hardware itself. An error in a design tool can introduce errors into the hardware. PLD design tool examples include synthesis and place and route tools. Electronics design tools examples include schematic design entry tools for printed circuit board design and circuit board layout software. Verification tool errors could fail to detect an error in the hardware or the hardware design. PLD verification tool examples include simulation tools, logic analyzers, and oscilloscopes. Electronics verification tools examples include analog and mixed-signal circuit simulator software. Rules for qualifying design tools are not provided in DO-254. If a design tool requires qualification, the applicant can follow strategies described in DO-254 Appendix B, or they may use the guidance for qualification of software development tools provided in RTCA DO-178B. Coordination with the certification authority is recommended for qualification of design tools. If the output of a design or verification tool is independently assessed, then no tool qualification is needed. Independent assessment includes any process or even another tool that verifies that the tool output is correct. The independent assessment can include review of the tool output or a check of the tool output with a separate, but dissimilar, tool. Verification tools used for measuring completion of elemental analysis do not require qualification.

The general intent behind DO-254 is that design tool outputs are checked by the verification processes including review, analysis, and test. If the design tool output is assessed, or verified to be correct, then qualification is not necessary. In most cases, design tool qualification would require cooperation from tool designers and also need ample schedule and budget resources. Most design and verification tools used by aerospace applications are common to the electronics industry. The emphasis in DO-254 is on assessing, or verifying, the tool's output or using relevant tool history.

Tools are assessed and fall into one of the categories shown in Table 14.4 if the output is not independently assessed.

For design tool qualification, a tool qualification plan must be written. The rigor of the plan is determined by the nature of the tool and the DAL of the hardware for which it is being used. A tool accomplishment summary is used to show compliance with the tool qualification plan.

TABLE 14.4 Tool Assessment and Qualification

Tool Type	Design Assurance Level	Relevant Tool History	Assessment	Qualification
Design	A, B	Yes	Qualification not required	
Design	A, B	No	Qualification required	Basic qualification Design tool qualification
Design	C	Yes	Qualification not required	
Design	C	No	Qualification required	Basic qualification
Design	D, E			
Verification	A, B	Yes	Qualification not required	
Verification	A, B	No	Qualification required	Basic qualification
Verification	C, D, E			
Verification— elemental analysis	A, B			

Note: Gray shading indicates that it is not applicable.

Level A, B, and C design tools and Level A and B verification tools without relevant history require a tool baseline in configuration management and problem reporting to track issues with the tool. These tools also require the "basic" tool qualification. "Basic" tool qualification consists of verifying that the tool produces correct outputs for the intended application. The basic qualification is performed with tests or analysis of the tool operation with respect to its requirements.

Tool assessment should be documented in the PHAC. Relevant tool history data and qualification of design tools should be discussed in the PHAC during the planning phase of the program and coordinated with the certification authority.

Practical suggestions for tool assessment and qualification include the following:

- Document tool usage on certification programs. Include project DAL, how tool was used, tool part number and version, computing environment that the tool was hosted on.
- Coordinate known problems or bugs with tools across departments or projects in a large organization.
- Ensure that the tool problem reporting spans the applications and groups.
- Contact tool vendor for any known problem list.
- Ensure that the tool is installed and configured properly.
- Be aware of automatic update features in tools.
- Archive a copy of the tool and a license for long-term maintenance.

14.11 FAA Order 8110-105

The FAA published Order 8110-105 originally in July 2008. Order 8110.105 Change 1 was subsequently published in September 2009. The Order documents FAA policy and is written for managers and staff of the FAA Aircraft Certification Service, designees of the FAA administrator, and organizations associated with the certification process. The Order explains FAA interpretation and application of DO-254 for simple and complex electronic hardware approval. The Order is specific to PLDs.

Topics covered in the Order are shown in Table 14.5.

While the Order is intended for FAA personnel and their designees, it behooves applicants and PLD developers to be familiar with the content and intent of the Order. The Order states specific topics of interest to the FAA that may go above and beyond content specific to DO-254. While not specifically referenced in the Order, the Job Aid[9] *Conducting Airborne Electronic Hardware Reviews* is used by the FAA and their designees to conduct stage of involvement (SOI) audits. As with the Order, it behooves applicants and PLD

TABLE 14.5   Order 8110.105 Topics

8110.105 Chapter	Topic	Comment
1	Introduction	
2	SEH/CEH Review Process	Introduces SOI audits for PLD development. Used in conjunction with the job aid *Conducting Airborne Electronic Hardware Reviews*
3	Determining FAA Involvement in Hardware Projects	Determines the level of FAA involvement or delegation to a designee for a program
4	Clarifying RTCA/DO-254 Topics Applicable to Both SEH and CEH	Clarifies and expands guidance for approval of simple and complex PLDs
5	Clarifying RTCA/DO-254 Topics Applicable Only to SEH	Clarifies and expands guidance for approval of simple PLDs
6	Clarifying RTCA/DO-254 Topics Applicable Only to CEH	Clarifies and expands guidance for approval of complex PLDs

developers to be familiar with the content of the Job Aid. The topics and questions in the Job Aid can be used by applicants and developers prior to formal audits to ensure readiness for a formal compliance audit.

To ensure program compliance with FAA guidance, the PHAC and hardware management plans and standards should be written to satisfy the topics covered in the Order. Examples to address include design standards for HDL, robustness testing, use of commercially available intellectual property (IP) cores, data to support claims of tool service history, tests and analyses required for DAL A and B simple devices, and provisions for measuring and recording test coverage of hardware requirements in the operational environment (i.e., flight hardware) for complex devices.

14.12 Certification Authorities Software Team Paper

The Certification Authorities Software Team (CAST) is an international group of certification and regulatory authority representatives from North America, South America, Asia, and Europe. CAST was formed to promote harmonization of certification and regulatory positions on software and complex electronic hardware aspects of safety.

CAST provides their findings in the form of position papers. CAST position papers are for education and information purposes and do not constitute official policy or guidance. Certification program use of information provided in CAST papers should be discussed with the appropriate certification authority.

Topics covered in the CAST papers are shown in Table 14.6.

TABLE 14.6   CAST Papers

Position Paper	Title	Date
27	CLARIFICATIONS ON THE USE OF RTCA DOCUMENT DO-254 AND EUROCAE DOCUMENT ED-80, DESIGN ASSURANCE GUIDANCE FOR AIRBORNE ELECTRONIC HARDWARE	June 2006
28	FREQUENTLY ASKED QUESTIONS (FAQs) ON THE USE OF RTCA DOCUMENT DO-254 AND EUROCAE DOCUMENT ED-80, DESIGN ASSURANCE GUIDANCE FOR AIRBORNE ELECTRONIC HARDWARE	December 2006
30	Simple Electronic Hardware and RTCA Document DO-254 and EUROCAE Document ED-80, Design Assurance Guidance for Airborne Electronic Hardware	August 2007
31	Technical Clarifications Identified for RTCA DO-254/EUROCAE ED-80	December 2012

14.13 What's Next

The European Aviation Safety Agency (EASA) has written a certification memoranda to require the use of DO-254 for all complex electronics within a system. The guidance is provided in EASA CM-SWCEH-001,[10] dated August 11, 2011. Section 7 of CM-SWCEH-001 states that all equipment and circuit board assemblies (CBA) with a design assurance classification of A, B, C, or D should meet Level D objectives for the equipment and CBA, regardless of the DAL of the system or aircraft function. This is the first regulatory step in using DO-254 in a manner consistent with its intended purpose.

Moving forward, the FAA will need to clarify how compliance to DO-254 will be delegated. A delegate such as a DER, authorized representative (AR), or a unit member (UM) with PLD compliance delegation may or may not have the background or experience for evaluating DO-254 for all electronics.

14.14 Conclusions

DO-254 provides objectives for hardware life cycle processes, activities to achieve these objectives, and outlines objective evidence for demonstrating that these objectives were accomplished. The purpose of compliance to DO-254 is to achieve the design assurance commensurate with the system safety requirements and functional hazards and provide evidence that the hardware performs its intended function, as expressed in the requirements. DO-254 provides guidance and does not strive to specify a specific life cycle.

Programs using DO-254 need the skills and competence to capture and express requirements and to produce and organize large volumes of data. Skills and competence are also needed for the testing, analysis and reviews of design data, test cases, procedures, and results to demonstrate verification coverage of requirements. Compliance data are intended to be a natural outcome of the development and verification process; it provides evidence that activities are complete, that processes complied with policies and standards, and that life cycle transition criteria were satisfied.

To date, DO-254 has only been required for PLDs. This narrow scope requires adapting DO-254 to a specific type of component. More programs and experience are needed with DO-254 applied to all complex electronics to appreciate its intent and benefits.

Further Reading

CAST Position Papers web page: www.faa.gov/aircraft/air_cert/design_approvals/air_software/cast/cast_papers/.
EASA web page: www.easa.europa.eu.
FAA Regulatory and Guidance Library (RGL) web page: rgl.faa.gov.
The Federal Aviation Administration Web Page: www.faa.gov.
The RTCA web page: www.rtca.org.
The SAE web page: www.sae.org.

References

1. RTCA DO-254, *Design assurance guidance for airborne electronic hardware,* RTCA Inc., Washington, DC, 2000. Copies of DO-254 may be obtained from RTCA, Inc., 1828 L St., NW, Suite 805, Washington, D.C. 20036-4001 U.S. (202) 833-9339. This document is also known as ED 80, *Design assurance guidance for airborne electronic hardware,* EUROCAE, Paris, 2. Copies of ED-80 may be obtained from EUROCAE, 17, rue Hamelin, 75783 PARIS CEDEX France, (331) 4505-7188.
2. FAA Order 8110.105 CHG 1, *SIMPLE AND COMPLEX ELECTRONIC HARDWARE APPROVAL GUIDANCE,* dated September 23, 2009.
3. SAE ARP4754A, *Guidelines for development of civil aircraft and systems,* SAE, Warrendale, PA, 2010.

4. SAE ARP4761, *Guidelines and methods for conducting the safety assessment process on civil airborne systems and equipment*, SAE, Warrendale, PA, 1996.

5. SAE ARP4754, *Certification considerations for highly-integrated or complex aircraft systems*, SAE, Warrendale, PA, 1996.

6. FAA Advisory Circular Number 20-152, *RTCA, INC., DOCUMENT RTCA/DO-254, DESIGN ASSURANCE GUIDANCE FOR AIRBORNE ELECTRONIC HARDWARE*, Federal Aviation Administration, June 2005.

7. RTCA DO-178B, *Software considerations in airborne systems and equipment certification*, RTCA Inc., Washington, DC, 1992. Copies of DO-178B may be obtained from RTCA, Inc., 1828 L St., NW, Suite 805, Washington, D.C. 20036-4001 U.S. (202) 833-9339. This document is also known as ED 12B, *Software considerations in airborne systems and equipment certification*, EUROCAE, Paris, 1992. Copies of ED-12B may be obtained from EUROCAE, 17, rue Hamelin, 75783 PARIS CEDEX France, (331) 4505-7188.

8. IEC TS 62239, *Process management for avionics—Preparation of an electronic components management plan*, International Electrotechnical Commission, Geneva, Switzerland, 2003.

9. Job Aid—Conducting airborne electronic hardware reviews, Aircraft Certification Service, Rev. (-) dated February 28, 2008.

10. EASA CM-SWCEH-001 Issue No.: 01, *Development assurance of airborne electronic hardware*, dated August 11, 2001.

II

Avionics Functions: Supporting Technology and Case Studies

This section deals with specific technology and use of different avionics, many of which are discussed in the previous section in a broad sense. Topics include:

1. Overview of functions accomplished by avionics and human–machine interface one chapter on human factors in general and other chapters on specific human machine interface technology.
2. Specific instances of avionics technologies that have contributed to situation awareness. There are two chapters on technologies that are in the background powering the avionics—wiring and batteries.
3. The final group of chapters is specific examples of different avionics suites where new technological strides have been made. These are illustrated using case studies of avionics architectures.

15

Human Factors Engineering and Flight Deck Design

Kathy H. Abbott
Federal Aviation
Administration

15.1 Introduction

This chapter briefly describes human factors engineering and considerations for civil aircraft flight deck design. The motivation for providing the emphasis on the human factor is that the operation of future aviation systems will continue to rely on humans in the system for effective, efficient, and safe operation. Pilots, mechanics, air traffic service personnel, designers, dispatchers, and many others are the basis for successful operations now and for the foreseeable future. There is ample evidence that failing to adequately consider humans in the design and operations of these systems is at best inefficient and at worst unsafe.

This becomes especially important with the continuing advance of technology. Technology advances have provided a basis for past improvements in operations and safety and will continue to do so in the future. New alerting systems for terrain and traffic avoidance, data link communication systems to augment voice-based radiotelephony, and new navigation systems based on Required Navigation Performance are just a few of the new technologies being introduced into flight decks.

Often such new technology is developed and introduced to address known problems or to provide some operational benefit. While introduction of new technology may solve some problems, it often introduces others. This has been true, for example, with the introduction of advanced automation.[1,2] Thus, while new technology can be part of a solution, it is important to remember that it will bring issues that may not have been anticipated and must be considered in the larger context (equipment design, training, integration into existing flight deck systems, procedures, operations, etc.). These issues are especially important to address with respect to the human operator.

The chapter is intended to help avoid vulnerabilities in the introduction of new technology and concepts through the appropriate application of human factors engineering in the design of flight decks. The chapter first introduces the fundamentals of human factors engineering, then discusses the flight deck design process. Different aspects of the design process are presented, with an emphasis on the incorporation of human factors in flight deck design and evaluation. To conclude the chapter, some additional considerations are raised.

15.2 Fundamentals

This section provides an overview of several topics that are fundamental to the application of Human Factors Engineering (HFE) in the design of flight decks. It begins with a brief overview of human factors, then discusses the design process. Following that discussion, several topics that are important to the application of HFE are presented: the design philosophy, the interfaces and interaction between pilots and flight decks, and the evaluation of the pilot/machine system.

15.2.1 Human Factors Engineering

It is not the purpose of this section to provide a complete tutorial on human factors. The area is quite broad and the scientific and engineering knowledge about human behavior and human performance, and the application of that knowledge to equipment design (among other areas), is much more extensive than could possibly be cited here.[3-8] Nonetheless, a brief discussion of certain aspects of human factors is desirable to provide the context for this chapter.

For the purposes of this chapter, human factors and its engineering aspects involve the application of knowledge about human capabilities and limitations to the design of technological systems.[9] Human factors engineering also applies to training, personnel selection, procedures, and other topics, but those topics will not be expanded here.

Human capabilities and limitations can be categorized in many ways, with one example being the SHEL model.[6] This conceptual model describes the components *Software, Hardware, Environment, and Liveware*. The SHEL model, as described in Ref. [6] is summarized below.

The center of the model is the human, or *Liveware*. This is the hub of human factors. It is the most valuable and most flexible component of the system. However, the human is subject to many limitations, which are now predictable in general terms. The "edges" of this component are not simple or straight, and it may be said that the other components must be carefully matched to them to avoid stress in the system and suboptimal performance. To achieve this matching, it is important to understand the characteristics of this component:

Physical size and shape—In the design of most equipment, body measurements and movement are important to consider at an early stage. There are significant differences among individuals, and the population to be considered must be defined. Data to make design decisions in this area can be found in anthropometry and biomechanics.

Fuel requirements—The human needs fuel (e.g., food, water, and oxygen) to function properly. Deficiencies can affect performance and well-being. This type of data is available from physiology and biology.

Input characteristics—The human has a variety of means for gathering input about the world around him or her. Light, sound, smell, taste, heat, movement, and touch are different forms of information perceived by the human operator; for effective communication between a system and the human operator this information must be understood to be adequately considered in design. This knowledge is available from biology and physiology.

Information processing—Understanding how the human operator processes the information received is another key aspect of successful design. Poor human–machine interface or system design that does not adequately consider the capabilities and limitations of the human information processing system can strongly affect the effectiveness of the system. Short- and long-term memory limitations are factors, as are the cognitive processing and decision-making processes used. Many human errors can be traced to this area. Psychology, especially cognitive psychology, is a major source of data for this area.

Output characteristics—Once information is sensed and processed, messages are sent to the muscles and a feedback system helps to control their actions. Information about the kinds of forces that can be applied and the acceptable direction of controls are important in design decisions. As another example, speech characteristics are important in the design of voice communication systems. Biomechanics and physiology provide this type of information.

Environmental tolerances—People, like equipment, are designed to function effectively only within a narrow range of environmental conditions such as temperature, pressure, noise, humidity, time of day, light, and darkness. Variations in these conditions can all be reflected in performance. A boring or stressful working environment can also affect performance. Physiology, biology, and psychology all provide relevant information on these environmental effects.

It must be remembered that humans can vary significantly in these characteristics. Once the effects of these differences are identified, some of them can be controlled in practice through selection, training, and standardized procedures. Others may be beyond practical control and the overall system must be designed to accommodate them safely. This *Liveware* is the hub of the conceptual model. For successful and effective design, the remaining components must be adapted and matched to this central component.

The first of the components that requires matching to the characteristics of the human is *Hardware*. This interface is the one most generally thought of when considering human–machine systems. An example is designing seats to fit the sitting characteristics of the human. More complex is the design of displays to match the human's information processing characteristics. Controls, too, must be designed to match the human's characteristics, or problems can arise from, for example, inappropriate movement or poor location. The user is often unaware of mismatches in this liveware–hardware interface. The natural human characteristic of adapting to such mismatches masks but does not remove their existence. Thus this mismatch represents a potential hazard to which designers should be alerted.

The second interface with which human factors engineering is concerned is that between *Liveware* and *Software*. This encompasses the nonphysical aspects of the systems such as procedures, manual and checklist layout, symbology, and computer programs. The problems are often less tangible than in the *Liveware-Hardware* interface and more difficult to resolve.

One of the earliest interfaces recognized in flying was between the human and the environment. Pilots were fitted with helmets against the noise, goggles against the airstream, and oxygen masks against the altitude. As aviation matured, the environment became more adapted to the human (e.g., through pressurized aircraft). Other aspects that have become more of an issue are disturbed biological rhythms and related sleep disturbances because of the increased economic need to keep aircraft, and the humans that operate them, flying 24 h a day. The growth in air traffic and the resulting complexities in operations are other aspects of the environment that are becoming increasingly significant now and in the future.

The last major interface described by the SHEL model is the human–human interface. Traditionally, questions of performance in flight have focused on individual performance. Increasingly, attention is being paid to the performance of the team or group. Pilots fly as a crew; flight attendants work as a team; maintainers, dispatchers, and others operate as groups; therefore, group dynamics and influences are important to consider in design.

The SHEL model is a useful conceptual model, but other perspectives are important in design as well. The reader is referred to the references cited for in-depth discussion of basic human behavioral considerations, but a few other topics are especially relevant to this chapter and are discussed here: usability, workload, and situation awareness.

15.2.1.1 Usability

The usability of a system is very pertinent to its acceptability by users; therefore, it is a key element to the success of a design. Nielsen[10] defines usability as having multiple components:

Learnability—The system should be easy to learn.
Efficiency—The system should be efficient to use.

Memorability—The system should be easy to remember.

Error—The system should be designed so that users make few errors during use of the system, and can easily recover from those they do make.

Satisfaction—The system should be pleasant to use so users are subjectively satisfied when using it.

This last component is indicated by subjective opinion and preference by the user. This is important for acceptability, but it is critical to understand that there is a difference between subjective preference and performance of the human–machine system. In some cases, the design that was preferred by the user was not the design that resulted in the best performance. This illustrates the importance of both subjective input from representative end users and objective performance evaluation.

15.2.1.2 Workload

In the context of the commercial flight deck, workload is a multidimensional concept consisting of: (1) the duties, amount of work, or number of tasks that a flight crew member must accomplish; (2) the duties of the flight crew member with respect to a particular time interval during which those duties must be accomplished; and/or (3) the subjective experience of the flight crew member while performing those duties in a particular mission context. Workload may be either physical or mental.[11]

Both overload (high workload, potentially resulting in actions being skipped or executed incorrectly or incompletely) and underload (low workload, leading to inattention and complacency) are worthy of attention when considering the effect of design on human–machine performance.

15.2.1.3 Situation Awareness

This can be viewed as the perception on the part of a flight crew member of all the relevant pieces of information in both the flight deck and the external environment, the comprehension of their effects on the current mission status, and the projection of the values of these pieces of information (and their effect on the mission) into the near future.[11]

Situation awareness has been cited as an issue in many incidents and accidents, and can be considered as important as workload. As part of the design process, the pilot's information requirements must be identified, and the information display must be designed to ensure adequate situation awareness. Although the information is available in the flight deck, it may not be in a form that is directly usable by the pilot, and therefore of little value.

Another area that is being increasingly recognized as important is the topic of organizational processes, policies, and practices.[12] It has become apparent that the influence of these organizational aspects is a significant, if latent, contributor to potential vulnerabilities in design and operations.

15.2.2 Flight Deck Design

The process by which commercial flight decks are designed is complex, largely unwritten, variable, and nonstandard.[11] That said, Figure 15.1 is an attempt to describe this process in a generic manner. It represents a composite flight deck design process based on various design process materials. The figure is not intended to exactly represent the accepted design process within any particular organization or program; however, it is meant to be descriptive of generally accepted design practice. (For more detailed discussion of design processes for pilot-system integration and integration of new systems into existing flight decks, see Refs. [13,14].)

The figure is purposely oversimplified. For example, the box labeled "Final Integrated Design" encompasses an enormous number of design and evaluation tasks, and can take years to accomplish. It could be expanded into a figure of its own that includes not only the conceptual and actual integration of flight deck components, but also analyses, simulations, flight tests, certification and integration based on these evaluations.

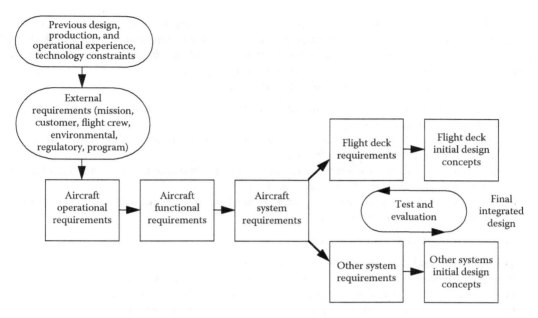

FIGURE 15.1 Simplified representation of the flight deck design process. (From Palmer, M.T. et al., NASA Technical Memorandum, 109171, January 1995.)

Flight deck design necessarily requires the application of several disciplines, and often requires trade-offs among those disciplines. Human factors engineering is only one of the disciplines that should be part of the process, but it is a key part of ensuring that the flight crew's capabilities and limitations are considered. Historically, this process tends to be very reliant on the knowledge and experiences of individuals involved in each program.

Human-centered or user-centered design has been cited as a desirable goal. That is, design should be focused on supporting the human operator of the system, much as discussed above on the importance of matching the hardware, software, and environment to the human component. A cornerstone of human-centered design is the design philosophy.

15.2.2.1 Flight Deck Design Philosophy

The design philosophy, as embodied in the top-level philosophy statements, guiding principles, and design guidelines, provides a core set of beliefs used to guide decisions concerning the interaction of the flight crew with the aircraft systems. It typically deals with issues such as allocation of functions between the flight crew and the automated systems, levels of automation, authority, responsibility, information access and formatting, and feedback, in the context of human use of complex, automated systems.[1,11]

The way pilots operate airplanes has changed as the amount of automation and its capabilities have increased. Automation has both provided alternate ways of accomplishing pilot tasks performed on previous generations of airplanes and created new tasks. The increased use of and flight crew reliance on flight deck automation makes it essential that the automation act predictably with actions that are well understood by the flight crew. The pilot has become, in some circumstances, a supervisor or manager of the automation.

Moreover, the automation must be designed to function in a manner that directly supports flight crews in performing their tasks. If these human-centered design objectives are not met, the flight crew's ability to properly control or supervise system operation is limited, leading to confusion, automation surprises, and unintended airplane responses.

Each airplane manufacturer has a different philosophy regarding the implementation and use of automation. Airbus and Boeing are probably the best-known for having different flight deck design philosophies. However, there is general agreement that the flight crew is and will remain ultimately responsible for the safety of the airplane they are operating.

Airbus has described its automation philosophy as

Automation must not reduce overall aircraft reliability; it should enhance aircraft and systems safety, efficiency, and economy.

Automation must not lead the aircraft out of the safe flight envelope and it should maintain the aircraft within the normal flight envelope.

Automation should allow the operator to use the safe flight envelope to its full extent, should this be necessary due to extraordinary circumstances.

Within the normal flight envelope, the automation must not work against operator inputs, except when absolutely necessary for safety.

Boeing has described its philosophy as follows:

The pilot is the final authority for the operation of the airplane.

Both crew members are ultimately responsible for the safe conduct of the flight.

Flight crew tasks, in order of priority, are safety, passenger comfort, and efficiency.

Design for crew operations based on pilot's past training and operational experience.

Design systems to be error tolerant.

The hierarchy of design alternatives is simplicity, redundancy, and automation.

Apply automation as a tool to aid, not replace, the pilot.

Address fundamental human strengths, limitations, and individual differences—for both normal and nonnormal operations.

Use new technologies and functional capabilities only when:

They result in clear and distinct operational or efficiency advantages.

There is no adverse effect to the human-machine interface.

One of the significant differences between the design philosophies of the two manufacturers is in the area of envelope protection. Airbus' philosophy has led to the implementation of what has been described as "hard" limits, where the pilot can provide whatever control inputs he or she desires, but the airplane will not exceed the flight envelope. In contrast, Boeing has "soft" limits, where the pilot will meet increasing resistance to control inputs that will take the airplane beyond the normal flight envelope, but can do so if he or she chooses. In either case, it is important for the pilot to understand what the design philosophy is for the airplane being flown.

Other manufacturers may have philosophies that differ from Boeing and Airbus. Different philosophies can be effective if each is consistently applied in design, training, and operations, and if each supports flight crew members in flying their aircraft safely. To ensure this effectiveness, it is critical that the design philosophy be documented explicitly and provided to the pilots who will be operating the aircraft, the trainers, and the procedure developers.

15.2.2.2 Pilot/Flight Deck Interfaces

The layout, controls, displays, and amount of automation in flight decks have evolved tremendously in commercial aviation.[15,16] What is sometimes termed the "classic" flight deck, which includes the B-727, the DC-10, and early series B-747, is typically characterized by dedicated displays, where one piece of data is generally shown on a dedicated gage or dial as the form of display. These aircraft are relatively lacking in automation. A representative "classic" flight deck is shown in Figure 15.2. All of these aircrafts are further characterized by the relative simplicity of their autopilot, which offers one or a few simple modes in each axis. In general, a single instrument indicates the parameter of a single sensor. In a few cases, such as the horizontal situation indicator, a single instrument indicates the "raw" output of

FIGURE 15.2 Representative "classic" flight deck (DC-10).

multiple sensors. Regardless, the crew is generally responsible for monitoring the various instruments and realizing when a parameter is out of range. A simple caution and warning system exists, but it covers only the most critical system failures.

The first generation of "glass cockpit" flight decks, which include the B-757/767, A-310, and MD-88, receive their nickname due to their use of cathode ray tubes (CRTs). A representative first-generation "glass cockpit" flight deck is shown in Figure 15.3. A mix of CRTs and instruments was used in this generation of flight deck, with instruments used for primary flight information such as airspeed and altitude. A key innovation in this flight deck was the "map display" and its coupling to the Flight Management System (FMS). This enabled the crew to program their flight plan into a computer and see their planned track along the ground, with associated waypoints, on the map display. Accompanying

FIGURE 15.3 Representative first-generation "glass cockpit" (B-757) flight deck.

the introduction of the map display and FMS were more complex autopilots (added modes from the FMS and other requirements). This generation of aircraft also featured the introduction of an integrated Caution and Warning System, usually displayed in a center CRT with engine information. A major feature of this Caution and Warning System was that it prioritized alerts according to a strict hierarchy of "warnings" (immediate crew action required), "cautions" (immediate crew awareness and future action required), and "advisories" (crew awareness and possible action required).[17]

The second generation of "glass cockpit" flight decks, which include the B-747-400, A-320/330/340, F-70/100, MD-11, and B-777, are characterized by the prevalence of CRTs (or LCDs in the case of the B-777) on the primary instrument panel. A representative second-generation "glass cockpit" flight deck is shown in Figure 15.4. CRT/LCDs are used for all primary flight information, which is integrated on a few displays. In this generation of flight deck, there is some integration of the FMS and autopilot—certain pilot commands can be input into either the FMS or autopilot and automatically routed to the other.

There are varying levels of aircraft systems automation in this generation of flight deck. For example, the MD-11 fuel system can suffer certain failures and take corrective action—the crew is only notified if they must take some action or if the failure affects aircraft performance. The caution and warning systems in this generation of flight decks are sometimes accompanied by synoptic displays that graphically indicate problems. Some of these flight decks feature fly-by-wire control systems—in the case of the A-320/330/340, this capability has allowed the manufacturer to tailor the control laws such that the flying qualities of these various size aircraft appear similar to pilots. The latest addition to this generation of flight deck, the B-777, has incorporated "cursor control" for certain displays, allowing the flight crew to use a touchpad to interact with "soft buttons" programmed on these displays.

Of note is the way that this flight deck design evolution affects the manner in which pilots access and manage information. Figure 15.2 illustrates the flight deck with dedicated gages and dials, with one display per piece of information. In contrast, the flight deck shown in Figure 15.4 has even more information available, and the pilot must access it in an entirely different manner. Some of the information is integrated in a form that the pilot can more readily interpret (e.g., moving map displays). Other information must be accessed through pages of menus. The point is that there has been a

FIGURE 15.4   Representative second-generation "glass cockpit" (Airbus A320) flight deck.

FIGURE 15.5 Gulfstream GV flight deck.

fundamental change in information management in the flight deck, not through intentional design but through introduction of technology, often for other purposes.

An example is shown in Figure 15.5 from the business aircraft community illustrating that the advanced technology discussed here is not restricted to large transport aircraft. In fact, new technology is quite likely to be more quickly introduced into these smaller, sophisticated aircraft.

Major changes in the flight crew interface with future flight decks are expected. While it is not known exactly what the flight decks of the future will contain or how they will function, some possible elements may include:

- Sidestick control inceptors, interconnected and with tailorable force/feel, preferably "backdriven" during autopilot engagement.
- Cursor control devices, which the military has used for many years, but the civil community is just starting to use (e.g., in the Boeing 777).
- Multifunction displays.
- Management of subsystems through displays and control-display units.
- "Mode-less" flight path management functions.
- Large, high-resolution displays having multiple signal sources (computer-generated and video).
- Graphical interfaces for managing certain flight deck systems.
- High-bandwidth, two-way datalink communication capability embedded in appropriate flight deck systems
- Replacement of paper with "electronic flight bags."
- Voice interfaces for certain flight deck systems.

These changes will continue to modify the manner in which pilots manage information within the flight deck, and the effect of such changes should be explicitly considered in the flight deck design process.

15.2.2.3 Pilot/Flight Deck Interaction

Although it is common to consider the pilot interfaces to be the only or primary consideration in human factors in flight deck design, the interaction between the pilot(s) and the flight deck must also be considered. Some of the most visible examples of the importance of this topic, and the consequences of vulnerabilities in this area, are in the implementation of advanced automation.

Advanced automation (sophisticated autopilots, autothrust, flight management systems, and associated displays and controls) has provided large improvements in safety (e.g., through reduced pilot workload in critical or long-range phases of flight) and efficiency (improved precision of flying certain flight paths). However, vulnerabilities have been identified in the interaction between the flight crews and modern systems.[2]

For example, on April 26, 1994, an Airbus A300–600 operated by China Airlines crashed at Nagoya, Japan killing 264 passengers and flight crew members. Contributing to the accident were conflicting actions taken by the flight crew and the autopilot. During complex circumstance, the flight crew attempted to stay on glide slope by commanding nose-down elevator. The autopilot was then engaged, and because it was still in go-around mode, commanded nose-up trim. A combination of an out-of-trim condition, high engine thrust, and retracting the flaps too far led to a stall. The crash provided a stark example of how a breakdown in the flight crew/automation interaction can affect flight safety. Although this particular accident involved an A300–600, other accidents, incidents, and safety indicators demonstrate that this problem is not confined to any one airplane type, airplane manufacturer, operator, or geographical region.

A lesson to be learned here is that design of the interaction between the pilot and the systems must consider human capabilities and limitations. A good human–machine interface is necessary but may not be sufficient to ensure that the system is usable and effective. The interaction between the pilot and the system, as well as the function of the system itself, must be carefully "human engineered."

15.2.3 Evaluation

Figure 15.1 showed test and evaluation (or just evaluation, for the remainder of the discussion) as an integral part of the design process. Because evaluation is (or should be) such an important part of design, some clarifying discussion is appropriate here. (See Ref. [18] for a more detailed discussion of the evaluation issues that are summarized below.)

Evaluation often is divided into verification (the process of demonstrating that the system works as designed) and validation (the process of assessing the degree to which the design achieves the system objectives of interest). Thus, validation goes beyond asking whether the system was built according to the plan or specifications; it determines whether the plan or specifications were correct for achieving the system objectives.

One common use of the term "evaluation" is as a synonym of "demonstration." That is, evaluation involves turning on the system and seeing if it basically resembles what the designer intended. This does not, however, provide definitive information on safety, economy, reliability, maintainability, or other concerns that are generally the motivation for evaluation.

It is not unusual for evaluation to be confused with demonstration, but they are not the same. In addition, there are several different types and levels of evaluation that are useful to understand. For example, *formative* evaluation is performed during the design process. It tends to be informal and subjective, and its results should be viewed as hypotheses, not definitive results. It is often used to evaluate requirements. In contrast, *formal* evaluation is planned during the design but performed with a prototype to assess the performance of the human/machine system. Both types of evaluations are required, but the rest of this discussion focuses on formal evaluation.

Another distinction of interest in understanding types of evaluation is the difference between *absolute* vs. *comparative* evaluations. *Absolute* evaluation is used when assessing against a standard of some kind. An example would be evaluating whether the pilot's response time using a particular system is less than some prespecified number. *Comparative* evaluation compares one design to another, typically an old design to a new one. Evaluating whether the workload for particular tasks in a new flight deck is equal to or less than in an older model is an example comparative evaluation. This type of evaluation is often used in the airworthiness certification of a new flight deck, to show its acceptability relative to an older, already certified flight deck. It may be advantageous for developers to expand an absolute evaluation into a comparative evaluation (through options within the new system) to assess system sensitivities.

Yet another important distinction is between *objective* vs. *subjective* evaluation. *Objective* evaluation measures the degree to which the objective criteria (based on system objectives) have been met. *Subjective* evaluation focuses on users' opinions and preferences. Subjective data are important but should be used to support the objective results, not replace them.

Planning for the evaluation should proceed in parallel with design rather than after the design is substantially completed. Evaluation should lead to design modification, and this is most effectively done in an iterative fashion.

Three basic issues, or levels of evaluation, are worth considering. The first is *compatibility*. That is, the physical presentation of the system must be compatible with human input and output characteristics. The pilot has to be able to read the displays, reach the controls, etc. Otherwise, it doesn't matter how good the system design is; it will not be usable.

Compatibility is important but not sufficient. A second issue is *understandability*. That is, just because the system is compatible with human input–output capabilities and limitations does not necessarily mean that it is understandable. The structure, format, and content of the pilot–machine dialogue must result in meaningful communication. The pilot must be able to interpret the information provided, and be able to "express" to the system what he or she wishes to communicate. For example, if the pilot can read the menu, but the options available are meaningless, that design is not satisfactory.

A designer must ensure that the design is both compatible and understandable. Only then should the third level of evaluation be addressed: that of *effectiveness*. A system is effective to the extent that it supports a pilot or crew in a manner that leads to improved performance, results in a difficult task being made less difficult, or enables accomplishing a task that otherwise could not have been accomplished. Assessing effectiveness depends on defining measures of performance based on the design objectives. Regardless of these measures, there is no use in attempting to evaluate effectiveness until compatibility and understandability are ensured.

Several different methods of evaluation can be used, ranging from static paper-based evaluations to in-service experience. The usefulness and efficiency of a particular method of evaluation naturally depends on what is being evaluated. Table 15.1 shows the usefulness and efficiency of several methods for each of the levels of evaluation.

As can be seen from this discussion, evaluation is an important and integral part of successful design.

TABLE 15.1 Methods of Evaluation

| Method | Levels of Evaluation | | |
	Compatibility	Understandability	Effectiveness
Paper evaluation: static	Useful and efficient	Somewhat useful but inefficient	Not useful
Paper evaluation: dynamic	Useful and efficient	Somewhat useful but inefficient	Not useful[a]
Part-task simulator: "Canned" scenarios	Useful but inefficient	Useful and efficient	Marginally useful but efficient[a]
Part-task simulator: model driven	Useful but inefficient	Useful and efficient	Somewhat useful and efficient
Full-task simulator	Useful but very inefficient	Useful but inefficient	Useful but somewhat inefficient
In-service evaluation	Useful but extremely inefficient	Useful but very inefficient	Useful but inefficient

Source: Electric Power Research Institute, Rep. NP-3701: Computer-generated display system guidelines, Vol. 2: Developing an evaluation plan, September 1984.

[a] Can be effective for formative evaluation.

15.3 Additional Considerations

15.3.1 Standardization

Generally, across manufacturers, there is a great deal of variation in existing flight deck systems design, training, and operation. Because pilots often operate different aircraft types, or similar aircraft with different equipage, at different points in time, another way to avoid or reduce errors is standardization of equipment, actions, and other areas.[19]

It is not realistic (or even desirable) to think that complete standardization of existing aircraft will occur. However, for the sake of the flight crews who fly these aircraft, appropriate standardization of new systems/technology/operational concepts should be pursued, as discussed below.

Appropriate standardization of procedures/actions, system layout, displays, color philosophy, etc. is generally desirable, because it has several potential advantages, including:

- Reducing potential for crew error/confusion due to negative transfer of learning from one aircraft to another;
- Reducing training costs, because you only need to train once; and
- Reducing equipment costs because of reduced part numbers, inventory, etc.

A clear example of standardization in design and operation is the Airbus A320/330/340 commonality of flight deck and handling qualities. This has advantages of reduced training and enabling pilots to easily fly more than one airplane type.

If standardization is so desirable, why is standardization not more prevalent? There are concerns that inappropriate standardization, rigidly applied, can be a barrier to innovation, product improvement, and product differentiation. In encouraging standardization, known issues should be recognized and addressed.

One potential pitfall of standardization that should be avoided is to standardize on the lowest common denominator. Another question is to what level of design prescription should standardization be done, and when does it take place? From a human performance perspective, consistency is a key factor. The actions and equipment may not be exactly the same, but should be consistent. An example where this has been successfully applied is in the standardization of alerting systems,[16] brought about by the use of industry-developed design guidelines. Several manufacturers have implemented those guidelines into designs that are very different in some ways, but are generally consistent from the pilot's perspective.

There are several other issues with standardization. One of them is related to the introduction of new systems into existing flight decks. The concern here is that the new system should have a consistent design/operating philosophy with the flight deck into which it is being installed. This point can be illustrated by the recent introduction of a warning system into modern flight decks. In introducing this new system, the question arose whether the display should automatically be brought up if an alert occurs (replacing the current display selected by the pilot). One manufacturer's philosophy is to bring the display up automatically when an alert occurs; another manufacturer's philosophy is to alert the pilot, then have the pilot select the display when desired. This is consistent with the philosophy of that flight deck of providing the pilot control over the management of displays. The trade-off between standardization across aircraft types (and manufacturers) and internal consistency with flight deck philosophy is very important to consider and should probably be done on a case-by-case basis.

The timing of standardization, especially with respect to introduction of new technology, is also critical.[4] It is desirable to deploy new technology early, because some problems are only found in the actual operating environment. However, if we standardize too early, then there is a risk of standardizing on a design that has not accounted for that critical early in-service experience. We may even unintentionally standardize a design that is error inducing. However, attempt to standardize too late and there may already be so many variations that no standard can be agreed upon. It is clear that standardization must be done carefully and wisely.

15.3.2 Error Management

Human error, especially flight crew error, is a recurring theme and continues to be cited as a primary factor in a majority of aviation accidents.[2,20] It is becoming increasingly recognized that this issue must be taken on in a systematic way or it may prove difficult to make advances in operations and safety improvements. However, it is also important to recognize that human error is also a normal by-product of human behavior, and most errors in aviation do not have safety consequences. Therefore, it is important for the aviation community to recognize that error cannot be completely prevented and that the focus should be on error management.

In many accidents where human error is cited, the human operator is blamed for making the error; in some countries the human operator is assigned criminal responsibility, and even some U.S. prosecutors seem willing to take similar views. While the issue of personal responsibility for the consequences of one's actions is important and relevant, it also is important to understand why the individual or crew made the error(s). In aviation, with very rare exceptions, flight crews (and other humans in the system) do not intend to make errors, especially errors with safety consequences. To improve safety through understanding of human error, it may be more useful to address errors as *symptoms* rather than *causes* of accidents. The next section discusses understanding of error and its management, then suggests some actions that might be constructive.

Human error can be distinguished into two basic categories: (1) those which presume the intention is correct, but the action is incorrect, (including *slips* and *lapses*), and (2) those in which the intention is wrong (including *mistakes* and *violations*).[21-23]

> *Slips* are where one or more incorrect actions are performed, such as in a substitution or insertion of an inappropriate action into a sequence that was otherwise good. An example would be setting the wrong altitude into the mode selector panel, even when the pilot knew the correct altitude and intended to enter it.[3]
>
> *Lapses* are the omission of one or more steps of a sequence. For example, missing one or more items in a checklist that has been interrupted by a radio call.
>
> *Mistakes* are errors where the human did what he or she intended, but the planned action was incorrect. Usually mistakes are the result of an incorrect diagnosis of a problem or a failure to understand the exact nature of the current situation. The plan of action thus derived may contain very inappropriate behaviors and may also totally fail to rectify a problem. For example, a mistake would be shutting down the wrong engine as a result of an incorrect diagnosis of a set of symptoms.
>
> *Violations* are the failure to follow established procedures or performance of actions that are generally forbidden. Violations are generally deliberate (and often well-meaning), though an argument can be made that some violation cases can be inadvertent. An example of a violation is continuing on with a landing even when weather minima have not been met before final approach. It should be mentioned that a "violation" error may not necessarily be in violation of a regulation or other legal requirement.

Understanding differences in the types of errors is valuable because management of different types may require different strategies. For example, training is often proposed as a strategy for preventing errors. However, errors are a normal by-product of human behavior. While training can help reduce some types of errors, they cannot be completely trained out. For that reason, errors should also be addressed by other means, and considering other factors, such as the consequences of the error or whether the effect of the error can be reversed. As an example of using design to address known potential errors, certain switches in the flight deck have guards on them to prevent inadvertent activation.

Error management can be viewed as involving the tasks of error avoidance, error detection, and error recovery.[23] Error avoidance is important, because it is certainly desirable to prevent as many errors as possible. Error detection and recovery are important, and in fact it is the safety consequences of errors that are most critical.

It seems clear that experienced pilots have developed skills for performing error management tasks. Therefore, it is possible that design, training, and procedures can directly support these tasks if we get a better understanding of those skills and tasks. However, the understanding of those skills and tasks is far from complete.

There are a number of actions that should be taken with respect to dealing with error, some of them in the design process:

Stop the blame that inhibits in-depth addressing of human error, while appropriately acknowledging the need for individual and organizational responsibility for safety consequences. The issue of blaming the pilot for errors has many consequences, and provides a disincentive to report errors.

Evaluate errors in accident and incident analyses. In many accident analyses, the reason an error is made is not addressed. This typically happens because the data are not available. However, to the extent possible with the data available, the types of errors and reasons for them should be addressed as part of the accident investigation.

Develop a better understanding of error management tasks and skills that can support better performance of those tasks. This includes:

Preventing as many errors as possible through design, training, procedures, proficiency, and any other intervention mechanism.

Recognizing that it is impossible to prevent all errors, although it is certainly important to prevent as many as possible.

Addressing the need for error management, with a goal of error tolerance in design, training, and procedures.

System design and associated flight crew interfaces can and should support the tasks of error avoidance, detection, and recovery. There are a number of ways of accomplishing this, some of which are mentioned here. One of these ways is through user-centered design processes that ensure that the design supports the human performing the desired task. An example commonly cited is the navigation display in modern flight decks, which integrates information into a display that provides information in a manner directly usable by the flight crew. This is also an example of a system that helps make certain errors more detectable, such as entering an incorrect waypoint. Another way of contributing to error resistance is designing systems that cannot be used or operated in an unintended way. An example of this is designing connectors between a cable and a computer such that the only place the cable connector fits is the correct place for it on the computer; it will not fit into any other connector on the computer.

15.3.3 Integration with Training/Qualification and Procedures

To conclude, it is important to point out that flight deck design should not occur in isolation. It is common to discuss the flight deck design separately from the flight crew qualification (training and recency of experience), considerations, and procedures. And yet, flight deck designs make many assumptions about the knowledge and skills of the pilots who are the intended operators of the vehicles. These assumptions should be explicitly identified as part of the design process, as should the assumptions about the procedures that will be used to operate the designed systems. Design should be conducted as part of an integrated, overall systems approach to ensuring safe, efficient, and effective operations.

References

1. Billings, C. E., *Aviation Automation: The Search for a Human-Centered Approach*, Mahwah, NJ: Lawrence Erlbaum Associates, 1997.
2. Federal Aviation Administration, The Human Factors Team report on the interfaces between flightcrews and modern flight deck systems, July 1996.

3. Sanders, M. S. and McCormick, E. J., *Human Factors in Engineering and Design*, 7th edn., New York: McGraw-Hill, 1993.

4. Norman, D. A., *The Psychology of Everyday Things*, also published as *The Design of Everyday Things*, New York: Doubleday, 1988.

5. Wickens, C. D., *Engineering Psychology and Human Performance*, 2nd edn., New York: Harper Collins College, 1991.

6. Hawkins, F., *Human Factors in Flight*, 2nd edn., Aldershot, U.K.: Avebury Aviation, 1987.

7. Bailey, R. W., *Human Performance Engineering: A Guide for System Designers*, Englewood Cliffs, NJ: Prentice-Hall, 1982.

8. Chapanis, A., *Human Factors in Systems Engineering*, New York: John Wiley & Sons, 1996.

9. Cardosi, K. and Murphy, E. (eds.), Human factors in the design and evaluation of air traffic control systems, DOT/FAA/RD-95/3, Cambridge, MA: Federal Aviation Administration, 1995.

10. Nielsen, J., *Usability Engineering*, New York: Academic Press, 1993.

11. Palmer, M. T., Roger, W. H., Press, H. N., Latorella, K. A., and Abbott, T. S., NASA Technical Memorandum, 109171, January 1995.

12. Reason, J., *Managing the Risks of Organizational Accidents*, Burlington, VT: Ashgate Publishing, 1997.

13. Society of Automotive Engineers, Pilot-system integration, Aerospace Recommended Practice (ARP) 4033, Warrendale, PA: SAE International, 1995.

14. Society of Automotive Engineers, Integration procedures for the introduction of new systems to the cockpit, ARP 4927, 1995.

15. Sexton, G., Cockpit: Crew-cockpit design and integration, in Wiener, E. L. and Nagel, D. C. (eds.), *Human Factors in Aviation*, San Diego, CA: Academic Press, pp. 495–526, 1988.

16. Arbuckle, P. D., Abbott, K. H., Abbott, T. S., and Schutte, P. C., Future flight decks, *21st Congress of the International Council of the Aeronautical Sciences*, Melbourne, Victoria, Australia, Paper Number 98-6.9.3, September, 1998.

17. Federal Aviation Administration, Aircraft Alerting Systems Standardization Study, Volume II: Aircraft Alerting Systems Design Guidelines, FAA Report No. DOT/FAA/RD/81-38, II, 1981.

18. Computer-generated display system guidelines. Volume 2: Developing an evaluation plan, Electric Power Research Institute EPRI, Palo, Alto, CA. Report No EPRI NP-3701, V2, 1984, 108 pp.

19. Abbott, K., Human error and aviation safety management, *Proceedings of the Flight Safety Foundations, 52nd Annual International Air Safety Seminar*, Rio de Janeiro, Brazil, November 8–11, 1999.

20. Boeing Commercial Airplane Group, Statistical summary of commercial jet aircraft accidents, World Wide Operations 1959–1995, April 1996.

21. Reason, J. T., *Human Error*, New York: Cambridge University Press, 1990.

22. Hudson, P. T. W. and Verschuur, W. L. G., Perceptions of procedures by operators and supervisors. *Paper SPE 46760 Proceedings of the 1998 International Conference on Health, Safety and Environment in Oil and Gas Exploration and Production*. Caracas. CD-ROM Society for Petroleum Engineers, Richardson, TX. p. 4, 1998.

23. Hudson, P. T. W., Bending the Rules. II. Why Do People Break Rules or Fail to Follow Procedures? and What Can You Do About It? *The Violation Manual*. P. T. W Hudson; R. Lawton; W. L. G. Verschuur, Leaden University, D. Parker, J. T. Reason, Manchester University.

24. Wiener, E. L., Intervention strategies for the management of human error, *Flight Safety Digest*, 116 pp., February 1995.

16

Head-Mounted Displays

James Melzer
Rockwell Collins Optronics

16.1 Introduction

Head-mounted displays (HMDs)* are personal information-viewing devices that can provide information in a way that no other display can. The information is always projected into the user's eyes and can be made reactive to head and body movements, replicating the way we view, navigate, and explore the world.[1] This unique capability lends itself to applications such as virtual reality for creating artificial environments,[2] to medical visualization as an aid in surgical procedures,[3,4] to military vehicles for viewing sensor imagery,[5] to airborne workstation applications reducing size, weight, and power over conventional displays,[6] to aircraft simulation and training,[7–9] to ground soldier applications,[10,11] and (central to this chapter) to avionics display applications.[12–15]

The goal of aircraft avionics is to help the pilot achieve situation awareness (SA), which Mica Endsley[16] describes as an active loop divided into three nested levels: "Level (1) *the perception of the elements in the environment within a volume of time and space*, Level (2) *the comprehension of their meaning*, and Level (3) *the projection of their status in the near future*." In achieving SA, the pilot gathers information from a variety of sources including cockpit displays, out-the-window visuals, radio communications, and even the tactile "feel" of the aircraft (Level 1), creates a mental model of what the information means (Level 2), and then predicts the implications of this (dynamic and accretionary) model to the future status of the aircraft or mission (Level 3) on a timescale that may be measured in hours, minutes, or seconds, repeating this cycle continuously throughout the flight. As in many situations in which an operator is under time constraints (air traffic control, battlefield management, medical procedures, firefighting, weather forecasting, or a football game) where a timely understanding of a dynamic situation is vital,[17–19] this loop can break down in the face of information overload, potentially causing the pilot to miss important cues (i.e., "failure to monitor"[20,21]).

---

* The term "Head-Mounted Display" is used in this chapter as a more generic term than Helmet-Mounted Display, which most often refers to military-oriented hardware. Other terms such as Helmet-Mounted Sight (HMS—which provides only a simple targeting reticle) along with Head Worn Display (HWD) are also sometimes used interchangeably.

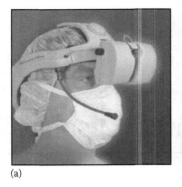

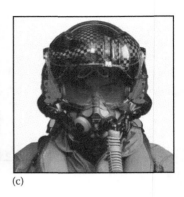

(a)   (b)   (c)

FIGURE 16.1   HMD designs for three different applications: (a) the CardioView® for minimally invasive cardiac surgery. (Photo courtesy of Vista Medical Technologies, Inc.) (b) A prototype of the U.S. Army's Land Warrior HMD. (Photo courtesy of Program Manager, Soldier, U.S. Army.) (c) The HMDS for the Joint Strike Fighter. (Photo courtesy of Vision Systems International, San Jose, CA.)

In some applications, such as with the medical and soldier's displays in Figure 16.1a and b, respectively, the HMD is used as a hands-free information source for viewing video, text, maps, or graphics. But to truly reap the benefits of the HMD as part of the aircraft's avionics (as in Figure 16.1c), it must be part of a visually coupled system (or VCS) that includes the HMD, and a head position and orientation tracker.[22,23] As the pilot turns his head, the tracker relays the orientation data to the mission computer, which updates the displayed information accordingly. This gives the pilot access to a myriad of real-time data that is *linked to head orientation*. Because the aircraft mission computer knows the pilot's head orientation, the HMD can display real-time data that is (1) screen-referenced (such as altitude, airspeed, or fuel status), (2) aircraft-referenced (such as the shape of the front of the aircraft), or (3) earth-referenced (either real objects such as runways or horizon lines or virtual objects such as safe pathway in the sky, threat/friendly locations engagement areas, waypoints, and adverse weather), the latter having been shown to provide a significant improvement on pilot performance.[24-28] In addition, it has been shown that the HMD can reduce pilot workload by converting cockpit data to useful information,[29] focus the pilot's attention,[30] stimulate both the ambient and focal modes of vision,[31-34] and combine the efficiencies of other sensory modalities such as auditory and haptic, thus enabling cross modal integration.[35,36]

In a fixed-wing fighter, a missile's sensor can be slaved to the pilot's line-of-sight, allowing him to designate targets away from the forward line-of-sight of the aircraft. In a helicopter, the pilot can point imaging sensors such as a forward-looking infrared (FLIR*) and fly in low light or adverse weather conditions. In this way, the HMD expands the pilot's field of regard to the full airspace around the aircraft, unlocking him from the interior of the cockpit. This reduces the time the pilot spends with head down in the cockpit, reduces perceptual switching time going from the cockpit to the outside world, and can direct the pilot to a target of interest and allow him track the target as it moves.[37]

The U.S. military introduced HMDs into fixed wing aircraft in the early 1970s for targeting air-to-air missiles. Several hundred Visual Targeting Acquisition Systems (VTAS) were fielded on F-4 Phantom fighter jets between 1973 and 1979.[38,39] This program was eventually abandoned because the HMD capabilities did not match the missile technology of the day.† HMDs were given new life when a Soviet MiG-29 was photographed in 1985 showing a simple helmet-mounted sight for off-axis targeting of the Vympel R-73 missile—also called the AA-11 Archer. In response, Israel initiated a

---

* Forward Looking Infra-Red (FLIR) is a sensor technology that views differences in black-body emissions—heat differences—from objects.
† There was also a Memorandum of Understanding signed in 1980 that relegated the development of short-range missile technology (and HMDs) to the Europeans.

fast-paced program that deployed the Elbit DASH HMD for off-axis targeting of the Rafael Python 4 missile in 1993–1994.[40]

Two domestic simulation studies—Vista Sabre[41] and Vista Sabre II[42]—demonstrated the clear advantage for a pilot equipped with an HMD when targeting missiles over a pilot using only a HUD. Encouraged by these studies and by a post-Berlin Wall examination of the close-combat capabilities of the HMD-equipped MiG-29,[43] the U.S. military initiated their own off-boresight missile targeting program. The result is the Joint Helmet-Mounted Cueing System (JHMCS, built by Vision Systems, International—VSI) currently deployed on the U.S. Navy's F/A-18, on the U.S. Air Force's F-15, and on both domestic and international versions of the F-16, with a total of 18 international countries using the JHMCS HMD on their fighter aircraft. The JHMCS gives pilots off-axis targeting symbology for the High Off-Boresight Missile, and aircraft status[44] and provides improved SA of the airspace around the aircraft (Figure 16.2).

The U.S. military's latest aircraft, the Joint Strike Fighter (JSF), is currently being outfitted with the latest version of the VSI HMD (see Figure 16.1c). Thales Visionics (formerly Gentex Visionics) provides their Scorpion HMD for the U.S. Air Force's A-10 aircraft (see Figure 16.13).[45] Other companies such as BAE (Striker HMD) and SAAB (Cobra HMD[46]) are providing HMDs for international fixed-wing aircraft.

The U.S. Army began installing HMDs on their rotary wing aircraft starting with the AH-1S Cobra gunship in the 1970s. A turreted machine gun is slaved to the pilot's head orientation via a mechanical linkage attached to his helmet. The pilot aims the weapon by superimposing a small helmet-mounted reticle on the target.[47] In the 1980s, the Army adopted the Integrated Helmet and Sighting System (IHADSS) for the AH-64 Apache helicopter. This monocular helmet-mounted display gives the pilot the ability—similar to the Cobra gunship—to target head-slaved weapons, with the added ability to display head-tracked FLIR

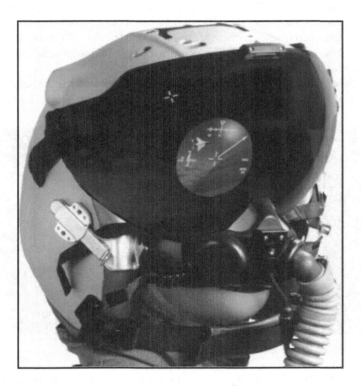

FIGURE 16.2 The U.S. Air Force and Navy's JHMCS helmet-mounted display. (Photo courtesy of Vision Systems International, San Jose, CA, used with permission.)

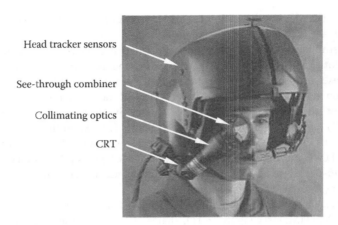

Head tracker sensors

See-through combiner

Collimating optics

CRT

FIGURE 16.3 The IHADSS is a monocular, monochrome, CRT-based, head-tracked, see-through helmet-mounted display used on the U.S. Army's AH-64 Apache helicopter. (Photo courtesy of Elbit Systems of America, Fort Worth, Texas.)

imagery for nighttime flying. Over 5000 of these CRT-based, monochrome systems have been delivered by Elbit Systems of America on this very successful program for the Army (Figure 16.3).

The U.S. Army also has extensive experience using helmet-mounted night vision goggles (NVGs) in aviation environments. These devices have their own unique set of performance, interface, and visual issues.[48–51] An interesting hybrid technique being used by the U.S. Army injects aircraft symbology into the front objective lens of the NVG. Though not head-tracked, it gives the pilot "eyes-up and out" access to aircraft status data. The so-called ANVIS HUD is a CRT-based design that has been successfully used on many of the Army's helicopters,[52] though these are being replaced by the ANVIS/HUD®- 24 which replaces the CRT with a small flat panel image source, reducing head supported weight and improving the center of gravity (CG). A similar device is shown in Figure 16.4.

In addition to these domestic applications, HMD-based pilotage systems are being adopted throughout the international aviation community on platforms such as Eurocopter's Tiger helicopter and the

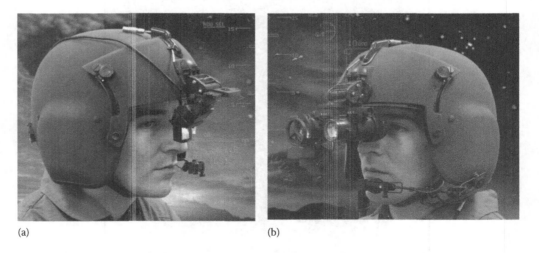

(a)   (b)

FIGURE 16.4 The EyeHUD®, a hybrid HMD and night vision goggle technology in which an HMD (shown in a direct-view configuration [a]) can also be used to inject imagery into the objective lens of the NVG (shown on the pilot's right eye in the photo [b]). (Photo courtesy of Rockwell Collins Display Products, Portland, Oregon.)

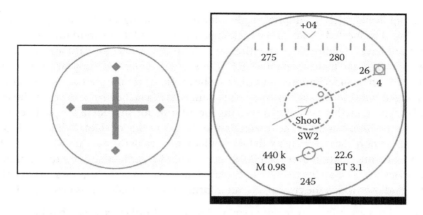

FIGURE 16.5 Comparison of early HMD reticle imagery (left) with more advanced HMD imagery.

South African Rooivalk.[53] NASA is currently investigating the use of HMDs for commercial pilots[54,55] to provide "better than visual" capability during Instrument Meteorological Conditions (IMC) as part of their NextGen program.

16.2 What Is an HMD?

In its simplest incarnation, an HMD consists of one or more image sources, collimating optics, and a means to mount the assembly on the pilot's head. In the IHADSS HMD shown in Figure 16.3, the monocular optics create and relay a virtual image of the image source (a cathode ray tube, or CRT), reflecting the imagery off the see-through combiner to the pilot's eye. This display module attaches to the right side of the aviator's protective helmet with adjustments that let the pilot position the display to see the entire image.

The early VTAS and Cobra helicopter HMDs used a simple targeting reticle to point weapons similar to the one shown in Figure 16.5. The JHMCS HMD has a more sophisticated targeting capability including "look-to" and shoot cues, as well as altitude, airspeed, compass heading, and artificial horizon data, though a more simplified approach is being implemented for fixed wing applications.[56–58] With the IHADSS, the pilot sees a similar symbology set superimposed over head-tracked FLIR data.

This collection of components, though deceptively simple, has at its core a complex interaction of system and hardware issues as well as visual, anthropometric, physical, and display issues. These in turn are viewed by an equally complex *human perceptual system*.[1] The designer's task is further complicated in the aircraft environment, because the HMD—now a *helmet*-mounted display—provides both display and protection for the pilot. Display performance issues such as luminance, contrast, alignment, and focus must be addressed but cannot impact pilotage or crash safety. For all these reasons, HMD design requires a careful balancing—a *suboptimization*—of both display and physical requirements.

16.3 Helmet-Mounted Display as Part of a Visually Coupled System

In an avionics application, the HMD—a helmet-mounted display or helmet-mounted sight—is part of a VCS consisting of the HMD, a head tracker, and mission computer.[22,23] As the pilot turns his head, the new orientation is communicated to the mission computer which updates the imagery as required. The information is always with the pilot, always ready for viewing.

Early cockpit-mounted displays—head-down displays—gave the pilot information on aircraft status but required him to return his attention to the interior of the cockpit for updates, reducing the time he spent looking outside the aircraft. As jets got faster and the allowable reaction time for pilots got shorter, head-up displays (HUD) improved pilot viewing by creating a virtual image that was projected on a combining glass located on top of the cockpit panel, in the pilot's forward line of sight. This meant the pilot did not have to redirect his attention away from the critical forward airspace or refocus his eyes to see the image, saving time. Because HUD imagery is *collimated*—appearing as though from a distant point—it can be superimposed on a distant object. This gives the pilot access to real-time geo- or aircraft-stabilized information such as compass headings, artificial horizons, or sensor imagery.

The HMD expands on this capability by placing the information in front of the pilot's eyes at all times and by linking the information to the orientation to the pilot's line of sight. While the HUD provides information about only the relatively small forward-looking area of the aircraft, with a head tracker, the HMD can provide information over the pilot's entire field of regard, all around the aircraft with eyes- and head-out viewing. This ability to link the displayed information with the pilot's line of sight increases the area of regard over which the critical aircraft information is available. This new capability can

- Cue the pilot's attention by providing a pointing reticle to where a sensor has located an object of interest
- Allow the pilot to slew head-tracked sensors such as a FLIR for flying at night or in adverse conditions
- Allow the pilot to aim weapons at targets that are off-boresight from the line of sight of the aircraft
- Allow the pilot to hand-off or receive target information (or location) from a remote platform, wingman, or other crewmember
- Provide the pilot with information that is either screen-referenced, aircraft-referenced, or earth-referenced[59]
- Reduce the time spent viewing the interior of the cockpit and increase the time spent viewing the outside world

And in general, it can provide increased SA to the pilot by giving him information about the entire space surrounding the aircraft.

One excellent example is the U.S. Army's AH-64 Apache helicopter equipped with the Integrated Helmet and Display Sighting System (IHADSS) HMD and head tracker. As the pilot moves his head in azimuth or elevation, the tracker communicates the head orientation to the servo system controlling the Pilot Night Vision System (PNVS) FLIR. The sensor follows his head movements, providing the pilot with a viewpoint as though his head was located on the nose of the aircraft. This gives the pilot the ability to "see" at night or in low light in a very intuitive and hands-off manner, similar to the way he would fly during daytime with the overlay of key flight data such as heading, altitude, and airspeed (Figure 16.6).

FIGURE 16.6 Linkage between the IHADSS helmet-mounted display and the Pilot Night Vision System (PNVS) in the AH-64 Apache helicopter. The PNVS is slaved to the pilot's head line of sight. As he turns his head, the PNVS turns to point in the same direction.

Studies are being conducted to find ways to squeeze even more out of the HMD in high-performance aircraft. A simulator study at the Naval Weapons Center explored using the HMD to provide "pathway in the sky" imagery to help pilots avoid threats and adverse weather.[60] Another experiment compensated for the loss of color and peripheral vision that accompanies g-induced loss of consciousness (g-loc). As the pilot began to "gray-out," the symbol set was reduced down to just a few critical items, positioned closer to the pilot's central area of vision. Another study provided helicopter pilots with earth-referenced navigation waypoints overlaid on terrain and battlefield engagement areas.[24,25] The results showed significant improvements in navigation, landing, the ability to maintain fire sectors, and an overall reduction in pilot workload.

16.4 HMD System Considerations and Trade-Offs

As discussed in the introduction, good HMD design relies on a suboptimization of requirements, trading off various performance parameters and requirements. The following sections will address some of these issues.

16.4.1 Ocularity

There are three optical configurations for HMDs: monocular, biocular, and binocular.

Monocular—A single image source viewed by a single eye. This is the lightest, least expensive, and simplest of all three approaches. Because of these advantages, most of the current HMD systems are monocular, such as the Elbit DASH, the Vision Systems International JHMCS (Figure 16.2), and the Elbit IHADSS (Figure 16.3). Some of the drawbacks include the potential for a laterally asymmetric CG and issues associated with focus, eye dominance, binocular rivalry, and ocular-motor instability.[61,62] There remain questions of how monocular HMDs interact with the dominant eye, though there is research[63] that defines the dominant eye simply as that which individuals use for monocular sighting tasks. This is the situation for Warfighters wearing the AN/PVS-14 NVG or the Land Warrior HMD, where soldiers place the HMD or NVG over their *nondominant* eye, leaving their dominant eye clear for weapon aiming.

Biocular—A single image source viewed by both eyes. The biocular approach is more complex than the monocular design, though it does stimulate both eyes, eliminating the potential for ocular-motor instability associated with monocular displays. Viewing imagery with two eyes has been shown to yield improvements in detection over one-eye viewing as well as providing a more comfortable viewing experience. However, two-eyed viewing is subject to much more stringent set of alignment, focus, and adjustment requirements. For *absolute* horizontal alignment, in *non-see-through* HMDs, the binocular alignment is not critical as long as it agrees with the focus to within ±1/4 diopter. For *relative* horizontal alignment, in *see-through* HMDs, the horizontal binocular alignment must be within 5–10 arcmin of the desired vergence distance and the focus should agree to within ±1/4 diopter (though temporary viewing can exceed this value[64]). For *absolute* vertical alignment, in *non-see-through* HMDs, the binocular alignment must be within 10 arcmin. For *relative* vertical alignment, in *see-through* HMDs, the binocular alignment must be within 3–6 arcmin.[65–67] The primary disadvantage of the biocular design is that the image source is usually located in the forehead region, making it more difficult to package. In addition, since the luminance from the single image source is split to both eyes, the brightness is cut in half.

Binocular—Each eye views an independent image source. This is the most complex, most expensive, and heaviest of all three options, but one which has all the advantages of a two-eyed system with the added benefit of providing partial binocular overlap (to enlarge the horizontal

TABLE 16.1 Comparison of the Advantages and Disadvantages of Monocular, Biocular, and Binocular HMDs

| Ocularity | Human Performance, Sensory, and Ergonomic Considerations | |
	Benefits	Disadvantages
Monocular (one image source viewed by one eye)	Minimum weight Simplest HMD; less stringent alignment Eye with no display remains dark adapted and continues to sample real world Least expensive	Possible visual rivalry problems, such as target suppression (involuntary), "cognitive switching," ocular-motor instability, eye dominance, and focus issues. Asymmetric center of mass. Smallest FOV; least information capability; more and larger head movements required. No stereoscopic depth information
Biocular (one image source viewed by both eyes)	No interocular rivalry, as with monocular Less complex to adjust than binocular Lighter weight than binocular Less expensive than binocular	Heavier than monocular. Possibly reduced luminance over monocular or binocular. No stereoscopic depth information. More complex alignment than monocular. Difficult to package, generally requires locating in the center of the forehead.
Binocular (two image sources viewed by both eyes)	Can provide stereo viewing Better depth information for mobility Improved target recognition over monocular Partial binocular overlap Symmetrical center of mass	Heaviest optics. Alignment and adjustments are more complex than monocular. Most expensive.

field of view [FOV] as discussed later in this chapter), stereoscopic imagery, and more packaging design freedom. A binocular HMD is subject to the same alignment, focus, and adjustment requirements as the biocular design, but the designer benefits by being able to move both the optics and the image sources *symmetrically away* from the face. Examples are the Helmet Integrated Display Sighting System (HIDSS; for the since-cancelled U.S. Army RAH-66 Comanche helicopter), and the F-35 HMD. A binocular system can also take advantage of some techniques for extending the horizontal FOV without compromising resolution (see Section 16.4.2; Table 16.1).

16.4.2 Field of View and Resolution

When asked for their initial HMD requirements, users will typically ask for more of both FOV *and* resolution. This is not surprising since the human visual system has a total FOV of 200° horizontal by 130° vertical,[68] with a grating acuity of 2 min of arc[69] in the central foveal region. However, for daytime air-to-air applications in a fixed wing aircraft, a large FOV is probably not necessary to display the symbology shown in Figure 16.5. This is because research[70] has shown that normal eye movements are limited to no more than ±15°. Constant movement outside of this range has the potential to cause eye strain, because the natural tendency is for an individual to keep their eyes in a "forward facing" orientation, so anything outside of the 15° range will elicit a head motion. For an HMD such as the JHMCS system where the pilot will receive aircraft and weapons status information, a 20° FOV was found to be sufficient. If the HMD is intended to display sensor imagery for night time pilotage such as with the IHADSS (a rectangular 30° by 40° FOV), the pilot will "paint" the sky with the HMD, creating a mental map of his surroundings. A larger FOV is advantageous, because it provides peripheral cues that contribute to the pilot's sense of self-stabilization, and it lowers pilot workload by reducing the range of head movements needed to fill in the mental map.[71-73] Most NVGs such as the ANVIS-6 have a FOV of 40° circular and the Quad-Eye provides up to 100° horizontal FOV using two image sensors per eye.

While display resolution contributes to overall image quality, there is also a direct relationship with performance.

If we look at the Johnson criteria for image recognition, we see that the amount of resolution required is (like most HMD-related issues) task dependent. For an object such as a tank, increased resolution will allow the pilot to detect ("something is there"), recognize ("it is a tank"), or identify ("it is a T-72 tank")[74] at a particular distance.

While more of each is desirable, FOV and resolution are linked by the following relationship:

$$H = F \times \tan \Theta$$

where F is the focal length of the collimating lens. Then if

- H is the size of the image source, Θ *is the FOV*, or the apparent size of the virtual image in space
- H is the pixel size, Θ *is the resolution* or apparent size of the pixel in image space

Thus, the focal length of the collimating lens *simultaneously* governs the FOV (which you want large) *and* the resolution (which you want small; Figure 16.7). For a display with a single image source, the result is either wide FOV *or* high resolution, but *not both* at the same time.

Given this $F \times \tan \Theta$ invariant, there are at least four ways to increase FOV of a display and still maintain resolution. These are (1) high-resolution area of interest, (2) partial binocular overlap, (3) optical tiling, and (4) dichoptic area of interest.[75,76] Of these, partial binocular overlap is more common for binocular flight applications, though there has been some investigation into dichoptic imaging techniques for HMDs,[77] and optical tiling is being used to expand the FOV of NVGs (Figure 16.8).[78]

Canting the two monocular channels in a binocular HMD provides partial binocular overlap (inward for convergent overlap, outward for divergent overlap). This enlarges the horizontal FOV, while maintaining the same resolution as the individual monocular channels. Partial overlap requires that two image sources and two video channels are available and that the optics and imagery are properly configured to compensate for any residual optical aberrations. Concerns have been voiced about the required minimum binocular overlap as well as the possibility that perceptual artifacts such as binocular rivalry—referred to as "luning"—may have an adverse impact on pilot performance. Although some studies of partial overlap found evidence of image fragmentation in some pilot/test subjects,[79,80] all tests

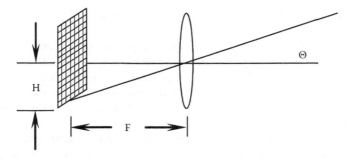

FIGURE 16.7 Illustration of how the focal length of the collimating lens determines the relationship between H, the size of the image source (or pixel size), and Θ, the FOV (or the resolution).

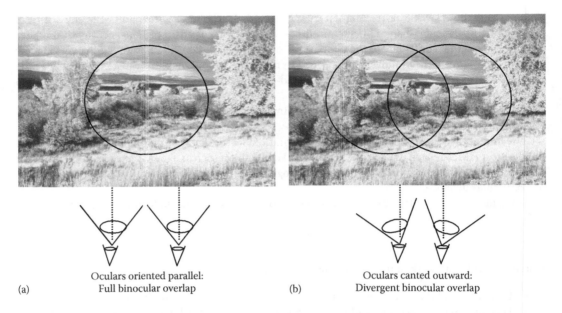

(a) Oculars oriented parallel: Full binocular overlap   (b) Oculars canted outward: Divergent binocular overlap

FIGURE 16.8 Comparison of (a) a full binocular overlap and (b) divergent partial binocular overlap. Note the increase in viewable imagery in the horizontal direction with the divergent overlap.

were conducted using static imagery, and interviews with pilots have shown that with a few minutes of practice, the artifacts are no longer noticeable. Several techniques have been shown effective in reducing the rivalry effects and their associated perceptual artifacts.[81]

16.4.3 Optical Design

The purpose of the optics in an HMD is threefold:

- Collimate the image source—creating a *virtual image*, which appears to be farther away than just a few inches from the face.
- Magnify the image source—making the imagery appear larger than the actual size of the image source.
- Relay the image source—creating the virtual image away from the image source, away from the front of the face.

Two of the most common optical design approaches common in HMDs are non–pupil forming and pupil forming.

The first is the *non-pupil-forming design*, similar to a simple magnifying lens—hence the term *simple magnifier*.[82,83] It is the easiest to design, the least expensive to fabricate, the lightest, and the smallest, though it does suffer from a short throw distance between the image source and the virtual image, putting whole assembly on the front of the head, close to the eyes. This approach is typically used for simple viewing applications such as the medical HMD (Figure 16.1a) and the Land Warrior display (Figures 16.1b and 16.9).

The second optical approach is a bit more complex, the *pupil-forming design*. This is more like the *compound microscope*, or a submarine periscope in which a first set of lenses creates an intermediate image of the image source, which is *relayed* by another set of lenses to where it creates a pupil (Figure 16.10).

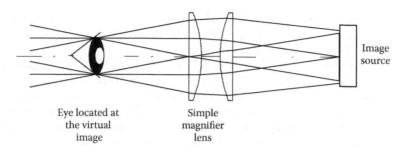

FIGURE 16.9 Diagram of a simple magnifier, or non-pupil-forming lens.

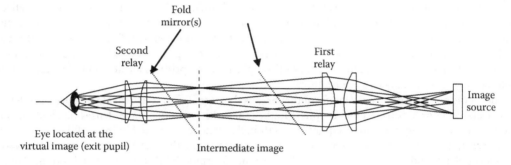

FIGURE 16.10 Pupil-forming optical design, similar to a compound microscope, binocular, or a periscope.

The advantage is that the pupil-forming design provides more path length from the image plane to the eye. This gives the designer more freedom to insert mirrors as required to fold the optical train away from the face to a more advantageous weight and CG location. The disadvantage is that the additional lenses increase the weight and cost of the HMD and that outside the exit pupil—the image of the stop—there is no imagery. This approach is typically used when the image source is large or when it is desirable to move the weight away from the front of the face such as in Figures 16.1c and 16.3 (Table 16.2).

In either case, the optical design must be capable of collimating, magnifying, and relaying the image with sufficiently small amounts of residual aberrations,[83] with manual focus (if required) and proper

TABLE 16.2 Summary of Some of the Advantages and Disadvantages of Pupil-Forming and Non-Pupil-Forming Optical Designs for HMDs

	Non-Pupil-Forming (Simple Magnifier)	Pupil-Forming (Relayed Lens Design)
Advantages	Simplest optical design. Fewer lenses and lighter weight. Does not "wipe" imagery outside of eye box. Less eyebox fit problems Mechanically the simplest and least expensive.	Longer path length means more packaging freedom. Can move some of the mass away from the front of face. More lenses provide better optical correction.
Disadvantages	Short path-length puts the entire display near the eyes/face. Short path-length means less packaging design freedom.	More complicated optical design. More lenses mean heavier design. Loss of imagery outside of pupil. Needs precision fitting, more and finer adjustments.

alignment (if a binocular system). In addition, the optical design must provide a sufficiently large exit pupil so that the user does not lose the image if the HMD shifts on the head as well as provide at least 25 mm of eye relief to allow the user to wear eyeglasses.

The exit pupil is found only in pupil-forming designs such as the JSF HMD (Figure 16.1c) and the IHADSS (Figure 16.3). In non-pupil-forming designs of Figure 16.1a and b, it is more correct to refer to a *viewing eyebox*, because there is a finite unvignetted viewing area. Classically, eye relief is defined as the distance along the optical axis from the last optical surface to the exit pupil. In an HMD with angled combining elements, eye relief should be measured from the eye to the closest point of the combiner (the so-called eye clearance distance), whether it is on the optical axis or not.

There have been new developments in HMD optical design, freeform optics, and waveguide optics, both of which offer the opportunity to reduce the size and weight of HMDs.

Improvements in precision fabrication of off-axis optical components have spurred the development of compact optical devices known as freeform prisms.[84-86] Imagery from a small microdisplay reflects from the first curved surface via total internal reflection to the second curved surface with a reflective coating (high reflectance for a non-see-through application, and partially reflecting for a see-through application), which is then reflected to the user's eye. The complex, off-axis optical surfaces allow numerous degrees of design freedom. For see-through, a second prism is located on the exterior of the collimating prism, as shown in Figure 16.11.

Another innovative optical design form that breaks the paradigm of classical HMD optics is referred to as waveguide or substrate-guided optics. Rather than the traditional large lenses or combiners in front of the eye, waveguide optics require only a thin, flat plate to be located in front of the eye.

Currently, there are two approaches, as shown in Figure 16.12. The first one uses a series of louvered output couplers[87] and is the basis for the Thales Visionics Scorpion HMD (using the Light-guide Optical Element, made by Lumus, also available commercially, see Figure 16.13). The second approach uses total internal reflection with either glass or holographic grating output couplers,[88-91] as shown on the right in Figure 16.12. This approach is the basis for the BAE Q-Sight HMD. The key to both designs is that a small input pupil is replicated numerous times, meaning that a smaller—and therefore a lighter and more compact—image source can be used. These designs are also starting to appear in small commercial HMDs because of their compact and lightweight packaging.

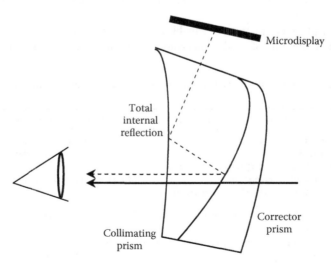

FIGURE 16.11 Schematic drawing of a see-through version of a freeform prism. (After Cheng, D. et al., *Appl. Opt.*, 48, 2655, 2009.)

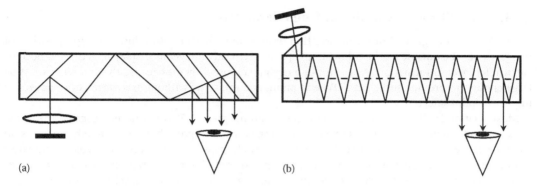

(a)   (b)

FIGURE 16.12 Two versions of waveguide optics: one that uses a louvered output coupling design (a, After Amitai, Y., United States Patent 6,829,095 B2, issued December 7, 2004) and the other (b, After Cameron, A., *Proc. SPIE, Head- and Helmet-Mounted Displays XIV: Design and Applications,* 7326, 73260H, 2009) that uses holographic grating output couplers. Each output ray is a replication of the small input pupil, creating an array of exit pupils to the eye.

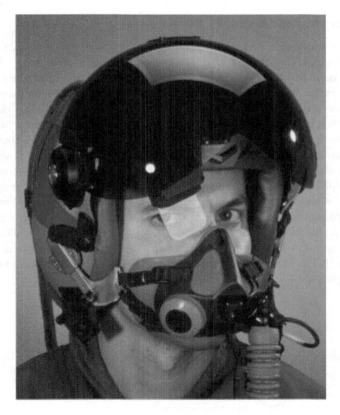

FIGURE 16.13 Scorpion HMD by Thales Visionics (formerly Gentex Visionics) using the louvered waveguide optic from Lumus. (Photo courtesy of Thales Visionics, Aurora, IL.)

16.4.4 See-Through Combiner Considerations

A see-through design is desired for aviation applications, but also to allow the superposition of imagery over the outside world, allowing the particularly advantageous display of geo-referenced or earth-referenced information. A see-through design is more difficult because the combiner must be large enough to provide sufficient FOV, exit pupil, and eye relief without excess weight or adversely impacting pilot safety.

Most aviation HMDs (currently) use only monochrome green (555 nm) imagery, because this is the peak daylight (photopic) visual sensitivity.[65] One of the ways to improve both see-through transmission and reflectance is to take advantage of high reflectance holographic notch filters and V-coats. (This latter term refers to an antireflective boundary layer coating designed to reduce reflections at a single wavelength.) The drawback is that while these special coatings reflect more of a specific display color, they transmit less of that *same* color, which makes the world look pink. With the advent of color displays in the cockpit, selectively reflecting green over another color may miscue the pilot. For these reasons, most aircraft HMD combiners have *spectrally neutral* reflective coatings.

Another consideration is see-through distortion. Pilots typically prefer to not have any optical combiners in front of their eyes (1) for ejection/crash safety and (2) to eliminate distortions and other viewing artifacts. Both of these concerns have driven fixed wing HMD designs to use the pilot's protective visor as the final optical collimating element (as on the JHMCS and JSF designs). Regardless of which configuration is chosen, distortionless see-through is important for the HMD and protective visor combination and is typically governed by military standards.[92]

16.4.5 Luminance and Contrast

In the high ambient luminance environment of an aircraft cockpit, daylight readability of displays is a critical issue. The combining element in an HMD—similar to the combiner of a HUD—reflects the projected imagery into the pilot's eyes. The pilot looks through the combining glass and sees the imagery superimposed on the outside world, so it cannot be 100% reflective—pilots always prefer to have as much see-through as possible. To view the HMD imagery against a bright background such as sun-lit clouds or snow, this less-than-perfect reflection efficiency means that the image source must be that much brighter. The challenge is to provide a combiner with good see-through transmission and still provide a high luminance image. There are limitations, though, because all image sources have a luminance maximum governed by the physics of the device as well as size, weight, and power of any ancillary illumination. In addition, other factors such as the transmission of the aircraft canopy and pilot's visor must be considered when determining the required image source luminance, as shown in Figure 16.14.

The image source luminance (B_I) is attenuated before entering the eye by the transmission of the collimating optics (T_O) and the reflectance of the combiner (R_C). The pilot views the distant object through the combiner (T_C or $1-R_C$), the protective visor (T_V), and the aircraft transparency (T_A) against the bright background (B_A). We can calculate the image source luminance for a desired contrast ratio (CR) of 1.3 using the expression[22]

$$CR = \frac{B_A + B_{Display}}{B_A}$$

where we know that the display luminance to the eye is given by

$$B_{Display} = B_I \times T_O \times R_C$$

and as observed by the pilot, the background is given by

$$B_O = T_C \times T_V \times T_A \times B_A$$

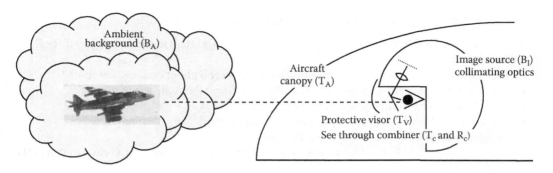

FIGURE 16.14   Contributions for determining image source luminance requirements for an HMD in an aircraft cockpit.

TABLE 16.3   Contributions for the Display Luminance Calculations for Four Different HMD Configurations

		Case 1—Clear Visor, 50% Combiner Transmission	Case 2—Dark Visor, 50% Combiner Transmission	Case 3—Clear Visor, 80% Combiner Transmission	Case 4—Dark Visor, 80% Combiner Transmission
Optics transmission	T_O	85%	85%	85%	85%
Combiner reflectance	R_C	50%	50%	20%	20%
Combiner transmission	T_C	50%	50%	80%	80%
Visor transmission	T_v	87%	12%	87%	12%
Aircraft canopy transmission	T_C	80%	80%	80%	80%
Ambient background luminance	B_C	10,000 fL	10,000 fL	10,000 fL	10,000 fL
Required image source luminance	B_I	2456 fL	339 fL	9826 fL	1355 fL

Rewriting, we can see that

$$CR = \frac{1 + \left(B_I \times T_O \times R_C\right)}{\left(T_C \times T_V \times T_A \times B_A\right)}$$

We can substitute some nominal values for the various contributions, as given in Table 16.3.

The first two cases compare the difference when the pilot is wearing a class 1 (clear) versus a class 2 (dark) visor.[92] The dark visor reduces the ambient background luminance, improving HMD image contrast against the bright clouds or snow. These first two cases are relatively simple, because they assume a combiner with 50% transmission and 50% reflectance (ignoring other losses). Since pilots need more see-through, this means a reduced reflectance. Cases 3 and 4 assume this more realistic combiner configuration with both clear and dark visors, resulting in a requirement for a much brighter image source.

16.4.6 Head-Mounting, Comfort, Biodynamics, and Safety

It is difficult to put a precise metric on the fit or comfort of an HMD, though it is always immediately evident to the wearer. Even if image quality is excellent, the user will reject the HMD if it does not fit well. Fitting and sizing is especially critical in the case of an HMD where in addition to being comfortable, it must provide a *precision* fit for the display relative to the pilot's eyes.

We can list the most important issues for achieving a good fit with an HMD:

- The user must be able to adjust the display (or have it properly fitted) to see the imagery throughout the mission.
- The HMD must be comfortable for a long duration of wear without causing "hot spots."
- The HMD must not slip with sweating or under g-loading, vibration, or buffeting.
- The HMD must be retained during crash or ejection.
- The weight of the head-borne equipment must be minimized.
- The mass moment of inertia must be minimized.
- The mass of the head-borne components should be distributed to keep the CG close to that of the head alone.

The human head weighs approximately 9–10 lb and sits atop the spinal column (Figure 16.15). The occipital condyles on the base of the skull mate to the superior articular facets of the first cervical vertebra (C1, or the Atlas).[93] These two small oblong mating surfaces on either side of the spinal column are the pivot points for the head.

The CG of the head is located at or about the tragion notch, the small cartilaginous flap in front of the ear. Because this is *up* and *forward* of the head/vertebra pivot point, there is a tendency for the head to tip downward, were it not for the strong counter force exerted by the muscles running down the back of the neck—hence when people fall asleep, they "nod off." Adding mass to the head in the form of an HMD can move the CG (now HMD + head) away from this ideal location. High vibration or buffeting, ejection, parachute opening or crash will greatly exacerbate the effect of this extra weight and displaced CG, with effects that can range from fatigue and neck strain to serious

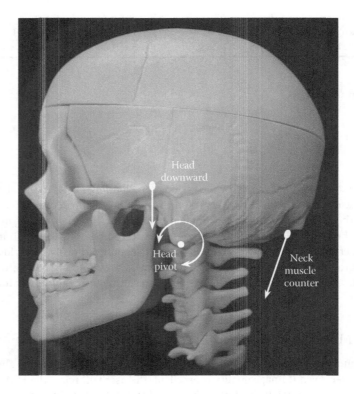

FIGURE 16.15 Human head and neck with the CG located near the tragion notch and the pivot point located at the occipital condyles.

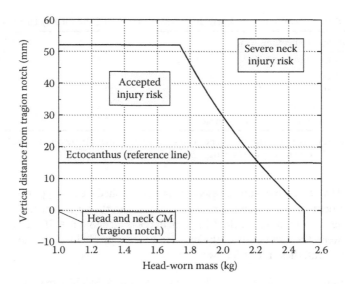

FIGURE 16.16 USAARL weight and *vertical* CG curve with the area under the curve considered crash safe in helicopter environments. (Data curve courtesy of U.S. Army Aeromedical Research Labs, Fort Rucker, AL.)

or mortal injury.[94] Designers can mitigate the impact of the added head-borne hardware by first minimizing the mass of the HMD, and then by optimizing its location to restore the head + HMD location to that of the head alone.

This is supported by the extensive biomechanics research at the U.S. Army's Aeromedical Research Laboratory (USAARL). Figure 16.16 gives a weight versus CG curve in the vertical direction, where the area under the curve is considered crash safe for a helicopter environment. The second graph (Figure 16.17) defines the weight/CG combination that will minimize fatigue.[23]

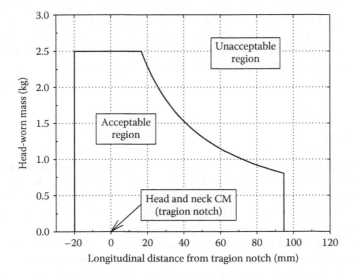

FIGURE 16.17 USAARL weight and *horizontal* CG curve with the area under the curve considered acceptable for fatigue in helicopter environments. (Data curve courtesy of U.S. Army Aeromedical Research Labs, Fort Rucker, AL.)

TABLE 16.4   Univariate (Uncorrelated) Anthropometric Data for Key Head Features

Critical Head Dimensions (cm)	5% Female	95% Male
Interpupillary distance (IPD)	5.66	7.10
Head length[a]	17.63	20.85
Head width	13.66	16.08
Head circumference	52.25	59.35
Head height (ectocanthus to top of head)[a]	10.21	12.77

Note the range of sizes for the 5th percentile female up to the 95th percentile male.[98]

[a] These data are head orientation dependent.

For fixed-wing HMDs, biomechanics research has established head-supported and center of mass boundaries as a refinement of a rectangular area located within the human head and neck known as the "Knox Box."[95] Using the occipital condyles (the pivot of the head about the C1 cervical vertebra—located approximately between the two points of the mastoid process just behind the ears, as shown in Figure 16.15) as the origin point, these are as follows:

- Maximum head-supported weight of 5 lb (2.5 kg, including MBU-20/P oxygen mask and 3 in. [7.6 cm] of hose, helmet, visor, and HMD components)
- Vertical CM limits (z-direction): between +0.5 in. (1.3 cm) and +1.5 in. (3.8 cm) above the occipital condyles
- *Threshold* CM horizontal limits (x-direction): between +0.5 in. (1.3 cm) forward and −0.8 in. (2 cm) posterior of the occipital condyles
- *Objective* CM horizontal limits (x-direction): between +0.2 in. (0.5 cm) forward and −0.8 in. (2 cm) posterior of the occipital condyles
- Lateral CM limits (y-direction): ±0.15 in. (±0.4 cm)

Anthropometry—"the measure of man"—is the compilation of data that define such things as the range of height for males and females, the size of our heads, and how far apart our eyes are. Used judiciously, these data can help the HMD designer achieve a proper fit, though an over-reliance can be equally problematical. One of the most common mistakes made by designers is to assume a correlation between various anthropometric measurements, because almost all sizing data are *univariate*—that is, they are completely uncorrelated with other data. For example, a person who has a 95th percentile head circumference will not necessarily have a 95th percentile interpupillary distance.[96] One bivariate study did try to correlate head length and head breadth for male and female aviators, resulting in a rather large spread of data.[97] Table 16.4 shows the univariate (uncorrelated) anthropometric data for key head features.

16.5 Summary

Head-mounted displays can provide a distinctly unique and personal viewing experience that no other display technology can. By providing the pilot with display information that is linked to head orientation, the pilot is freed from having to return his attention to the cockpit interior and is able to navigate and fly the aircraft in a more intuitive and natural manner. This is an effective means of providing a pilot aircraft status as well as information about the surrounding airspace.

But these capabilities are not without a price. HMDs require careful attention to the complex interactions between hardware and human perceptual issues, made only more complex by the need for the HMD to provide life support in an aviation environment. Only when all factors are considered and the requirements successfully suboptimized with an understanding of the aviator's tasks and environment, can this be done.

Recommended Reading

Barfield, W. and Furness, T.A. *Virtual Environments and Advanced Interface Design*. New York: Oxford University Press, 1995.

Boff, K.R. and Lincoln, J.E. *Engineering Data Compendium, Human Perception and Performance*. Wright-Patterson Air Force Base, OH: Human Engineering Division, Harry G. Armstrong Aerospace Medical Research Laboratory, 1988.

Kalawsky, R.A. *The Science of Virtual Reality and Virtual Environments*. New York: Addison-Wesley Publishing Company, 1993.

Karim, M.A. (Ed.) *Electro-Optical Displays*. New York: Marcel Dekker, Inc., 1992.

Lewandowski, R.J. *Helmet- and Head-Mounted Displays, Selected SPIE Papers on CD-ROM*. SPIE Press, Vol. 11, 2000, Bellingham, Washington.

Melzer, J.E. and Moffitt, K.W. (Eds.) *Head-Mounted Displays: Designing for the User*. Createspace Independent Publishing Platform, 2011.

Rash, C.E. (Ed.) *Helmet-Mounted Displays: Design Issues for Rotary-Wing Aircraft*. Washington, DC: U.S. Government Printing Office, 1999.

Rash, C.E., Russo, M.B., Letowski, T.R., and Schmeisser, E.T. (Eds.) *Helmet-Mounted Displays: Sensation, Perception and Cognition Issues*. Fort Rucker, AL: U.S. Army Aeromedical Research Laboratory, 2009. Available for download from: http://www.usaarl.army.mil/publications/HMD_Book09/index.htm.

Velger, M. *Helmet-Mounted Displays and Sights*. Norwood, MA: Artech House, Inc., 1998.

References

1. Gibson, J.J. *The Ecological Approach to Visual Perception*. Hillsdale, NJ: Lawrence Erlbaum Associates, 1986.
2. Kalawsky, R.S. *The Science of Virtual Reality and Virtual Environments: A Technical, Scientific and Engineering Reference on Virtual Environments*. Wokingham, England: Addison-Wesley, 1996.
3. Schmidt, G.W. and Osborn, D.B. Head-mounted display system for surgical visualization. *Proc. SPIE, Biomedical Optoelectronic Instrumentation*, 2396, 345–350, 1995.
4. Pankratov, M.M. New surgical three-dimensional visualization system. *Proc. SPIE, Lasers in Surgery: Advanced Characterization, Therapeutics, and Systems*, 2395, 143–144, 1995.
5. Casey, C.J. Helmet-mounted displays on the modern battlefield. *Proc. SPIE, Helmet- and Head-Mounted Displays IV*, 3689, 270–277, 1999.
6. Browne, M.P. Head-mounted workstation displays for airborne reconnaissance applications. *Proc. SPIE, Cockpit Displays V: Displays for Defense Applications*, 3363, 348–354, 1998.
7. Lacroix, M. and Melzer, J. Helmet-mounted displays for flight simulators. *Proceedings of the IMAGE VII Conference*, Tucson, AZ, June 12–17, 1994.
8. Casey, C.J. and Melzer, J.E. Part-task training with a helmet integrated display simulator system. *Proc. SPIE, Large-Screen Projection, Avionic and Helmet-Mounted Displays*, 1456, 175–179, 1991
9. Simons, R. and Melzer, J.E. HMD-based training for the U.S. Army's AVCATT—A collective aviation training simulator. *Proc. SPIE, Large Screen Projection, Avionic and Helmet-Mounted Displays*. 5079, 1–6, 2003.
10. Melzer, J.E. Integrated headgear for the Future Force Warrior and beyond (Invited Paper). *Proc. SPIE*, 5801, *Cockpit and Future Displays for Defense and Security*, 173–178, 2005.
11. Schuyler, W.J. and Melzer, J.E. Integrated headgear for the Future Force Warrior: Results of first field evaluation. *Proc. SPIE, Head and Helmet-Mounted Displays XII: Design and Applications*, 6557, 20, 2007.
12. Foote, B. Design guidelines for advanced air-to-air helmet-mounted display systems. *Proc. SPIE, Helmet- and Head-Mounted Displays III*, 3362, 94–103, 1998.

13. Belt, R.A., Kelley, K., and Lewandowski, R. Evolution of helmet-mounted display requirements and Honeywell HMD/HMS systems. *Proc. SPIE, Helmet- and Head-Mounted Displays III*, 3362, 373–385, 1998.

14. Bayer, M.M., Rash, C.E., and Brindle, J.H. Introduction to helmet-mounted displays, Chapter 3, in: Rash, C.E., Russo, M.B., Letowski, T.R., and Schmeisser, E.T. (Eds.), *Helmet-Mounted Displays: Sensation, Perception and Cognition Issues*. Fort Rucker, AL: U.S. Army Aeromedical Research Laboratory, pp. 47–108, 2009.

15. Melzer, J.E. and Rash, C.E. The potential of an interactive HMD, Chapter 19, in: Rash, C.E., Russo, M.B., Letowski, T.R., and Schmeisser, E.T. (Eds.), *Helmet-Mounted Displays: Sensory, Perception and Cognition Issues*. Fort Rucker, AL: U.S. Army Aeromedical Research Laboratory, pp. 887–898, 2009.

16. Endsley, M.R. Toward a theory of situation awareness in dynamic systems. *Human Factors*, 37(1), 32–64, 1995.

17. Endsley, M.R. Theoretical underpinnings of situation awareness: A critical review, in: Endsley, M.R. and Garland, D.J. (Eds.), *Situation Awareness Analysis and Measurement*. Mahway, NJ: Lawrence Erlbaum Associates, pp. 3–32, 2000.

18. Endsley, M.R. and Hoffman, R.R. The Sacagawea principal. *Proceedings of the IEEE: Intelligent Systems*, 17(6), 80–85, November 2002.

19. Uhlarik, J. and Comerford, D.A. A review of situation awareness literature relevant to pilot surveillance functions. Final Report: DOT/FAA/AM-02/3. Washington, DC: Office of Aerospace Medicine, 2002.

20. Endsley, M.R. Measurement of situation awareness in dynamic systems. *Human Factors*, 37(1), 65–84, 1995.

21. Smith, D.J. Situation(al) awareness (SA) in effective command and control. Retrieved July 12, 2008, http://www.smithsrisca.demon.co.uk/situational-awareness.html, 2006.

22. Kocian, D.F. Design considerations for virtual panoramic display (VPD) helmet systems. *AGARD Conference Proceedings No. 425, the Man-Machine Interface in Tactical Aircraft Design and Combat Automation*, 22-1, 1987.

23. Rash, C.E. (Ed.) *Helmet-Mounted Displays: Design Issues for Rotary-Wing Aircraft*. Washington, DC: U.S. Government Printing Office, 1999.

24. Rogers, S.P., Asbury, C.N., and Haworth, L.A. Evaluation of earth-fixed HMD symbols using the PRISMS helicopter flight simulator. *Proc. SPIE: Helmet- and Head-Mounted Displays IV*, 3689, 54–65, 1999.

25. Rogers, S.P., Asbury, C.N., and Szoboszlay, Z.P. Enhanced flight symbology for wide field-of-view helmet-mounted displays. *Proc. SPIE: Helmet- and Head-Mounted Displays VIII: Technologies and Applications*, 5079, 321–332, 2003.

26. Jenkins, J.C. Development of helmet-mounted display symbology for use as a primary flight reference. *Proc. SPIE: Helmet- and Head-Mounted Displays VIII: Technologies and Applications*, 5079, 333–345, 2003.

27. Jenkins, J.C., Thurling, A.J., and Brown, B.D. Ownship status helmet-mounted display symbology for off-boresight tactical applications. *Proc. SPIE: Helmet- and Head-Mounted Displays VIII: Technologies and Applications*, 5079, 346–360, 2003.

28. Jenkins, J.C., Sheesley, D.G., and Bivetto, F.C. Helmet-mounted display symbology for enhanced trend and attitude awareness. *Proc. SPIE: Helmet- and Head-Mounted Displays IX: Technologies and Applications*, 5442, 164–178, 2004.

29. Albery, W.B. Multisensory cueing for enhancing orientation information during flight. *Aviation Space and Environmental Medicine*, 78, B186–B190, 2007.

30. Wickens, C.D. and McCarley, J.S. *Applied Attention Theory*. Boca Raton, FL: CRC Press/Taylor & Francis Group, 2008.

31. Wickens, C.D. Multiple resources and performance prediction. *Theoretical Issues in Ergonomics Science*, 3, 159–177, 2002.

32. Leibowitz, H.W., Shupert, C.L., and Post, R.B. Emergent techniques for assessment of visual performance. Retrieved from http://www.nap.edu/openbook/POD227/html, 1985.

33. Wickens, C.D. Spatial awareness biases. Technical Report AHFD-02-6/NASA-02-4. Moffett Field, CA: NASA Ames Research Center, 2002.

34. Uhlarik, J. and Comerford, D.A. A review of situation awareness literature relevant to pilot surveillance functions. Final Report: DOT/FAA/AM-02/3. Washington, DC: Office of Aerospace Medicine, 2002.

35. Spence, C. and Driver, J. Cross-modal links in attention between audition, vision and touch: Implications for interface design. *International Journal of Cognitive Ergonomics*, 1, 351–373, 1997.

36. Driver, J. and Spence, C. Cross-modal links in spatial attention. *Philosophical Transactions of the Royal Society of London B*, 353, 1319–1331, 1998.

37. Yeh, M., Wickens, C.D., and Seagull, F.J. Effects of frame of reference and viewing condition attentional issues with helmet-mounted displays. Technical Report ARL-98-1/ARMY-FED-LAB1. Champaign, IL: Aviation Research Lab, Institute of Aviation, University of Illinois at Urbana-Champaign, 1998.

38. Belt, R.A., Kelley, K., and Lewandowski, R. Evolution of helmet-mounted display requirements and Honeywell HMD/HMS systems. *Proc. SPIE, Helmet- and Head-Mounted Displays III*, 3362, 373, 1998.

39. Dornheim, M. VTAS sight fielded, shelved in 1970s. *Aviation Week & Space Technology*, October 23, 1995, p. 51.

40. Dornheim, M.A. and Hughes, D. U.S. intensifies efforts to meet missile threats. *Aviation Week & Space Technology*, October 16, 1995, p. 36.

41. Arbak, C. Utility evaluation of a helmet-mounted display and sight. *Proc. SPIE, Helmet-Mounted Displays*, 1116, 138, 1989.

42. Merryman, R.F.K. Vista Sabre II: Integration of helmet-mounted tracker/display and high off-boresight missile seeker into F-15 aircraft. *Proc. SPIE, Helmet-and Head-Mounted Displays and Symbology Design Requirements*, 2218, 173–185, 1994.

43. Lake, J. NATO's best fighter is made in Russia. *The Daily Telegraph*, August 26, 1991, p. 22.

44. Goodman, G.W., Jr. First look, first kill. *Armed Forces Journal International*, 32, July 2000.

45. Atac, R. Applications of the Scorpion color helmet-mounted cueing system. *Proc. SPIE Head- and Helmet-Mounted Displays XV: Design and Applications*, 7688, 768–803, 2010.

46. Larsson, J. and Blomqvist, T. The Cobra helmet mounted display system for Gripen. *Proc. SPIE 6955, Head- and Helmet-Mounted Displays XIII: Design and Applications*, 695505, 2008.

47. Braybrook, R. Looks can kill. *Armada International*, 4, 44, 1998.

48. Sheehy, J.B. and Wilkinson, M. Depth perception after prolonged usage of night vision goggles. *Aviation, Space and Environmental Medicine*, 60, 573, 1989.

49. Donohue-Perry, M.M., Task, H.L., and Dixon, S.A. Visual acuity versus field of view and light level for night vision goggles (NVGs). *Proc. SPIE, Helmet- and Head-Mounted Displays and Symbology Design Requirements*, 2218, 71–82, 1994.

50. Crowley, J.S., Rash, C.E., and Stephens, R.L. Visual illusions and other effects with night vision devices. *Proc. SPIE, Helmet-Mounted Displays III*, 1695, 166, 1992.

51. DeVilbiss, C.A., Ercoline, W.R., and Antonio, J.C. Visual performance with night vision goggles (NVGs) measured in U.S. Air Force aircrew members. *Proc. SPIE Helmet- and Head-Mounted Displays and Symbology Design Requirements*, 2218, 64–71, 1994.

52. Yona, Z., Weiser, B., and Hamburger, O. Day/night ANVIS/HUD-24 (day HUD) flight test and pilot evaluations. *Proc. SPIE, Helmet- and Head-Mounted Displays IX: Technologies and Applications*, 5442, 225–236, 2004.

53. Mace, T.K., Van Zyl, P.H., and Cross, T. Integration, development, and qualification of the helmet-mounted sight and display on the Rooivalk Attack Helicopter. *Proc. SPIE, Helmet- and Head-Mounted Displays VI*, 4361, 12–24, 2001.

54. Arthur, J.J., Prinzel, L.J., Williams, S.P., Bailey, R.E., Shelton, K.J., and Norman, R.M. Enhanced/ synthetic vision and head-worn display technologies for terminal maneuvering area NextGen operations. *Proc. SPIE, Display Technologies and Applications for Defense, Security, and Avionics V*, 8042, 80420T (16 pages), 2011.

55. Bailey, R.E., Shelton, K.J., and Arthur, J.J. Head-worn displays for NextGen. *Proc. SPIE, Head- and Helmet-Mounted Displays XVI: Design and Applications*, 8041, 80410G (15 pages), 2011.

56. Jenkins, J.C. Development of helmet-mounted display symbology for use as a primary flight reference. *Proc. SPIE, Helmet- and Head-Mounted Displays VIII: Technologies and Applications*, 5079, 333–345, 2003.

57. Jenkins, J.C., Thurling, A.J., and Brown, B.D. Ownship status helmet-mounted display symbology for off-boresight tactical applications. *Proc. SPIE, Helmet- and Head-Mounted Displays VIII: Technologies and Applications*, 5079, 346–360, 2003.

58. Jenkins, J.C., Sheesley, D.G., and Bivetto, F.C. Helmet-mounted display symbology for enhanced trend and attitude awareness. *Proc. SPIE, Helmet- and Head-Mounted Displays IX: Technologies and Applications*, 5442, 164–178, 2004.

59. Melzer, J.E. HMDs as enablers for situation awareness, the OODA loop and sensemaking. *Proc. SPIE, Head- and Helmet-Mounted Displays XVII: And Display Technologies and Applications for Defense, Security, and Avionics VI*, 8383, 83830M (9 pages), 2012.

60. Procter, P. Helmet displays boost safety and lethality. *Aviation Week & Space Technology*, February 1, 1999, p. 81.

61. Rash, C.E. and Verona, R.W. The human factor considerations of image intensification and thermal imaging systems, Chapter 16, in: Karim, M.A. (Ed.), *Electro-Optical Displays*. New York: Marcel Dekker, Inc., pp. 653–710, 1992.

62. Moffitt, K.W. Ocular responses to monocular and binocular helmet-mounted display configurations. *Proc. SPIE, Helmet-Mounted Displays*, 1116, 142–149, 1989.

63. Mapp, A.P., Ono, H., and Barbeito, R. What does the dominant eye dominate? A brief and somewhat contentious review. *Perception and Psychophysics*, 65, 310–317, 2003.

64. Wang, B. and Ciuffreda, K.J. Depth-of-focus of the human eye: Theory and clinical implications. *Survey of Ophthalmology*, 51(1), 75–85, 2006.

65. Boff, K.R. and Lincoln, J.E. *Engineering Data Compendium, Human Perception and Performance*. Wright-Patterson AFB, OH: Harry G. Armstrong Aerospace Medical Research Laboratory, 1988.

66. Moffitt, K.W. Designing HMDs for viewing comfort, in: Melzer, J.E. and Moffitt, K.W. (Eds.), *Head-Mounted Displays: Designing for the User*. Createspace Independent Publishing Platform, pp. 117–146, 2011.

67. Self, H.C. Critical tolerances for alignment and image differences for binocular helmet-mounted displays, Technical Report AAMRL-TR-86-019. Wright-Patterson AFB, OH: Armstrong Aerospace Medical Research Laboratory, 1986.

68. U.S. Department of Defense. MIL-HDBK-141 Optical Design. 1962.

69. Smith, G., and Atchison, D.A. *The Eye and Visual Optical Instruments*. New York: Cambridge University Press, 1997.

70. Bahill, A.T., Adler, D., and Stark, L. Most naturally occurring human saccades have magnitudes of 15 degrees or less. *Investigative Ophthalmology*, 14, 468–469, 1975.

71. Wells, M.J., Venturino, M., and Osgood, R.K. Effect of field of view size on performance at a simple simulated air-to-air mission. *Proc. SPIE, Helmet-Mounted Displays*, 1116, 126–138, 1989.

72. Kasper, E.F., Haworth, L.A., Szoboszlay, Z.P., King, R.D., and Halmos, Z.L. Effects of in-flight field-of-view restriction on rotorcraft pilot head movement. *Proc. SPIE, Head-Mounted Displays II*, 3058, 34–46, 1997.

73. Szoboszlay, Z.P., Haworth, L.A., Reynolds, T.L., Lee, A.G., and Halmos, Z.L. Effect of field-of-view restriction on rotorcraft pilot workload and performance: Preliminary results. *Proc. SPIE, Helmet- and Head-Mounted Displays and Symbology Design Requirements II*, 2465, 142–154, 1995.

74. Lloyd, J.M. *Thermal Imaging Systems*. New York: Plenum Press, 1975.

75. Melzer, J.E. Overcoming the field of view: Resolution invariant in head-mounted displays. *Proc. SPIE, Helmet- and Head-Mounted Displays III*, 3362, 284–293, 1998.

76. Hoppe, M.J. and Melzer, J.E. Optical tiling for wide-field-of-view head-mounted displays. *Proc. SPIE, Current Developments in Optical Design and Optical Engineering VIII*, 3779, 146–154, 1999.

77. Browne, M.P., Moffitt, K., Hopper, D.G., and Fath, B. Preliminary experimental results from a dichoptic vision system (DiVS). *Proc. SPIE, Head- and Helmet-Mounted Displays XVI: Design and Applications*, 8041, 804105, 2011.

78. Jackson, T.W. and Craig, J.L. Design, development, fabrication, and safety-of-flight testing of a panoramic night vision goggle. *Proc. SPIE, Head- and Helmet-Mounted Displays IV*, 3689, 98, 1999.

79. Klymenko, V., Verona, R.W., Beasley, H.H., and Martin, J.S. Convergent and divergent viewing affect luning, visual thresholds, and field-of-view fragmentation in partial binocular overlap helmet-mounted displays. *Proc. SPIE, Helmet- and Head-Mounted Displays and Symbology Design Requirements*, 2218, 2, 1994.

80. Klymenko, V., Harding, T.H., Beasley, H.H., Martin, J.S., and Rash, C.E. Investigation of helmet-mounted display configuration influences on target acquisition. *Proc. SPIE, Head- and Helmet-Mounted Displays*, 4021, 316, 2000.

81. Melzer, J.E. and Moffitt, K. An ecological approach to partial binocular-overlap. *Proc. SPIE, Large Screen, Projection and Helmet-Mounted Displays*, 1456, 124, 1991.

82. Task, H.L. HMD image sources, optics and visual interface, in: Melzer, J.E. and Moffitt, K.W. (Eds.), *Head-Mounted Displays: Designing for the User*. Createspace Independent Publishing Platform, pp. 55–82, 2011.

83. Fischer, R.E. Fundamentals of HMD optics, in: Melzer, J.E. and Moffitt, K.W. (Eds.), *Head-Mounted Displays: Designing for the User*. Createspace Independent Publishing Platform, pp. 83–116, 2011.

84. Hua, H. Sunglass-like displays become a reality with free-form optical technology. SPIE Newsroom, doi:10.1117/2.1201208.004375, 2012.

85. Cheng, D., Wang, Y., Hua, H., and Talha, M.M. Design of an optical see-through head-mounted display with a low f-number and large field of view using a freeform prism. *Applied Optics*, 48, 2655–2668, 2009.

86. Rolland, J.P., Kaya, I., Thompson, K.P., and Cakmakci, O. Invited paper: Head-worn displays—Lens design. *SID Symposium Digest*, 855–858, 2010.

87. Amitai, Y. Substrate-guided optical beam expander. United States Patent 6,829,095 B2, issued December 7, 2004.

88 DeJong, C.D. Full-color, see-through, daylight-readable, goggle-mounted display. *Proc. SPIE, Head- and Helmet-Mounted Displays XVI: Design and Applications*, 8041, 80410E, 2011.

89. Mukawa, H., Akutsu, K., Matsumura, I., Nakano, S., Yoshida, T., Kuwahara, M., Aiki, K., and Ogawa, M. Distinguished paper: A full color eyewear display using holographic planar waveguides, *SID Symposium Digest*, 39, 89–92, 2008.

90. Cameron, A. The application of holographic optical waveguide technology to Q-Sight™ family of helmet mounted displays. *Proc. SPIE, Head- and Helmet-Mounted Displays XIV: Design and Applications*, 7326, 73260H, 2009.

91. Cameron, A.A. Optical waveguide technology and its application in head-mounted displays. *Proc. SPIE 8383, Head- and Helmet-Mounted Displays XVII; and Display Technologies and Applications for Defense, Security, and Avionics VI*, 83830E, 2012.

92. U.S. Department of Defense. MIL-V-85374, Military specification, visors, shatter resistant. 1979.

93. Perry, C.E. and Buhrman, J.R. HMD head and neck biomechanics, in: Melzer, J.E. and Moffitt, K.W. (Eds.), *Head-Mounted Displays: Designing for the User*. Createspace Independent Publishing Platform, pp. 147–174, 2011.

94. Guill, F.C. and Herd, G.R. An evaluation of proposed causal mechanisms for "ejection associated" neck injuries. *Aviation, Space and Environmental Medicine*, 60, A26, July 1989.

95. Knox, F.S., Buhrman, J.R., Perry, C.E, and Kaleps, I. Interim head/neck criteria. Consultation Report. Wright-Patterson Air Force Base, OH: Armstrong Laboratories, Escape and Impact Protection Branch, 1991.

96. Whitestone, J.J. and Robinette, K.M. Fitting to maximize performance of HMD systems, in: Melzer, J.E. and Moffitt, K.W. (Eds.), *Head-Mounted Displays: Designing for the User*. Createspace Independent Publishing Platform, pp. 175–206, 2011.

97. Barnaba, J.M. Human factors issues in the development of helmet mounted displays for tactical, fixed-wing aircraft. *Proc. SPIE, Head-Mounted Displays II*, 3058, 2–14, 1997.

98. Gordon, C.C., Churchill, T., Clauser, C.E., Bradtmiller, B., McConville, J.T., Tebbetts, I., and Walker, R.A. 1988 anthropometric survey of U.S. army personnel: Summary statistics interim report. Technical Report TR-89/027. Natick, MA: U.S. Army Natick, 1989.

17

Head-Up Display

Robert B. Wood
*Rockwell Collins Head-Up
Guidance Systems*

Peter J. Howells
*Rockwell Collins Head-Up
Guidance Systems*

17.1 Introduction

During early military head-up display (HUD) development, it was found that pilots using HUDs could operate their aircraft with greater precision and accuracy than they could with conventional flight instrument systems.[1,2] This realization eventually led to the development of the first HUD systems intended specifically to aid the pilot during commercial landing operations. This was first accomplished by Sextant Avionique for the Dassault Mercure aircraft in 1975 and then by Sundstrand and Douglas Aircraft Company for the MD80 series aircraft in the late 1970s (see Figure 17.1).

In the early 1980s, Flight Dynamics developed a holographic optical system to display an inertially derived aircraft flight path with precision landing guidance, thus providing the first wide field-of-view (FOV) head-up guidance system (HGS). Subsequently, Alaska Airlines became the first airline to adopt this technology and perform routine fleet-wide manually flown CAT IIIa operations on B-727-100/200 aircraft using the flight dynamics system. Once low-visibility operations were successfully demonstrated using a HUD, regional airlines opted for this technology (see Figure 17.2), as an alternative to a fail passive autoland system, to help maintain their schedules when the weather fell below CAT II minimums, and to improve situational awareness.

By the end of the century, many airlines had installed HGSs, and thousands of pilots were fully trained in their use. HUD-equipped aircraft had logged more than 6,000,000 flight hours and completed over 30,000 low-visibility operations. HUDs are now well-established additions to aircraft cockpits, providing additional operational capabilities while enhancing situational awareness, resulting in improved aircraft safety. Aircraft manufacturers are now offering HUDs on almost all new flight decks in the air transport and regional and large business aircraft markets. Boeing's 787 Dreamliner was the first commercial aircraft to certify the standard flight deck with a HUD for each pilot in all phases of flight.

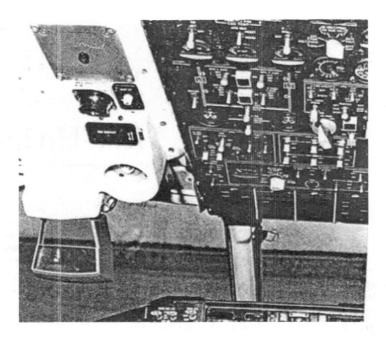

FIGURE 17.1 Early commercial HUD.[3]

FIGURE 17.2 Commercial manually flown CAT IIIa HUD installed in a B-737-800.

In 2009, the Flight Safety Foundation issued a special report[15] on the safety impact of adding Head-up Guidance System Technology (HGST) to a flight deck and concluded that "the HGST might have positively influenced 38% of the accidents overall" and also concluded that HGST would have prevented or positively influenced 69% of the takeoff and landing accidents analyzed. This special report was noticed by governments around the world including the Civil Aviation Authority of China, which has plans to require the installation of HUDs in aircraft flying in Chinese airspace.

17.2 HUD Fundamentals

HUD systems are comprised of two major subsystems: the pilot display unit(s) (PDU) and the HUD computer. The PDU is installed in the flight deck either in the glareshield or above the pilot's head. The PDU is attached to precisely aligned brackets that are part of the aircraft structure so that that the collimated HUD image seen by the pilot overlays the outside scene. The HUD requires a very bright image source that is usually monochrome green. In early systems, this was a cathode ray tube (CRT) with the symbols drawn using analog signals that move a bright spot around the CRT screen. The latest digital HUD designs use a monochrome active-matrix liquid-crystal display (AMLCD) with a very bright backlight made up of an array of green light-emitting diodes (LED). Attached to the HUD image source is an optical system that projects the light from the image source onto the HUD combiner in front of the pilot's eyes. The combiner glass has a special coating that only reflects the color of light emitted by the image source and allows all the other colors in the real-world scene to pass through, enabling the pilot to view the collimated outside world overlaid with the collimated HUD image. The PDU is precisely located in the cockpit so a pilot positioned at the cockpit design eye position (DEP) can view the HUD information accurately aligned with the outside world. This allows, for example, the computer-generated zero pitch line to overlay the real-world horizon when the aircraft is on the ground.

The cockpit DEP is defined as the optimum cockpit location that meets the requirements of 14 CFR 25.773[4] and 27.775.[5] From this location, the pilot can easily view all relevant head-down instruments and the outside world scene through the aircraft windshield while being able to access all required cockpit controls. The HUD "eyebox" always includes the cockpit DEP, allowing pilots to fly the aircraft using the HUD from the same physical location as an aircraft without a HUD.

The PDU is connected to the HUD computer that is installed in the electronics bay of the aircraft. The HUD computer can either be a stand-alone HUD computer unit or a HUD module that is part of an integrated modular avionics (IMA) cabinet system such as the common computing system on the Boeing 787 or the Pro Line Fusion system on the Bombardier Global Express. In IMA architectures, the HUD software function shares computing and interface resources with other avionics functions. The HUD computer receives flight information from aircraft sensors and systems, runs algorithms to verify and format the data, and then generates the characters and symbols that make up the HUD image. Some HUD computers are also capable of generating high-integrity guidance commands and cues for precision low-visibility takeoff, approach, landing (flare), and rollout. Some HUD computers can also add a video overlay to the symbology. The interface between the HUD computer and the PDU is usually a digital video fiber-optic link based on the Aeronautical Radio, Incorporated (ARINC) 818 standard.

17.2.1 Optical Configurations

The optics in HUD systems are used to "collimate" the HUD image so that essential flight parameters, navigational information, and guidance are superimposed on the collimated outside world scene. Each HUD design has a vertical and horizontal FOV that is the angular region over which the HUD image is visible to the pilot.

The four distinct FOV definitions used to describe the FOV characteristics of a HUD are illustrated in Figure 17.3 and summarized as follows:

1. *Total FOV (TFOV)*—The maximum angular extent over which symbology from the image source can be viewed by the pilot with *either eye* allowing vertical and horizontal head movement within the HUD eyebox.
2. *Instantaneous FOV (IFOV)*—The union of the two solid angles subtended at each eye by the clear apertures of the HUD optics from a fixed head position within the HUD eyebox. Thus, the instantaneous FOV is comprised of what the left eye sees plus what the right eye sees from a fixed head position within the HUD eyebox.

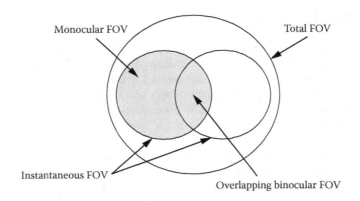

FIGURE 17.3   HUD FOV defined.

3. *Binocular overlapping FOV*—The intersection of the two solid angles subtended at each eye by the clear apertures of the HUD optics from a fixed head position within the HUD eyebox. The binocular overlapping FOV thus defines the maximum angular extent of the HUD display that is visible to both eyes simultaneously.

4. *Monocular FOV*—The solid angle subtended at the eye by the clear apertures of the HUD optics from a fixed eye position. Note that the monocular FOV size and shape may change as a function of eye position within the HUD eyebox.

The FOV characteristics are designed and optimized for a specific cockpit geometric configuration based on the intended function of the HUD. For example, if the HUD is intended to support landing of the aircraft, then the FOV will be centered on where the pilot will be looking during a normal approach. In some cases, the cockpit geometry may impact the maximum available FOV.

One of the most significant advances in HUD optical design in the last 30 years is the change from optical systems that collimate by refraction to systems that collimate by reflection or, in some cases, by diffraction. The move toward more complex reflective collimation systems has resulted in larger display FOVs that expand the usefulness of HUDs as full-time primary flight references. A more recent optical design is substrate-guided optics that use internal reflection within a special combiner to relay the collimated HUD image to the pilot's eyes in a very compact package.

17.2.1.1 Refractive Optical Systems

Figure 17.4 shows the optical configuration of a refractive HUD system. This configuration is similar to the basic HUD optical systems in use since the 1950s.[6] In this optical configuration, the image source is collimated by a combination of refractive lens elements designed to provide a highly accurate display over a moderate display FOV. Note that an internal mirror is used to fold the optical system to reduce the physical size of the packaging envelope of the HUD. Also shown in Figure 17.4 is the HUD combiner glass, a flat semitransparent plate designed to reflect approximately 25% of the collimated light from the CRT and transmit approximately 70% of the real-world luminance.

Note that the vertical instantaneous FOV can be increased by adding a second flat combiner glass, displaced vertically above and parallel with the first.

17.2.1.2 Reflective Optical Systems

In the late 1970s, HUD optical designers looked for ways to increase the HUD display total and instantaneous FOVs.[7,8] Figure 17.5 shows a typical overhead-mounted reflective HUD optical system designed specifically for a commercial transport cockpit.

The reflective optics can be thought of as two distinct optical subsystems. The first is a relay lens assembly designed to reimage and preaberrate the HUD image source to an intermediate aerial image,

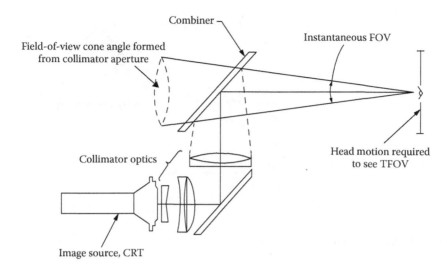

FIGURE 17.4 Refractive optical systems.

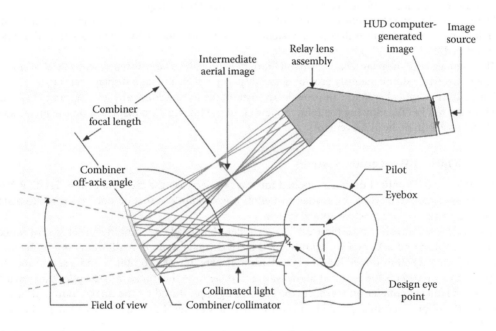

FIGURE 17.5 **(See color insert.)** Reflective optical systems (overhead mounted).

located at one focal length from the optically powered combiner/collimator element. If needed, the relay lens may include a prism element to "bend" the optical path over the pilot's head. This design allows a HUD to be installed in aircraft with restricted headroom.

The second optical subsystem is the combiner/collimator element that reimages and collimates the intermediate aerial image for viewing by the pilot. As in the refractive systems, the pilot's eyes focus at optical infinity, looking through the combiner to see the virtual image. The combiner is tilted off-axis with respect to the axial chief ray from the relay lens assembly to reflect the HUD image into the pilot's eyes.

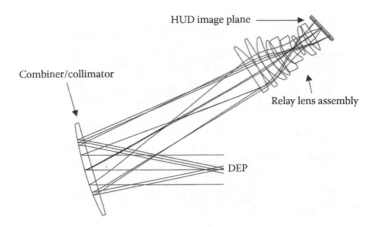

FIGURE 17.6   Reflective optical system raytrace.

The combiner off-axis angle, although required for image viewing reasons, significantly increases the optical aberrations within the system, which must be compensated in the relay lens and with image predistortion to ensure the pilot sees a well-correlated, accurate virtual HUD image.

Figure 17.6 illustrates the optical ray trace of a typical reflective HUD system showing the complexity of the relay lens assembly. (This is the optical system used on the first manually flown CAT IIIa HUD system ever certified.)[9]

The complexity of the relay lens, shown in Figure 17.6, provides a large instantaneous FOV over a fairly large eyebox, while simultaneously providing low display parallax and high display accuracy.

The reflective optical system can provide an instantaneous and binocular overlapping FOV that is equal to the total FOV, allowing the pilot to view all of the information displayed on the image source with each eye with no head movement.

17.2.1.3 Optical Waveguide Systems

In the 1990s, HUD optical designers looked for ways to decrease the size and cost of HUD PDUs so the benefits of the HUD would be available to pilots of small aircraft that do not have room to install a traditional refractive or reflective optical system.

The optical system shown in Figure 17.7 uses a flat plate of glass (combiner) that is positioned in front of the pilot and acts as a waveguide for the HUD image source. An input coupling bends the light from an LED-illuminated liquid-crystal microdisplay into the combiner and a semitransparent output coupler redirects the light out of the waveguide, sending the image into the pilot's forward field of view. The output coupler only reflects the color of light produced by the image source illuminator allowing the pilot to see an expanded and collimated view of the HUD imager overlaid on the outside world scene.

17.2.1.4 Optical System Comparison

Table 17.1 summarizes typical FOV performance characteristics for different optical systems.

All commercially certified HUD systems in airline operation today use reflective optical systems because of the superior display FOV characteristics. Refractive optics are used in glareshield-mounted HUDs on fighter aircraft where overhead-mounted optics are impractical. Waveguide optics is used in compact HUD designs for aircraft where there is not enough room for overhead-mounted optics and there is a willingness to accept lower optical performance for lower cost.

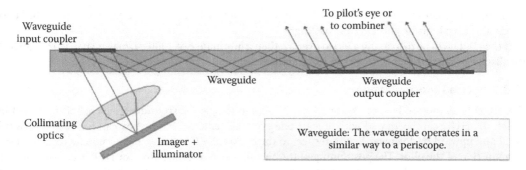

FIGURE 17.7 Waveguide optical system.

TABLE 17.1 Typical HUD Fields of View

	Refractive Optics (Dual Combiners)	Reflective Optics	Waveguide Optics
Total field of view	25°–30° diameter	30°V × 42° H	21° V × 30° H
Instantaneous FOV	16° V × 17.8° H	30°V × 42° H	21° V × 30° H
Overlapping binocular FOV	16° V × 6° H	30°V × 42° H	21° V × 15° H
Monocular FOV	16° V × 12° H	30°V × 42° H	21° V × 22.5° H

17.2.2 Significant Optical Performance Characteristics

This section summarizes other important optical characteristics associated with conformal HUD systems. It is clear that the HUD FOV, luminance, and display line width characteristics must meet basic performance requirements.[10] However, optical system complexity and cost are driven by HUD eyebox size, combiner off-axis angle, display accuracy requirements, and optical parallax error compensation. Without a well-corrected optical system, conformal symbology will not properly overlay the outside world view and symbology will not remain fixed with respect to the real-world view as the head is moved around within the HUD eyebox.

17.2.2.1 Display Luminance and Contrast Ratio

The HUD should be capable of providing a usable display under all foreseeable ambient lighting conditions, ranging from a sunlit cloud with a luminance of 10,000 foot-lamberts (ftL; or 34,000 cd/m²) to a night approach into a sparsely lit runway. HUD contrast ratio is a measure of the relative luminance of the display with respect to the real-world background and is defined as follows:

$$\text{HUD contrast ratio} = \frac{\text{Display luminance} + \text{Real-world luminance}}{\text{Real-world luminance}}$$

The display luminance is the photopically weighted image source light output that reaches the pilot's eyes. Real-world luminance is the luminance of the real world as seen through the HUD combiner. (By convention, the transmission of the aircraft windshield is left out of the real-world luminance calculation.)

It is generally agreed that a HUD contrast ratio (CR) of >1.2 is the minimum needed for display viewing and that a CR of >1.3 is preferable. A HUD contrast ratio of 1.3 against a 10,000 ftL cloud seen through a HUD combiner with an 80% photopic transmission requires a display luminance at the pilot's eye of 2,400 ftL, a luminance about 10 times higher than most head-down displays. This minimum luminance translates to an image source brightness of >9000 ftL, a luminance easily met with the latest AMLCD with a high-brightness LED backlight.

If the HUD will be displaying a video overlay, then a luminance of 4000 ftL may be needed to support the display of more than one or two video grayshades.

17.2.2.2 Head Motion Eyebox

The HUD head motion box, or "eyebox," is a 3D region in space surrounding the cockpit DEP within which the HUD can be viewed with at least one eye. The center of the eyebox can be located forward or aft and upward or downward, with respect to the cockpit DEP to better accommodate the actual sitting position of the pilot. The positioning of the cockpit eye reference point[11] or DEP is dependent on a number of ergonomically related cockpit issues such as head-down display visibility, the over-the-nose down-look angle, and the physical location of various controls such as the control yoke and the landing gear handle.

The HUD eyebox should be as large as possible to allow maximum head motion without losing display information. The relay lens exit aperture, the spacing between the relay lens and combiner, the combiner to DEP distance, and the combiner focal length all impact the eyebox size. Modern HUD eyebox dimensions are typically 5.2 in. lateral, 3.0 in. vertical, and 6.0 in. longitudinal.

In all HUDs, the monocular instantaneous FOV is reduced (or vignettes) with lateral or vertical eye displacement, particularly near the edge of the eyebox. Establishing a minimum monocular FOV from the edge of the eyebox thus ensures that even when the pilot's head is decentered so that one eye is at the edge of the eyebox, useful display FOV is still available. A 10° horizontal by 10° vertical monocular FOV generally can be used to define the eyebox limits. In reflective HUDs, relatively small head movements (>1.5 in. laterally) will cause one eye to be outside of the eyebox and see no display; however, the other eye will see the total FOV, so no information is lost to the pilot.

17.2.2.3 HUD Display Accuracy

Display accuracy is a measure of how precisely the projected HUD image overlays the real-world view seen through the combiner and windshield from any eye position within the eyebox. Display accuracy is a monocular measurement and, for a fixed display location, is numerically equal to the angular difference between a HUD-projected symbol element and the corresponding real-world feature as seen through the combiner and windshield. The total HUD system display accuracy error budget includes optical errors, overhead to combiner misalignment errors, windshield variations, environmental conditions (including temperature), assembly tolerances, and installation errors. Optical errors are both head position and field angle dependent.

The following display accuracy values are achievable in commercial HUDs when all the error sources are accounted for:

Boresight	± 3.0 milliradians (mrad)
Total display accuracy	± 7.0 milliradians (mrad)

The boresight direction is used as the calibration direction for zeroing all electronic errors. Boresight errors include the mechanical installation of the HUD hardpoints to the airframe, electronic drift due to thermal variations, and manufacturing tolerances for positioning the combiner element. Refractive HUDs with integrated combiners (i.e., F-16) are capable of achieving display accuracies of about half of the errors identified previously.

17.2.2.4 HUD Parallax Errors

Within the binocular overlapping portion of the FOV, the left and right eyes view the same location on the HUD image source. These slight angular errors between what the two eyes see are called binocular parallax errors or collimation errors. The binocular parallax error for a fixed field point within the total

FOV is the angular difference in rays entering two eyes separated horizontally by the interpupillary distance (assumed to be 2.5 in.). If the projected virtual display image were perfectly collimated at infinity from all eyebox positions, the two ray directions would be identical, and the parallax errors would be zero. Parallax errors consist of both horizontal and vertical components.

Parallax errors in refractive HUDs are generally less than about 1.0 mrad due to the rotational symmetry of the optics and because of the relatively small overlapping binocular FOV.

17.2.2.5 Display Line Width

The HUD line width is the angular dimension of displayed symbology elements. Acceptable HUD line widths are between 0.7 and 1.2 mrad when measured at the 50% intensity points. The displayed line width is dependent on the effective focal length of the optical system and the physical line width on the image source. A typical wide FOV reflective HUD optical system with a focal length of 5 in. will provide an antialiased display line width of about 1.2 mrad using 3–4 pixels on the image source. Any HUD optical system aberrations will adversely affect apparent display line width. These aberrations include uncorrected chromatic aberrations (lateral color) and residual uncompensated coma and astigmatism. Minimizing these optical errors during the optimization of the HUD relay lens design will also help meet the parallax error requirements.

Table 17.2 summarizes the optical performance characteristics of a commercial wide-angle reflective HUD optical system.

TABLE 17.2   HUD Optical System Summary (Reflective HUD)

1.	Combiner design	Wide FOV wavelength selective, stowable, inertial breakaway (HIC[a] compliant)
2.	DEP to combiner distance	9.5″–13.5″ (cockpit geometry dependent)
3.	Display FOVs	
	Total display FOV	24°–28° vertical × 30°–42° horizontal
	Instantaneous FOV	24°–28° vertical × 30°–42° horizontal
	Overlapping binocular FOV	22°–24° vertical × 24°–40° horizontal
4.	Head motion box or eyebox	Typical dimensions
	Horizontal	4.7″–5.4″
	Vertical	2.5″–3.0″
	Depth (fore/aft)	4.0″–7.0″
5.	Head motion to view TFOV	None
6.	Display parallax errors (typical)	
	Convergence	95% of data points <2.5 mrad
	Divergence	95% of data points <1.0 mrad
	Dipvergence	93% of data points <1.5 mrad
7.	Display accuracy (2 sigma)	
	Boresight	<2.5–4.0 mrad
	TFOV	<5.0–9.0 mrad
8.	Combiner transmission and coloration	78%–82% photopic (day-adapted eye)
		84% scotopic (night-adapted eye)
		<0.03 color shift u′v′ coordinates
9.	Display luminance and contrast ratio	
	Symbology and video	Up to 4000 foot-lambert (ftL)
	Display contrast ratio	1.2–1.4:1 (10,000 ftL background)
10.	Display line width	1.2 mrads
11.	Secondary display image intensity	<0.5% of the primary image from eyebox

[a] Head injury criteria.

17.2.3 Video Display on the HUD

Early HUDs only had to display the symbols and text of the flight symbology but customers are now requesting that the HUD also display the conformal video from new situational awareness tools such as Enhanced Vision System (EVS) and Synthetic Vision System (SVS). Displaying symbology and video on an analog CRT HUD is very challenging[16] because the CRT uses beam steering "stroke writing" to draw each symbol on screen in turn. To draw video, the beam has to scan repeatedly across the CRT varying the intensity of the beam to form the image. Drawing both the symbology and then the video limits maximum brightness and increases the heat generated in the projection unit. The market demand for video overlays on the HUD accelerated the transition from CRT to a high-resolution, digital image source where the entire HUD image is refreshed every frame.

17.2.3.1 Video Processing and Alignment

Video processing is used to optimize the received video image for display to the pilot on the HUD. This may include any or all of the following:

- Converting from a raster scan to a digital image to eliminate the overscan and flyback
- Converting analog video to a monochrome digital pixel image
- Applying image improvement algorithms to enhance image contrast and reduce image noise
- Upscaling from the raw image resolution to the resolution of the HUD image source

The HUD includes a video image alignment capability that uses stored settings to shift the position of the video on the HUD image source to ensure that information in the video is accurately aligned with the outside world. For example, a mechanic or pilot can display an EVS camera image on the HUD and notice that a distant object displayed in the HUD video does not overlay the object as seen through its combiner. The operator can then use control panel buttons to shift the image slightly so that it directly overlays the real world. These offsets can then be saved for use in flight.

17.2.3.2 Displaying Video Grayshades

When the HUD is in normal use, the pilot adjusts the luminance of the HUD symbology to a comfortable level that is visible against the outside scene as viewed through the combiner. An automatic luminance algorithm then modulates the luminance, based on a forward-looking light sensor, to maintain the selected contrast ratio. The pilot normally selects a low-contrast ratio so the symbology is only just visible. This low contrast between the symbology and the outside scene allows the pilot to more easily maintain a balance between attention to the outside visual scene and the visual scan of the flight information. The fact that the imagery on the HUD is always viewed against a background of the outside scene limits the range of luminance displayed on the HUD that can be detected by pilot. The low limit is the minimum luminance that is noticed by the pilot against the background. The high limit in low background luminance conditions is the HUD luminance above which the pilot's night vision sensitivity will be compromised and the outside scene becomes difficult to see behind the bright HUD image. The high limit in high background luminance conditions is the maximum luminance of the HUD image source. This limited range of image luminance also limits the number of grayshades that can be displayed in the video image that is strongly dependent on ambient luminance and maximum HUD luminance capability.

When a video display capability is added to the HUD, controls are added that allow the pilot to adjust the overall brightness of the video and adjust the video contrast to enhance detail in the image. There is a danger that the pilot will be distracted by constantly adjusting the controls to improve the image, and it is recommended that the video contrast is controlled by an automatic algorithm that maximizes the number of grayshades at the selected luminance level.

17.2.3.3 Maintaining Symbology Contrast

The primary use of the HUD is the display of flight information symbology. The display of video on the HUD from an EVS or SVS can interfere with that primary use if the video obscures or reduces the

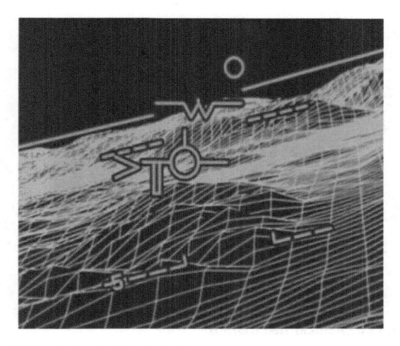

FIGURE 17.8 HUD symbology detail showing haloing.

readability of the symbology. The HUD designer must ensure that the symbology is clear and readable regardless of the luminance settings selected by the pilot and the content of the video image. This challenge is particularly difficult because the range of grayshades available for display of the video is limited, and there will be a tendency for the pilot to raise the luminance of the video to compensate. The obscuration occurs when symbology overlays a bright area of video and there is insufficient contrast between the symbology and the video, making it difficult to read.

Solutions to this challenge include the following:

- Artificially limiting the luminance of the video to ensure it is always less bright than the symbology.
- Raising the luminance of the symbology above the level selected by the pilot to be brighter than the video in areas when the video is bright to maintain the contrast ratio. This is an automatic feature of CRT HUDs where the symbology and video are "drawn" separately because the area of phosphor is hit twice by the electron beam increasing the luminance of that spot.
- Adding a symbology mask to the video to ensure that the area around the symbols is clear. This technique—called haloing—maintains the contrast between the symbology and the background but is only really practical on HUDs with a digital image source (Figure 17.8).

17.2.3.4 Video Interface Standards

Early EVS sensors used the RS-170A standard for transferring video to the HUD. This standard, used for television transmission before the transition to digital TV in 2009, transmits an analog waveform on coaxial cable. In 2007, ARINC issued ARINC 818—Avionics Digital Video Bus (ADVB), a standard that defined a point-to-point, serial protocol for transmission of video, audio, and data using the Fibre Channel protocol. The video is normally transmitted over a fiber-optic cable that ensures a high-resolution, low-noise signal is received by the HUD.

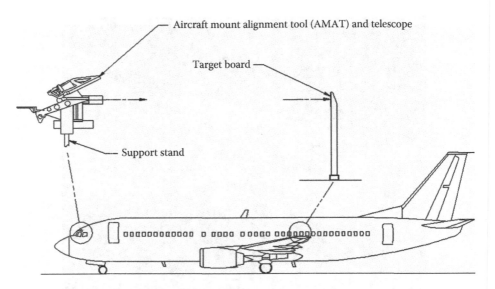

FIGURE 17.9   Boresighting the HUD hardpoints.

17.2.4 HUD Mechanical Installation

The intent of the HUD is to display symbolic information that overlays the real world as seen by the pilot. To accomplish this, the HUD PDU must be very accurately aligned to the pitch, roll, and heading axis of the aircraft. For this reason, the angular relationship of the HUD PDU with respect to the cockpit coordinates is crucial. The process of installing and aligning the HUD attachment bushings or hardpoints into the aircraft is referred to as "boresighting." Boresighting is usually only done once when the HUD is first installed on the aircraft. (Although the alignment of the HUD may be checked occasionally, once installed, the hardpoints are permanent and rarely need adjustment.)

Some reflective HUDs utilize mating bushings for the PDU hardware that are installed directly to the aircraft structure. Once the bushings are aligned and boresighted to the aircraft axis, they are fixed in place using a structural epoxy. Figure 17.9 illustrates a typical installation method for HUD boresighting. In this case, the longitudinal axis of the aircraft is used as the boresight reference direction. Using special tooling, the overhead unit (OHU) and combiner bushings are aligned with a precisely positioned target board located near the rear of the fuselage. This boresighting method does not require the aircraft to be jacked and leveled.

Other HUD installations use a tray that attaches to the aircraft structure and provides a stable interface to the HUD LRUs. The PDU tray must still be installed and boresighted to the aircraft axis.

17.2.5 HUD System Hardware Components

A typical commercial reflective HUD system includes three principal line replaceable units (LRUs). HUD LRUs can be interchanged on the flight deck without requiring any alignment or recalibration. The cockpit-mounted LRUs include the OHU and combiner (the PDU). The HUD computer is located in the electronics bay or another convenient location. A HUD interconnect diagram is shown in Figure 17.10.

17.2.5.1 HUD Overhead Unit

The OHU, positioned directly above the pilot's head, interfaces with the HUD computer receiving either analog signals (to drive a CRT) or a digital video stream. The OHU electronics converts the

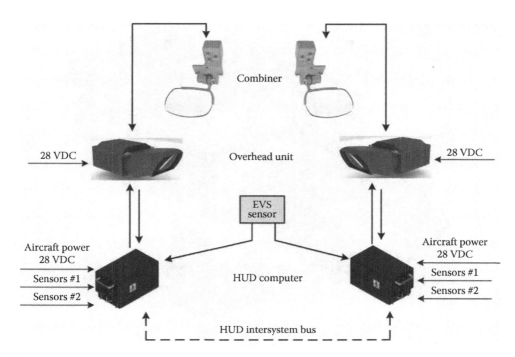

FIGURE 17.10   Typical HUD interconnect diagram.

video data to an image on a high-brightness AMLCD, and the control data embedded in the video is used to control the brightness of the LED backlight. The AMLCD is optically coupled to the relay lens assembly that reimages the HUD image to an intermediate aerial image one focal length away from the combiner LRU, as illustrated in the optical schematic in Figure 17.5. The combiner reimages the intermediate image at optical infinity for viewing by the pilot. The OHU includes all of the electronics necessary to drive the image source and monitor the built-in-test (BIT) status of the LRU that is then transmitted to the HUD computer using an ARINC 429 data bus. The OHU also provides the electronic interfaces to the combiner LRU.

A typical OHU is illustrated in Figure 17.11. This LRU contains all electronic circuitry required to drive the image source and all BIT-related functions. The following are the major OHU subsystems:

- Relay lens assembly
- Desiccant assembly (prevents condensation within the relay lens)
- Monochrome AMLCD
- LED backlight with heatsink
- ARINC 818 interface
- Backlight control and AMLCD driver board
- Low-voltage power supplies and energy storage
- BIT and monitoring circuits
- OHU chassis

In some HUD systems, the PDU may provide data back to the HUD computer as part of the "wrap-around" critical symbol monitor feature.[12] Real-time monitoring of certain critical symbol elements (i.e., horizon line) provides the high-integrity levels required for certifying a HUD as a primary flight reference display. Other monitored critical data on the HUD may include instrument landing system (ILS) data, airspeed, flight path vector, and low-visibility guidance symbology.

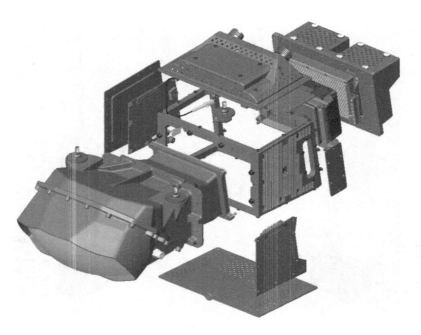

FIGURE 17.11   HUD OHU chassis (WFOV reflective optics).

17.2.5.2 HUD Combiner

The combiner is an optical–mechanical LRU consisting of a precision support structure for the wavelength-selective semireflective combiner element and a mechanism allowing the combiner to be stowed and to break away. The combiner LRU attaches to a precision prealigned mating interface permanently mounted to the aircraft structure. The support structure positions the combiner glass with respect to the cockpit DEP and the OHU. The combiner mechanism allows the glass to be stowed upward when not in use and to break away during a rapid aircraft deceleration, thus meeting the cockpit "head injury criteria" (HIC) requirements.[13] The combiner locks into both the stowed and breakaway positions and requires positive actions by the pilot to return it to the deployed position. Many HUD combiner assemblies include a built-in alignment detector that monitors the glass position in real time. Figure 17.12 shows a commercial HUD PDU installed in aircraft structure. The combiner usually includes the HUD optical controls (brightness and contrast).

17.2.5.3 HUD Computer

The HUD computer interfaces with the aircraft sensors and systems, performs data conversions, validates data, computes command guidance (if applicable), positions and formats symbols, generates the display list, and converts the display list into a HUD image for display by the PDU. In some commercial HUD systems, the HUD computer performs all computations associated with low-visibility takeoff, approach, landing, and rollout guidance and all safety-related performance and failure monitoring. Because of the critical functions performed by these systems, the displayed data must meet the highest integrity requirements. The HUD computer architecture and software are designed specifically to meet these requirements.

   One of the key safety requirements for a full flight regime HUD is that the display of unannunciated, hazardously misleading attitude on the HUD must be improbable and that the display of unannunciated hazardously misleading low-visibility guidance must be extremely improbable. An analysis of these requirements leads to the system architecture shown in Figure 17.13.

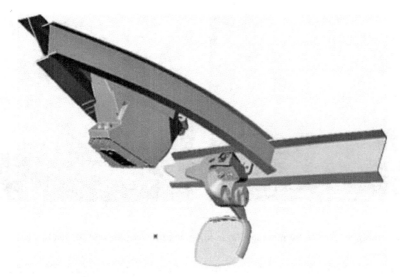

FIGURE 17.12 HUD Overhead Unit and Combiner installed in aircraft structure.

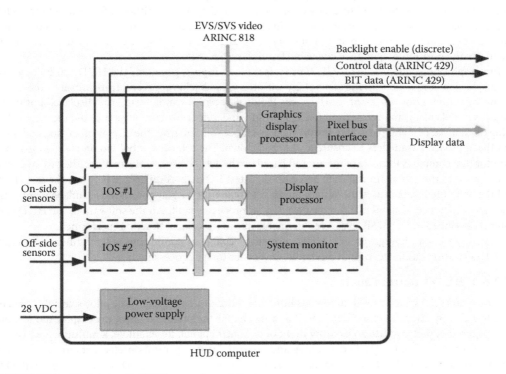

FIGURE 17.13 High-integrity HUD computer architecture.

In this architecture, primary data are brought into the HUD computer via dual independent input/ output (I/O) subsystems from the primary sensors and systems on the aircraft. The avionics interface for a specific HUD computer depends on the avionics suite and can include a combination of any of the following interfaces: ARINC 429, ARINC 629, ASCB, AFDX, or MIL-STD-1553B. Older aircraft will often include analog or synchro interfaces to sensors. The I/O subsystem also includes the interfaces to the OHU and combiner and will often include outputs to the flight data recorder and central maintenance computer.

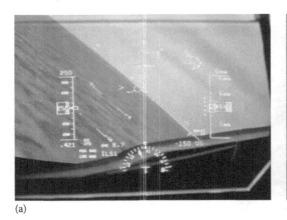 

(a)   (b)

FIGURE 17.14   Commercial HUD symbology: (a) in flight, holding altitude and (b) just before touchdown.

Figure 17.14 are photographs of a typical commercial HUD symbology set. The aircraft sensor data needed to generate these displays are listed in Table 17.3. In general, two sources of the critical data are required to meet the safety and integrity requirements for displaying attitude information to the pilot.

The display processor (DP) verifies the validity of the data, compares equivalent data from the redundant sensors, computes the display symbol locations, and generates a symbol display list. The DP also calculates the position of guidance cue using landing system data.

The graphics display processor (GDP) processes video inputs, generates the HUD symbology from the display list, and combines the symbology with the video background and transmits it to the OHU.

The system monitor processor (SM) verifies the display path by monitoring the displayed position of critical symbols using an inverse function algorithm,[12] independently computes the guidance algorithms using off-side data for comparison to the primary guidance solution from the display processor, and monitors the approach parameters to ensure a safe touchdown. The critical symbol monitor is a wraparound monitor that computes the state of the aircraft based on the actual display information. The displayed state is compared to the actual aircraft state based on the latest I/O data. A difference between the actual state and the computed state causes the system monitor to blank the display through two independent paths, since any difference in states could indicate a display processor fault. All the software in the HUD computer is developed to DO-178B level A requirements due to the critical functions performed.

Figure 17.15 is a photograph of a HUD computer capable of computing takeoff guidance, manual CAT IIIa landing guidance, rollout guidance, and raster image processing.

17.2.5.4 HUD Control Panel

Commercial HUD systems used for low-visibility operations often require some pilot-selectable data not available on any aircraft system bus as well as a means for the pilot to control the display mode. Some HUD operators prefer to use an existing flight deck control panel, for example, a multipurpose control display unit (MCDU), for HUD data entry and control. Other operators prefer a stand-alone control panel, dedicated to the HUD function. Figure 17.16 shows a stand-alone HUD control panel certified for use in CAT IIIa HUD systems.

17.2.6 Aspects of HUD Certification

Certification requirements for a HUD system depend on the functions performed. As the role of HUDs have expanded from CAT IIIa landing devices (including takeoff and rollout guidance) to full flight regime primary flight references, the certification requirements have become more complex. It is beyond the scope of this chapter to describe all the certification issues and requirements for a primary

TABLE 17.3   Sensor Data Required for Full Flight Regime Operation

Input Data	Data Source
Attitude	Pitch and roll angles—two independent sources
Airspeed	Calibrated airspeed
	Low-speed awareness (e.g., Vstall)
	High-speed awareness (e.g., Vmo)
Altitude	Barometric altitude (pressure altitude corrected with altimeter setting)
	Radio altitude
Vertical speed	Vertical speed (inertial if available, otherwise raw air data)
Slip/skid	Lateral acceleration
Heading	Magnetic heading
	True heading or other heading (if selectable)
	Heading source selection (if other than magnetic selectable)
Navigation	Selected course
	VHF omni-directional radio (VOR) bearing/deviation
	Distance measuring equipment (DME) distance
	Localizer deviation
	Glideslope deviation
	Marker beacons
	Bearings/deviations/distances for any other desired nav signals (e.g., ADF, TACAN, RNAV/FMS)
Reference	Selected airspeed
Information	Selected altitude
	Selected heading
	Other reference speed information (e.g., V_1, V_R, Vapch)
	Other reference altitude information (e.g., landing minimums [decision height (DH)/MDA], altimeter setting)
Flight path	Pitch angle
	Roll angle
	Heading (magnetic or true, same as track)
	Ground speed (inertial or equivalent)
	Track angle (magnetic or true, same as heading)
	Vertical speed (inertial or equivalent)
	Pitch rate, yaw rate
Flight path acceleration	Longitudinal acceleration
	Lateral acceleration
	Normal acceleration
	Pitch angle
	Roll angle
	Heading (magnetic or true, same as track)
	Ground speed (inertial or equivalent)
	Track angle (magnetic or true, same as heading)
	Vertical speed (inertial or equivalent)
Automatic flight control system	Flight director guidance commands
	Autopilot/flight director modes
	Autothrottle modes
Miscellaneous	Wind speed
	Wind direction (and appropriate heading reference)
	Mach
	Windshear warning(s)
	Ground proximity warning(s)
	Traffic collision avoidance system (TCAS) resolution advisory information

FIGURE 17.15   High-integrity HUD computer.

FIGURE 17.16   HUD control and data entry panel.

flight display HUD; however, the basic requirements are not significantly different from primary flight display (PFD) head-down display certification requirements.

The Federal Aviation Administration (FAA) has documented the requirements for systems providing guidance in low-visibility conditions in Advisory Circular AC 120-28, "Criteria for Approval of Category III Weather Minima for Takeoff, Landing, and Rollout." The certification of the landing guidance aspects of the HUD is fundamentally different from automatic landing systems because the human pilot is in the active control loop during the beam tracking and flare. The following summarizes the unique aspects of the certification process for a manual category III system:

1. *Control law development*—The guidance control laws are developed and optimized based on the pilot's ability to react and respond. These "pilot-centered" control laws must be tailored for a pilot of average ability. The monitors must be designed and tuned to detect approaches that will be outside the footprint requirement, yet they cannot cause a go-around rate greater than about 4%.

2. *Motion-based simulator campaign*—Historically, approximately 1400 manned approaches in an approved motion-based simulator, with at least 12 certification authority pilots, are required for performance verification for a FAA/European Aviation Safety Agency (EASA) certification. The Monte Carlo test case ensemble is designed to verify the system performance throughout the flight envelope expected in field operation. The full range of environment conditions must be sampled

(head winds, cross winds, tail winds, turbulence, etc.) along with variations in the airfield conditions (sloping runways, ILS beam offsets, beam bends, etc.). Finally, the sensor data used by the HUD must be varied according to the manufacturer's specified performance tolerances. Failure cases must also be simulated. Time history data for each approach, landing, and rollout are recorded to perform the required data reduction analysis. A detailed statistical analysis is required to demonstrate, among other characteristics, the longitudinal, lateral, and vertical rate touchdown footprint. Finally, the analysis must project out the landing footprint to a one-in-a-million (10-6) probability.

3. *Aircraft flight test*—Following a successful simulator performance verification campaign, the HUD must be demonstrated in actual flight trials on a fully equipped aircraft. As in the simulator case, representative head winds, cross winds, and tail winds must be sampled for certification. Failure conditions are also run to demonstrate system performance and functionality. This methodology has been used to certify HUD systems providing manual guidance for takeoff, landing, and rollout on a variety of different aircraft types.

17.3 Applications and Examples

This section describes how the HUD is used on a typical aircraft using typical symbology sets that are displayed to a pilot in specific phases of flight. The symbology examples used in this section are taken from a Rockwell Collins HGS® installed on an in-service aircraft.

In addition to symbology, this section also discusses the pilot-in-the-loop optimized guidance algorithms that are provided as part of an HGS. Another feature of some HUDs is the display of video images on the HUD and the uses of this feature—where the HUD is only a display device—are discussed.

17.3.1 Symbology Sets and Modes

To optimize the presentation of information, the HUD has different symbology sets that present only the information needed by the pilot in that phase of flight. For example, the aircraft pitch information is not important when the aircraft is on the ground.

These symbology sets either are selected as modes by the pilot or are displayed automatically when a certain condition is detected.

17.3.1.1 Primary Mode

The HGS primary or "full symbology" mode can be used during all phases of flight from takeoff to landing. This mode supports low-visibility takeoff operations, all en route operations, and approaches to CAT I or II minimums using the flight director guidance. The HGS primary mode display is very similar to the head-down PFD to ease the pilot's transition from head-down instruments to head-up symbology. Figure 17.17 shows a typical in-flight primary mode display symbols.

17.3.1.2 Approach Mode

The HGS approach or "decluttered" mode shown in Figure 17.18 supports approach and landing operations. The display has been decluttered to maximize visibility by removing the altitude and airspeed tape data and replacing them with digital values. The horizontal situation indicator (HSI) is also removed, with ILS raw data (localizer and glideslope deviation) now being displayed near the center of the display.

The HUD displays the flight director guidance as a circular symbol called a guidance cue. The pilot follows the guidance by maneuvering the aircraft so the flight path symbol overlays the guidance cue as shown in Figure 17.19(a)

On some aircraft, the HGS calculates the position of the guidance cue to provide pilot-in-the-loop guidance for low-visibility operations (see Section 17.3.3).

The speed error tape (symbol on left wing of flight path) shows the difference between the selected airspeed and the current airspeed. The flight path acceleration caret shows current acceleration state

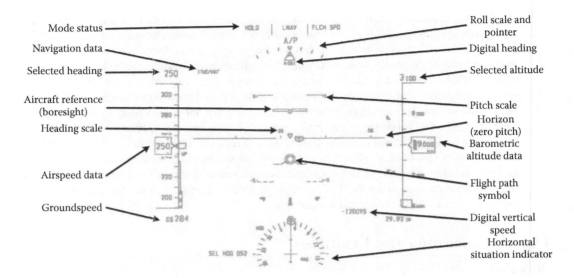

FIGURE 17.17   **(See color insert.)** HGS primary mode symbology: in-flight.

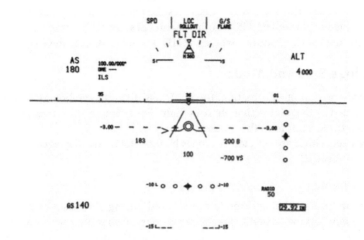

FIGURE 17.18   HGS approach mode symbology.

with a position below the left wing of the flight path indicating a deceleration. The aircraft is on-speed and stable when there is no speed error displayed and the acceleration caret is level with the left wing. If a speed error is displayed, then the pilot adjusts thrust (or pitch) so the caret is an equal distance on the opposite side of the wing, continuing to adjust thrust as the speed error is reduced. Figure 17.19 shows some examples of this energy management feature.

Following touchdown, the display changes to a ground symbology mode that removes unnecessary symbology to assist with the landing rollout. The guidance cue helps the pilot stay on centerline while the aircraft is decelerated to exit the runway.

17.3.1.3 Unusual Attitude

The HGS unusual attitude (UA) display is designed to aid the pilot in the recognition and recovery from UA situations. The UA symbology is automatically activated whenever the aircraft exceeds normal operational roll or pitch values and deactivated once the aircraft is restored to controlled flight. When the

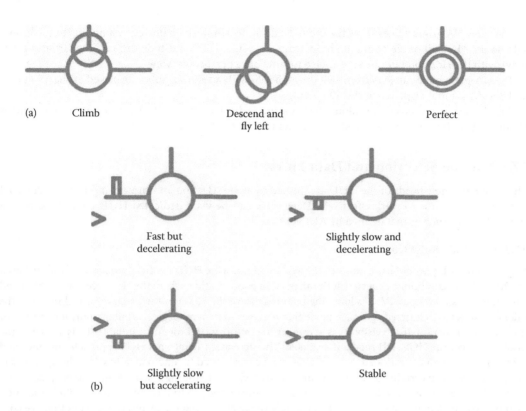

FIGURE 17.19 HGS cues for (a) guidance, and (b) energy management.

UA symbology is deactivated, the HGS returns to displaying the symbology for the selected operational mode. When activated, the UA display replaces the selected operational mode symbology.

The UA symbology includes a large circle (UA attitude display outline) centered on the combiner (see Figure 17.20). The circle is intended to display the UA attitude symbology in a manner similar to an attitude direction indicator (ADI) "ball." The UA horizon line represents zero degrees pitch attitude and is parallel to the actual horizon. The UA horizon line always remains within the outline to provide a sufficient sky/ground indication and always shows the closest direction to and the roll orientation of the actual horizon. The aircraft reference symbol is displayed on top of a portion of the UA horizon line and UA ground lines whenever the symbols coincide.

FIGURE 17.20 Unusual attitude symbology display.

The three UA ground lines show the ground side of the UA horizon line corresponding to the brown side on an ADI ball or electronic flight instrument system (EFIS) attitude display. The Ground Lines move with the Horizon Line and are angled to simulate a perspective view.

The UA pitch scale range is from −90 through +90 with a zenith symbol displayed at the +90 point and a nadir symbol displayed at the −90 point.

The UA roll scale is positioned along the UA attitude display outline. The UA roll scale pointer rotates about the UA aircraft reference symbol to always point straight up in the Earth frame.

17.3.2 Mode Selection and Data Entry

The pilot data entry needs of the HUD are limited to mode selection and runway input for the internal guidance algorithms to work effectively. Data entry can be via a dedicated HUD control panel or a shared MCDU, such as that defined by ARINC 739.

17.3.2.1 Mode Selection

On some aircraft, the pilot has a number of ways to configure the HUD for an approach and landing based on the visibility conditions expected at the airport. In good weather, where the cloud "ceiling" is high and the runway visual range (RVR) is long, the pilot may leave the HUD in the primary or full mode or the pilot may select a decluttered approach mode that removes some symbol groups but has only flight director guidance information. If the visibility is at or near the category III limit and a suitable ILS is available, the pilot will select the HGS AIII mode, if available. This activates the internal HGS approach guidance and monitoring. To reduce workload, some HUDs can be configured to automatically select the appropriate landing mode when certain conditions are met, such as when the landing system deviations become active.

Another mode that is available for selection, but only on the ground, is the test mode where the pilot, or more usually a maintenance person, can verify the health of the HUD and the sensors that are connected to the system.

17.3.2.2 Data Entry

To make use of the HUD-based guidance, the pilot must enter the following information:

- Runway elevation—the altitude of the runway threshold
- Runway length—official length of the runway in feet or meters
- Reference glideslope—the published descent angle to the runway, for example, 3°

On some aircraft, these data may be sent from the flight management system (FMS) and confirmed by the pilot.

17.3.3 HUD Guidance

On some aircraft, the HUD can provide a head-up guidance capability. An HGS provides pilot-in-the-loop, low-visibility takeoff and landing capability that is more cost effective than that provided by an autoland system.

HGS computes guidance to touchdown using deviations from the ILS or global navigation satellite system (GNSS) landing system (GLS) to keep the pilot at the center of the optimum landing path. The method for guiding the pilot is to drive the flight director guidance cue, with the internal guidance algorithm positioning the guidance cue horizontally and vertically as shown in Section 17.3.1.2. The movement of the guidance cue is optimized for pilot-in-the-loop flying. This optimization includes

- Limiting the movement of the cue to rates that are achievable by a normal pilot
- Anticipating the natural delay between the movement of the cue and reaction of the pilot/aircraft
- Filtering out short-term cue movements that may be seen in turbulent air

In addition to approach guidance where the goal is to keep the aircraft in the center of the approach path beam, guidance is also provided for other phases of the approach and landing.

During the flare phase—a pitch-up maneuver prior to touchdown—the guidance cue emulates the normal rate and magnitude of pullback that the pilot would use during a visual landing. During the rollout phase—where the goal is to guide the aircraft down the centerline of the runway—the pilot is given smooth horizontal commands that are easy to follow.

All these algorithms have to work safely for all normal wind and turbulence conditions. Because accurately following the guidance is critical to the safety of the aircraft, the algorithms include monitors to ensure that the information is not misleading and monitors to ensure that the pilot is following the commands. If the system detects the pilot is significantly deviating from the path or speed target, the system will display an approach warning message to both pilots that requires the pilot to abort the landing.

The pilot can also select AIII mode prior to takeoff and the HGS will provide guidance to runway centerline and initial lift off.

17.3.4 HUD Annunciations

An important element of any system is the annunciations that inform or alert the pilots to problems that require their action. In a well-managed flight deck, the role of each of the pilots is designed to be complementary. The pilot flying (PF) is responsible for control of the aircraft. The pilot not flying (PNF) is responsible for navigation and communication as well as monitoring the performance of the PF.

All the status information needed to safely fly the aircraft is displayed on the HGS for the pilot including the following:

- Mode status—Modes of the HGS guidance or the guidance source.
- Cautions—Approaching operating limitations or loss of a sensor.
- Warnings—Loss of a critical sensor requiring immediate action.
- System failure—HGS has failed and the pilot should not use the system.

Because of the optical technology used in a HUD, the PNF cannot see the annunciations displayed to the PF. To support PNF monitoring, the HGS outputs some or all of these annunciations either to a flight deck central warning system or to a dedicated annunciator panel in front of the other pilot.

17.3.5 HUD Vision Systems

For a HUD that has video display capability, there are several available or emerging vision systems that provide a conformal representation of the view ahead of the aircraft. This conformal view is an enhancement or replacement for a clear weather daylight view and gives the pilot the equivalent of a visual flight rules (VFR) condition. Two types of HUD vision systems have been certified for use, each with different characteristics and advantages.

17.3.5.1 Enhanced Vision

The FAA describes an Enhanced Flight Vision System as the conformal display of video from an EVS sensor on a HUD displaying flight path information. The EVS sensor uses one or more detectors of infrared and visible light to enhance the pilot's natural vision view ahead of the aircraft (see Figure 17.21). The FAA allows pilots who can see the runway environment with an EFVS to descend below minimum descent altitude (MDA) and continue the approach until the pilot can see the runway with natural vision or until the aircraft descends to decision height of 100 ft height above terrain.

The EVS display is also helpful during taxi and provides improved situational awareness during takeoff.

FIGURE 17.21   (See color insert.) HUD symbology with enhanced vision video overlay.

The EVS sensor does not see through all environmental conditions and may not always be able to display a useful image on the HUD.

The interface between the EVS sensor and the HUD can be an analog video format (i.e., RS-170A), or some sensors can provide digital video using the ARINC 818 standard.

Sensor technologies that are candidates for EVS include

- Forward-looking infrared, either cooled (InSb) or uncooled (InGaAs or microbolometer)
- Millimeter wave radar (mechanical or electronic scan)
- Millimeter wave radiometers (passive camera)
- Ultraviolet sensors

Although the concept of interfacing a sensor with a HUD to achieve additional operational credit is straightforward, there are a number of technical and certification issues that must be overcome including pilot workload, combiner see-through with a video image, sensor boresighting, integrity of the sensor, and potential failure modes. In addition, the location of the sensor on the aircraft needs to be selected with care to minimize parallax between the sensor image and the real world as seen by the pilot through the HUD combiner and the impact on aircraft aerodynamic characteristics.

17.3.5.2 Synthetic Vision

Synthetic Vision is an alternative approach to improving the pilot's situational awareness. In this HUD vision system, an onboard system generates a "real-world view" of the outside scene based on a terrain database using GNSS position, track, heading, and altitude information. The 3D graphics engine generates the terrain view using shadowing and highlighting to make the view look more realistic. The view, shown in Figure 17.22, is covered with a grid and a random texture to enhance the sense of movement across the terrain. The SVS can also add geolocated features on the texture such as obstacles and runways to help the pilot maintain situational awareness.

One advantage of the synthetic vision view is that it is available in all weather conditions; however, the view does not show the real world and cannot be used to confirm that the runway is free of obstacles.

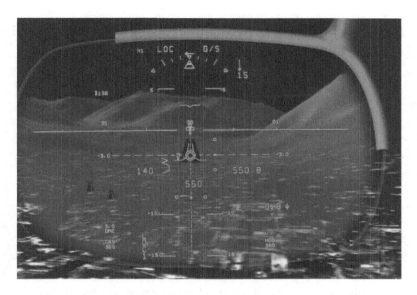

FIGURE 17.22   HUD symbology with synthetic vision video overlay.

17.3.5.3 Combined Vision

If both EVS and SVS video have been routed to the HUD computer, then sophisticated video processing can be used to merge the EVS to the SVS image to provide a combined vision view to the pilot. The video processing extracts the real-world details from the EVS image and adds them to the all-weather SVS image for a "best of both worlds" view ahead of the aircraft.

17.3.6 Recent Developments

HUD suppliers continue to develop new features for the HUD that will improve safety while reducing size and cost so this valuable safety tool is available to more pilots. Some of these are described in the following text.

17.3.6.1 Compact HUD

The waveguide optical system described earlier has enabled the design of a "compact" HUD that is significantly smaller than reflective HUDs. The compact HUD can be installed on small aircraft where the restricted head room will not allow the installation of an overhead projection unit. The compact HUD displays a collimated image and is accurately installed so the symbology is conformal. The compact HUD optical system cannot meet the same optical performance characteristics as a typical reflective HUD, with practical limitations on FOV, luminance, and eyebox, restricting use to smaller aircraft.

17.3.6.2 Surface Guidance Symbology

A natural extension of HUD use is to provide situational awareness and guidance when the aircraft is travelling to and from the runway to the parking area or terminal gate. The development of special symbology that helps the pilot navigate while taxing helps eliminate runway incursions and can keep the traffic moving safely even in low-visibility conditions and at night.

The addition of an EVS or SVS video overlay adds to the safety margin.

17.3.6.3 Color HUD

Due to the complexity of wide FOV reflective HUD optical systems, the optical designer must use all means available to meet display accuracy and parallax error requirements. All certified reflective HUDs

FIGURE 17.23   (**See color insert.**) HUD symbology perceived color varies with background color.

today are monochromatic, generally using a narrowband green-emitting phosphor. The addition of a second color to the HUD is a desirable natural progression in HUD technology; however, one of the technical challenges associated with adding a second (or third) display color is maintaining the performance standards available in monochrome displays. One method for solving this problem uses a collimator with two independent embedded curvatures, one optimized for green symbology and the other optimized for red symbology, each with a wavelength-selective coating.[14]

One fundamental issue associated with color symbology on HUDs is the effects of the real-world background color "adding" to the display color (green), resulting in an unintended perceived display color, as shown in Figure 17.23.

Definitions

Boresight: The aircraft longitudinal axis, used to position the HGS during initial installation and as a reference for symbol positioning. Also is the process of aligning the HUD precisely with respect to the aircraft reference frame.

Collimation: The optical process needed to produce parallel rays of light. The viewer's eyes are then focused as if viewing a distant object.

Conformal: Symbols that are positioned so that they accurately overlay the real world.

Eyebox: The HUD eyebox is a 3D area around the flight deck DEP where all of the data shown on the combiner can be seen.

References

1. Naish, J. (1972) Michael, applications of the head-up display (HUD) to a commercial jet transport. *J. Aircraft*, 9(8), 530–536.
2. Naish, J.M. (1964) Combination of information in superimposed visual fields. *Nature*, 202(4933), 641–646.
3. Sundstrand Data Control, Inc. (1979) Head up display system.
4. Part 25—Airworthiness standards: Transport category airplanes, Subpart D-Design and Construction, Sec. 25.773 Pilot Compartment View.

5. Part 25—Airworthiness standards: Transport category airplanes, Subpart D-Design and Construction, Sec. 25.775 Windshield and Windows.
6. Vallance, C.H. (1983) The approach to optical system design for aircraft head up display. *Proc. SPIE*, 399: 15–25.
7. Hughes. (1976) Optical display systems utilizing holographic lenses, U.S. Patent 3,940,204.
8. Marconi. (1981) Head up displays, U.S. Patent 4,261,647.
9. Wood, R.B. and Hayford, M.J. (1988) Holographic and classical head up display technology for commercial and fighter aircraft, *Proc. SPIE 0883, Holographic Optics: Design and Applications*, 36(April 12, 1988); doi:10.1117/12.944123; http://dx.doi.org/10.1117/12.944123
10. SAE. (1998) AS8055 minimum performance standard for airborne head up display (HUD). Warrendale, PA: Society of Automotive Engineers.
11. Stone, G. (1987) The design eye reference point, *SAE 6th Aerospace Behavioral Engineering Technology Conference Proceedings, Human/Computer Technology: Who's in Charge?* pp. 51–57.
12. Desmond, J. (1997) Method and apparatus for detecting control system data processing errors, U.S. Patent 4,698,785.
13. Part 25—Airworthiness standards: Transport category airplanes, Subpart C-Structure, Sec. 25.562 Emergency Landing Dynamic Conditions.
14. Gohman, J.A., et al. (1988) Multi-color head-up display system, U.S. Patent 5,710,668.
15. Flight Safety Foundation Special Report: Head-Up Guidance System Technology—A Clear Path to Increasing Flight Safety. November 2009. http://flightsafety.org/files/hgs_nov09.pdf
16. Howells, P.J. and Brown, R. (2007) Challenges with displaying enhanced and synthetic vision video on a head-up display. *Proceedings of the SPIE Enhanced and Synthetic Vision 2007*, Orlando, FL, April 27, 2007, Vol. 6559, 65590F; doi:10.1117/12.720445.

18

Display Devices: RSD™ (Retinal Scanning Display)

Thomas M. Lippert
Microvision, Inc. *

18.1 Introduction

This chapter relates performance, safety, and utility attributes of the retinal scanning display (RSD) as employed in a helmet-mounted pilot-vehicle interface, and by association, in panel-mounted head up display (HUD) and head down display (HDD) applications. Because RSD component technologies are advancing so rapidly, quantitative analyses and design aspects are referenced to permit a more complete description here of the first high-performance RSD system developed for helicopters.

Visual displays differ markedly in how they package light to form an image. The RSD depicted in Figure 18.1, is a relatively new optomechatronic device based initially on red, green, and blue diffraction-limited laser light sources. The laser beams are intensity modulated with video information, optically combined into a single, full-color pixel beam, then scanned into a raster pattern by a roster optical scanning engine (ROSE) comprised of miniature oscillating mirrors, much as the deflection yoke of a cathode-ray tube (CRT) writes an electron beam onto a phosphor screen. RSDs are unlike CRTs in that conversion of electrons to photons occurs prior to beam scanning, thus eliminating the phosphor screen altogether along with its re-radiation, halation, saturation, and other brightness- and contrast-limiting factors. This means that the RSD is fundamentally different from other existing display technologies in that there is no planar emission or reflection surface—the ROSE creates an optical pupil directly. Like the CRT, an RSD may scan out spatially continuous (nonmatrix-addressed)

---

* This is a reprinted chapter with minor editorial changes and changes due to technology or regulatory evolution by technical editors since the original author could not be reached. The author's affiliation is from the original contribution; current affiliation of the author is unknown.

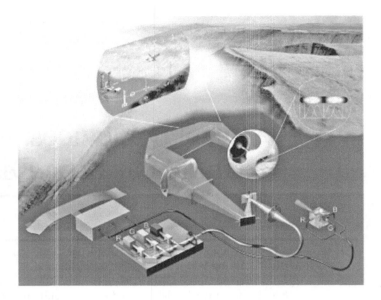

FIGURE 18.1   Functional component diagram of the RSD HMD.

information along each horizontal scan line, while the scan lines form discrete information samples in the vertical image dimension.*

18.2 An Example Avionic HMD Challenge

Consider the display engineering problem posed by Figure 18.1. An aircraft flying the contour of the earth will transit valleys as well as man-made artifacts: towers, power lines, buildings, and other aircraft. On this flight the pilot is faced with a serious visual obscurant in the form of ground fog, rendered highly opaque by glare from the sun.

The pilot's situational awareness and navigation performance are best when flying "eyes-out" the wind-shield, in turn requiring "eyes-out" electronic display of his own aircraft attitude and status information. Particularly under degraded visual conditions additional imagery of obstacles (towers, the earth, etc.) synthesized from terrain data bases and mapped into the pilot's ever-changing direction of gaze via Global Positioning System data reduce the hazards of flight. The question has been, which technology can provide a display of adequate brightness, color, and resolution to adequately support pilotage as viewed against the harsh real-world conditions described.

For over 30 years, researchers and designers have improved the safety and effectiveness of HMDs so that mission-critical information would always be available "eyes-out" where the action is, unlike "eyes-in" traditional HDDs.[1] U.S. Army AH-64 Apache helicopter pilots are equipped with such an HMD, enabling nap-of-the-earth navigation and combat at night with video from a visually coupled infrared imager and data computer. This particular pilot-vehicle interface has proven its reliability and effectiveness in over 1 million hours of flight and was employed with great success in the Desert Storm Campaign. Still, it lacks the luminance required for optimal grayscale display during typical daylight missions, much less the degraded conditions illustrated above.

The low luminance and contrast required for nighttime readability is relatively easy to achieve, but it is far more difficult to develop an HMD bright enough and of sufficient contrast for daylight use. The information must be displayed as a dynamic luminous transparency overlaying the real-world's complex features, colors, and motion. In order to display an image against a typical real-world daytime

---

* At the time this volume was published EUROCAE has not approved the publication of ED-124 pending internal review.

scene luminance of 3000 fL, the virtual display peak luminance must be about 1500 fL at the pilot's eye. And depending on the efficiency of the specific optics employed, the luminance at the display light source may need to be many times greater. The display technology that provides the best HMD solution might also provide the optimal HUD and HDD approaches.

18.3 CRTs and MFPs

Army aviation is the U.S. military leader in deployed operational HMD systems. The Apache helicopter's monochrome green CRT Helmet Display Unit (HDU) presents pilotage FLIR (forward-looking infrared) imagery overlaid with flight symbology in a 40° (H) × 30° (V) monocular field of view (FOV). The Apache HDU was developed in the late 1970s and early 1980s using the most advanced display technology then available. The RAH-66 Comanche helicopter program expanded the display's performance requirements to include night and day operability of a monochrome green display with a binocular 52° H × 30° V FOV and at least 30° of left/right image overlap.

The Comanche's Early Operational Capability Helmet Integrated Display Sighting System (EOC HIDSS) prototype employed dual miniature CRTs. The addition of a second CRT pushed the total head-supported weight for the system above the Army's recommended safety limit. Weight could not be removed from the helmet itself without compromising safety, so even though the image quality of the dual-CRT system was good, the resulting reduction in safety margins was unacceptable.

The U.S. Army Aircrew Integrated Systems (ACIS) office initiated a program to explore alternate display technologies for use with the proven Aircrew Integrated Helmet System Program (AIHS, also known as the HGU-56/P helmet) that would meet both the Comanche's display requirements and the Army's safety requirements.

Active-matrix liquid-crystal displays (AMLCD), active-matrix electroluminescent (AMEL) displays, field-emission displays (FEDs), and organic light-emitting diodes (OLEDs) are some of the alternative technologies that have shown progress. These postage-stamp size miniature flat-panel (MFP) displays weigh only a fraction as much as the miniature CRTs they seek to replace.

AMLCD is the heir apparent to the CRT, given its improved luminance performance. Future luminance requirements will likely be even higher, and there are growing needs for greater displayable pixel counts to increase effective range resolution or FOV, and for color to improve legibility and enhance information encoding. It is not clear that AMLCD technology can keep pace with these demands.

18.4 Laser Advantages, Eye Safety

The RSD offers distinct advantages over other display technologies because image quality and color gamut are maintained at high luminances limited only by eye-safety considerations.[2,3] The light-concentrating aspect of the diffraction-limited laser beam can routinely produce source luminances that exceed that of the solar disc. Strict engineering controls, reliable safeguards, and careful certification are mandatory to minimize the risk of damage to the operator's vision.[4] Of course, these safety concerns are not limited to laser displays; any system capable of displaying extremely high luminances should be controlled, safeguarded, and certified.

Microvision's products are routinely tested and classified according to the recognized eye safety standard—the maximum permissible exposure (MPE)—for the specific display in the country of delivery. In the U.S. the applicable agency is the Center for Devices and Radiological Health (CDRH) Division of the Food and Drug Administration (FDA). The American National Standards Institute's Z136.1 reference, "The Safe Use of Lasers," provides MPE standards and the required computational procedures to assess compliance. In most of Europe the IEC 60825-1 provides the standards.

Compliance is assessed across a range of retinal exposures to the display, including single-pixel, single scan line, single video frame, 10 s, and extended-duration continuous retinal exposures. For most scanned laser displays, the worst-case exposure leading to the most conservative operational usage is found to be the extended-duration continuous display MPE. Thus, the MPE helps define laser power and scan-mirror operation-monitoring techniques implemented to ensure safe operation. Examples include shutting down the laser(s) if the active feedback signal from either scanner is interrupted and automatically attenuating the premodulated laser beam for luminance control independent of displayed contrast or grayscale.

18.5 Light Source Availability and Power Requirements

Another challenge to manufacturers of laser HMD products centers on access to efficient, low-cost lasers or diodes of appropriate collectible power (1–100 mW), suitable wavelengths (430–470, 532–580, and 607–660 nm), low video-frequency noise content (<3%), and long operating life (10,000 h). Diodes present the most cost-effective means because they may be directly modulated up from black, while lasers are externally modulated down from maximum beam power.

Except for red, diodes still face significant development hurdles, as do blue lasers. Operational military-aviation HMDs presently require only a monochrome green, G, display which can be obtained by using a 532-nm diode-pumped solid-state (DPSS) laser with an acoustic-optic modulator (AOM). Given available AOM and optical fiber coupling efficiencies, the 1500-fL G RSD requires about 50 mW of laser beam power. Future requirements will likely include red + green, RG, and full color, RGB, display capability.

18.6 Microvision's Laser Scanning Concept

Microvision has developed a flexible component architecture for display systems (Figure 18.1). RGB video drives AOMs to impress information on Gaussian laser beams, which are combined to form full-color pixels with luminance and chromaticity determined by traditional color-management techniques. The aircraft-mounted photonics module is connected by single-mode optical fiber to the helmet, where the beam is air propagated to a lens, deflected by a pair of oscillating scanning mirrors (one horizontal and one vertical), and brought to focus as a raster format intermediate image. Finally, the image is optically collimated and combined with the viewer's visual field to achieve a spatially stabilized virtual image presentation.

The AIHS Program requires a production display system to be installed and maintained as a helicopter subsystem—designated Aircraft Retained Unit (ARU)—plus each pilot's individually fitted protective helmet, or Pilot Retained Unit (PRU). Microvision's initial concept-demonstration HMD components meet these requirements (Figure 18.2).

Microvision's displays currently employ one horizontal line-rate scanner—the Mechanical Resonant Scanner (MRS)—and a vertical refresh galvanometer. Approaches using a bi-axial microelectro-mechanical system (MEMS) scanner are under development. Also, as miniature green laser diodes become available, Microvision expects to further reduce ARU size, weight, and power consumption by transitioning to a small diode module (Figure 18.1, lower-right) embedded in the head-worn scanning engine, which would also eliminate the cost and inefficiency of the fiber optic link.

For the ACIS project, a four-beam concurrent writing architecture was incorporated to multiply by 4 the effective line rate achievable with the 16-kHz MRS employed in unidirectional horizontal writing mode. The vertical refresh scanner was of the 60-Hz saw-tooth-driven servo type for progressive line scanning. The f/40 writing beams, forming a narrow optical exit pupil (Figure 18.3), are diffraction-multiplied to form a 15-mm circular matrix of exit pupils.

The displayed resolution of a scanned-light-beam display[5] is limited by three parameters: (1) spot size and distribution as determined by cascaded scan-mirror apertures (D), (2) total scan-mirror deflection angles in the horizontal or vertical raster domains (*Theta*), and (3) dynamic scan-mirror flatness under normal operating conditions. Microvision typically designs to the full-width/half-maximum Gaussian spot overlap criterion, thus determining the spot count per raster line. Horizontal and vertical displayable spatial resolutions, limited by (D)*(*Theta*), must be supported by adequate scan-mirror dynamic flatness for the projection engine to perform at its diffraction limit. Beyond these parameters,

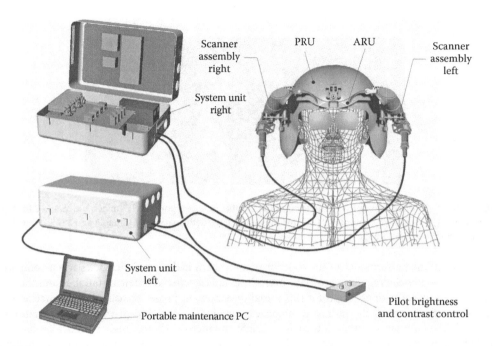

FIGURE 18.2 Microvision's RSD components meet the requirements of the AIHS HIDSS program for an HMD.

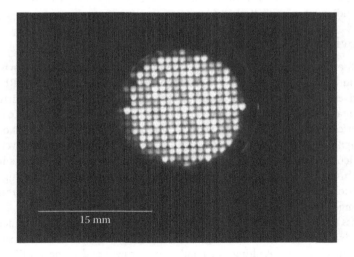

FIGURE 18.3 The far-field beamlet structure of a spot-multiplied (expanded) RSD OEP. The unexpanded 1 mm exit pupil is represented by a single central spot.

image quality is affected by all the components common to any video projection display. Electronics, photonics, optics, and packaging tolerances are the most significant.

18.6.1 Government Testing of the RSD HMD Concept

Under the ACIS program, the concept version of the Microvision RSD HMD was delivered to the U.S. Army Aeromedical Research Laboratory (USAARL) for testing and evaluation in February 1999.[6]

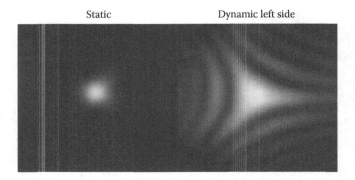

FIGURE 18.4 The effect of improved mirror design is visible in these spot (pixel) images, normalized for size but not for intensity, for scanned spots at $\sim\lambda/4$ P-P mechanical mirror deformation (left image) and $\sim2\lambda$ P-P mechanical mirror deformation (right image).

As expected, the performance of the concept-phase system had some deficiencies when compared to the RAH-66 Comanche requirements. However, these deficiencies were few in number and the overall performance was surprisingly good for this initial development phase. Measured performance for exit pupil, eye relief, alignment, aberrations, luminance transmittance, and field-of-view met the requirements completely. The luminance output of the left and right channels—although high, with peak values of 808 and 1111 fL, respectively—did not provide the contrast values required by Comanche in all combinations of ambient luminance and protective visor. Of greatest concern was the modulation transfer function (MTF)—and the analogous Contrast Transfer Function (CTF)—exhibiting excessive rolloff at high spatial frequencies, and indicating a "soft" displayed image.

18.6.2 Improving RSD Image Quality

The second AIHS program phase concentrated on improving image quality. Microvision identified the sources of the luminance, contrast, and MTF/CTF deficiencies found by USAARL. A few relatively straightforward fixes such as better fiber coupling, stray light baffling, and scan-mirror edge treatment provided the luminance and low-spatial-frequency contrast improvements required to meet specification, but MTF/CTF performance at high spatial frequencies presented a more complex set of issues.

Each image-signal-handling component in the system contributes to the overall system MTF. Although the video electronics and AOM-controller frequency responses were inadequate, they were easily remedied through redesign and component selection. Inappropriate mounting of fixed fold mirrors in the projection path led to the accumulation of several wavelengths of wave-front error and resultant image blurring. This problem, too, is readily solved.

The second class of problems pertains to the figure of the scan mirrors. Interferometer analyses of the flying spot under dynamic horizontal scanning conditions indicated excessive mirror surface deformation (~2 peak-to-peak mechanical), resulting in irregular spot growth and reduced MTF/CTF performance (Figure 18.4).

Three fast-prototyping iterations brought the mirror surface under control ($\sim\lambda/4$) to achieve acceptable spot profiles at the raster edge. Thus, the component improvements described above are expected to result in MTF/CTF performance meeting U.S. Army specification.

18.7 Next Step

The next step in the evolution of the helicopter pilot's laser HMD is the introduction of daylight-readable color. Microvision first demonstrated a color VGA format RSD HMD in 1996, followed by SVGA in 1998. Development of a 1280 × 1024-color-pixel (SXGA) binocular HMD project is being made possible by ACIS's Virtual Cockpit Optimization Program (VCOP), which began with software-reconfigurable virtual flight simulations in 2000 and proceeded to in-flight virtual cockpit demonstrations in 2001. For these demonstrations, the aircraft's traditional control-panel instrumentation served only as an emergency backup function. Figure 18.1, with which this chapter began, represents the VCOP RGB application concept.

One configuration of the VCOP simulation/operation HMD acknowledges the limited ability of the blue component to generate effective contrast against white clouds or blue sky. Because the helmet tracker used in any visually-coupled system will "know" when the pilot is "eyes out" or "head down," the HMD may employ graphics and imaging sensor formats in daylight readable greenscale, combined with red, for "eyes out" information display across established green/yellow/red caution advisory color codes, switching to full color formats at lower luminances for "head down" displays of maps, etc.

The fundamental capabilities of the human visual system, along with ever increasing imaging sensor and digital image generation bandwidths, require HMD spatial resolutions greater than SXGA. For this reason, the U.S. Air Force Research Laboratory has contracted Microvision Inc. to build the first known HDTV HMD (1920 × 1080 pixels in a noninterlaced 60 Hz frame refresh digital video format). The initial system will be a monocular 100-fL monochrome green fighter pilot training HMD with growth-to-daylight readable binocular color operation.

An effort of 30 years has only scratched the surface of the HMD's pilot vehicle interfacing potential. It is expected that the RSD will open new avenues of pilot-in-the-loop research and enable safer, more effective air and ground operations.

Definitions

Helmet-mounted display (HMD): Head-Up Display (HUD); Head-Down Display (HDD).

Image viewing zone (IVZ): The range of locations from which an entire virtual image is visible while fixating any of the image's boundaries.

Optical Exit Pupil (OEP): The aerial image formed by all compound magnifiers, which defines the IVZ.

Optomechatronic: Application of integrated optical, mechanical, and electronic elements for imaging and display.

Retinal Scanning Display (RSD): A virtual image projection display which scans a beam of light to form a visible pattern on the retina. The typical 15-mm OEP of a helmet-mounted RSD OEP permits normal helmet shifting in operational helicopter environments without loss of image. Higher-*g* environments may require larger OEPs.

ROSE: Raster Optical Scanning Engine.

Virtual image projection (VIP): An optical display image comprised of parallel or convergent light bundles.

Virtual retinal display (VRD): A subcategory of RSD specifically characterized by an optical exit pupil less t han 2 mm, for Low Vision Aiding (LVA), vision testing, narrow field of view, or "agile" eye-following OEP display systems. This is the most light-efficient form of RSD.

Acknowledgments

This work was partially funded by U.S. Army Contract No. DAAH23-99-C-0072, Program Manager, Aircrew Integrated Systems, Redstone Arsenal, AL. The author wishes to express appreciation for the outstanding efforts of the Microvision, Inc. design and development team, and for the guidance and support provided by the U.S. Army Aviation community, whose vision and determination have made these advances in high-performance pilotage HMD systems possible.

Further Reading

Microvision, Inc. Website: www.microvision.com. Accessed on April 28, 2014.

References

1. Rash, C. E., Ed., Helmet-mounted displays: Design issues for rotary-wing aircraft, U.S. Army Medical Research and Materiel Command, Fort Detrick, MD, 1999.
2. Kollin, J., A retinal display for virtual environment applications, *SID International Symposium, Digest of Technical Papers*, Vol. XXIV, pp. 827–828. Playa del Rey, CA: Society for Information Display, May 1993.
3. de Wit, G. C., A virtual retinal display for virtual reality, Doctoral dissertation, Ponsen & Looijen BV, Wageningen, the Netherlands, 1997.
4. Gross, A., Lorenson, C., and Golich, D., Eye-safety analysis of scanning-beam displays, Society for Information Display International Symposium Digest of Technical Papers, pp. 343–345, May 1999, Wiley Online Library.
5. Urey, H., Nestorovic, N., Ng, B., and Gross, A., Optics designs and system MTF for laser scanning displays, Helmet and Head Mounted Displays IV, *Proc. SPIE*, 3689, 238–248, 1999.
6. Rash, C. E., Harding, T. H., Martin, J. S., and Beasley, H. H., Concept phase evaluation of the Microvision, Inc., Aircrew Integrated Helmet System, HGU-56/P, virtual retinal display, USAARL Report No. 99-18. Fort Rucker, AL: U.S. Army Aeromedical Research Laboratory, 1999.

19

Vision Systems

Steven D. Young
Langley Research Center
National Aeronautics and
Space Administration

Lynda J. Kramer
Langley Research Center
National Aeronautics and
Space Administration

Randall E. Bailey
Langley Research Center
National Aeronautics and
Space Administration

19.1 Introduction

In 1929, "Jimmy" Doolittle ushered in a new era of aviation by making the first completely blind takeoff and landing, demonstrating the promise of flight where pilots refer primarily, or exclusively, to their instruments. Following Doolittle's lead, many of the avionics systems introduced since then have sought to enable and improve instrument-based flying. These include, for example, attitude indicators, radio navigation systems, and the instrument landing systems (ILS). Further, a set of processes and procedures called instrument flight rules (IFR) were developed to assure safe use of these systems. Yet today, limited visibility remains a critical factor affecting both the safety and capacity of aviation operations. For example, in commercial aviation, controlled flight into terrain (CFIT), runway incursion/excursion (RI/RE), and loss of control (LOC) due to loss of attitude awareness are leading accident categories that frequently result when external visibility is reduced. In general aviation, an additional accident category is inadvertent flight into instrument meteorological conditions (IMC) (i.e., low visibility), wherein pilots with little IMC experience or IFR training fly into deteriorating weather/visibility conditions and either collide with unexpected terrain or lose control of the vehicle because of the lack of external references. In terms of capacity, instrument-based flying (i.e., IMC/IFR) significantly constrains operations. Low visibilities (and ceilings) impose decision altitudes/decision heights (DA/DH), increase requirements for separation and in-trail spacing on final approach and on takeoff, add requirements for specialized crew training, levy increased airport infrastructure needs (e.g., lighting systems), and prescribe requirements for aircraft equipage (e.g., autoland systems). In addition, low visibility can contribute to unexpected delays (e.g., due to lower taxi speeds and a higher likelihood of wrong turns on the airport surface). All of these issues have led to research, development, and deployment of vision systems, which are defined here as an electronic means of providing pilots with intuitive, visual-like, flight references that seek to transform instrument-based flight into the equivalent of a clear-weather daytime operation, or better.

Initial attempts to solve the visibility problem with vision system technology employed imaging sensors that could "see through" obscurations along and ahead of the flight path. Such systems, now termed enhanced vision systems (EVSs), improve visual acquisition by providing sensor-based video imagery to pilots regarding external scene topography (e.g., terrain and vertical structures) and airport features (e.g., runway edge lights while on approach to landing). Several types of sensors have been developed for this purpose, such as forward-looking infrared (FLIR), millimeter-wave radar (MMWR), and low-light-level image intensifying. The sensor type employed by a particular system largely determines the visual advantage in overcoming obscurations such as darkness, fog, haze, rain, and snow. Most current EVS designs provide sensor imagery on a head-up display (HUD) to allow pilots to cross-check images with any external visual cues that might be available or become available as the aircraft transitions through various visibility levels or atmospheric conditions. The most common current sensor technology in use is FLIR, primarily due to maturity level (i.e., proven performance in several domains), relatively small installation footprint, and a tremendous visual advantage, especially in conditions of smoke, haze, and darkness. EVS has been the subject of research and development for over five decades with the military leading the way but with interesting spin-off applications in the commercial aviation sector. The most significant feature of an EVS design and installation is that the camera is rigidly mounted to the aircraft and operates independently of navigation or other aircraft systems that may provide information on the aircraft position and attitude. On the other hand, the quality, content, and field of regard of the associated display is critically dependent upon the sensor and any associated image processing.

In contrast, a synthetic vision system (SVS) provides computer-generated imagery (CGI) of the external topography and man-made features based upon estimates of aircraft attitude and position and one or more georeferenced databases or models (e.g., terrain, obstacles, and airport features). The required databases are created prior to flight by mapping or survey techniques and updated as needed to account for changes that may occur. Scenes derived from these databases are depicted relative to the aircraft's estimated position and attitude, which come from onboard navigation and attitude/heading reference systems. An SVS has unlimited field of regard and is unaffected by atmospheric conditions. As such, it can be particularly advantageous in conditions for which EVS sensors cannot penetrate. For some designs, SVS may also include traffic (e.g., from the traffic alerting and collision avoidance system [TCAS]), terrain proximity indications (e.g., from the terrain awareness and warning system [TAWS]), and flight path depiction (e.g., highway in the sky or tunnel). Weather-penetrating sensors are not required for SVS, although they can be used to supplement the CGI and/or provide a level of integrity checking. The rapid emergence of reliable global positioning system (GPS) information and initiatives to globally map terrain and airport features and to maintain these data has made SVS feasible, with products entering the commercial and business aviation industry around 2006.

In summary, vision systems are intended to improve a pilot's situation and spatial awareness during low-visibility conditions and at night. More than 30 years of research and development have shown that such systems can help to mitigate several accident precursors, including

- Loss of vertical and lateral spatial awareness with respect to flight path
- Loss of terrain and traffic awareness during terminal area operations
- Unclear escape or go-around path even after recognition of problem
- Loss of attitude awareness in cases where there is no visible horizon
- Loss of situation awareness relating to the runway operations
- Unclear path guidance on the airport surface

Along with these safety-related benefits, studies have also shown efficiency-related benefits, such as reduced taxi time and reduced takeoff and landing minima.

19.2 Operating Concepts and Functions

There are several application domains for which vision systems are utilized. These include approach and landing, runway operations, surface/taxi operations, departure/go-around operations, and even some low-altitude en route operations (e.g., for general aviation). In addition to these domains, rotorcraft operators have made use of these systems for low-altitude obstacle avoidance and "see-through" capability during brownout landings. Regardless of the domain, the operating concept remains to provide a capability that enables visual flight rules (VFR)-like procedures to be used in reduced visibility conditions by providing the missing out-the-window cues via an electronic means.

In order to achieve this operating concept for the aforementioned domains, SVS technologies, in particular, allow for additional pilot decision support functions to be developed. These functions are not necessarily required to enable VFR-like operations, but they can provide other benefits. Several such functions have been developed, tested, and in some cases certified and approved for use on aircraft. Examples include

- Taxi route display and guidance
- Route deviation detection and alerting (in the air and on the surface)
- Runway incursion detection and alerting
- Runway excursion prediction
- Taxi hold-short overrun detection and alerting
- Terrain proximity warnings (e.g., TAWS)
- Obstacle proximity warnings
- Collision avoidance guidance (traffic, terrain, obstacles)
- Landing rollout and turnoff guidance
- Mission rehearsal (e.g., previewing approach paths)
- Detection and warning regarding Notices to Airmen (NOTAMs) (e.g., closed runways)
- Detection and warning regarding violating airspace constraints or boundaries

19.3 Enabling Technologies

Most EVSs are comprised of relatively few components, typically an imaging sensor, a processor, a display, and pilot controls. Passive sensors operating in the infrared band are the most common and have been demonstrated to perform very well in nighttime conditions. Typically, FLIR sensor and processing technologies spanning multiple bandwidths are tuned to capture thermal emissions (e.g., from runway surfaces) as well as radiance from incandescent lighting (e.g., airport and runway lights) to produce an enhanced image optimized for airport operations. Example images are shown in Figures 19.1 and 19.2 for the EVS II system manufactured by Kollsman. Fluctuations in performance for some weather conditions have led to testing of alternate sensing technologies by several manufacturers and researchers, most prominently sensors operating in the millimeter wave band. Regardless of the sensor utilized, video imagery is produced that is processed to optimize the image and remove artifacts and noise before passing to a cockpit display.

FIGURE 19.1 **(See color insert.)** Example of EVS imagery (right) as compared with the corresponding out-the-window view (left) while in fog and cloud on final approach. (Courtesy of Kollsman Inc., Wichita, KS.)

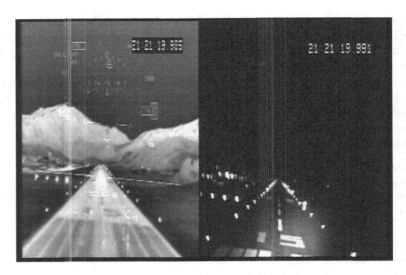

FIGURE 19.2   Example of EVS imagery (left) as compared with the corresponding out-the-window view (right) in darkness near touchdown. (Courtesy of Kollsman Inc., Wichita, KS.)

EVS imagery may be provided on head-down displays, but typically HUDs are employed to keep pilots' "eyes out," with the video imagery more intuitively overlaying any available out-the-window cues. The application of EVS imagery on a HUD in this way has led to certification and operation approval of multiple commercial products under a U.S. regulatory term defined by the FAA as "enhanced flight vision systems (EFVSs)." Under this definition, an EFVS provides a means by which the pilot may meet an enhanced flight visibility requirement where enhanced flight visibility is defined as "the average forward horizontal distance, from the cockpit of an aircraft in flight, at which prominent topographical objects may be clearly distinguished and identified by day or night by a pilot using an enhanced flight vision system." Critical elements of an EFVS are the EVS sensor performance characteristics and the method of display of this conformal "visual-like" EVS imagery with symbology. EFVS symbology typically includes an inertial flight path marker and a flight path angle reference cue. These are used to indicate flight path tracking and projection to the point of intended landing. Such applications of "scene-linked symbology" (i.e., symbolic cues positioned to overlay real-world positions and translated in flight as though they were actual objects in the external scene) facilitate more efficient cognitive processing and mitigate issues such as attention tunneling and symbology fixation. Per current regulations, an approved EFVS may be used in lieu of natural vision and as the primary reference for maneuvering the airplane in some conditions and as long as the required visual references are continuously and distinctly visible and identifiable by the pilot. A typical EFVS installation is shown in Figure 19.3.

In contrast, SVS provides a virtual visual environment by presenting a view of the external environment based primarily upon models/databases combined with positioning and attitude/heading reference systems. Because the content and representation of SVS can vary greatly across designs, careful attention is required when defining the SVS intended function (i.e., describing the intended use of the system). For example, terrain presentations must address color, texture, and shadowing, whether gridding or other properties are needed to promote vection and splay, and whether cultural features should be depicted (e.g., bodies of water, major landmarks and highways, and buildings within close proximity to airport). Similarly, when operating on the surface, are runway/taxiway edge lines sufficient, or should centerlines and markings also be modeled and displayed? In addition to stored information, SVS may also include other information regarding the external environment, as long as systems or functions that can provide this information are installed and integrated.

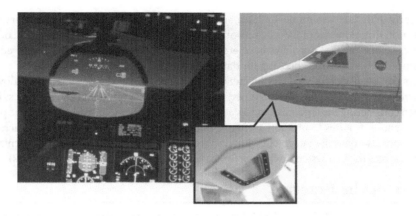

FIGURE 19.3   (See color insert.) Example of an EVS installation.

Figure 19.4 provides a typical generalized architecture for an SVS. As shown in the figure, SVS enabling technology begins with a robust source of aircraft state information. This is most commonly provided by GPS, integrated with inertial navigation system (INS) if available. Performance can be improved using the wide-area augmentation system (WAAS) and/or satellite-based augmentation system (SBAS). Accurate models of features in the external environment are also required—these are features that do not change frequently such as terrain, airport features, obstacles, and published trajectories or flight paths. A processing resource uses the reported position and attitude to provide a perspective, or view, of these models in real time. SVS takes advantages of advances in geospatial modeling, remote sensing, and computational power and is analogous to GPS moving map systems in the automotive and personal electronic device industry.

For some designs, SVSs may also represent traffic (e.g., from the TCAS), terrain proximity (e.g., from the TAWS), and flight path (e.g., highway in the sky, or tunnel). To account for features that may change

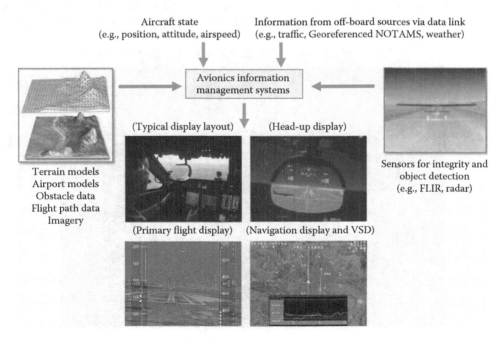

FIGURE 19.4   (See color insert.) SVS elements. (Note: A subset of these elements may be used in particular designs.)

more frequently than database/model updates, some SVS architectures also include imagery overlays, or feature detection functions, from EV-like sensors (e.g., sensors that can detect the location of temporary obstacles or obstructions that may have been erected). Added sensors may also be employed to support integrity assurance (i.e., on-line checking for database or positioning system errors).

SVS can make use of several displays including HUDs, head-worn displays, and head-down primary flight and navigation displays. The effect of display minification on interpretation and awareness is a primary concern when considering the type, size, and location of display. A second consideration is the potential use of visual momentum features to promote reference frame and information understanding across different displays. SVSs are also typically integrated with TAWS to assure there are no conflicting warnings or guidance provided across the two systems, which do make use of similar data (e.g., terrain models).

19.4 State of the Practice

Vision systems such as EVS/EFVS and SVS have penetrated the marketplace at a remarkably fast pace. Examples include the previously mentioned Kollsman EVS/EFVS (Figures 19.1 and 19.2), as well as Honeywell's Head-Down Primary Flight SVS (Figures 19.5 and 19.6), and Rockwell Collins' HUD SVS (Figure 19.7). Nearly all prominent avionics manufacturers have systems such as these: either already certified and operationally approved or well on their way (e.g., Garmin, GE, Universal Avionics, and Thales). Much of the initial motivation for equipage has been for safety reasons (e.g., improvements in pilot awareness of terrain and airport environments). As operational experience continues to grow with the early adopters, technology maturity and confidence in performance is increasing. As a result, the development of compelling business cases that reinforce the safety benefits are emerging. For example, National Aeronautics and Space Administration (NASA) and others have shown that vision system technologies can provide significant improvement in flight technical error to aid in meeting required navigation performance (RNP) criteria.

Another more specific example involves obtaining operational credit through EVS/EFVS capability (i.e., using a HUD and FLIR sensor). In 2004, Chapter 14 of the U.S. Code of Federal Regulations (CFR) Section §91.175 was amended such that operators conducting straight-in instrument approaches (in other than Category II or Category III operations) could operate below the published DA, DH, or minimum descent altitude (MDA) when using an approved EFVS. The key concept under the revisions to §91.175

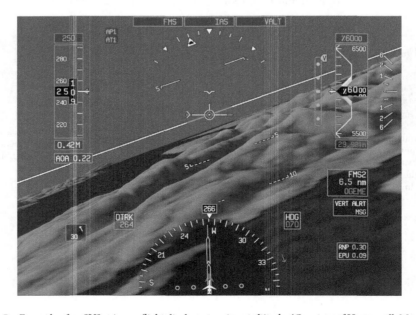

FIGURE 19.5   Example of an SVS primary flight display—turning at altitude. (Courtesy of Honeywell, Morristown, NJ.)

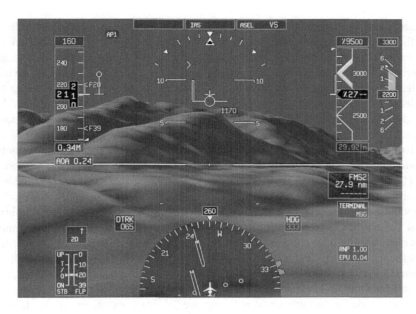

FIGURE 19.6   Example of an SVS primary flight display—descending. (Courtesy of Honeywell, Morristown, NJ.)

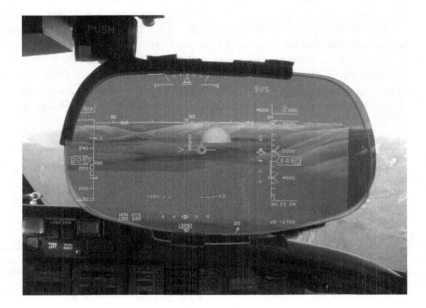

FIGURE 19.7   Example of an SVS HUD. (Courtesy of Rockwell Collins, Cedar Rapids, IA.)

is that an EFVS can be used in lieu of the required natural vision from the DA/DH/MDA to 100 feet (ft) height above the touchdown zone elevation (TDZE). Minimum aviation system performance standards (MASPS) are also now available (RTCA DO-315) and an FAA Advisory Circular (AC 90-106) provides guidance for design and certification criteria of EFVS for operational credit. DO-315 also provides performance standards for SVS but with no additional operational credit implied. In other words, installing SVS does not change the airplane's existing operational capability or certification basis, although it does provide the improvements to situation awareness and flight technical error previously mentioned.

With regard to the databases required for SVSs, from 1998 to the present, a joint committee under the auspices of the RTCA and the European Organisation for Civil Aviation Equipment (RTCA/EUROCAE) has developed industry standards for terrain, obstacle, and airport databases when used in aviation systems. The aviation systems for which these standards were developed include not only vision systems but also other systems that require such data (e.g., TAWS and instrument procedure design tools). Since these standards were published, the industry has by and large adopted them as part of their development process. In addition, significant investments have been made by both data originators (e.g., satellite-based and airborne terrain mapping service providers) and data integrators (e.g., organizations that provide aviation charting and mapping products) to collect, distribute, and maintain global datasets that meet these standards. Further, the International Civil Aviation Organization (ICAO) has adopted the majority of these U.S. and European standards as international standards that must be complied with globally. Tables 19.1 and 19.2 provide some of the quality standards specified for terrain and obstacle databases. ICAO has required that this quality of data be made available by member states (i.e., countries) via the Aeronautical Information Service (AIS) in the very near future. At present, multiple commercial and government organizations have collected and made available databases that meet, or exceed, these standards. As for the standards for airport feature databases, over 30 features are defined as "shall be captured," including runway/taxiway edges and centerlines, landing threshold points, gate locations, and building locations. Accuracy requirements range from 5 to 0.5 m, depending on the type of feature.

For SVS, one criteria for deciding whether to approve SV-based operations will be the quality of the data(bases) used by the system. Quality in this case can include many metrics, but the most relevant are accuracy, integrity, timeliness (e.g., age), and post-spacing (for terrain models). Of these, integrity is the most difficult to assess and confirm. Integrity addresses two problems: (1) has the data been corrupted or altered since it was collected and (2) can the data provide misleading information. To quantify integrity, at least two independent observations are required. In good visibility, the independent observation can come from the pilot, but as the visibility goes down, so does the ability of the pilot to provide this check. A partial solution comes from the fact that as more flights occur, it can be argued that integrity will constantly improve (i.e., every clear daytime approach and landing will be confirmation of the SV presentation for that flight path). However, this does not completely cover all cases (e.g., flights in new areas or just

TABLE 19.1   Standards for Terrain Database Quality for Aviation Uses

Attributes	Area 1 (State)	Area 2 (Terminal)	Area 3 (Surface)	Area 4 (Cat II/III)
Post-spacing	3 arc-sec (~90 m)	1 arc-sec (~30 m)	0.6 arc-sec (~20 m)	0.3 arc-sec (~9 m)
Vertical accuracy (m)	30	3	0.5	1
Horizontal accuracy (m)	50	5	0.5	2.5
Confidence level (%)	90	90	90	90
Integrity level	10^{-3}	10^{-5}	10^{-5}	10^{-5}

Sources: RTCA DO-276B/EUROCAE ED-98B, User requirements for terrain and obstacle data, RTCA, Inc., Washington, DC, September 2012; Annex 15, *Aeronautical Information Services*, 13th edn., International Civil Aviation Organization (ICAO), Montreal, Quebec, Canada, July 2010.

TABLE 19.2   Standards for Obstacle Database Quality for Aviation Uses

Attributes	Area 1 (State)	Area 2 (Terminal)	Area 3 (Surface)	Area 4 (Cat II/III)
Vertical accuracy (m)	30	3	0.5	1
Horizontal accuracy (m)	50	5	0.5	2.5
Confidence level (%)	90	90	90	90
Integrity level	10^{-3}	10^{-5}	10^{-5}	10^{-5}

Sources: RTCA DO-276B/EUROCAE ED-98B, User requirements for terrain and obstacle data, RTCA, Inc., Washington, DC, September 2012; Annex 15, *Aeronautical Information Services*, 13th edn., International Civil Aviation Organization (ICAO), Montreal, Quebec, Canada, July 2010.

after database updates). A complete solution requires an independent measurement in-flight and research has shown this can be done using existing onboard radar or other sensors. Of course, with each sensor comes limitations such as range, accuracy, and weather conditions in which they can operate. The integrity of the positioning and attitude/heading reference system must also be addressed. However, there are mature designs already in place for these systems (e.g., receiver autonomous integrity monitoring [RAIM] for GPS).

In addition, data completeness is an important criteria. A complete minimum set of data for SVS would include aircraft position and attitude, as well as terrain, obstacles, and airport features for the anticipated flight operation. Additional data required for some SVSs would be the flight plan (e.g., for tunnel display) and traffic positions. Most concerns today are associated with the completeness of obstacle data. However, as previously discussed, the RTCA/EUROCAE and ICAO activities have reduced these concerns to simply the question of "What about changes that occur between database updates?" and "How can we be sure there is no missing data?" The latter can be mitigated to some extent as is done for integrity—by using onboard radar, or other sensors, to detect obstacles that may not be in the databases, and imaging sensors may also be added to augment the synthetic scene. This concept has led to research and development on "combined vision systems" (CVSs) (to be discussed in the next section).

19.5 What's Next: Operational Considerations

The revision to 14 CFR §91.175 created an operating credit for EFVS adopters but also introduced a potential distraction during the approach and landing maneuver (at as low as 100 ft). By rule, at no lower than 100 ft above the touchdown zone, one of the required visual landing references must be distinctly visible and identifiable by the pilot *without* reliance on the vision system. To mitigate this issue and avoid this potential attention distraction, minimum performance standards were published to support EFVS operations throughout the approach, and to touchdown, in visibility as low as 1000 ft runway visual range (RVR) by *sole use* of an approved EFVS (i.e., in lieu of natural vision). Simply stated, standards are now in place where the visual segment of the approach may be accomplished by using either "enhanced flight visibility" or natural vision. The precedence that this change sets is potentially quite significant with respect to future uses of vision systems.

With the framework for this capability established, MASPS to safely use an EFVS to operate in this regime down to 300 ft RVR have been published (RTCA DO-341). The basic principles for this operation are almost identical to that established under 14 CFR §91.175: (1) An EFVS provides the enhanced flight visibility sufficient to conduct the operation; (2) the EFVS uses a conformal display of EVS imagery and symbology whereby the combination of an inertial flight path marker and a flight path angle reference cue is shown on a HUD or equivalent display; (3) the required visual references for the operation are continuously in view; and (4) the pilot uses the EFVS to maneuver the aircraft to safely conduct the operation. Operations, whether flown in 1000 ft RVR or 300 ft RVR, are fundamentally the same. The most significant difference is that natural vision is not sufficient in 300 ft RVR. Natural vision cannot be relied upon to mitigate an EFVS failure. Therefore, the integrity and availability requirements for the EFVS increase tremendously. The performance of the EFVS sensor system must be sufficient to penetrate almost all atmospheric conditions, unaffected by the runway type, width, and surrounding surfaces, to provide integrity and availability levels rivaling Category III ILS systems. To meet these requirements, designs that make use of redundancy are likely essential, including the cockpit displays.

EFVS, to date, is being applied as an alternative means of vision—a capability in lieu of, or equivalent to, natural vision to satisfy requirements in the visual segment of an operation. Conversely, SVS, due to its inherent technologies, is seen as a way to improve instrument flight segment requirements. As such, the joint RTCA SC-213/EUROCAE WG 79 committee published DO-315B to establish minimum performance standards for possible operational credit for SVS, leveraging FAA Order 8400.13 (*Procedures for the Evaluation and Approval of Facilities for Special Authorization Category I Operations and All Category II and III Operations*). Specifically, DO-315B establishes performance standards for an SVS enabling lower than standard Category I minima or a reduction in the required minimum visibility. The current performance standards for SVS operational credit do not require the use of a HUD.

The objectives of EVS/EFVS and SVS technology mirror each other as they both attempt to obviate low-visibility conditions as a causal factor in aviation accidents while replicating the operational benefits of clear-day flight operations regardless of the actual outside visibility. The approaches that enable this capability, however, are significantly different yet complementary. SV provides weather independence, potentially unlimited field of regard, navigation, and cultural feature information that can be tailored by the system designer. In comparison, EVS operations are based on a real-time sensor image whose performance is related to the sensor type. Under many low-visibility conditions, an EVS can provide improvement over the pilot's natural vision. Most importantly, EVS provides a direct view of the vehicle external environment.

A CVS attempts to provide "the best of both worlds"—bringing the advantages of SVS while using EVS to offset its limitations, and vice versa. How these technologies are combined and used on the flight deck depend upon the intended operational use and environment. Candidate designs have been tested wherein both are combined on the same display, while others make use of separate independent displays that can be cross-checked by the pilot. Spatial and/or temporal separation of information has also been evaluated. For integrated displays, designs may fuse the information from both sources or, more commonly, use an insert for the EVS/EFVS portion within a larger field-of-regard SVS portion. Integration in this way assures that the pilot can discern the source of the information (SVS or EVS) from the image. An example of a CVS is shown in Figure 19.8.

Spatial separation of SV/EV in CVSs has been shown to be effective, but workload (cognitive integration and scanning) can be high. However, designs that colocate displays may be problematic as well. For instance, the pilot flying while landing may use head-down SV on approach and then switch to head-up EV on final. This transition can be eased by applying "visual momentum" methods to make sure correlation can be maintained when switching between the two sources and viewpoints.

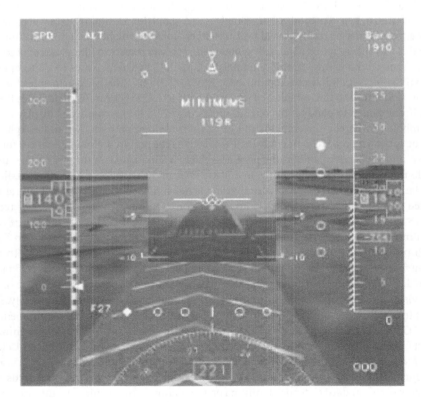

FIGURE 19.8   Example of an CVS primary flight display.

19.6 The Future: EVO and BTV

As currently envisioned, EFVSs, SVSs, and CVSs are not intended to change the roles and responsibilities of air navigation service providers (ANSPs) for spacing and separation from other traffic. Vision systems are also not intended to change the roles and responsibilities of the airport service providers for foreign object damage control, wildlife control, or other airport infrastructure and service functions. However, the future growth and usage of such systems may enable the flight crew to self-separate from other traffic, terrain, and obstacles; or reduce requirements for airport infrastructure.

One of the goals for the future air transportation system is "equivalent visual operations (EVOs)." EVOs can have two interpretations: (1) equivalent traffic flow rates and patterns such that capacity becomes independent of visibility and (2) equivalent visual cues such that there is a less abrupt change from VFR operations to IFR operations. In terms of vision systems, it is this latter interpretation that is driving much vision system research and development. Can vision systems provide what is required to use VFR-like rules regardless of the actual visibility and regardless of traffic flow rates and patterns?

In one form of EVO, separation authority is delegated to the flight crew (as it is under VFR), although the flight operation may be supervised by air traffic management (ATM). Visual displays will provide an electronic means of providing sufficient "visibility" (with commensurate accuracy, integrity, and availability) of the terrain, obstacles, airport features, and other required flight references to enable the flight crew to "see and avoid," "see to follow," and "self-navigate." In addition, the vision systems should provide this information in an intuitive fashion, analogous to visual flight today, enabling minimal training or transitional impacts from VFR to IFR flight operations.

The principle of "see and avoid" is defined under the U.S. CFR Part 91.113, across all classes of airspace, as "when weather conditions permit, regardless of whether an operation is conducted under instrument flight rules or visual flight rules, vigilance shall be maintained by each person operating an aircraft so as to see and avoid other aircraft." To enable see and avoid by an electronic means (i.e., via vision systems), "see and avoid" or maybe more appropriately "sense and avoid" support must be provided. This likely will comprise onboard sensors in addition to wireless data link communications (e.g., TCAS and ADS-B) to sense the presence of other vehicles and provide sufficient fidelity to allow the pilot (or system) to steer clear. This capability dovetails closely with work being performed by the unmanned aircraft systems (UAS) community for "sense and avoid."

"Self-navigation," in this context, is the ability to identify and safely fly with respect to visual flight references (e.g., navigation with respect to cultural features such as roads, rivers, and large man-made structures) and the ability to safely conduct "visual" approaches, landings, and takeoff operations, without collision with the terrain or obstacles. "See to follow," within the EVO context, alleviates the controllers' responsibility to provide vectors for an aircraft under its control. Because the controller is no longer responsible for separation or spacing, the controller workload significantly drops. For example, when spacing off a lead aircraft during final approach, once the pilot acknowledges traffic awareness, he/she becomes responsible (as in VFR flight) to maintain the aircraft in-trail and maintain appropriate separation. Vision system technology must provide sufficient information to perform this function safely. Onboard merging and spacing decision support functions and/or guidance can streamline this operation to optimize spacing and situation awareness. These capabilities are not to imply that displays must replicate the external world references. Instead, these capabilities are designed to achieve equivalent levels of performance and safety whereby displayed information is an analog of the real world, with symbolic or imagery augmenting this analog.

In addition, we know that there are many nonphysical features in the external environment that pilots cannot see, regardless of the visibility, but are of operational import. These include the 4D flight plan, wake vortices, closed runways/taxiways, airspace boundaries, and regions where conflicts can be predicted (e.g., using "conflict probes"). However, we know that these features can be measured, defined, and/or data-linked and then represented on displays in a geospatial frame of reference. Highway-in-the-sky (aka tunnel) displays are the most mature example of this. Acknowledging that such nonvisible nonphysical features can be provided to aid situation awareness, a new flight deck system concept emerges that has been referred to as better-than-visual (BTV) systems. Reaching a mature and comprehensive BTV capability is seen as an incremental process from today's EVS/EFVS and SVS technologies.

The first incremental step to BTV is a transition to EVO capability. The next step would be to provide the capability for the crew to see and avoid or mitigate those hazards or features that are otherwise "invisible" to the flight crew. Third, and finally, capabilities are developed to support new ATM operating concepts such as trajectory-based operations, in particular those that present unique challenges on the flight deck.

At this point, it is believed that five technical challenges should be overcome to fully enable the BTV concept: (1) an unobtrusive lightweight head-worn display that can provide unlimited field of regard and easy retrofit, (2) comprehensive high-integrity information processing techniques that can provide the crew with the required information and with quality commensurate with its use, (3) decision support functions such as those described previously, (4) display format and symbology improvements for traditional head-up and head-down displays that can mitigate the trend to increasing information and operational complexity, and (5) robust communications interfaces to the ubiquitous "wireless" communications infrastructures planned for the future.

References

Annex 15. (July 2010). *Aeronautical Information Services*, 13th edn. International Civil Aviation Organization (ICAO), Montreal, Quebec, Canada.

RTCA DO-276B/EUROCAE ED-98B. (September 2012). User requirements for terrain and obstacle data. RTCA, Inc., Washington, DC.

Further Reading

Alexander, A., Prinzel, L. J., Wickens, C. D., Kramer, L. J., Arthur, J. J., and Bailey, R. E. (April 2009). Evaluating the effects of dimensionality in advanced avionic display concepts for synthetic vision systems. *International Journal of Aviation Psychology*, 19(2), 105–130.

Alexander, A. L., Wickens, C. D., and Hardy, T. J. (2005). Synthetic vision systems: The effects of guidance symbology, display size, and field of view. *Human Factors: The Journal of the Human Factors and Ergonomics Society*, 47(4), 693–707.

Annex 4. (July 2009). *Aeronautical Charts*, 11th edn. International Civil Aviation Organization (ICAO), Montreal, Quebec, Canada.

Annex 14. (July 2013). *Aerodromes*. Vol. I—*Aerodrome Design and Operations*, 6th edn. and Vol. II—*Heliports*, 3rd edn. International Civil Aviation Organization (ICAO), Montreal, Quebec, Canada.

Arthur, J. J., III, Prinzel, L. J., III, Kramer, L. J., Bailey, R. E., and Parrish, R. V. (2003). CFIT prevention using synthetic vision. *Proceedings of SPIE, Enhanced and Synthetic Vision 2003*, Orlando, FL, Vol. 5018, pp. 146–157.

Arthur, J. J., III, Prinzel, L. J., III, Shelton, K. J., Kramer, L. J., Williams, S. P., Bailey, R. E., and Norman, R. M. (April 2009). Synthetic vision enhanced surface operations with head-worn display for commercial aircraft. *International Journal of Aviation Psychology*, 19(2), 158–181.

Arthur, J. J., Williams, S. P., Prinzel, L. P., Kramer, L. J., and Bailey, R. E. (April 2004). Flight simulator evaluation of display media devices for synthetic vision concepts. In C. E. Rash and C. E. Reese (eds.), *Proceedings of SPIE, Helmet- and Head-Mounted Displays IX: Technologies and Applications*, Orlando, FL, Vol. 5442, pp. 213–224. SPIE, Orlando, FL.

Atkins, M. L., Foyle, D. C., Hooey, B. L., and McCann, R. S. (1999). Head-up display symbology for surface operations: Comparisons among scene-linked symbology sets for optimum turn navigation. In R. S. Jensen, B. Cox, J. D. Callister, and R. Lavis (eds.), *Proceedings of the 10th International Symposium on Aviation Psychology*, Columbus, OH, Vol. 2, pp. 784–790. Ohio State University, Columbus, OH.

Bailey, R., Prinzel, L., and Young, S. (2011). Concept of operations for integrated intelligent flight deck displays and decision support. NASA technical memorandum TM-2011-217-081. National Aeronautics and Space Administration, Washington, DC.

Bailey, R. E., Kramer, L. J., and Williams, S. P. (2010). Enhanced vision for all-weather operations under NextGen. *Proceedings of SPIE, Enhanced and Synthetic Vision Conference 2010*, Orlando, FL, Vol. 7689, pp. 768903-1–768903-18.

Campbell, J. L., UijtdeHaag, M., Vadlamani, A., and Young, S. (2003). The application of Lidar to synthetic vision system integrity. *Proceedings of the 22st AIAA/IEEE Digital Avionics Systems Conference*, Indianapolis, IN, October 14–16, pp. 9C2-1–9C2-7, 2003.

Döhler, H.-U. and Korn, B. (2006). EVS based approach procedures: IR-image analysis and image fusion to support pilots in low visibility. *ICAS 2006 25th International Congress of the Aeronautical Sciences*, Hamburg, Germany, September 3–8, pp. 1–10, 2006.

FAA/AC 20-167. (June 2010). Airworthiness approval of enhanced vision system, synthetic vision system, combined vision system, and enhanced flight vision system equipment. Federal Aviation Administration, Washington, DC.

FAA/AC 90-106. (June 2010). Enhanced flight vision systems. Washington, DC: FAA.

Foyle, D. C., Andre, A. D., McCann, R. S., Wenzel, E., Begault, D., and Battiste, V. (1996). Taxiway navigation and situation awareness (T-NASA) system: Problem, design philosophy and description of an integrated display suite for low-visibility airport surface operations. *SAE Transactions: Journal of Aerospace*, 105, 1411–1418.

French, G. and Schnell, T. (2003). Terrain awareness and pathway guidance for head-up displays (TAPGUIDE): A simulator study of pilot performance. *Proceedings of 22nd IEEE/AIAA Digital Avionics Systems Conference*, Indianapolis, IN, Vol. 2, pp. 9.C.4–9.1–7.

Jones, D. R., Prinzel, L. J., III, Otero, S. D., and Barker, G. D. (2009). Collision avoidance for airport traffic concept evaluation. *Proceedings of the 28th AIAA/IEEE Digital Avionics Systems Conference*, Orlando, FL, October 25–29, pp. 4C4-1–4C4-15, 2009.

Korn, B., Biella, M., and Lenz, H. (2008). Operational landing credit with EVS head down display: Crew procedure and human factors evaluation. *Proc. of SPIE. Enhanced and Synthetic Vision 2008, SPIE Defense and Security Symposium*, Orlando, FL, Vol. 6957, March 16–20, pp. 695707-1–695707-11, 2008.

Korn, B. and Döhler, H.-U. (2006). A system is more than the sum of its parts—Conclusions of DLRs EVS Project ADVISE-PRO. *Proceedings of the 25th AIAA/IEEE Digital Avionics Systems Conference*, Portland, OR, October 15–19, pp. 4B4-1–4B4-8, 2006.

Korn, B. and Hecker, P. (2002). Enhanced and synthetic vision: Increasing pilot's situation awareness under adverse weather conditions. Air traffic management for commercial and military systems. *Proceedings of the 21st AIAA/IEEE Digital Avionics Systems Conference*, Irvine, CA, October 27–31, pp. 11C2-1–11C2-10, 2002, Omnipress, Madison, WI.

Korn, B., Lorenz, B., Döhler, H.-U., Többen, H., and Hecker, P. (2005). Passive "Radar-PAPI" landing aids for precision straight-in approach and landing in low visibility. *International Journal of Applied Aviation Studies*, 4(2), 13–38.

Korn, B., Schmerwitz, S., Lorenz, B., and Döhler, H.-U. (2009). Combining enhanced and synthetic vision for autonomous all weather approach and landing. *The International Journal of Aviation Psychology*, 19(1), 49–75 (Taylor & Francis Group).

Kramer, L. J., Arthur, J. J., III, Bailey, R. E., and Prinzel, L. J., III. (2005). Flight testing an integrated synthetic vision system. *Proceedings of SPIE, Enhanced and Synthetic Vision 2005*, Orlando, FL, Vol. 5802, pp. 1–12.

Kramer, L. J., Bailey, R. E., and Prinzel, L. J. (April 2009). Commercial flight crew decision-making during low-visibility approach operations using fused synthetic/enhanced vision systems. *International Journal of Aviation Psychology*, 19(2), 131–157.

Kramer, L. J., Prinzel, L. J., III, Bailey, R. E., and Arthur, J. J., III (2003). Synthetic vision enhances situation awareness and RNP capabilities for terrain-challenged approaches. *Proceedings of the American Institute of Aeronautics and Astronautics Third Aviation Technology, Integration, and Operations Technical Forum*, Denver, CO, Vol. 6814, pp. 1–11. AIAA, Reston VA.

Kramer, L. J., Williams, S. P., and Bailey, R. E. (2008). Simulation evaluation of synthetic vision as an enabling technology for equivalent visual operations. *Proceedings of SPIE, Enhanced and Synthetic Vision Conference 2008*, Orlando, FL, Vol. 6957, pp. 1–15.

Lemos, K. and Schnell, T. (2003). Synthetic vision systems: Human performance assessment of the influence of terrain density and texture. *Proceedings of 22nd IEEE/AIAA Digital Avionics Systems Conference*, Indianapolis, IN, Vol. 2, pp. 9.E.3–9.1–10.

Parrish, R. V., Kramer, L. J., Bailey, R. E., Jones, D. R., Young, S. D., Arthur, J. J., III, Prinzel, L. J., III, Harrah, S., and Glaab, L. J. (May 2008). Aspects of synthetic vision display systems and the best practices of the NASA's SVS Project. NASA TP-215130. NASA, Washington, DC.

Powell, J. D., Jennings, C., and Holforty, W. (2005). Use of ADS-B and perspective displays to enhance airport capacity. Paper presented at the *24th Digital Avionics Systems Conference*, Washington, DC, October 30–November 3, 2005, Vol. 1, pp. 4.D.4–4.1–9.

RTCA DO-272C/EUROCAE ED-99C. (September 2011). User requirements for aerodrome mapping information. RTCA, Inc., Washington, DC.

RTCA DO-291B/EUROCAE ED-119B. (September 2011). Interchange standards for terrain, obstacle, and aerodrome mapping data. RTCA, Inc., Washington, DC.

RTCA/DO-315B. (March 2011). Minimum Aviation System Performance Standards (MASPS) for enhanced vision systems, synthetic vision systems, combined vision systems and enhanced flight vision systems. RTCA Inc., Washington, DC.

RTCA/DO-341. (September 2012). Minimum Aviation System Performance Standards (MASPS) for an enhanced flight vision system to enable all-weather approach, landing and roll-out to a safe taxi speed. RTCA Inc., Washington, DC.

Schiefele, J., Howland, D., Maris, J., Pschierer, C., Wipplinger, P., and Meuter, M. (2005). Human factors flight trial analysis for 3D SVS: Part II. *Proceedings of SPIE, Enhanced and Synthetic Vision 2005*, Orlando, FL, Vol. 5802, pp. 195–206.

Schnell, T., Kwon, Y., Merchant, S., Etherington, T., and Vogl, T. (2004). Improved flight technical performance in flight decks equipped with synthetic vision information system displays. *International Journal of Aviation Psychology*, 4, 79–102.

Schnell, T., Theunissen, E., and Rademaker, R. (2005). Human factors test & evaluation of an integrated synthetic vision and sensor-based flight display system for commercial and military applications. Paper presented at the NATO Research and Technology Organization, Human Factors and Medicine panel workshop entitled *Toward Recommended Methods for Testing and Evaluation of EV and E/SV-Based Visionic Devices*, Williamsburg, VA, April 26–27, pp. 18-1–18-32, 2005.

Suarez, B., Kirk, K., and Theunissen, E. (2012). Development, integration and testing of a stand-alone CDTI with conflict probing support. *Proceedings of the Infotech@Aerospace Conference*, Garden Grove, CA. AIAA 2012-2487, Reston, VA, pp. 1005–1016.

Tadema, J., *Unmanned Aircraft Systems HMI and Automation—Tackling Control, Integrity, and Integration Issues*. Shaker Publishing B.V., Maastricht, the Netherlands.

Tadema, J. and Theunissen, E. (2007). A display concept for UAV autoland monitoring: Rationale, design and evaluation. *Proceedings of the 26th AIAA/IEEE Digital Avionics Systems Conference*, Dallas, TX, October 21–25, pp. 5B4-1–5B4-12.

Theunissen, E. (1997). *Integrated Design of a Man-Machine Interface for 4-D Navigation*. Delft University Press, Delft, the Netherlands.

Theunissen, E. and Etherington, T. J. (2011). Reinventing the past: Avionics systems that didn't make it. *Proceedings of the 30th AIAA/IEEE Digital Avionics Systems Conference*, Seattle, WA, October 16–20, pp. 8B3-1–8B3-18.

Theunissen, E., Roefs, F. D., and Etherington, T. J. (2009). Synthetic vision: Application areas, rationale, implementation and results. *International Journal of Aviation Psychology*, 19(1), 1–24.

Theunissen, E. and Uijt de Haag, M. (2011). Towards a seamless integration of awareness support and alerting systems: Why and how. *Proceedings of the 30th AIAA/IEEE Digital Avionics Systems Conference*, Seattle, WA, October 16–20, pp. 6E1-1–6E1-16.

Uijt De Haag, M., Sayre, J., Campbell, J., Young, S., and Gray, R. (April 2005). Terrain database integrity monitoring for synthetic vision systems. *IEEE Transactions on Aerospace and Electronic Systems*, 41(2), 386–406 (Institute for Electronics and Electrical Engineers).

Vadlamani, A. and Uijt de Haag, M. (2008). Runway obstacle detection using onboard sensors: Modeling and simulation analysis. *Proceedings of the 27th IEEE/AIAA Digital Avionics Systems Conference (DASC)*, St. Paul, MN, October 26–30, pp. 5A4-1–5A4-12, 2008.

Young, S. and Bailey, R. (2008). Equivalent visual flight deck technologies. ARMD technical seminar. National Aeronautics and Space Administration, Washington, DC, January 31, 2008.

Young, S. and Jones, D. (2001). Runway incursion prevention: A technology solution. *International Air Safety Seminar*, Athens, Greece, November 5–9, pp. 1–22, 2001.

Young, S. D., Kakarlapudi, S., and Uijt de Haag, M. (2005). A shadow detection and extraction algorithm using digital elevation models and x-band weather radar measurements. *International Journal of Remote Sensing*, 26(8), 1531–1549 (Taylor & Francis Group Publishing).

Young, S. and Uijt de Haag, M. (August 2005). Detection of digital elevation model errors using x-band weather radar. *AIAA Journal of Aerospace Computing, Information, and Communication*, 2, 309–326.

20

Speech Recognition and Synthesis

Douglas W. Beeks
Beeks Engineering and Design

20.1 Introduction

The application of speech recognition (SR) in aviation is rapidly evolving and perhaps moving toward more common use on future flightdecks. The concept of using SR in aviation is not new. The use of speech recognition and voice control (VC) has been researched for more than 20 years, and many of the proposed benefits have been demonstrated in varied applications. Continuing advances in computer hardware and software are making the use of voice control applications on the flightdeck more practical, flexible, and reliable. There is little argument that the easiest and most natural and ideal way for a human to interact with a computer is by direct voice input (DVI).

While speech recognition has improved over the past several years, it has not reached the level of capability and reliability of one person talking to another. Using SR and DVI in a flightdeck atmosphere likely brings to mind thoughts of the computer on board the starship Enterprise from the science fiction classic *Star Trek*, or possibly of the HAL9000 computer from the movie *2001: A Space Odyssey*. The expectation of a voice control system like the computer on the Enterprise and the HAL9000 computer is that it be highly reliable, work in adverse and stressful conditions, be transparent to the user, and understand its users accurately without having to tailor their individual speech and vocabulary to suit the system. Current speech recognition and voice control systems are not able to achieve this level of performance expectations, although the ability and flexibility of speech recognition and its application to voice control has increased over the past few years. Whether or not a speech recognition system will ever be able to function to the level of one person speaking to another remains to be seen.

The current accuracy rate of speech recognition is in the lower to mid 90% range. Some speaker-dependent systems, and generally those with small vocabularies, have shown accuracy rates into the upper 90% range. While at first glance that might sound good, consider that with a 90% accuracy rate, 1 in 10 words will be incorrectly recognized. Also consider that this 90% and greater accuracy may be under ideal conditions; many times this high accuracy rate is achieved in a controlled and sterile lab environment. Under actual operating conditions, including cockpit noise, random noises, bumps and thumps, multiple people talking at once, etc., the accuracy rate of speech recognition systems can erode significantly.

Currently, a few military applications are planning on using SR to provide additional methods to support the Man-Machine Interface (MMI) to reduce the workload on the pilot in advanced aircraft. The Eurofighter Typhoon has SR capabilities. Numerous aviation companies worldwide are conducting research and studies into how the available SR technology can be incorporated into current equipment designs and designs of the future for both the civilian and military marketplace.

20.2 How Speech Recognition Works: A Simplistic View

Speech recognition is based on statistical pattern matching. One of the more common methods of speech recognition based on pattern matching uses Hidden Markov Modeling (HMM) comprising two types of pattern models, the acoustical model and the language model. Which of the two models will be used, and in some cases both will be required, depends on the complexity of the application. Complex speech recognition applications, such as those supporting continuous or connected speech recognition, will use a combination of the acoustical and language models.

In a simple application using only the acoustical model, the application will process the uttered word into phonemes, which are the fundamental parts of speech. These phonemes are converted to a digital format. This digital format, or pattern, is then matched against stored patterns by the speech processor in search of a match from a stored database of word patterns. From the match, the phoneme and word can be identified.

In a more complex method, the speech processor will convert the utterance to a digital signal by sampling the voice input at some rate, commonly 16 kHz. The required acoustical signal processing can be accomplished using several techniques. Some commonly used techniques are Linear Predictive Coding (LPC) cochlea modeling, Mel Frequency Cepstral Coefficients (MFCC), and others. For this example, the sampled data is converted to the frequency domain using a fast-Fourier transformation. The transformation will analyze the stored data at 1/30–1/100 of a second (3.3–100 ms) intervals, and convert the value into the frequency domain. The resulting graph from the converted digital input will be compared against a database of known sounds. From these comparisons, a value known as a feature number will be determined.

The feature numbers will be used to reference a phoneme found using that feature number. This, ideally, would be all that is required to identify a particular phoneme, however, this will not work for a number of reasons. Background noises, the user not pronouncing a word the same way every time, and the sound of a phoneme will vary, depending on the surrounding phonemes that may add variance to the sound being processed. To overcome problems of variability of the different phonemes, the phonemes are assigned to more than one feature number. Since the speech input was analyzed at an interval of 1/30–1/100 of a second and a phoneme or sound may last from 500 ms to 2 s, many feature numbers may be assigned to a particular sound. By using statistical analysis of these feature numbers and the probability that any one sound may contain those feature numbers, the probability of that sound being a particular phoneme can be determined.

To be able to recognize words and complete utterances, the speech recognizer must also be able to determine the beginning and the end of a phoneme. The most common method to determine the beginning and endpoint is by using the Hidden Markov Models (HMM) technique. The HMM is a state transition model and will use probabilities of feature numbers to determine the likelihood of transitioning from one state to another. Each phoneme is represented by a HMM. The English language is made up of 45–50 phonemes. A sequence of HMM will represent a word. This would be repeated for each word in the vocabulary. While the system can now recognize phonemes, phonemes do not always sound the same, depending on the phoneme preceding and following it. To address this problem, phonemes are placed in groups of three, called tri-phones, and as an aid in searching, similar sounding tri-phones are grouped together.

From the information obtained from the HMM state transitions, the recognizer is able to hypothesize and determine which phoneme likely was spoken, and then by referring this to a lexicon, the recognizer is able to determine the word that likely was spoken.

This is an overly simplified definition of the speech recognition process. There are numerous adaptations of the HMM technique and other modeling techniques. Some of these techniques are neural networks (NNs), dynamic time warping (DTW), and combinations of techniques.

20.2.1 Types of Speech Recognizers

There are two types of speech recognizers: speaker-dependent and speaker-independent.

20.2.1.1 Speaker-Dependent Systems

Speaker-dependent recognition is exactly that, speaker dependent. The system is designed to be used by one person. To operate accurately, the system will need to be "trained" to the user's individual speech patterns. This is sometimes referred to as "enrollment" of the speaker with the system. The speech patterns for the user will be recorded and patterned from which a template will be created for use by the speech recognizer. Because of the required training and storage of specific speech templates, the performance and accuracy of the speaker-dependent speech recognition engine will be tied to the voice patterns of a specific registered user. Speaker-dependent recognition, while being the most restrictive, is the most accurate, with accuracy rates in the mid to upper 90% range. For this reason, past research and applications for cockpit applications have opted to use speaker-dependent recognition.

The major drawback of this system is that it is dedicated to a single user, and that it must be trained prior to its use. Many applications will allow the speech template to be created elsewhere prior to use on the hosting system. This can be done at separate training stations prior to using the target system by transferring the created user voice template to the target system. If more than one user is anticipated, or if the training of the system is not desirable, a speaker-independent system might be an option.

20.2.1.2 Speaker-Independent Recognizers

Speaker-independent recognition systems are independent of the user. This type of system is intended to allow multiple users to access a system using voice input. Examples of speaker-independent systems are directory assistance programs and an airline reservation system with a voice input driven menu system. Major drawbacks with a speaker-independent system, in addition to increased complexity and difficult implementation, are its lower overall accuracy rate, higher system overhead, and slower response time. The impact of these drawbacks continues to lessen with increased processor speeds, faster hardware, and increased data storage capabilities.

A variation of the speaker-independent system is the speaker-adaptive system. The speaker-adaptive system will adapt to the speech pattern, vocabulary, and style of the user. Over time, as the system adapts to the users' speech characteristics, the error rate of the system will improve, exceeding that of the independent recognizer.

20.2.2 Vocabularies

A vocabulary is a list of words that are valid for the recognizer. The size of a vocabulary for a given speech recognition system affects the complexity, processing requirements, and the accuracy of that system. There are no established definitions for how large a vocabulary should be, but systems using smaller vocabularies can result in better recognizer accuracy. As a general rule, a small vocabulary may contain up to 100 words, a medium vocabulary may contain up to 1,000 words, a large vocabulary may contain up to 10,000 words, and a very large vocabulary may contain up to 64,000 words, and above that the vocabulary is considered unlimited. Again, this is a general rule and may not be true in all cases.

The size of a vocabulary will be dependent upon the purpose and intended function of the application. A very specific application may require only a few words and make use of a small vocabulary, while an application that would allow dictation or setting up airline reservations would require a very large vocabulary.

How can the size and contents of a vocabulary be determined? The words used by pilots are generally specific enough to require a small to medium vocabulary. Words that can or should be in the vocabulary could be determined in a number of ways. Drawing from the knowledge of how pilots would engage a desired function or task is one way. This could be done using a questionnaire or some similar survey method.

Another way to gather words for the vocabulary is to set up a lab situation and use the "Wizard of Oz" technique. This technique would have a test evaluator behind the scenes acting upon the commands given by a test subject. The test subject would have various tasks and scenarios to complete. While the test subject runs through the tasks, the words and phrases used by the subject are collected for evaluation. After running this process numerous times, the recorded spoken words and phrases will be used to construct a vocabulary list and command syntax, commonly referred to as a grammar. The vocabulary could be refined in further tests by only allowing those contained words and phrases to be valid, and have test subjects again run through a suite of tasks. Observations would be made as to how well the test subjects were able to complete the tasks using the defined vocabulary and syntax. Based on these tests, and the evaluation results, the vocabulary is modified as required.

A paper version of the evaluation process could be administered by giving the pilot a list of tasks, and then asking them to write out what commands they would use to perform the task. Following this data collection step, a second test could be generated having the pilot choose from a selected list of words and commands what he would likely say to complete the task. As a rule, pilots will tend to operate in a predictable manner, and this lends itself to a reduced vocabulary size and structured grammar.

20.2.3 Modes of Operation for Speech Recognizers

There are two modes of operation for a speech recognizer: continuous recognition, and discrete or isolated word recognition.

20.2.3.1 Continuous Recognition

Continuous speech recognition systems are able to operate on a continuous spoken stream of input in which the words are connected together. This type of recognition is more difficult to implement due to several inherent problems such as determining start and stop points in the stream and the rate of the spoken input.

The system must be able to determine the start and endpoint of a spoken stream of continuous speech. Words will have varied starting and ending phonemes depending on the surrounding phonemes. This is called "co-articulation." The rate of the spoken speech has a significant impact on the accuracy of the recognition system. The accuracy will degrade with rapid speech.

20.2.3.2 Discrete Word Recognition

Discrete or isolated word recognition systems operate on single words at a time. The system requires a pause between saying each word. The pause length will vary, and on some systems the pause length can be set to determined lengths. This type of recognition system is the simplest to perform because the endpoints are easier for the system to locate, and the pronunciation of a word is less likely to affect the pronunciation of other words (co-articulation effects are reduced). A user of this type of system will speak in a broken fashion. This system is the type most people think of in terms of a voice recognition system.

20.2.4 Methods of Error Reduction

There are no real standards by which error rates of various speech recognizers are measured and defined. Many systems claim accuracy rates in the high 90% range, but under actual usage with surrounding noise conditions, the real accuracy level may be much less. Many factors can impact the accuracy of SR systems. Some of these factors include the individual speech characteristics of the user, the operating environment, and the design of the SR system itself.

There are four general error types impacting the performance of a SR system; these are substitution errors, insertion errors, rejection errors, and operator errors:

- Substitution errors occur when the SR system incorrectly identifies a word from the vocabulary. An example might be the pilot calling out "Tune COM one to one two four point seven" and the SR system incorrectly recognizes that the pilot spoke "Tune NAV one to one two four point seven." The SR system substituted NAV in place of COM. Both words may be defined and valid in the vocabulary, but the system selected the wrong word.
- Insertion errors may occur when some source of sound other than a spoken word is interpreted by the system as valid speech. Random cockpit noise might at some time be identified as a valid word to the SR system. The use of noise-canceling microphones and PTT can help to reduce this type of error.
- Rejection errors occur when the SR system fails to respond to the user's speech, even if the word or phrase was valid.
- Operator errors occur when the user is attempting to use words or phrases that are not identifiable to the SR system. A simple example might be calling out "change the radio frequency to one one eight point six" instead of "Tune COM one to one one point eight six," which is recognized by the vocabulary.

When designing a speech recognition application, several design goals and objectives should be kept in mind:

- *Limitations of the hardware and the software*—Keep in mind the limitations of the hardware and the software being used for the application. Will the system need to have continuous recognition and discrete word recognition? Will the system need to be speaker independent, or will the reduced accuracy in using a speaker-independent recognizer be acceptable. Will the system be able to handle the required processing in an acceptable period of time? Will the system operate acceptably in the target environment?
- *Safety*—Will using SR to interface with a piece of equipment compromise safety? Will an error in recognition have a serious impact on the safety of flight? If the SR system should fail, is there an alternate method of control for that application?
- *Train the system in the environment in which it is intended to be used*—As discussed earlier, a SR system that has a 99% accuracy in the lab, may be frustrating and unusable in actual cockpit conditions. The speech templates or the training of the SR system needs to be done in the actual environment, or in as similar an environment as possible.
- *Don't try to use SR for tasks that don't really fit*—The problem with a new tool, like a new hammer, is that everything becomes a nail to try out that new hammer. Some tasks are natural candidates for using SR, many are not. Do not force SR onto a task if it is not appropriate for use of SR. Doing so will add significant risk and liability. Good target applications for SR include radio tuning functions, navigation functions, FMS functions, and display mode changes. Bad target applications for SR would be things that can affect the safety of flight, in short, anything that will kill you.
- *Incorporate error correction mechanisms*—Have the system repeat, using either voice synthesis or through a visual display, what it interprets, and allow the pilot to accept or reject this recognition. Allow the system to be able to recognize invalid recognition. If the recognizer interprets that it

heard the pilot call out an invalid frequency, it should recognize it as invalid and possibly query the pilot to repeat, or prompt the pilot by saying or displaying that the frequency is invalid.

- *Provide feedback of the SR system's activities*—Allow the user to interact with the SR system. Have the system speak, using voice synthesis, or display what it is doing. This will allow the user to either accept or reject the recognizer interpretation. This may also serve as a way to prompt a user for more data that may have been left out of the utterance. "Tune COM 1 to...." After a delay, the system might query the user for a frequency: "Please select frequency for COM1." If the user selects some repeated command, the system may repeat back the command as it is executed: "Tuning COM 1 to ...."

20.2.4.1 Reduced Vocabulary

One way to dramatically increase the accuracy of a SR system is to reduce the number of words in a vocabulary. In addition to the reduction in words, the words should be carefully chosen to weed out words that sound similar.

Use a trigger phrase to gain the attention of the recognizer. The trigger phrase might be as simple as "computer..." followed by some command. In this example, "computer" is the trigger phrase and alerts the recognizer that a command is likely to follow. This can be used with a system that is always on-line and listening.

Speech recognition errors can be reduced using a noise-canceling microphone. The flightdeck is not the quiet, sterile place a lab or a desktop might be. There are any number of noises and chatter that could interfere with the operation of speech recognition. Like humans, a recognizer can have increased difficulty in understanding commands in a noisy environment. In addition to the use of noise-canceling microphones, the use of high-quality omnidirectional microphones will offer further reduction in recognition errors. Using push-to-talk (PTT) microphones will help to reduce the occurrence of insertion errors as well as recognition errors.

20.2.4.2 Grammar

Grammar definition plays an important role in how accurate a SR application may be. It is used to not only define which words are valid to the system, but what the command syntax will be. A grammar notation frequently used in speech recognition is Context Free Grammar (CFG). A sample of a valid command in CFG is

$$\langle start \rangle = tune(COM \mid NAV) radio$$

This definition would allow valid commands of "tune COM radio," and "tune NAV radio." Word order is required, and words cannot be omitted. However, the grammar can be defined to allow for word order and omitted words.

20.3 Recent Applications

Though speech recognition has been applied to various flightdeck applications over the past 20 years, limitations in both hardware and software capability have kept the use of speech recognition from serious contention as a flightdeck tool. Even though there have been several notable applications of speech recognition in the recent past, and there are several current applications of speech recognition in the cockpit of military aircraft, it will likely be several more years before the civilian market will see such applications reach the level of reliability and pilot acceptance to see them commonly available.

In the mid-1990s, NASA performed experiments using speech recognition and voice control on an OV-10A aircraft. The experiment involved 12 pilots. The speech recognizer used for this study was an ITT VRS-1290 speaker-dependent system. The vocabulary used in this study was small, containing 54 words.

The SR system was tested under three separate conditions: on the ground, 1g conditions, and 3g conditions. There was no significant difference in SR system performance found between the three conditions. The accuracy rates for the SR system under these three test conditions was 97.27% in hangar conditions, 97.72 under 1g conditions, and 97.11% under 3g conditions.[1]

A recent installation that is now in production is a military fighter, the Eurofighter Typhoon. This aircraft will be the first production aircraft with voice interaction as a standard OEM configuration with speech recognition modules (SRMs). The speech recognizer is speaker dependent, and sophisticated enough to recognize continuous speech. The supplier of the voice recognition system for this aircraft is Smiths Industries. In addition, the system has received general pilot acceptance. Since the system is speaker dependent, the pilot must train the speech recognizer to his unique voice patterns prior to its use. This is done at ground-based, personal computer (PC) support stations. The PC is used to create a voice template for a specific pilot. The created voice template is then transferred to the aircraft prior to flight, via a data loader. Specifications for the recognizer include a 250-word vocabulary, a 200 ms response time, continuous speech recognition, and an accuracy rate of 95%–98%.[2]

20.4 Flightdeck Applications

The use of speech recognition, the enabling technology for voice control, should not be relied on as the sole means of control or entering data and commands. Speech recognition is more correctly defined as an assisted method of control and should have reversionary controls in place if the operation and performance of the SR system is no longer acceptable. It is not a question of whether voice control will find its way into mainstream aviation cockpits, but a question of when and to what degree. As the technology of SR continues to evolve, care must be exercised so that SR does not become a solution looking for a problem to solve. Not all situations will be good choices for the application of SR. In a high workload atmosphere, such as the flightdeck, the use of SR could be a logical choice for use in many operations, leading to a reduction in workload and head-down time.

Current speech recognition systems are best assigned to tasks that are not in themselves critical to the safety of flight. In time, this will change as the technology evolves. The thought of allowing the speech recognition system to gain the ability to directly impact flight safety brings to mind an example that occurred at a speech recognition conference several years ago. While a speech recognition interface on a PC was being discussed and demonstrated before an audience, a member of the audience spoke out "format C: return," or something to that effect. The result was the main drive on the computer was formatted, erasing its contents. Normally an event such as this impacts no one's safety, however, if such unrestricted control were allowed on an aircraft, there would be serious results.

Some likely applications for voice control on the flightdeck are navigation functions, communications functions such as frequency selection, toggling of display modes, checklist functions, etc.

20.4.1 Navigation Functions

For navigation functions, SR could be used as a method of entering waypoints and inputting FMS data. Generally, most tasks requiring the keyboard to be used to enter data into the FMS would make good use of a SR system. This would allow time and labor savings in what is a repetitive and time consuming task. Another advantage of using SR is that the system is able to reduce confusion and guide the user by requesting required data. The use of SR with FMS systems is being evaluated and studied by both military and civilian aviation.

20.4.2 Communication Functions

For communication functions, voice control could be used to tune radio frequencies by calling out that frequency. For example, "Tune COM1 to one one eight point seven." The SR system would interpret this utterance, and would place the frequency into stand-by. The system may be designed to have the SR system repeat the recognized frequency back through a voice synthesizer to the pilot for confirmation prior to the frequency being placed into standby. The pilot would then accept the frequency and make it active or reject it. This would be done with a button press to activate the frequency. Another possible method of making a frequency active would be to do this by voice alone. This does bring about some added risk, as the pilot will no longer be physically making the selection. This could be done by a simple, "COM one Accept" to accept the frequency, but leave it in pre-select. Reject the frequency by saying, "COM one Reject," and to activate the frequency by saying, "COM one activate."

The use of SR would also allow a pilot to query systems, such as by requesting a current frequency setting; "What is COM one?" The ASR system could then respond with the current active frequency and make possible pre-select. This response could be by voice or by display. Other possible options would be to have the SR respond to ATC commands by moving the command frequency change to the pre-select automatically. Having done this, the pilot would only have to command "Accept," "Activate," or "Reject." The radio would never on its own, place a frequency from standby to active mode.

With the use of a GPS position-referenced database, a pilot might only have to call out "Tune COM one Phoenix Sky Harbor Approach." By referencing the current aircraft location to a database, the SR systems could look up the appropriate frequency and place it into pre-select. The system might respond back with, "COM one Phoenix Sky Harbor Approach at one two oh point seven." The pilot would then be able to accept and activate the frequency without having to know the correct frequency numbers or having to dial the frequency into the radio. Clearly a time-saving operation. Possible drawbacks are out-of-date radio frequencies in the database or no frequency listing. This can be overcome by being able to call out specific frequencies if required. "Tune COM one to one two oh point seven."

20.4.3 Checklist

The use of speech recognition is almost a natural for checklist operations. The pilot may be able to command the system with "configure for take-off." This could lead to the system bringing up an appropriate checklist for take-off configuration. The speech system could call out the checklist items as they occur and the pilot, having completed and verified the task, could press a button to accept and move on to the next task. It may be possible to allow a pilot to verbally check-off a task vs. a button selection; however, that does bring about an opportunity for a recognition error.

Definitions

Accuracy: Generally, accuracy refers to the percentage of times that a speech recognizer will correctly recognize a word. This accuracy value is determined by dividing the number of times that the recognizer correctly identifies a word by the number of words input into the SR system.

Continuous speech recognition: The ability of the speech recognition system to accept a continuous, unbroken stream of words and recognize it as a valid phrase.

Discrete word recognition: This refers to the ability of a speech recognizer to recognize a discrete word. The words must be separated by a gap or pause between the previous word and successive words. The pause will typically be 150 ms or longer. The use of such a system is characterized by "choppy" speech to ensure the required break between words.

Grammar: This is a set of syntax rules determining valid commands and vocabulary for the SR system. The grammar will define how words may be ordered and what commands are valid. The grammar definition structure most commonly used is known as "context free grammar" or CFG.

Isolated word recognition: The ability of the SR system to recognize a specific word in a stream of words. Isolated word recognition can be used as a "trigger" to place the SR system into an active standby mode, ready to accept input.

Phonemes: Phonemes are the fundamental parts of speech. The English language is made up from 45 to 50 individual phonemes.

Speaker dependent: This type of system is dependent upon the speaker for operation. The system will be trained to recognize one person's speech patterns and acoustical properties. This type of system will have a higher accuracy rate than a speaker-independent system, but is limited to one user.

Speaker independent: A speaker-independent system will operate regardless of the speaker. This type of system is the most desirable for a general use application, however the accuracy rate and response rate will be lower than the speaker-dependent system.

Speech synthesis: The use of an artificial means to create speech-like sounds.

Text to speech: A mechanism or process in which text is transformed into digital audio form and output as "spoken" text. Speech synthesis can be used to allow a system to respond to a user verbally.

Tri-phones: These are groupings of three phonemes. The sound a phoneme makes can vary depending on the phoneme ahead of it and after it. Speech recognizers use tri-phones to better determine which phoneme has been spoken based upon the sounds preceding and following it.

Verbal artifacts: These are words or phrases, spoken with the intended command that have no value content to the command. This is sometimes referred to simply as garbage when defining a specific grammar. Grammars may be written to allow for this by disregarding and ignoring these utterances, for example, the pilot utterance, "uhhhhhmmmmmmm, select north up mode." The "uhhhhhmmmmmmm" would be ignored as garbage.

Vocabulary: The vocabulary a speech recognition system is made up of the words or phrases that the system is to recognize. Vocabulary size is generally broken into four sizes; small, with tens of words, medium with a few hundred words, large with a few thousand words, very large with up to 64,000 words, and unlimited. When a vocabulary is defined, it will contain words that are relative, and specific to the application.

Bibliography

Anderson, T. R., Applications of speech-based control, in *Proceedings of the Alternative Control Technologies: Human Factors Issues*, Wright-Patterson AFB, Dayton, OH, October 14–15, 1998.

Anderson, T. R., The technology of speech-based control, in *Proceedings of the Alternative Control Technologies: Human Factors Issues*, Wright-Patterson AFB, Dayton, OH, October 14–15, 1998.

Bekker, M. M., A comparison of mouse and speech input control of a text-annotation system, *Behaviour & Information Technology*, 14(1), 1995.

Eurofighter Typhoon Speech Recognition Module. Available: http://www.eurofighter.com/news-and-events/2008/05/direct-voice-input-technology. Accessed on May 15, 2014.

Hart, S. G., Helicopter human factors, in *Human Factors in Aviation*, Wiener, E. L. and Nagel, D. C. (eds.), Academic Press, San Diego, CA, pp. 591–638, 1988.

Hopkin, V. D., Air traffic control, in *Human Factors in Aviation*, Wiener, E. L. and Nagel, D. C. (eds.), Academic Press, San Diego, CA, 1988, Chapter 19.

Jones, D. M., Frankish, C. R., and Hapeshi, K., Automatic speech recognition in practice, *Behaviour and Information Technology*, 2, 109–122, 1992.

Leger, A., Synthesis and expected benefits analysis, in *Proceedings of the Alternative Control Technologies: Human Factors Issues*, Wright-Patterson AFB, Dayton, OH, October 14–15, 1998.

Rood, G. M., Operational rationale and related issues for alternative control technologies, in *Proceedings of the Alternative Control Technologies: Human Factors Issues*, Wright-Patterson AFB, Dayton, OH, October 14–15, 1998.

Rudnicky, A. I. and Hauptmann, A. G., Models for evaluating interaction protocols in speech recognition, School of Computer Science, Carnegie Mellon University, Pittsburgh, PA, 1991.

Wickens, C. D. and Flach, J. M., Information processing, in *Human Factors in Aviation*, Wiener, E. L. and Nagel, D. C. (eds.), Academic Press, San Diego, CA, 1988, Chapter 5.

Williamson, D. T., Barry, T. P., and Liggett, K. K., Flight test results of ITT VRS-1290 in NASA OV10A. Pilot-Vehicle Interface Branch (WL/FIGP), WPAFB, Dayton, OH.

Williges, R. C., Williges, B. H., and Fainter, R. G., Software interfaces for aviation systems, in *Human Factors in Aviation*, Wiener, E. L. and Nagel, D. C. (eds.), Academic Press, San Diego, CA, 1988, Chapter 14.

References

1. Williamson, D. T., Barry, T. P., and Liggett, K. K., Flight test results of ITT VRS-1290 in NASA OV10A. Pilot-Vehicle Interface Branch (WL/FIGP), WPAFB, Dayton, OH.

2. The Eurofighter Typhoon Speech Recognition Module [online]. Available: http://www.eurofighter. com/news-and-events/2008/05/direct-voice-input-technology. Accessed on May 15, 2014.

21

Terrain Awareness

Barry C. Breen
Honeywell

21.1 Enhanced Ground Proximity Warning System

The Enhanced Ground Proximity Warning System (EGPWS)* is one of the newest systems becoming standard on all military and civilian aircraft. Its purpose is to help provide situational awareness to terrain and to provide predictive alerts for flight into terrain. This system has a long history of development and its various modes of operation and warning/advisory functionality reflect that history:

- Controlled Flight Into Terrain (CFIT) is the act of flying a perfectly operating aircraft into the ground, water, or a man-made obstruction. Historically, CFIT is the most common type of fatal accident in worldwide flying operations.
- Analysis of the conditions surrounding CFIT accidents, as evidenced by flight recorder data, Air Traffic Control (ATC) records, and experiences of pilots in Controlled Flight Towards Terrain (CFTT) incidents, have identified common conditions which tend to precede this type of accident.
- Utilizing various onboard sensor determinations of the aircraft current state, and projecting that state dynamically into the near future, the EGPWS makes comparisons to the hazardous conditions known to precede a CFIT accident. If the conditions exceed the boundaries of safe operation, an aural and/or visual warning/advisory is given to alert the flight crew to take corrective action.

---

* There are other synonymous terms used by various government/industry facets to describe basically the same equipment. The military (historically at least) and at least one non-U.S. manufacturer refer to GPWS and EGPWS as Ground Collision Avoidance Systems (GCAS), although the military is starting to use the term EGPWS more frequently. The FAA, in its latest regulations concerning EGPWS functionality, have adopted the term Terrain Awareness Warning System (TAWS).

21.2 Fundamentals of Terrain Avoidance Warning

The current state of the aircraft is indicated by its position relative to the ground and surrounding terrain, attitude, motion vector, accelerations vector, configuration, current navigation data, and phase of flight. Depending upon operating modes (see next section) required or desired, and EGPWS model and complexity, the input set can be as simple as GPS position and pressure altitude or fairly large including altimeters, air data, flight management data, instrument navigation data, accelerometers, inertial references, etc. (see Figure 21.1).

The primary input to the "classic" GPWS (nonenhanced) is the Low Range Radio (or Radar) Altimeter (LRRA), which calculates the height of the aircraft above the ground level (AGL) by measuring the time it takes a radio or radar beam directed at the ground to be reflected back to the aircraft. Imminent danger of ground collision is inferred by the relationship of other aircraft performance data relative to a safe height above the ground. With this type of system, level flight toward terrain can only be implied by detecting rising terrain under the aircraft; for flight towards steeply rising terrain, this may not allow enough time for corrective action by the flight crew.

The EGPWS augments the classic GPWS modes by including in its computer memory a model of the earth's terrain and man-made objects, including airport locations and runway details. With this digital terrain elevation and airports database, the computer can continuously compare the aircraft state vector to a virtual three-dimensional map of the real world, thus predicting an evolving hazardous situation much in advance of the LRRA-based GPWS algorithms.

The EGPWS usually features a colored or monochrome display of terrain safely below the aircraft (shades of *green* for terrain and *blue* for water is standard). When a potentially hazardous situation

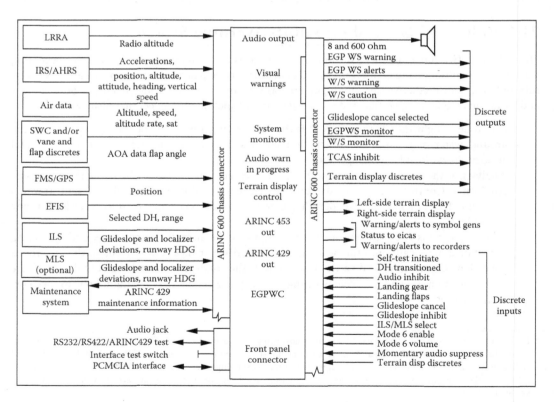

FIGURE 21.1　Typical air transport EGPWS installation.

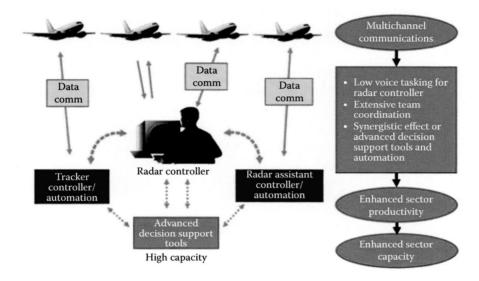

FIGURE 1.2 Tomorrow: Evolved sector with data comm. (Courtesy of Federal Aviation Administration, Washington, DC, Available online at: https://www.faa.gov/about/office_org/field_offices/fsdo/orl/local_more/media/fy13summit/NEXTGEN_MCO_Safety_Summit.pdf, accessed on April 27, 2014.)

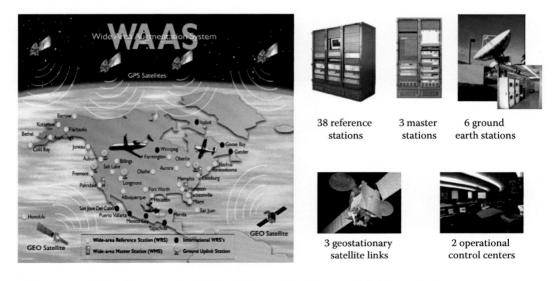

FIGURE 4.3 WAAS. (Courtesy of the Federal Aviation Administration, Washington, DC.)

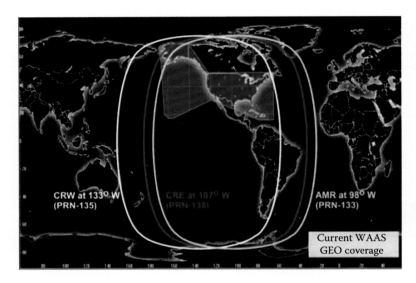

FIGURE 4.4 WAAS GEO coverage. (Courtesy of the Federal Aviation Administration, Washington, DC.)

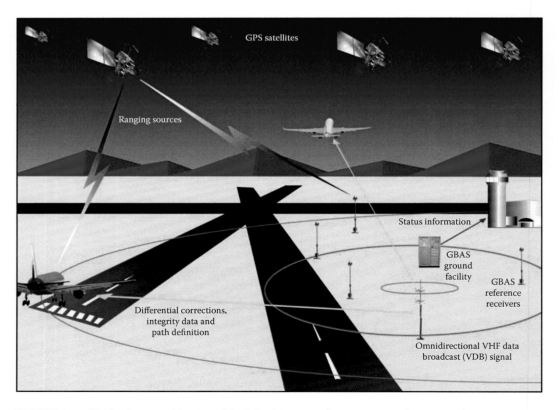

FIGURE 4.5 GBAS architecture. (Courtesy of the Federal Aviation Administration, Washington, DC.)

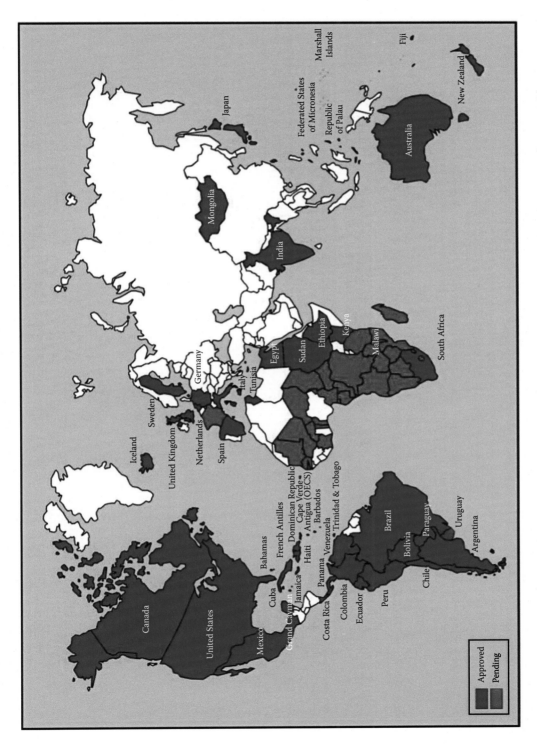

FIGURE 4.14 Nations that have approved GPS for aircraft navigation in IMC. (Courtesy of the Federal Aviation Administration, Washington, DC.)

FIGURE 8.1 Cockpit voice recorders.

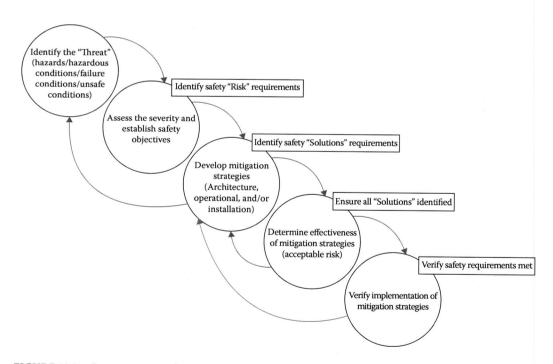

FIGURE 10.3 Generic system safety process model.

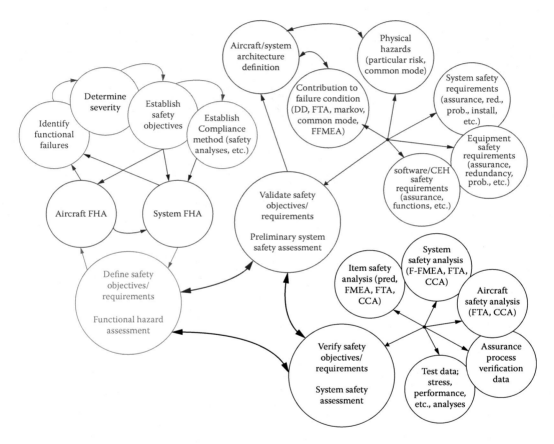

FIGURE 10.4 Safety assessment process.

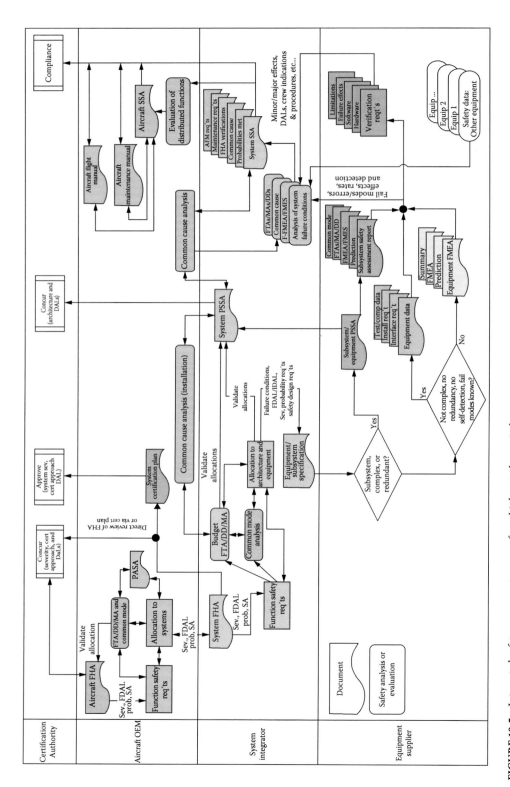

FIGURE 10.5 Integrated safety assessment process from highest to lowest indenture.

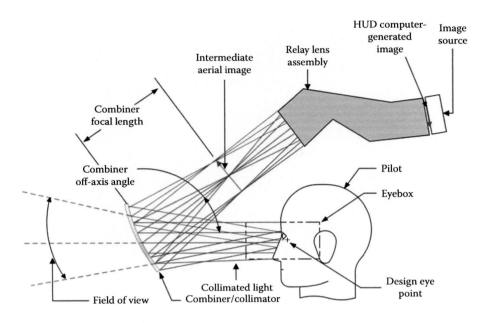

FIGURE 17.5 Reflective optical systems (overhead mounted).

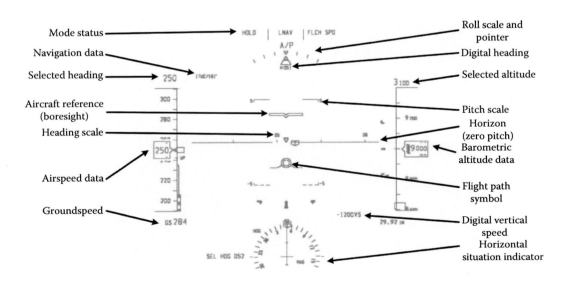

FIGURE 17.17 HGS primary mode symbology: in-flight.

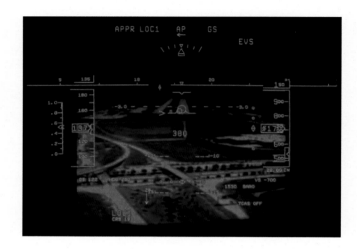

FIGURE 17.21   HUD symbology with enhanced vision video overlay.

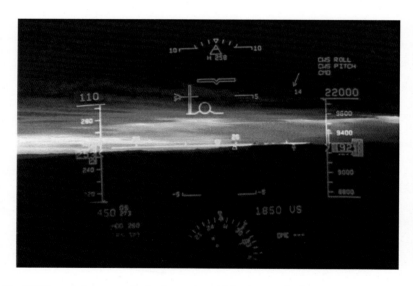

FIGURE 17.23   HUD symbology perceived color varies with background color.

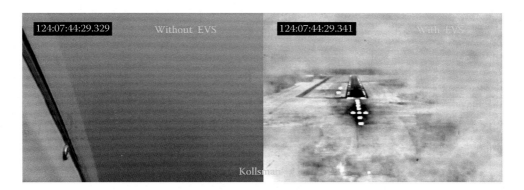

FIGURE 19.1 Example of EVS imagery (right) as compared with the corresponding out-the-window view (left) while in fog and cloud on final approach. (Courtesy of Kollsman Inc., Wichita, KS.)

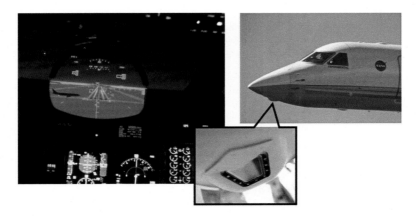

FIGURE 19.3 Example of an EVS installation.

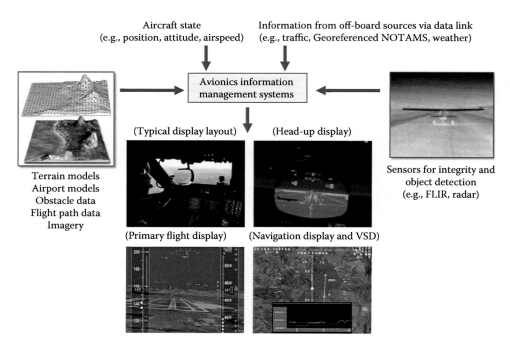

FIGURE 19.4 SVS elements. (Note: A subset of these elements may be used in particular designs.)

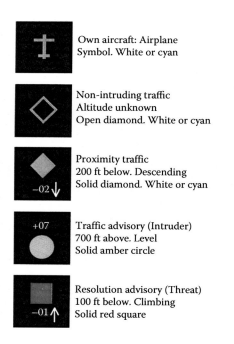

FIGURE 22.7 Standardized symbology for TA display.

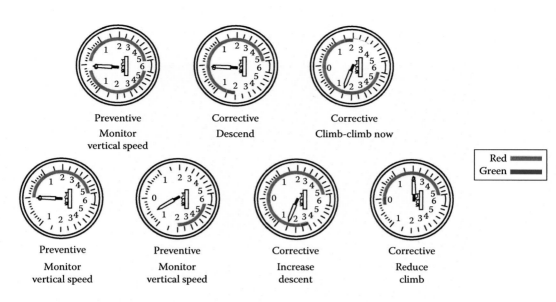

Preventive	Corrective
Monitor	Descend
vertical speed	
	Corrective
	Climb-climb now

Red
Green

Preventive	Preventive	Corrective	Corrective
Monitor	Monitor	Increase	Reduce
vertical speed	vertical speed	descent	climb

FIGURE 22.8 Typical RA indications.

FIGURE 22.9 Combined TA/RA display.

FIGURE 22.10 Joint-use weather radar and traffic display.

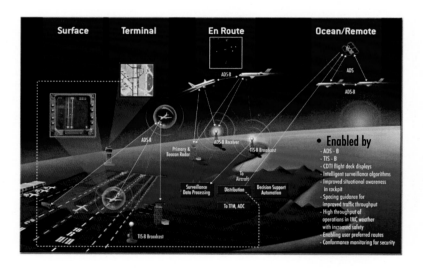

FIGURE 23.2 ADS-B supports operations during all flight phases.

FIGURE 23.6 CDTI.

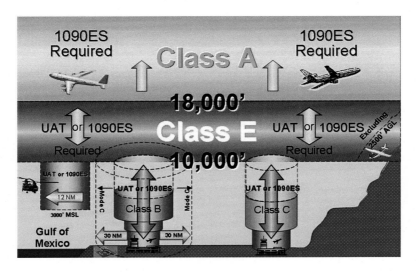

FIGURE 23.9 FAA ADS-B OUT rule airspace requirements. (From Advisory Circular (AC) 90-114 (Change 1), Automatic dependent surveillance-broadcast (ADS-B) operations, U.S. Department of Transportation, Federal Aviation Administration, Washington, DC, dated September 21, 2012.)

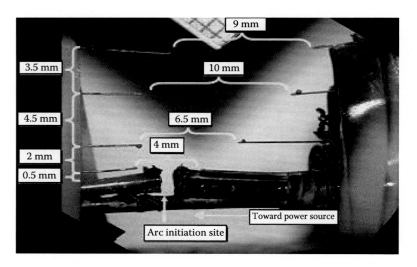

FIGURE 25.1 This is an example of the arc damage to copper strands suspended over an electrical arc. The top strand was nearly 0.5 in. from the arc initiation site. (Courtesy of Lectromec, Chantilly, VA.)

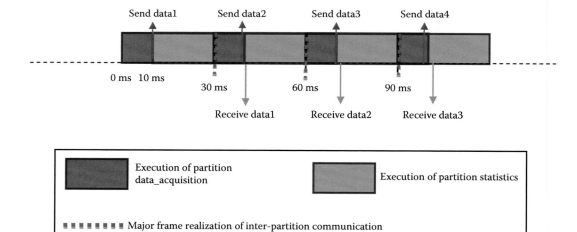

FIGURE 39.4 Scheduling of partitions.

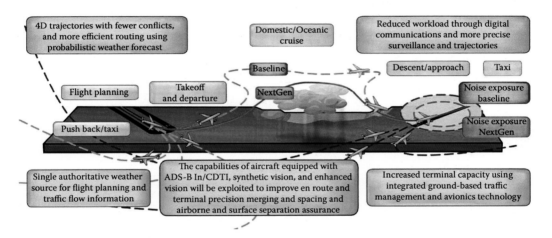

FIGURE 45.1 Targeted NextGen 2025 capabilities for a typical flight. (From Joint Planning and Development Office, Targeted NextGen capabilities for 2025, November 2011.)

FIGURE 45.3   Safe coexistence of many airspace users will be critical in the future.

exists, the EGPWS alerts the flight crew with aural and/or visual warnings. Advisory (informative) situational awareness information may consist simply of an aural statement, for example, "One Thousand," as the aircraft height AGL passes from above to below 1000 ft. Cautionary alerts combine an informative aural warning, for example, "Too Low Flaps" for flying too low and slow without yet deploying landing flaps, with a *yellow* visual alert. The cautionary visual alert can be an individual lamp with a legend such as "GPWS" or "TERRAIN"; it can be a yellow text message displayed on an Electronic Flight Instrument System (EFIS) display; or, in the case of the enhanced alert with a display of surrounding terrain, the aural "CAUTION TERRAIN" or "TERRAIN AHEAD," accompanied by both a *yellow* lamp and a rendering of the hazardous terrain on the display in *bright yellow*.

When collision with terrain is imminent and immediate drastic recovery action must be taken by the flight crew to avert disaster, the standard aural alert is a loud, commanding "PULL UP" accompanied by a *red* visual alert. Older aircraft with no terrain display utilize a single red "pull up" lamp; modern EFIS-equipped aircraft put up the words PULL UP in bright red on the Primary Flight Display (PFD). On the display of surrounding terrain, usually integrated on the EFIS Horizontal Situation Indicator, the location of hazardous terrain turns *bright red*.*

21.3 Operating Modes

The various sets of hazardous conditions that the EGPWS monitors and provides alerts for are commonly referred to as **Modes**.† These are described in detail in the following paragraphs.

Modes 1 through 4 are the original classic GPWS modes, first developed to alert the pilots to unsafe trajectory with respect to the terrain. The original analogue computer model had a single red visual lamp and a continuous siren tone as an aural alert for all modes. Aircraft manufacturer requirements caused refinement to the original modes, and added the voice "Pull Up" for Modes 1 through 4 and a new Mode 5 "Glideslope." Mode 6 was added with the first digital computer models about the time of Boeing 757/767 aircraft introduction; and Mode 7 was added when windshear detection became a requirement in about 1985.‡

The latest addition to the EGPWS are the Enhanced Modes: Terrain Proximity Display, Terrain Ahead Detection, and Terrain Clearance Floor. For many years, pilot advocates of GPWS requested that Mode 2 be augmented with a display of approaching terrain. Advances in memory density, lower costs, increased computing power, and the availability of high-resolution maps and Digital Terrain Elevation Databases (DTED) enabled this advancement. Once displayable terrain elevation database became a technical and economic reality, the obvious next step was to use the data to *look ahead* of the aircraft path and predict terrain conflict well before it happened, rather than waiting for the downward-looking sensors.

Combining the DTED with a database of airport runway locations, heights, and headings allows the final improvement—warnings for normal landing attempts where there is no runway.

---

* Note that all of the EGPWS visual indication examples in this overview discussion are consistent with the requirements of FAR 25.1322.

† The EGPWS modes described here are the most common for commercial and military transport applications. Not discussed here are more specialized warning algorithms, closely related to terrain-following technology, that have been developed for military high-speed low-altitude operations. These are more related to advanced terrain database guidance, which is outside the scope of enhanced situation awareness function.

‡ Though not considered CFIT, analysis of windshear-related accidents has resulted in the development of reactive windshear detection algorithms. At the request of Boeing, their specific reactive windshear detection algorithm was hosted in the standard commercial GPWS, about the same time the 737-3/4/500 series aircraft was developed. By convention this became Mode 7 in the GPWS. The most common commercially available EGPWS computer contains a Mode 7 consisting of both Boeing and non-Boeing reactive windshear detection algorithms, although not all aircraft installations will use Mode 7. There also exist "standalone" reactive windshear detection computers; and some aircraft use only predictive wind shear detection, which is a function of weather radar.

21.3.1 Mode 1: Excessive Descent Rate

The first ground proximity mode warns of an excessive descending barometric altitude rate near the ground, regardless of terrain profile. The original warning was a straight line at 4000 ft/min barometric sinkrate, enabled at 2400 ft AGL, just below the altitude at which the standard commercial radio altimeters came into track (2500 ft AGL). This has been refined over the years to a current standard for Mode 1 consisting of two curves, an outer cautionary alert and a more stringent inner warning boundary. Exceeding the limits of the outer curve results in the voice alert "Sinkrate"; exceeding the inner curve results in the voice alert "Pull Up."

Figure 21.2 illustrates the various Mode 1 curves, including the current standard air transport warnings, the DO-161A minimum warning requirement, the original curve and the Class B TSO C151 curves for 6–9 passenger aircraft and general aviation. Note that the Class B curves, which use GPS height above the terrain database instead of radio altitude, are not limited to the standard commercial radio altimeter range of 2500 ft AGL.

21.3.2 Mode 2: Excessive Closure Rate

Rate-of-change of radio altitude is termed the closure rate, with the positive sense meaning that the aircraft and the ground are coming closer together. When the closure rate initially exceeds the Mode 2 warning boundary, the alert "Terrain Terrain" is given. If the warning condition persists, the voice is changed to a "Pull Up" alert.

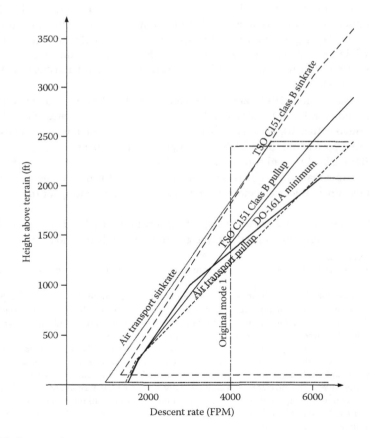

FIGURE 21.2 Mode 1 warning curves.

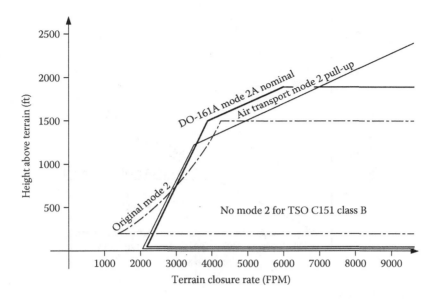

FIGURE 21.3 Mode 2 curves.

Closure rate detection curves are the most difficult of the classic GPWS algorithms to design. Tall buildings, towers, trees, and rock escarpments in the area of final approach can cause sharp spikes in the computed closure rate. Modern Mode 2 algorithms employ complex filtering of the computed rate, with varying dynamic response dependent upon phase of flight and aircraft configuration. The Mode 2 detection algorithm is also modified by specific problem areas by using latitude, longitude, heading, and selected runway course—a technique in the EGPWS termed "Envelope Modulation."

Landing configuration closure rate warnings are termed Mode 2B; cruise and approach configurations are termed Mode 2A. Figure 21.3 illustrates some of the various Mode 2A curves, including the original first Mode 2 curve, the current standard air transport Mode 2A "Terrain-Terrain-Pullup" warning curve, and the DO-161A nominal Mode 2A warning requirement. Note that the Class B TSO C151 EGPWS does not use a radio altimeter and therefore has no Mode 2.

21.3.3 Mode 3: Accelerating Flight Path Back into the Terrain after Take-Off

Mode 3 (Figure 21.4) is active from liftoff until a safe altitude is reached. This mode warns for failure to continue to gain altitude. The original Mode 3, still specified in DO-161A as Mode 3A, produced warnings for any negative sinkrate after take-off until 700 ft of ground clearance was reached. The mode has since been redesigned (designated 3B in DO-161A) to allow short-term sink after take-off but detect a trend to a lack of climb situation. The voice callout for Mode 3 is "Don't Sink." This take-off mode now remains active until a time-integrated ground clearance value is exceeded; thus allowing for a longer protection time with low-altitude noise abatement maneuvering before climb-out.

Altitude loss is computed by either sampling and differentiating altitude MSL or integrating altitude rate during loss of altitude. Because a loss is being measured, the altitude can be a corrected or uncorrected pressure altitude, or an inertial or GPS height. Typical Mode 3 curves are linear, with warnings for an 8-ft loss at 30 ft AGL, increasing to a 143-ft loss at 1500 ft AGL.

21.3.4 Mode 4: Unsafe Terrain Clearance Based on Aircraft Configuration

The earliest version of Mode 4 was a simple alert for descent below 500 ft with the landing gear up. Second generations of Mode 4 added additional alerting at lower altitudes for flaps not in landing

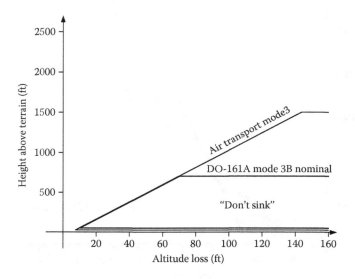

FIGURE 21.4   Mode 3 curves.

position. The warning altitude for flaps was raised to the 500-ft level for higher descent rates. There are three of these types of Mode 4 curves still specified as alternate minimum performance requirements in DO-161A (see Figure 21.5).

Modern Mode 4 curves are airspeed-enhanced, rather than descent rate alone, and for high airspeeds will give alerts at altitudes up to 1000 ft AGL.

Currently, EGPWS Mode 4 has three types of alerts based upon height AGL, mach/airspeed, and aircraft configuration, termed Modes 4A, 4B, and 4C (Figure 21.6). Two of the curves (4A, 4B) are active during cruise until full landing configuration is achieved with a descent "close to the ground"—typically 700 ft for a transport aircraft. Mode 4C is active on take-off in conjunction with the previously described Mode 3. All three alerts are designed with the intent to warn of flight "too close to the ground" for the current speed/configuration combination. At higher speeds, the alert commences at higher AGL and the voice alert is always "Too Low Terrain." At lower speeds, Mode 4A warning is "Too Low Gear" and the Mode 4B warning is "Too Low Flaps."

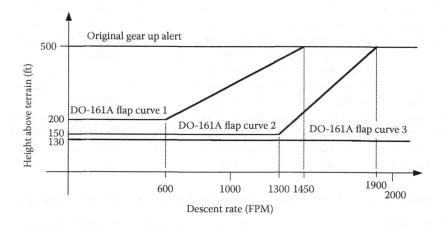

FIGURE 21.5   Old GPWS Mode 4 curves.

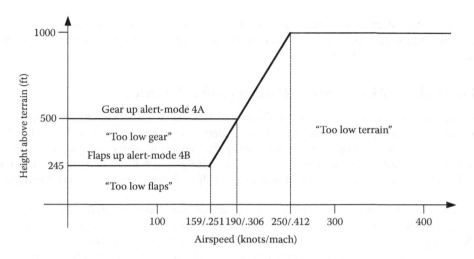

FIGURE 21.6 EGPWS Mode 4.

Mode 4C compliments Mode 3, which warns on an absolute loss of altitude on climb-out, by requiring a continuous gain in height above the terrain. If the aircraft is rising, but the terrain under is also rising, Mode 4C will alert "Too Low Terrain" on take-off if sufficient terrain clearance is not achieved prior to Mode 3 switching out.

21.3.5 Mode 5: Significant Descent below the ILS Landing Glide Path Approach Aid

This mode warns for failure to remain on an instrument glide path on approach. Typical warning curves alert for 1.5–2.0 dots below the beam, with a wider divergence allowed at lower altitudes. The alerts and warnings are only enabled when the crew is flying an ILS approach, as determined by radio frequency selections and switch selection. Most installations also include separate enable switch and a warning cancel for crew use when flying some combination of visual and or other landing aids and deviation from the ILS glide path is intentional. Although the mode is typically active from 1000 ft AGL down to about 30 ft, allowance in the design of the alerts must also be made for beam capture from below, and level maneuvering between 500 and 1000 ft without nuisance alerting.

Figure 21.7 illustrates the Mode 5 warnings for a typical jet air transport. When the outer curve is penetrated, the voice message "Glideslope" is repeated at a low volume. If the deviation below the beam

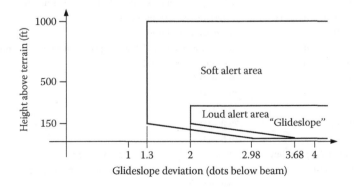

FIGURE 21.7 EGPWS Mode 5.

increases or altitude decreases, the repetition rate of the voice is increased. If the altitude/deviation combination falls within the inner curve, the voice volume increases to the equivalent of a warning message and the repetition rate is at maximum.

21.3.6 Mode 6: Miscellaneous Callouts and Advisories

The first application of this mode consisted of a voice alert added to the activation of the decision height discrete on older analog radio altimeters. This voice alert is "Minimums" or "Decision Height," which adds an extra level of awareness during the landing decision point in the final approach procedure. Traditionally, this callout would be made by the pilot not flying (PNF). Automating the callout frees the PNF from one small task enabling him to more easily monitor other parameters during the final approach.

This mode has since been expanded as a "catch all" of miscellaneous aural callouts requested by air transport manufacturers and operators, many of which also were normally an operational duty of the PNF (see Table 21.1). In addition to the radio altitude decision height, callouts are now available at barometric minimums, at an altitude approaching the decision height or barometric minimums, or at various combinations of specific altitudes. There are also "smart callouts" available that only call the altitude for nonprecision approaches (ILS not tuned). The EGPWS model used on Boeing aircraft will also callout for V_1 on take-off and give aural "engine out" warnings. Finally, included in the set of Mode 6 callouts are warnings of overbanking (excessive roll angle) (Figure 21.8).

21.3.7 Mode 7: Flight into Windshear Conditions

Windshear is a sudden change in wind direction and/or windspeed over a relatively short distance in the atmosphere and can have a detrimental effect on the performance of an aircraft. The magnitude of a windshear is defined precisely in engineering terms by the sum of the rate of change of horizontal wind, and the vertical wind divided by the true airspeed of the aircraft:

$$F = -\left(\frac{w_{\text{wind}}}{V_A} + \frac{\dot{u}_{\text{wind}}}{g} \right)$$

TABLE 21.1   Examples of EGPWS Mode 6 Callouts

Voice Callout	Description
Radio altimeter	Activates at 2500 ft as radio altimeter comes into track
Twenty five hundred	(Alternate to radio altimeter)
One thousand	Activates at 1000 ft AGL
Five hundred (smart)	Activates at 500 ft AGL for non-precision approaches only
One hundred	Activates at 100 ft AGL
Fifty	50 ft AGL
Forty	40 ft AGL
Thirty	30 ft AGL
Twenty	20 ft AGL
Ten	10 ft AGL
Approaching minimums	100 ft above the selected decision height
Minimums	At pilot selected decision height—may be AGL or barometric
Decision height	(Alternate to minimums)

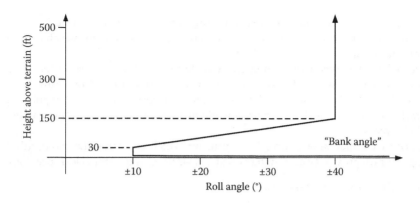

FIGURE 21.8 EGPWS Mode 6 overbank (excessive roll) alert.

where

F is expressed in units of g and is positive for increasing energy windshears

w_{wind} is the vertical wind velocity (fps), positive for downdrafts

$\dot{u}_{wind} = \dfrac{du_{wind}}{dt}$ is the rate of change of horizontal wind velocity

V_A is the true airspeed (fps)

g is the gravitational acceleration, 32.178 fps^2

There are various techniques for computing this windshear factor from onboard aircraft sensors (air data, inertial accelerations, etc.). The EGPWS performs this computation and alerts the crew with the aural "Windshear, Windshear, Windshear," when the factor exceeds predefined limits as required by TSO C117a.

21.3.8 Envelope Modulation

Early GPWS equipment was plagued by false and nuisance warnings, causing pilots to distrust the equipment when actual hazardous conditions existed. Many approach profiles and radar vectoring situations violated the best-selected warning curve designs. Even as GPWS algorithms were improved, there still existed some approaches that required close proximity to terrain prior to landing.

Modern GPWS equipment adapts to this problem by storing a table of known problem locations and providing specialized warning envelope changes when the aircraft is operating in these areas. This technique is known as GPWS envelope modulation.

An example exists in the southerly directed approaches to Glasgow Scotland, Runway 23. The standard approach procedures allow an aircraft flying level at 3000 ft barometric altitude to pass over mountain peaks with heights above 1700 ft when approaching this runway. At nominal airspeeds the difference in surrounding terrain height will generate closure rates well within the nominal curve of Figure 21.3. With the envelope modulation feature the GPWS, using latitude, longitude, and heading, notes that the aircraft is flying over this specific area and temporarily lowers the maximum warning altitude for Mode 2 from 2450 ft to the minimum 1250 ft AGL. This eliminates the nuisance warning while at the same time providing the minimum required DO-161A protection for inadvertent flight closer to the mountain peaks on the approach path.

21.3.9 Enhanced Modes

The enhanced modes provide terrain and obstacle awareness beyond the normal sensor-derived capabilities of the standard GPWS. Standard GPWS warning curves are deficient in two areas, even with the best designs. One area is immediately surrounding the airport; which is where a large majority of CFIT accidents occur. The other is flight directly into precipitous terrain, for which little or no Mode 2 warning time may occur.

The enhanced modes solve these problems by making use of a database of terrain and obstacle spot heights and airport runway locations arranged in a grid addressed by latitude and longitude. This combined terrain/airports/obstacle database—a virtual world within the computer—provides the ability to track the aircraft position in the real world given accurate *x-y-z* position combined with the aircraft velocity vector.

This database technique allows three improvements which overcome the standard GPWS modes shortcomings: terrain proximity display, terrain ahead alerting, and terrain clearance floor.

21.3.9.1 Terrain Proximity Display

The terrain proximity display is a particular case of a horizontal (plan view) moving map designed to enhance vertical and horizontal situational awareness. The basic display is based upon human factors studies recommending a minimum of contours and coloring. The display is purposely compatible with existing three-color weather radar displays, allowing economical upgrade of existing equipment.

Terrain well below the flight path of the aircraft is depicted in shades of green, brighter green being closer to the aircraft and sparse green-to-black for terrain far below the aircraft. Some displays additionally allow water areas to be shown in cyan (blue). Terrain in the proximity of the aircraft flight path, but posing no immediate danger (it can be easily flown over or around) is depicted in shades of yellow. Terrain well above the aircraft (nominally more than 2000 ft above flight level), toward which continued safe flight is not possible, is shown in shades of red.

21.3.9.2 Terrain Ahead Alerting

Terrain (and/or obstruction) alerting algorithms continually compare the state of the aircraft flight to the virtual world and provide visual and/or aural alerts if impact is possible or probable. Two levels of alerting are provided, a cautionary alert and a hard warning. The alerting algorithm design is such that, for a steady approach to hazardous terrain, the cautionary alert is given much in advance of the warning alert. Typical design criteria may try to issue caution up to 60 s in advance of a problem and a warning within 30 s.

Voice alerts for the cautionary alert are "Caution, Terrain" or "Terrain Ahead." For the warnings on turboprop and turbojet aircraft, the warning aural is "Terrain Terrain Pullup" or "Terrain Ahead Pullup," with the pullups being repeated continuously until the aircraft flight path is altered to avoid the terrain.

In conjunction with the aural alerts, yellow and red lamps may be illuminated, such as with the standard GPWS alerts. The more compelling visual alerts are given by means of the Terrain Awareness Display. Those areas that meet the criteria for the cautionary alert are illuminated in a bright yellow on the display. If the pullup alert occurs, those areas of terrain where an immediate impact hazard exists are illuminated in bright red. When the aircraft flight path is altered to avoid the terrain, the display returns to the normal terrain proximity depictions as the aural alerts cease.

21.3.9.3 Terrain Clearance Floor

The standard Modes 2 and 4 are desensitized when the aircraft is put in landing configuration (flaps down and/or gear lowered) and thus fail to alert for attempts at landing where there is no airport. Since the EGPWS database contains the exact position of all allowable airport runways, it is possible to define an additional alert, a terrain clearance floor, at all areas where there are no runways. When the aircraft descends below this floor value, the voice alert "Too Low Terrain" is given. This enhanced mode alert is also referred to as premature descent alert.

21.4 EGPWS Standards

ARINC 594—Ground Proximity Warning System: This is the first ARINC characteristic for Ground Proximity Warning Systems and defines the original analog interfaced system. It applies to the original model (MkI and MkII) GPWS systems featuring Modes 1–5, manufactured by Sundstrand Data Control, Bendix, Collins, Litton and others. It also applies to the AlliedSignal (Honeywell) MkVII digital GPWS, which featured Modes 1–7 and a primarily analog interface for upgrading older models.

ARINC 743—Ground Proximity Warning System: This characteristic applies to primarily digital (per ARINC 429 DITS) interfaced Ground Proximity Warning Systems, such as the AlliedSignal/Honeywell MkV series, which was standard on all newer Boeing aircraft from the 757/767 up through the introduction of the 777.

ARINC 762—Terrain Avoidance and Warning System: This characteristic is an update of ARINC 743 applicable to the primarily digital interfaced (MkV) Enhanced GPWS.

ARINC 562—Terrain Avoidance and Warning System: This proposed ARINC characteristic will be an update of ARINC 594, applicable to the primarily analog interfaced (MkVII) Enhanced GPWS.

RTCA DO-161A—Minimum Performance Standards, Airborne Ground Proximity Warning System: This 1976 document still provides the minimum standards for the classic GPWS Modes 1–5. It is required by both TSO C92c and the new TSO C151 for EGPWS (TAWS).

TSO C92c—Ground Proximity Warning, Glideslope Deviation Alerting Equipment: This TSO covers the classic Modes 1–6 minimum performance standards. It basically references DO-161A and customizes and adds features of the classic GPWS which were added subsequent to DO-161A, including voice callouts signifying the reason for the alert/warnings, Mode 6 callouts, and bank angle alerting.

CAA Specification 14 (U.K.)—Ground Proximity Warning Systems: This is the United Kingdom CAA standard for Modes 1–5 and also specifies some installation requirements. As with the U.S. TSOs, Spec 14 references DO-161A and customizes and augments features of the classic GPWS which are still required for U.K. approvals. Most notably, the U.K. version of Mode 5 is less stringent and requires a visual indication of Mode 5 cancellation. Spec 14 also requires that a stall warning inhibit the GPWS voice callouts, a feature which is found only on U.K.-certified installations.

TSO C117a—Airborne Windshear Warning and Escape Guidance Systems for Transport Airplanes: This TSO defines the requirements for EGPWS Mode 7, reactive low level windshear detection.

TSO C151a—Terrain Awareness and Warning System (TAWS): This TSO supersedes TSO C92c for certain classes of aircraft and requires the system to include the enhanced modes. It also extends coverage down to smaller aircraft, including a non-required functionality for small Part 91 piston powered aircraft. It describes three classes of TAWS equipment. Class A, the standard EGPWS, contains all the modes previously described here including the terrain display. Class B is for intermediate sized turbine powered aircraft. Class B requirements include the enhanced modes but do not require the old TSO C92c modes that depend upon a radio altimeter—altitude above ground level is instead computed from MSL altitude and the terrain display is optional. Class C is similar to Class B but the "must warn" and "must not warn" requirements are more appropriate for small general aviation operations. Class C is entirely voluntary but includes a requirement that a higher accuracy vertical source be used (e.g., GPS altitude)—as such the installation does not require an air data computer typically used on larger aircraft (although air data can be used if available).

AC 23-18—Installation of Terrain Awareness and Warning System (TAWS) Approved for Part 23 Airplanes.

AC 25-23—Airworthiness Criteria for the Installation Approval of a Terrain Awareness and Warning System (TAWS) for Part 25 Airplanes.

FAR 91.223—Terrain Awareness and Warning System: Requires that at least Class B equipment is installed on all turbine-powered U.S.-registered aircraft with six passenger seats or more.

FAR 121.354—Terrain Awareness and Warning System: Requires that Class A equipment is installed, including a terrain display, for all aircraft used in domestic, flag, and supplemental operations.

FAR 135.154—Terrain Awareness and Warning System: For all aircraft used in commuter and on-demand operations, requires Class A equipment is installed, including a terrain display, for turbine-powered aircraft with ten passenger seats or more; and, requires at least Class B equipment is installed for all turbine-powered aircraft with six to nine passenger seats.

RTCA DO-200A—Standards for Processing Aeronautical Data: This standard is required by TSO C151c to be used for processing the TAWS terrain database.

Further Reading

1. *Controlled Flight Into Terrain, Education and Training Aid*—This joint publication of ICAO, Flight Safety Foundation, and DOT/FAA consists of two loose-leaf volumes and an accompanying video tape. It is targeted toward the air transport industry, containing management, operations, and crew training information, including GPWS. Copies may be obtained by contacting the Flight Safety Foundation, Alexandria, Virginia.

2. *DOT Volpe NTSC Reports on CFIT and GPWS*—These may be obtained from the USDOT and contain accident analyses, statistics, and studies of the effectivity of both the classic and enhanced GPWS warning modes. There are a number of these reports which were developed in response to NTSB requests. Of the two most recent reports, the second one pertains to the Enhanced GPWS in particular:

   a. Spiller, David—Investigation of Controlled Flight Into Terrain (CFIT) Accidents Involving Multi-engine Fixed-wing Aircraft Operating Under Part 135 and the Potential Application of a Ground Proximity Warning System (Cambridge, MA: U.S. Department of Transportation, Volpe National Transportation Systems Center) March 1989.

   b. Phillips, Robert O.—Investigation of Controlled Flight Into Terrain Aircraft Accidents Involving Turbine Powered Aircraft with Six or More Passenger Seats Flying Under FAR Part 91 Flight Rules and the Potential for Their Prevention by Ground Proximity Warning Systems (Cambridge, MA: U.S. Department of Transportation, Volpe National Transportation Systems Center) March 1996.

22

Traffic Alert and Collision Avoidance System II (TCAS II)

Steve Henely
Rockwell Collins

22.1 Introduction

The traffic alert and collision avoidance system (TCAS) provides a solution to the problem of reducing the risk of midair collisions between aircraft. TCAS is a family of airborne systems that function independently of ground-based air traffic control (ATC) to provide collision avoidance protection. The TCAS concept makes use of the radar beacon transponders carried by aircraft for ground ATC purposes and provides no protection against aircraft that do not have an operating transponder.

TCAS I provides proximity alerts only, to aid the pilot in the visual acquisition of potential threat aircraft. TCAS II provides traffic advisories (TAs) and resolution advisories (recommended evasive maneuvers) in a vertical direction to avoid conflicting traffic. Development of TCAS III, which was to provide TAs and resolution advisories in the horizontal as well as the vertical direction, was discontinued in favor of emerging systems. This chapter will focus on TCAS II.

Based on a congressional mandate (U.S. Public Law 100-223), the Federal Aviation Administration (FAA) issued a rule effective February 9, 1989, that required the equipage of TCAS II on airline aircraft with more than 30 seats by December 30, 1991. Public Law 100-223 was later amended (U.S. Public Law 101-236) to permit the FAA to extend the deadline for TCAS II fleet-wide implementation to December 30, 1993. In December of 1998, the FAA released a technical standard order (TSO) that approved Change 7, resulting in the DO-185A TCAS II requirement. Change 7 incorporates software enhancements to reduce the number of false alerts. Based on extensive analysis of TCAS II Change 7 performance since 2000, additional changes were identified to improve resolution advisory (RA) logic. Validation of these logic changes was performed in Europe and the United States, resulting in the publication of Change 7.1 of the minimum operational performance standard (MOPS), DO-185B.

22.2 Components

TCAS II consists of the Mode S/TCAS control panel, the Mode S transponder, the TCAS computer, antennas, traffic and RA displays, and an aural annunciator. Figure 22.1 is a block diagram of TCAS II. Control information from the Mode S/TCAS control panel is provided to the TCAS computer via the Mode S transponder. TCAS II uses a directional antenna, mounted on top of the aircraft. In addition to receiving range and altitude data on targets above the aircraft, this directional antenna is used to transmit interrogations at varying power levels in each of four 908 azimuth segments. An omnidirectional or directional transmitting and receiving antenna is mounted at the bottom of the aircraft to provide TCAS with range and altitude data from traffic that is below the aircraft. TCAS II transmits transponder interrogations on 1030 MHz and receives transponder replies on 1090 MHz.

The TA display depicts the position of the traffic relative to the TCAS aircraft to assist the pilot in visually acquiring threatening aircraft. The RA can be displayed on a standard vertical speed indicator (VSI), modified to indicate the vertical rate that must be achieved to maintain safe separation from threatening aircraft. When an RA is generated, the TCAS II computer lights up the appropriate display segments and RA compliance is accomplished by flying to keep the VSI needle out of the red segments. On newer aircraft, the RA display function is integrated into the primary flight display (PFD). Displayed traffic and resolution advisories are supplemented by synthetic voice advisories generated by the TCAS II computer.

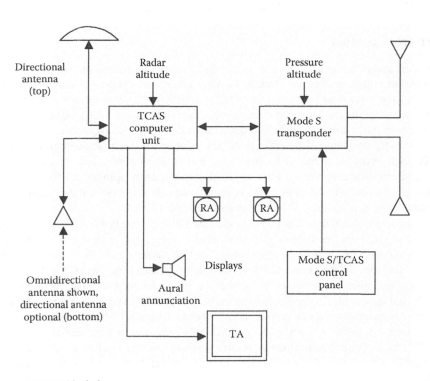

FIGURE 22.1 TCAS II block diagram.

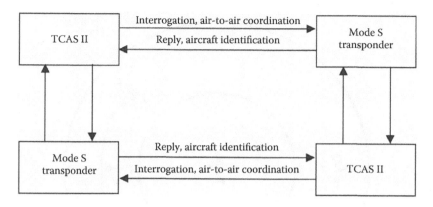

FIGURE 22.2 Interrogation/reply between TCAS systems.

22.3 Surveillance

TCAS listens for the broadcast transmission (squitters), which is generated once per second by the Mode S transponder and contains the discrete Mode S address of the sending aircraft. Upon receipt of a valid squitter message, the transmitting aircraft identification is added to a list of aircraft the TCAS aircraft will interrogate. Figure 22.2 shows the interrogation/reply communications between TCAS systems. TCAS sends an interrogation to the Mode S transponder with the discrete Mode S address contained in the squitter message. From the reply, TCAS can determine the range and the altitude of the interrogated aircraft.

There is no selective addressing capability with Mode A/C transponders, so TCAS uses the Mode C only all-call message to interrogate these types of Mode A/C transponders at a nominal rate of once per second. Mode C transponders reply with altitude data while Mode A transponders reply with no data in the altitude field. All Mode A/C transponders that receive a Mode C all-call interrogation from TCAS will reply. Since the length of the reply is 21 μs, Mode A/C-equipped aircraft within a range difference of 1.7 nmi from the TCAS will generate replies that overlap each other, as shown in Figure 22.3. These overlapping Mode A/C replies are known as synchronous garble.

Hardware degarblers can reliably decode up to three overlapping replies. The whisper–shout technique and directional transmissions can be used to reduce the number of transponders that reply to a single interrogation. A low power level is used for the first interrogation step in a whisper–shout sequence. In the second whisper–shout step, a suppression pulse is first transmitted at a slightly lower level than the first interrogation, followed 2 ms later by an interrogation at a slightly higher power level than the first interrogation. The whisper–shout procedure shown in Figure 22.4 reduces the possibility of garble by suppressing most of the transponders that had replied to the previous interrogation but eliciting replies from an additional group of transponders that did not reply to the previous interrogation. Directional interrogation transmissions further reduce the number of potential overlapping replies.

22.4 Protected Airspace

One of the most important milestones in the quest for an effective collision avoidance system is the development of the range/range rate (tau). This concept is based on time to go, rather than distance to go, to the closest point of approach. Effective collision avoidance logic involves a trade-off between providing the necessary protection with the detection of valid threats while avoiding false alarms. This trade-off is accomplished by controlling the sensitivity level, which determines the tau, and therefore the dimensions of the protected airspace around each TCAS-equipped aircraft.

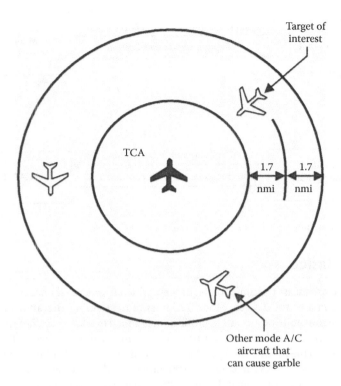

FIGURE 22.3   Synchronous garble area.

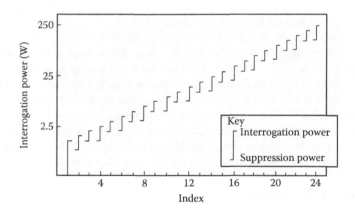

FIGURE 22.4   Whisper–shout interrogation.

The pilot can select three modes of TCAS operation: *standby*, *TA-only*, and *automatic*. These modes are used by the TCAS logic to determine the sensitivity level. When the *STANDBY* mode is selected, the TCAS equipment does not transmit interrogations. Normally, the *STANDBY* mode is used when the aircraft is on the ground. In *TA-ONLY* mode, the equipment performs all of the surveillance functions and provides TAs but not RAs. The *TA-ONLY* mode is used to avoid unnecessary distractions while at low altitudes and on final approach to an airport. When the pilot selects *AUTOMATIC* mode, the TCAS logic selects the sensitivity level based on the current altitude of the aircraft. Table 22.1 shows the altitude thresholds at which TCAS automatically changes its sensitivity level selection and the associated tau values for altitude-reporting aircraft.

The boundary lines depicted in Figure 22.5 show the combinations of range and range rate that would trigger a TA with a 40 s tau and an RA with a 25 s tau. These TA and RA values correspond to sensitivity level 5 from Table 22.1. As shown in Figure 22.5, the boundary lines are modified at close range to provide added protection against slow closure encounters.

TABLE 22.1   Sensitivity Level Selection Based on Altitude

Altitude (ft)	Sensitivity Level	Tau Values (s) TA	RA
0–1,000 AGL	2	20	NA
1,000–2,350 AGL	3	25	15
2,350–5,000 MSL	4	30	20
5,000–10,000 MSL	5	40	25
10,000–20,000 MSL	6	45	30
20,000–42,000 MSL	7	48	35
Greater than 42,000 MSL	7	48	35

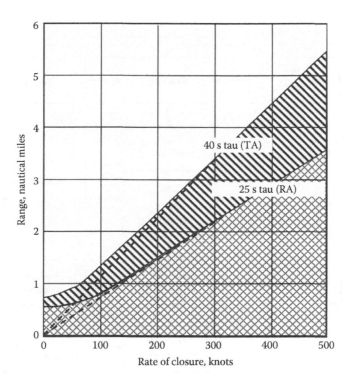

FIGURE 22.5   TA/RA tau values for sensitivity level 5.

22.5 Collision Avoidance Logic

The collision avoidance logic functions are shown in Figure 22.6. This description of the collision avoidance logic is meant to provide a general overview. There are many special conditions relating to particular geometry, thresholds, and equipment configurations that are not covered in this description. Using surveillance reports, the collision avoidance logic tracks the slant range and closing speed of each target to determine the time in seconds until the closest point of approach. If the target is equipped with an altitude-encoding transponder, collision avoidance logic can project the altitude of the target at the closest point of approach.

A range test must be met and the vertical separation at the closest point of approach must be within 850 ft for an altitude-reporting target to be declared a potential threat and a TA to be generated. The range test is based on the RA tau plus approximately 15 s. A non-altitude-reporting target is declared a potential threat if the range test alone shows that the calculated tau is within the RA tau threshold associated with the sensitivity level being used.

A two-step process is used to determine the type of RA to be selected when a threat is declared. The first step is to select the sense (upward or downward) of the RA. Based on the range and altitude tracks of the potential threat, the collision avoidance logic models the potential threat's path to the closest point of approach and selects the RA sense that provides the greater vertical separation. The second RA step is to select the strength of the RA. The least disruptive vertical rate maneuver that will achieve safe separation is selected. Possible resolution advisories are listed in Table 22.2.

In a TCAS/TCAS encounter, each aircraft transmits Mode S coordination interrogations to the other to ensure the selection of complementary resolution advisories. Coordination interrogations contain information about an aircraft's intended vertical maneuver.

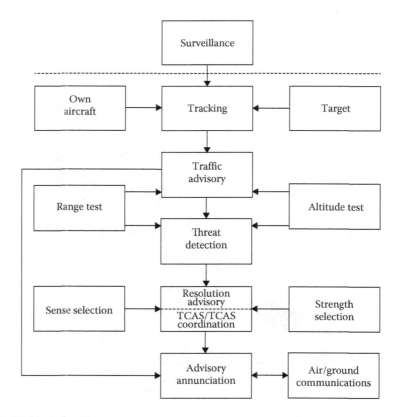

FIGURE 22.6 CAS logic functions.

TABLE 22.2   Resolution Advisories

Upward Sense	Type	Downward Sense
Increase climb to 2500 fpm	Positive	Increase descent to 2500 fpm
Reversal to climb	Positive	Reversal to descend
Maintain climb	Positive	Maintain descent
Crossover climb	Positive	Crossover descend
Climb	Positive	Descend
Don't descend	Negative vsl	Don't climb
Don't descend >500 fpm	Negative vsl	Don't climb >500 fpm
Don't descend >1000 fpm	Negative vsl	Don't climb >1000 fpm
Don't descend >2000 fpm	Negative vsl	Don't climb >2000 fpm

Note: Any combination of climb and descent restrictions may be given simultaneously (normally in multiaircraft encounters); fpm, feet per minute; vsl, vertical speed limit.

22.6 Cockpit Presentation

The TA display can either be a dedicated TCAS display or a joint-use weather radar and traffic display (see Figure 22.10). In some aircraft, the TA display will be an electronic flight instrument system (EFIS) or flat-panel display that combines TA and RA information on the same display. Targets of interest on the TA display are depicted in various shapes and colors as shown in Figure 22.7.

The pilot uses the RA display to determine whether an adjustment in aircraft vertical rate is necessary to comply with the RA determined by TCAS. This determination is based on the position of the VSI needle with respect to the lighted segments. If the needle is in the red segments, the pilot should change

Own aircraft: Airplane Symbol. White or cyan

Non-intruding traffic Altitude unknown Open diamond. White or cyan

Proximity traffic 200 ft below. Descending Solid diamond. White or cyan

Traffic advisory (Intruder) 700 ft above. Level Solid amber circle

Resolution advisory (Threat) 100 ft below. Climbing Solid red square

FIGURE 22.7   (See color insert.) Standardized symbology for TA display.

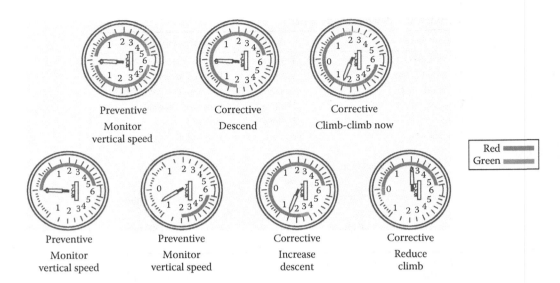

FIGURE 22.8　(**See color insert.**) Typical RA indications.

the aircraft vertical rate until the needle falls within the green "fly-to" segment. This type of indication is called a corrective RA. A preventive RA is when the needle is outside the red segments and the pilot should simply maintain the current vertical rate. The green segment is lit only for corrective resolution advisories. RA display indications corresponding to typical encounters are shown in Figure 22.8.

Figure 22.9 shows a combined TA/RA display indicating a TA (potential threat 200 ft below), RA (threat 100 ft above), and nonthreatening traffic (1200 ft above). The airplane symbol on the lower

FIGURE 22.9　(**See color insert.**) Combined TA/RA display.

FIGURE 22.10 (See color insert.) Joint-use weather radar and traffic display.

middle section of the display indicates the location of the aircraft relative to traffic. Figure 22.10 shows an example of a traffic display with a path cue showing the path constraint given by the RA.

22.7 Hybrid Surveillance and Airborne Collision Avoidance Systems

Hybrid surveillance is a new feature that may be included as an optional capability in TCAS II. TCAS II units equipped with hybrid surveillance may use passive surveillance to track intruders that meet validation criteria and are not projected to be near-term collision threats. Active surveillance uses the standard TCAS transponder interrogation as described in Section 22.3 and passive surveillance uses position data, typically based on GPS, that has been broadcast from the intruder's transponder. An intruder is tracked with active surveillance when it comes close to being a collision threat. The intent of hybrid surveillance is to reduce the number of required TCAS interrogations, through the use of ADS-B (discussed in Chapter 23) data, without any degradation of the safety and effectiveness of TCAS, to assist in reducing frequency spectrum congestion. The MOPS for TCAS II Hybrid Surveillance was approved on December 13, 2006.

Studies are being conducted on the next generation of collision avoidance systems that will provide the same role as TCAS II in the future NextGen airspace. Airborne collision avoidance systems (ACAS X) is a family of aircraft collision avoidance systems that will leverage optimized threat logic and ADS-B (along with traditional TCAS range, bearing, and range rate measurements) to provide functional and performance improvements to today's TCAS II systems. ACAS X is being designed to support NextGen procedures that will safely bring aircraft in closer proximity to one another that, in some scenarios, would cause existing TCAS II alerting logic to provide TAs. ACAS X alerting and advisory logic is being developed to accommodate a broader range of aircraft vertical maneuvering performance capabilities than the TCAS II alerting and advisory logic. This means that ACAS X will support general aviation aircraft, small unmanned aircraft, and helicopters capabilities, in addition to air transport fixed wing aircraft.

23

Automatic Dependent Surveillance—Broadcast

Joel M. Wichgers
Rockwell Collins

23.1 Introduction: What Is ADS-B?

Automatic dependent surveillance—broadcast (ADS-B) is an emerging air traffic surveillance technology that enables suitably equipped aircraft and airport ground vehicles to be tracked by (1) air traffic controllers without the need for conventional radar and (2) pilots of other aircraft that are equipped with ADS-B receive equipment. ADS-B traffic surveillance information transmitted by equipped aircraft and airport ground vehicles (referred to as ADS-B OUT) is expected to replace radar as the primary source of traffic surveillance used by air traffic controllers to control aircraft worldwide. Equally important, ADS-B will enable a broad range of on-aircraft applications (referred to as ADS-B IN) that will allow pilots to more safely and efficiently operate their aircraft at reduced distances from other traffic.

ADS-B is an integral system in plans for upgrading the aviation infrastructure around the world to support enhanced aircraft operations. For example, in the United States, ADS-B is an important part of the Federal Aviation Administration's (FAA) plan to overhaul the National Airspace Air Transportation System, which is referred to as NextGen (short for next generation). NextGen has been architected to improve the safety, efficiency, capacity, security, and environmental friendliness of the air transportation system through the use of advanced operational procedures enabled by combining ADS-B with better navigation, communications, and information management systems. As part of the transformation to NextGen, the FAA plans to transition from using radar to using ADS-B as its primary means of air traffic control (ATC) surveillance post-2020 and approve ADS-B-enabled on-aircraft applications

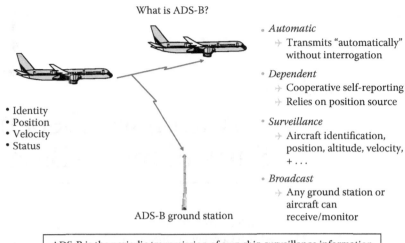

What is ADS-B?

- **Identity**
- **Position**
- **Velocity**
- **Status**

- *Automatic*
  - Transmits "automatically" without interrogation
- *Dependent*
  - Cooperative self-reporting
  - Relies on position source
- *Surveillance*
  - Aircraft identification, position, altitude, velocity, + . . .
- *Broadcast*
  - Any ground station or aircraft can receive/monitor

ADS-B ground station

ADS-B is the periodic transmission of own ship surveillance information by cooperative aircraft and airport ground vehicles.

FIGURE 23.1   ADS-B overview.

that support reduced aircraft spacing and delegated separation. Similarly in Europe, ADS-B is viewed as an integral system to enable the Single European Sky Air Traffic Management Research (SESAR) initiatives for improving the air transportation system in Europe.

More accurate than radar, ADS-B systems determine their own ship surveillance information very precisely using global navigation satellite system (GNSS) receivers installed on the aircraft or vehicle. With a more precise understanding of the location of aircraft and operationally relevant airport ground vehicles, the air transportation system can be designed to make better use of the airspace. ADS-B is a lower-cost surveillance technology than radar, and it enables both pilots and air traffic controllers to "see" and control aircraft with more precision over a far larger portion of the Earth than has ever been possible before. As illustrated in Figure 23.1, ADS-B is a surveillance technology that has the following features:

Automatic—Transmits surveillance information "automatically" without interrogation

Dependent—Depends on equipped aircraft/vehicles to cooperatively self-report their surveillance information relying on the availability of a suitable onboard position source (e.g., GNSS receiver)*

Surveillance—Provides surveillance information for accurate aircraft/vehicle tracking to ATC and other users

Broadcast—Aircraft/vehicles broadcast position and other data to all aircraft and ground stations equipped to receive ADS-B simultaneously

23.2 ADS-B Concept of Operation

At its core, ADS-B is based on aircraft and airport ground vehicles cooperatively transmitting traffic surveillance information that is updated frequently. The ADS-B surveillance information is typically significantly better than today's traffic surveillance information and, as such, enables a family of ground-based and aircraft-based applications that will improve the safety and operational efficiency of the air transportation system. The concept of ADS-B had its early beginnings in the 1990s as the concept of "free flight" was being developed by the aviation industry to provide more flexible

---

* ADS-B dependence on the availability of a position source is similar to that of secondary surveillance radar that requires equipped aircraft to cooperatively self-report their altitude.

and efficient airspace operations. Much of the early focus and vision for ADS-B operations was to enable aircraft with advanced traffic applications to support a range of capabilities to enable "free flight" and now NextGen and SESAR. While considerable initial focus for ADS-B was on the aircraft-based traffic applications and technologies, it was recognized early on that ADS-B could also provide significant benefits to air traffic service providers with reduced surveillance system costs, improved surveillance system coverage, and better surveillance performance to support ATC services. Based upon the improved surveillance information broadcast via ADS-B to both ATC and other aircraft, a broad range of new operational procedures are being developed and starting to be deployed that will improve the efficiency and safety of aircraft operations.

ADS-B uses two fundamental components to support its operation including GNSS to determine own ship position and velocity and a broadcast communications link to share this surveillance information with other users. This is far different from primary surveillance radars, which operate by transmitting radio waves from fixed terrestrial antennas and then determining the range and bearing of aircraft based upon the reflected signals. It is also different from secondary surveillance radars (SSR) that depend on aircraft altitude reports and on active replies from aircraft transponders to SSR interrogations for measuring range and bearing to each aircraft. Unlike radar, the accuracy of ADS-B surveillance information does not significantly degrade with range from the radar ground station or the target's altitude. Furthermore, the update intervals for ADS-B surveillance information do not depend on the rotational speed of radar's mechanical antennas. In contrast, a system using ADS-B creates and listens for periodic surveillance reports from aircraft/vehicles that are updated frequently (typically every second) and are quite accurate as they use GNSS as the source of position and velocity. The quality of the ADS-B surveillance data is typically much better than radar, since the position is typically much more accurate, the update rate is more frequent, and the velocity information is much better. For radar, detecting changes in aircraft velocity and direction requires several radar sweeps that are often spaced multiple seconds apart.

Figure 23.2 illustrates the concept of operations for ADS-B, which supports operations during all flight phases. ADS-B-equipped aircraft and airport ground vehicles use GNSS receivers to derive precise position and velocity state information, which is augmented with other aircraft/vehicle parameters and transmitted during all phases of operation including oceanic/remote, en route, terminal, and airport surface operations. This surveillance information is broadcast periodically such that ADS-B ground system receivers can utilize the information for ATC services and ADS-B receivers installed on other aircraft and airport ground vehicles can use it for on-aircraft/vehicle traffic applications.

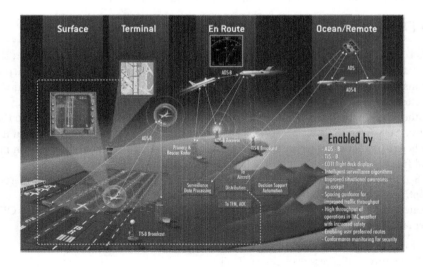

FIGURE 23.2 **(See color insert.)** ADS-B supports operations during all flight phases.

ADS-B aircraft determines their own state information, which includes 3D position and velocity, using a GNSS receiver as the source for their reported horizontal position and horizontal velocity and an air data system for the altitude and altitude rate. ADS-B ground vehicles also use a GNSS receiver to establish horizontal position and velocity, but they do not need an altitude source since they indicate in their surveillance broadcast that they are on the ground. This state information is transmitted (broadcast) along with a unique 24-bit vehicle identification code at frequent intervals via a data link. This is referred to as "ADS-B OUT."

ADS-B ground stations receive the broadcasts and relay the information to ATC for the precise tracking of ADS-B-equipped aircraft and vehicles. ATC can then utilize this information to provide aircraft separation assurance, traffic flow organization and management, and other ATC services. With better surveillance information than is available with conventional radar, ADS-B surveillance enables performance-based aircraft operations that make more effective use of available airspace in the air transportation system.

ADS-B receivers can also be installed on aircraft and other airport ground vehicles for receiving traffic surveillance information that is broadcast by surrounding aircraft/vehicles and other ground surveillance systems (including automatic dependent surveillance—rebroadcast [ADS-R] and Traffic Information Services—Broadcast [TIS-B] that are described later), such that when the information is processed, it can provide pilots/vehicle operators with information regarding the surrounding traffic. The surveillance information can be processed and used to improve the safety and efficiency of aircraft operations. ADS-B-enabled on-aircraft safety applications typically identify potential traffic conflicts and provide pilots with appropriate indications and alerts, like the Traffic Situational Awareness with Alerts (TSAA) application. Some ADS-B-enabled safety applications also provide guidance for resolving the traffic conflict (e.g., conflict detection and resolution). Applications that improve efficiency are typically the ones that achieve predictable and reduced aircraft spacing between aircraft, including the flight deck interval management (FIM) application. Such on-aircraft applications that use ADS-B surveillance information are referred to as "ADS-B IN" applications.

ADS-B surveillance information may be transmitted by airport ground vehicles to improve operations on the airport surface. Some concepts of operation call for equipping only vehicles that operate on taxiways and runways (excluding gate, ramp, parking stands, and maintenance areas) to include ground vehicles like snowplows, emergency vehicles, maintenance vehicles, and service vehicles. A more expanded concept of operation calls for equipping virtually all vehicles that operate on the airport surface to also include fuel trucks, catering vehicles, baggage movement vehicles, and grass cutters.

ADS-B includes two core processes (1) ADS-B OUT and (2) ADS-B IN as overviewed earlier. These two core processes are explained in greater detail in the following sections.

23.3 ADS-B OUT

ADS-B OUT refers to an aircraft/vehicle transmitting a periodic self-report of their own ship ADS-B surveillance information. This is illustrated in Figure 23.3. Equipped aircraft and airport ground vehicles transmit their vehicle state vector (horizontal and vertical position, horizontal and vertical velocity), state quality (indications of position accuracy and integrity and velocity accuracy), identification, and other information over a data link approved by the aviation authority for use in the airspace. In the future, ADS-B surveillance broadcasts are anticipated to also include aircraft intent information to support better prediction of where the aircraft will be in the future. Intent information may include current and future waypoints programmed into the aircraft's flight management system.

The ADS-B OUT traffic surveillance information has been designed to support numerous applications, for both ATC and ADS-B IN on-aircraft applications.

As indicated in Table 23.1, three data links have been defined for broadcasting ADS-B OUT surveillance information including

1. 1090 MHz extended squitter (1090ES)
2. 978 MHz universal access transceiver (UAT)
3. VHF data link—mode 4 (VDL-M4)

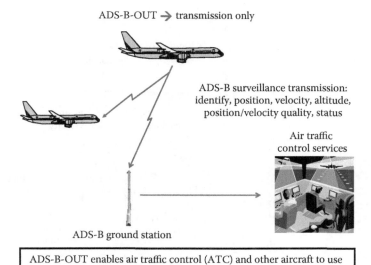

ADS-B-OUT → transmission only

ADS-B surveillance transmission:
identify, position, velocity, altitude,
position/velocity quality, status

Air traffic
control services

ADS-B ground station

ADS-B-OUT enables air traffic control (ATC) and other aircraft to use
the cooperatively transmitted ADS-B surveillance information.

FIGURE 23.3 ADS-B OUT: self-reporting of own aircraft/vehicle surveillance information.

TABLE 23.1 ADS-B OUT Data Links

Link	Applicable Standard	Description	Applicability
ADS-B 1090 MHz extended squitter (1090ES)	RTCA DO-260B	An extension of the Mode-S technology in which 1090ES avionics periodically broadcast short ADS-B messages at 1090 MHz that provide their identity (24-bit address), vehicle state (position, velocity), and other aircraft status information	1090ES has been internationally coordinated as the globally harmonized interoperable data link for ADS-B that is being used throughout most of the world including the United States, Canada, Central America, Europe, Asia, and the Pacific. In the United States, which allows either 1090ES or UAT, 1090ES is applicable to aircraft that fly anywhere in the U.S. national airspace, including high-altitude airspace above FL180.
Universal access transceiver (UAT)—978 MHz	RTCA DO-282B	ADS-B technology in which UAT avionics periodically transmit messages at 978 MHz that provide their identity, vehicle state, and other status information	Applicable to aircraft flying in the U.S. national airspace below FL180 (mainly general aviation aircraft).
ADS-B VHF data link mode 4 (VDL-M4)	EUROCAE ED-108A	ADS-B technology in which VDL-M4 avionics periodically transmit messages in the allocated ADS-B OUT 25 kHz channel in the VHF band (117.975–137 MHz) that provides their identity, vehicle state, and other status information	ADS-B link used by a few regional implementations in the world, for example, in the Scandinavian countries of Northern Europe.

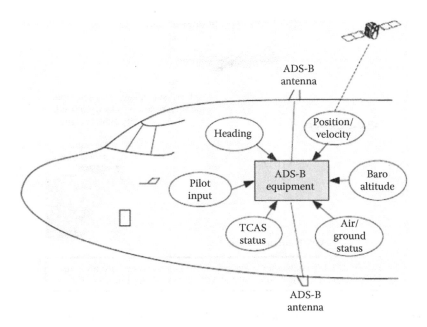

FIGURE 23.4 ADS-B OUT aircraft functional system diagram. (From Advisory Circular (AC) 20-165A, Airworthiness approval of automatic dependent surveillance—Broadcast (ADS-B) out systems, U.S. Department of Transportation, Federal Aviation Administration, Washington, DC, dated November 7, 2012.)

For complete and detailed specification of these three ADS-B links including the broadcast messages and content, refer to the latest revisions of RTCA DO-260 for 1090ES [5], RTCA DO-282 for UAT [6], and EUROCAE ED-108 for VDL-M4 [3]. The International Civil Aviation Organization has recommended 1090ES for international ADS-B use, and as a result, 1090ES is the ADS-B link used operationally in most of the world. In the United States, in addition to the 1090ES ADS-B link, the FAA has also approved the use of UAT to satisfy their ADS-B OUT rule (described in Section 23.8) in limited portions of the national airspace for aircraft that fly below flight level (FL) 180. As of this writing, the UAT ADS-B link has been operationally approved only in the United States, and it is mainly applicable to general aviation aircraft because of the airspace where it satisfies the FAA's ADS-B OUT rule. The use of VDL-M4 for ADS-B has seen a few regional implementations, for example, in the Scandinavian countries of Northern Europe.

A depiction of the typical on-aircraft functional information sources used to generate ADS-B OUT is illustrated in Figure 23.4. The information sources include a source of position/velocity. To date, only GNSS-based systems have been approved by civil aviation authorities for ADS-B OUT surveillance broadcasts, but in the future, alternative position sources may also meet the data quality (e.g., accuracy and integrity) requirements for ADS-B OUT. A barometric altitude source is required as well as a means to determine whether the aircraft is on the ground or in the air. The source for the flight identifier information (e.g., AA-123, which is separate from the unique 24-bit aircraft identifier) can be a pilot input device or can be automatically identified using data link information. A source of heading is required for most aircraft with an exception for small aircraft without electronic heading sensors. A source identifying the traffic alert and collision avoidance system (TCAS) status is required for aircraft equipped with TCAS II.

23.4 ADS-B IN

ADS-B IN (see Figure 23.5) refers to an appropriately equipped aircraft's ability to receive, process, and display information obtained via ADS-B OUT transmissions by other aircraft/vehicles as well as to receive, process, and display information provided by ground-based surveillance services

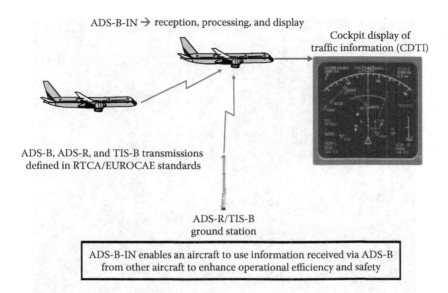

ADS-B-IN → reception, processing, and display

Cockpit display of traffic information (CDTI)

ADS-B, ADS-R, and TIS-B transmissions defined in RTCA/EUROCAE standards

ADS-R/TIS-B
ground station

> ADS-B-IN enables an aircraft to use information received via ADS-B from other aircraft to enhance operational efficiency and safety

FIGURE 23.5   ADS-B IN: on-aircraft traffic surveillance.

(described later) including ADS-R,* TIS-B,† and Flight Information Services—Broadcast (FIS-B).‡ The display of the received surveillance information is processed appropriately for the pilot and is provided on a display commonly referred to as a cockpit display of traffic information (CDTI).

23.4.1 ADS-B IN Cockpit Display

A CDTI is the generic name for a display that provides pilots with surveillance information about other traffic, including the traffic's relative position to own ship, and may also include application-specific information, such as traffic indications, alerts, and spacing guidance. Traffic information displayed on a CDTI may be based upon information obtained from one or more traffic information sources, including ADS-B, ADS-R, TIS-B, and TCAS. In addition to traffic surveillance information and any associated application-specific information, a display may also provide other information including navigation waypoints, weather, terrain, airspace structure and limitations, obstacles, airport maps, and other flight-relevant information. An example CDTI is shown in Figure 23.6, with own ship position indicated by the large triangle near the bottom center and traffic depicted with other smaller triangles or chevrons.

23.4.2 ADS-B IN Applications

Numerous ADS-B IN-enabled traffic applications have been envisioned by the aviation industry and are at various stages of development. Some of the traffic applications have been standardized, developed, and fielded, while others are currently being developed and standardized by aviation standards groups including RTCA and European Organization for Civil Aviation Equipment (EUROCAE). In addition to

---

* ADS-R: Ground service that relays ADS-B surveillance information received from transmitting aircraft/vehicles on one ADS-B data link technology onto a second ADS-B data link technology for ADS-B IN applications use on aircraft that can only receive on the second ADS-B link.

† TIS-B: Ground service that provides traffic surveillance information that is obtained from non-ADS-B surveillance systems for use by ADS-B IN applications.

‡ FIS-B: Ground service that broadcasts meteorological and aeronautical data. As of this writing, FIS-B is only available on the UAT data link.

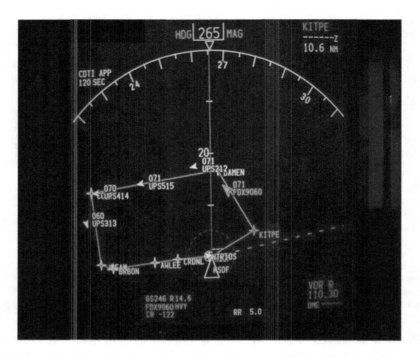

FIGURE 23.6   (See color insert.) CDTI.

those that have been already standardized, many more ADS-B IN applications have been proposed. The ADS-B IN applications are being grouped into five broad application categories including (1) situational awareness, (2) extended situational awareness, (3) spacing, (4) delegated separation, and (5) self-separation. A description of each of these five application categories is provided in the following:

1. *Situational awareness* applications are those that are intended to enhance the pilot's knowledge of surrounding traffic that are in the air as well as those on the airport surface. The improved situational awareness may improve pilot decision making that is expected to result in safer and more efficient flights. There are no changes to the pilot or controller responsibilities for these applications.
2. *Extended situational awareness* applications add provisions to the basic situational awareness applications such as cueing the pilot to traffic conditions through indications and alerts or providing information that may support a reduced aircraft separation standard during an operational procedure.
3. *Spacing applications* require pilots to achieve and maintain a given longitudinal spacing with designated aircraft as specified by ATC instruction. While pilots are given new tasks associated with conducting spacing applications, separation responsibility remains with the controller.
4. *Delegated separation* applications are those where the controller delegates separation responsibility and transfers the corresponding separation task to pilots, who ensure that the applicable separation requirements are met. The separation responsibility delegated to the pilots is limited to designated aircraft within the limitations of the clearance, which is limited in time, space, and scope. Except for the specific limited delegation, separation responsibility for all other aircraft remains the controller's responsibility.
5. *Self-separation* applications are those that require pilots to separate their aircraft from all surrounding traffic in accordance with the applicable separation requirements and flight rules.

Table 23.2 identifies a number of example ADS-B IN-enabled applications that have been developed or proposed, including applications from each of the five application categories.

TABLE 23.2 ADS-B IN Airborne Applications

Application Category	Name	Acronym	System Requirements	Avionics Requirements	Application Description
Situational awareness	Airborne situational awareness	AIRB	DO-319	DO-317	Provides situational awareness of airborne traffic
	Airport surface situational awareness	SURF	DO-322	DO-317	Provides situational awareness of traffic on the airport surface and airborne traffic near the runways
Extended situational awareness	Visual separation on approach	VSA	DO-314	DO-317	Assists pilots in acquiring and maintaining visual contact with preceding aircraft during visual separation on approach
	Oceanic in-trail procedure	ITP	DO-312	DO-317	Assists pilots in determining whether the initiation criteria for oceanic climb or descend through are satisfied
	Airport surface situational awareness with indications and alerts	SURF IA	DO-323	To be established	Provides situational awareness of traffic on the airport surface and airborne traffic near the runways with indications and alerts
	Traffic situational awareness with alerts	TSAA	DO-338	To be established	Airborne traffic situational awareness with advisories and alerts to support visual acquisition and avoidance of traffic
Spacing	Flight deck interval management—spacing	FIM-S	DO-328	To be established	Flight deck–based interval management for achieving or maintaining longitudinal spacing from one or more designated aircraft
Delegated separation	Independent closely spaced parallel approaches	ICSPA	DO-289	To be established	Airborne application to support conducting independent, simultaneous approaches to closely spaced parallel runways in instrument meteorological conditions
	CDTI-enabled delegated separation	CEDS	N/A	To be established	Airborne application to support pilots with safely separating from designated aircraft
Self-separation	Airborne conflict management	ACM	DO-289	To be established	Application to prevent loss of separation (conflict detection and resolution) and provide advisory information for trajectories that may cause a conflict (conflict prediction)

23.5 Ground-Based Information Services

There are three ground-based information services associated with ADS-B including (1) ADS-R, (2) TIS-B, and (3) FIS-B. Each of these services is described in the following.

23.5.1 ADS-R

ADS-R is a ground-based traffic information service that relays ADS-B information transmitted by an aircraft or vehicle using one ADS-B link technology (e.g., UAT) and received by the ground station

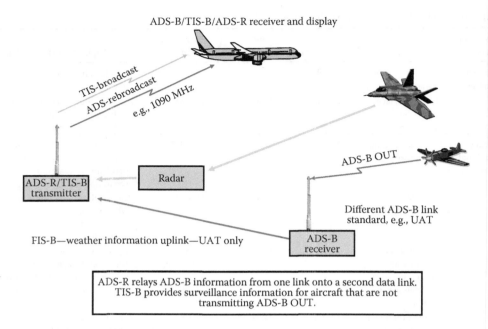

FIGURE 23.7   ADS-R and TIS-B services.

for subsequent rebroadcast for use by an aircraft or vehicle using another ADS-B link technology (e.g., 1090ES). As shown in Figure 23.7, the ADS-R system receives ADS-B transmissions by active ADS-B-equipped aircraft and continuously monitors the presence of proximate aircraft with differing ADS-B links. When such aircraft are in proximity of each other, the ADS-R system instructs ADS-B ground stations within range of both aircraft to rebroadcast surveillance information received on one link frequency to aircraft on the other link frequency (e.g., UAT to 1090ES and vice versa). The ADS-R multilink gateway service is a companion to the TIS-B service (described in Section 23.5.2) for providing ADS-B IN aircraft with a complete set of traffic surveillance information for all aircraft.

23.5.2 Traffic Information Services: Broadcast

Not all aircraft will be broadcasting their position via ADS-B. Such conditions may occur for aircraft that are either not ADS-B equipped or in conditions where the ADS-B OUT equipment installed on an aircraft is not operational. TIS-B is a ground-based traffic information service that fills this traffic surveillance information gap by broadcasting traffic surveillance information for those aircraft/vehicles that are not broadcasting ADS-B surveillance information and for which ground surveillance information is available, such that ADS-B IN-equipped aircraft have a complete set of traffic surveillance information for aircraft in their vicinity.

TIS-B receives traffic surveillance information from available non-ADS-B surveillance systems, including radar, Airport Surface Detection System—Model X (ASDE-X),* and multilateration† systems. This surveillance information is processed and correlated with traffic surveillance information that is received

---

* ASDE-X is a surveillance system that allows air traffic controllers to track surface movement of aircraft and vehicles. The FAA is deploying it at major airports in the United States. ASDE-X uses a variety of surveillance sources including radar, multilateration, and ADS-B.
† Multilateration is a ground surveillance technology that determines aircraft and vehicle positions by measuring the difference in the time of reception between two or more ground stations at known locations for signals transmitted by the aircraft or vehicle.

via ADS-B. The TIS-B system uses this information to transmit traffic surveillance information for non-ADS-B-equipped aircraft/vehicles to active ADS-B IN users. The TIS-B service is complementary to the ADS-R service and the ADS-B OUT surveillance information provided by other aircraft in order to allow ADS-B IN users to have a complete picture of the nearby traffic without duplication. As of this writing, TIS-B is being implemented by the FAA as described therein. In the future, the TIS-B system may provide one or more other modes of operation, including (1) a full traffic mode whereby all traffic known to the TIS-B system in a region is broadcast or (2) a best available traffic surveillance mode whereby TIS-B will transmit known traffic surveillance information in the conditions where either (a) no ADS-B OUT information is being broadcast directly by a given traffic vehicle or (b) higher-quality traffic surveillance information is known by the TIS-B system than is being broadcast by the traffic vehicle on ADS-B OUT.

23.5.3 Flight Information Service: Broadcast

FIS-B is a ground-based service that provides meteorological and aeronautical data to suitably equipped aircraft. While not directly considered to be part of ADS-B, the FAA is providing the FIS-B service using the UAT data link (978 MHz), which is the link typically being used by general aviation aircraft for ADS-B in the United States. FIS-B ground stations receive weather and aeronautical data from a variety of information sources and generate sets of products specific to their location and region of interest for broadcast to aircraft users. These products are broadcast over the UAT link so that pilots of aircraft that receive the FIS broadcast have timely information of regional weather and National Airspace System (NAS) status and changes that might impact their flight.

Current FIS-B products include the following: Airmen's Meteorological Information (AIRMET), Significant Meteorological Information (SIGMET), Convective SIGMET, Meteorological Aviation Routine Weather Report (METAR), Continental United States Next-Generation Radar (CONUS NEXRAD), Regional NEXRAD, Notice to Airmen (NOTAM), pilot report (PIREP), special use airspace (SUA) status, terminal aerodrome forecast (TAF), winds and temperatures aloft, and TIS-B service status.

Additional FIS-B products may be offered in the future, including Echo Tops, Cloud Tops, Icing NowCast, One-Minute Observations (OMO), Lightning, and Digital Automated Terminal Information System (D-ATIS).

23.6 ADS-B Ground Receiver Network

Networks of ADS-B ground stations and associated traffic surveillance receivers are being deployed by air traffic service providers to provide surveillance coverage to receive ADS-B OUT transmissions throughout the airspace to support the needs of ATC. ADS-B ground station networks are being deployed in many countries for achieving improved traffic surveillance coverage over airspace regions where surveillance information was limited or for operational cost-effectiveness (e.g., replacement of SSR with ADS-B ground stations to obtain improved surveillance coverage at a lower cost). ADS-B ground stations provide aircraft surveillance information that typically has better accuracy and updates rates than SSR for use in ATC automation systems and aircraft control services, which will provide the opportunity for safely reducing the minimum aircraft separation standards to enable more efficient flight operations.

ADS-B ground stations receive and process the ADS-B OUT surveillance broadcasts by aircraft and airport ground vehicles for use by the ATC automation systems and for presentation on controller displays.

As part of the FAA's airspace improvement initiatives and path to NextGen, the FAA contracted the development of a network of ADS-B ground system transceivers that receive ADS-B broadcasts from equipped aircraft/vehicles and provide services for broadcasting ADS-R, TIS-B, and FIS-B information. The ADS-B ground network being deployed in the United States is expected to be fully operational in 2014 and is planned to contain approximately 800 strategically located ground stations to receive ADS-B

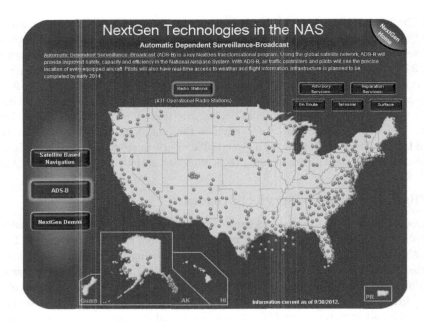

FIGURE 23.8   ADS-B ground network in process of development. (From U.S. Department of Transportation, Federal Aviation Administration, http://www.faa.gov/nextgen/flashmap.)

OUT transmissions by aircraft/vehicles in the NAS. A significant portion of this ADS-B ground station network is already operational at this time. These ground stations and their network computers process and provide ADS-B information to the FAA at designated service delivery points. The FAA will utilize this information to provide ATC services. The FAA and its designated service providers supply weather information to the ground network for uplink on FIS-B and traffic information gathered through SSR and multilateration systems for uplink on TIS-B. Figure 23.8 illustrates the location of ADS-B ground network stations deployed as of September 30, 2012.

23.7 ADS Space-Based Satellites

Some air traffic service providers have a desire and need to receive accurate and timely aircraft surveillance information to control aircraft in regions where it is difficult to install and maintain a ground network of ADS-B ground stations including oceanic, remote, and polar regions. In such regions, satellites may be used to receive aircraft surveillance information and relay it to ATC. There are two primary techniques being considered for utilizing satellites to obtain ADS-B surveillance information from aircraft outside the coverage region of an ADS-B ground station network: (1) installing Mode-S 1090ES and UAT ADS-B receivers directly onto the satellites and (2) installing a converter function on the aircraft that utilizes the ADS-B OUT surveillance information from the on-aircraft systems and appropriately convert the surveillance information for transmission to the satellites. For the former technique, the satellite systems would receive and consolidate the ADS-B signals prior to transmission to the ground, whereas in the latter technique, the satellite would function as a "bent pipe" to retransmit the surveillance information received from the aircraft such that it can be received by the ground. With both of these techniques, the satellite ground stations would receive and consolidate the traffic surveillance information transmitted by the satellites and provide the information to ATC and other users. ADS-capable satellites can be used to augment the ADS-B ground receiver networks to provide worldwide surveillance coverage.

Another means to receive traffic surveillance information in remote regions is called automatic dependent surveillance—contract (ADS-C) (as opposed to ADS-B), which is described in Section 23.10.

23.8 ADS-B OUT Mandates

ADS-B OUT is being mandated for aircraft entering designated airspace by civil aviation authorities and air traffic service providers around the world including those in Australia, Canada, Europe, the United States and others. While this section will not go into detail on all the ADS-B OUT mandates throughout the world as such information is subject to change, it will describe the FAA mandate for ADS-B as an example of airspace mandates in place at the time of this writing.

ADS-B is an important part of the FAA's planned NextGen airspace upgrade such that ADS-B can be utilized as the primary means of aircraft surveillance for ATC. To ensure that all aircraft provide this surveillance, the FAA has published a final rule on May 28, 2010 [4], requiring ADS-B OUT equipment to be installed on virtually all aircraft operating in designated airspaces (including airspace classes A, B, C, and E) as identified in Table 23.3 and depicted in Figure 23.9. The FAA rule will take effect on January 1, 2020. As of this writing, the FAA does not have plans for mandating ADS-B IN equipage; although, various means to encourage ADS-B IN equipage are being considered, which could potentially result in a future mandate.

Per FAA rule for ADS-B OUT that is current at the time of this writing, the ADS-B OUT equipment installed on aircraft must meet one of two sets of FAA standards, either (1) Technical Standard Order (TSO)-166c [8], which references RTCA/DO-260B for 1090ES, or (2) TSO-154c [7], which references RTCA/DO-282B for UAT. 1090ES is required for all aircraft that operate at and above 18,000 ft (class A airspace) and either 1090 MHz or UAT for aircraft that operate below 18,000 ft (in class B, C and specified portions of class E airspaces).

TABLE 23.3   Airspace Where ADS-B OUT Is Required by the FAA ADS-B OUT Rule (Dated May 28, 2010)

Class of Airspace	Airspace Description (per Designations in the United States)	FAA ADS-B OUT Requirements Effective on January 1, 2020
Class A	Airspace from 18,000 ft MSL up to and including FL600, including the airspace overlying the waters within 12 NM of the cost of the 48 contiguous states and Alaska.	Required for all aircraft (except aircraft without electrical system).
Class B	Airspace from the surface to 10,000 ft surrounding the nation's busiest airports. The configuration of class B airspace is tailored around each airport and consists of several layers.	Required for all aircraft. In addition to class B airspace, the rule requires equipage for all aircraft operating within 30 NM of FAA-specified airports, which are among the busiest in the nation.
Class C	Airspace from the surface to 4000 ft above the airport elevation surrounding specified airports.	Required for all aircraft.
Class D	Airspace from the surface to 2500 ft above the airport elevation surrounding those airports that have an operational control tower.	Not required.
Class E	Controlled airspace that is not in Class A, B, C, or D is classified as Class E. Class E airspace extends upward from the surface or a designated altitude to the overlying or adjacent controlled airspace. Unless otherwise designated at a lower altitude, Class E airspace begins at 14,500 ft MSL over the United States, including that airspace overlying the waters within 12 NM of the coast of the 48 contiguous states and Alaska. Class E airspace does not include the airspace at 18,000 ft MSL up to FL600 (as this is class A airspace) but does include airspace above FL600.	In portions of the Class E airspace, ADS-B OUT is required for all aircraft (expect aircraft without electrical systems) as specified: • Required for aircraft operating above 10,000 ft MSL in the lower 48 states and the District of Columbia, excluding the airspace at and below 2500 ft above the ground. • Required in the Gulf of Mexico for aircraft at and above 3000 ft within 12 NM of the U.S. coastline.
Class G	Uncontrolled airspace. Class G airspace is airspace that is not designated as Class A, B, C, D, or E.	Not required.

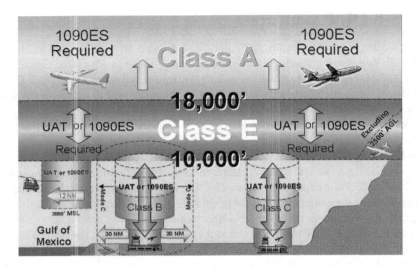

FIGURE 23.9   **(See color insert.)** FAA ADS-B OUT rule airspace requirements. (From Advisory Circular (AC) 90-114 (Change 1), Automatic dependent surveillance-broadcast (ADS-B) operations, U.S. Department of Transportation, Federal Aviation Administration, Washington, DC, dated September 21, 2012.)

Note that virtually all aircraft that operate in the airspace classes identified in the table are required to equip with ADS-B OUT, with a few exceptions for aircraft that were not originally or subsequently certified with an electrical system installed, for example, some balloons and gliders.

23.9 ADS-B in Relation to TCAS

ADS-B is a traffic surveillance technology developed to provide surveillance information to both ground-based and aircraft-based traffic surveillance applications in support of safe aircraft flight operations. The TCAS, known internationally as the airborne collision avoidance system (ACAS), is a "last resort" safety system intended to reduce the probability of occurrence of midair collisions between aircraft in the event where aircraft loss of separation has occurred.

ADS-B surveillance differs from TCAS surveillance in that ADS-B broadcasts position and velocity information, while TCAS derives relative position information through an interrogate–reply protocol. ADS-B supports a larger operating range (potentially 90–120 NM or more) for air-to-air reception, whereas TCAS is only intended to provide surveillance to approximately 15 NM. ADS-B surveillance typically has better accuracy than TCAS. ADS-B surveillance is based upon highly accurate GNSS position measurements; while TCAS measures range with great accuracy, it has relatively poor bearing measurement accuracy due to the limitations of available and cost-effective antennas that can accurately resolve bearing and be installed on aircraft.

ADS-B is seen as a valuable technology that can enhance the operation of TCAS via a concept that is referred to as "hybrid surveillance," which is described in Chapter 22. By making use of ADS-B IN surveillance information, hybrid surveillance can potentially improve TCAS by

- Decreasing the number of active interrogations required by TCAS, thus reducing frequency congestion and effectively increasing its operational range in high-density traffic airspace
- Reducing unnecessary TCAS advisories and alerts by incorporating the typically more accurate ADS-B state information as well as other information from ADS-B into the TCAS surveillance processing

- Integrating TCAS and ADS-B IN traffic displays, thereby providing a single display with the best available surveillance information for all known traffic to the pilots
- Extending the collision avoidance to below 1000 ft above ground level to include the detection of runway incursions

23.10 ADS-B in Relation to ADS-C

There are two commonly recognized types of ADS for aircraft including:

- ADS-B
- ADS-C, which is also known as ADS-addressed (ADS-A)

ADS-B has been described in this chapter and it is different than ADS-C. While ADS-B is a broadcast intended for all to hear, ADS-C is based on a negotiated one-to-one peer relationship between an aircraft providing automatic dependent surveillance (ADS) information and a ground facility requiring receipt of ADS messages. ADS-C is used, for example, to support free flight using the Future Air Navigation System (FANS). During flight over areas without radar coverage (e.g., oceanic remote and polar), ADS-C reports are periodically sent by an aircraft to the controlling ATC using the Aircraft Communication Addressing and Reporting System (ACARS) as the communication protocol.

23.11 ADS-B Challenges

While ADS-B is being deployed, there are a few concerns being voiced about its implementation, including (1) frequency congestion on 1090 MHz, (2) what happens when GNSS is unavailable, (3) privacy of aircraft operations, and (4) the potential disruption of the air transportation system from malicious spoofing. Each of these concerns is briefly touched on in the following texts.

The 1090 MHz frequency being used by ADS-B 1090ES is the same frequency that is also used by aircraft transponders for replying to interrogations from both TCAS and SSR systems. With three systems sharing one frequency, there are concerns that in very high traffic density airspace with dense traffic volumes, that it may significantly reduce the usable range for each of these systems. Various mitigations are being explored by the regulatory agencies to address this spectrum congestion issue, including a next generation of TCAS that will utilize ADS-B surveillance information to reduce the number of interrogations and subsequent replies, known as hybrid surveillance.

Another concern often expressed is what happens to ADS-B when the GNSS source used to determine the own ship position and velocity information is not available due to, for example, GNSS outages, interference, or intentional jamming. Various backup strategies are being considered, which include the following: (1) retaining a backup surveillance system of primary and secondary surveillance radars to maintain ATC surveillance in the situation where ADS-B surveillance information becomes unavailable, (2) using surveillance determined from multilateration systems to track aircraft positions when ADS-B surveillance information is not available, and (3) equipping aircraft with one or more alternative sources of position and velocity information suitable to support ADS-B OUT (the latter of which is commonly referred to as alternative position, navigation, and timing or APNT).

In the area of privacy, concerns have been voiced about the potential to track aircraft movements and use that information in ways that were not intended. This might include tracking the movements of special aircraft or aircraft owned by very important people (VIPs) such as celebrities or corporate executives. The ADS-B standards have been developed with a means to mitigate such privacy concerns through the use of anonymous aircraft identifiers.

"Spoofing" is a concern often voiced regarding malicious attempts to disrupt the air traffic system by introducing many false ADS-B traffic targets. Various mitigations to this concern are being deployed to confirm ADS-B surveillance information using alternate sources of air traffic surveillance, including multilateration and radar.

23.12 Concluding Summary

ADS-B is an integral element of air traffic service provider strategies for upgrading the aviation infrastructure to support enhanced airspace operations that will improve the safety, efficiency, capacity, and environmental friendliness of the air transportation system. ADS-B OUT is the cooperative self-reporting of own ship surveillance information, and it is being mandated for aircraft that operate in much of the airspace around the world. ADS-B and related technologies provide better traffic surveillance information for use on the aircraft and on the ground. With ADS-B IN, many on-aircraft applications that were previously not possible are being enabled. As of this writing, ADS-B IN is not being mandated; although, this may change in the future. As ADS-B OUT equipage becomes prevalent, there is more incentive for ADS-B IN equipage as it enables many on-aircraft applications that improve aircraft operational efficiency and safety. ADS-B is expected to become the primary means of surveillance for ATC around the world.

References

1. Advisory Circular (AC) 20-165A, Airworthiness approval of automatic dependent surveillance—Broadcast (ADS-B) out systems, U.S. Department of Transportation, Federal Aviation Administration, Washington, DC, dated November 7, 2012.
2. Advisory Circular (AC) 90-114 (Change 1), Automatic dependent surveillance-broadcast (ADS-B) operations, U.S. Department of Transportation, Federal Aviation Administration, Washington, DC, dated September 21, 2012.
3. Minimum operational performance standard for VDL Mode 4 aircraft transceiver, ED–108A, EUROCAE, September 2005.
4. FAA ADS-B OUT Rule, Automatic dependent surveillance–broadcast (ADS–B) out performance requirements to support air traffic control (ATC) service; Final Rule, U.S. Department of Transportation, Federal Aviation Administration, Washington, DC, published in the *Federal Register*, 75(103), pp. 30160–30195, dated May 28, 2010.
5. Minimum operational performance standards for 1090 MHz extended squitter automatic dependent surveillance—Broadcast (ADS-B) and traffic information services—Broadcast (TIS-B) [including changes], DO-260B, RTCA, December 2, 2009.
6. Minimum operational performance standards for universal access transceiver (UAT) automatic dependent surveillance—Broadcast (ADS-B) [including changes], DO-282B, RTCA, December 2, 2009.
7. Technical Standard Order 154c, Universal access transceiver (UAT) automatic dependent surveillance—Broadcast (ADS-B) equipment operating on the frequency of 978 MHz, U.S. Department of Transportation, Federal Aviation Administration, Washington, DC, effective date December 2, 2009.
8. Technical Standard Order 166b, Extended squitter automatic dependent surveillance—Broadcast (ADS-B) and traffic information service—Broadcast (TIS-B) equipment operating on the radio frequency of 1090 megahertz (MHz), U.S. Department of Transportation, Federal Aviation Administration, Washington, DC, effective date December 2, 2009.
9. U.S. Department of Transportation, Federal Aviation Administration, http://www.faa.gov/nextgen/flashmap.

24

Flight Management Systems

Randy Walter
GE Aviation Systems

24.1 Introduction

The flight management system typically consists of two units, a computer unit and a control display unit. The computer unit can be a standalone unit providing both the computing platform and various interfaces to other avionics or it can be integrated as a function on a hardware platform such as an Integrated Modular Avionics (IMA) cabinet. The Control Display Unit (CDU or MCDU) provides the primary human/machine interface for data entry and information display. Since hardware and interface implementations of flight management systems can vary substantially, this discussion will focus on the functional aspects of the flight management system.

The flight management system provides the primary navigation, flight planning, and optimized route determination and en route guidance for the aircraft and is typically comprised of the following interrelated functions: navigation, flight planning, trajectory prediction, performance computations, and guidance.

To accomplish these functions the flight management system must interface with several other avionics systems. As mentioned above, the implementations of these interfaces can vary widely depending upon the vintage of equipment on the aircraft but generally will fall into the following generic categories.

- Navigation sensors and radios
  - Inertial/attitude reference systems
  - Navigation radios
  - Air data systems
- Displays
  - Primary flight and navigation
  - Multifunction
  - Engine
- Flight control system
- Engine and fuel system
- Data link system
- Surveillance systems

Figure 24.1 depicts a typical interface block diagram.

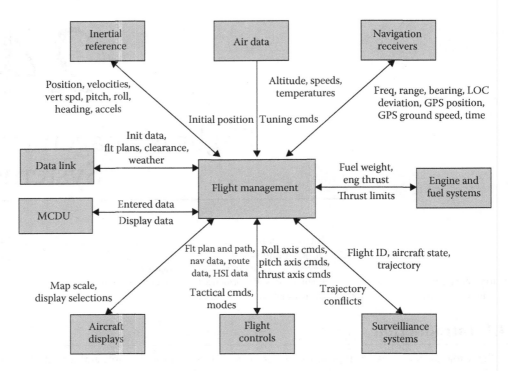

FIGURE 24.1   Typical interface block diagram.

Today, flight management systems can vary significantly in levels of capability because of the various aviation markets they are intended to serve. These range from simple point to point lateral navigators to the more sophisticated multisensor navigation, optimized four-dimensional flight planning/ guidance systems. The flight management system in its simplest form will slowly diminish as reduced separation airspace standards place more demands on the aircraft's ability to manage its trajectory more accurately, even though lateral-only navigators will continue to have a place in recreational general aviation.

With its current role in the aircraft, the flight management system becomes a primary player in the current and future communications navigation surveillance for air traffic management (CNS/ATM) environment. Navigation within required navigation performance (RNP) airspace, data-linked clearances and weather, aircraft trajectory-based traffic management, time navigation for aircraft flow control, and seamless low-visibility approach guidance all are enabled through advanced flight management functionality.

24.2 Fundamentals

At the center of the FMS functionality is the flight plan construction and subsequent construction of the four-dimensional aircraft trajectory defined by the specified flight plan legs and constraints and the aircraft performance. Flight plan and trajectory prediction work together to produce the four-dimensional trajectory and consolidate all the relevant trajectory information into a flight plan/profile buffer. The navigation function provides the dynamic current aircraft state to the other functions. The vertical, lateral steering, and performance advisory functions use the current aircraft state from navigation and the information in the flight plan/profile buffer to provide guidance, reference, and advisory information relative to the defined trajectory and aircraft state.

- The navigation function—responsible for determining the best estimate of the current state of the aircraft.
- The flight planning function—allows the crew to establish a specific routing for the aircraft.
- The trajectory prediction function—responsible for computing the predicted aircraft profile along the entire specified routing.
- The performance function—provides the crew with aircraft unique performance information such as takeoff speeds, altitude capability, and profile optimization advisories.
- The guidance functions—responsible for producing commands to guide the aircraft along both the lateral and vertical computed profiles.

Depending on the particular implementation, the ancillary input/output (I/O), built-in test equipment (BITE), and control display functions may be included as well. Since the ancillary functions can vary significantly, this discussion will focus on the core flight management functions.

There are typically two loadable databases that support the core flight management functions. These are the navigation database which must be updated on a 28-day cycle and the performance database that only gets updated if there's been a change in the aircraft performance characteristics (i.e., engine variants or structural variants affecting the drag of the aircraft).

The navigation database contains published data relating to airports, navaids, named waypoints, airways and terminal area procedures along with RNP values specified for the associated airspace. The purpose of the navigation data base is twofold. It provides the navigation function location, frequency, elevation, and class information for the various ground-based radio navigation systems. This information is necessary to select, auto-tune, and process the data from the navigation radios (distance, bearing, or path deviation) into an aircraft position. It also provides the flight plan function with airport, airport-specific arrival, departure, and approach procedures (predefined strings of terminal area waypoints), airways (predefined enroute waypoint strings), and named waypoint information that allows for rapid route construction. A detailed description of the actual data content and format can be found in ARINC 424.

The performance database contains aircraft/engine model data consisting of drag, thrust, fuel flow, speed/altitude envelope, thrust limits, and a variety of optimized and tactical speed schedules that are unique to the aircraft. Figure 24.2 shows the interrelationships between the core functions and the databases.

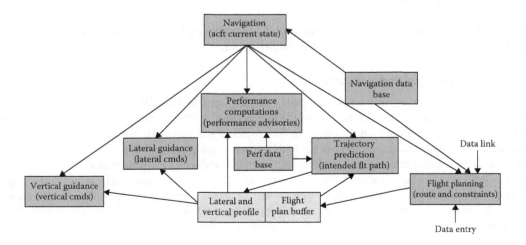

FIGURE 24.2 Flight management functional block diagram.

24.2.1 Navigation

The navigation function within the FMS computes the aircraft's current state (generally WGS-84 geodetic coordinates) based on a statistical blending of multisensor position and velocity data. The aircraft's current state data usually consists of:

- Three-dimensional position (latitude, longitude, altitude)
- Velocity vector
- Altitude rate
- Track angle, heading, and drift angle
- Wind vector
- Estimated Position Uncertainty (EPU)
- Time

The navigation function is designed to operate with various combinations of autonomous sensors and navigation receivers. The position update information from the navigation receivers is used to calibrate the position and velocity data from the autonomous sensors, in effect providing an error model for the autonomous sensors. This error model allows for navigation coasting based on the autonomous sensors while maintaining a very slow growth in the EPU. If the updating from navigation aids such as distance measurement equipment (DME), very high frequency omni range (VOR), or global positioning system (GPS) is temporarily interrupted, navigation accuracy is reasonably maintained, resulting in seamless operations. This capability becomes very important for operational uses such as RNAV approach guidance where the coasting capability allows completion of the approach even if a primary updating source such as GPS is lost once the approach is commenced. A typical navigation sensor complement consists of:

- Autonomous sensors
  - Inertial reference
  - Air data
- Navigation receivers
  - DME receivers
  - VOR/LOC receivers
  - GPS receivers

The use of several navigation data sources also allows cross-checks of raw navigation data to be performed to ensure the integrity of the FMS position solution.

24.2.1.1 Navigation Performance

The navigation function, to be RNP airspace compliant per DO-236, must compute an Estimated Position Uncertainty (EPU) that represents the 95% accuracy performance of the navigation solution. The EPU is computed based on the error characteristics of the particular sensors being used and the variance of the individual sensors position with respect to other sensors. The RNP for the airspace is defined as the minimum navigation performance required for operation within that airspace. It is specified by default values based on the flight phase retrieved from the navigation data base for selected flight legs or crew-entered in response to ATC-determined airspace usage. A warning is issued to the crew if the EPU grows larger than the RNP required for operation within the airspace. The table below shows the current default RNP values for the various airspace categories.

Airspace Definition	Default RNP (nm)
Oceanic—no VHF navaids within 200 nm	12.0
Enroute—above 15,000 ft	2.0
Terminal	1.0
Approach	0.5

A pictorial depiction of the EPU computation is shown below for a VOR/VOR position solution. A similar representation could be drawn for other sensors.

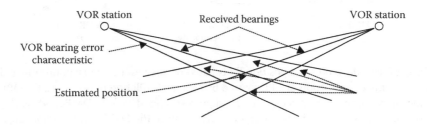

As can be seen from the diagram, the estimated position uncertainty (EPU) is dependent on the error characteristics of the particular navigation system being used as well as the geometric positioning of the navaids themselves. Other navigation sensors such as an inertial reference system have error characteristics that are time-dependent. More information pertaining to EPU and various navigation navaid system error characteristics can be found in RTCA DO-236.

24.2.1.2 Navigation Receiver Management

The various navigation receivers require different levels of FMS management to obtain a position update solution.

GPS—The GPS receiver is self-managing in that the FMS receives position, velocity, and time information without any particular FMS commands or processing. Typically, the FMS will provide an initial position interface to reduce the satellite acquire time of the receiver and some FMSs may provide an estimated time of arrival associated with a final approach fix waypoint to support the Predictive Receiver Autonomous Integrity Monitor (PRAIM) function in the GPS. More information on the GPS interface and function can be found in ARINC 743.

VHF navaids (DME/VOR/ILS)—The DME/VOR/ILS receivers must be tuned to an appropriate station to receive data. The crew may manually tune these receivers but the FMS navigation function will also auto-tune the receivers by selecting an appropriate set of stations from its stored navigation database and sending tuning commands to the receiver(s). The selection criteria for which stations to tune are:

○ Navaids specified within a selected flight plan procedure, while the procedure is active.
○ The closest DME navaids to the current aircraft position of the proper altitude class that are within range (typically 200 nm).
○ Collocated DME/VORs within reasonable range (typically 25 nm).
○ ILS facilities if an ILS or localizer (LOC) approach has been selected into the flight plan and is active.

Since DMEs receive ranging data and VORs receive bearing data from the fixed station location, the stations must be paired to determine a position solution as shown below:

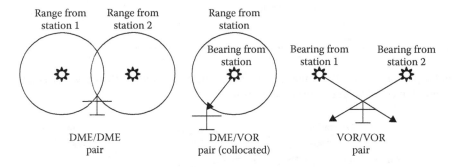

The pairing of navaids to obtain a position fix is based on the best geometry to minimize the position uncertainty (minimize the portion of EPU caused by geometric dilution of precision, GDOP). As can be seen from the figure above, the FMS navigation must process range data from DMEs and bearing data from VORs to compute an estimated aircraft position. Further, since the DME receives slant range data from ground station to aircraft, the FMS must first correct the slant range data for station elevation and aircraft altitude to compute the actual ground-projected range used to determine position. The station position, elevation, declination, and class are all stored as part of the FMS navigation data base. There are variations in the station-tuning capabilities of DME receivers. A standard DME can only accept one tuning command at a time, an agility-capable DME can accept two tuning commands at a time, and a scanning DME can accept up to five tuning commands at a time. VOR receivers can only accept one tuning command at a time.

An ILS or LOC receiver works somewhat differently in that it receives cross-track deviation information referenced to a known path into a ground station position. These facilities are utilized as landing aids and therefore are located near runways. The FMS navigation function processes the cross-track information to update the cross-track component of its estimated position. More information about DME/VOR/ILS can be found in ARINC 709, 711, and 710, respectively.

24.2.2 Flight Planning

The basis of the FMC flight profile is the route that the aircraft is to fly from the departure airport to the destination airport. The FMS flight planning function provides for the assembly, modification, and activation of this route data known as a flight plan. Route data are typically extracted from the FMC navigation data base and typically consists of a departure airport and runway, a standard instrument departure (SID) procedure, enroute waypoints and airways, a standard arrival (STAR) procedure, and an approach procedure with a specific destination runway. Often the destination arrival (or approach transition) and approach procedure are not selected until the destination terminal area control is contacted. Once the routing, along with any route constraints and performance selections, are established by the crew, the flight plan is assembled into a "buffer" that is used predominantly by the trajectory predictions in computing the lateral and vertical profile the aircraft is intended to fly from the departure airport to the destination airport.

The selection of flight planning data is done by the crew through menu selections either on the MCDU or navigation display or by data link from the airline's operational control. Facilities are also provided for the crew to define additional navigation/route data by means of a supplemental navigation data base. Some of the methods for the crew to create new fixes (waypoints) are listed below.

PBD waypoints—Specified as bearing/distance off existing named waypoints, navaids, or airports.
PB/PB waypoints—Specified as the intersections of bearings from two defined waypoints.

ATO waypoints—Specified by an along-track offset (ATO) from an existing flight plan waypoint. The waypoint that is created is located at the distance entered and along the current flight plan path from the waypoint used as the fix. A positive distance results in a waypoint after the fix point in the flight plan while a negative distance results in a waypoint before the fix point.

Lat/Lon waypoints—Specified by entering in the latitude/longitude coordinates of the desired waypoint.

Lat/Lon crossing waypoints—Created by specifying a latitude or longitude. A waypoint will be created where the active flight plan crosses that latitude or longitude. Latitude or longitude increments can also be specified, in which case several waypoints are created where the flight plan crosses the specified increments of latitude or longitude.

Intersection of airways—Created by specifying two airways. A waypoint will be created at the first point where the airways cross.

Fix waypoints—Created by specifying a "fix" reference. Reference information includes creation of abeam waypoints and creation of waypoints where the intersections of a specified radial or distance from the "fix" intersects the current flight plan.

Runway extension waypoints—Created by specifying a distance from a given runway. The new waypoint will be located that distance from the runway threshold along the runway heading.

Abeam waypoints—If a direct-to is performed, selection of abeam points results in waypoints being created at their abeam position on the direct-to path. Any waypoint information associated with the original waypoint is transferred to the newly created waypoints.

FIR/SUA intersection waypoints—Creates waypoints where the current flight plan crosses flight information region (FIR) boundaries and special use areas (SUA) that are stored in the navigation data base.

The forward field of view display system shows a presentation of the selected segments of the flight plan as the flight plan is being constructed and flown.

The crew can modify the flight plan at any time. The flight plan modification can come from crew selections or via data link from the airline operational communications or air traffic control in response to a tactical situation. An edit to the flight plan creates a modified (or temporary) version of the flight plan that is a copy of the active flight plan plus any accrued changes made to it. Trajectory predictions are performed on the modified flight plan with each edit and periodically updated, which allows the crew to evaluate the impact of the flight plan changes prior to acceptance. When the desired changes have been made to the crew's satisfaction this modified flight plan is activated by the crew.

24.2.2.1 Flight Plan Construction

Flight plans are normally constructed by linking data stored in the navigation data base. The data may include any combination of the following items:

- SID/STAR/approach procedures
- Airways
- Prestored company routes
- Fixes (en route waypoints, navaids, nondirectional beacons, terminal waypoints, airport reference points, runway thresholds)
- Crew-defined fixes (as referenced above)

These selections may be strung together using clearance language, by menu selection from the navigation data base, by specific edit actions, or data link.

Terminal area procedures (SIDs, STARs, and approaches) consist of a variety of special procedure legs and waypoints. Procedure legs are generally defined by a leg heading, course or track, and a leg termination type. The termination type can be specified in many ways such as an altitude, a distance, or intercept of another leg. More detail on the path construction for these leg types and terminators will be discussed in Section 24.2.3. Refer to ARINC 424 specification for further detail about what data and format are contained in the NDB to represent these leg types and terminations.

AF	DME Arc to a Fix
CA	Course to an Altitude
CD	Course to a Distance
CF*	Course to a Fix
CI	Course to an Intercept
CR	Course to Intercept a Radial
DF*	Direct to a Fix
FA*	Course from Fix to Altitude
FC	Course from Fix to Distance
FD	Course from Fix to DME Distance
FM	Course from Fix to Manual Term
HA*	Hold to an Altitude
HF*	Hold, Terminate at Fix after 1 Circuit
HM*	Hold, Manual Termination
IF*	Initial Fix
PI	Procedure Turn
RF*	Constant Radius to a Fix
TF*	Track to Fix
VA	Heading to Altitude
VD	Heading to Distance
VI	Heading to Intercept next leg
VM	Heading to Manual Termination
VR	Heading to Intercept Radial

Many of these leg types and terminations have appeared because of the evolution of equipment and instrumentation available on the aircraft and do not lend themselves to producing repeatable, deterministic ground tracks. For example, the ground track for a heading to an altitude will not only be dependent on the current wind conditions but also the climb performance of each individual aircraft. One can readily see that to fly this sort of leg without an FMS, the crew would follow the specified heading using the compass until the specified altitude is achieved, as determined by the aircraft's altimeter. Unfortunately, every aircraft will fly a different ground track and in some cases be unable to make a reasonable maneuver to capture the following leg. For the FMS, the termination of the leg is "floating" in that the lat/lon associated with the leg termination must be computed. These nondeterministic-type legs present problems for the air traffic separation concept of RNP airspace and for this reason RTCA DO-236 does not recommend the use of these legs in terminal area airspace, where they are frequently used today. These leg types also present added complexity in the FMS path construction algorithms since the path computation becomes a function of aircraft performance. With the advent of FMS and RNAV systems, in general, the need for non-deterministic legs simply disappears along with the problems and complexities associated with them.

Waypoints may also be specified as either "flyover" or "nonflyover." A flyover waypoint is a waypoint whose lat/lon position must be flown over before the turn onto the next leg can be initiated whereas a nonflyover waypoint does not need to be overflown before beginning the turn onto the next leg.

---

* These leg types are recommended in DO-236 as the set that produces consistent ground tracks and the only types that should be used within RNP airspace.

24.2.2.2 Lateral Flight Planning

To meet the tactical and strategic flight planning requirements of today's airspace, the flight planning function provides various ways to modify the flight plan at the crew's discretion.

Direct-to—The crew can perform a direct-to to any fix. If the selected fix is a downtrack fix in the flight plan, then prior flight plan fixes are deleted from the flight plan. If the selected fix is not a downtrack fix in the flight plan, then a discontinuity is inserted after the fix and existing flight plan data are preserved.

Direct/intercept—The direct/intercept facility allows the crew to select any fixed waypoint as the active waypoint and to select the desired course into this waypoint. This function is equivalent to a direct-to except the inbound course to the specified fix which may be specified by the crew. The inbound course may be specified by entering a course angle, or if the specified fix is a flight plan fix, the crew may also select the prior flight plan-specified course to the fix.

Holding pattern—Holding patterns may be created at any fix or at current position. All parameters for the holding pattern are editable including entry course, leg time/length, etc.

Fixes—Fixes may be inserted or deleted as desired. A duplicate waypoint page will automatically be displayed if there is more than one occurrence of the fix identifier in the navigation database. Duplicate fixes are arranged starting from the closest waypoint to the previous waypoint in the flight plan.

Procedures—Procedures (SIDs, STARs, and approaches including missed approach procedures) may be inserted or replaced as desired. If a procedure is selected to replace a procedure that is in the flight plan, the existing procedure is removed and replaced with the new selection.

Airway segments—Airway segments may be inserted as desired.

Missed approach procedures—The flight planning function also allows missed approach procedures to be included in the flight plan. These missed approach procedures can either come from the navigation database where the missed approach is part of a published procedure, in which case they will be automatically included in the flight plan, or they can be manually constructed by entry through the MCDU. In either case, automatic guidance will be available upon activation of the missed approach.

Lateral offset—The crew can create a parallel flight plan by specifying a direction (left or right of path) and distance (up to 99 nm) and optionally selecting a start and/or end waypoint for the offset flight plan. The flight planning function constructs an offset flight plan, which may include transition legs to and from the offset path.

24.2.2.3 Vertical Flight Planning

Waypoints can have associated speed, altitude, and time constraints. A waypoint speed constraint is interpreted as a "cannot exceed" speed limit, which applies at the waypoint and all waypoints preceding the waypoint if the waypoint is in the climb phase, or all waypoints after it if the waypoint is in the descent phase. A waypoint altitude constraint can be of four types—"at," "at or above," "at or below," or "between." A waypoint time constraint can be of three types—"at," "after," "before," "after," and "before" types are used for en route track-crossings and the "at" type is planned to be used for terminal area flow control.

Vertical flight planning consists of selection of speed, altitude, time constraints at waypoints (if required or desired), cruise altitude selection, aircraft weight, forecast winds, temperatures, and destination barometric pressure as well as altitude bands for planned use of aircraft anti-icing. A variety of optimized speed schedules for the various flight phases are typically available. Several aircraft performance-related crew selections may also be provided. All these selections affect the predicted aircraft trajectory and guidance.

24.2.2.4 Atmospheric Models

Part of the flight planning process is to specify forecast conditions for temperatures and winds that will be encountered during the flight. These forecast conditions help the FMS to refine the trajectory predictions to provide more accurate determination of estimated times of arrival (ETAs), fuel burn, rates of climb/descent, and leg transition construction.

The wind model for the climb segment is typically based on an entered wind magnitude and direction at specified altitudes. The value at any altitude is interpolated between the specified altitudes to zero on the ground and merged with the current sensed wind. Wind models for use in the cruise segment usually allow for the entry of wind (magnitude and direction) for multiple altitudes at en route waypoints. Future implementation of en route winds may be via a data link of a geographical current wind grid database maintained on the ground. The method of computing winds between waypoints is accomplished by interpolating between entries or by propagating an entry forward until the next waypoint entry is encountered. Forecast winds are merged with current winds obtained from sensor data in a method which gives a heavier weighting to sensed winds close to the aircraft and converges to sensed winds as each waypoint-related forecast wind is sequenced. The wind model used for the descent segment is a set of altitudes with associated wind vector entered for different altitudes. The value at any altitude is interpolated from these values, and blended with the current sensed wind.

Forecast temperature used for extrapolating the temperature profile is based on the International Standard Atmosphere (ISA) with an offset (ISA deviation) obtained from pilot entries and/or the actual sensed temperature.

Forecast temperature = 15 + ISA dev − 0.001983 × altitude   altitude < 36,089
Forecast temperature = −56.5   altitude > 36,089

Air pressure is also used in converting speed between calibrated airspeed, mach, and true airspeed.

δ (Pressure ratio) = $(1 - 0.0000068753 \times \text{altitude})^{5.2561}$ altitude < 36,089
δ (Pressure ratio) = $0.22336 \times e^{(4.8063 \times (36,089 - \text{altitude})/100,000)}$

24.2.3 Trajectory Predictions

Given the flight plan, the trajectory prediction function computes the predicted four-dimensional flight profile (both lateral and vertical) of the aircraft within the specified flight plan constraints and aircraft performance limitations based on entered atmospheric data and the crew-selected modes of operation. The lateral path and predicted fuel, time, distance, altitude, and speed are obtained for each point in the flight plan (waypoints as well as inserted vertical breakpoints such as speed change, cross-over, level off, top of climb [T/C], top of descent [T/D] points). The flight profile is continuously updated to account for nonforecasted conditions and tactical diversions from the specified flight plan.

To simplify this discussion, the flight path trajectory is broken into two parts—the lateral profile (the flight profile as seen from overhead) and the vertical profile (the flight profile as seen from the side). However, the lateral path and vertical path are interdependent in that they are coupled to each other through the ground speed parameter. Since the speed schedules that are flown are typically constant CAS/mach speeds for climb and descent phases, the TAS (or ground speed) increases with altitude for the constant CAS portion and mildly decreases with altitude for the constant mach portion, as shown in the following equations.

$$\text{Mach} = \text{sqrt}\left[\left(\frac{1}{\delta}\left\{\left[1 + 0.2\left(\frac{\text{CAS}}{661.5}\right)^2\right]^{3.5} - 1\right\} + 1\right)^{0.286} - 1\right]$$

$\text{TAS} = 661.5 \times \text{mach} \times \text{sqrt}[\theta]$

CAS = calibrated airspeed in knots

TAS = true airspeed in knots

   δ = atmospheric pressure ratio (actual temperature/S.L. std. temperature)

   θ = atmospheric temperature ratio (actual temperature/S.L. std. temperature)

The significance of the change in airspeed with altitude will become apparent in the construction of the lateral and vertical profile during ascending and descending flights as described in the next section. Further, since the basic energy balance equations used to compute the vertical profile use TAS, these speed conversion formulas are utilized to convert selected speed schedule values to true airspeed values.

24.2.3.1 Lateral Profile

Fundamentally, the lateral flight profile is the specified route (composed of procedure legs, waypoints, hold patterns, etc.), with all the turns and leg termination points computed by the FMS according to how the aircraft should fly them. The entire lateral path is defined in terms of straight segments and turn segments which begin and end at either fixed or floating geographical points. Computing these segments can be difficult because the turn transition distance and certain leg termination points are a function of predicted aircraft speed (as noted in the equations below), wind, and altitude, which, unfortunately, are dependent on how much distance is available to climb and descend. For example, the turn transition at a waypoint requires a different turn radius and therefore a different distance when computed with different speeds. The altitude (and therefore speed of the aircraft) that can be obtained at a waypoint is dependent upon how much distance is available to climb or desend. So, the interdependency between speed and leg distance presents a special problem in formulating a deterministic set of algorithms for computing the trajectory. This effect becomes significant for course changes greater than 45°, with the largest effect for legs such as procedure turns which require a 180° turn maneuver.

Lateral turn construction is based on the required course change and the aircraft's predicted ground speed during the turn. If the maximum ground speed that the aircraft will acquire during the required course change is known, a turn can be constructed as follows:

$$\text{Turn radius (ft)} = \frac{GS^2}{g \times \tan \phi}$$

$$\text{Turn arclength (ft)} = \Delta \text{Course} \times \text{turn radius}$$

$$GS = \text{maximum aircraft ground speed during the turn}$$

$$g = \text{acceleration due to gravity}$$

$$\phi = \text{nominal aircraft bank angle used to compute a turn}$$

For legs such as constant radius to a fix (RF) where the turn radius is specified, a different form of the equation is used to compute the nominal bank angle that must be used to perform the maneuver.

$$\phi = \arctan \left[\frac{GS^2}{\text{Turn radius} \times g} \right]$$

To determine the maximum aircraft ground speed during the turn the FMC must first compute the altitude at which the turn will take place, and then the aircraft's planned speed based on the selected speed schedule and any applicable wind at that altitude. The desired bank angle required for a turn is predetermined based on a trade-off between passenger comfort and airspace required to perform a lateral maneuver.

The basis for the lateral profile construction is the leg and termination types mentioned in the flight plan section. There are four general leg types:

- Heading (V)—aircraft heading
- Course (C)—fixed magnetic course
- Track (T)—computed great circle path (slowly changing course)
- Arc (A or R)—an arc defined by a center (fix) and a radius

There are six leg terminator types:

1. Fix (F)—terminates at geographic location
2. Altitude (A)—terminates at a specific altitude
3. Intercept next leg (I)—terminates where leg intercepts the next leg
4. Intercept radial (R)—terminates where leg intercepts a specific VOR radial
5. Intercept distance (D or C)—terminates where leg intercepts a specific DME distance or distance from a fix
6. Manual (M)—leg terminates with crew action

Not all terminator types can be used with all leg types. For example, a track leg can only be terminated by a fix since the definition of a track is the great circle path between two geographic locations (fixes). Likewise, arc legs are only terminated by a fix. In a general sense, heading and course legs can be graphically depicted in the same manner understanding that the difference in the computation is the drift angle (or aircraft yaw).

Figure 24.3 depicts a graphical construction for the various leg and terminator types. The basic construction is straightforward. The complexity arises from the possible leg combinations and formulating proper curved transition paths between them. For example, if a TF leg is followed by a CF leg where the specified course to the fix does not pass through the terminating fix for the prior TF leg, then a transition path must be constructed to complete a continuous path between the legs.

In summary, the lateral flight path computed by the FMC contains much more data than straight lines connecting fixed waypoints. It is a complete prediction of the actual lateral path that the aircraft will fly under FMS control. The constructed lateral path is critical because the FMC will actually control the aircraft to it by monitoring cross-track error and track angle error, and issuing roll commands to the autopilot as appropriate.

24.2.3.2 Vertical Profile

The fundamental basis for the trajectory predictor is the numerical integration of the aircraft energy balance equations including variable weight, speed, and altitude. Several forms of the energy balance equation are used to accommodate unrestricted climb/descent, fixed gradient climb/descent, speed change, and level flight. The integration steps are constrained by flight plan-imposed altitude and speed restrictions as well as aircraft performance limitations such as speed and buffet limits, maximum altitudes, and thrust limits. The data that drives the energy balance equations come from the airframe/engine-dependent thrust, fuel flow, drag, and speed schedule models stored in the performance data base. Special construction problems are encountered for certain leg types such as an altitude-terminated leg because the terminator has a floating location. The location is dependent upon where the trajectory integration computes the termination of the leg. This also determines the starting point for the next leg.

The trajectory is predicted based on profile integration steps—the smaller the integration step the more accurate the computed trajectory. For each step the aircraft's vertical speed, horizontal speed, distance traveled, time used, altitude change, and fuel burned is determined based on the projected aircraft target speed, wind, drag, and engine thrust for the required maneuver. The aircraft's vertical state is computed for the end of the step and the next step is initialized with those values. Termination of an integration step can occur when a new maneuver type must be used due to encountering an altitude or speed constraint, flight phase change, or special segments such as turn transitions where finer integration steps may be prudent. The vertical profile is comprised of the following maneuver types: unrestricted ascending and descending segments, restricted ascending and descending segments, level segments, and speed change segments. Several forms of the energy balance equation are used depending on the maneuver type for a given segment of the vertical profile. Assumptions for the thrust parameter are maneuver type and flight phase dependent.

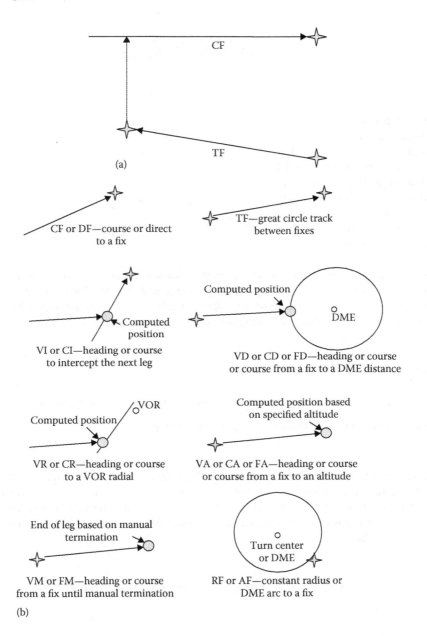

FIGURE 24.3 Basic lateral leg construction. (a) TF leg followed by a CF leg. (b) Computation of transition path.

24.2.3.3 Maneuver Types

Unrestricted ascending and descending segments—The following form of the equation is typically used to compute the average vertical speed for fixed altitude steps (*dh* is set integration step). Using fixed altitude steps for this type of segment allows for deterministic step termination at altitude constraints. For ascending flight the thrust is generally assumed to be the take-off, go-around, or climb thrust limit. For descending flight the thrust is generally assumed to be at or a little above flight idle.

$$\frac{V}{S} = \frac{(T - D)V_{ave} / GW}{(T_{act} / T_{std}) + (V_{ave} / g)(dV_{true} / dh)}$$

where

    T is the avg. thrust (lb)

    D is the avg. drag (lb)

    GW is the A/C gross weight (lb)

    T_{act} is the ambient temperature (K)

    T_{std} is the std. day temperature (K)

    V_{ave} is the average true airspeed (ft/s)

    $g = 32.174$ ft/s^2

    $dV_{true} =$ Delta V_{true} (ft/s)

    dh is the desired altitude step (ft)

The projected aircraft true airspeed is derived from the pilot-selected speed schedules and any applicable airport or waypoint-related speed restrictions. Drag is computed as a function of aircraft configuration, speed, and bank angle. Fuel flow and therefore weight change is a function of the engine thrust. Once V/S is computed for the step the other prediction parameters can be computed for the step.

    $dt = \dfrac{dh}{V/S}$, where $dt =$ delta time for step

    $ds = dt(V_{true} +$ average along track wind for segment$)$, where $ds =$ delta distance for step

    $dw = dt \times$ fuel flow(T), where $w =$ delta weight for step

Restricted ascending and descending segments—The following form of the equation is typically used to compute the average thrust for fixed altitude steps (dh and V/S are predetermined). Using fixed altitude steps for this type of segment allows for deterministic step termination at altitude constraints. The average V/S is either specified or computed based on a fixed flight path angle (FPA).

$$\frac{V}{S_{ave}} = GS_{ave}\tan FPA, \quad \text{where } GS_{ave} = \text{segment ground speed (ft/s)}$$

The fixed FPA can in turn be computed based on a point to point vertical flight path determined by altitude constraints, which is known as a geometric path. With a specified V/S or FPA segment the thrust required to fly this profile is computed.

$$T = \frac{W \times V/S_{ave}}{V_{ave}} - \left(1 + \frac{V_{ave}}{g}\frac{dV_{true}}{dh}\right) + D$$

The other predicted parameters are computed as stated for the unrestricted ascending and descending segment.

Level segments—Constant-speed-level segments are a special case of the above equation. Since dV_{true} and V/S_{ave} are by definition zero for level segments, the equation simplifies to $T = D$. Level segments are typically integrated based on fixed time or distance steps so the other predicted parameters are computed as follows:

dt = set integration step
and
$ds = dt(V_{true} +$ average along track wind for segment) $\qquad ds$ = delta distance for step
or
ds = set integration step
and
$dt = ds/(V_{true} +$ average along track wind for segment) $\qquad dt$ = delta time for step
$dw = dt \times$ fuel flow (T) $\qquad dw$ = delta weight for step

Speed change segments—The following form of the equation is typically used for speed change segments to compute the average time for a fixed dV_{true} step. The V/S_{ave} used is predetermined based on ascending, descending, or level flight along with the operational characteristics of the flight controls or as for the case of geometric paths computed based on the required FPA. The thrust is assumed to be flight idle for descending flight, take-off or climb thrust limit for ascending flight, or cruise thrust limit for level flight.

$$dt = \frac{dV_{true}}{g}\left\{ \frac{(T-D)}{GW} - \left(\frac{T_{act}}{T_{std}} \frac{V/S_{ave}}{V_{ave}} \right) \right\}$$

$$dh = \frac{V}{S_{ave}} \times dt$$

For all maneuver types the altitude rate, speed change, or thrust must be corrected for bank angle effects if the maneuver is performed during a turn transition. The vertical flight profile that the FMC computes along the lateral path is divided into three phases of flight: climb, cruise, and descent.

The climb phase—The climb phase vertical path, computed along the lateral path, is typically composed of the segments shown in Figure 24.4. In addition to these climb segments there can also be altitude level-off segments created by altitude restrictions at climb waypoints and additional target speed acceleration segments created by speed restrictions at climb waypoints.

The cruise phase—The cruise phase vertical path, computed along the lateral path, is very simple (Figure 24.5). It's typically composed of a climb speed to cruise speed acceleration or deceleration segment followed by a segment going to the FMC-computed top of descent. The cruise phase typically is predicted level at cruise altitude via several distance- or time-based integration steps. Unlike the climb and descent phase, the optimal cruise speeds slowly change with the changing weight of the aircraft, caused by fuel burn. If step climb or descents are required during the cruise phase, these are treated as unrestricted ascending flight and fixed V/S or FPA descents. At each step the FMC computes the aircraft's along-path speed, along-path distance traveled, and fuel burned based on the projected aircraft target speed, wind, drag, and engine thrust. The projected aircraft true airspeed is derived from the pilot-selected cruise speed schedule and applicable airport-related speed restrictions. Drag is computed as a function of aircraft speed and bank angle. For level flight, thrust must be equal to drag. Given the required thrust, the engine power setting can be computed, which becomes the basis for computing fuel burn and throttle control guidance.

The descent phase—The descent phase vertical path, computed along the lateral path, can be composed of several vertical leg types as shown in Figure 24.6.

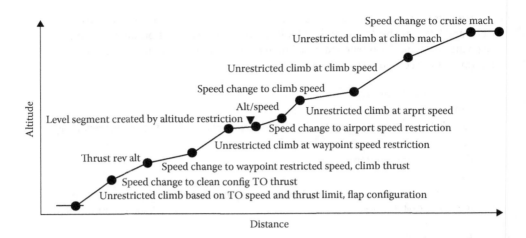

FIGURE 24.4   Typical climb profile.

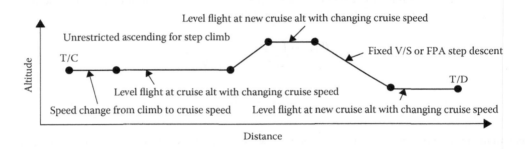

FIGURE 24.5   Typical cruise profile.

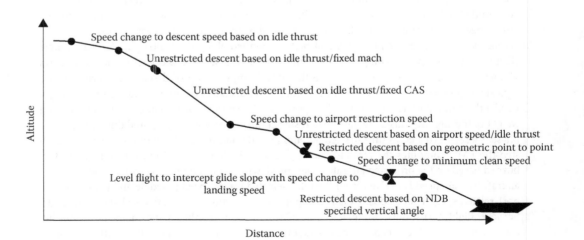

FIGURE 24.6   Typical descent profile.

In addition to these descent segments, there can also be altitude level-off segments created by altitude restrictions at descent waypoints and additional targets speed deceleration segments created by speed restrictions at descent waypoints as well as eventual deceleration to the landing speed for the selected flaps configuration.

24.2.3.4 NDB Vertical Angles

These leg types are generally used in the approach. The desired approach glide slope angle that assures obstacle clearance is packed as part of the waypoint record for the approach in the Navigation Data Base (NDB). The angle is used to compute the descent path between the waypoint associated with the angle and the first of the following to be encountered (looking backwards):

1. Next lateral waypoint with an NDB vertical angle record
2. Next "at" constraint
3. First approach waypoint

A new NDB gradient can be specified on any waypoint. This allows the flexibility to specify multiple FPAs for the approach if desired. The integration basis for this leg assumes a thrust level compatible with maintaining the selected speed schedule at the descent rate specified by the NDB angle. Decelerations that can occur along these legs because of various restrictions (both regulatory and airframe) assume performing the speed change at idle thrust at the vertical speed specified by the NDB angle. If within the region where flaps are anticipated, then the deceleration model is based on a flaps configuration performance model.

Default approach vertical angle—Generally, this leg is used in lieu of a specified NDB angle to construct a stabilized nominal glide slope between the glide slope intercept altitude (typically 1500 ft above the runway) to the selected runway. The integration basis for this leg is the same as the NDB angle.

Direct to vertical angle—This leg type provides a vertical "Dir to" capability for use in tactical situations. The path constructed is the angle defined by the current 3-D position of the aircraft and the next appropriate reference point (usually the next altitude constraint). For a pending vertical "direct to" the direct to angle is updated on a periodic basis to account for the movement of the aircraft. In determining the direct to angle the aircraft 3-D position is extrapolated to account for the amount of time required to compute the trajectory for VNAV guidance to avoid path overshoots when the trajectory is available. The integration basis for this leg assumes a thrust level compatible with maintaining the selected speed schedule at the descent rate specified by the direct to gradient. Decelerations that can occur along these descent legs because of various restrictions (both regulatory and aircraft) assume performing the speed change at idle thrust for the anticipated flaps/landing gear configuration.

Computed vertical angle—This leg type provides constant angle vertical paths between constraints that are part of the vertical flight plan. These geometric paths provide for repeatable, stabilized, partial power descent paths at lower altitudes in the terminal area. The general rules for proper construction of these paths are:

o Vertical maneuvering should be minimized. This implies that a single angle to satisfy a string of altitude constraints is preferred. This can occur when "At or above" and "At or below" altitude constraints are contained in the flight plan.

o If a string of "At or above" and/or "At or below" constraints can be satisfied with an unrestricted, idle power path, then that path is preferred.

o Computed gradient paths should be checked for flyability (steeper than idle). The computation of the idle path (for the anticipated and idle with drag devices deployed) should account for a minimum deceleration rate if one is contained within the computed gradient leg.

The integration basis for this leg assumes a thrust level compatible with maintaining the selected speed schedule at the descent rate specified by the computed vertical angle. Decelerations that can occur along these descent legs because of various restrictions (both regulatory and airframe) assume performing the speed change at idle thrust limited for a maximum deceleration rate.

> *Constant V/S*—This leg type provides a strategic, shallower-than-idle initial descent path if desired. The construction of this path is dependent on a vertical speed and intercept altitude being requested. The integration basis for this leg assumes a thrust level compatible with maintaining the selected speed schedule at the descent rate specified by the commanded v/s. Decelerations that can occur along these descent legs because of various restrictions (both regulatory and airframe) assume performing the speed change at idle thrust limited for a maximum deceleration rate.
>
> *Unrestricted descent*—The unrestricted descent uses performance data to construct an energy-balanced idle descent path when not constrained by altitude constraints. The integration basis for this leg assumes maintaining the selected speed schedule at idle thrust. This results in a changing vertical speed profile. Decelerations that can occur along these descent legs because of various restrictions (both regulatory and aircraft) assume performing the speed change at a minimum vertical speed rate and idle thrust limited for a maximum deceleration rate. The minimum vertical speed can be based on energy sharing or a precomputed model. An idle thrust factor allows the operator to create some margin (shallower or steeper) in the idle path construction.

24.2.4 Performance Computations

The performance function provides the crew information to help optimize the flight or provide performance information that would otherwise have to be ascertained from the aircraft performance manual. FMSs implement a variety of these workload reduction features; only the most common functions are discussed here.

24.2.4.1 Speed Schedule Computation

Part of the vertical flight planning process is the crew selection of performance modes for each flight phase based on specific mission requirements. These performance modes provide flight profile optimization through computation of flight phase-dependent, optimized speed schedules that are used as a basis for both the trajectory prediction, generation of guidance speed targets, and other performance advisories.

The selection of a specific performance mode for each flight phase results in the computation of an optimized speed schedule, which is a constant CAS, constant mach pair, which becomes the planned speed profile for each flight phase. The altitude where the CAS and mach are equivalent is known as the crossover altitude. Below the crossover altitude the CAS portion of the speed schedule is the controlling speed parameter and above the crossover altitude the mach portion is the controlling speed. The performance parameter that is optimized is different for each performance mode selection.

Climb

- Economy (based on Cost Index)—speed that optimizes overall cost of operation (lowest cost).
- Maximum angle of climb—speed that produces maximum climb rate with respect to distance.
- Maximum rate of climb—speed that produces maximum climb rate with respect to time.
- Required time of arrival speed (RTA)—speed that optimizes overall cost of operation, while still achieving a required time of arrival at a specific waypoint.

Cruise

- Economy (based on Cost Index)—speed that optimizes overall cost of operation (lowest cost).
- Maximum endurance—speed that produces lowest fuel burn rate, maximizing endurance time.
- Long range cruise—speed that produces best fuel mileage, maximizing range.
- Required time of arrival (RTA)—speed that optimizes overall cost of operation, while still achieving a required time of arrival at a specific waypoint.

Descent

- Economy (based on Cost Index)—speed that optimizes overall cost of operation (lowest cost).
- Maximum descent rate—speed that produces maximum descent rate with respect to time.
- Required time of arrival (RTA)—speed that optimizes overall cost of operation, while still achieving a required time of arrival at a specific waypoint.

All flight phases allow a manually entered CAS/mach pair as well.

It may be noted that one performance mode that is common to all flight phases is the "economy" speed mode which minimizes the total cost of operating the airplane on a given flight. This performance mode uses a Cost Index, which is the ratio of time-related costs (crew salaries, maintenance, etc.) to fuel cost as one of the independent variables in the speed schedule computation.

$$\text{Cost Index (CI)} = \frac{\text{Flight time-related cost}}{\text{Fuel cost}}$$

The cost index allows airlines to weight time and fuel costs based on their daily operations.

24.2.4.2 Maximum and Optimum Altitudes

An important parameter for the flight crew is the optimum and maximum altitude for the aircraft/engine type, weight, atmospheric conditions, bleed air settings, and the other vertical flight planning parameters. The optimum altitude algorithm computes the most cost-effective operational altitude based solely on aircraft performance and forecasted environmental conditions. Fundamentally, the algorithm searches for the altitude that provides the best fuel mileage.

$$\text{Altitude that maximizes the ratio:} \quad \frac{\text{Ground speed}}{\text{Fuel burn rate}}$$

The maximum altitude algorithm computes the highest attainable altitude based solely on aircraft performance and forecasted environmental conditions, while allowing for a specified rate of climb margin.

$$\text{Altitude that satisfies the equality:} \quad \text{Min climb rate} = \text{TAS} \times \frac{(\text{Thrust} - \text{drag})}{\text{Weight}}$$

Optimum altitude is always limited by maximum altitude. The algorithms for these parameters account for the weight reduction caused by the fuel burn in achieving the altitudes. The speeds assumed are the selected performance modes.

> *Trip altitude*—Another important computation that allows the crew to request an altitude clearance to optimize the flight is the recommended cruise altitude for a specified route known as trip altitude. This altitude may be different from the optimum altitude in that for short trips the optimum altitude may not be achievable because of the trip distance. This algorithm searches for the altitude that satisfies the climb and descent while preserving a minimum cruise time.
>
> *Alternate destinations*—To help reduce crew workload during flight diversion operations the FMS typically provides alternate destination information. This computation provides the crew with distance, fuel, and ETA for selected alternate destinations. The best trip cruise altitude may be computed as well. The computations are based either on a direct route from the current position to the alternate or continuing to the current destination, execution of a missed approach at the destination, and then direct to the alternate. Also computed for these alternate destinations are available holding times at the present position and current fuel state vs. fuel required to alternates. Usually included for the crew convenience is the CDU/MCDU retrieval of suitable airports that are nearest the aircraft.

Step climb/descent—For longer-range flights often the achievable cruise altitude is initially lower than the optimum because of the heavy weight of the aircraft. As fuel is burned off and the aircraft weight reduced, it becomes advantageous to step climb to a higher altitude for more efficient operation. The FMS typically provides a prediction of the optimum point(s) at which a step climb/descent maneuver may be initiated to provide for more cost-effective operation. This algorithm considers all the vertical flight planning parameters, particularly the downstream weight of the aircraft, as well as entered wind data. The time and distance to the optimum step point for the specified step altitude is displayed to the crew, as well as the percent savings/penalty for the step climb/descent vs. the current flight plan. For transoceanic aircraft it is typical for the trajectory prediction function to assume that these steps will be performed as part of the vertical profile, so that the fuel predictions are more aligned with what the aircraft will fly.

Thrust limit data—To prevent premature engine maintenance/failure and continued validation of engine manufacturer's warrantees, it becomes important not to overboost the aircraft engines. The engine manufacturers specify flight phase-dependent thrust limits that the engines are designed to operate reliably within. These engine limits allow higher thrust levels when required (take-off, go-around, engine out) but lower limits for non-emergency sustained operation (climb and cruise). The thrust limits for take-off, climb, cruise, go around, and continuous modes of operation are computed based on the current temperature, altitude, speed, and type of engine/aircraft and engine bleed settings. Thrust limit data are usually represented by "curve sets" in terms of either engine RPM (N1) or engine pressure ratio (EPR), depending on the preferred engine instrumentation package used to display the actual engine thrust. The "curve sets" typically have a temperature-dependent curve and an altitude-dependent curve along with several correction curves for various engine bleed conditions. The algorithms used to compute the thrust limits vary among engine manufacturers.

Take-off reference data—The performance function provides for the computation, or entry, of V_1, V_R and V_2 take-off speeds for selected flap settings and runway, atmospheric, and weight/CG conditions. These speeds are made available for crew selection for display on the flight instruments. In addition, take-off configuration speeds are typically computed. The take-off speeds and configuration speeds are stored as data sets or supporting data sets in the performance database.

Approach reference data—Landing configuration selection is usually provided for each configuration appropriate for the operation of the specific aircraft. The crew can select the desired approach configuration and the state of that selection is made available for other systems. Selection of an approach configuration also results in the computation of a landing speed based on a manually entered wind correction for the destination runway. In addition, approach configuration speeds are computed and displayed for reference and selection for display on the flight instruments. The approach and landing speeds are stored as data sets in the performance database.

Engine-out performance—The performance function usually provides engine-out performance predictions for the loss of at least one engine. These predictions typically include:

- ○ Climb at engine-out climb speed
- ○ Cruise at engine-out cruise speed
- ○ Driftdown to engine-out maximum altitude at driftdown speed
- ○ Use of maximum continuous thrust

The engine out speed schedules are retrieved from the performance data base and the trajectory predictions are computed based on the thrust available from the remaining engines and the increased aircraft drag created by engine windmilling and aircraft yaw caused by asymmetrical thrust.

24.2.5 Guidance

The FMS typically computes roll axis, pitch axis, and thrust axis commands to guide the aircraft to the computed lateral and vertical profiles as discussed in Section 24.2.3. These commands may change forms depending on the particular flight controls equipment installed on a given aircraft. Other guidance information is sent to the forward field of view displays in the form of lateral and vertical path information, path deviations, target speeds, thrust limits and targets, and command mode information.

24.2.5.1 Lateral Guidance

The lateral guidance function typically computes dynamic guidance data based on the predicted lateral profile described in Section 24.2.3. The data are comprised of the classic horizontal situation information:

- Distance to go to the active lateral waypoint (DTG)
- Desired track (DTRK)
- Track angle error (TRKERR)
- Cross-track error (XTRK)
- Drift angle (DA)
- Bearing to the go to waypoint (BRG)
- Lateral track change alert (LNAV alert)

A common mathematical method to compute the above data is to convert the lateral path lat/lon point representation and aircraft current position to earth-centered unit vectors using the following relationships:

P = earth centered unit position vector with x, y, z components
$X = COS$ (lat) COS (lon)
$Y = COS$ (lat) SIN (lon)
$Z = SIN$ (lat)

For the following vector expressions $\times$ is the vector cross product and $\cdot$ is the vector dot product. For any two position vectors that define a lateral path segment:

$N = Pst \times Pgt$ $\qquad$ N = *unit vector normal to* Pst *and* Pgt
$Pap = N \times (Ppos \times N)$ $\qquad$ Pgt = *go to point unit position vector*
$\qquad$ Pst = *start point unit position vector*
$DTGap = earth\ radius \times arcCOS\ (Pgt{\cdot}Pap)$ $\qquad$ Pap = *along path position unit vector*
$DTGpos = earth\ radius \times arcCOS\ (Pgt{\cdot}Ppos)$ $\qquad$ $Ppos$ = *current position unit vector*
$XTRK = -earth\ radius \times arcCOS(Pap{\cdot}Ppos)$ $\qquad$ *(full expression)*
$XTRK = -earth\ radius \times N{\cdot}Ppos$ $\qquad$ *(good approximation)*
$\quad Est = Z \times P$ $\qquad$ Est = *East-pointing local level unit vector*
$\quad Nth = P \times Est$ $\qquad$ Nth = *North-pointing local level unit vector*

$$Z = \begin{vmatrix} 0 \\ 0 \\ 1 \end{vmatrix} \quad Z\ axis\ unit\ vector$$

$DTRK = arcTAN\ [(-N{\cdot}Nth_{ap})/(-N{\cdot}Est_{ap})$
$BRG = arcTAN\ [(-N{\cdot}Nth_{pos})/(-N{\cdot}Est_{pos})$
$TRKERR = DTRK - Current\ Track$
$DA = Current\ Track - Current\ Heading$
$LNAV\ Alert\ is\ set\ when\ the\ DTG/ground\ speed\ <10\ s\ from\ turn\ initiation$

The above expressions can also be used to compute the distance and course information between points that are displayed to the crew for the flight plan presentation. The course information is generally displayed as magnetic courses, due to the fact that for many years a magnetic compass was the primary heading sensor and therefore all navigation information was published as magnetic courses. This historical-based standard requires the installation of a worldwide magnetic variation model in the FMS since most of the internal computations are performed in a true course reference frame. Conversion to magnetic is typically performed just prior to crew presentation.

The lateral function also supplies data for a graphical representation of the lateral path to the navigation display, if the aircraft is so equipped, such that the entire lateral path can be displayed in an aircraft-centered reference format or a selected waypoint center reference format. The data for this display are typically formatted as lat/lon points with identifiers and lat/lon points with straight and curved vector data connecting the points. Refer to ARINC 702A for format details. In the future the FMS may construct a bit map image of the lateral path to transmit to the navigation display instead of the above format.

Lateral leg switching and waypoint sequencing—As can be seen in Section 24.2.3.1, the lateral path is composed of several segments. Most lateral course changes are performed as "flyby" transitions. Therefore anticipation of the activation of the next vertical leg is required, such that a smooth capture of that segment is performed without path overshoot. The turn initiation criteria are based on the extent of the course change, the planned bank angle for the turn maneuver, and the ground speed of the aircraft.

$$\text{Turn radius} = \frac{\text{Ground speed}^2}{\left[g \times \text{TAN}\left(\phi_{\text{nominal}} \right) \right]}$$

$$\text{Turn initiation distance} = \frac{\text{Turn radius}}{\text{TAN(Course change}/2) + \text{Roll in distance}}$$

The roll in distance is selected based on how quickly the aircraft responds to a change in the aileron position. Transitions that are flyby but require a large course change (>135°) typically are constructed for a planned overshoot because of airspace considerations. Turn initiation and waypoint sequence follow the same algorithms except the course change used in the above equations is reduced from the actual course change to delay the leg transition and create the overshoot. The amount of course change reduction is determined by a balance in the airspace utilized to perform the overall maneuver. For "flyover" transitions, the activation of the next leg occurs at the time the "flyover" waypoint is sequenced.

The initiation of the turn transition and the actual sequence point for the waypoint are not the same for "flyby" transitions. The waypoint is usually sequenced at the turn bisector point during the leg transition.

Roll control—Based on the aircraft current state provided by the navigation function and the stored lateral profile provided by the trajectory prediction function, lateral guidance produces a roll steering command that can be engaged by the flight controls. This command is both magnitude and rate limited based on aircraft limitations, passenger comfort, and airspace considerations. The roll command is computed to track the straight and curved path segments that comprise the lateral profile. The roll control is typically a simple control law driven by the lateral cross-track error and track error as discussed in the prior section as well as a nominal roll angle for the planned turn transitions. The nominal roll angle is zero for straight segments but corresponds to the planned roll angle used to compute lateral transition paths to follow the curved segments.

$$\text{Roll} = \text{XTRK gain} \times \text{XTRK} + \text{TRK gain} \times \text{TRK error} + \phi_{\text{nominal}}$$

where ϕ_{nominal} is the nominal planned roll angle.

The gain values used in this control loop are characteristic of the desired aircraft performance for a given airframe and flight controls system.

Lateral capture path construction—At the time LNAV engagement with the flight controls occurs, a capture path is typically constructed that guides the airplane to the active lateral leg. This capture path is usually constructed based on the current position and track of the aircraft if it intersects the active lateral leg. If the current aircraft track does not intersect the active lateral leg, then LNAV typically goes into an armed state waiting for the crew to steer the aircraft into a capture geometry before fully engaging to automatically steer the aircraft. Capture of the active guidance leg, is usually anticipated to prevent overshoot of the lateral path.

24.2.5.2 Vertical Guidance

The vertical guidance function provides commands of pitch, pitch rate, and thrust control to the parameters of target speeds, target thrusts, target altitudes, and target vertical speeds (some FMS provide only the targets depending on the flight management/flight control architecture of the particular aircraft). Much like the lateral guidance function, the vertical guidance function provides dynamic guidance parameters for the active vertical leg to provide the crew with vertical situation awareness. Unlike the lateral guidance parameters, the vertical guidance parameters are somewhat flight phase dependent.

Flight Phase	Vertical Guidance Data
Takeoff	Take-off speeds V_1, V_2, V_R
	Take-off thrust limit
Climb	Target speed based on selected climb speed schedule, flight plan speed restriction, and airframe limitations
	Target altitude intercept
	Altitude constraint violation message
	Distance to top of climb
	Climb thrust limits
Cruise	Target speed based on selected cruise speed schedule, flight plan speed restriction, and airframe limitations
	Maximum and optimum altitude distance to step climb point
	Distance to top of descent
	Cruise thrust limit
	Cruise thrust target
Descent	Target speed based on selected descent speed schedule, flight plan speed restriction, and airframe limitations
	Target altitude intercept
	Vertical deviation
	Desired V/S
	Energy bleed-off message
Approach	Target speed based on dynamic flap configuration
	Vertical deviation
	Desired V/S
Missed approach	Target speed based on selected climb speed schedule, flight plan speed restriction, and airframe limitations
	Target altitude intercept
	Altitude constraint violation message
	Distance to top of climb
	Go-around thrust limit

Vertical guidance is based on the vertical profile computed by the trajectory prediction function as described in a previous section as well as performance algorithms driven by data from the performance data base.

The mathematical representation of the vertical profile is the point type identifier, distance between points, which includes both lateral and vertical points, speed, altitude, and time at the point. Given this information, data for any position along the computed vertical profile can be computed.

$$\text{Path gradient} = \frac{\left(\text{alt}_{\text{start}} - \text{alt}_{\text{end}}\right)}{\text{Distance between points}}$$

Therefore the path reference altitude and desired *V/S* at any point is given by:

$$\text{Path altitude} = \text{alt}_{\text{end}} + \text{Path gradient} \times \text{DTGap}$$

$$\text{Vertical deviation} = \text{Current altitude} - \text{Path altitude}$$

$$\text{Desired } V/S = \text{Path gradient} \times \text{Current ground speed}$$

In the same manner time and distance data to any point or altitude can be computed as well. The target speed data are usually not interpolated from the predicted vertical profile since it is only valid for on-path flight conditions. Instead, it is computed based on the current flight phase, aircraft altitude, relative position with respect to flight plan speed restrictions, flaps configuration, and airframe speed envelope limitations. This applies to thrust limit computations as well.

Auto flight phase transitions—The vertical guidance function controls switching of the flight phase during flight based on specific criteria. The active flight phase becomes the basis for selecting the controlling parameters to guide the aircraft along the vertical profile. The selected altitude is used as a limiter in that the vertical guidance will not allow the aircraft to fly through that altitude (except during approach operations where the selected altitude may be pre-set for a missed approach if required). When on the ground with the flight plan and performance parameters initialized, the flight phase is set to take-off. After liftoff, the phase will switch to climb when the thrust revision altitude is achieved. The switch from climb to cruise (level flight) phase usually occurs when the aircraft is within an altitude acquire band of the target altitude.

$$|\text{Cruise altitude} - \text{current altitude}| < \text{Capture gain} \times \text{Current vertical speed}$$

The capture gain is selected for aircraft performance characteristics and passenger comfort. The switch from cruise to descent can occur in various ways. If the crew has armed the descent phase by lowering the preselected altitude below cruise altitude, then descent will automatically initiate at an appropriate distance before the computed T/D to allow for sufficient time for the engine to spool down to descent thrust levels so that the aircraft speed is coordinated with the initial pitch-over maneuvers. If the crew has not armed the descent by setting the selected altitude to a lower level, then cruise is continued past the computed T/D until the selected altitude is lowered to initiate the descent. Facilities are usually provided for the crew to initiate a descent before the computed T/D in response to ATC instructions to start descending.

Vertical leg switching—As can be seen in Section 24.2.3.2, the vertical path is composed of several segments. Just as in the lateral domain it is desirable to anticipate the activation of the next vertical leg such that a smooth capture of that segment is performed without path overshoot. It therefore becomes necessary to have an appropriate criteria for vertical leg activation. This criteria is typically in the form of an inequality involving path altitude difference and path altitude rate difference.

- $|\text{Path altitude } (n) - \text{Path altitude } (n+1)| \times \text{Capture gain} < |\text{Desired } V/S \ (n) - \text{Desired } V/S \ (n+1)|$

The capture gain is determined based on airframe performance and passenger comfort.

Pitch axis and thrust axis control—The pitch command produced by vertical guidance is based on tracking the speed target, FMS path, or acquiring and holding a target altitude depending on the flight phase and situation. If VNAV is engaged to the flight controls an annunciation of the parameter controlling pitch is usually displayed in the crew's forward field of view.

Flight Phase	Pitch Axis Control	Thrust Axis Control	Pitch/Thrust Mode Annunciation
Take-off	None until safely off ground then same as climb	Take-off thrust limit	Vspd/TO limit
Climb and cruise climb	Capture and track speed target	Climb thrust limit	Vspd/CLB limit
Level flight	Capture and maintain altitude	Maintain speed target	Valt/CRZ limit
Unrestricted descent	Capture and track vertical path	Set to flight idle	Vpath/CRZ limit
Restricted descent and approach	Capture and track vertical path	Set to computed thrust required, then maintain speed	Vpath/CRZ limit
Descent path capture from below and cruise descent	Capture and track fixed V/S capture path	Set to computed thrust required, then maintain speed	Vpath/CRZ limit
Descent path capture from above	Capture and track upper speed limit	Set to flight idle	Vspd/CRZ limit
Missed approach	Capture and track speed target	Go-around thrust limit	Vspd/GA limit

Control strategy may vary with specific implementations of FMSs. Based on the logic in the above table, the following outer loop control algorithms are typically used to compute the desired control parameters.

Pitch axis control—The control algorithms below are representative of control loop equations that could be used and are by no means the only forms that apply. Both simpler and more complex variations may be used.

Vspd

    Capture

    Delta pitch = Speed rate gain × (Airspeed rate – Capture rate)

    Track

    Delta pitch = (Airspeed gain × Airspeed error + Speed rate gain × Airspeed rate)/ V_{true}

Vpath

    Capture

    V/S error = Fixed capture V/S – Current V/S

    Delta pitch = Path capture gain × arcSIN (V/S error/V_{true})

    Track

    Delta pitch = (VS gain × V/S error + Alt error gain × Alt error)/V_{true}

Valt

    Capture

    Capture V/S = Alt capture gain × Alt error

    V/S error = Capture V/S – Current V/S

    Delta pitch = V/S gain ×× arcSIN (V/S error/V_{true})

    Track

    Delta pitch = (VS gain × Current V/S + Alt error gain × Alt error)/V_{true}

Proper aircraft pitch rates and limits are typically applied before final formulation of the pitch command. Once again, the various gain values are selected based on the aircraft performance and passenger comfort.

Thrust axis control—The algorithms below are representative of those that could be utilized to determine thrust settings. Quite often the thrust setting for maintaining a speed is only used for an initial throttle setting. Thereafter the speed error is used to control the throttles.

Thrust Limit

Thrust limit = f(temp, alt, spd, engine bleed air): *stored as data sets in the performance DB*

Flight Idle

Idle thrust = f(temp, alt, spd, engine bleed air): *stored as data sets in the performance DB*

Thrust Required

$$T = \frac{W \times V/S_{\text{ave}}}{V_{\text{ave}}} \left(1 + \frac{V_{\text{ave}}}{g} \frac{dV_{\text{true}}}{dh} \right) + D$$

RTA (required time of arrival)—The required time of arrival or time navigation is generally treated as a dynamic phase-dependent speed schedule selection (refer to Section 24.2.4). From this standpoint the only unique guidance requirements are the determination of when to recompute the phase-dependent speed schedules based on time error at the specified point and perhaps the computation of the earliest and latest times achievable at the specified point.

RNAV approach with VNAV guidance—The only unique guidance requirement is the increased scale in the display of vertical deviation when the approach is initiated. The vertical profile for the approach is constructed as part of the vertical path by trajectory predictions, complete with deceleration segments to the selected landing speed (refer to Section 24.2.4).

24.3 Summary

This chapter is an introduction to the functions that comprise a flight management system and has focused on the basic functionality and relationships that are fundamental to understanding the flight management system and its role in the operations of the aircraft. Clearly, there is a myriad of complexity in implementing each function that is beyond the scope of this publication.

The future evolution of the flight management system is expected to focus not on the core functions as described herein, but on the use within the aircraft and on the ground of the fundamental information produced by the flight management system today. The use of the FMS aircraft state and trajectory intent within the aircraft and on the ground to provide strategic conflict awareness is a significant step toward better management of the airspace. Communication of the optimized user-preferred trajectories will lead to more efficient aircraft operation. The full utilization of RNP-based navigation will increase the capacity of the airspace. Innovative methods to communicate FMS information and specify flight plan construction with the crew to make flight management easier to use are expected as well. Clearly, the FMS is a key system in moving toward the concepts embodied in CNS future airspace.

25

Electrical Wiring Interconnect System

Michael Traskos
Lectromec

25.1 Introduction

This chapter provides a summary of electrical wiring interconnection system (EWIS) design considerations, regulatory requirements, and trends associated with EWIS. Although the consideration of EWIS at the system level is comparatively new in aerospace design and maintenance, the industry has made significant progress in the design, assessment, and maintenance over the last 15 years.

25.1.1 What Is EWIS?

The capability of all aircraft to maintain airworthiness relies on reliable transmission of information and electrical energy through the aircraft. A major factor affecting aircraft availability is the integrity of individual systems and the equipment supporting the flow of data and electrical energy to those aircraft systems. A well-designed and maintained EWIS interconnects aircraft subsystems and enables reliable aircraft operation.

Historically in aircraft construction and maintenance, the wiring system was often looked at as a "fit-and-forget" system. The wiring system was often that last place maintenance actions would be performed and structured wire inspection programs were nearly nonexistent. The examination of wiring has evolved over the last three decades, typically spurred by a major aircraft incident. Starting in the mid-1980s, attention was generated when a major hazard (arc tracking) was identified with a widely used wire insulator. It was not until the late 1990s when the idea of addressing wire and power delivery components as a system started to be considered.

The watershed moment for the EWIS occurred in response to two major commercial airliner crashes in the 1990s in which wire failures were identified as the root cause. Following the failure investigations, the Federal Aviation Administration (FAA) worked with the aerospace industry to develop new rules regarding EWIS. Among the FAA programs was one to establish the condition of aging wiring system components and validate the adequacy of visual inspections on recently retired transport aircraft. The FAA's Aging Transport Systems Rulemaking Advisory Committee (ATSRAC) was tasked to characterize commercial EWIS integrity and provide recommended actions for assessing EWIS air airworthiness. The ATSRAC found evidence of aging wiring, materials degradation, inadequate installations, and lacking maintenance practices. Implementation of the ATSRAC recommendations was a major part of the FAA's Enhanced Airworthiness Program for Airplane Systems (EAPAS) (FAA *Aging Nonstructural Systems Research*, Christopher D. Smith, Manager, Aging Nonstructural Systems Research FAA, and William J. Hughes Technical Center).

25.1.2 EWIS Components

The EWIS is comprised of the following components:

- Wires and cables
- Harness protective sleeving
- Connectors
- Circuit breakers
- Relays
- Harness clamps
- Spot ties/lacing tape
- Terminals
- Power buses
- Splices

Typically, the consideration of EWIS and an independent system stops at the LRU allowing for the individual components to be assessed separately.

Fiber-optic cables are not considered as part of the FAA's definition of EWIS. It is however included in the U.S. military's definition of EWIS.

25.2 Governing Regulatory Requirements/ Mandates (Military/Civil)

This section covers the regulatory guidelines for civilian and military aircraft.

25.2.1 FAA Regulations

There are a number of FAA rules and regulations pertaining to EWIS. Many of these rules were based on the recommendations made by ATSRAC. These regulations cover a wide range of EWIS-related topics:

- §25.1703—Function and installation
- §25.1705—Systems and functions
- §25.1707—System separation
- §25.1709—System safety
- §25.1711—Component identification
- §25.1713—Fire protection
- §25.1715—Electrical bonding and protection against static electricity
- §25.1717—Circuit protective devices
- §25.1719—Accessibility provisions
- §25.1721—Protection of EWIS
- §25.1723—Flammable fluid fire protection

- §25.1725—Powerplants
- §25.1727—Flammable fluid shutoff means
- §25.1729—Instructions for continued airworthiness
- §25.1731—Powerplant and APU fire detector system
- §25.1733—Fire detector systems, general
- SFAR 88—Fuel tank system fault tolerance evaluation requirements

Of these, regulation §25.1709 has perhaps the largest impact on EWIS system design and certification requirements.

Each EWIS must be designed and installed so that

1. Each catastrophic failure condition
   a. Is extremely improbable
   b. Does not result from a single failure
2. Each hazardous failure condition is extremely remote

This rule, introduced as part of the FAA's EWIS safety initiative, requires that EWIS failure conditions be investigated much in the same way as other aircraft systems (the wording was such that it would closely follow §25.1309 for system safety assessments). The main guidance for assessing wiring system on new platforms is Advisory Circular (AC) 25.1701. The guidance in this document suggests that both the functional impacts of wire failure be considered as well as the physical damage. AC 25.1701 states that "regardless of probability, any single arcing failure should be assumed for any power-carrying wire."

The inter- and intrabundle arcing damage must be considered. For interharness arcing damage, this should include the consideration of damage to other wires in the harnesses, as well as the impact if there is an uncommanded electrical transfer of energy from one circuit to another. To do this, the wire harness contents must be known as well as the failure impacts if there is an open, short, or uncommanded application of energy to the circuit.

For interharness damage, an assessment should include estimation of direct and indirect damage to nearby systems and components. Figure 25.1 shows that damage is possible to nearby systems even if there is physical separation. Recent research has shown that with as little as 0.2 s of arcing, it is possible to breach a pressurized hydraulic line with a separation distance of 1.0 in. (Traskos et al., 2011).

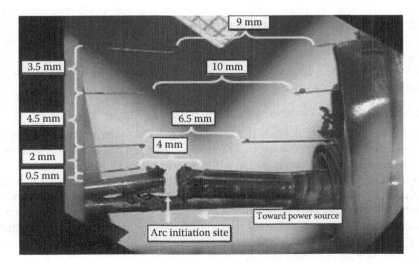

FIGURE 25.1 **(See color insert.)** This is an example of the arc damage to copper strands suspended over an electrical arc. The top strand was nearly 0.5 in. from the arc initiation site. (Courtesy of Lectromec, Chantilly, VA.)

The main contributing factors for electrical arc damage include the following:

- *Fault current*: The fault current (maximum available current available) is the largest factor in the arc damage level. This can be affected by the distance from the power source and/or resistance in the circuit.
- *Circuit protection*: Depending on the circuit breaker type and the fault current level, circuit breaker activation may take several seconds before opening (as in the case of thermal circuit breakers) or open within a few arcing cycles (in the case of arc fault circuit interrupters). The circuit protection does not impact the fault current, only the duration of the arc event.
- *Wire type*: Some wire types are more prone to producing greater arc damage than others, but none can fully eliminate the risk of electrical arcing.
- *Separation or segregation*: If it is not possible to increase the separation distance between the wire harness and the component, then segregation must be considered and can be accomplished with the addition of a physical barrier that limits the effective harness arcing range.

Unfortunately, there is no general rule-of-thumb safe separation distance as the physical, electrical, and environmental conditions can make each configuration unique.

25.2.1.1 SFAR 88

SFAR 88 stands for Special Federal Aviation Regulation 88. This regulation addresses fuel system safety by requiring an analysis of wiring and other electrical components (pumps, fuel quantity indicators, etc.), not only inside the tanks but in circuitry and items located outside the tanks including adjacent dry bays. Effective December 2004, operators must include FAA-approved provisions in their inspection and maintenance programs to assure the safety of their aircraft fuel systems.

All turbine-powered aircraft with a capacity of 30 or more passengers or a payload capacity of 7500 lb or more are required to show compliance. SFAR 88 affects manufacturers (the type certificate, or TC, holders), those companies holding supplemental type certificates (STCs) involving modifications and installations affecting fuel systems (mostly manufacturers), operators, repair stations, and FAA inspectors.

The following is a selective listing of what is expected to be addressed when showing compliance with SFAR 88:

- *Fuel pumps*: Ground fault interrupters (GFIs) for all tanks, replacing existing pump power relays. GFIs *and* automatic shutoff at low fuel levels may be required.
- *Fuel quantity indication systems (FQISs)*: Since 115 V ac current could potentially short to FQIS signal wiring, barrier device protection may be required for all CWTs.
- *Lightning protection*: Improved bonding of fasteners, clamps, and pipes attached to (or in) the tank structure will be necessary.
- *Fault current protection*: K-type fasteners are deemed an unreliable bond, and in-tank bonding jumpers may have to be installed in some aircraft.

25.2.1.2 Enhanced Zonal Awareness Program

The enhanced zonal awareness procedure (EZAP) was introduced to the aviation industry in 2005 and covered requirements for the assessments of new and existing aircraft (the final rule was enacted in 2007). This required development of plans for regular EWIS assessment and inspection.

Along with this regulation, the FAA released AC 25-27 entitled "Development of Transport Category Airplane Electrical Wiring Interconnection Systems Instructions for Continued Airworthiness Using an Enhanced Zonal Analysis Procedure." This AC provides "… guidance for developing maintenance and inspection instructions for EWIS using an enhanced zonal analysis procedure." Furthermore, this AC provided a description of the type and interval of inspections necessary to maintain airworthiness. An EZAP process overview is shown in Figure 25.2.

A fully developed EZAP must cover the inspection interval and specific cleaning tasks for each zone to minimize the presence of combustible materials. Two levels of visual inspection are identified within

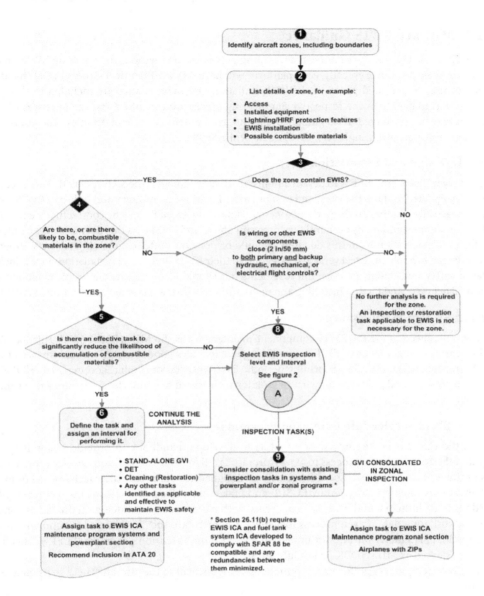

FIGURE 25.2 FAA high level flow diagram guidance for generating EWIS EZAP instructions for continued airworthiness. (From AC 25-27A, *Development of Transport Category Airplane Electrical Wiring Interconnection Systems Instructions for Continued Airworthiness Using and Enhanced Zonal Analysis Procedure*, issued May 4, 2010, U.S. Department of Transportation Federal Aviation Administration.)

the AC: general visual inspection (GVI) and detailed visual inspection (DET). The GVI is described as a limited inspection within touching distance of the wire/harnesses within a zone. The goal is to identify any obvious irregularities of EWIS components. The GVI is typically limited to areas that are limited access areas and are in benign environments. The DET is an intensive wire bundle examination that may include mirrors, magnifying lens, or other tools to assist in physical EWIS examination.

The selection of the particular inspection to be used is based on the physical, environmental, and mechanical stresses to which the wire bundle may be subjected. The AC provides a robust process to determine the frequency and type of inspection to be used.

25.2.2 Military EWIS Guidance

Like the FAA, the U.S. military has been developing processes and methods for assessing EWIS on both new and existing aircraft. Typically, each platform will have its own Technical Orders (TOs) that direct the platform as to accepted practices. EWIS installation and maintenance are included in this. If no guidance is available, then the respective engineering offices are consulted and the practices of other platforms may be considered. If no guidance is found, recommendations will default to the guidance in the Society of Automotive Engineers (SAE) document AS50881.

25.2.2.1 Design and Installation

The primary military aircraft EWIS design and installation document is the SAE AS50881. This document is regularly updated and has the latest and best practices identified by military and industry EWIS subject matter experts. Depending on the particular system, following AS50881 may be contractually enforced or used as a guidance standard by both military program offices and OEMs. As this document is an industry-accepted guide, deviation from this document should be done only after engineering analysis. Programs are encouraged to apply the latest version of AS50881 to their EWIS design and installation since it includes the latest safety and design practice improvements but should also consider the implementation cost. Additional detailed guidance on installing and maintaining EWIS components can be found in ARP 4404.

25.2.2.2 System Maintenance

An often referenced guide for EWIS component maintenance is the MIL-HDBK-522.* This handbook provides inspection guidance of EWIS issues. The guide includes good descriptions and images of common discrepancies found while performing periodic EWIS inspects or troubleshooting. MIL-HDBK-522 covers information and guidance on what to look for with regard to connectors, routing, wire crimping, shielding, etc.

25.2.2.3 EWIS Service Life Extension Programs

In 2012, the U.S. Air Force undertook an effort to develop a handbook to provide weapons systems program offices a systematic process to assess aircraft EWIS for overall condition, service life extension, and continued airworthiness. The document was designed to align with the Mechanical Equipment and Subsystems Integrity Program (MECSIP) and makes extensive use of lessons learned from EWIS-related military, industry, and FAA ACs concerned with maintaining aircraft airworthiness. It contains the framework for a data-driven process to achieve and maintain the physical and functional integrity of the EWIS, particularly for aircraft undergoing service life extension programs (SLEPs). This document was released as MIL-HDBK-525 in July 2013.

This seven-step process performs a comprehensive assessment of the aircraft. These steps include the following:

- *Task 1*: Document aircraft EWIS and identify critical circuits and functions.
- *Task 2*: Analyze EWIS failure and maintenance data.
- *Task 3*: Conduct an aircraft physical and electrical inspection and document overall condition of the aircraft EWIS.
- *Task 4*: Conduct a comprehensive materials/aging analysis of wiring and electrical components removed from the aircraft.
- *Task 5*: Analyze and provide an overall aircraft electrical system risk and life assessment.
- *Task 6*: Develop action plan.
- *Task 7*: Develop follow-up plans to reassess EWIS to achieve desired outcomes.

The process flowchart for the EWIS SLEP risk assessment is shown in Figure 25.3.

---

* http://quicksearch.dla.mil/basic_profile.cfm?ident_number=277535&method=basic, accessed on May 27, 2014.

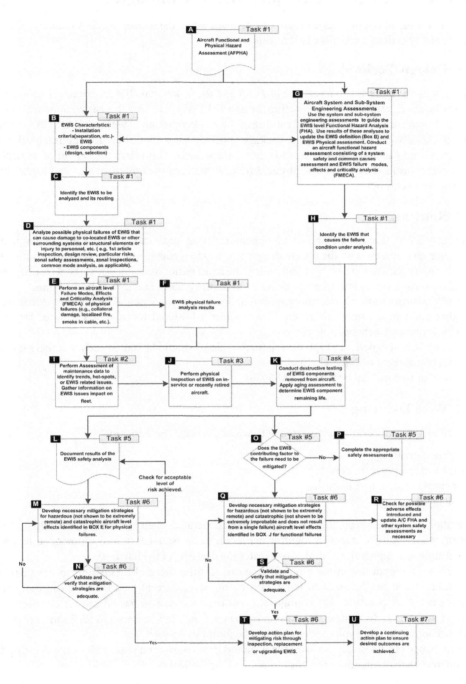

FIGURE 25.3 U.S. military program flow diagram for step-by-step guidance to assess EWIS on an aging platform and determine actions for service life extension. (From MIL-HDBK-525, http://quicksearch.dla.mil/basic_profile. cfm?ident_number=279725&method=basic, accessed on May 27, 2014.)

25.3 EWIS Design Techniques and Technologies

There are a number of considerations that must be made when designing EWIS. A couple of the unique challenges of EWIS design are discussed in this section.

25.3.1 Design Technology

The EWIS design is done at two levels: the physical and the functional. The functional is addressed with the development of circuit designs supporting the non-EWIS systems, and this can be done with any modern circuit design software. The physical routing of the EWIS is designed with 3D CAD/CAE software.

Some of the advanced system design packages provide tools that allow for the integration of the circuit design and the physical routing of system components. This offers the benefit of providing the supplemental tools to ensure that the physical routing matches the circuit design and highlight areas where they are in conflict.

25.3.2 Routing

The routing of EWIS through an aircraft is a difficult task that impacts most of the aircraft design groups. This is an iterative process with the system design, space management, and safety groups. Routing should be designed to minimize the weight, ensure adequate bend radius as to not stress the wires, and provide sufficient separation from system components, redundant systems, and hydraulic/fuel lines.

The EWIS routing must also consider maintenance and environmental conditions. Areas that may have higher ambient temperature profiles may require additional thermal shielding or larger gauge wires to limit resistive wire heating.

Further, areas that experience high flexing (such as door) should be routed such that there is sufficient slack to not place strain or cause chafing.

Guidance for routing can be found in AS50881.

25.3.3 Wire Derating

There are four types of derating associated with wire selection. These include

1. Altitude derating
2. Ambient temperature derating
3. Harness construction derating
4. Voltage derating

The operational altitude affects the dissipation of heat generated when electrical current flows down the wire. If there are multiple stages of flight or operation to be assessed, it is recommended to select the highest altitude (the higher the altitude, the greater the derating of the harness).

The ambient temperature in the aircraft zone impacts the thermal dissipation to the surrounding environment. The lower the ambient temperature, the larger current load the wire can carry without passing the rated temperature. The configuration of the harness also has an impact on this; the more wires, the less thermal dissipation. Harness derating guidance can be found in SAE standard AS50881.

Guidance for calculating voltage derating can be found in AS5088.

The wire gauge (and possibly the wire type) should be reconsidered if after all of the derating factors are considered, the wire cannot meet the circuit load requirements.

25.3.4 Electromagnetic Interference

This may require additional electromagnetic interference (EMI) shielding of the harness. For equipment sensitive to EMI, the wire routing should be designed to maximize the separation distance from EMI-generating devices. Additional guidance on installation of EMI shielded wires can be found in AIR4465.

25.4 EWIS Sustainment

An aircraft's EWIS will, over time, degrade in performance. To address this, the FAA requires that new and existing Part 25 aircraft have an Enhanced Zonal Assessment Program (EZAP) that defines the maintenance activities and intervals for wiring.

25.4.1 EZAP

As identified in the earlier part of this chapter, the typical means that is prescribed for the inspection of EWIS components for commercial aircraft is based on the analysis done as part of the EZAP.

The inspection intervals identified here can range from an inspection every couple of flight hours to once every D-check. The factors that affect the determination of inspection frequency are the hostility of the environment and the likelihood of accidental damage. Those areas with more severe conditions are to be inspected more regularly; those with more benign are inspected less frequently.

As part of the U.S. Navy's approach to EWIS reliability, dedicated teams of individuals go on-site to maintenance locations to perform hands-on reviews with the maintenance crews. These reviews will typically include a team reviewing an aircraft after it has been serviced, identify and document all of the EWIS-related discrepancies, and then review with the maintenance team the issues and the proper means to resolve the discrepancy.

25.4.2 Degradation of Components

An item to note here is that the EZAP process does not identify the longevity or service life of particular components. If additional data are available at the MTTF or other reliability metrics, these should be integrated with the maintenance and inspection plan.

As the EWIS can be found throughout an aircraft, the rate of degradation of the components will vary accordingly. Those exposed to uncontrolled environmental conditions will undoubtedly experience high degradation and higher failure rates than those in benign environmental conditions. Areas such as the wheel wells and wing leading/trailing edges are particular areas that degrade quickly. The degradation can be exacerbated by the exposure to chemicals such as deicers. As some commercially available deicers have an electrical conductivity greater than saltwater, these can contribute to shorting or electrical arcing events.

25.5 Failure of EWIS Components

As EWIS constitutes a wide range of system components, technologies, and materials, there are a number of failure/degradation modes associated with each. The following provides a brief overview of many of these failure/degradation modes as to what should be considered when EWIS is inspected.

25.5.1 Wires

Wire is perhaps the most critical and most susceptible to damage by various forms of mechanical, electrical, and chemical stresses. The majority of aircraft wire insulation is of a thin-walled polymer construction (in the case of some standard constructions, this may be as thin as 5 mils on a 20 AWG wire). While aerospace-grade wire undergoes a rigorous testing prior to being permitted on an aircraft installation, over time, due to degradation or poor installation, the wire may fail:

1. A wire's primary insulation and cable jacket may be damaged or degrade, showing one or more of the following (some of these are due to degradation, while some are due to maintenance actions):
   a. Chafing, fraying, cuts, cracking, thermal damage, softening, cold flow, unraveling/layer separation, recession, thinning, and other forms of insulation layer deformation/separation/breaches

    b.   Discoloration/charring from stress exposure, aging, or overheating
    c.   Loss of dielectric or insulation resistance
  2. The primary conductor and/or shield braid may show one or more of the following:
    a.   Broken or damaged strands
    b.   Corrosion
    c.   Red plague style corrosion
    d.   Discoloration from internal or external high-temperature exposure
    e.   Short between the shield and primary conductor

As mentioned earlier in this chapter, the probability of failure and deterioration increases in severe wind and moisture prone (SWAMP) areas or areas with high vibration (e.g., near engines), high temperature (e.g., near engines), high moisture (e.g., wheel well), fluid contamination (e.g., under passenger cabin), or high maintenance (e.g., cargo area). If degraded or failed wiring is found, it should be removed and analyzed to determine failure cause. After assessment, the proper maintenance action can be taken.

25.5.2 Splices

A splice is a connection of two or more conductors or cables and provides good mechanical strength as well as good connectivity. The most common splice failure mode is high resistance due to an issue with the interconnection between the splice barrel and wire. This high resistance may be a result of overheating due to improper crimping, corrosion, or aging. Environmentally sealed splices should be used in accordance with standard wiring practices or SAE AS81824/1 or equivalent specification, particularly in unpressurized and SWAMP areas. The possibility of fluid contamination in any installation should be always considered. Guidance can also be found in AS50881.

25.5.3 Cables

An electrical cable is where two or more insulated conductors are contained in a common covering (e.g., jacket, shielding, and molding). The failure modes of cables are the same as for wires, with the addition of shorting between the shield and primary conductor. Controlled impedance cable (e.g., twisted pair and coaxial cable) should be accessed for impedance characteristics such as impedance, velocity of propagation, and voltage standing wave ratio.

25.5.4 Connectors

Like wire, connectors are also susceptible to damage caused by various forms of exposure. The connector is susceptible to the following:

1. External damage caused by corrosion and mechanical stresses (repeated mating during maintenance actions).
2. Internal damage caused by handling, often found in areas with frequent maintenance activity.
3. Corrosion of the worn pins and sockets.
4. Exposure to petroleum-based fluids, moisture, saltwater, and cleaning and deicing solutions.
5. High-temperature zones such as engine and auxiliary power unit bays. The failure probability increases in SWAMP areas.

Connector failures are often associated with intermittent faults or "cannot duplicate" type issues. These can be difficult to troubleshoot and can lead to repeat maintenance activities attempting to identify the root cause of a problem. Guidance for evaluation and inspection of connectors can be found in MIL-HDBK-522.

25.5.5 Circuit Breakers

Circuit breakers are critical to protecting EWIS components from over current conditions. Most circuit breakers used on legacy aircraft are calibrated mechanical devices that activate when current exceeds the trip curve characteristic. Circuit breakers are susceptible to wear from repeated use. Like other EWIS components, circuit breakers are live limited. This "wear-out" can be damage caused by various forms of exposure, as well as "cycle wear." The breaker is susceptible to arcing damage, wear-out of the trip mechanism, or degradation and corrosion from various fluids and heat exposure. It is possible to measure contact resistance and use the value to assess the overall condition of the breaker. Probability of failure increases in SWAMP areas. Guidance for circuit breaker evaluation and inspection can be found in MIL-HDBK-522. One particular item to note is the cycling guidelines in the handbook.

Arc fault circuit breakers (AFCBs) are a technology found on several platforms. An AFCB device replaces a thermal circuit breaker and is another method of reducing the risk of arc track wire damage. The device detects the current and voltage signature of an electrical arc and rapidly trips the circuit breaker. On some models, there is a visual indication if an arc fault has been detected.

25.6 Future Trends

Of emerging EWIS technologies, the two that are likely to have the greatest impact in system design are distributed power applications and high-voltage systems.

25.6.1 Distributed Power Applications

The classic aircraft power system design was to run wire from the engines, through the fuselage, to a power bus near the flight deck, to a circuit breaker, then back through the aircraft to the destination. For large aircraft, the "overlapping" wire run can be hundreds of feet. In a distributed power system, generated power is delivered to nodes throughout the aircraft. These nodes then turn on/off power to components based on flight crew or system commands.

The weight benefit to distributed power is dependent upon the application. If retrofitted to an existing platform, there may be little or no weight savings. For new platforms integrating distributed power into the system design, there is much more weight savings. Further, this reduces the amount of wire, thereby reducing the potential locations for failures to be generated.

An additional benefit distributed power allows for in system design is to limit electrical potential on inactive circuits (i.e., only power circuits if the end device needs power). By reducing the number of energized circuits, this can reduce the available energy in the case of electrical fire or arcing.

Lastly, depending on the system used, data logging and circuit monitoring are possible. This information can be used for troubleshooting or actively used by the system to open circuits that generate failure signatures.

25.6.2 Higher-Voltage Applications

For a long time, aircraft have typically run on voltages less that 115 V ac phase to ground. Recent aircraft design is focused on using higher-voltage power generators for the following:

The use of higher voltages has a number of impacts onto the wiring system:

1. By running at higher voltages, less current is needed to deliver the same power. This in turn means that smaller gauge wires can be used, reducing the wire and cable weight.
2. The insulating materials and wire specifications used have to be further examined. In 2011, the SAE wire and cable group (8D) removed the 600 V rating from many wire specifications because it

was realized that there was no defined test specification to qualify wires to this level. This has not been much of a concern in the past, because most systems operated at or below 115 V/208 V. But as the systems are starting to use high voltage for distribution through the aircraft and start pushing the rating for the wire, this can become more of a concern. It may be necessary to increase the thickness of the insulating materials to provide adequate dielectric strength and to pass accepted damage tolerance requirements.

3. Depending on the voltage used, it may be necessary to increase the separation distance at connection (connectors, terminal boards, etc.). The increase in size may be difficult in systems with already limited available space.

Reference

Traskos, M.G., Linzey, W.G., and Traskos, S.R. Examination of EWIS and pressurized hydraulic lines, *Proceedings 2nd Aircraft Airworthiness, and Sustainment Conference 2011*, San Diego, CA, April 2011, http://quicksearch.dla.mil/, accessed on May 27, 2014.

26

Batteries

David G. Vutetakis
Concorde Battery
Corporation

26.1 Introduction

The battery is an essential component of almost all aircraft electrical systems. Batteries are used to start engines and auxiliary power units (APUs), to provide emergency backup power for essential avionics equipment, to assure no-break power for navigation units and fly-by-wire computers, and to provide ground power capability for maintenance and preflight checkouts. Many of these functions are mission critical, so the performance and reliability of an aircraft battery is of considerable importance. Other important requirements include environmental ruggedness, a wide operating temperature range, ease of maintenance, rapid recharge capability, and tolerance to abuse.

Historically, only a few types of batteries have been found to be suitable for aircraft applications. Up until the 1950s, vented lead–acid (VLA) batteries were used exclusively (Earwicker, 1956). In the late 1950s, military aircraft began converting to vented nickel–cadmium (VNC) batteries, primarily because of their superior performance at low temperature. The VNC battery subsequently found widespread use in both military and commercial aircraft (Fleischer, 1956; Falk and Salkind, 1969). The only other type of battery used during this era was the vented silver–zinc battery, which provided an energy density about three times higher than VLA and VNC batteries (Miller and Schiffer, 1971). This battery type was applied to several types of U.S. Air Force fighters (F-84, F-105, and F-106) and U.S. Navy helicopters (H-2, H-13, and H-43) in the 1950s and 1960s. Although silver–zinc aircraft batteries were attractive for reducing weight and size, their use has been discontinued due to poor reliability and high cost of ownership.

In the late 1960s and early 1970s, an extensive development program was conducted by the U.S. Air Force and Gulton Industries to qualify sealed nickel–cadmium (SNC) aircraft batteries for military and commercial applications (McWhorter and Bishop, 1972). This battery technology was successfully demonstrated on a Boeing KC-135, a Boeing 727, and a UH-1F helicopter. Before the technology could be transitioned into production, however, Gulton Industries was taken over by SAFT and a decision was made to terminate the program.

In the late 1970s and early 1980s, the U.S. Navy pioneered the development of sealed lead–acid (SLA) batteries for aircraft applications (Senderak and Goodman, 1981). SLA batteries were initially applied to the AV-8B and F/A-18, resulting in a significant reliability and maintainability (R&M) improvement compared with VLA and VNC batteries. The Navy subsequently converted the C-130, H-46, and P-3 to SLA batteries. The U.S. Air Force followed the Navy's lead, converting numerous aircraft to SLA batteries, including the A-7, B-1B, C-5, C-130, C-141, KC-135, F-4, F-16, and F-117 (Vutetakis, 1994). The term "high-reliability, maintenance-free battery" or (HRMFB) was coined to emphasize the improved R&M capability of sealed-cell aircraft batteries. The use of HRMFBs soon spun off into the commercial sector, and numerous commercial and general aviation aircraft today have been retrofitted with SLA batteries.

In the mid-1980s, spurred by increasing demands for HRMFB technology, a renewed interest in SNC batteries took place. A program to develop advanced SNC batteries was initiated by the U.S. Air Force, and Eagle-Picher Industries was contracted for this effort (Flake, 1988; Johnson et al., 1994). The B-52 bomber was the only aircraft to retrofit this technology, and Eagle-Picher subsequently discontinued the SNC battery. SNC batteries were also developed by ACME Aerospace for several aircraft applications, including the F-16 fighter, Apache AH-64 helicopter, MD-90, and Boeing 777 (Anderman, 1994). Only the Apache AH-64 and Boeing 777 continue to use ACME SNC batteries.

In the 1990s, as an alternative to the sealed technologies, "low-maintenance" or "ultralow-maintenance" nickel–cadmium batteries were introduced (Scardaville and Newman, 1993). These types of batteries are intended to be direct replacements of conventional VNC batteries, avoiding the need to replace or modify the charging system. Although the battery still requires scheduled maintenance for electrolyte filling, the maintenance frequency can be decreased significantly. This type of battery was originally developed by SAFT and more recently by Marathon. Flight tests were successfully performed by the U.S. Navy on the H-1 helicopter, and most VNC batteries used by the Navy have been converted to the low-maintenance technology. This technology has also been applied to various commercial aircraft.

Most recently, lithium-ion technology has been adapted to aircraft batteries (Vutetakis and Timmons, 2008). Lithium-ion batteries are much lighter than lead–acid or nickel–cadmium batteries, which makes them very attractive for aircraft applications. However, the weight savings must be balanced with higher cost and increased safety concerns. So far, lithium-ion batteries are being used in just a few military (B-2 and F-35) and commercial (B787) aircraft.

Determining the most suitable battery type and size for a given aircraft type requires detailed knowledge of the application requirements (load profile, duty cycle, environmental factors, and physical constraints) and the characteristics of available batteries (performance capabilities, charging requirements, life expectancy, and cost of ownership). With the various battery types available today, considerable expertise is required to select the best type and size of battery for a given aircraft application. The information contained in this chapter will provide general guidance for original equipment design and for upgrading existing aircraft batteries. More detailed information can be found in the sources listed at the end of the chapter.

26.2 General Principles

26.2.1 Battery Fundamentals

Batteries operate by converting chemical energy into electrical energy through electrochemical discharge reactions. Batteries are composed of one or more cells, each containing a *positive electrode*, a *negative electrode*, a *separator*, and an *electrolyte*. Cells can be divided into two major classes: primary and secondary. Primary cells are not rechargeable and must be replaced once the reactants are depleted. Secondary cells are rechargeable and require a dc charging source to restore reactants to their fully charged state. Examples of primary cells include carbon–zinc (Leclanche or dry cell), alkaline–manganese, mercury–zinc, silver–zinc, and lithium cells (e.g., lithium–manganese dioxide, lithium–sulfur dioxide, and lithium–thionyl chloride). Examples of secondary cells include lead–lead dioxide (lead–acid), nickel–cadmium, nickel–iron, nickel–hydrogen, nickel–metal hydride, silver–zinc, silver–cadmium, and lithium–ion. For aircraft applications, secondary cells are the most prominent, but primary cells are sometimes used for powering critical avionics equipment (e.g., flight data recorders).

Batteries are rated in terms of their *nominal voltage* and *ampere-hour (Ah) capacity*. The voltage rating is based on the number of cells connected in series and the nominal voltage of each cell (2.0 V for lead–acid, 1.2 V for nickel–cadmium, and 3.2–3.7 V for lithium–ion). The most common voltage rating for aircraft batteries is 24 V. A 24 V, lead–acid battery contains 12 cells, while a 24 V, nickel–cadmium battery contains either 19 or 20 cells (the U.S. military rates 19-cell batteries at 24 V). Voltage ratings of 22.8, 25.2, and 26.4 V are also common with nickel–cadmium batteries, consisting of 19, 20, or 22 cells, respectively. Twelve-volt lead–acid batteries, consisting of six cells in series, are also used in many general aviation aircraft. For lithium-ion batteries, seven or eight cells are connected in series to make a 24 V battery.

The Ah capacity available from a fully charged battery depends on its temperature, rate of discharge, and age. Normally, aircraft batteries are rated at room temperature (25°C), the *C-rate* (1 h rate), and beginning of life. Military batteries, however, often are rated based on the end-of-life capacity, that is, the minimum capacity before the battery is considered unserviceable. Capacity ratings of aircraft batteries vary widely, generally ranging from as low as 3 Ah to about 65 Ah.

The maximum power available from a battery depends on its internal construction. High-rate cells, for example, are designed specifically to have very low internal impedance as required for starting turbine engines and APUs. For lead–acid batteries, the peak power has traditionally been defined in terms of the cold-cranking ampere (*CCA*) rating. For nickel–cadmium batteries, the peak power rating has traditionally been defined in terms of the current at maximum power, or *Imp* rating. These ratings are based on different temperatures (−18°C for CCA, 23°C for Imp), making it difficult to compare different battery types. Furthermore, neither rating adequately characterizes the battery's initial peak current capability, which is especially important for engine start applications. A standard method of rating the peak power capability of an aircraft battery is defined in IEC 60952-1 (2004), RTCA DO-293 (2004) and RTCA DO-311 (2008). These standards define two peak power ratings, the *Ipp* and the *Ipr*, at three different temperatures (23°C, −18°C, and −30°C). The Ipp rating is the current at 0.3 s when discharged at half the nominal voltage. The Ipr rating is the current at 15 s when discharged at half the nominal voltage. Other peak power specifications have been included in some military standards. For example, MIL-B-8565/15 specifies the initial peak current, the current after 15 s, and the capacity after 60 s during a 14 V, constant voltage discharge at two different temperatures (24°C and −26°C) (Vutetakis and Viswanathan, 1996).

The *state of charge* of a battery is the percentage of its capacity available relative to the capacity when it is fully charged. By this definition, a fully charged battery has a state of charge of 100%, and a battery with 20% of its capacity removed has a state of charge of 80%. The *state of health* of a battery is the percentage of its capacity available when fully charged relative to its rated capacity. For example, a battery rated at 30 Ah, but only capable of delivering 24 Ah when fully charged, will have a state of health of $24/30 \times 100 = 80\%$. Thus, the state of health takes into account the loss of capacity as the battery ages.

26.2.2 Lead–Acid Batteries

26.2.2.1 Theory of Operation

The chemical reactions that occur in a lead–acid battery are represented by the following equations:

$$\text{Positive electrode:} \quad PbO_2 + H_2SO_4 + 2H^+ + 2e^- \underset{\text{charge}}{\overset{\text{discharge}}{\rightleftharpoons}} PbSO_4 + 2H_2O \tag{26.1}$$

$$\text{Negative electrode:} \quad Pb + H_2SO_4 \underset{\text{charge}}{\overset{\text{discharge}}{\rightleftharpoons}} PbSO_4 + 2H^+ + 2e^- \tag{26.2}$$

$$\text{Overall reaction:} \quad PbO_2 + Pb + 2H_2SO_4 \underset{\text{charge}}{\overset{\text{discharge}}{\rightleftharpoons}} 2PbSO_4 + 2H_2O \tag{26.3}$$

As the cell is charged, the sulfuric acid (H_2SO_4) concentration increases and becomes highest when the cell is fully charged. Likewise, when the cell is discharged, the acid concentration decreases and becomes most dilute when the cell is fully discharged. The acid concentration generally is expressed in terms of specific gravity (SG), which is the weight of the electrolyte compared to the weight of an equal volume of pure water. The cell's SG can be estimated from its open-circuit voltage (OCV) using the following equation:

$$SG = OCV - 0.84 \qquad (26.4)$$

There are two basic cell types: vented and recombinant. Vented cells have a flooded electrolyte, and the hydrogen and oxygen gases generated during charging are vented from the cell container. Recombinant cells have a starved or gelled electrolyte, and the oxygen generated from the positive electrode during charging diffuses to the negative electrode where it recombines to form water by the following reaction:

$$Pb + H_2SO_4 + \frac{1}{2}O_2 \rightarrow PbSO_4 + H_2O \qquad (26.5)$$

The recombination reaction suppresses hydrogen evolution at the negative electrode, thereby allowing the cell to be sealed. In practice, the recombination efficiency is not 100% and a resealable valve regulates the internal pressure at a relatively low value, generally below 5 psig. For this reason, SLA cells are often called "valve-regulated lead–acid" (VRLA) cells.

26.2.2.2 Cell Construction

Lead–acid cells are composed of alternating positive and negative plates, interleaved with single or multiple layers of separator material. Plates are made by pasting active material onto a grid structure made of lead or lead alloy. The electrolyte is a mixture of sulfuric acid and water. In flooded cells, the separator material is porous rubber, cellulose fiber, or microporous plastic. In recombinant cells with starved electrolyte technology, a glass fiber mat separator is used, sometimes with an added layer of microporous polypropylene. Gel cells, the other type of recombinant cell technology, are made by adding powdered silica to the electrolyte to form a gelatinous structure that surrounds the electrodes and separators.

26.2.2.3 Battery Construction

Lead–acid aircraft batteries are constructed using injection-molded, plastic *monoblocs* that contain a group of cells connected in series. Monoblocs typically are made of polypropylene, but ABS is used by at least one manufacturer. Normally, the monobloc serves as the battery case, similar to a conventional automotive battery. For more robust designs, monoblocs are assembled into a separate outer container made of steel, aluminum, or fiberglass-reinforced epoxy. Cases usually incorporate an electrical receptacle for connecting to the external circuit with a quick connect/disconnect plug. Two generic styles of receptacles are common: the "Elcon style" and the "Cannon style." The Elcon style is equivalent to military type MS3509. The Cannon style has no military equivalent but is produced by Cannon and other connector manufacturers. Batteries sometimes incorporate thermostatically controlled heaters to improve low-temperature performance. The heater is powered by the aircraft's AC or DC bus. Figure 26.1 shows an assembly drawing of a typical lead–acid aircraft battery; this particular example does not incorporate a heater.

26.2.2.4 Discharge Performance

Battery performance characteristics usually are described by plotting voltage, current, or power versus discharge time, starting from a fully charged condition. Typical discharge performance data for SLA aircraft batteries are illustrated in Figures 26.2 and 26.3. Figure 26.4 shows the effect of temperature on the capacity when discharged at the C-rate. Manufacturers' data should be obtained for current information on specific batteries of interest.

26.2.2.5 Charge Methods

Constant voltage charging at 2.3–2.4 V per cell is the preferred method of charging lead–acid aircraft batteries. For a 12-cell battery, this equates to 27.6–28.8 V, which generally is compatible with the voltage available from the aircraft's 28 V DC bus. Thus, lead–acid aircraft batteries normally can be charged by

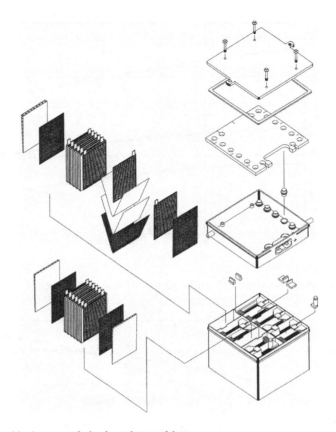

FIGURE 26.1   Assembly drawing of a lead–acid aircraft battery.

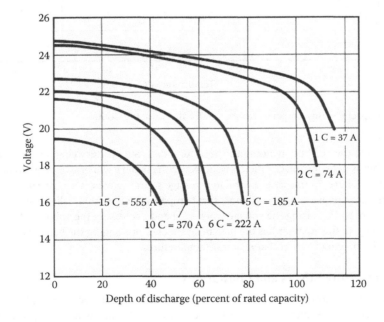

FIGURE 26.2   Discharge curves at 25°C for a 24 V/37 Ah SLA aircraft battery.

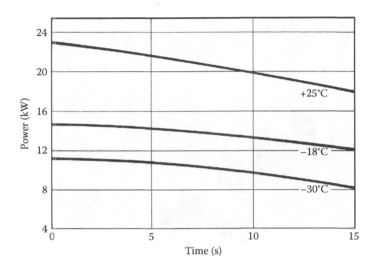

FIGURE 26.3   Maximum power curves (12 V discharge) for a 24 V/37 Ah SLA battery.

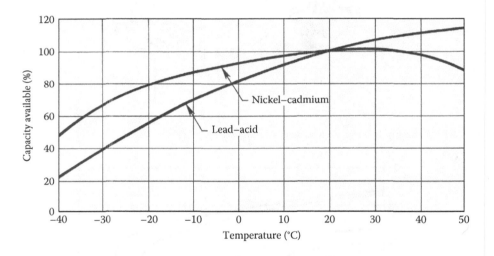

FIGURE 26.4   Capacity versus temperature for aircraft batteries at the C-rate.

direct connection to the DC bus, avoiding the need for a dedicated battery charger. If the voltage regulation on the DC bus is not controlled sufficiently, however, the battery will be overcharged or undercharged causing premature failure. In this case, a regulated voltage source may be necessary to achieve acceptable battery life. Some aircraft use voltage regulators that compensate, either manually or automatically, for the battery temperature by increasing the voltage when cold and decreasing the voltage when hot. Adjusting the charging voltage in this manner has the beneficial effect of prolonging the battery's service life at high temperature and achieving faster recharge at low temperatures.

26.2.2.6 Temperature Effects and Limitations

Lead–acid batteries generally are rated at 25°C (77°F) and operate best around this temperature. Exposure to low ambient temperatures results in performance decline, whereas exposure to high ambient temperatures results in shortened life.

TABLE 26.1 Freezing Points of Sulfuric Acid–Water Mixtures

Specific Gravity at 15°C	Cell OCV (V)	Battery OCV (V)	Freezing Point	
			(°C)	(°F)
1.000	1.84	22.08	0	+32
1.050	1.89	22.68	−3	+26
1.100	1.94	23.28	−8	+18
1.150	1.99	23.88	−15	+5
1.200	2.04	24.48	−27	−17
1.250	2.09	25.08	−52	−62
1.300	2.14	25.68	−70	−95
1.350	2.19	26.28	−49	−56
1.400	2.24	26.88	−36	−33

The lower temperature limit is dictated by the freezing point of the electrolyte. The electrolyte freezing point varies with acid concentration, as shown in Table 26.1. The minimum freezing point is a chilly −70°C (−95°F) at an SG of 1.300. Since fully charged batteries have SGs in the range of 1.28–1.33, they are not generally susceptible to freezing even under extreme cold conditions. However, when the battery is discharged, the SG drops and the freezing point rises. At low SG, the electrolyte first will turn to slush as the temperature drops. This is because the water content freezes first, gradually raising the SG of the remaining liquid so that it remains unfrozen. Solid freezing of the electrolyte in a discharged battery requires temperatures well below the slush point; a practical lower limit of −30°C is often specified. Solid freezing can damage the battery permanently (i.e., by cracking cell containers), so precautions should be taken to keep the battery charged or heated when exposed to temperatures below −30°C.

The upper temperature limit is generally in the range of 60°C–70°C. Capacity loss is accelerated greatly when charged above this temperature range due to vigorous gassing and/or rapid grid corrosion. The capacity loss generally is irreversible when the battery is cooled.

26.2.2.7 Service Life

The service life of a lead–acid aircraft battery depends on the type of use it experiences (e.g., rate, frequency, and depth of discharge), environmental conditions (e.g., temperature and vibration), charging method, and the care with which it is maintained. Service lives can range from 1 to 10 years, depending on the application. Table 26.2 shows representative cycle life data as a function of the depth of discharge. Manufacturers' data should be consulted for specific batteries of interest.

26.2.2.8 Storage Characteristics

Lead–acid batteries always should be stored in the charged state. If allowed to remain in the discharged state for a prolonged time period, the battery becomes damaged by "sulfation." Sulfation occurs when lead sulfate forms into large, hard crystals, blocking the pores in the active material. The sulfation creates a high-impedance condition that makes it difficult for the battery to accept recharge. The sulfation may or may not

TABLE 26.2 Cycle Life Data for SLA Aircraft Batteries

Depth of Discharge (% of Rated Capacity)	Number of Cycles to End of Life
10	2000
30	670
50	400
80	250
100	200

be reversible, depending on the discharge conditions and specific cell design. The ability to recovery from deep discharge has been improved in recent years by electrolyte additives, such as sodium sulfate.

VLA batteries normally are supplied in a dry, charged state (i.e., without electrolyte), which allows them to be stored almost indefinitely (i.e., 5 years or more). Once activated with electrolyte, periodic charging is required to overcome the effect of self-discharge and to prevent sulfation. The necessary charging frequency depends on the storage temperature. At room temperature (25°C), charging every 30 days is typically recommended. More frequent charging is necessary at higher temperatures (e.g., every 15 days at 35°C), and less frequent charging is necessary at low temperatures (e.g., every 120 days at 10°C).

SLA batteries can be supplied only in the activated state (i.e., with electrolyte), so storage provisions are more demanding compared with dry charged batteries. As in the case of activated VLA batteries, periodic charging is necessary to overcome the effects of self-discharge and to prevent sulfation. The rate of self-discharge of SLA batteries varies widely from manufacturer to manufacturer, so the necessary charging frequency also varies widely. For example, recommended charging frequencies can range from 3 to 12 months.

26.2.2.9 Maintenance Requirements

Routine maintenance of lead–acid aircraft batteries is required to assure airworthiness and to maximize service life. For vented-cell batteries, electrolyte topping must be performed on a regular basis to replenish the water loss that occurs during charging. Maintenance intervals are typically 2–4 months. A capacity test or load test usually is included as part of the servicing procedure. For sealed-cell batteries, water replenishment obviously is unnecessary, but periodic capacity measurements generally are recommended. Capacity check intervals can be based either on calendar time (e.g., every 6–12 months after the first year) or operating hours (e.g., every 500 h after the first 1000 h). Refer to the manufacturers' maintenance instructions for specific batteries of interest.

26.2.2.10 Failure Modes and Fault Detection

The predominant failure modes of lead–acid cells are summarized as follows:

- Shorts caused by growth of the positive grid, shedding or mossing of active material, or mechanical defects protruding from the grid. Manifested by inability of the battery to hold a charge (rapid decline in OCV).
- Loss of electrode capacity due to active material shedding, excessive grid corrosion, sulfation, or passivation. Manifested by low capacity and/or inability to hold voltage under load.
- Water loss and resulting cell dry-out due to leaking seal, repeated cell reversals, or excessive overcharge (this mode applies to sealed cells or to vented cells that are improperly maintained). Manifested by low capacity and/or inability to hold voltage under load.

The detection of these failure modes is straightforward if the battery can be removed from the aircraft. The battery capacity and load capability can be measured directly, and the ability to hold a charge can be inferred by checking the OCV over time. However, the detection of these failure modes while the battery is in service is more difficult. The more critical the battery is to the safety of the aircraft, the more important it becomes to detect battery faults accurately. A number of onboard detection schemes have been developed for critical applications, mainly for military aircraft (Vutetakis and Viswanathan, 1995).

26.2.2.11 Disposal

Lead, the major constituent of the lead–acid battery, is a toxic (poisonous) chemical. As long as the lead remains inside the battery container, no health hazard exists. Improper disposal of spent batteries can result in exposure to lead, however. Environmental regulations in the United States and abroad prohibit the disposal of lead–acid batteries in landfills or incinerators. Fortunately, an infrastructure exists for recycling the lead from lead–acid batteries. The same processes used to recycle automotive batteries are used to recycle aircraft batteries. Federal, state, and local regulations should be followed for proper disposal procedures.

26.2.3 Nickel–Cadmium Batteries

26.2.3.1 Theory of Operation

The chemical reactions that occur in a nickel–cadmium battery are represented by the following equations:

$$\text{Positive electrode:} \quad 2NiOOH + 2H_2O + 2e^- \underset{\text{charge}}{\overset{\text{discharge}}{\rightleftarrows}} 2Ni(OH)_2 + 2(OH)^- \qquad (26.6)$$

$$\text{Negative electrode:} \quad Cd + 2(OH)^- \underset{\text{charge}}{\overset{\text{discharge}}{\rightleftarrows}} Cd(OH)_2 + 2e^- \qquad (26.7)$$

$$\text{Overall reaction:} \quad 2NiOOH + Cd + 2H_2O \underset{\text{charge}}{\overset{\text{discharge}}{\rightleftarrows}} 2PbSO_4 + 2H_2O \qquad (26.8)$$

There are two basic cell types: vented and recombinant. Vented cells have a flooded electrolyte, and the hydrogen and oxygen gases generated during charging are vented from the cell container. Recombinant cells have a starved electrolyte, and the oxygen generated from the positive electrode during charging diffuses to the negative electrode where it recombines to form cadmium hydroxide by the following reaction:

$$Cd + H_2O + \frac{1}{2}O_2 \rightarrow Cd(OH)_2 \qquad (26.9)$$

The recombination reaction suppresses hydrogen evolution at the negative electrode, thereby allowing the cell to be sealed. Unlike VRLA cells, recombinant nickel–cadmium cells are sealed with a high-pressure vent that releases only during abusive conditions. Thus, these cells remain sealed under normal charging conditions. However, provisions for gas escape must still be provided when designing battery cases since abnormal conditions may be encountered periodically (e.g., in the event of a charger failure that causes an overcurrent condition).

26.2.3.2 Cell Construction

The construction of nickel–cadmium cells varies significantly, depending on the manufacturer. In general, cells feature alternating positive and negative plates with separator layers interleaved between them, a potassium hydroxide (KOH) electrolyte of approximately 31% concentration by weight (SG 1.300), and a prismatic cell container with the cell terminals extending through the cover. The positive plate is impregnated with nickel hydroxide, and the negative plate is impregnated with cadmium hydroxide. The plates differ according to manufacturer with respect to the type of the substrate, type of plaque, impregnation process, formation process, and termination technique. The most common plate structure is made of nickel powder sintered onto a substrate of perforated nickel foil or woven screens. At least one manufacturer (ACME) uses nickel-coated polymeric fibers to form the plate structure. Cell containers typically are made of nylon, polyamide, or steel. One main difference between vented cells and sealed (recombinant) cells is the type of separator. Vented cells use a gas barrier layer to prevent gases from diffusing between adjacent plates. Recombinant cells feature a porous separator system that permits gas diffusion between plates.

26.2.3.3 Battery Construction

Nickel–cadmium aircraft batteries generally consist of a steel case containing identical, individual cells connected in series. The number of cells depends on the particular application, but generally 19 or 20 cells are used. The end cells of the series are connected to the battery receptacle located on the

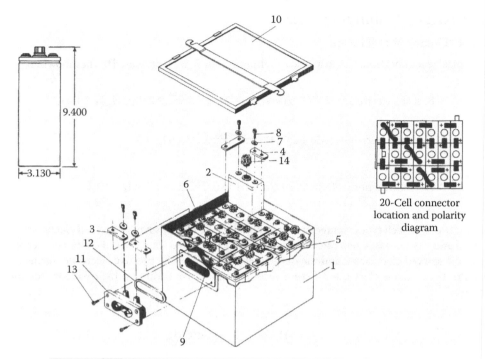

Item	Description	Quantity	Item	Description	Quantity
1	Can assembly	1	8	Socket head cap screw	42
2	Cell assembly	20	9	Spacer	1
3	Connector	12	10	Cover assembly	1
4	Connector	5	11	Receptacle assembly	1
5	Connector	3	12	Rectangular ring	1
6	Connector	1	13	Phillips head screw	4
7	Belleville spring	42	14	Filler cap and vent assembly	20

FIGURE 26.5 Assembly drawing of a nickel–cadmium aircraft battery.

outside of the case. The receptacle is usually a two-pin, quick disconnect type; both Cannon and Elcon styles commonly are used. Cases are vented by means of vent tubes or louvers to allow escape of gases produced during overcharge. Some battery designs have provisions for forced air cooling, particularly for engine start applications. Thermostatically controlled heating pads sometimes are employed on the inside or outside of the battery case to improve low-temperature performance. Power for energizing the heaters normally is provided by the aircraft's AC or DC bus. Temperature sensors often are included inside the case to allow regulation of the charging voltage. In addition, many batteries are equipped with a thermal switch that protects the battery from overheating if a fault develops or if battery is exposed to excessively high temperatures. A typical aircraft battery assembly is shown in Figure 26.5.

26.2.3.4 Discharge Performance

Typical discharge performance data for VNC aircraft batteries are illustrated in Figures 26.6 and 26.7. Discharge characteristics of SNC batteries are similar to VNC batteries. Figure 26.4 shows the effect of temperature on discharge capacity at the C-rate. Compared with lead–acid batteries, nickel–cadmium batteries tend to have more available capacity at low temperature, but less available capacity at high temperature. Manufacturers' data should be consulted for current information on specific batteries of interest.

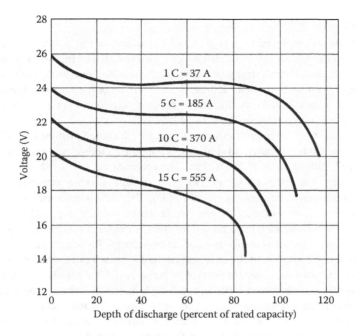

FIGURE 26.6 Discharge curves at 25°C for a 24 V/37 Ah VNC aircraft battery.

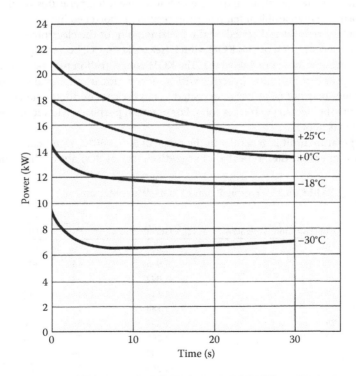

FIGURE 26.7 Maximum power curves (12 V discharge) for a 24 V/37 Ah VNC aircraft battery.

26.2.3.5 Charge Methods

A variety of methods are employed to charge nickel–cadmium aircraft batteries. The key requirement is to strike an optimum balance between overcharging and undercharging, while achieving full charge in the required time frame. Overcharging results in excessive water loss (vented cells) or heating (sealed cells). Undercharging results in capacity fading. Some overcharge is necessary, however, to overcome coulombic inefficiencies associated with the electrochemical reactions. In practice, recharge percentages on the aircraft generally range between 105% and 120%.

For vented-cell batteries, common methods of charging include constant potential, constant current, or pulse current. Constant potential charging is the oldest method and normally is accomplished by floating a 19-cell battery on a 28 V DC bus. The constant current method requires a dedicated charger and typically uses a 0.5–1.5 C-rate charging current. Charge termination is accomplished using a temperature-compensated voltage cutoff (VCO). The VCO temperature coefficient is typically (–) 4mV/°C. In some cases, two constant current steps are used: the first step at a higher rate (e.g., C-rate) and the second step at a lower rate (e.g., 1/3–1/5 of the C-rate). This method is more complicated but results in less gassing and electrolyte spewage during overcharge. Pulse current methods are similar to the constant current methods, except the charging current is pulsed rather than constant.

For sealed-cell batteries, only constant current or pulse current methods should be used. Constant potential charging can cause excessive heating, resulting in thermal runaway. Special attention must be given to the charge termination technique in sealed-cell batteries, because the voltage profile is relatively flat as the battery becomes fully charged. For example, it may be necessary to rely on the battery's temperature rise rather than voltage rise as the signal for charge termination.

26.2.3.6 Temperature Effects and Limitations

Nickel–cadmium batteries, like lead–acid batteries, normally are rated at room temperature (25°C) and operate best around this temperature. Exposure to low ambient temperatures results in performance decline, and exposure to high ambient temperatures results in shortened life.

The lower temperature limit is dictated by the freezing point of the electrolyte. Most cells are filled with an electrolyte concentration of 31% KOH, which freezes at –66°C. Lower concentrations will freeze at higher temperatures, as shown in Table 26.3. The KOH concentration may become diluted over time as a result of spillage or carbonization (reacting with atmospheric carbon dioxide), so the freezing point of a battery in service may not be as low as expected. As in the case of dilute acid electrolytes, slush ice will form well before the electrolyte freezes solid. For practical purposes, a lower operating temperature limit of –40°C often is quoted.

The upper temperature limit is generally in the range of 50°C–60°C; significant capacity loss occurs when batteries are operated (i.e., repeated charge/discharge cycles) above this temperature range.

TABLE 26.3   Freezing Points of KOH–Water Mixtures

Concentration (Weight%)	Specific Gravity at 15°C	Freezing Point (°C)	(°F)
0	1.000	0	+32
5	1.045	−3	+27
10	1.092	−8	+18
15	1.140	−15	+5
20	1.188	−24	−11
25	1.239	−38	−36
30	1.290	−59	−74
31	1.300	−66	−87
35	1.344	−50	−58

The battery capacity often is recoverable, however, when the battery is cooled to room temperature and subjected to several deep discharge cycles.

26.2.3.7 Service Life

The service life of a nickel–cadmium aircraft battery depends on many factors, including the type of use it experiences (e.g., rate, frequency, and depth of discharge), environmental conditions (e.g., temperature and vibration), charging method, and the care with which it is maintained and reconditioned. Thus, it is difficult to generalize the service life that can be expected. All things being equal, the service life of a nickel–cadmium battery is typically longer than that of a lead–acid battery.

26.2.3.8 Storage Characteristics

Nickel–cadmium batteries can be stored in any state of charge and over a broad temperature range (i.e., −65°C to 60°C). For maximum shelf life, however, it is best to store batteries between 0°C and 30°C. Vented-cell batteries normally are stored with the terminals shorted together. Shorting of sealed-cell batteries during storage is not recommended, however, since it may cause cell venting and/or cell reversal.

When left on open circuit during periods of nonoperation, nickel–cadmium batteries will self-discharge at a relatively fast rate. As a rule of thumb, the self-discharge rate of sealed cells is approximately 1% per day at 20°C (when averaged over 30 days), and the rate increases by 1% per day for every 10°C rise in temperature (e.g., 2%/day at 30°C and 3%/day at 40°C). The self-discharge rate is somewhat less for vented cells. The capacity lost by self-discharge usually is recoverable when charged in the normal fashion.

26.2.3.9 Maintenance Requirements

Routine maintenance of nickel–cadmium aircraft batteries is required to assure airworthiness and to maximize service life. Maintenance intervals for vented-cell batteries in military aircraft are typically 60–120 days. Maintenance intervals for commercial aircraft can be as low as 100 and as high as 1000 flight hours, depending on the operating conditions. Maintenance procedures include capacity checks, cell equalization (deep discharge followed by shorting cell terminals for at least 8 h), isolating and replacing faulty cells (only if permitted; this practice generally is not recommended), cleaning to remove corrosion and carbonate buildup, and electrolyte adjustment. For sealed-cell batteries, maintenance requirements are much less demanding. Electrolyte adjustment is unnecessary, and the extent of corrosion is greatly reduced. However, some means of assuring airworthiness is still necessary, such as periodic capacity measurement. Manufacturers' recommendations should be followed for specific batteries of interest.

26.2.3.10 Failure Modes and Fault Detection

The predominant failure modes of nickel–cadmium cells are summarized as follows:

- Shorts caused by cadmium migration through the separator, swelling of the positive electrode, degradation of the separator, or mechanical defects protruding from the electrode. Manifested by inability of the battery to hold a charge (soft shorts) or dead cells (hard shorts)
- Water loss and resulting cell dry-out due to leaking seal, repeated cell reversal, or excessive overcharge (this mode applies to sealed cells or to vented cells that are improperly maintained). Manifested by low capacity and/or inability to hold voltage under load
- Loss of negative (cadmium) electrode capacity due to passivation or active material degradation. Manifested by low capacity and/or inability to hold voltage under load. Usually reversible by deep discharge followed by shorting cell terminals or by "reflex" charging (pulse charging with momentary discharge between pulses)
- Loss of positive (nickel) electrode capacity due to swelling or active material degradation. Manifested by low capacity that is nonrestorable

As discussed under lead–acid batteries, the detection of these failure modes is relatively straightforward if the battery can be removed from the aircraft. For example, the battery capacity and load capability can be directly measured and compared against pass/fail criteria. The occurrence of soft shorts (i.e., a high-impedance short between adjacent plates) is more difficult to detect but often can be identified by monitoring the end-of-charge voltage of individual cells.

The detection of these failure modes while the battery is in service is more difficult. As in the case of lead–acid batteries, a number of onboard detection schemes have been developed for critical applications (Vutetakis and Viswanathan, 1995). The more critical the battery is to the safety of the aircraft, the more important it becomes to detect battery faults accurately.

26.2.3.11 Disposal

Proper disposal of nickel–cadmium batteries is essential because cadmium is a toxic (carcinogenic) chemical. In the United States and abroad, spent nickel–cadmium batteries are considered to be hazardous waste, and their disposal is strictly regulated. Several metallurgical processes have been developed for reclaiming and recycling the nickel and cadmium from nickel–cadmium batteries. These processes can be used for both vented and sealed cells. Federal, state, and local regulations should be followed for proper disposal procedures.

26.2.4 Lithium-Ion Batteries

26.2.4.1 Theory of Operation

The chemical reactions that occur in a lithium-ion battery are represented by the following equations:

$$\text{Positive electrode:}\quad \text{Li}_{1-x}\text{FePO}_4 + x\text{Li} + x\text{e}^- \underset{\text{charge}}{\overset{\text{discharge}}{\rightleftarrows}} \text{LiFePO}_4 \tag{26.10}$$

$$\text{Negative electrode:}\quad \text{Li}_x\text{C} \underset{\text{charge}}{\overset{\text{discharge}}{\rightleftarrows}} \text{C} + x\text{Li} + x\text{e}^- \tag{26.11}$$

$$\text{Overall reaction:}\quad \text{Li}_x\text{C} + \text{Li}_{1-x}\text{FePO}_4 \underset{\text{charge}}{\overset{\text{discharge}}{\rightleftarrows}} \text{LiFePO}_4 + \text{C} \tag{26.12}$$

These equations assume the cathode material is made of LiFePO_4. Similar equations can be written based on other cathode materials, such as LiCoO_2, LiNiO_2, and LiMn_2O_4. On discharge, lithium ions are removed from the carbon layers of the negative electrode and accumulate in the positive electrode. On charge, lithium ions leave the positive electrode and are intercalated in the carbon layers. Unlike lead–acid and nickel–cadmium cells, the electrolyte does not participate in the electrochemical reactions.

26.2.4.2 Cell Construction

Lithium-ion cells are available in three basic designs: cylindrical, prismatic, and pouch. The cylindrical design consists of a spiral wound coil of two composite electrodes separated by a microporous film, fitted into a cylindrical-shaped metal case. The prismatic design consists of flat electrodes and microporous separators, fitted into a prismatic-shaped metal case. The pouch design also has flat electrodes and separators, but they are fitted into a vacuum-sealed polymeric pouch. The prismatic and pouch designs are generally more costly but have better packing efficiency when assembled into a battery, resulting in higher energy density.

The electrolyte used in lithium-ion cells is a flammable organic solvent that poses an inherent safety hazard. In all designs, the cell is hermetically sealed to prevent escape of electrolyte. Cylindrical and prismatic designs generally have a nonresealable venting mechanism to safely vent gases if the internal pressure gets abnormally high.

Other safety features of some lithium-ion cells include shutdown separators (separators become highly resistive if the cell gets too hot), internal fuses, and current interrupt devices (disconnects the cell terminal internally in response to swelling of the cell).

26.2.4.3 Battery Construction

There are two basic approaches to the design and construction of lithium-ion aircraft batteries. One approach is to connect a sizable number of small cylindrical cells in parallel to make a cell pack, then connect these cell packs in series to make a 24 V battery. This approach does not have the best packaging efficiency and requires many cell connections but does take advantage of low-cost cylindrical cells that are available in high-volume production (i.e., 18650 and 26650 cells). The other approach is to use larger-capacity cells that do not need to be paralleled and connect these in series to make a 24 V battery. This approach minimizes the number of cell connections, improves packaging efficiency, and increases the battery reliability by reducing the probability of having a bad cell.

Lithium-ion aircraft batteries normally include some form of a battery management system (BMS). The BMS may include a cell balancing circuit and charge control circuit, temperature monitoring, and/or a disconnect switch. The battery needs to be carefully integrated with the aircraft electrical system to assure optimum reliability and safety. Other battery design features, such as container, cover, and receptacles, are similar to that of lead–acid and nickel–cadmium aircraft batteries.

26.2.4.4 Charge Methods

Constant voltage charging is the preferred method of charging lithium-ion aircraft batteries. Depending on the cathode chemistry, the charge voltage can range from 3.6 to 4.2 V per cell. With most lithium-ion cells, the inrush current needs to be controlled, so generally it is not feasible to float the battery on the 28 V DC bus. Thus, a dedicated battery charger is normally necessary. The charger also needs to monitor individual cells or paralleled cell groups to assure they do not become out of balance with the rest of the battery. Excessively high cell voltages can cause the cells to pressurize, resulting in venting of flammable gases that can ignite or even explode. To prevent cells from getting out of balance, electronic balancing circuits within the battery are generally used.

26.2.4.5 Temperature Effects and Limitations

Most lithium-ion batteries have a narrower temperature range compared to lead–acid and nickel–cadmium cells. The lower operating temperature limit is typically in the range of −10°C to −20°C, and the upper operating temperature limit is typically 45°C–60°C, depending on the cell type. At low temperatures, the internal resistance increases substantially, so the high-rate discharge capability is generally poor. At high temperatures, the battery life is shortened because the charging process rapidly degrades the active material. Research is ongoing to extend the operating temperature range of lithium-ion technology.

26.2.4.6 Storage Characteristics

Lithium-ion batteries have a self-discharge rate much lower than lead–acid and nickel–cadmium cells. Batteries are typically shipped and stored around 50% state of charge and can be stored for up to 2 years when the ambient temperature is 25°C or below. Batteries should not be allowed to self-discharge to an OCV less than approximately 2.5 V per cell (20 V for an eight-cell battery), or irreversible capacity loss can occur.

26.2.4.7 Maintenance Requirements

Because lithium-ion batteries have only recently been adapted for aircraft applications, their maintenance requirements have not been established. Since the cells are sealed, there obviously is no need to maintain electrolyte levels. The frequency of capacity tests to assure airworthiness will need to be established as operational experience is gained but is expected to be similar to SLA and SNC batteries.

26.2.4.8 Service Life

Lithium-ion cells have very long cycle life, comparable to that of nickel–cadmium batteries (e.g., 1000–2000 cycles at 80% depth of discharge per cycle). Predictions of 5–10-year service life have been made. However, unlike nickel–cadmium batteries, lithium-ion batteries are not tolerant of being left in the deeply discharged condition, such as when the power switch is inadvertently left on overnight or for a long weekend. This type of condition results in rapid corrosion of the copper anode, causing irreversible capacity loss. To prevent this occurrence, it is advisable to include a means of disconnecting the battery from the load, such as a timer circuit when the aircraft is powered down. Otherwise, the service life of the lithium-ion battery could be very short. Lithium-ion cells also suffer irreversible capacity loss when maintained at 100% state of charge for a prolonged time period, so the service life of batteries that stay fully charged (i.e., emergency batteries) may also be limited.

26.2.4.9 Disposal

Lithium-ion batteries are classified as nonhazardous waste and disposal is therefore not regulated. However, these batteries do contain recyclable materials and recycling options should be considered when the battery reaches its end of service life.

26.3 Applications

Designing a battery for a new aircraft application or for retrofit requires a careful systems engineering approach. To function well, the battery must be interfaced carefully with the aircraft's electrical system. The battery's R&M depends heavily on the type of charging system to which it is connected; there is a fine line between undercharging and overcharging the battery. Many airframe manufacturers have realized that it is better to prepare specifications for a "battery system" rather than to have separate specifications for the battery and the charger. This approach assures that the charging profile is tuned correctly to the specific characteristics of the battery and to the aircraft's operational requirements.

26.3.1 Commercial Aircraft

In general, aircraft batteries must be sized to provide sufficient emergency power to support flight essential loads in the event of failure of the primary power system. FAA regulations impose a minimum emergency power requirement of 30 min on all commercial airplanes. Some airlines impose a longer emergency requirement, such as 40 or 60 min, due to frequent bad weather on their routes or for other reasons. The emergency requirement for extended twin operations (ETOPS) imposed on two-engine aircraft operating over water is a full 90 min, although 60 min is allowed with operating restrictions. The specified emergency power requirement may be satisfied by batteries or other backup power sources, such as a ram air turbine. If a ram air turbine is used, a battery still is required for transient fill-in. Specific requirements pertaining to aircraft batteries can be found in the Federal Aviation Regulations (FARs), Sections 25.1309, 25.1333, 25.1351, and 25.1353. FAA Advisory Circular No. 25.1333-1 describes specific methods to achieve compliance with applicable FAR sections. For international applications, Civil Aviation Authority (CAA) and Joint Airworthiness Authority (JAA) regulations should be consulted for additional requirements.

When used for APU or engine starting, the battery must be sized to deliver short bursts of high power, as opposed to the lower rates required for emergency loads. APU start requirements on large commercial aircraft can be particularly demanding; for instance, the APU used on the Boeing 757 and 767 airplanes has a peak current requirement of 1200 A (Gross, 1991). The load on the battery starts out very high to deliver the inrush current to the motor, then falls rapidly as the motor develops back electromotive force (EMF). Within 30–60 s, the load drops to zero as the APU ignites and the starter cutoff point is reached. The worst-case condition is starting at altitude with a cold APU and a cold battery; normally, a lower temperature limit of −18°C is used as a design point. A rigorous design methodology for optimizing aircraft starter batteries was developed by Evjen and Miller (1971).

When nickel–cadmium batteries are used for APU or engine starting applications, FAA regulations require the battery to be protected against overheating. Suitable means must be provided to sense the battery temperature and to disconnect the battery from the charging source if the battery overheats.

TABLE 26.4 Military Aircraft Batteries

Military Part No.	Type	Rating[a] (Ah)	Max. Wt. (lb)	Applications	Notes
MIL-PRF-8565 series					
D8565/1-1	SNC	2.0 (26 V)	8.6	AV-8A/C, CH-53D/E, MH-53E	Superseded by D8565/1-2
D8565/1-2	SLA	1.5	6.8	Same as D8565/1-1	Contains integral charger
D8565/2-1	VNC	30	88.0	OV-10D	Superseded by M81757/12-1
D8565/3-3	SLA	15	47.4	V-22(EMD)	MS3509 connector
D8565/4-1	SLA	7.5	26.0	F/A-18A/B/C/D, CH-46D/E, HH-46A, UH-46A/D, F-117A	MS27466T17B6S connector
D8565/5-1	SLA	30	80.2	C-1A, SP-2H, A-3B, KA-3B, RA-3B, ERA-3B, NRA-3B, UA-3B, P-3A/B/C, EP-3A/B/E, RP-3A, VP-3A, AC-130A/H/U, C-130A/B/E/F/H, DC-130A, EC-130E/H/G/Q, HC-130H/N/P, KC-130F/R/T, LC-130F/H/R, LC-130F/H/R, MC-130E/H, NC-130A/B/H, WC-130E/H, C-18A/B, EC-18B/D, C-137B/C, EC-137D, E-8A, TS-2A, US-2A/B, T-28B/C, QT-33A, MH-53J, MH-60G	Equivalent to D8565/5-2, except the use of MS3509 connector
D8565/5-2	SLA	30	80.2	Same as D8565/5-1 (for aircraft equipped with Cannon style mating connector)	Equivalent to D8565/5-1, except the use of Cannon connector
D8565/6-1	SLA	1.5	6.4	V-22A, CV-22A, CH-47E, E-2C, S-3B	MS27466715B5S connector
D8565/7-1	SLA	24	63.9	AV-8B, TAV-8B, VH-60A, V-22A, CV-22A, MV-22B	MS3509 connector
D8565/7-2	SLA	24	63.9	Same as D8565/7-1	Replacement for D8565/7-1 with higher rate capability
D8565/8-1	SLA	15	43.0	T-45A	Cannon connector
D8565/9-1	SLA	24	63.0	T-34B/C, U-6A	MS3509 connector
D8565/9-2	SLA	24	63.0	None identified	Cannon connector
D8565/10-1	VNC	35	85.0	AH-1W	MS3509 connector. Equipped with temperature sensor. Spec superseded by MIL-PRF-8157/17
D8565/11-1	SLA	10	34.8	F-4D/E/G, C-141B, MH-60E, NC-141A, YF-22A	Equivalent to D8565/11-2, except the use of MS3509 connector
D8565/11-2	SLA	10	34.8	None identified	Equivalent to D8565/11-1, except the use of Cannon connector
D8565/12-1	SLA	35	90.0	Developed for P-7, which got cancelled	MS3509 connector. Includes heater circuit
D8565/13-1	SLA	10	31.0	Carousel IV, LTN-72 Inertial Navigation Systems (INS)	ARINC, 1/2 ATR case
D8565/14-1	SLA	15	45.2	F/A-18E/F	D38999/24YG11SN connector
D8565/15-1	SLA	35	90.0	C/KC-135 series	MS3509 connector
D8565/16-1	SLA	5	14.6	H-60	

(Continued)

TABLE 26.4 (*Continued*) Military Aircraft Batteries

Military Part No.	Type	Rating[a] (Ah)	Max. Wt. (lb)	Applications	Notes
D8565/17-1	SLA	0.33	3.5	EA-6B	
D8565/18-1	SLA	10	33.0	F-5, T-38	
D8565/19-1	SLA	15	50.0	F-16 (pre-Block 40)	Contains integral heater
D8565/20-1	SLA	17	50.0	F-16 (Block 40-50)	No heater
MIL-PRF-8565 specials					
MS3319-1	VNC	0.75	3.5	HH-2D,SH-2D/F	MS3106-12S-3P connector
MS3337-2	SNC	0.40	4.0	F-4S	Obsolete
MS3346-1	VNC	2.5	10.0	A-7D/E, TA-7C	Obsolete
MS3487-1	VNC	18	50.0	AH-1G	Equivalent to BB-649A/A
MS7334-2	SNC	0.33	3.5	E-1B, EA-6B, US-2D	MS3106R14S-7P connector
MIL-PRF-81757 series (tri-service)					
M81757/7-2	VNC	10	34.0	CH-46A/D/E/F, HH-46A, UH-46A/D, U-8D/F, F-5E/F	Replaceable cells. Supersedes MS24496-1 and MS24496-2
M81757/7-3	VNC	10	34.0	Same as M81757/7-2	Nonreplaceable cells. Supersedes MS18045-44, MS18045-48, and MS90221-66W
M81757/8-4	VNC	20	55.0	C-2A, T-2C, T-39A/B/D, OV-10A	Replaceable cells. Supersedes MS24497-3, MS24497-5, and M81757/8-2
M81757/8-5	VNC	20	55.0	Same as M81757/8-4	Nonreplaceable cells. Supersedes MS90365-1, MS90365-2, MS90321-68W, MS90321-77, MS90321-78W, MS18045-45, MS18048-49, and M81757/8-3
M81757/9-2	VNC	30	80.0	CT-39A/E/G, NT-39A, TC-4C, HH-1K, TH-1L, UH-1E/H/L/N, AH-1J/T, LC-130F/R, OV-1B/C/D	Replaceable cells. Supersedes MS24498-1 and MS24498-2
M81757/9-3	VNC	30	80.0	Same as M81757/9-2	Nonreplaceable cells. Supersedes MS18045-46, MS18045-50, MS90321-75W, and MS90321-69W
M81757/10-1	VNC	6 (23 V)		A-6E, EA-6A, KA-6D	Nonreplaceable cells. Supersedes MS90447-2 and MS90321-84W
M81757/11-3	VNC	20	55.0	HH-2D, SH-2D/F/G, HH-3A/E, SH-3D/G/H, UH-3A/H, VH-3A/D	Nonreplaceable cells. Supersedes MS90377-1, MS90321-79W, and M81757/11-1
M81757/11-4	VNC	20	55.0	None identified	Nonreplaceable cells with temperature sensor. Supersedes MS90377-1, MS90321-79W, and M81757/11-2
M81757/12-1	VNC	30	88.0	OV-10D	Nonreplaceable cells, air-cooled. Supersedes D8565/2-1
M81757/12-2	VNC	30	88.0	C-2A (REPRO), OV-10D	Nonreplaceable cells, air-cooled, with temperature sensor
M81757/13-1	VNC	30	80.0	EA-3B, ERA-3B, UA-3B	Nonreplaceable cells. Supersedes MS18045-75
M81757/14-1	VNC	5.5	17.5	SH-60, MH-60	Low maintenance
M81757/15-1	VNC	25	55.0	H-2, H-3	Low maintenance

Designation	Type	Capacity	Voltage	Application	Notes
M81757/15-2	VNC	25	52.0	T-2	Low maintenance
M81757/15-3	VNC	25	56.0	AH-1Z, UH-1Y, VH-3D	Low maintenance
M81757/16-1	VNC	35	78.0	A-10, UH-1N	Low maintenance
D8565/10-1	VNC	35	85.0	AH-1W	Spec that is MIL-PRF-81757/17
M81757/18-1	VNC	55 (12 V)	55.0	Unknown	
M81757/19-2	VNC	10	34.0	Same as BB-432B/A	
M81757/21-1	VNC	13	27.0	Same as BB-664/A	
BB-series (U.S. Army)					
BB-432A/A	VNC	10	34.0	CH-47A/B/C, U-8F	Equivalent to M81757/7-2
BB-432B/A	VNC	10	34.0	CH-47D	Equivalent to BB-432A/A, except that it includes a temperature sensor
BB-433A/A	VNC	30	80.0	C-12C/D/F/L, OV-1D, EH-1H/X, UH-1H/V, RU-21A/B/C/H	Equivalent to M81757/9-2
BB-434/A	VNC	20	55.0	CH-54	Equivalent to M81757/8-4
BB-476/A	VNC	13	27.6	OH-58A/B/C	
BB-558/A	VNC	17	38.5	OH-58D	
BB-564/A	VNC	13	25.0	AH-64A	Superseded by BB-664/A
BB-638/U	VLA	31	80.0	None identified	Equivalent to M83769/1-1
BB-638A/U	VLA	31	80.0	None identified	Equivalent to M83769/6-1
BB-639/U	VLA	18	56.0	None identified	Equivalent to M83769/2-1
BB-640/U	VLA	8.4	34.0	None identified	Equivalent to M83769/3-1
BB-649A/A	VNC	18	50.0	AH-1E/F/P/S	Equivalent to MS3487-1
BB-664/A	VNC	13	27.0	AH-64A	
BB-678A/A	VNC	13	24.8	OH-6A	
BB-693A/U	VNC	30	83.0	Vulcan	
BB-708/U	VNC	5.5	15.0	OV-1D (Mission Gear Equipment)	
BB-716/A	VNC	5.5	17.5	EH-60A, HH-60H/J, MH-60S, SH-60B/F, UH-60A	
MIL-PRF-29595 series (rechargeable lithium batteries)					
M29595/1-1	LIB	25	26.0	Same as D8565/4-1	Not yet approved for acquisition
M29595/2-1	LIB	50	53.0	Same as D8565/10-1	Not yet approved for acquisition
M29595/3-1	LIB	30	28.2	Same as D8565/14-1	Not yet approved for acquisition
M29595/4-1	LIB	55	45.0	Same as D8565/7-2, M81757/15-3	Not yet approved for acquisition
M29595/5-1	LIB	15	15.0	Same as M81757/14-1	Not yet approved for acquisition

[a] Capacity rating is based on the 1 h rate unless otherwise noted. Voltage rating is 24 V unless otherwise noted.

This requirement originated in response to numerous instances of battery thermal runaway, which usually occurred when 19-cell batteries were charged from the 28 V DC bus. Most instances of thermal runaway were caused by degradation of the cellophane gas barrier, thus allowing gas recombination and resultant cell heating during charging. Modern separator materials (e.g., Celgard) have greatly reduced the occurrence of thermal runaway as a failure mode of nickel–cadmium batteries, but the possibility still exists if the electrolyte level is not properly maintained.

Application of lithium-ion batteries to commercial aircraft should be approached very cautiously. Even though there is a weight advantage compared to SLA or VNC batteries, there are additional safety hazards because of the higher energy density and the use of flammable and toxic electrolyte. There have already been several battery fires on aircraft certified with lithium-ion batteries that forced the FAA to issue emergency airworthiness directives (e.g., Boeing 787, Cessna CJ4 Model 525C). The future of lithium-ion batteries will depend on whether they can be designed to achieve an equivalent level of safety compared with SLA or VNC batteries.

Commercially available aircraft batteries come in 12 and 24 V versions with capacities ranging from about 1 Ah to about 65 Ah. Detailed specifications for aircraft batteries are generally available on the website of those companies offering their products for sale (see listing at the end of this chapter).

26.3.2 Military Aircraft

A listing of commonly used military aircraft batteries is provided in Table 26.4. This listing includes only those batteries that have been assigned a military part number based on an approved military specification; nonstandard batteries are not included. Detailed characteristics and performance capabilities can be found by referring to the applicable military specifications. A number of nonstandard battery designs have been proliferated in the military due to the unique form, fit, and/or functional requirements of certain aircraft. Specifications for these batteries normally are obtainable only from the aircraft manufacturer. Specific examples of battery systems used in present-day military aircraft were described by Vutetakis (1994).

Definitions

Ampere-Hour Capacity: The quantity of stored electrical energy, measured in Ah, which the battery can deliver from its completely charged state to its discharged state. The dischargeable capacity depends on the rate at which the battery is discharged; at higher discharge rates, the available capacity is reduced.

C-Rate: The discharge rate, in amperes, at which a battery can deliver 1 h of capacity to a fixed voltage endpoint (typically 18 or 20 V for a 24 V battery). Fractions or multiples of the C-rate also are used. C/2 refers to the rate at which a battery will discharge its capacity in 2 h; 2C is twice the C-rate or that rate at which the battery will discharge its capacity in half an hour. This rating system helps to compare the performance of different sizes of cells.

CCA: The numerical value of the current, in amperes, that a fully charged lead–acid battery can deliver at −18°C (0°F) for 30 s to a voltage of 1.2 V per cell (i.e., 14.4 V for a 24 V battery). In some cases, 60 s is used instead of 30 s.

Electrolyte: An ionically conductive, liquid medium that allows ions to flow between the positive and negative plates of a cell. In lead–acid cells, the electrolyte is a mixture of sulfuric acid (H_2SO_4) and deionized water. In nickel–cadmium cells, the electrolyte is a mixture of potassium hydroxide (KOH) dissolved in deionized water.

Imp: The current, in amperes, delivered at 15 s during a constant voltage discharge at half of the nominal voltage of the battery (i.e., at 12 V for a 24 V battery). The Imp rating normally is based on a battery temperature of 23°C (75°F), but manufacturers generally can supply Imp data at lower temperatures as well.

Ipp: Similar to Imp and Ipr, except it is the current delivered at 0.3 s during the constant voltage discharge.

Ipr: Same definition as Imp.

Monobloc: A group of two or more cells connected in series and housed in a one-piece enclosure with suitable dividing walls between cell compartments. Typical monoblocs come in 6, 12, or 24 V configurations. Monoblocs are commonly used in lead–acid batteries but rarely used in nickel–cadmium aircraft batteries.

Negative Electrode: The electrode from which electrons flow when the battery is discharging into an external circuit. Reactants are electrochemically oxidized at the negative electrode. In the lead–acid cell, the negative electrode contains spongy lead and lead sulfate ($PbSO_4$) as the active materials. In the nickel–cadmium cell, the negative electrode contains cadmium and cadmium hydroxide ($Cd(OH)_2$) as the active materials.

Nominal Voltage: The characteristic operating voltage of a cell or battery. The nominal voltage is 2.0 V for lead–acid cells and 1.2 V for nickel–cadmium cells. These voltage levels represent the approximate cell voltage during discharge at the C-rate under room temperature conditions. The actual discharge voltage depends on the state of charge, state of health, discharge time, rate, and temperature.

Positive Electrode: The electrode to which electrons flow when the battery is discharging into an external circuit. Reactants are electrochemically reduced at the positive electrode. In the lead–acid cell, the positive electrode contains lead dioxide (PbO_2) and lead sulfate ($PbSO_4$) as the active materials. In the nickel–cadmium cell, the positive electrode contains nickel oxyhydroxide ($NiOOH$) and nickel hydroxide ($Ni(OH)_2$) as the active materials.

Separator: An electrically insulating material that is used to prevent metallic contact between the positive and negative plates in a cell but permits the flow of ions between the plates. In flooded cells, the separator includes a gas barrier to prevent gas diffusion and recombination of oxygen. In sealed cells, the separator is intended to allow gas diffusion to promote high recombination efficiency.

State of Charge: The available capacity of a battery divided by the capacity available when fully charged, normally expressed on a percentage basis. Sometimes referred to as "true state of charge."

State of Health: The available capacity of a fully charged battery divided by the rated capacity of the battery, normally expressed on a percentage basis. Sometimes referred to as "apparent state of charge." Can also be used in a more qualitative sense to indicate the general condition of the battery.

Further Reading

The following reference material contains further information on various aspects of aircraft battery design, operation, testing, maintenance, and disposal:

IEC 60952-1, 2nd edn. 2004. Aircraft batteries—Part 1: General test requirements and performance levels.

IEC 60952-2, 2nd edn. 2004. Aircraft batteries—Part 2: Design and construction requirements.

IEC 60952-3, 2nd edn. 2004. Aircraft batteries—Part 3: Product specification and declaration of design and performance (DDP).

NAVAIR 17-15BAD-1, Naval Aircraft and Naval Aircraft Support Equipment Storage Batteries. Request for this document should be referred to Commanding Officer, Naval Air Technical Services Facility, 700 Robbins Avenue, Philadelphia, PA 19111.

Rand, D. A. J., Moseley, P. T., Garche, J., and Parker, C. D. (eds.). 2004. *Valve-Regulated Lead-Acid Batteries*. Elsevier, Amsterdam, the Netherlands.

Reddy, T.B. (ed.). 2011. *Linden's Handbook of Batteries*, 4th edn. McGraw-Hill, New York.

RTCA DO-293. 2004. Minimum operational performance standards for nickel-cadmium and lead-acid batteries.

RTCA DO-311. 2008. Minimum operational performance standards for rechargeable lithium battery systems.

SAE Aerospace Standard AS8033. 1981. Nickel-cadmium vented rechargeable aircraft batteries (Non-sealed, maintainable type).

The following companies manufacture aircraft batteries and can be contacted for technical assistance and pricing information:

Nickel–Cadmium Batteries

Acme Electric Corporation
Aerospace Division
528 W. 21st Street
Tempe, Arizona 85282
Phone (480) 894-6864
www.acmeelec.com

MarathonNorco Aerospace, Inc.
8301 Imperial Drive
Waco, Texas 76712
Phone (817) 776-0650
www.mptc.com

SAFT America Inc.
711 Industrial Boulevard
Valdosta, Georgia 31601
Phone (912) 247-2331
www.saftbatteries.com

Lithium-Ion Batteries

Concorde Battery Corporation
2009 San Bernardino Road
West Covina, California 91790
Phone (800) 757-0303
www.concordebattery.com

SAFT America Inc.
711 Industrial Boulevard
Valdosta, Georgia 31601
Phone (912) 247-2331
www.saftbatteries.com

Lead–Acid Batteries

Concorde Battery Corporation
2009 San Bernardino Road
West Covina, California 91790
Phone (800) 757-0303
www.concordebattery.com

Enersys Energy Products Inc.
(Hawker Batteries)
617 N. Ridgeview Drive
Warrensburg, MO 64093
Phone (800) 964-2837
www.enersysinc.com

Teledyne Battery Products
840 West Brockton Avenue
Redlands, California 92375
Phone (800) 456-0070
www.gillbatteries.com

References

Anderman, M. 1994. Ni-Cd Battery for aircraft; battery design and charging options. *Proceedings of the Ninth Annual Battery Conference on Applications and Advances*, California State University, Long Beach, CA, pp. 12–19.

Earwicker, G. A. 1956. Aircraft batteries and their behavior on constant-potential charge. In *Aircraft Electrical Engineering*, G. G. Wakefield (ed.), pp. 196–224. Regel Aeronautical Society, London, U.K.

Evjen, J. M. and Miller, L. D., Jr. 1971. Optimizing the design of the battery-starter/generator system. SAE Paper 710392.

Falk, S. U. and Salkind, A. J. 1969. *Alkaline Storage Batteries*, pp. 466–472. John Wiley & Sons, New York.

Flake, R. A. 1988. Overview on the evolution of aircraft battery systems used in air force aircraft. SAE Paper 881411.

Fleischer, A. 1956. Nickel-Cadmium batteries. *Proceedings of the 10th Annual Battery Research and Development Conference*, Power Sources Division, U.S. Signal Corps Engineering Laboratories, Fort Monmouth, NJ, pp. 37–41.

Gross, S. 1991. Requirements for rechargeable airplane batteries. *Proceedings of the Sixth Annual Battery Conference on Applications and Advances*, California State University, Long Beach, CA.

Johnson, Z., Roberts, J., and Scoles, D. 1994. Electrical characterization of the negative electrode of the USAF 20-year-life maintenance-free sealed nickel-cadmium aircraft battery over the temperature range -40°C to +70°C. *Proceedings of the 36th Power Sources Conference*, Cherry Hill, NJ, pp. 292–295.

McWhorter, T. A. and Bishop, W. S. 1972. Sealed aircraft battery with integral power conditioner. *Proceedings of the 25th Power Sources Symposium*, Cherry Hill, NJ, pp. 89–91.

Miller, G. H. and Schiffer, S. F. 1971. Aircraft zinc-silver oxide batteries. In *Zinc-Silver Oxide Batteries*, A. Fleischer (ed.), pp. 375–391. John Wiley & Sons, New York.

Scardaville, P. A. and Newman, B. C. 1993. High power vented nickel-cadmium cells designed for ultra low maintenance. *Proceedings of the Eigth Annual Battery Conference on Applications and Advances*, California State University, Long Beach, CA.

Senderak, K. L. and Goodman, A. W. 1981. Sealed lead-acid batteries for aircraft applications. *Proceedings of the 16th IECEC*, Atlanta, GA, pp. 117–122.

Vutetakis, D. G. 1994. Current status of aircraft batteries in the U.S. Air Force. *Proceedings of the Ninth Annual Battery Conference on Applications and Advances*, California State University, Long Beach, CA, pp. 1–6.

Vutetakis, D. G. and Timmons, J. B. 2008. A comparison of lithium-ion and lead-acid aircraft batteries. *Proceedings of the 2008 SAE Power Sources Conference*, Paper No. 2008-01-2875, Society of Automotive Engineers, New York.

Vutetakis, D. G. and Viswanathan, V. V. 1995. Determining the state-of-health of maintenance-free aircraft batteries. *Proceedings of the 10th Annual Battery Conference on Applications and Advances*, California State University, Long Beach, CA, pp. 13–18.

Vutetakis, D. G. and Viswanathan, V. V. 1996. Qualification of a 24-Volt, 35-Ah sealed lead-acid aircraft battery. *Proceedings of the 11th Annual Battery Conference on Applications and Advances*, California State University, Long Beach, CA, pp. 33–38.

27
Genesis: An IMA Architecture for Boeing B-787 and Beyond

Randy Walter
GE Aviation Systems

**Christopher
B. Watkins**
*Gulfstream Aerospace
Corporation*

27.1 Genesis Platform Concept

Avionics architectures have evolved over time as avionics have become more and more integrated. Within traditional architectures, an aircraft function was typically allocated to a redundant set of line-replaceable units (LRUs) where the function was wholly contained by an LRU. This functional independence between LRUs provided clearly defined boundaries where functional responsibility was wholly owned by a single organization (the LRU supplier). Modern avionics architectures have leveraged integrated avionics platforms to provide cost and weight savings to the aircraft as well as new integrated capabilities. To accomplish this, the aircraft functions are allocated to multiple LRUs, which can optionally be provided by separate system suppliers. Since an aircraft function's boundary now spans multiple systems, the function's responsibility is not fully owned by any single system supplier organization. The role of a system integrator has grown to own and manage the aircraft functions as implemented across multiple systems and LRUs. The Boeing 787 includes a good example of an integrated modular avionics (IMA) architecture. The Genesis platform represents the foundational architecture used by the B-787 and other aircraft and is described within this chapter.

The Genesis platform is a hardware/software platform that provides computing, communication, and input–output (I/O) services for implementing real-time embedded systems, known as hosted functions.

Multiple systems can be architected and overlaid on the partitioned platform resources to form a highly integrated system with the unique characteristic of full isolation and independence of each individual system. As such, the Genesis platform provides the starting point for the synthesis of a highly integrated real-time system, hence the name (Generic Networked Elements for the Synthesis of Integrated Systems). The platform elements are architected to maintain a high-integrity, fault-tolerant environment, necessary for hosting critical system functionality.

Genesis is a new class of integrated computing platforms that encompasses IMA as well as other real-time computing platforms. It is targeted for supporting highly critical applications but can be scaled to appropriately support lower levels of application integrity. Genesis is characterized by the following features and architectural advantages:

Significant Features	Architectural Advantages
Integrated system architecture	Reduced system size
Computing, communication, and I/O	Reduced system weight
Network-centric communications	Reduced system power
Composable and extensible architecture	Reduced system cost
Robust partitioning	Reduced part numbers
Real-time deterministic system	Minimized interconnect wiring
High-integrity platform	Supported efficient sensor and effector placement
Fault containment	Minimized wire lengths for passive cooling, which
Fail-passive design	in turn improves electrical signal quality
Asynchronous component clocking	Life-critical applications hosting capability
Change containment	
Open system environment	
Legacy LRU compatibility	

The Genesis platform is an avionics infrastructure component and does not provide an aircraft function by its own right. Aircraft functions (i.e., air data system, landing gear system, flight management system [FMS]) can be hosted within the Genesis platform and thus are referred to as "hosted functions." Hosted functions are allocated to the platform resources to form a "functional" architecture specific to each system to meet the availability, operational, safety, and topological requirements for each function. Hosted functions can "own" unique sensors, effectors, devices, and nonplatform LRUs, which become part of the functional system architecture. Multiple hosted functions share the platform resources within a virtual system environment enforced by partitioning mechanisms that are implemented as part of the platform design. The virtual system partitioning environment guarantees that hosted functions are isolated from each other; therefore, they cannot interfere with each other regardless of faults that may occur within the hosted functions or the platform resources. The platform design guarantees each hosted function its allocated share of the computing, network, and I/O resources. Those resource allocations are predetermined and communicated to the platform components via loadable configuration files, which become the source of information for the run time enforcement of the hosted function resource guarantees. The configuration files also define the data flow between systems in the context of data publishers and subscribers. Data signals can be added or rerouted on the shared backbone simply by loading an updated (and aircraft approved) configuration file during a maintenance action. The need for physical wiring changes is reduced. As such, the configuration files become an important artifact for the overall system integration effort.

The Genesis platform is a scalable platform that allows the integrated system architecture to be sized for each platform element. The number of computing elements, I/O elements, and communication links can be scaled to meet the needs of a specific implementation. This ability, along with diverse interface types offered with the I/O element, allows the Genesis platform to be used at any level of system integration deemed appropriate, without the need to redesign the platform.

The Genesis platform enables the implementation of highly integrated system architectures, which benefit aerospace and other industries with significantly reduced system weight, part numbers, and equipment cost while maintaining lower cost-of-change characteristics associated with federated systems.

27.1.1 Comparison to Traditional Federated Architectures

The Genesis architecture is provided in contrast to the traditional architecture characterized by federated systems. Federated systems are architected to provide the following services in each system:

- Separate processing
- Separate infrastructure
- Separate I/O
- Internal system bus

In addition, I/O is routed point to point between sensors/effectors and systems in a federated system architecture. A federated system architecture diagram is presented in Figure 27.1.

In contrast to federated systems, Genesis is architected to provide the following services for an integrated set of systems:

- Common infrastructure, which standardizes the development environment for the aircraft
- Common processing via shared general processing modules (GPMs), with robustly partitioned application software
- Specific I/O via shared remote interface units (RIUs), also known as remote data concentrators (RDCs)
- Distributed system bus (avionics full-duplex switched Ethernet network ARINC 664 Part 7 [A664-P7])

The Genesis architecture utilizes network-centric communications. I/O is routed using the A664-P7 network backbone between the GPMs, aircraft LRUs, and RIUs to sensors/effectors and non-A664-P7 buses: ARINC 429 and controller area network (CAN) (ISO 11898). A diagram of the Genesis system architecture is presented in Figure 27.2.

As shown in Figure 27.2, the Genesis architecture presents a "virtual system" concept to replace the *physical* systems as packaged in a federated architecture. Figure 27.2 portrays four *virtual systems* that

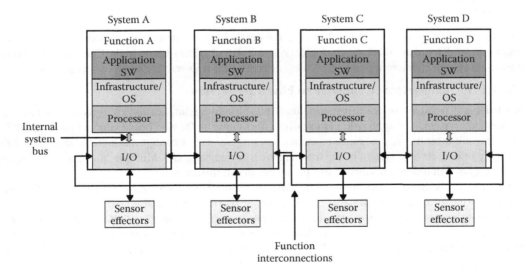

FIGURE 27.1   Federated system architecture.

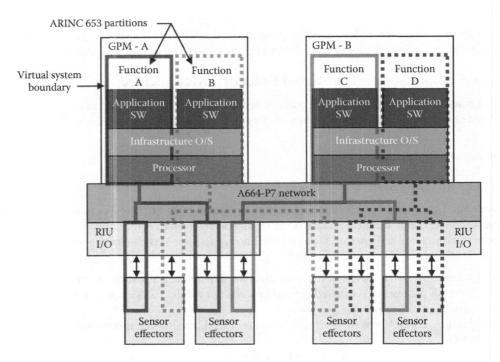

FIGURE 27.2   Genesis architecture identifying the "virtual systems."

are intended to be equivalent to the four physical systems shown in Figure 27.1. The "virtual system" consists of the same logical groupings of components as contained by a physical system:

- Application software
- Infrastructure/operating system (OS)
- Processor
- System bus
- I/O

Therefore, a key difference between Genesis and federated architectures is the definition of the logical system. In a federated architecture, the logical system is the physical system. In the Genesis architecture, the logical system is different than the physical system and is thus referred to as a "virtual system."

27.1.1.1 Virtual Systems Compared to Physical Systems

In a federated architecture, the target computer and the software application are typically packaged as a "physical" system. The application is usually linked with the OS and board support package (BSP), and the resultant software executable is verified as a single software configuration item. The application normally utilizes its own system bus to interface with its dedicated I/O. Multiple "physical" systems are then integrated in order to perform a specific set of aircraft functions.

In contrast, the major components of the Genesis architecture (GPM, A664-P7 network, RIU) provide a "virtual system" environment. The platform hosts the software application on a GPM, which is a computing resource shared between many software applications. The GPM hardware and platform software, along with configuration data developed by the system integrator, forms the equivalent of a target computer for RTCA DO-178B/C* purposes. When a software application is integrated with the

---

* Although at the time of writing this chapter DO-178C has been published, actual application of DO-178C and approval of software for its compliance have not yet occurred.

target computer, it forms a "virtual system." Multiple "virtual systems" can be provided by a single GPM (see Figure 27.2). The distinction between the application "virtual system" in the GPM and an application system in the federated environment is that the application "virtual system" in the GPM is a software configuration item without hardware. This allows the system to be more easily updated or upgraded in the field through a software update. Hardware modifications are not needed to update or add functionality given sufficient GPM resource availability. Each GPM is comparable to one physical system.

The "virtual system" concept extends to the A664-P7 network. The federated system's internal data bus is represented by a network-centric data communications environment. However, with the Genesis architecture, the "virtual systems" share the A664-P7 network as a data transport mechanism. The ARINC 664 P7 concept of virtual links (VLs) provide network transport partitioning for the data messages that originate from the GPM, an A664-P7 interface or a networked RIU I/O element. Each VL is allocated guaranteed network bandwidth (data size and rate) and maximum network delivery latency and jitter.

The "virtual system" concept also extends to the RIU, which is configured to provide I/O services for multiple "virtual systems." The RIU employs temporal partitioning mechanisms through scheduled read/write operations and allows for physical separation between I/O contained within multiple independent fault zones (IFZs) to segregate functional signals as determined by the system integrator. The IFZ boundaries ensure that RIU faults do not affect I/O interfaces outside of the faulted IFZ.

27.1.2 Platform Architecture

The platform consists of the following major components:

- GPMs to support functional processing needs
- RIUs to support system analog and legacy digital interfaces (the working plant)
- Avionics full-duplex (A664-P7) switched Ethernet network for communication between platform elements

These elements, including the quantity of each element, can be packaged in a variety of ways to form a specific physical implementation of the platform. Elements can be packaged as LRUs (sealed units of replaceable equipment) or in module or card form. Modules and cards can then be grouped within cabinets or integrated LRUs that may share common resources such as power supplies and cooling.

It should be noted that a platform could be implemented without GPMs if the intent was to integrate a set of federated LRUs that don't require a shared processing element. For example, when upgrading/architecting a flight deck using systems already developed with dedicated processing, the Genesis architecture can still be leveraged for its shared data I/O and transport. As compared to point-to-point wiring, Genesis's shared data backbone provides a significant wire weight savings. Additionally Genesis's ability to gateway I/O between different protocols/formats eases the integration effort and reduces the need for signal translation boxes.

The Common Core System (CCS) was developed by Smiths Aerospace (company later acquired by GE Aviation) for the Boeing 787 Dreamliner and is a specific implementation of the Genesis platform that groups GPMs and some A664-P7 switches within a cabinet structure. The CCS implements the RIU as an RDC and the remaining A664-P7 switches as LRUs that are distributed throughout locations within the airplane to facilitate separation and minimize wiring to subsystems, sensors, and effectors.

The platform uses an "open system" environment that enables independent suppliers to architect and implement their systems on the platform by complying with industry standard interfaces at all levels within the platform. The key open standards utilized are ARINC 653, ARINC 664 P7, ARINC 429, ARINC 665, and ISO 11898 (CAN bus).

The Genesis platform is an asynchronous system, meaning that component schedules are independent of each other. Each unit internally controls when data are produced; there is no attempt to order operations between units at the platform level. This decouples the elements at the network interface, helping to prevent individual unit behavior from propagating through the system and perturbing the operation of other units. This unit-level independence emulates the federated system environment, producing the same system-level characteristics. Applications must account for this environment in their design, avoiding implementations that depend on synchronous data behavior. Asynchronicity makes the total system more robust by eliminating any dependency on synchronization mechanisms.

Genesis is a configurable resource platform. Functions are allocated with the resources they require to perform their task in the form of sufficient processing time and memory, network I/O communication, and interface resources for analog signals, legacy digital signals, and other digital bus types. These resource allocations are mechanized within the platform through specific predefined configuration tables loaded into each platform unit. The configuration tables represent the resource allocations that are guaranteed to each function to perform its task. These resource guarantees, along with the platform's system partitioning characteristics, form the cornerstone of hosted system independence and, therefore, change containment within the system. These properties allow individual functions to change (i.e., new/modified data traffic or hosted function software) without collateral impact to other functions. This platform characteristic becomes the certification basis for incremental change, enabling unit acceptance for certification at the individual function level instead of the entire integrated system level (all functions hosted on the platform). The certification concept of "incremental acceptance" is documented in RTCA DO-297 "Integrated Modular Avionics (IMA) Development Guidance and Certification Considerations." This guidance document describes one way for a Genesis architecture to be certified.

The Genesis platform architecture represented in Figure 27.3 depicts the building-block approach utilized to scale the platform for a specific implementation. The major building blocks consist of computing elements, network elements, and I/O elements. The platform is scalable by adding (or subtracting) building blocks for each of the resources depending upon the specific needs of the hosted functions for a given set of systems. Utilization of this scalability attribute does not alter the fundamental architecture or operation of the platform and does not impact the existing hosted functions as long as their associated resource allocation guarantees remain intact.

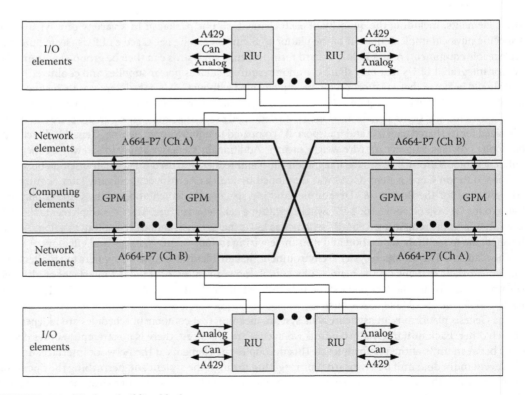

FIGURE 27.3   Platform building blocks.

Figure 27.4 shows a generic architecture for a fully integrated system, and Figure 27.5 shows the architectural implementation for the 787 CCS. The main difference between these two architectural representations is that the 787 CCS implementation minimizes the interconnect wiring between network switches. From a strategic standpoint, the system integrator must weigh the cost of the switch interconnects to the level of flexibility and redundancy afforded to the network communication paths.

The Genesis platform readily accepts integration of other "specialized" or legacy LRUs into the overall integrated system by providing interfaces for the network and other serial digital interfaces for ARINC 429 and CAN bus. This allows the architectural flexibility to integrate "specialized" functions that may require unique hardware.

The platform components are located on the aircraft to provide physical segregation for redundant system elements as well as the network and I/O topology to minimize interconnect wiring. Sensors and effectors can be positioned where they perform best or are needed. The RIU is a remote, "bolt-on" LRU designed for passive cooling. This allows signal conversion and concentration onto the platform communication media to be done in a manner local to the source, negating the need to run several long signal wires to a central location. This not only reduces installation weight but also improves signal quality for analog device interconnect.

27.1.3 Genesis Platform Major Component Characteristics

Certain attributes are required for each major component for the platform to provide system independence and an environment that simplifies implementations of all hosted functions. These attributes dictate integrity, availability, partitioning, and fault-containment requirements for each component. The platform must stay "function neutral" to minimize the cost of change and to maintain the system flexibility. This concept drives certain system architectural decisions concerning how Genesis accomplishes its platform mission.

27.1.3.1 GPM Characteristics

GPMs help to standardize the computing environment through the use of common development processes across multiple systems and suppliers. GPMs also enable a common (central) data load process that enables system functionality to be updated/upgraded through a software update. This can serve to relieve the need to add/modify hardware in the field.

The GPM is an independent computing platform that hosts core software and provides the hosted applications a robust partitioned environment and infrastructure services including I/O, health monitor, and nonvolatile file storage and retrieval based on the ARINC 653 standard. Execution time window, cyclic period, and memory needs, as well as remedial recovery action for process and partition level faults, are allocated to each application and conveyed to the core software through configuration files specific to each GPM. All these allocations are enforced by the partitioning mechanisms designed into the core software and hardware. Likewise, each application specifies its I/O data needs in the form of messages, which form a logical data group of one to many parameters. These groupings represent ARINC 653 data ports, which form the direct communication link with an application.

Application communication can be internal, A653 to A653, or external A653 to network. Internal A653 to A653 communication is memory-to-memory transfer. It is a facility used for GPM-specific application program interfaces (APIs) and when necessary to meet demanding latency requirements for multiple partition applications residing in the same GPM. When an application requests access to an assigned A653 port mapped to the network, the core software executes network end system (ES) interface drivers to read or write the requested message from or to the network ES message buffers. These network message buffers are referred to as network communication ports (com port).

There is a direct correspondence between an A653 port and a specific com port (Figure 27.6). The core software has knowledge of the A653/A653 and A653/com port mapping from the application-specific portion of the configuration file. Both A653 and com ports can be sampling or queuing. Sampling ports

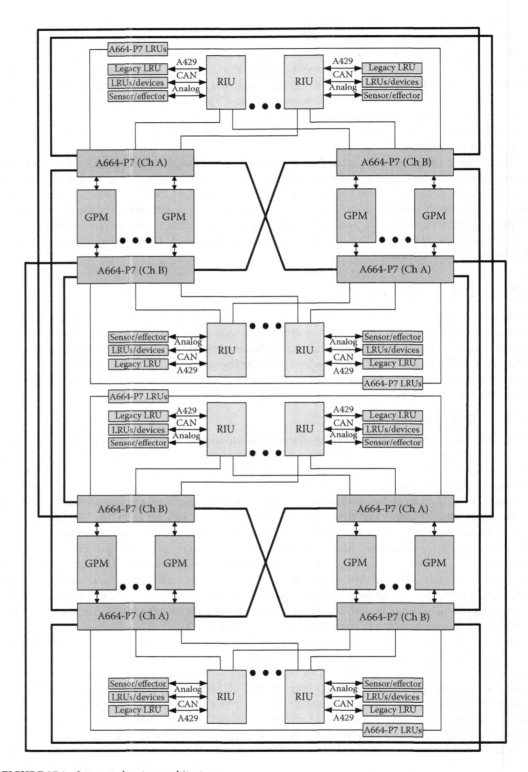

FIGURE 27.4   Integrated system architecture.

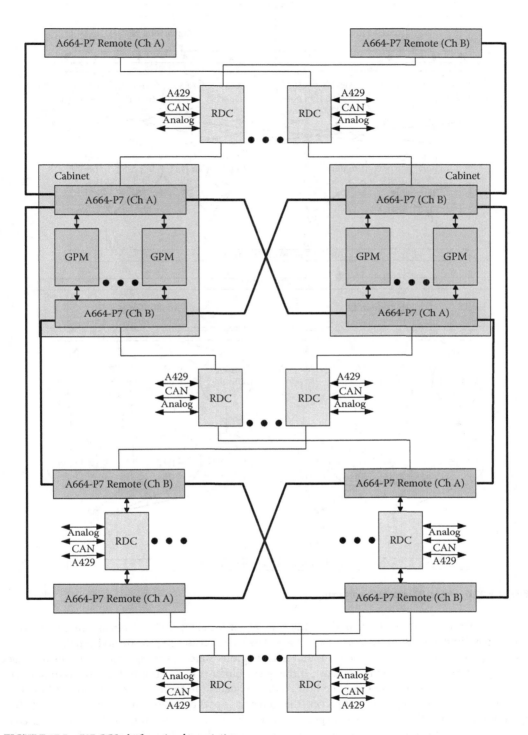

FIGURE 27.5 787 CCS platform implementation.

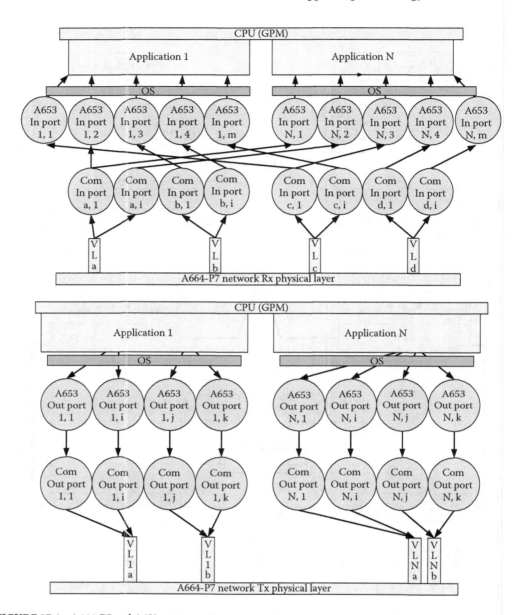

FIGURE 27.6   A664-P7 and A653 port mapping.

overwrite old data with the latest update. The contents of the sampling port remain after being read, allowing unlimited reads of the port. If a periodic data source ceases to provide updated data, then the last receipt of sampling port data remains and is left to grow "stale." Queuing ports aggregate messages in the queue buffer until read by the host. If the queue buffer is filled prior to the host reading it, any intervening messages are lost. Once the host reads the queue, the contents are emptied until the next message arrives. The time required to access the com port is part of the execution time allocated to that application. Most of these network com ports are actually user datagram protocol (UDP) ports, used in conjunction with the Internet protocol (IP) from a protocol point of view. However, applications do not have to be aware of the underlying network protocol. The platform does provide a trivial file transfer protocol (TFTP) library function that can be included in any partition.

One or more com ports from a single application are assigned to a VL, which provides the network transport partitioning for the applications data messages. When data are written to a com port, it is transmitted on the network when the VL is eligible for transmission. Eligibility is managed through the communications mechanism for bandwidth regulation, referred to as the bandwidth allocation gap (BAG) by the A-664 standard. The data message is immediately transmitted if it is BAG eligible and the transmission queue depth is zero. The transmission is delayed if it is not BAG eligible or the transmission queue depth is not zero. Com port data messages are defined by their source application in logical data groupings by virtue of their associated ARINC 653 port definitions. Assignment of com ports to VLs are made by the integrator based on identifying both targeted and general consumers of particular data flows and upon transmission performance constraints.

The GPM hardware is designed to be fail passive, which means there are no single faults that can cause erroneous behavior and that undetected fault sequences resulting in erroneous behavior are extremely improbable (consistent with critical function category requirements). This means that each GPM has autonomous integrity properties, allowing critical functions to be implemented without the need for cross-channel consolidation schemes (voting) between redundant computing elements to achieve the required integrity. Redundant applications may choose to synchronize their functional states using communication links provided by the A664-P7 network. Redundant applications for the computing domain are only necessary to achieve function availability requirements. If a GPM fault occurs, the GPM will fail "silent," ceasing network communication. Data that are sourced by applications contained within the GPM will quickly go "stale" on the network, conveying the application failure to any subscriber of that data. Likewise, a debilitating failure contained within application-specific system resources will result in no data transmission for that application. This is enforced via core software health monitor mechanisms. The fail-silent behavior embodies the cornerstone for system redundancy management and fault-tolerant system behavior. A faulted GPM will reset and attempt recovery, and data transmission will resume only after successful recovery to normal operation.

27.1.3.2 A664-P7 Network Characteristics

The communication backbone of the platform is an ARINC 664 part 7 network (A664-P7) comprised of the network ES hosted in each connecting end node and multiple network switches. The network is arranged in a dual-channel switched star topology with each end node having a redundant, full-duplex point-to-point connection with two independent communication pathways (A and B channels). ESs may only employ a single connection to the network through a single switch if their availability requirement allows. The A/B dual-channel connection allows redundant transmission and reception of data through two independent network paths, ensuring that a loss of communication is extremely improbable. The channel redundancy is managed in the ES, which is contained in each end node host; therefore, the ES only presents a single data stream to the consuming host and only requires a single data stream from the source host. If there is a loss of a redundant data path, the ES continues to stream data from the alternate channel. The loss is essentially a transparent event for the data publishers and subscribers; however, a separate indication for the loss of channel redundancy is provided to the virtual systems that depend on network redundancy for their functional availability.

More than two independent paths are possible through the switched network if the networked unit contains multiple ESs or if there are multiple, redundant networked units. In such a case, each ES could be set up to route traffic through physically separate ARINC 664 switches. Additionally, channel redundancy could be turned off in the ES such that the A and/or B channels are used for independent data transmission. This option may be selected by the system architect if data transport redundancy was not needed for system availability or if a high-integrity system benefited from independent data routes for comparison/voting at the destination.

Figure 27.7 depicts the suggested A664-P7 network switch interconnection topology. Notice that there are two separate redundant channels of communication (labeled as A and B channels). Each switch on a channel is directly connected to all other switches on that channel. This topology of interconnected

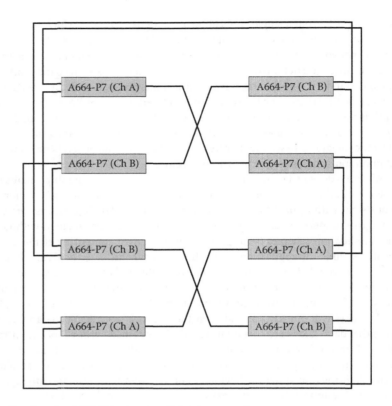

FIGURE 27.7   Topology for A664-P7 network switch interconnection.

switches maximizes the capabilities for redundant network paths and minimizes the impact when a switch is lost. The topology allows the system integrator to best optimize the network routes during initial installation and provides the greatest flexibility for future growth of the virtual systems by minimizing bottlenecks in the network paths.

Each network channel is designed to contain all faults that may occur between Tx and Rx UDP ports, meaning that there are no single faults that can cause erroneous behavior and undetected fault sequences and that erroneous behavior is extremely improbable (consistent with critical function category requirements). This fault-containment behavior can be mechanized using an ES-to-ES integrity algorithm implemented in the network ES, which allows the redundant channels to be used for high availability and ensures that function integrity is not compromised by the network, eliminating the need for application cross comparisons of multiple communication paths to achieve communication integrity. If an integrity algorithm is not implemented in the network ES, then signal integrity can still be achieved through signal cross comparisons at the destination. A dual ES in a synchronized mode can be implemented for full fail-passive ES behavior.

Network data partitioning is enforced using ARINC 664-compliant VLs as depicted in Figure 27.8. The VL is the network transport "pipe" for data packets from one publisher to one or more subscribers, analogous to an ARINC 429 point-to-multipoint bus. VLs have guaranteed allocated network bandwidth (data size and rate) and guaranteed maximum network delivery latency and jitter. Further, the ordering of data packets within the VL is maintained from publisher to subscriber. The network switches route VLs from a configured input physical port to a configured output physical port or ports. During this process, the switch checks to ensure that the size and transmission rate of each VL does not exceed its allocation. If these limits are exceeded, the messages are dropped, thereby containing any data flow

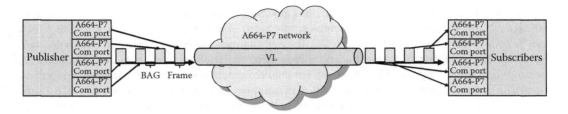

FIGURE 27.8 VLs.

misbehavior and enforcing the network partitioning. The switch has knowledge of these allocated VL attributes through network configuration files.

As shown in Figure 27.9, VLs can be comprised of up to four sub-VLs, which carry messages from their associated com port buffers. The sub-VLs are utilized in order to regulate data flow within the VL by "prioritizing" VL messages through sub-VL groupings. The sub-VL transmissions are serviced round-robin within each VL. The com port buffer represents messages of one or more parameters from a single source. Com ports are the basic data element and are statically configured to be updated at different rates or aperiodically. Messages will only be transmitted when they are written by the source publisher and the allocated VL is BAG eligible, which represents the predefined transmit interval for the group of ports assigned to a VL. If a message is larger than the maximum payload size for a network frame, then the message is segmented into multiple network frames for transmittal. A buffered network frame is limited in size according to the VL-defined maximum frame size, and only one network frame can be transmitted during each BAG interval. BAG is the primary means for regulating data flow through the network and protects the network from babbling sources.

VL priorities can be configured at the switch ports to increase performance for performance-critical messages. In such a configuration, higher-priority messages are routed through the switch port before lower-priority messages. The message delivery performance is exchanged between the VL priority levels, thereby decreasing the guaranteed transport time for higher-priority VL messages and increasing the guaranteed transport time for lower-priority VL messages. The system configuration tools include traffic analysis to ensure that high-priority VLs won't be able to "starve" out low-priority VLs by consuming too much of the switch port bandwidth.

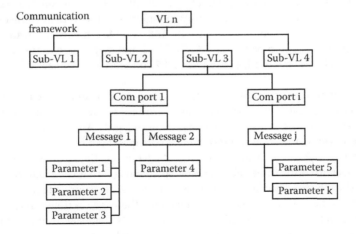

FIGURE 27.9 Communication framework.

The A664-P7 ES can be configured for various modes of operation depending on the host's particular needs. It supports dual-channel communications (referred to as network channels A and B), which can be configured independently for each VL. The A/B channels can be configured for dual redundant operation (identical data on both channels), independent operation (different data on both channels), or single operation (only one channel active). For dual redundant operation (redundancy management turned on), the publishing ES sends the identical message over both the A and B networks and the subscribing ES passes the first valid message received and drops the redundant copy. In this configuration, a loss of either network channel is transparent to the host.

For independent, high-integrity operation (redundancy management turned off), a pair of ESs can be connected such that each ES only connects to one network (one to A and the second to B). A pair of subscribing ESs receive both copies of the message so that a downstream data compare can verify signal integrity. A pair of ESs is required so that a single fault in the ES cannot compromise the comparison. This dual-lane architecture reduces signal availability since a loss of either network channel would result in signal loss (through miscompare); however, it increases signal integrity since erroneous data on a channel would be detected through the downstream compare operation. If an integrity algorithm is not utilized in the network, then downstream compares may be required to achieve signal integrity requirements.

Finally, the network channels can be configured for single operation to support nonessential functions. In this case, only one of the A/B channels is utilized to send a single thread of data, or both channels are utilized to send two separate single threads of data. This configuration reduces wiring but also minimizes the signal availability and integrity and should only be used for nonessential data.

27.1.3.3 RIU Characteristics

An RIU is sometimes referred to as an RDC in industry. The RIU is the gateway between the A664-P7 network, analog and digital devices, legacy A429 buses, and linear CAN subnets. As such, the RIU provides analog-to-digital and digital-to-analog conversion services along with network formatting, range checking, scaling, offset, linearization, threshold, and filter services that are specific for each signal. An analog device interface such as a pressure sensor, valve, motor, or synchro can be comprised of one or many analog signals. Devices such as sensors may require excitation by the RIU. The RIU provides conversions at the analog signal level (also known as electrical primitive) but treats the appropriate grouping of signals as an interface. It has knowledge of the electrical primitives (by pin) assigned to specific interfaces along with the other selected services through configuration files.

The RIU provides a digital-to-digital gateway for A429 and CAN linear subnets. Fundamentally, the RIU accumulates bytes received from these buses in buffers specific to each connected bus and maps these data to specific A664-P7 com ports within its A664-P7 ES. Similarly for transmissions, the RIU retrieves bytes from specific com ports in its A664-P7 ES and maps these to transmit buffers for each connected bus. The RIU has knowledge of the A664-P7 com port and connected bus mapping through configuration files.

The primary partitioning method for the RIU I/O complement is to provide multiple I/O IFZs using physical separation to segregate functional signals as determined by the system integrator. To that end, each RIU contains several I/O IFZs that share a common high-integrity A664-P7 ES interface. The A664-P7 ES interface is designed in a manner to prevent cross corruption between IFZs that share the interface, allowing the system integrator to ensure segregation between functional I/O signals by grouping function-specific signals within the same IFZ. Gateway operations (both analog and digital) are performed on a repetitive cyclic schedule specified in the configuration file. The allocated schedule (or update rate) is specified for each interface and each connected digital bus. There is a maximum rate for any gateway service defined for the architecture. The interface schedule consists of a major frame within which there are several minor frames. The duration of both the major frame and the minor frames is a configurable fixed period. Each minor frame consists of a number of scheduling blocks (SBs) that provide I/O processing, which means that all SBs and associated I/O will have a guaranteed period within which they will be run.

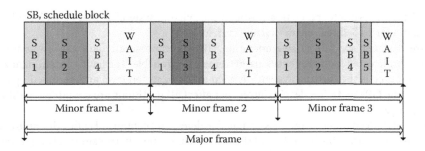

SB, schedule block

FIGURE 27.10 Major/minor frame scheduling policy.

Each minor frame will be implemented in such a way that a degree of free time (wait function will loop until a given time) will be allocated at the end (see Figure 27.10). This has a twofold advantage. First, it ensures that SBs cannot overrun into the next minor frame. Second, depending on the amount of spare time that has been allocated, the free time allows for a certain amount of future expansion.

The RIU maps between interfaces and A664-P7 com ports. The mapping granularity can be allocated to parameters within the com ports or the entire com port depending on the specific interface. Analog signals always map at the parameter level. ARINC 429 interfaces always map at the com port level and CAN interfaces map at either level. The RIU has knowledge of this mapping through unique configuration files loaded into each RIU.

The RIU architecture is comprised of two distinct integrity regions, one that is shared across all resource users and a second that provides independent resources that can be used on an individual user basis. The shared region is designed such that either there are no single faults or fault sequences that can cause undetected erroneous behavior are extremely improbable. The independent resource region is designed such that undetected erroneous behavior is improbable. The importance of this architecture is that it prevents faults in the common resource from causing simultaneous malfunction of the independent resources and the functions that use them. This architecture eases the safety considerations associated with resource allocation of the RIU complement and cross corruption of functions. Each individual RIU output is forced to a safe state in the presence of a fault or absence of a valid command for that interface. Likewise, a detected fault in any analog or gateway input interface is tagged as invalid data in the transmitted network packet. RIU service guarantees are enforced by an independent safety monitor that uses a separate time base from the processor. The safety monitor must be "kicked" by a keyword from the processor within a time window; otherwise, the outputs will be put in a safe state via a hardware mechanism.

27.1.4 Platform Integrity and Fault Containment

Platform integrity is characterized by the following properties:

- Redundancy
- Fault isolation mechanisms
- "Virtual system" partitioning
- Fault-containment zones

As part of the Genesis architecture, redundancy is provided for all physical components of the system to meet the fault tolerance objectives for the hosted function systems. The fault isolation philosophy is to contain faults at the Genesis "virtual system" boundaries, allowing redundant elements to continue with a seamless flow of data. The concept of a "virtual system" describes the architectural feature in which a function is contained within a portion of a larger physical system. Multiple "virtual systems" can reside within one physical system. The "virtual system" is isolated through temporal and spatial partitioning mechanisms as opposed to physical partitioning mechanisms that define the boundaries of a physical system. An integral characteristic of the "virtual system" is that faults must be contained within the boundaries defined for the "virtual system."

GPMs are designed for fail-passive operation. There are no single fault exposures for erroneous behavior, and undetected fault sequences are extremely improbable. The fault-containment zone for the computing resource is the "virtual system" defined at the partition level and bounded by the physical boundaries of the GPM. Module-level failures that affect all partitions are contained within a GPM as shown in Figure 27.11.

Since each network channel (A and B) is designed to detect and contain all faults, the fault-containment zone for the network forms the boundary of the A664-P7 UDP port in transmitting and receiving ES for each channel. Most VL transmission faults are isolated at the switch boundary, and all faults are isolated at the receiving A664-P7 ES port level boundary. This prevents data corrupted during transport within the network from propagating to the host, as shown in Figure 27.12. To accomplish this, the ES manages the redundancy between the A and B network channels and only presents the subscriber with a single piece of data.

However, if the level of network fault detection provided by the A664-p7 implementation is not sufficient to meet a signal integrity requirement (i.e., integrity algorithm is not implemented in ES), then "redundancy management" can be turned off and a pair of ESs can be used to deliver the signal through independent paths for comparison at the destination. This form of signal integrity (integrity at the destination) reduces signal availability since a loss of data at either network channel would cause the comparator to fail at the destination. Therefore, integrity at the destination should only be utilized if the A664-P7 fault detection (integrity at the source) is not implemented to fully support the signal integrity requirements. As with any system, the architect must trade integrity with availability since they can be opposing design goals.

The I/O domain (RIU) leverages the fact that sensor/effectors and A429 buses are not (generally) high-integrity devices and require DC to achieve high integrity, so any major essential or life-critical system requires multiple independent channels for I/O. The RIU architecture provides multiple IFZs within each RIU. Each IFZ provides fault detection for internal faults, which is generally an order of magnitude better than the sensors/effectors they interface to the system. However, the design of the shared resources within the RIU prevents fault propagation across multiple IFZs. Containment of all I/O faults, as in a federated system architecture, is still incumbent upon providing physically separated signals through IFZs. It should also be noted that even though there are multiple IFZs within a single RIU, the signals may need to be connected to physically separate RIUs to support redundancy across multiple safety zones of the aircraft. For example, the architect may need to guard against events such as blade or bird strike that could take out a single RIU and thus all of its IFZs simultaneously.

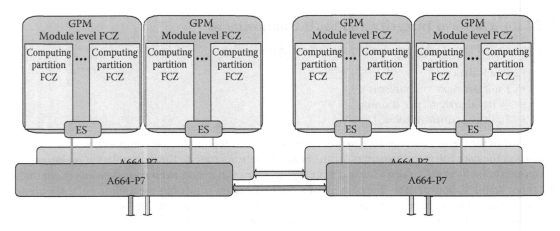

FIGURE 27.11 Computing FCZs.

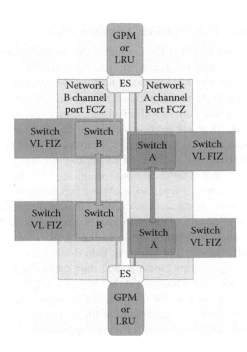

FIGURE 27.12 Network FCZ.

Consuming applications can use downstream consolidation (DC) methods to detect I/O faults, including the remaining small percentage attributable to the RIU. This is shown in Figure 27.13.

This fault-containment scheme has implications for those hosted functions (essential systems) that do not require multiple-channel I/O. These systems require sufficient system segregation at the I/O level to preclude misbehavior of multiple essential systems due to an undetected fault in a single RIU. To that end, the RIU complement provides IFZs to achieve segregation of essential system I/O.

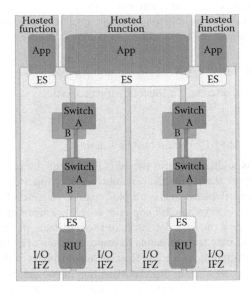

FIGURE 27.13 I/O FCZ.

27.1.4.1 How a Hosted Function Identifies a System Fault

The platform fault-containment zones form the foundation for communicating platform faults to hosted functions. Computing, network, and I/O faults are "seen" directly by hosted functions through platform component "fail-silent" attributes. If a platform fault is detected, published data will stop updating within the A664-P7 com ports or dataset functional status will indicate a dataset-specific fault. Each hosted function can only recognize faults applicable to its subscribed data. This allows each function to determine its own functional response to the data failure. The A664-P7 ES reports network A and B channel status directly to hosted functions as header information associated with each com port. For hosted function effector outputs, the RIU returns an output status message that includes output monitor values and network redundancy status for the output command. This allows hosted functions to monitor their analog/digital output interfaces and a view of their communication redundancy.

Undetected erroneous behavior must also be addressed for critical signals that are required by hosted functions to avoid hazardous or catastrophic failure conditions as defined by SAE ARP-4761. If a fault within the Genesis platform can go undetected (i.e., network Integrity algorithm not implemented), then the subscribing system couldn't rely on the fail-silent behavior for the offending component. Instead, the system would need to set up multiple, independent data paths through the Genesis platform and then complete a data comparison operation within the subscribing LRUs (integrity at the destination). Example hazards that may drive such signal integrity requirements include erroneous display of air data on all primary displays or erroneous command of flight controls. Common mode faults must be addressed when critical hosted functions such as these are allocated to the Genesis platform. Introducing dissimilarity in the hardware architecture is the most common method of addressing common mode failures. An alternate approach would be to demonstrate that each Genesis component is processing a different set of signals according to asynchronous schedules; therefore, a common fault couldn't affect all units simultaneously (similar to approach of hosting dissimilar software on common hardware).

The majority of hosted functions should be able to rely on the Genesis platform's fail-silent attributes for fault detection. The example platform fault scenario shown in Figure 27.14 depicts the platform fault indications used by hosted functions to determine their functional health if a GPM fails:

- All data published by partitions within the GPM fail silent (stops transmitting).
- This is seen by subscribers of data as A664-P7 Rx com ports going stale as determined by the difference in ES receive time compared to current ES time.
- Each subscriber determines how long to wait before declaring data too old.
- ES receive time is contained in each A664-P7 Rx com port header.

27.1.5 Platform Fault Tolerance

Platform fault tolerance is characterized by the following properties:

- Redundancy
- Fail-silent behaviors
- Data consolidation (at A664-P7 ES and hosted function)

Functional fault tolerance is architected into the "virtual system" by allocating sufficient redundant functional elements to support the required availability of that function. This can take the form of multiple copies for processing elements and multiple "virtual" channels (IFZs) for I/O elements. The consuming functions are provided redundant copies of input data from multiple independent sources. As mentioned in earlier sections, the fault detection and containment philosophy used by the platform is for the source to cease valid data transmissions in response to uncorrectable hard or soft faults. This characteristic becomes a built-in data validity indication to consumers of that specific data because the receiving port stops receiving "new" data. Receive-time tagging of the message ports allows the consumer to determine the freshness of data and decide when a source has gone "invalid." This allows the consuming function to select a source or sources based on its view of received data validity.

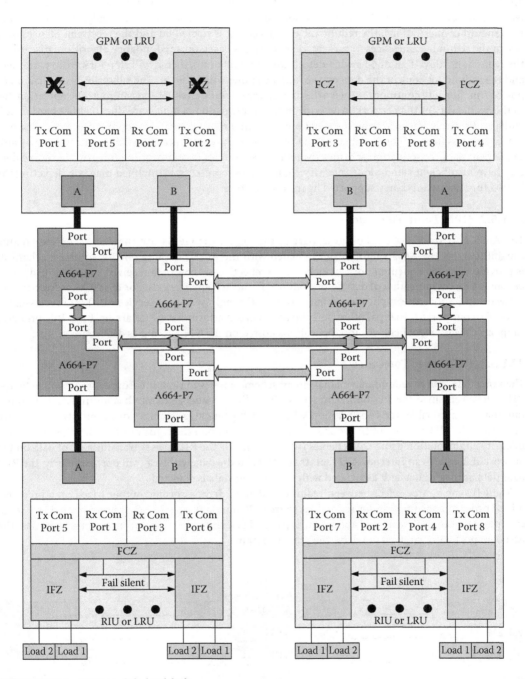

FIGURE 27.14 GPM module-level fault.

27.1.5.1 ARINC 664 P7 Network Fault Tolerance

The interconnecting platform network is a dual-channel arrangement that the network ESs manage. Network messages are typically configured for redundant transmission on each independent network. In standard configurations, the redundancy management is turned on and the receiving ES uses only one of the redundant messages to present to applications based on a redundancy algorithm, discarding the redundant copy. If an A664 switch detects an internal failure, then it will stop forwarding messages and the redundant copy of the data on the failed channel will go silent. The subscribing application or hosted function will continue to receive the data on the working channel and can continue operation due to the availability of the network. Any error in the transmission of a message due to either hard or soft faults is effectively screened from the functional system. From a systems standpoint, the data consumer sees a seamless flow of "good" data until the aggregate redundancy is exhausted. Applications may also implement data consolidation schemes for redundant sources to augment the integrity of source data that have insufficient stand-alone integrity. Figure 27.15 represents a simplified diagram depicting the hosted function redundancy supported by the architecture.

27.1.5.2 GPM Fault Tolerance

The A653 port read services hosted as part of the OS on the GPMs use the A664-P7 time tagging capability to set the A653 validity flag associated with the A653 sampling port. A similar mechanism is provided for A653 queuing ports, which will be reflective of "new data received since last read." This becomes a direct indication of data source failures for hosted applications. For the GPM, infrastructure software will only write application-requested data (through A653 port service calls) into the assigned A664-P7 com ports if that partition and the associated platform hardware are healthy. This property supports the fail-silent behavior for applications hosted on the GPM acting as data sources.

27.1.5.3 RIU Fault Tolerance

The same data source redundancy management scheme applies to signaling sources represented by the RIU. However, since the I/O hardware string (including sensor) generally has an undetected failure rate that is improbable, the consumption scheme for higher-integrity functions typically is to receive data from all valid redundant I/O sources. It is incumbent upon the consuming application to provide integrity augmentation using DC schemes (such as voting). The RIU ceases transmission of data on the network if it detects an internal RIU fault that affects all the data within a com port. Similarly, the RIU flags data as invalid if a fault associated with that specific data is detected.

The RIU can source-select between redundant data to drive a specific output based on a predetermined order of consumption for redundant sources. Redundant source data are represented as a set of com ports that the RIU will read and use in a specified priority order. The consumption scheme uses the first com port with valid data to drive the specific output.

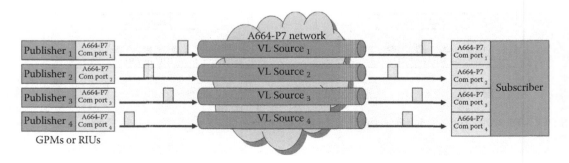

FIGURE 27.15   Generic redundancy architecture.

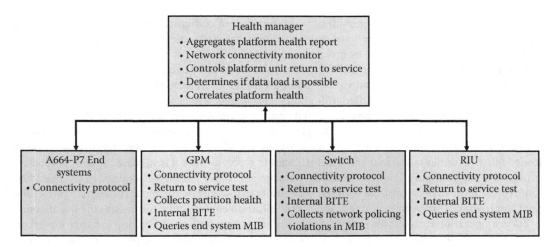

FIGURE 27.16   Genesis health management.

27.1.6 Platform Health Management and Fault Isolation

The Genesis platform health manager provides the following services:

- Provides platform fault isolation
- Monitors connectivity of all platform and nonplatform LRUs
- Communicates configuration status to a higher-level health manager
- Controls return to service tests
- Determines when data load of line-replaceable modules (LRMs) and LRUs is possible

The platform has a built-in health manager function, which is implemented as an A653 application on the GPMs. This function aggregates and reports the health of the platform and its components (Figure 27.16). Based on its internal built-in test equipment (BITE), each platform LRU periodically reports its health to the platform health manager. The platform health manager also monitors the connectivity of all LRUs interfaced to the network using a simple request and response protocol. The health manager controls return to service tests (initiated BITE) and data load of platform units to ensure the integrated system can remain operational during these maintenance actions. The platform health data are correlated by the Genesis health manager to provide definitive fault isolation for the platform, supporting both production build and operational maintenance. The health manager also indicates the dispatch readiness of the platform resources based on a predefined minimum equipment list for the platform unit complement, indicating any platform failures that would prevent safe dispatch.

27.1.7 Platform Configuration Management

The Genesis platform configuration management function provides the following services:

- Validates hardware and software compatibility
- Provides consistency check for hosted function software
- Communicates configuration status to a higher-level configuration manager
- Generates a part-number report for maintenance personnel

The platform has a built-in configuration manager, which is implemented as an A653 application on the GPMs. This function ensures that the installed platform unit-level hardware and software are compatible with the installation and with each other. The compatibility levels that are validated by the configuration manager are unit-level and system-level compatibility as well as hosted function software consistency.

The configuration manager also provides communication of configuration status to a higher-level aircraft configuration manager as a means to provide an "all go" indication or identify specific configuration problems for correction by maintenance action.

Unit-level compatibility is performed internal to a platform unit by comparing unit-specific hardware and software version numbers. The version numbers do not reflect specific part numbers but rather groups of elements that are compatible. This allows interoperability of compatible part numbers within the platform so that benign unit modifications do not force massive retrofit of platform elements.

The primary means used to confirm the platform system configuration is a platform manifest. A manifest is a list of software numbers intended for a specific installation. The platform manifest is loaded into the configuration manager and broadcast to each unit after it is validated against hardwired installation program pins. Using the loadable software part (LSP) numbers contained in the manifest, units confirm they are compatible with the other units that comprise the platform and are proper for their installation. This ensures interface and communication compatibility for the specific installation.

The configuration manager also provides a consistency check for hosted function software loaded on each GPM, ensuring that each of the redundant copies is consistent across the platform. This is accomplished by instrumenting GPMs to report the LSP description and associated LSP number for each hosted function application they contain. The configuration manager then compares LSP numbers for the same LSP description.

All platform units periodically report all hardware and LSP numbers. The platform configuration manager removes redundancies and consolidates an LSP numbers report for maintenance display. This report communicates any existing configuration faults in a manner that allows easy correction of the configuration problem by maintenance personnel.

27.2 Functional System Implementation on the Platform

27.2.1 Architecting Systems on the Platform

The process for architecting a system on the Genesis platform is as follows:

- Define the aircraft functions that comprise the system.
- Define the system elements required to implement each aircraft function.
- Define the resource usage demands for each system element.
- Define the data exchange between aircraft system elements and functions.
- Optimize the logical architecture for the platform.
- Allocate the aircraft system elements to the platform to define the physical architecture.

The first step in synthesizing an integrated system to be hosted on the platform is to define the aircraft functions that comprise the system. From this function list, system elements emerge as processing applications, I/O needs, and special equipment (non-Genesis platform elements). Attributes are then defined for each of these system elements such as availability, integrity, and performance needs. From these attributes, a "logical" system architecture is formulated that is used to scale the platform to determine the number of platform elements required to host the integrated system. The next step is to define the data exchange between system elements and other aircraft systems. The definition of the functional communication forms the basic logical communication structure within which actual network messages and datasets are organized. The logical communication links define the VL framework for the platform network (Figure 27.17). This framework is associated with functional system elements, and when those elements are allocated to physical resources, the associated communication link structure follows the functional assignments to form the physical system architecture. By necessity, the communication link structure must be defined by the interfacing functions because they best know how their system exchanges data with other systems and the sensor and effector complement of the system. The optimization of the network resource is extremely dependent upon how this structure is defined, thereby

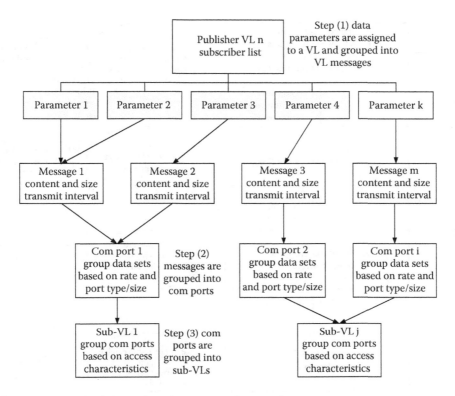

FIGURE 27.17 Process for defining the communication framework.

keeping system-specific targeted communication isolated from more general purpose multiple-user data. The task of structuring the communication framework is a top-down process that identifies the primary data exchanges between publishers and subscribers, the data content required for the exchange, and the port structure that best mechanizes the framework (see Figure 27.18).

The network configuration process begins with definition of the communication link structure, which is performed from a systems view, much like defining A429 connections between LRUs in a federated system approach. Each system must identify its associated publish communication link structure with a view of the subscribers and the data they need. The integrated system architect starts by identifying targeted user links, multiple-user links, gateway in links, and gateway out links. Every system should start with these four categories, adding within a category or removing categories as the system need dictates. A maximum latency requirement is defined and associated with each link.

For each communication link defined in the communication structure, the message elements need to be identified at a high level. For example, a VL connecting an aircraft FMS to a display unit may contain some periodic parametric data, such as position, velocity, time, wind vector, and a periodic file, such as an A661 navigation display file. As part of this high-level definition, update rates and estimated data sizes must be defined.

The actual com ports and characteristics can be defined from the message framework. In order to determine optimal bandwidth usage, the system architect separates periodic data into sampling and queuing update rate groupings, combining groups only if it is a lower bandwidth requirement to combine a group. This analysis compares the com port overhead penalty to the additional bandwidth usage that is created by adding lower-rate parameters to a higher-rate com port. In general, file transfers and A429 should be set up as queuing ports, with parametric data being set up as sampling ports. This process will result in a set of sampling and queuing ports defined for each VL structure.

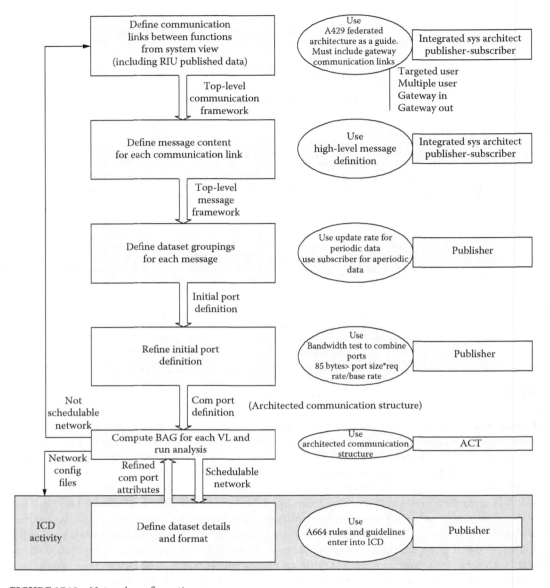

FIGURE 27.18 Network configuration process.

27.2.2 Integrating Federated Systems Using Genesis Platform

A Genesis platform can be implemented without the use of GPM elements. If an aircraft is being architected using federated systems that contain their own dedicated processing function, then GPMs may not be required. In this case, the shared data transport and interface gateway functions provided by the A664-networked RIUs remain a powerful platform for integrating federated system LRUs with dissimilar interfaces. The shared network would replace the dedicated point-to-point wiring that is traditionally used between federated systems. The shared data network greatly reduces wiring within the aircraft, thus saving weight and installation complexity. In this way, the Genesis architecture remains relevant for cockpit upgrade programs or new aircraft programs that utilize traditional federated LRUs instead of porting their processing over to an A653 environment within a GPM.

The same process is used to implement systems in a Genesis architecture that doesn't include GPMs. Functionality would not be allocated to a GPM application, but the data-level integration tasks would remain as described. Since the Genesis platform health management and configuration management applications were allocated to a GPM, an alternate solution would need to be provided in the absence of GPMs. The Genesis components would continue to participate as a member system in whatever approach is selected for the health and configuration managers.

27.2.3 Architecture Configuration Toolset

The Genesis platform is a configurable set of infrastructure resources to support the implementation of aircraft functions. An unconfigured platform provides no functionality; each platform component would sit idle and no resources could be utilized. The configuration of the platform is the "glue" that binds the system together. The configuration is not only responsible for defining the system interfaces but also orchestrates the system performance. The system integrator's resource allocation guarantees are established by the platform configuration. The task of optimizing the system resources and the pledge of assurance for the validity of generated configuration files are integral responsibilities that should not be underestimated. The Genesis architecture relies heavily on DO-178B/C qualified tools to perform the configuration and associated system analysis processes. The configuration and analyses of an integrated avionics environment are large and complex. If this process was not supported by a qualified set of tools, then extensive manual verification checks and manual analyses would be required for both the initial configuration and subsequent changes.

Once an integrated system logical architecture is defined, the logical architecture is allocated to the scaled platform architecture to form the physical architecture for the integrated system. To support this system allocation, an architecture configuration toolset (ACT) is provided that allows graphical representations of the architecture resources to be created (see Figure 27.19). The integrator graphically structures the physical instantiation of the architecture using building-block system elements and allocates these resources to specific system elements in the form of applications and their associated communication

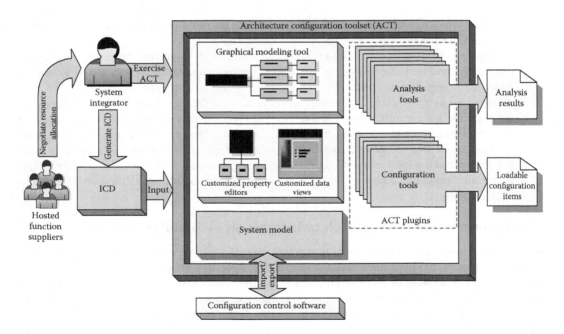

FIGURE 27.19   ACT.

structure. Since this can consist of a lot of data, the tools can also be interfaced to an interface control document (ICD) file format to allow mass input of data for later viewing through the graphical interface. The graphical data view is especially helpful in the data validation and debug activities.

Definition of the hosted functions and allocation of the platform resources to those functions is necessary to achieve successful functional integration while ensuring efficient use of the available resources. The ACT is designed to aid in defining, configuring, modeling, and analyzing the integrated system. Hosted function resource allocation model elements are updated, through the use of the toolset, to reflect the current state and understanding of the functions, ensuring a complete and updated system model is maintained for comprehensive allocation of platform resources. From this system model, the appropriate platform element configuration files are produced to implement the integrated system.

The basic functions of the ACT include the following:

- *Defining the physical architecture*—This allows the user to graphically construct the physical architecture including platform elements and system unique elements and allows overlay of the logical architecture onto specific resources.
- *Defining I/O messages*—The toolset allows the user to define and organize the I/O messages used in the communication of data between the system elements and other systems.
- *Analyzing Genesis performance*—For each hosted function, the toolset calculates the system performance relative to the maximum end-to-end response time and compares this to the end-to-end requirements for latency of the I/O signals as configured by the integrator.
- *Generating a model of integrated system*—Use analysis tools to evaluate the physical architecture definition and generate an integrated system model that simultaneously satisfies the guaranteed resource definitions (resource quantity and performance) for all hosted function systems.
- *Defining the Genesis configuration*—The toolset allows the user to generate configuration files for all platform components: GPMs, A664 network switches, A664 ESs, and RIUs. A GPM configuration file defines its software partitions and communications ports. Network configuration files define how the network data flow is controlled between publishers and subscribers and how bandwidth is allocated on the shared network. An RIU configuration file defines the conversion operations the RIU will perform on the I/O signals and the gateway interconnect for serial digital buses (A429, CAN).
- *Managing configuration*—The toolset has a configuration management capability for storing, cataloging, and securing versions of the system configurations, I/O messages, and associated analysis and requirements data.
- *Assisting the user in making incremental Genesis changes*—Changes to the system configuration are expected not only after entry into service but even during the integration and test phases of a development program. Many changes will be limited in scope where only relatively small changes are needed in one or a small set of hosted functions. The toolset is designed to assist the user in making incremental Genesis changes and performs a "difference" of all the configuration files, providing a change-impact assessment to validate limited verification for the new configuration.

27.2.4 Contract-Based Approach to Integration of Modular Systems

The Genesis platform is a modular architecture that utilizes an "open system" environment to allow independent suppliers to implement their aircraft functions on the platform. This developmental independence and partitioning of responsibilities provides the system integrator, platform provider, and hosted function supplier with an efficient project organization to separately complete their designs, integration, verification, and certification efforts. While a degree of developmental independence exists, the dependencies that remain must be properly recognized and managed accordingly to ensure a successful integration.

The hosted function uses the platform architecture and, as a result, is ultimately dependent upon the resources provided by the platform. The dependencies can be characterized through a contract-based approach for integrating the hosted functions with the platform.

The "contract" between suppliers of the platform and the hosted functions forms a formal mode of communication that describes what resources or services can be guaranteed to the hosted functions. The objective of the contract is twofold. First, it communicates (and guarantees) the platform properties to the developers of the hosted functions. The contract provides a formal record of platform properties that the developers can rely upon when developing their own systems that depend upon the system or architecture. The contract needs to document the safety aspects of the platform such as the availability and integrity of the platform components (likely in the form of SAE ARP 4761 compliant FHA and PSSA/SSA). The contract also needs to document the resource allocation guarantees (likely in the form of an ICD). Second, the contract forms a piece of certification evidence. Since the hosted functions are allocated to the platform resources to form a "functional" architecture, each hosted function must identify their dependence upon the specific claims of guarantees in their safety and certification cases submitted to the certification authorities. For example, this can be achieved through references to the platform safety analyses within the individual system safety analyses and through system requirement traceability to the Genesis platform ICD. The point is that certain documentation must be shared between supplier organizations so that proper references can be made in certification deliverables. This "sharing" of certification artifacts is a necessary action within an integrated avionics architecture and can affect how documents are structured so that proprietary data are not needlessly shared between organizations.

Multisystem integration methods are not a new concept, but due to the heightened degree of integration in the Genesis architecture, it is more important to formalize detailed communications between system providers than in the integration efforts for traditional federated systems.

27.2.4.1 Contract Roles and Responsibilities

The basic roles and responsibilities for the contract are split between the system integrator, the platform provider, and the hosted function supplier. The system integrator documents the platform resource allocation guarantees within a platform ICD, for example, processor time allocation, memory allocation, communication bandwidth, and I/O allocation. These allocations form guarantees for the hosted function supplier. The platform provider documents the platform safety guarantees (regardless of specific ICD allocation) in their contract formed by their safety documents (SAE ARP 4761-compliant FHA and PSSA/SSA). The platform provider is then responsible for verifying their safety guarantees and the qualified platform tools (ACT) that are used to validate the resource allocation guarantees defined by the system integrator in the platform ICD. The platform provider must provide their verification evidence to the certification authorities, but delivery to the hosted function supplier is not necessary. The hosted function supplier accepts the contracts listing the claims of guarantees (i.e., platform safety documents and ICD) and retains it as a formal record on which to base their platform assumptions and certification arguments. In lieu of providing verification evidence of the platform behavior, the hosted function supplier is able to reference the platform contract data in their certification and safety arguments. The hosted function supplier only has to verify the function and performance of their hosted function per their function requirements. The system integrator maintains the burden of providing evidence for the integrated system that includes all hosted functions that were verified independently. This will include top-level safety documentation (SAE ARP 4761 compliant FHA and PSSA/SSA), which considers cumulative loss of (and undetected erroneous behavior for) each Genesis platform component that hosts multiple hosted functions. Even though individual hosted function safety analyses show loss of their function as only a major event, the combinatorial loss of that function along with other cohosted functions may lead to an elevated safety event such as hazardous or catastrophic. For example, consider if the aircraft's primary displays and standby displays were hosted by the Genesis platform. The individual loss of either system would be less of a safety effect than if both were simultaneously lost due to a single platform failure (which would be catastrophic). This catastrophic condition needs to be mitigated through appropriate platform allocation activities, and the proof needs to be documented in the top-level safety analyses that consider all hosted functions together on the platform. While the system integrator technically owns the platform ICD, each individual hosted function supplier is responsible for verifying their hosted function meets all requirements in the context of the platform ICD. The verification evidence for an individual hosted function is valid since the platform provider provides separate evidence to support the partitioning mechanisms that keep a system from being unintentionally impacted by another hosted function. The system integrator is likely not responsible for any formal verification of the ICD, other than that embodied by the top-level safety analyses. Some "feel good" integration tests will likely be performed by the system integrator, but those are not submitted for certification evidence. A comprehensive test at the fully integrated system level is not feasible due to the combinatorial explosion of test cases related to the large number of inputs and outputs that are simultaneously possible across all hosted function systems. That said, this responsibility should be negotiated and clearly defined with your certification authority.

27.2.4.2 Using Contractual Guarantees to Form Safety and Certification Arguments

In order to minimize the costs of change for the integrated system, it is important to form separate safety and certification arguments for each hosted function. This does not alleviate the need for a top-level argument for the integration of the individually argued systems, but the complexity of the top-level argument is reduced. If the safety and certification arguments made for the integrated system do not reason at the individual hosted function system level, then the system-wide arguments will be very complex and a change to any hosted function would require the entire argument to be reevaluated and revalidated.

An important aspect of the Genesis architecture is that the platform is provided independent of the hosted functions that will eventually utilize it. Therefore, the arguments for platform safety and certification are independent of any specific functionality that is formed when a hosted function is integrated with the platform. While the platform can be developed in isolation, the hosted function supplier must also consider its dependencies upon the platform. The system integrator is then left to reason about the integration of the all the hosted functions together upon the platform.

It can be shown that the approach to segmenting the safety and certification evidence into separate, individual sources of evidence leads to a complete certification argument when considered as a whole. The logic used to form these safety and certification arguments is based upon a compositional reasoning approach referred to as assume–guarantee or rely–guarantee reasoning that is commonly utilized for modular systems in the domain of computer science. This type of reasoning can be visualized as shown in Figure 27.20, where X represents a hosted function system that is integrated with platform Y.*

System Y (platform) must guarantee its resources and services, and system X (hosted function) must assume that the guaranteed resources and services hold true. Since system X depends on system Y, system X can only hold true when the guaranteed properties of Y hold true. However, the converse is not true; system Y does not maintain any dependence upon system X.

FIGURE 27.20 Visualization of the concept of compositional reasoning.

* J. Rushby. Modular Certification. NASA Contractor Report. CR-2002-212130, NASA Langley Research Center, Dec. 2002.

$$\frac{\left\langle P_1 \right\rangle Y \left\langle P_2 \right\rangle \quad \left\langle P_2 \wedge P_3 \right\rangle X \left\langle P_4 \right\rangle}{\left\langle \text{true} \right\rangle Y \parallel X \left\langle P_1 \wedge P_2 \wedge P_3 \right\rangle}$$

FIGURE 27.21 A formal representation of compositional reasoning for two systems.

This reasoning is formalized in Figure 27.21, which denotes the logic as <p>Z<q>, asserting that if Z is a system where *p* holds true, then the system must satisfy q. This notation is applied to our example, thus stating that if we assume the properties of P_1, then Y guarantees P_2. Likewise, by assuming the properties of P_2 and the additional properties P_3, X guarantees P_4. This form of reasoning allows the system providers to reason separately about their systems Y and X, and then the system integrator is able to deduce properties about the composition of Y and X, denoted Y||X, given the properties < P_1 and P_2 and P_3 >. The properties P_1, P_2, and P_3 represent guarantees that must be made (and shared) between system providers so that the arguments for safety and certification at the integrated system level can be completed.

27.3 Certification Aspects

27.3.1 Certification Approach

The Genesis architecture presents some philosophical differences for the certification approach as compared to traditional aircraft certification approaches. The first difference is that the integrated system is organized into modules (including platform and hosted function modules) that are incrementally accepted for certification per the guidance defined in RTCA DO-297. Each module requires its own set of design assurance, safety, and certification data. The hosted function and platform modules are submitted for acceptance by the certification authorities prior to the system integrator's system-level and aircraft-level submittals. Once all modules have been submitted for acceptance, the integrated system certification can be sought under a system-level certification argument that is based upon the individual arguments for module acceptance. The certification process of incremental acceptance can be characterized by six distinct phases*:

- Phase 1: Platform acceptance
- Phase 2: Hosted function acceptance
- Phase 3: Integrated system acceptance
- Phase 4: Aircraft integration and certification with Genesis system
- Phase 5: Incremental change of platform or hosted functions
- Phase 6: Reuse of platform or hosted functions

The second philosophical difference under the Genesis architecture is that hosted functions are certified according to a logical architecture rather than a physical architecture. The technical merit of this approach is justified through the contract-based integration mechanism provided by the platform guarantees. A hosted function's arguments supporting acceptance for certification are based upon a guaranteed set of resources provided by the platform elements. The guarantees describe the logical architecture provided to the hosted function. The physical architecture can be manipulated without disturbing the certification case as long as the modified physical architecture is shown to satisfy the existing platform guarantees made for the logical system definition of the previously verified hosted function. For example, an application can be moved from one GPM to another without necessarily affecting the existing certification artifacts for that hosted application. The system integrator would need to update their

---

* Similar phases are referenced RTCA DO-297, entitled "Integrated modular avionics (IMA) development guidance and certification considerations."

certification artifacts to document the new system integration solution. This would include updates to the safety analysis and confirmation that the functional performance guarantees (i.e., data transport latency) are still met by the new system definition. However, the certification artifacts at the individual hosted function level could remain unchanged.

27.3.2 Minimal Cost of Change

Cost of change is the cost (time, effort, and money) incurred when there is a change in a hosted function or a change in a platform component. The cost-of-change concept is especially important in the Genesis architecture due to the high degree of integration between systems. Without the appropriate system design and certification approach, the changes in a hosted function could cause dramatic and unnecessary costs for the suppliers of hosted functions sharing the platform resources, the system integrator, the end users, and even the changed hosted function. Similarly, updates to platform components (hardware or software) due to technology refresh or obsolescence could cause significant cost impact to hosted functions. It is important to acknowledge that the benefits realized by the Genesis architecture can be outweighed by high costs of change if change is not managed appropriately in the architecture implementation.

The following are targets for minimum cost of change:

- Eliminating costs associated with unaffected hosted functions when changes are made to affected hosted functions (the primary objective of reducing cost of change)
- Limiting costs related to scaling, extending, and upgrading the platform
- Limiting impacts on unaffected platform elements when technology refreshes occur
- Incrementally accepting platform or hosted function components for certification

In terms of design, the platform implementation must support computing robust partitioning within the GPMs, abstraction of software applications from computing hardware within the GPM, network partitioning, I/O update rate segmentation, and qualified configuration mechanisms and tools.

In terms of certification, the platform certification arguments must support incremental acceptance and logical architectures. If the certification arguments are based upon a physical architecture, then any changes to the physical architecture (even those that do not disrupt the logical architecture) will require the certification arguments to be revalidated. Consider the example of moving an application between GPMs. In this case, the cost of change would essentially be limited to (1) the qualified ACT report that provides evidence of all resource guarantees being met and (2) updated safety analyses at the system integrator level that document the hosting of the app on a different physical GPM and the associated impact to the aircraft if that GPM failed and all installed apps were lost.

27.3.3 Platform Acceptance for Certification

The Genesis platform is independently accepted for certification apart from the hosted functions. While the hosted functions rely upon the platform for their arguments for certification acceptance, the platform arguments are made in isolation. To adhere to the incremental approach to certification, the platform can be broken down into modules that can be incrementally accepted for certification. The platform developer can decide on the appropriate breakdown based upon the distributed roles and responsibilities for the program. Typical platform module breakdowns may individually seek certification acceptance for the processing module hardware (GPMs), OS (for GPMs), infrastructure applications, network elements (A664 P7), and I/O elements (RIU).

A set of module acceptance data that establishes the module's physical characteristics, functional characteristics, performance characteristics, and interfaces (physical, electrical, and software as applicable) should be provided to the certification authorities with each module. In order to maximize reuse, the usage domain can be described, including usage limitations and guidance for implementation (usage domain described in DO-297). As in all certification approaches, the module's validation and verification evidence must be provided to the certification authorities. A portion of the platform evidence, if not the whole, will be required by the hosted function suppliers when developing their own evidence. As a platform-dependent system, the hosted function suppliers will need to reference the platform evidence when they apply for incremental acceptance.

27.3.4 Incremental Change and Module Reuse

A great benefit of a modular structure of incrementally accepted certification arguments is that this organization aids in providing a minimal cost of change during an incremental change in the platform or a hosted function. Traditional certification techniques require a single, complex argument that embraces all certification details. Under the Genesis approach, rather than revalidating the entire complex argument, only the smaller, less complex, modified argument module must be revalidated along with the top-level integrated system argument module. For example, if a GPM hosted application is changed, then its certification module would be updated along with the integrated-system-level module, but the GPM hardware and OS modules would remain unchanged. Other modules that rely on the modified application module would also need to be revalidated if any of its guarantees are changed (i.e., data latency, integrity, redundancy). If the module guarantees remain unchanged, then other modules that depend upon it do not require revalidation or reverification.

Module reuse is also enabled by the Genesis architectural approach. Similar to incremental change, the Genesis approach provides incrementally accepted modules to be reused in alternate instances of the platform at minimal cost. The claim for "minimal cost" is predicated on the assumption that the original arguments for module certification acceptance can be shown to be satisfied by the new platform. The top-level integrated system arguments for the alternate platform would need to be formed, but most of the underlying logic that supported a module's acceptance in the original certification effort can likely be reused. For example, a GPM module from one aircraft certification activity can be reused in another aircraft certification activity with minimal impact to the existing module certification artifacts (requirements, validation, verification) if the same guarantees can be made with respect to that certification module. This would require the same functional and safety requirements to be levied on the GPM, and existing input data latency maximums would need to be met for GPM infrastructure messages. A majority of the module data would not change, and the work would be minimized to a new requirement traceability analysis that demonstrates the GPM module relies on the same input guarantees and satisfies the same output guarantees.

27.3.5 Regulatory Guidance

When the Genesis architecture was conceived, the common regulatory documents for the certification of complex aircraft systems did not specifically address the incremental approach to certification nor an approach to certifying systems according to a logical architecture (traditional guidance is focused on the physical architecture). Due to these deficiencies, a minimal cost of change was not fostered by these traditional sources of regulatory guidance:

- RTCA DO-178B Software considerations in airborne systems and equipment certification
- RTCA DO-254 Design assurance guidance for airborne electronic hardware
- SAE ARP-4754 Certification considerations for highly-integrated or complex aircraft systems

Fortunately, recent developments have generated new guidance that more appropriately addresses the unique properties of an IMA platform. New guidance is likely to increase as architectures such as the Genesis platform continue to grow in popularity. At the time of this publication, the following guidance was available:

- RTCA DO-178C, Software Considerations in Airborne Systems and Equipment Certification
- SAE ARP-4754A, Guidelines for Development of Civil Aircraft and Systems
- SAE ARP 4761, Guidelines and Methods for Conducting the Safety Assessment Process on Civil Airborne Systems
- Federal Aviation Administration (FAA) AC 20-174, Development of Civil Aircraft and Systems

- FAA AC 20-171, Alternatives to RTCA DO-178B for Software in Airborne Systems and Equipment
- FAA AC 20-115C, Airborne Software Assurance
- FAA AC 20-152 RTCA, Inc., Document RTCA DO-254, Design Assurance Guidance for Airborne Electronic Hardware
- FAA TSO C-153, Integrated Modular Avionics Hardware Elements
    - Addresses reusability of certification credit for hardware elements (but not software)
    - Typical interpretations still focus on integrated system and do not allow for stand-alone architectural elements
- FAA AC20-145, Guidance for Integrated Modular Avionics (IMA) that Implement TSO-C153 Authorized Hardware Elements
    - Supplement to TSO C-153 that offers no additional guidance for incremental acceptance or reuse
- RTCA DO-297, Integrated Modular Avionics (IMA) Development Guidance and Certification Considerations
    - Supports most of the certification approach for the Genesis platform
- FAA AC20-170, Integrated Modular Avionics Development, Verification, Integration, and Approval Using RTCA DO-297 and Technical Standard Order-C153
    - Sets forth an acceptable means of compliance for aircraft and engines that utilize IMA systems. Supplements guidance of DO-297 and TSO C-153
- FAA AC20–148, Reusable Software Components
    - Supports incremental certification of software components

27.4 Conclusions and Summary

An avionics architecture has been presented that demonstrates how avionics have evolved into a highly integrated environment. Technical details of the architecture have been described as well as the roles and responsibilities for the integration and certification processes. This should give the reader an appreciation of the benefits and complexities relating to a highly integrated avionics environment.

In summary, the Genesis platform is a hardware/software platform that provides computing, communication, and I/O services for implementing real-time embedded systems. This IMA architecture contrasts with the traditional federated architecture by utilizing shared resources to host multiple "virtual systems" that were traditionally hosted as separate "physical systems." The Genesis architecture employs a network-centric environment of computing, I/O, and communication elements and provides connections for other nonplatform systems (connected via A664-P7, A429, CAN, or analog/digital I/O). The architecture uses robust partitioning mechanisms to establish fault-containment capabilities and to ensure a high degree of "virtual system" integrity.

As compared to traditional certification strategies, the main differences for the suggested certification approach include the processes of incremental acceptance and certification by logical system (instead of by physical system). A contract-based approach to the integration of modular systems is proposed to support these certification methodologies.

Genesis is a composable and extensible architecture that is targeted for supporting highly critical applications but can be scaled to appropriately support lower levels of application integrity. As compared to the federated architectures, the Genesis architecture offers advantages of a reduced system size, weight, power, and cost. The platform provides system integrators with a foundation that allows them to quickly and efficiently build up an integrated system that takes advantage of a common development effort (open environment), common set of resources (less resource waste), and a common set of integration objectives for each system developed (increases coordination between system suppliers).

The Boeing 787 and other aircraft have already chosen to utilize the Genesis architecture to achieve an IMA solution. The trend of integrating avionics together is expected to continue, and therefore the Genesis concepts presented in this chapter are only the starting point for a new age of avionics system architectures.

28

Boeing B-777 Avionics Architecture

Michael J. Morgan
Honeywell

28.1 Introduction

The avionics industry has long recognized the substantial cost benefits that could be realized using a large-scale integrated computing architecture for airborne avionics. Technology achievements by airframe, avionics, and semiconductor manufacturers allow implementation of these integrated avionics architectures resulting in substantial life cycle cost benefits. The Boeing 777 Airplane Information Management System (AIMS) represents the first application of an integrated computing architecture in a commercial air transport.

28.2 Background

Since 1988, the avionics industry has made a significant effort to develop the requirements and goals for a next-generation integrated avionics architecture. This work is documented in ARINC Project Paper 651 [1]. Top-level goals of the Integrated Modular Avionics (IMA) architecture are to reduce overall cost of ownership through reduced spares requirements (includes reduction in cost of spare Line Replaceable Modules [LRM] and reduction in number of LRMs required), reduce equipment removal rate, and reduce weight and volume in both avionics and wiring. In addition, IMA addresses the airlines' demand for better MTBUR/MTBF (Mean Time Between Unscheduled Removals as a fraction of Mean Time Between Failures), improved system performance (response time), increased airborne functionality, better fault isolation and test, and maintenance-free dispatch for extended intervals.

Technology trends in microprocessor and memory technology demand that airborne computing architectures evolve if the avionics industry is to meet the goals of IMA. By exploiting these developments in the microprocessor and memory industries very highly integrated architectures previously not technologically feasible or cost-effective may now be realized. These functionally integrated architectures minimize life cycle cost by minimizing the duplication of hardware and software elements (see Figure 28.1).

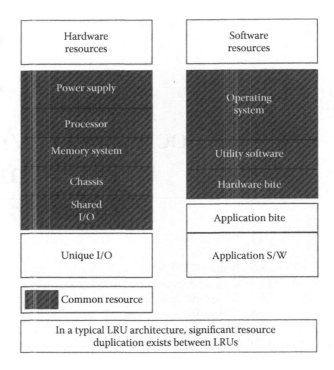

FIGURE 28.1　Components of a typical LRU.

High levels of functional integration dictate availability and integrity requirements far exceeding the requirements for federated architectures. Resource availability requirements must be sufficient to probabilistically preclude the simultaneous loss of multiple functions utilizing shared resources. These availability requirements imply application of fault-tolerant technology. Although fault tolerance is required to meet the integrity and availability goals of IMA, it is also directly compatible with the airline goal for deferred maintenance. Furthermore, since fault-tolerant technology requires high-integrity monitoring, it also is compatible with airlines' desires for improved fault isolation, better maintenance diagnostics, and reduced unconfirmed removal rate (MTBUR). Current IMA implementations are realizing a more than six times improvement in unconfirmed equipment removals over a typical federated Line Replaceable Unit-based (LRU-based) architecture.

High functional integration also implies the requirement to maintain functional independence for software using any shared resource. Strict CPU separation is not sufficient to ensure that functions will not adversely affect each other. Input-output (I/O) resource sharing demands a backplane bus architecture that has extremely high integrity and enforces rigid partitioning between all users. Processor resource sharing requires a robust software partitioning system in which all partition protection elements are monitored to ensure isolation integrity.

Robust partitioning protection must be performed as an integral part of the architecture, and isolation must not be dependent upon the integrity of the application software. In this environment, the robust partitioning architecture would be certified as a standalone element allowing functional software to be updated and certified independently of other functions sharing the same computational or I/O resources. Since it is anticipated that airborne functionality will continue to increase and that the majority of this increase will be accommodated via software changes alone, this partitioned environment will provide flexibility in responding to evolving system requirements (e.g., CNS/ATM).

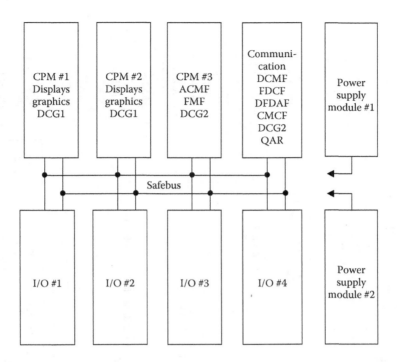

FIGURE 28.2 AIMS baseline functional distribution.

28.3 Boeing 777 AIMS

Now in its second-generation implementation, the Boeing 777 AIMS implements the IMA concept in an architecture supporting a high degree of functional integration and reducing duplicated resources to a minimum. In this architecture, the conventional LRUs, which typically contain a single function, are replaced with dual integrated cabinets, which provide the processing and the I/O hardware and software required to perform the following functions (see Figure 28.2):

- Flight Management
- Display
- Central Maintenance
- Airplane Condition Monitoring
- Communication Management (including flight deck communication)
- Data Conversion Gateway (ARINC 429/629 Conversion)

The integrated cabinets are connected to the airplane interfaces via a combination of ARINC 429, ARINC 629, and discrete I/O channels (see Figure 28.3; note that for clarity the 429 and discrete channels are not shown).

28.4 Cabinet Architecture Overview

The heart of the AIMS system consists of dual cabinets located in the electronics bay, each containing four core processor modules (CPMs), four input/output modules (IOMs), and two power supply modules. Additional space is reserved in the cabinet to add CPMs and IOMs as may be required for future growth (see Figure 28.4). The following are shared platform resources provided by AIMS:

- Common processor and mechanical housing
- Common input/output ports, aircraft-level power conditioning, and mechanical housing

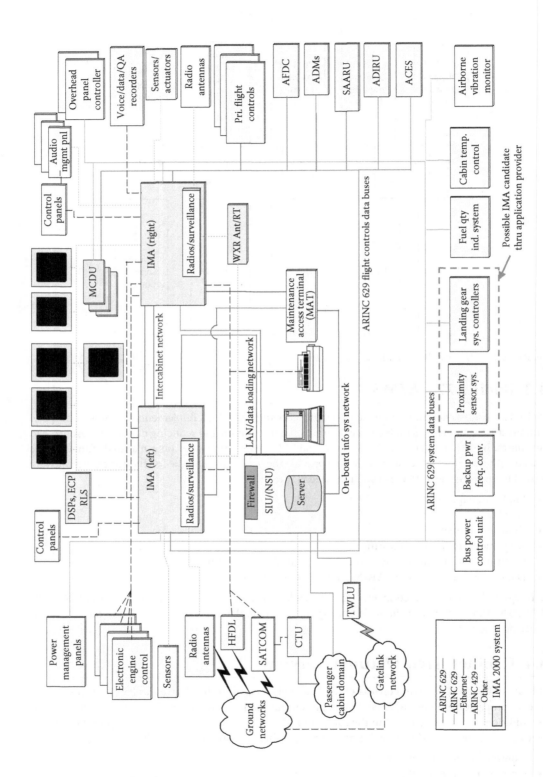

FIGURE 28.3   Airplane interface schematic.

FIGURE 28.4 AIMS cabinet.

- Common backplane bus (SAFEbus®) to move data between CPMs and between CPMs and IOMs
- Common operating system and built-in test (BIT) and utility software

Instead of individual applications residing in a separate LRU, applications are integrated on common CPMs. The IOMs transmit data from the CPMs to other systems on the airplane, and receive data from these other systems for use by the CPM applications. A high-speed backplane bus, called SAFEbus, provides a 60-Mbit/s data pipe between any of the CPMs and IOMs in a cabinet. Communication between AIMS cabinets is accomplished through four ARINC 629 serial buses.

The robust partitioning provided by the architecture allows applications to use common resources without any adverse interactions. This is achieved through a combination of memory management and deterministic scheduling of application software execution. Memory is allocated before run time, and only one application partition is given write-access to any given page of memory. Scheduling of processor resources for each application is also done before run time and is controlled by a set of tables loaded onto each CPM and IOM in the cabinet. This set of tables operates synchronously and controls application scheduling on the CPMs as well as data movement between modules across the SAFEbus.

Hardware fault detection and isolation is achieved via a lock-step design of the CPMs, IOMs, and the SAFEbus. Each machine cycle on the CPMs and IOMs is performed in lock-step by two separate processing channels, and comparison hardware ensures that each channel is performing identically. If a miscompare occurs, the system will attempt retries where possible before invoking the fault-handling and logging software in the operation system. The SAFEbus has four redundant data channels that are compared in real time to detect and isolate bus faults. The applications hosted on AIMS are listed below, along with the number of redundant copies of each application per shipset in parentheses:

- Displays (4)
- Flight Management/Thrust Management (2)
- Central Maintenance (2)
- Data Communication Management (2)
- Flight Deck Communication (2)
- Airplane Condition Monitoring (1)
- Digital Flight Data Acquisition (2)
- Data Conversion Gateway (4)

All of the IOMs in the two AIMS cabinets are identical. The CPMs have common hardware for processor, memory, power, and SAFEbus interface, but have the capability to include a custom I/O card to provide specific hardware for an application "client." The client hardware in AIMS includes the displays graphics generator, the digital flight data acquisition interface to the data recorder, Aircraft Communications

Addressing and Reporting System (ACARS) modem interface, and the airplane condition monitoring memory. The following are other flight deck hardware elements that make up the AIMS system:

- Six flat panel display units
- Three control and display units
- Two Electronic Flight Instrument System (EFIS) display control panels
- Display select panel
- Cursor control devices
- Display remote light sensors

28.5 Backplane Bus

As stated, the cabinet LRMs are interconnected via dual high-speed serial buses called SAFEbus (see Figure 28.5) that provide the only communication mechanism between the processing and I/O elements of the integrated functions. As such, extremely high availability and integrity requirements are necessary to preclude the simultaneous loss of multiple functions and to preserve robust partitioning of I/O resources. In addition, SAFEbus itself is required to provide and enforce the integrity of this key shared resource. Absolute data integrity must be ensured independent of hardware and software failures within any of the CPMs. In this environment, SAFEbus behaves as a generic and virtual resource capable of supporting high levels of I/O integration.

The SAFEbus protocol is driven by a sequence of commands stored in the internal table memory of each Bus Interface Unit (BIU). Each command corresponds to a single message transmission. All BIUs are synchronized so that at any given point in time all BIUs "know" the state of the bus and are at equivalent points in their tables. Because buffer addresses are stored in tables, they do not need to be transmitted over the bus, and since all transactions are scheduled deterministically, there is no need

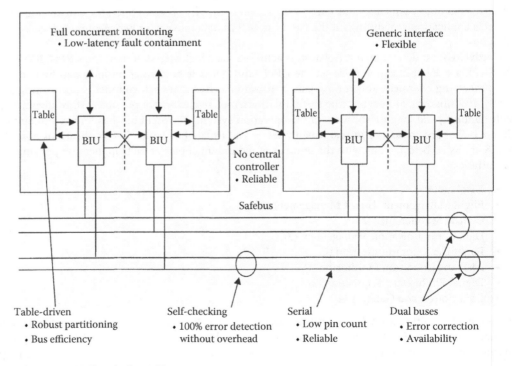

FIGURE 28.5 SAFEbus dual serial buses.

to arbitrate the bus. This allows for extremely high bus efficiency (94%) with no bits required to be dedicated to address control and minimal bits required to control data. A more detailed description of SAFEbus operation can be found in ARINC Project Paper 659 and also in Ref. [2].

28.6 Maintenance

The requirements for fault tolerance allow increased design flexibility and capability for deferred maintenance operation. By taking advantage of the high-integrity hardware monitoring, which fault-tolerant design mandates, the AIMS cabinets are capable of instantaneous fault detection and confinement. This increased fault visibility allows the cabinet to suppress most faults prior to producing a flight deck effect. This is an important step in reducing the Mean Time Between Removal (MTBR) of the equipment. In addition, fault tolerance provides the capability for deferring maintenance to regular (and thus schedulable) intervals. Depending upon the "fail-to-dispatch" probability that a particular airline may be willing to "endure," dispatch can continue for 10–30 days without maintenance following any first failure in the AIMS.

28.7 Growth

Functional growth is provided in the cabinets through two paths: (1) installed growth in the form of excess computing and backplane resources provided as part of the baseline AIMS and (2) spare LRM slots provided in each cabinet. Spare computing and backplane resources may be used by any function (new or existing) that requires additional throughput or I/O. Existing spare I/O hardware—for example 629 terminals, 429 terminals, and discrete I/O—are also available for use by any function integrated into the cabinet. Spare LRM slots may be used for additional processing, I/O, or additional unique hardware that may be required for a specific function. Additional processing modules may be added as required without changes to existing cabinet hardware. Addition of I/O may require wiring changes if new airplane interfaces are needed.

Further Reading

This chapter is substantially a reprint of material originally presented in the following sources:

Morgan, M.J., Honeywell, Inc., Integrated modular avionics for next generation airplanes, *IEEE Aerospace and Electronic Systems Magazine*, 6:9–12, August 1991.
Witwer, R., Honeywell, Inc., Developing the 777 airplane information management system (AIMS): A view from program start to one year of service, *IEEE Transactions on Aerospace and Electronic System*, 33:637–641, August 1996.

References

1. ARINC. Design guidance for integrated modular avionics, ARINC 651-1, Aeronautical Radio Inc., 1999.
2. Hoyme, K., Driscoll, K., Herrlin J., and Radke, K., Honeywell, Inc., ARINC 629 and SAFEbus®: Data buses for commercial aircraft, *Scientific Honeyweller*, Fall, 57–70, 1991.

29

Boeing B-777:
Fly-by-Wire Flight Controls

Gregg F. Bartley
*Federal Aviation
Administration**

29.1 Introduction

Fly-By-Wire (FBW) primary flight controls have been used in military applications such as fighter airplanes for a number of years. It has been a rather recent development to employ them in a commercial transport application. The 777 is the first commercial transport manufactured by Boeing which employs a FBW primary flight control system. This chapter will examine a FBW primary flight control system using the system on the 777 as an example (Figure 29.1). It must be kept in mind while reading this chapter that this is only a single example of what is currently in service in the airline industry. There are several other airplanes in commercial service made by other manufacturers that employ a different architecture for their FBW flight control system than described here.

A FBW flight control system has several advantages over a mechanical system. These include:

- Overall reduction in airframe weight.
- Integration of several federated systems into a single system.

---

* This is a reprinted chapter with minor editorial changes and changes due to technology or regulatory evolution by technical editors since the original author could not be reached. The author's affiliation is from the original contribution; current affiliation of the author is unknown.

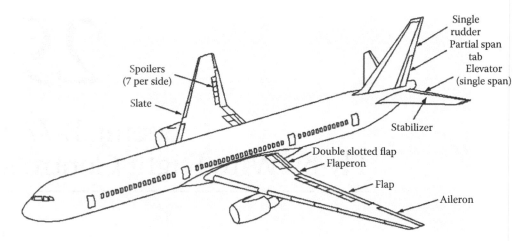

FIGURE 29.1 The primary flight control system on the Boeing 777 is comprised of the outboard ailerons, flaperons, elevator, rudder, horizontal stabilizer, and the spoiler/speedbrakes.

- Superior airplane handling characteristics.
- Ease of maintenance.
- Ease of manufacture.
- Greater flexibility for including new functionality or changes after initial design and production.

29.2 System Overview

Conventional primary flight controls systems employ hydraulic actuators and control valves controlled by cables that are driven by the pilot controls. These cables run the length of the airframe from the cockpit area to the surfaces to be controlled. This type of system, while providing full airplane control over the entire flight regime, does have some distinct drawbacks. The cable-controlled system comes with a weight penalty due to the long cable runs, pulleys, brackets, and supports needed. The system requires periodic maintenance, such as lubrication and adjustments due to cable stretch over time. In addition, systems such as the yaw damper that provide enhanced control of the flight control surfaces require dedicated actuation, wiring, and electronic controllers. This adds to the overall system weight and increases the number of components in the system.

In a FBW flight control system, the cable control of the primary flight control surfaces has been removed. Rather, the actuators are controlled electrically. At the heart of the FBW system are electronic computers. These computers convert electrical signals sent from position transducers attached to the pilot controls into commands that are transmitted to the actuators. Because of these changes to the system, the following design features have been made possible:

- Full-time surface control utilizing advanced control laws. The aerodynamic surfaces of the 777 have been sized to afford the required airplane response during critical flight conditions. The reaction time of the control laws is much faster than that of an alert pilot. Therefore, the size of the flight control surfaces could be made smaller than those required for a conventionally controlled airplane. This results in an overall reduction in the weight of the system.
- Retention of the desirable flight control characteristics of a conventionally controlled system and the removal of the undesirable characteristics. This aspect is discussed further in the section on control laws and system functionality.
- Integration of functions such as the yaw damper into the basic surface control. This allows the separate components normally used for these functions to be removed.
- Improved system reliability and maintainability.

29.3 Design Philosophy

The philosophy employed during the design of the 777 primary flight control system maintains a system operation that is consistent with a pilot's past training and experience. What is meant by this is that however different the actual system architecture is from previous Boeing airplanes, the presentation to the pilot is that of a conventionally controlled mechanical system. The 777 retains the conventional control column, wheel, and rudder pedals, whose operations are identical to the controls employed on other Boeing transport aircrafts. The flight deck controls of the 777 are very similar to those of the Boeing 747-400, which employs a traditional mechanically controlled primary flight control system.

Because the system is controlled electronically, there is an opportunity to include system control augmentation and envelope protection features that would have been difficult to provide in a conventional mechanical system. The 777 primary flight control system has made full use of the capabilities of this architecture by including such features as

- Bank angle protection
- Turn compensation
- Stall and overspeed protection
- Pitch control and stability augmentation
- Thrust asymmetry compensation

More will be said of these specific features later. What should be noted, however, is that none of these features limit the action of the pilot. The 777 design utilizes *envelope protection* in all of its functionality rather than *envelope limiting*. Envelope *protection* deters pilot inputs from exceeding certain predefined limits but does not prohibit it. Envelope *limiting* prevents the pilot from commanding the airplane beyond set limits. For example, the 777 bank angle protection feature will significantly increase the wheel force a pilot encounters when attempting to roll the airplane past a predefined bank angle. This acts as a prompt to the pilot that the airplane is approaching the bank angle limit. However, if deemed necessary, the pilot may override this protection by exerting a greater force on the wheel than is being exerted by the backdrive actuator. The intent is to inform the pilot that the command being given would put the airplane outside of its normal operating envelope, but the ability to do so is not precluded. This concept is central to the design philosophy of the 777 primary flight control system.

29.4 System Architecture

29.4.1 Flight Deck Controls

As noted previously, the 777 flight deck uses standard flight deck controls; a control column, wheel, and rudder pedals that are mechanically linked between the Captain's and First Officer's controls. This precludes any conflicting input between the Captain and First Officer into the primary flight control system. Instead of the pilot controls driving quadrants and cables, as in a conventional system, they are attached to electrical transducers that convert mechanical displacement into electrical signals.

A gradient control actuator is attached to the two control column feel units. These units provide the tactile feel of the control column by proportionally increasing the amount of force the pilot experiences during a maneuver with an increase in airspeed. This is consistent with a pilot's experience in conventional commercial jet transports.

Additionally, the flight deck controls are fitted with what are referred to as "backdrive actuators." As the name implies, these actuators backdrive the flight deck controls during autopilot operation. This feature is also consistent with what a pilot is used to in conventionally controlled aircraft and allows the pilot to monitor the operation of the autopilot via immediate visual feedback of the pilot controls that is easily recognizable.

29.4.2 System Electronics

There are two types of electronic computers used in the 777 primary flight control system: the Actuator Control Electronics (ACE), which is primarily an analog device, and the Primary Flight Computer (PFC), which utilizes digital technology. There are four ACEs and three PFCs employed in the system. The function of the ACE is to interface with the pilot control transducers and to control the primary flight control system actuation with analog servo loops. The role of the PFC is the calculation of control laws by converting the pilot control position into actuation commands, which are then transmitted to the ACE. The PFC also contains ancillary functions, such as system monitoring, crew annunciation, and all the primary flight control system onboard maintenance capabilities.

Four identical ACEs are used in the system, referred to as L1, L2, C, and R. These designations correspond roughly to the left, center, and right hydraulic systems on the airplane. The flight control functions are distributed among the four ACEs. The ACEs decode the signals received from the transducers used on the flight deck controls and the primary surface actuation. The ACEs convert the transducer position into a digital value and then transmit that value over the ARINC 629 data busses for use by the PFCs. There are three PFCs in the system, referred to as L, C, and R. The PFCs use these pilot control and surface positions to calculate the required surface commands. At this time, the command of the automatic functions, such as the yaw damper rudder commands, are summed with the flight deck control commands, and are then transmitted back to the ACEs via the same ARINC 629 data busses. The ACEs then convert these commands into analog commands for each individual actuator.

29.4.3 ARINC 629 Data Bus

The ACEs and PFCs communicate with each other, as well as with all other systems on the airplane, via triplex, bi-directional ARINC 629 flight controls data busses, referred to as L, C, and R. The connection from these electronic units to each of the data busses is via a stub cable and an ARINC 629 coupler. Each coupler may be removed and replaced without disturbing the integrity of the data bus itself.

29.4.4 Interface to Other Airplane Systems

The primary flight control system transmits and receives data from other airplane systems by two different pathways. The Air Data and Inertial Reference Unit (ADIRU), Standby Attitude and Air Data Reference Unit (SAARU), and the Autopilot Flight Director Computers (AFDC) transmit and receive data on the ARINC 629 flight controls data busses, which is a direct interface to the primary flight computers. Other systems, such as the Flap Slat Electronics Unit (FSEU), Proximity Switch Electronics Unit (PSEU), and Engine Data Interface Unit (EDIU) transmit and receive their data on the ARINC 629 systems data busses. The PFCs receive data from these systems through the Airplane Information Management System (AIMS) Data Conversion Gateway (DCG) function. The DCG supplies data from the systems data busses onto the flight controls data busses. This gateway between the two main sets of ARINC 629 busses maintains separation between the critical flight controls busses and the essential systems busses but still allows data to be passed back and forth (Figure 29.2).

29.4.5 Electrical Power

There are three individual power systems dedicated to the primary flight control system, which are collectively referred to as the Flight Controls Direct Current (FCDC) power system. An FCDC Power Supply Assembly (PSA) powers each of the three power systems. Two dedicated Permanent Magnet Generators (PMG) on each engine generate AC power for the FCDC power system. Each PSA converts the PMG alternating current into 28 V DC for use by the electronic modules in the Primary Flight Control System. Alternative power sources for the PSAs include the airplane Ram Air Turbine (RAT),

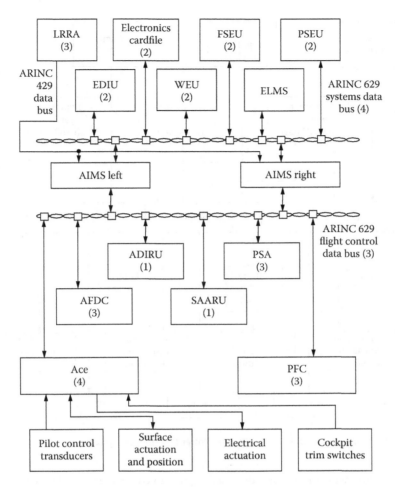

FIGURE 29.2 Block diagram of the electronic components of the 777 Primary Flight Control System, as well as the interfaces to other airplane systems.

the 28-V DC main airplane busses, the airplane hot battery buss, and dedicated 5 Ah FCDC batteries. During flight, the PSAs draw power from the PMGs. For on-ground engines-off operation or for in-flight failures of the PMGs, the PSAs draw power from any available source (Figure 29.3).

29.5 Control Surface Actuation

29.5.1 Fly-by-Wire Actuation

The control surfaces on the wing and tail of the 777 are controlled by hydraulically powered, electrically signaled actuators. The elevators, ailerons, and flaperons are controlled by two actuators per surface, the rudder is controlled by three. Each spoiler panel is powered by a single actuator. The horizontal stabilizer is positioned by two parallel hydraulic motors driving the stabilizer jackscrew.

The actuation powering the elevators, ailerons, flaperons, and rudder have several operational modes. These modes, and the surfaces that each are applicable to, are defined below.

Active—Normally, all the actuators on the elevators, ailerons, flaperons, and rudder receive commands from their respective ACEs and position the surfaces accordingly. The actuators will remain in the active mode until commanded into another mode by the ACEs.

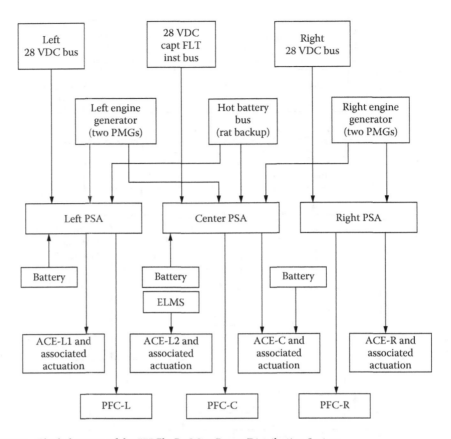

FIGURE 29.3   Block diagram of the 777 Fly-By-Wire Power Distribution System.

Bypassed—In this mode, the actuator does not respond to commands from its ACE. The actuator is allowed to move freely, so that the redundant actuator(s) on a given surface may position the surface without any loss of authority, i.e., the actuator in the active mode does not have to overpower the bypassed actuator. This mode is present on the aileron, flaperon, and rudder actuators.

Damped—In this mode, the actuator does not respond to the commands from the ACE. The actuator is allowed to move, but at a restricted rate which provides flutter damping for that surface. This mode allows the other actuator(s) on the surface to continue to operate the surface at a rate sufficient for airplane control. This mode is present on elevator and rudder actuators.

Blocked—In this mode, the actuator does not respond to commands from the ACE, and it is not allowed to move. When both actuators on a surface (which is controlled by two actuators) have failed, they both enter the "Blocked" mode. This provides a hydraulic *lock* on the surface. This mode is present on the elevator and aileron actuators.

An example using the elevator surface illustrates how these modes are used. If the inboard actuator on an elevator surface fails, the ACE controlling that actuator will place the actuator in the "Damped" mode. This allows the surface to move at a limited rate under the control of the remaining operative outboard actuator. Concurrent with this action, the ACE also arms the "Blocking" mode on the outboard actuator on the same surface. If a subsequent failure occurs that will cause the outboard actuator to be placed in the "Damped" mode by its ACE, both actuators will then be in the "Damped" mode and have their "Blocking" modes armed. An elevator actuator in this configuration enters the "Blocking" mode, which hydraulically locks the surface in place for flutter protection.

29.5.2 Mechanical Control

Spoiler panel 4 and 11 and the alternate stabilizer pitch trim system are controlled mechanically rather than electrically. Spoilers 4 and 11 are driven directly from control wheel deflections via a control cable. The alternate horizontal stabilizer control is accomplished by using the pitch trim levers on the flight deck aisle stand. Electrical switches actuated by the alternate trim levers allow the PFCs to determine when alternate trim is being commanded so that appropriate commands can be given to the pitch control laws.

Spoiler panels 4 and 11 are also used as speedbrakes, both in the air and on the ground. The speedbrake function for this spoiler pair only has two positions: stowed and fully extended. The speedbrake commands for spoilers 4 and 11 are electrical in nature, with an ACE giving an *extend* or *retract* command to a solenoid-operated valve in each of the actuators. Once that spoiler pair has been deployed by a speedbrake command, there is no control wheel speedbrake command mixing, as there is on all the other fly-by-wire spoiler surfaces (Figure 29.4).

29.6 Fault Tolerance

"Fault tolerance" is a term that is used to define the ability of any system to withstand single or multiple failures which results in either no loss of functionality or a known loss of functionality or reduced level of redundancy while maintaining the required level of safety. It does not, however, define any particular method that is used for this purpose. There are two major classes of faults that any system design must deal with. These are

1. A failure which results in some particular component becoming totally inoperative. An example of this would be a loss of power to some electronic component, such that it no longer performs its intended function.
2. A failure which results in some particular component remaining active, but the functionality it provides is in error. An example of this failure would be a Low Range Radio Altimeter whose output is indicating the airplane is at an altitude 500 ft above the ground when the airplane is actually 200 ft above the ground.

One method that is used to address the first class of faults is the use of redundant elements. For example, there are three PFCs in the 777 primary flight control system, each with three identical computing "lanes" within each PFC. This results in nine identical computing channels. Any of the three PFCs themselves can fail totally due to loss of power or some other failure which affects all three computing lanes, but the primary flight control system loses no functionality. All four ACEs will continue to receive all their surface position commands from the remaining PFCs. All that is affected is the level of available redundancy. Likewise, any single computing lane within a PFC can fail, and that PFC itself will continue to operate with no loss of functionality. The only thing that is affected is the amount of redundancy of the system. The 777 is certified to be dispatched on a revenue flight, per the Minimum Equipment List (MEL), with two computing lanes out of the nine total (as long as they are not within the same PFC channel) for 10 days and for a single day with one total PFC channel inoperative.

Likewise, there is fault tolerance in the ACE architecture. The flight control functions are distributed among the four ACEs such that a total failure of a single ACE will leave the major functionality of the system intact. A single actuator on several of the primary control surfaces may become inoperative due to this failure, and a certain number of spoiler symmetrical panel pairs will be lost. However, the pilot flying the airplane will notice little or no difference in handling characteristics with this failure. A total ACE failure of this nature will have much the same impact to the Primary Flight Control System as that of a hydraulic system failure.

The second class of faults is one that results in erroneous operation of a specific component of the system. The normal design practice to account for failures of this type is to have multiple elements doing

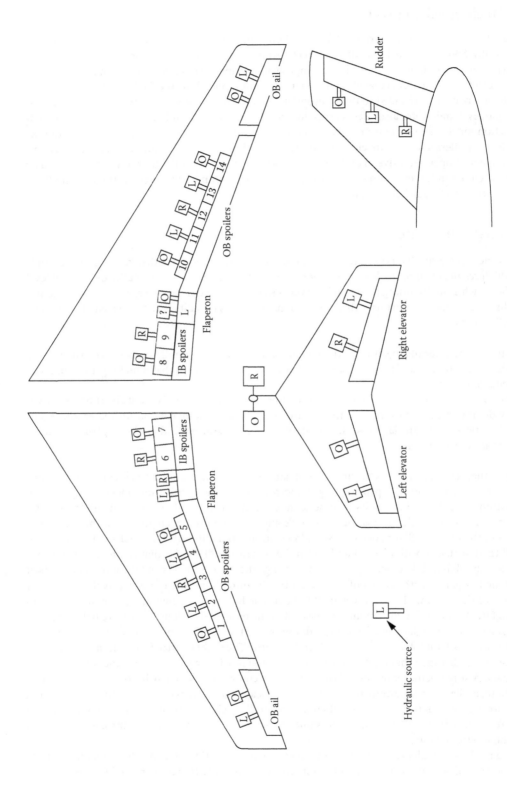

FIGURE 29.4 Schematic representation of the Boeing 777 Primary Flight Control System's hydraulic power and electronic control functional distribution. *Note:* Spoilers 4 and 11 are commanded via cables from the control wheel and vial the aces from the speedbrake lever. The stabilizer is commanded via the cables through the aisle stand levers only and otherwise is commanded through the aces.

the same task and their outputs voted or compared in some manner. This is sometimes referred to as a "voting plane." All critical interfaces into the 777 FBW Primary Flight Control System use multiple inputs which are compared by a voting plane. For interfaces that are required to remain operable after a first failure, at least three inputs must be used. For example, there are three individual Low Range Radio Altimeter (LRRA) inputs used by the PFCs. The PFCs compare all three inputs and calculates a mid-value select on the three values; i.e., the middle value LRRA input is used in all calculations which require radio altitude. In this manner, any single failure of an LRRA that results in an erroneous value will be discarded. If a subsequent failure occurs which causes the remaining two LRRA signals to disagree by a preset amount, the PFCs will throw out both values and take appropriate action in those functions which use these data.

Additionally, a voting plane scheme is used by the PFCs on themselves. Normally, a single computing lane within a PFC channel is declared as the "master" lane, and that lane is responsible for transmitting all data onto the data busses for use by the ACEs and other airplane systems. However, all three lanes are simultaneously computing the same control laws. The outputs of all three computing lanes within a single PFC channel are compared against each other. Any failure of a lane that will cause an erroneous output from that lane will cause that lane to be condemned as "failed" by the other two lanes.

Likewise, the outputs from all three PFC channels themselves are compared. Each PFC looks at its own calculated command output for any particular actuator, and compares it with the same command that was calculated by the other two PFC channels. Each PFC channel then does a mid-value select on the three commands, and that value (whether it was the one calculated by itself or by one of the other PFC channels) is then output to the ACEs for the individual actuator commands. In this manner, it is assured that each ACE receives identical commands from each of the PFC channels.

By employing methods such as those described above, it is assured that the 777 primary flight control system is able to withstand single or multiple failures and be able to contain those failures in such a manner that the system remains safe and does not take inappropriate action due to those failures.

29.7 System Operating Modes

The 777 FBW primary flight control system has three operating modes: normal, secondary, and direct. These modes are defined below

Normal—In the "Normal" mode, the PFCs supply actuator position commands to the ACEs, which convert them into an analog servo command. Full functionality is provided, including all enhanced performance, envelope protection, and ride quality features.

Secondary—In the "Secondary" mode, the PFCs supply actuator position commands to the ACEs, just as in the "Normal" mode. However, functionality of the system is reduced. For example, the envelope protection functions are not active in the "Secondary" mode. The PFCs enter this mode automatically from the "Normal" mode when there are sufficient failures in the system or interfacing systems such that the "Normal" mode is no longer supported. An example of a set of failures that will automatically drop the system into the "Secondary" mode is total loss of airplane air data from the ADIRU and SAARU. The airplane is quite capable of being flown for a long period of time in the "Secondary" mode. It cannot, however, be dispatched in this condition.

Direct—In the "Direct" mode, the ACEs do not process commands from the PFCs. Instead, each ACE decodes pilot commands directly from the pilot controller transducers and uses them for the closed loop servo control of the actuators. This mode will automatically be entered due to total failure of all three PFCs, failures internal to the ACEs, loss of the flight controls ARINC 629 data busses, or some combination of these failures. It may also be selected manually via the PFC disconnect switch on the overhead panel in the flight deck. The airplane handling characteristics in the "Direct" mode closely match those of the "Secondary" mode.

29.8　Control Laws and System Functionality

The design philosophy employed in the development of the 777 primary flight control system control laws stresses aircraft operation consistent with a pilot's past training and experience. The combination of electronic control of the system and this philosophy provides for the feel of a conventional airplane, but with improved handling characteristics and reduced pilot workload.

29.8.1　Pitch Control

Pitch control is accomplished through what is known as a *maneuver demand* control law, which is also referred to as a C*U control law. C* (pronounced "C-Star") is a term that is used to describe the blending of the airplane pitch rate and the load factor (the amount of acceleration felt by an occupant of the airplane during a maneuver). At low airspeeds, the pitch rate is the controlling factor. That is, a specific push or pull of the column by the pilot will result in some given pitch rate of the airplane. The harder the pilot pushes or pulls on the column, the faster the airplane will pitch nose up or nose down. At high airspeeds, the load factor dominates. This means that, at high airspeeds, a specific push or pull of the column by the pilot will result in some given load factor.

The "U" term in C*U refers to the feature in the control law which will, for any change in the airspeed away from a referenced trim speed, cause a pitch change to return to that referenced airspeed. For an increase in airspeed, the control law will command the airplane nose up, which tends to slow the airplane down. For a decrease in airspeed, the control law causes a corresponding speed increase by commanding the airplane nose down. This introduces an element of speed stability into the airplane pitch control. However, airplane configuration changes, such as a change in the trailing edge flap setting or lowering the landing gear, will NOT result in airplane pitch changes, which would require the pilot to re-trim the airplane to the new configuration. Thus, the major advantage of this type of control law is that the nuisance-handling characteristics found in a conventional, mechanically controlled flight control system which increase the pilot workload are minimized or eliminated, while the desirable characteristics are maintained.

While in flight, the pitch trim switches on the Captain's and First Officer's control wheels do not directly control the horizontal stabilizer as they normally do on conventionally controlled airplanes. When the trim switches are used in flight, the pilot is actually requesting a new referenced trim speed. The airplane will pitch nose up or nose down, using the elevator surfaces, in response to that reference airspeed change to achieve that new airspeed. The stabilizer will automatically trim, when necessary, to offload the elevator surface and allow it to return to its neutral surface when the airplane is in a trimmed condition. When the airplane is on the ground, the pitch trim switches do trim the horizontal stabilizer directly. While the alternate trim levers (described previously) move the stabilizer directly, even in flight, the act of doing so will also change the C*U referenced trim speed such that the net effect is the same as would have been achieved if the pitch trim switches on the control wheels had been used. As on a conventional airplane, trimming is required to reduce any column forces that are being held by the pilot.

The pitch control law incorporates several additional features. One is called landing flare compensation. This function provides handling characteristics during the flare and landing maneuvers consistent with that of a conventional airplane, which would have otherwise been significantly altered by the C*U control law. The pitch control law also incorporates stall and overspeed protection. These functions will not allow the referenced trim speed to be set below a predefined minimum value or above the maximum operating speed of the airplane. They also significantly increase the column force that the pilot must hold to fly above or below those speeds. An additional feature incorporated into the pitch control law is turn compensation, which enables the pilot to maintain a constant altitude with minimal column input during a banked turn.

The unique 777 implementation of maneuver demand and speed stability in the pitch control laws means that:

- An established flight path remains unchanged unless the pilot changes it through a control column input, or if the airspeed changes and the speed stability function takes effect.
- Trimming is required only for airspeed changes and not for airplane configuration changes.

29.8.2 Yaw Control

The yaw control law contains the usual functionality employed on other Boeing jetliners, such as the yaw damper and rudder ratio changer (which compensates a rudder command as a function of airspeed). However, the 777 FBW rudder control system has no separate actuators, linkages, and wiring for these functions, as have been used in previous airplane models. Rather, the command for these functions are calculated in the PFCs and included as part of the normal rudder command to the main rudder actuators. This reduces weight, complexity, maintenance, and spares required to be stocked.

The yaw control law also incorporates several additional features. The gust suppression system reduces airplane tag wag by sensing wind gusts via pressure transducers mounted on the vertical tail fin and applying a rudder command to oppose the movement that would have otherwise been generated by the gust. Another feature is the wheel-rudder crosstie function, which reduces sideslip by using small amounts of rudder during banked turns.

One important feature in the yaw control is Thrust Asymmetry Compensation, or TAC. This function automatically applies a rudder input for any thrust asymmetry between the two engines which exceed approximately 10% of the rated thrust. This is intended to cancel the yawing moment associated with an engine failure. TAC operates at all airspeeds above 80 kn even on the ground during the take-off roll. It will not operate when the engine thrust reversers are deployed.

29.8.3 Roll Control

The roll control law utilized by the 777 primary flight control system is fairly conventional. The outboard ailerons and spoiler panels 5 and 10 are locked out in the faired position when the airspeed exceeds a value that is dependent upon airspeed and altitude. It roughly corresponds to the airplane "flaps up" maneuvering speed. As with the yaw damper function described previously, this function does not have a separate actuator, but is part of the normal aileron and spoiler commands. The bank angle protection feature in the roll control law has been discussed previously.

29.8.4 757 Test Bed

The control laws and features discussed here were incorporated into a modified 757 and flown in the summer of 1992, prior to full-scale design and development of the 777 primary flight control system. The Captain's controls remained connected to the normal mechanical system utilized on the 757. The 777 control laws were flown through the First Officer's controls. This flying testbed was used to validate the flight characteristics of the 777 fly-by-wire system, as was flown by Boeing, customer, and regulatory agency pilots. When the 777 entered into its flight test program, its handling characteristics were extremely close to those that had been demonstrated with the 757 flying testbed.

29.8.5 Actuator Force-Flight Elimination

One unique aspect of the FBW flight control system used on the 777 is that the actuators on any given surface are all fully powered at all times. There are two full-time actuators driving each of the elevator, aileron, and flaperon surfaces, just as there are three full-time actuators on the rudder. The benefit of this particular implementation is that each individual actuator was able to be sized smaller than it would have had to have been if each surface was going to be powered by a single actuator through the entire flight regime. In addition, there is not a need for any redundancy management of an active/standby actuation system. However, this does cause a concern in another area. This is a possible actuator force-fight condition between the multiple actuators on a single flight control surface.

Actuator force-fight is caused by the fact that no two actuators, position transducers, or set of controlling servo loop electronics are identical. In addition, there always will be some rigging differences of the multiple actuators as they are installed on the airplane. These differences will result in one actuator attempting to position a flight control surface in a slightly different position than its neighboring actuator. Unless addressed, this would result in a twisting moment upon the surface as the two actuators fight each other to place the surface in different positions. In order to remove this unnecessary stress on the flight control surfaces, the primary flight computer control laws include a feature which "nulls out" these forces on the surfaces.

Each actuator on the 777 primary flight control system includes what is referred to as a Delta Pressure, or Delta P, pressure transducer. These transducer readings are transmitted via the ACEs to the PFCs, which are used in the individual surface control laws to remove the force-fight condition on each surface. The PFCs add an additional positive or negative component to each of the individual elevator actuator commands, which results in the difference between the two Delta P transducers being zero. In this way, the possibility of any force-fight condition between multiple actuators on a single surface is removed. The surface itself, therefore, does not need to be designed to withstand these stresses, which would have added a significant amount of weight to the airplane.

29.9 Primary Flight Controls System Displays and Annunciations

The primary displays for the primary flight control system on the 777 are the Engine Indication and Crew Alerting System (EICAS) display and the Multi-Function Display (MFD) in the flight deck. Any failures that require flight crew knowledge or action are displayed on these displays in the form of an English language message. These messages have several different levels associated with them, depending upon the level of severity of the failure.

> *Warning (Red with accompanying aural alert):* A nonnormal operational or airplane system condition that requires immediate crew awareness and immediate pilot corrective compensatory action.
> *Caution (Amber with accompanying aural alert):* A nonnormal or airplane system condition that requires immediate crew awareness. Compensatory or corrective action may be required.
> *Advisory (Amber with no accompanying aural alert):* A nonnormal operational or airplane system condition which requires crew awareness. Compensatory or corrective action may be required.
> *Status (White):* No dispatch or Minimum Equipment List (MEL) related items requiring crew awareness prior to dispatch.

Also available on the MFD, but not normally displayed in flight, is the flight control synoptic page, which shows the position of all the flight control surfaces.

29.10 System Maintenance

The 777 primary flight control system has been designed to keep line maintenance to a minimum, but when tasks do need to be accomplished, they are straightforward and easy to understand.

29.10.1 Central Maintenance Computer

The main interface to the primary flight control system for the line mechanic is the Central Maintenance Computer (CMC) function of AIMS. The CMC uses the Maintenance Access Terminal (MAT) as its primary display and control. The role of the CMC in the maintenance of the primary flight control system is to identify failures present in the system and to assist in their repair. The two features utilized by the CMC that accomplish these tasks are maintenance messages and ground maintenance tests. Maintenance messages describe to the mechanic, in simplified English, what failures are present in the system and the components possibly at fault. The ground maintenance tests exercise the system, test for active and latent failures, and confirm any repair action taken. They are also used to unlatch any EICAS and maintenance messages that may have become latched due to failures.

The PFCs are able to be loaded with new software through the data loader function on the MAT. This allows the PFCs to be updated to a new software configuration without having to take them out of service.

29.10.2 Line Replaceable Units

All the major components of the system are Line Replaceable Units (LRU). This includes all electronics modules, ARINC 629 data bus couplers, hydraulic and electrical actuators, and all position, force, and pressure transducers. The installation of each LRU has been designed such that a mechanic has ample space for component removal and replacement, as well as space for the manipulation of any required tools.

Each LRU, when replaced, must be tested to assure that the installation was accomplished correctly. The major LRUs of the system (transducers, actuators, and electronics modules) have LRU replacement tests that are able to be selected via a MAT pull-down menu and are run by the PFCs. These tests are user-friendly and take a minimum amount of time to accomplish. Any failures found in an LRU replacement test will result in a maintenance message, which details the failures that are present.

29.10.3 Component Adjustment

The primary surface actuators on the 777 are replaced in the same manner as on conventional airplanes. The difference is how they are adjusted. Each elevator, aileron, flaperon, and rudder actuator has what is referred to as a null adjust transducer, which is rotated by the mechanic until the actuator is positioned correctly. For example, when a rudder actuator is replaced, all hydraulic systems are depressurized except for the one that supplies power to the actuator that has just been replaced. The null adjust transducer is then adjusted until the rudder surface aligns itself with a mark on the empennage, showing that the actuator has centered the rudder correctly.

The transducers used on the pilot controls are, for the most part, individual LRUs. However, there are some packages, such as the speedbrake lever position transducers and the column force transducers, which have multiple transducers in a single package. When a transducer is replaced, the primary flight controls EICAS maintenance pages are used to adjust the transducer to a certain value at the system rig point. There are CMC-initiated LRU replacement tests which check that the component has been installed correctly and that all electrical connections have been properly mated.

29.11 Summary

The Boeing 777 fly-by-wire primary flight control system uses new technology to provide significant benefits over that of a conventional system. These benefits include a reduction in the overall weight of the airplane, superior handling characteristics, and improved maintainability of the system. At the same time, the control of the airplane is accomplished using traditional flight deck controls, thereby allowing the pilot to fly the airplane without any specialized training when transferring from a more conventional commercial jet aircraft. The technology utilized by the 777 primary flight control system has earned its way onto the airplane, and is not just technology for technology's sake.

Abbreviations

ACE	Actuator control electronics
ADIRU	Air data inertial reference unit
ADM	Air data module (static and total pressure)
AFDC	Autopilot flight director computer
AIMS	Airplane Information Management System
ARINC	Aeronautical Radio Inc. (industry standard)
C	Center
C*U	Pitch control law utilized in the primary flight computer
CMC	Central maintenance computer function in AIMS
DCGF	Data conversion gateway function of AIMS
EDIU	Engine data interface unit
EICAS	Engine indication and crew alerting system
ELMS	Electrical load management system
FBW	Fly-by-wire
FCDC	Flight controls direct current (power system)
FSEU	Flap slat electronic unit
L	Left
L1	Left 1
L2	Left 2
LRRA	Low range radio altimeter
LRU	Line replaceable unit
MAT	Maintenance access terminal
MEL	Minimum equipment list
MFD	Multi-function display
MOV	Motor-operated valve
PCU	Power control unit (hydraulic actuator)
PFC	Primary flight computer
PMG	Permanent magnet generator
PSA	Power supply assembly
R	Right
RAT	Ram air turbine
SAARU	Standby attitude and air data reference unit
TAC	Thrust asymmetry compensation
WEU	Warning electronics unit

30

New Avionics Systems: Airbus A330/A340

Peter Potocki
de Montalk
Airbus Industrie*

30.1 Overview

The A330/A340 project is a twin program—the first time that an aircraft has been designed from the outset both with four engines and also with two engines. Both aircraft types have essentially the same passenger and freight capacity. The four-engined A340 is optimized for long-range missions, but is also efficient at shorter ranges. With two engines, the A330 offers even better operating economics for the missions where an airline does not need the very long range of the A340.

The key to obtaining substantial commonality between the two products was the realization that on the two different aircraft many features could, in fact, be engineered the same way without a penalty. This approach has provided very substantial advantages for the operators, the airframe manufacturer, and for the equipment vendors. In effect, by designing for both sister aircraft from the outset, the requirements were engineered in common, and any added features for either of the two aircrafts could be introduced at a point in the design where they cost very little extra in terms of price, weight, reliability/maintainability or fuel burn. As a result, the two aircrafts can use the same parts (except the engine-related parts), the same aircrews, and the same airport and maintenance environment, and cost almost the same to develop as a single aircraft. Also, both are supremely efficient.

The A340 is offered in two configurations, allowing operators to tailor capacity and capability to demand. The larger A340-300 has the same fuselage length as the A330 and, while seating 300–350

---

* This is a reprinted chapter with minor editorial changes and changes due to technology or regulatory evolution by technical editors since the original author could not be reached. The author's affiliation is from the original contribution; current affiliation of the author is unknown.

passengers, has seat-mile costs close to those of the latest 747, making it an economical alternative on long-range routes with lower traffic densities.

The A340-200, seating 250–300 passengers, has the longest range capability of any commercial airliner available. Its low trip costs, coupled with the operating flexibility of four engines, make it an ideal aircraft for taking over when long-range twins become uneconomic.

While the A340 serves very long routes, the A330 is designed to serve high-growth, high-density regional routes. At the same time, it has the capability to operate economically on extended-range international routes. With typically 335 seats in a two-class arrangement, the A330 has a range of 4500 nmi with a full complement of passengers and baggage and 3200 nmi with maximum payload, making it ideal as a direct replacement for the costlier trijets and as a growth replacement for earlier twinjets.

30.2 Highlights

The A330 and A340 are built on the technological background established by two previous, complementary, product lines. The A300/A310 series is the world's best-established twin-jet twin-aisle aircraft program, with a very large number of technologically advanced features that transfer to larger, longer-range aircraft. The A319/A320/A321 series is the world's most advanced single-aisle aircraft program, again offering a large number of features that are found on much larger aircraft.

During the entire development process, there has been an insistence on securing the maximum commonality that could be achieved with the other programs without loss of efficiency. Using selected features from each of these product lines and updating them as needed resulted in an all-new A330/A340 aircraft program remarkably free of teething troubles, while at the same time providing a new benchmark for aircraft in this size category. As an added benefit, the technological features of the A330/A340 can, in many cases, be used to improve the established older product lines.

30.3 Systems

Before the A330/A340 entered into service, the world's most technologically advanced airliner, in any category, was the A320. Its design formed the basis for the A330/A340 systems.

30.4 Cockpit

The A330/A340 cockpit is designed to be identical to that of the A320 from the point of view of the crew. The exceptions to this rule are associated with the size of the aircraft and the added needs of the long-range mission, such as improved dispatchability, polar navigation capability, and of course, engine-related features. The result is that the 130-seat-capacity short/medium haul A319 up to the 340-plus seat capacity very long-haul A340 have the most advanced flight deck of any airliner, enabling the same crews to fly any of these aircraft with minimum additional training.

30.5 User Involvement

The design of the A330/A340 cockpit has evolved from the same methods that were used successfully on the first Airbus A300. The initial design of the cockpit (and the systems) was based on three features:

1. The existing cockpit from the previous aircraft (the A320, in this case)
2. The geometry of the A330/A340 nose section (which is based on the geometry of the A300, A310, and 300–600)
3. Applicable new research and development work carried out since the A320 had been designed

This initial design was reviewed by a task force consisting of pilots and engineers of each of the launch airlines in light of their experience with the A320 or with other aircraft operating on the intended routes for the A330 and A340.

The task force met a number of times over a period of over a year. At each of these steps, the design of the A330/A340 was refined, and certain features were mocked up for the next iteration in the review. The final design of the aircraft system and cockpit is essentially the one that the airline task force experienced and "flew" in the simulators during their final sessions.

30.6 Avionics

The avionics of the A330/A340 are highly integrated for optimal crew use and for optimal maintenance. As with all previous new and derivative aircraft since the A300FF of 1981, the primary data bus standard is ARINC 429 with ARINC 600 packaging. Other industry bus standards are used in specific applications in which ARINC 429 is not suitable.

30.7 Instruments

The six Cathode Ray Tubes (CRTs) on the main instrument panel display present flight and systems information to the pilots. This arrangement provides excellent visibility of all CRTs. Flight information is provided by the Electronic Flight Instrument System (EFIS) consisting of a Primary Flight Display (PFD) and a Navigation Display (ND) in front of each pilot. The Electronic Centralized Aircraft Monitor (ECAM), consisting of the engine/warning display on the upper screen and aircraft systems display on the lower screen, provides the systems information. Sensors throughout the aircraft continuously monitor the systems, and if a parameter moves out of the normal range, they automatically warn the pilot.

During normal flight, the ECAM presents systems displays according to the phase of flight, showing the systems in which the pilot is interested, such as some secondary engine data, pressurization, and cabin temperature. The pilot can, by manual selection, interrogate any system at any time. Should another system require attention, the ECAM will automatically present it to the flight crew for action. Should a system fault occur that results in a cascade of other system faults, ECAM identifies the originating fault and presents the operational checklists without any need for added crew actions. The information display formats currently in use enable the pilots to assimilate the operational situation of the aircraft much more easily than on the previous generation of aircraft.

There are substantial advantages on the maintenance side as well; the entire Electronic Instrument System (EIS) consists of only three Line Replaceable Unit (LRU) types, enabling significant dispatchability and spare stocks availability. In fact, all the flight information (including standby) is presented on only 11 instruments of 6 types. A new EIS using liquid crystal displays is installed on the A330/A340 and A320 family of aircraft, offering improved capabilities and cost of ownership.

30.8 Navigation

Dual Flight Management Systems (FMS; integrated with the Flight Guidance and Flight Envelope computing functions) combine the data from the aircraft navigation sensors, including the GPS installation. Backup navigation facilities are included in each pilot's multipurpose control/display unit (MCDU), allowing the aircraft to be dispatched with an inoperative FMS.

The FMS permits the crew to select an optimal flight plan for their route from a selection in the airline navigation data base, allowing the aircraft to fly automatically, through the autopilot or flight director, from just after take-off until the crew elects to carry out a precision approach and automatic landing. The "canned" flight plan captures the data needed for flight from the specifications entered by the crew prior to departure, as well as along the route as conditions change and more current information on weather and routing becomes available. New FMSs, with improved cost of ownership and capability, are installed on aircraft delivered from mid-2000. The same new FMSs are being installed on the A320 family.

30.9 Flight Controls

The flight control system of the A330/A340 is essentially the same as that of the A320, with five computers of two different types allowing the pilot to control the aircraft in pitch, roll, and yaw. The layout of the pilot controls is essentially the same as that of the A320 series, as are the handling qualities. The technology features are also essentially the same, with extensive use of dissimilarity in the hardware and in the software, and extensive segregation in the hydraulic and electrical power supplies and signaling lanes. As with the A320 series, mechanical signaling is used for the rudder and for the horizontal stabilizer trim backup.

Detail changes have been introduced reflecting the longer mission times, especially of the A340, to provide better access to the system, and the opportunity has been taken to reduce the variety of backup submodes that the crew must use, making the aircraft even easier to fly.

Like the A320 series, the A330/A340 is a conventional, naturally stable airliner. The electronic flight controls offer a number of advantages to the pilot. There is a large reduction in manually operated mechanical parts, easier troubleshooting, and no need for rigging. Optimal use of the control surfaces is facilitated, as is the use of maneuver load alleviation. The passengers and crew benefit as well, since the aircraft is more comfortable and easier to fly with precision in turbulence, while the flight envelope and structural protection features allow the crew to immediately use the whole capability of the aircraft should it be needed in an extreme situation.

30.10 Central Maintenance System

The A320, with its Central Fault Display System (CFDS), pioneered the industry standard for Central Maintenance Systems (CMS). This industry background of experience has been built into the A330/A340 CMS. It enables troubleshooting and return-to-service testing to be carried out rapidly and with a high degree of confidence from the cockpit. Much of the CMS information may also be accessed remotely, via Aircraft Communications Addressing and Reporting System (ACARS), allowing the aircraft to be greeted upon arrival by a maintenance technician who already has a good idea of the exact nature of any defect and has likely been able to procure from stores the proper spare LRU required to resolve the fault.

Compared with the previous generation of CMSs, such as the A320 CFDS, the A330/A340 CMS has been improved in a number of respects, allowing troubleshooting to take place on more than one system at a time and with even clearer data available for the job. There has been a significant improvement in dependability as well. The systems designers and the equipment vendors have paid great attention to maintainability standards, and a maintenance message filter facility has been incorporated that enables known false messages to be eliminated by the CMS, so that the mechanic does not apply the maxim "Falsus in Uno, Falsus in Omnibus."

30.11 Communications

There is a quiet revolution going on in the way that the crew communicates with the ground. This has been taking place in two ways. The A330/A340 uses the same full-capability standardized flight crew audio and frequency selections system as used on the A320 series, which is also largely used on other recent derivative aircraft. This is a break from the traditional highly customized lower-capability systems.

The other aspect of the revolution is more far-reaching: voice communication is giving way to data communications, with the advantages of lower error rates, more timely service, and lower costs. This started with highly customized ACARS systems for company communications, using very high frequencies (VHFs). The A330/A340 is equipped with a standardized ACARS system that can be used by any customer, with allowance for customers to easily introduce their own custom features to reflect their own needs.

These initial ACARS systems have been extended to offer worldwide coverage, even in mid-ocean and sparsely inhabited areas, using the Inmarsat facilities and high-frequency (HF) data link, and to cover not only company communications but also Air Traffic Control (ATC) services, starting with predeparture and oceanic clearances. On aircraft delivered since 1998, the ACARS unit has been replaced by the Air Traffic Services Unit (ATSU), which is designed to also accommodate safety-related ATC functions using the Aeronautical Telecommunications Network (ATN), offering the majority of ATC and other communication services now using voice and, more importantly, offering profitable migration to the ATN. The ATSU is the first unit to host software from a number of different vendors. The same ATSU is also used on the A320 family of aircraft. The ATN upgrade is being implemented to be available when the corresponding communication and ATC services are in service.

30.12 Flexibility and In-Service Updates

The first generation of aircraft with widespread digital systems, such as the A300FF, A310, and B767, suffered from some of the same disadvantages as their analog-system predecessors because their avionics were not designed to accommodate unplanned changes. Once a design change was made, equipment had to be removed from the aircraft, program memories had to be reloaded in the avionics workshops (sometimes by physically changing parts), and the equipment reinstalled. At some point, the airframe manufacturer usually got involved to certify the change. There was an advantage in the avionics shop, because reloading a program and retesting is a faster and cheaper activity than installing a kit of new electronic parts, but the major cost of carrying out the change on the aircraft stayed the same.

The A330/A340 systems have, to a large extent, overcome this disadvantage, at greatly reduced cost, by having *in situ* facilities on the aircraft for those digital LRUs that have been identified as requiring in-service change. Two techniques are used, depending on the criticality of the LRU concerned and other practicalities:

1. On-board replaceable memories (OBRMs)—memory modules located on the (accessible) front panel of an LRU—come in industry-standard sizes, cost much less than the LRU itself, and can be "recycled" many times. The visible part of the OBRM contains the LRU's software part number section. OBRMs comply with the toughest criticality criteria, enable classic configuration control of the LRU, and require no tools to change. They have been in use on the A320 since 1988.
2. On-board data loading, using 3.5-in. diskettes and other media, is a little slower. Although it is even less expensive than OBRMs, it does require a data loader to be carried to or installed on board the aircraft and an adaptation to the aircraft's classic configuration control techniques. The same data loader is used for the FMS data base.

Both techniques enable software updates to be carried out overnight on the whole fleet.

Another aspect is flexibility in dealing with airlines' changing needs. The basic equipment for the aircraft is designed with a number of pin-programmable features that correspond to frequently requested airline changes and other systems like the FMS, in which the airline loads a database that specifies its own preferences. These features allow airlines to pool databases and standard spares at outstations and still obtain the kind of operation that they need. Another feature is partitioned software, in which heavily customized systems like ACARS can be certified just once for all users, with one set of "core" software. The airline may load its own additional operational software on top of this core to reflect its own needs.

Lastly, certain systems like the optional Aircraft Condition Monitoring System (ACMS), which used to be heavily customized, use a combination of these techniques to enable an airline to select the features that it needs out of a very extensive selection that forms a superset of the needs of all the customers.

30.13 Development Environment

The development of each Airbus aircraft has been supported by an Iron Bird whole-aircraft systems rig and by supporting systems rigs that enable work to proceed simultaneously without mutual interference. The A330/A340 model is no exception, and a number of facilities have been constructed specifically for this program. These methods are now being used by other airframe manufacturers.

Proper software development is an essential part of systems development throughout the aircraft, and a number of software tools have been developed, notably in the areas of formal methods, rapid prototyping, automatic coding, and rapid data recovery and analysis. These are supplemented by large, fast data recording and telemetry facilities on the test aircraft fleet, associated with real-time and rapid-playback test data displays for the benefit of the flight test observers on board the test aircraft and for the test and systems engineers on the ground.

The result of this environment, the proper use of features from previous programs, and the proper management of test data flow and the resulting decision process has created an aircraft that has had a remarkable trouble-free period of introduction into service. This is true both in terms of customer satisfaction and in terms of measurable parameters such as delay rate, which have been up to an order of magnitude better for A340 than for the previous derivative long-range aircraft that entered service.

30.14 Support Environment

The A330/A340 airplanes have a number of unique support features, apart from those previously described. As with other Airbus aircraft, an airframe-wide Automatic Test Equipment unit is available to customers, along with an airframe-wide test program suite. No other airframe manufacturer offers this facility for avionics. The Aircraft Maintenance Manual and Trouble Shooting Manual have been carefully designed to integrate with the CMS for easier, faster fault rectification. For those airlines that wish to use it, a software package for a PC-compatible laptop is available to further speed fault-finding. The documentation is also compatible with open industry computer text and graphics standards to facilitate the introduction of intelligent maintenance documentation systems.

31

Airbus Electrical Flight Controls: A Family of Fault-Tolerant Systems

Pascal Traverse
Airbus SAS

31.1 Introduction

The first electrical flight control system for a civil aircraft was designed by Aerospatiale (now fully integrated in Airbus) and installed on the Concorde. This is an analog, full-authority system for all control surfaces. The commanded control surface positions are directly proportional to the stick inputs. A mechanical back-up system is provided on the three axes.

The first generation of electrical flight control systems with digital technology appeared on several civil aircraft at the start of the 1980s with the Airbus A310 program. These systems control the slats, flaps, and spoilers. These systems were designed with very stringent safety requirements (control surface runaway must be extremely improbable). As the loss of these functions results in a supportable increase in the crew's workload, it is possible to lose the system in some circumstances.

The Airbus A320 (certified in early 1988) is the first example of a second generation of civil electrical flight control aircraft, rapidly followed by the A330/A340 aircraft (certified at the end of 1992). These aircraft benefit from the significant experience gained in the technologies used for a

TABLE 31.1   Incremental Introduction of New Technologies

First Flight In:	1955	1969	1972	1978–1983	1983	1987
Servo-controls and artificial	x	x	x	x	x	→x
Feel						
Electro-hydraulic actuators		x	x	x	x	→x
Command and monitoring		x	x	x	x	→x
Computers						
Digital computers			x	x	x	→x
Trim, yaw damper, protection	x	x	x	x	x	→x
Electrical flight controls		x		x	x	→x
Side-stick, control laws				x		→x
Servoed aircraft (auto-pilot)	x	x	x	x	x	→x
Formal system safety		x	x	x	x	→x
Assessment						
System integration testing	x	x	x	x	x	→x
	Carevelle	Concorde	A300	Flight test Concorde A300	A310, A300–600	A320

fly-by-wire system (see Table 31.1). The distinctive feature of these aircrafts is that all control surfaces are electrically controlled and that the system is designed to be available under all circumstances.

This system was built to very stringent dependability requirements both in terms of safety (the system may generate no erroneous signals) and availability (the complete loss of the system is extremely improbable).

The overall dependability of the aircraft fly-by-wire system relies in particular on the computer arrangement (the so-called control/monitor architecture), the system tolerance to both hardware and software failures, the servo-control and power supply arrangement, the failure monitoring, and the system protection against external aggressions. It does this without forgetting the flight control laws which minimize the crew workload, the flight envelope protections which allow fast reactions while keeping the aircraft in the safe part of the flight envelope, and finally the system design and validation methods.

The aircraft safety is demonstrated by using both qualitative and quantitative assessments; this approach is consistent with the airworthiness regulation. Qualitative assessment is used to deal with design faults, interaction (maintenance, crew) faults, and external environmental hazard. For physical ("hardware") faults, both qualitative and quantitative assessments are used (Traverse et al., 2010). The quantitative assessment covers the FAR /CS 25.1309 requirement, and links the failure condition classification (minor to catastrophic) to its probability target.

This chapter describes the Airbus fly-by-wire systems from a fault-tolerant standpoint. The fly-by-wire basic principles are common to all Airbus currently in production: the A320 (and its family: A318, A319, A321), the A330/A340 (including the significant variant of the A340-600), the A380, the A400M and the A350. These principles are presented first, followed by the description of the main system features common to these aircraft, the failure detection and reconfiguration procedures, the specificities of these aircraft, and the design, development, and validation procedures. Future trends in terms of fly-by-wire fault-tolerance conclude this overview.

31.2 Fly-by-Wire Principles

On aircraft of the A300 and A310 type, the pilot orders are transmitted to the servo-controls by an arrangement of mechanical components (rods, cables, pulleys, etc.). In addition, specific computers and actuators driving the mechanical linkages restore the pilot feels on the controls and transmit the auto-pilot commands (see Figure 31.1).

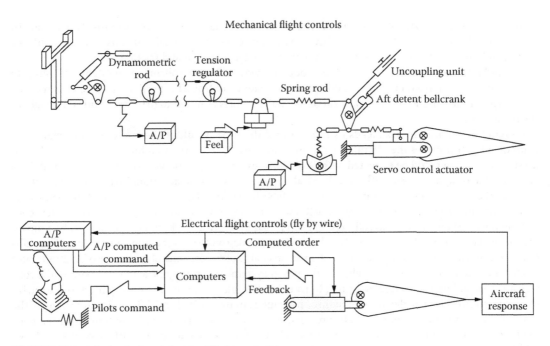

FIGURE 31.1 Mechanical and electrical flight control.

The term fly-by-wire has been adopted to describe the use of electrical rather than mechanical signalling of the pilot's commands to the flying control actuators. One can imagine a basic form of fly-by-wire in which an airplane retained conventional pilot's control columns and wheels, hydraulic actuators (but electrically controlled), and artificial feel as experienced in the 1970s with the Concorde program. The fly-by-wire system would simply provide electrical signals to the control actuators that were directly proportional to the angular displacement of the pilot's controls, without any form of enhancement.

In fact, the design of the A320 et al. flight control systems takes advantage of the potential of fly-by-wire for the incorporation of control laws that provide extensive stability augmentation and flight envelope limiting (Favre, 1993). The positioning of the control surfaces is no longer a simple reflection of the pilot's control inputs and conversely, the natural aerodynamic characteristics of the aircraft are not fed back directly to the pilot (see Figure 31.2).

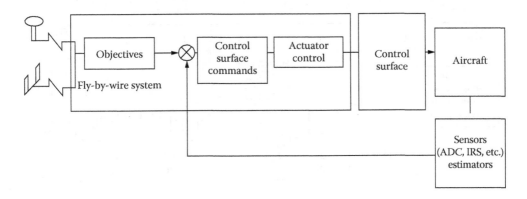

FIGURE 31.2 Flight control laws.

The sidesticks, now part of a modern cockpit design with a large visual access to instrument panels, can be considered as the natural issue of fly-by-wire, since the mechanical transmissions with pulleys, cables, and linkages can be suppressed with their associated backlash and friction.

The induced roll characteristics of the rudder provide sufficient roll manoeuvrability enabling to design a mechanical back-up on the rudder alone for lateral control. This permitted the retention of the advantages of the sidestick design, now rid of the higher efforts required to drive mechanical linkages to the roll surfaces.

Looking for minimum drag leads us to minimize the negative lift of the horizontal tail plane and consequently diminishes the aircraft longitudinal stability. It was estimated for the Airbus family that no significant gain could be expected with rear centre-of-gravity positions beyond a certain limit. This allowed us to design a system with a mechanical back-up requiring no additional artificial stabilization.

These choices were obviously fundamental to establish the now-classical architecture of the Airbus fly-by-wire systems (Figures 31.3 and 31.4), namely a set of five (or more) full-authority digital computers controlling the three pitch, yaw, and roll axes and completed by an ultimate back-up (mechanical back-up on the trimmable horizontal stabilizer and on the rudder or analog on elevator, aileron and rudder). (Two additional computers as part of the auto pilot system are in charge of rudder control in the case of A320 aircraft.)

Of course, a fly-by-wire system relies on the power systems energizing the actuators to move the control surfaces and on the computer system to transmit the pilot controls. The energy used to pressurize the servo-controls is provided by a set of three hydraulic circuits, one of which is sufficient to control the aircraft. One of the three circuits can be pressurized by a Ram air turbine, which automatically extends in case of an all-engine flame-out. From A380, electrical actuators are used. Airbus standard (A380, A400M, and A350) is thus now to use two hydraulic circuits plus two electrical circuits to power the actuators, Ram Air Turbine providing electrical power.

The electrical power is normally supplied by two segregated networks, each driven by one or two generators, depending on the number of engines. In case of loss of the normal electrical generation, an emergency generator supplies power to a limited number of fly-by-wire computers (among others). These computers can also be powered by the two batteries.

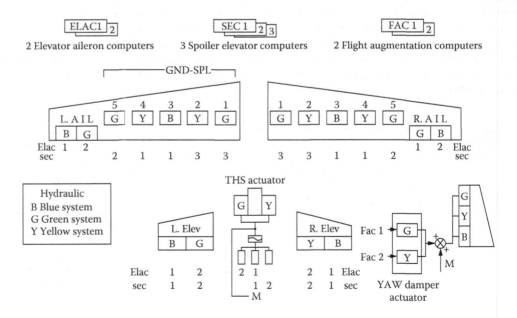

FIGURE 31.3   A320/A321 flight control system architecture.

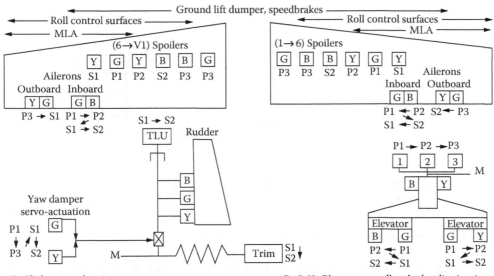

FIGURE 31.4 A330/A340 flight control system architecture.

31.3 Main System Features

31.3.1 Computer Arrangement

31.3.1.1 Redundancy

The five fly-by-wire computers are simultaneously active. They are in charge of control law computation as a function of the pilot inputs as well as individual actuator control. The system incorporates sufficient redundancies to provide the nominal performance and safety levels with one failed computer, while it is still possible to fly the aircraft safely with one single computer active.

As a control surface runaway may affect the aircraft safety (elevators for example), each computer is divided into two physically separated channels (Figure 31.5). The first one, the control channel, is permanently monitored by the second one, the monitor channel. In case of disagreement between control and monitor, the computer affected by the failure is passivated, while the computer with the next highest priority takes control. The repartition of computers, servo-controls, hydraulic circuit, and electrical bus bars and priorities between the computers are dictated by the safety analysis including the engine burst analysis and a set of particular risks.

31.3.1.2 Dissimilarity

Despite the nonrecurring costs induced by dissimilarity, it is fundamental that the five computers all be of different natures to avoid common mode failures. These failures could lead to the total loss of the electrical flight control system.

Consequently, two types of computers may be distinguished:

1. 1.2 ELAC (elevator and aileron computers) and 3 SEC (spoiler and elevator computers) on A320 family
2. 2.3 FCPC (flight control primary computers) and 2 FCSC (flight control secondary computers) on A330/A340 and the others

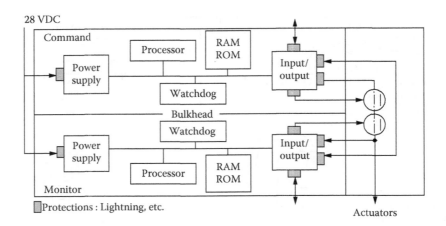

FIGURE 31.5   Command and monitoring computer architecture.

Taking the A320 as an example, the ELACs have been designed around 68,010 microprocessors and the SECs with 80,186 microprocessors. We therefore have two different design and manufacturing teams with different microprocessors (and associated circuits), different computer architectures, and different functional specifications. At the software level, the architecture of the system leads to the use of four software packages (ELAC control channel, ELAC monitor channel, SEC control channel, and SEC monitor channel) when, functionally, one would suffice.

31.3.1.3 Serve-Control Arrangement

Ailerons and elevators can be positioned by two servo-controls in parallel. As it is possible to lose control of one surface, a damping mode was integrated into each servo-control to prevent flutter in this failure case. Generally, one servo-control is active and the other one is damped. In case of loss of electrical control, the elevator actuators are centered by a mechanical feedback to increase the horizontal stabilizer efficiency.

Rudder and horizontal stabilizer controls are designed to receive both mechanical and electrical inputs (or electrical input only, depending on the A/C model). One servo-control per spoiler surface is sufficient. The spoiler servo-controls are pressurized in the retracted position in case of loss of electrical control.

31.3.1.4 Flight Control Laws

The general objective of the flight control laws integrated in a fly-by-wire system is to improve the natural flying qualities of the aircraft, in particular in the fields of stability, control, and flight domain protections. In a fly-by-wire system, the computers can easily process the anemometric and inertial information as well as any information describing the aircraft state. Consequently, control laws corresponding to simple control objectives could be designed. The stick inputs are transformed by the computers into pilot control objectives which are compared to the aircraft's actual state measured by the inertial and anemometric sensors. Thus, as far as longitudinal control is concerned, the sidestick position is translated into vertical load factor demands, while lateral control is achieved through roll rate, sideslip, and bank angle objectives.

The stability augmentation provided by the flight control laws improves the aircraft flying qualities and contributes to aircraft safety. As a matter of fact, the aircraft remains stable in case of perturbations such as gusts or engine failure due to very strong spin stability, unlike conventional aircraft. Aircraft control through objectives significantly reduces the crew workload; the fly-by-wire system acts as the inner loop of an autopilot system, while the pilot represents the outer loop in charge of objective management.

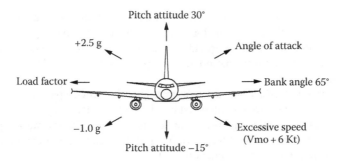

FIGURE 31.6 A320 flight envelope protections.

Finally, protections forbidding potentially dangerous excursions out of the normal flight domain can be integrated in the system (Figure 31.6). The main advantage of such protections is to allow the pilot to react rapidly without hesitation, since he knows that this action will not result in a critical situation.

31.3.1.5 Computer Architecture

Each computer can be considered as being two different and independent computers placed side by side (see Figure 31.5). These two (sub) computers have different functions and are placed adjacent to each other to make aircraft maintenance easier. Both command and monitoring channels of the computer are simultaneously active or simultaneously passive, ready to take control.

Each channel includes one or more processors, their associated memories, input/output circuits, a power supply unit, and specific software. When the results of these two channels diverge significantly, the links between the computer and the exterior world are cut by the channel or channels which detected the failure. The system is designed so that the computer outputs are then in a dependable state (signal interrupt via relays). Failure detection is mainly achieved by comparing the difference between the control and monitoring commands with a predetermined threshold. As a result, all consequences of a single computer fault are detected and passivated, which prevents the resulting error from propagating outside of the computer. This detection method is completed by permanently monitoring the program sequencing and the program correct execution.

Flight control computers must be robust. In particular, they must be especially protected against overvoltages and undervoltages, electromagnetic aggressions, and indirect effects of lightning. They are cooled by a ventilation system but must operate correctly even if ventilation is lost.

31.3.1.6 Installation

The electrical installation, in particular the many electrical connections, also comprises a common-point risk. This is avoided by extensive segregation. In normal operation, two electrical generation systems exist without a single common point. The links between computers are limited; the links used for monitoring are not routed with those used for control. The destruction of a part of the aircraft is also taken into account; the computers are placed at three different locations, certain links to the actuators run under the floor, others overhead, and others in the cargo compartment.

31.4 Failure Detection and Reconfiguration

31.4.1 Flight Control Laws

The control laws implemented in the flight control system computers have full authority and must be elaborated as a function of consolidated information provided by at least two independent sources in agreement.

Consequently, the availability of control laws using aircraft feedback (the so-called normal laws) is closely related to the availability of the sensors. The Airbus aircraft fly-by-wire systems use the information of three air data and inertial reference units (ADIRUs), as well as specific accelerometers and rate gyros. Moreover, in the case of the longitudinal normal law, analytical redundancy is used to validate the pitch rate information when provided by a single inertial reference unit. The load factor is estimated through the pitch rate information and compared to the available accelerometric measurements to validate the IRS data.

After double or triple failures, when it becomes impossible to compare the data of independent sources, the normal control laws are reconfigured into laws of the direct type where the control surface deflection is proportional to the stick input. To enhance the dissimilarity, the more sophisticated control laws with aircraft feedback (the normal laws) are integrated in one type of computer, while the other type of computer incorporates the direct laws only.

31.4.2 Actuator Control and Monitor

The general idea is to compare the actual surface position to the theoretical surface position computed by the monitoring channel. When needed, the control and monitor channels use dedicated sensors to perform these comparisons. Specific sensors are installed on the servovalve spools to provide an early detection capability for the elevators. Both channels can make the actuator passive. A detected runaway will result in the servo-control deactivation or computer passivation, depending on the failure source.

31.4.3 Comparison and Robustness

Specific variables are permanently compared in the two channels. The difference between the results of the control and monitoring channels are compared with a threshold. This must be confirmed before the computer is disconnected. The confirmation consists of checking that the detected failure lasts for a sufficiently long period of time. The detection parameters (threshold, temporization) must be sufficiently "wide" to avoid unwanted disconnections and sufficiently "tight" so that undetected failures are tolerated by the computer's environment (the aircraft). More precisely, all systems tolerance (most notably sensor inaccuracy, rigging tolerances, computer asynchronism) are taken into account to prevent undue failure detection, and errors which are not detectable (within the signal and timing thresholds) are assessed with respect to their handling quality and structural loads effect.

Note that to detect an undue oscillation, detection threshold becomes fairly sophisticated, as a function of the frequency of the signal to be detected (Goupil, 2011; Cazes et al., 2012).

31.4.4 Latent Failures

Certain failures may remain masked a long time after their occurrence. A typical case is a monitoring channel affected by a failure resulting in a passive state and detected only when the monitored channel itself fails. Tests are conducted periodically so that the probability of the occurrence of an undesirable event remains sufficiently low (i.e., to fulfil [FAR/CS 25] § 25.1309 quantitative requirement). Typically, a computer runs its self-test and tests its peripherals during the power-on of the aircraft, and therefore at least once a day or at each flight.

31.4.5 Reconfiguration

As soon as the active computer interrupts its operation relative to any function (control law or actuator control), one of the standby computers almost instantly changes to active mode with no or limited jerk on the control surfaces. Typically, duplex computers are designed so that they permanently transmit healthy signals which are interrupted as soon as the "functional" outputs (to an actuator, for example) are lost.

31.4.6 System Safety Assessment

The aircraft safety is demonstrated using qualitative and quantitative assessments. Qualitative assessment is used to deal with design faults, interaction (maintenance, crew) faults, and external environmental hazard. For physical ("hardware") faults, both qualitative and quantitative assessments are done. In particular, this quantitative assessment covers the link between failure condition classification (Minor to Catastrophic) and probability target.

31.4.7 Warning and Caution

It is deemed useful for a limited number of failure cases to advise the crew of the situation, and possibly that the crew act as a consequence of the failure. Nevertheless, attention has to be paid to keep the level of crew workload acceptable. The basic rule is to get the crews attention only when an action is necessary to cope with a failure or to cope with a possible future failure. On the other hand, maintenance personnel must get all the failure information.

The warnings and cautions for the pilots are in one of the following three categories:

1. Red warning with continuous sound when an immediate action is necessary (e.g., to reduce airplane speed).
2. Amber caution with a simple sound, such that the pilot be informed although no immediate action is needed (e.g., in case of loss of flight envelope protections an airplane speed should not be exceeded).
3. Simple caution (no sound), such that no action is needed (e.g., a loss of redundancy).

Priority rules among these warnings and cautions are defined to present the most important message first (see also Traverse et al. [1994]).

31.5 Aircraft Specificities

Although basic principles are the same from A320 to the latest A350, each Aircraft has specificities, either due to the size and the mission of the plane, or to an incremental introduction of new technologies.

31.5.1 A330/A340

The general design objective relative to the A330/A340 fly-by-wire system was to reproduce the architecture and principles chosen for the A320 as much as possible for the sake of commonality and efficiency, taking account of the A340 particularities (long-range four-engine aircraft).

As is now common for each new program, the computer functional density was increased between the A320 and A330/A340 programs: The number of computers was reduced to perform more functions and control an increased number of control surfaces (Figure 31.3).

In term of dissimilarity between the two types of computers (PRIM and SEC), a major decision was to force the choice of technology used by these computers, instead of relying on the independent choice of each development team. This is coming from the reduced number of potential technologies; a consequence of this choice is to develop both computers within Airbus.

The general concept of the A320 flight control laws was maintained, adapted to the aircraft characteristics, and used to optimize the aircraft performance, as follows:

- The angle of attack protection was reinforced to better cope with the aerodynamic characteristics of the aircraft.
- The Dutch roll damping system was designed to survive against rudder command blocking, thanks to an additional damping term through the ailerons, and to survive against an extremely

improbable complete electrical failure thanks to an additional autonomous damper. The outcome of this was that the existing A300 fin could be used on the A330 and A340 aircraft with the associated industrial benefits.

- The take-off performance could be optimized by designing a specific law that controls the aircraft pitch attitude during the rotation.
- The flexibility of fly-by-wire was used to optimize the minimum control speed on the ground (VMCG). In fact, the rudder efficiency was increased on the ground by fully and asymmetrically deploying the inner and outer ailerons on the side of the pedal action as a function of the rudder travel: the inner aileron is commanded downwards, and the outer aileron (complemented by one spoiler) is commanded upwards.
- A first step in the direction of structural mode control through fly-by-wire was made on the A340 program through the so-called "turbulence damping function" destined to improve passenger comfort by damping the structural modes excited by turbulence.

31.5.2 A340-600

A340-600 is derived from the basic A340 model by increasing the length of the fuselage. This characteristic led to:

Specific control laws, able to manage both rigid and flexible body modes (Kubica et al., 1995).

A full electrical control of the rudder to get a very precise control without getting the maintenance cost needed to fight the wear of a mechanical control. Accordingly an ultimate back-up was designed with an analog computer linked to the pedals in the cockpit and to one of the rudder servocontrol. This device has been subsequently installed on all A330 and A340.

The auto-pilot inner loop has been integrated in the flight control computer to reduce the level of asynchronism between the functions.

31.5.3 A380

The size of the A380 is an obvious challenge, requiring adaptation of control laws, of the architecture of the system to cope with multiple control surfaces, of the actuators and the power systems (Lelaie, 2012).

However, the most striking evolution is certainly the use of electrical actuation. In addition to hydraulic servocontrol, Electro-Hydrostatic Actuators and Electrical Back-up Hydraulic Actuators are used (Figure 31.7—Todeschi, 2007). They are powered by the two independent electrical circuits that were on previous aircraft devoted to avionics only. The power architecture has thus evolved from three hydraulic circuits for actuation and two electrical circuits for avionics (typical A320/A330/A340) to three hydraulic circuits used for actuation and two electrical circuits used for both actuation and avionics.

The primary reason for the use of electrical actuation was to improve the survivability of the A380 against particular risks such that an engine rotor burst. Indeed, electrical cables can easily be routed in passenger area (main and upper decks of A380), which is not usually done for hydraulic pipes.

Additionally, electrical actuation allowed to simplify the hydraulic system and to change the ram air turbine from a hydraulic generator to an electrical one. A key enabler was the availability of the technology: prototypes of EHA and EBHA had been flown on A320 years before the A380 launch.

On top of electrical actuation, the A340-600 experience has been prolonged by integrating the complete auto-pilot and the flight controls together and by having an ultimate back-up fully electrical, with an analog computer linked to the side-sticks and pedals and to a rudder, ailerons and elevators.

New functions have also been introduced on A380.

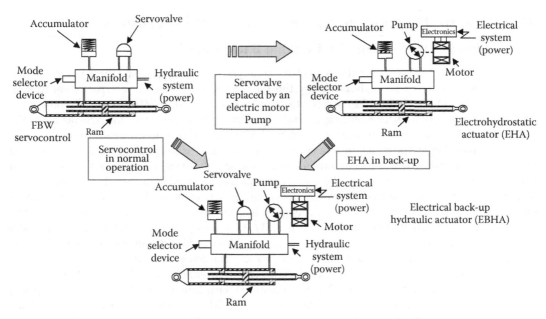

FIGURE 31.7   Electrical actuation.

Brake to Vacate to control automatically the braking and spoiler systems In order to reach a pre-selected exit from the runway (Villaumé, 2009).

Runway Overrun Protection: both by warning the crew if the landing conditions may be unsuitable and by monitoring the available landing distance in real time (Jacob et al., 2009).

AP/FD TCAS: a support to the pilot to answer a Resolution Advisory from the Traffic Collision Avoidance System, either automatically or with the Flight Director (Botargues, 2009); TCAS Alert Prevention (TCAP)to avoid nuisance Resolution Advisory (Botargues, 2012).

31.5.4 A400M

A400M is a multi-role airlifter with architecture very close to the A380 one. The main difference is related to the functions to be performed and the manoeuvrability requirements needed for the European NATO nations.

The safety objectives for the FBW part of the system (PRIM plus SEC) have been defined on civil Airbus without benefiting from the ultimate back-up. However, on the A400M, the back-up is fully capable of continued safe flight and landing and this must be demonstrated to the customer. This results in a decreased need for dissimilarity between PRIM and SEC; they share a common type of hardware (but different functions and software). Tolerance to a design or hardware manufacturing error is ensured by the functional dissimilarity between PRIM and SEC on the one hand, and the total hardware dissimilarity between PRIM/SEC and the ultimate back-up redundancy, on the other.

31.5.5 A350

A350 is also very close to the A380 architecture. One major difference is the introduction of a variable camber function with a tight synchronization of the spoilers to flaps move.

A350 flight control system is also the first to be distributed in the aircraft, with actuator control being decentralized in electronics module attached to the actuators themselves. Data between these actuator modules and the main flight control computers are exchanged through bidirectional data buses, to MIL-STD-1553 standard.

31.6 Design, Development, and Validation Procedures

31.6.1 Fly-by-Wire System Certification Background

An airline can fly an airplane only if this airplane has a type certificate issued by the aviation authorities of the airline country. For a given country, this type certificate is granted when the demonstration has been made and accepted by the appropriate organization (Federal Aviation Administration in the U.S., European Aviation Safety Agency in several European countries, etc.) that the airplane meets the country's aviation rules and consequently a high level of safety. Each country has its own set of regulatory materials although the common core is very large. They are basically composed of two parts: the requirements on one part, and a set of interpretations and acceptable means of compliance in a second part. An example of requirement is "The aeroplane systems must be designed so that the occurrence of any failure condition which would prevent the continued safe flight and landing of the aeroplane is extremely improbable" (in Federal and Joint Aviation Requirements 25.1309, [FAR/CS 25]). An associated part of the regulation (Advisory Circular from FAA, Advisory Material—Joint from CS 25.1309) gives the meaning and discuss such terms as "failure condition," and "extremely improbable." In addition, guidance is given on how to demonstrate compliance (Traverse et al., 2010).

The aviation regulatory materials are evolving to be able to cover new technologies (such as the use of fly-by-wire systems). This is done through special conditions targeting specific issues of a given airplane, and later on by modifying the general regulatory materials. With respect to Airbus fly-by-wire airplane, the following innovative topics were addressed for certification (note: some of these topics were also addressing other airplane systems):

- Flight envelope protections
- Side-stick controller
- Static stability
- Interaction of systems and structure
- System safety assessment
- Lightning indirect effect and electromagnetic interference
- Integrity of control signal transmission
- Electrical power
- Software verification and documentation, automatic code generation
- System validation
- Application-specific integrated circuit

31.6.2 The A320 Experience

31.6.2.1 Design

The basic element developed on the occasion of the A320 program is the so-called SAO specification (Spécification Assistée par Ordinateur), the Aerospatiale graphic language defined to clearly specify control laws and system logics. One of the benefits of this method is that each symbol used has a formal definition with strict rules governing its interconnections. The specification is under the control of a configuration management tool and its syntax is partially checked automatically.

31.6.2.2 Software

The software is produced with the essential constraint that it must be verified and validated. Also, it must meet the world's most severe civil aviation standards (level 1 software to DO178A [1985]—see also Barbaste and Desmons [1988]). The functional specification acts as the interface between the aircraft manufacturer's world and the software designer's world. The major part of the A320 flight control software specification is a copy of the functional specification. This avoids creating errors when translating the functional specification into the software specification. For this "functional" part of the software, validation is not required as it is covered by the work carried out on the functional specification. The only part of the software specification to be validated concerns the interface between the hardware and the software (task sequencer, management of self-test software inputs/outputs). This part is only slightly modified during aircraft development.

To make software validation easier, the various tasks are sequenced in a predetermined order with periodic scanning of the inputs. Only the clock can generate interrupts used to control task sequencing. This sequencing is deterministic. A part of the task sequencer validation consists in methodically evaluating the margin between the maximum execution time for each task (worst case) and the time allocated to this task. An important task is to check the conformity of the software with its specification. This is performed by means of tests and inspections. The result of each step in the development process is checked against its specification. For example, a code module is tested according to its specification. This test is, first of all, functional (black box), then structural (white box).

Adequate coverage must be obtained for the internal structure and input range. The term "adequate" does not mean that the tests are assumed as being exhaustive. For example, for the structural test of a module, the equivalence classes are defined for each input. The tests must cover the module input range taking these equivalence classes and all module branches (among other things) as a basis. These equivalence classes and a possible additional test effort have the approval of the various parties involved (aircraft manufacturer, equipment manufacturer, airworthiness authorities, designer, and quality control).

The software of the control channel is different from that of the monitoring channel. Likewise, the software of the ELAC computer is different from that of the SEC computer (the same applies to the FCPC and FCSC on later A/C models). The aim of this is to minimize the risk of a common error which could cause control surface runaway (control/monitoring dissimilarity) or complete shutdown of all computers (ELAC/SEC dissimilarity).

The basic rule to be retained is that the software is made in the best possible way. This has been recognized by several experts in the software field both from industry and from the airworthiness authorities. Dissimilarity is an additional precaution which is not used to reduce the required software quality effort.

31.6.2.3 System Validation

Simulation codes, full-scale simulators and flight tests were extensively used in a complementary way to design, develop, and validate the A320 flight control system (see also Chatrenet [1989]), in addition to analysis and peer review.

A "batch" type simulation code called OSMA (Outil de Simulation des Mouvements Avion) was used to initially design the flight control laws and protections, including the nonlinear domains and for general handling quality studies.

A development simulator was then used to test the control laws with a pilot in the loop as soon as possible in the development process. This simulator is fitted with a fixed-base faithful replica of the A320 cockpit and controls and a visual system; it was in service in 1984, as soon as a set of provisional A320 aero data, based on wind tunnel tests, was made available. The development simulator was used to develop and initially tune all flight control laws in a closed-loop cooperation process with flight test pilots.

Three "integration" simulators were put into service in 1986. They include the fixed replica of the A320 cockpit, a visual system for two of them, and actual aircraft equipment including computers, displays, control panels, and warning and maintenance equipment. One simulator can be coupled to the "iron bird" which is a full-scale replica of the hydraulic and electrical supplies and generation, and is fitted with all the actual flight control system components including servojacks. The main purpose of these simulators is to test the operation, integration, and compatibility of all the elements of the system in an environment closely akin to that of an actual aircraft.

Finally, flight testing remains the ultimate and indispensable way of validating a flight control system. Even with the current state of the art in simulation, simulators cannot yet fully take the place of flight testing for handling quality assessment. On this occasion a specific system called SPATIALL (Système Pour Acquisition et Traitement d'Informations Analogiques ARINC et Logiques) was developed to facilitate the flight test. This system allows the flight engineer to:

- Record any computer internal parameter
- Select several preprogrammed configurations to be tested (gains, limits, thresholds, etc.)
- Inject calibrated solicitations to the controls, control surfaces, or any intermediate point.

The integration phase complemented by flight testing can be considered as the final step of the validation side of the now-classical V-shaped development/validation process of the system.

31.6.3 The A340 Experience

31.6.3.1 Design

The definition of the system requires that a certain number of actuators be allocated to each control surface and a power source and computers assigned to each actuator. Such an arrangement implies checking that the system safety objectives are met. A high number of failure combinations must therefore be envisaged. A study has been conducted with the aim of automating this process.

It was seen that a tool which could evaluate a high number of failure cases, allowing the use of capacity functions, would be useful and that the possibility of modelling the static dependencies was not absolutely necessary even though this may sometimes lead to a pessimistic result. This study gave rise to a data processing tool which accepts as input an arrangement of computers, actuators, hydraulic and electrical power sources, and also specific events such as simultaneous shutdown of all engines and, therefore, a high number of power sources. The availability of a control surface depends on the availability of a certain number of these resources. This description was made using a fault tree-type support as input to the tool.

The capacity function used allows the aircraft roll controllability to be defined with regard to the degraded state of the flight control system. This controllability can be approached by a function which measures the roll rate available by a linear function of the roll rate of the available control surfaces. It is then possible to divide the degraded states of the system into success or failure states and thus calculate the probability of failure of the system with regards to the target roll controllability.

The tool automatically creates failure combinations and evaluates the availability of the control surfaces and, therefore, a roll controllability function. It compares the results to the targets. These targets are, on the one hand, the controllability (availability of the pitch control surfaces, available roll rate, etc.) and, on the other hand, the reliability (a controllability target must be met for all failure combinations where probability is greater than a given reliability target). The tool gives the list of failure combinations which do not meet the targets (if any) and gives, for each target of controllability, the probability of no satisfaction. The tool also takes into account a dispatch with one computer failed.

31.6.3.2 Automatic Programming

The use of automatic programming tools is becoming widespread. This tendency appeared on the A320 and is being confirmed on the A340 (in particular, the FCPC is, in part, programmed automatically). Such a tool has SAO sheets as inputs, and uses a library of software packages, one package being allocated to each symbol. The automatic programming tool links together the symbol's packages.

The use of such tools has a positive impact on safety. An automatic tool ensures that a modification to the specification will be coded without stress even if this modification is to be embodied rapidly (situation encountered during the flight test phase for example). Also, automatic programming, through the use of a formal specification language, allows onboard code from one aircraft program to be used on another.

Note that the functional specification validation tools (simulators) use an automatic programming tool. This tool has parts in common with the automatic programming tool used to generate codes for the flight control computers. This increases the validation power of the simulations. For dissimilarity reasons, only the FCPC computer is coded automatically (the FCSC being coded manually). The FCPC automatic coding tool has two different code translators, one for the control channel and one for the monitoring channel.

31.6.3.3 System Validation

The A320 experience showed the necessity of being capable of detecting errors as early as possible in the design process, to minimize the debugging effort along the development phase. Consequently, it was decided to develop tools that would enable the engineers to actually fly the aircraft in its environment to check that the specification fulfils the performance and safety objectives before the computer code exists.

The basic element of this project is the so-called SCADE specification (the SCADE language having replaced the SAO one used on A320 [Polchi, 2005]), the graphic language defined to clearly specify control laws and system logics. The specification is then automatically coded for engineering simulation purposes in both control law and system areas.

In the control law area, OCAS (Outil de Conception Assistée par Simulation) is a real-time simulation tool that links the SAO definition of the control laws to the already-mentioned aircraft movement simulation (OSMA). Pilot orders are entered through simplified controls including side-stick and engine thrust levels. A simplified PFD (primary flight display) visualizes the outputs of the control law. The engineer is then in a position to judge the quality of the control law that he has just produced, in particular with respect to law transition and nonlinear effects. In the early development phase, this same simulation was used in the full-scale A340 development simulator with a pilot in the loop.

In the system area, OSIME (Outil de SImulation Multi Equipement) is an expanded time simulation that links the SAO definition of the whole system (control law and system logic) to the complete servo-control modes and to the simulation of aircraft movement (OSMA). The objective was to simulate the whole fly-by-wire system including the three primary computers (FCPC), the two secondary computers (FCSC), and the servo-controls in an aircraft environment.

This tool contributed to the functional definition of the fly-by-wire system, to the system validation, and to the failure analysis. In addition, the behavior of the system at the limit of validity of each parameter, including time delays, could be checked to define robust monitoring algorithms. Non-regression tests have been integrated very early into the design process to check the validity of each new specification standard.

Once validated, both in the control law and system areas using the OCAS and OSIME tools, a new specification standard is considered to be ready to be implemented in the real computers (automatic coding) to be further validated on a test bench, simulator, and on the aircraft (Figure 31.8).

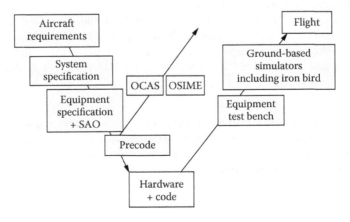

FIGURE 31.8 Validation methodology.

31.6.4 The A340-600 Experience

In the 1990s an integration of regulatory materials resulted in a set of four documents:

1. A document on system design, verification and validation, configuration management, quality assurance (ARP 4754, 1994; Landi and Nicholson, 2011).
2. A document on software design, verification, configuration management, quality assurance (DO178B, 1992)
3. A document on hardware design, verification, configuration management, quality assurance (DO254, 1995)
4. A document on the system safety assessment process (ARP 4761, 1994)

It is noteworthy that the A340-600 was certainly the first aircraft to be certified with a system (the flight control system) demonstrated compliant to the ARP4754.

31.6.5 The A380 Experience

The A380 has seen the introduction of model-checking techniques at several points in the development.

First of all, an automatic formal technique (proof tool) has been used to verify low-level requirements (Souyris et al., 2011). Formal verification consists of using an evidence tool to demonstrate that all possible executions of a component code meet its low-level formal requirements.

Beyond Unitary Verification, the following verification activities now benefit from equipped formal techniques (automated): calculation of a guaranteed upper bound marker of the amount of memory used for the program stack and calculation of a guaranteed upper marker of all execution time for each functional task. All these activities have been recognized by Certification Authorities.

In addition, model checking is used on some limited part of the functional specification of the software ("SCADE specification").

31.7 Future Trends

The fly-by-wire systems developed on the occasion of the A320 has demonstrated it's adaptability to very different airplanes (most notably the A380 and the A400M) and to innovation. If we consider the future trends that were mentioned in the 1996 Issue of this book,

1. Electrical actuation has been introduced on A380
2. Ultimate back-up is now fully electrical (first step on A340-600, concluded on A380)
3. Structural mode control has been significantly introduced with A340-600
4. Integrated Modular Avionics appeared on A380; some flight controls function are integrated (maintenance and warning) and the design of the PRIM computer is shared with the IMA modules' one.

So, what's next?

On one hand, we can certainly expect new types of electrical actuators, namely Electro-Mechanical Actuators that wouldn't need any hydraulic transmission inside (Todeschi, 2010).

On another hand, we have seen new functions on A380 that are providing increased safety margins. We can certainly expect more to come in the future.

References

ARP 4754. 1994. *System Integration Requirements.* Society of Automotive Engineers (SAE) and European Organization for Civil Aviation Electronics (EUROCAE). Revision A published in 2010. SAE: Warrendale, PA.

ARP 4761. 1994. *Guidelines and Tools for Conducting the Safety Assessment Process on Civil Airborne Systems and Equipment.* Society of Automotive Engineers (SAE) and European Organization for Civil Aviation Electronics (EUROCAE). EUROCAE: Malakoff, France.

Barbaste, L. and Desmons, J.P. 1988. Assurance qualité du logiciel et la certification des aéronefs/ Expérience A320. 1er séminaire EOQC sur la qualité des logiciels, Brussels, Belgium, pp. 135–146, April 1988.

Botargues, P. 2009. P. Airbus AP/FD TCAS mode: A new step towards safety improvement. *Safety First— The Airbus Safety Magazine,* February 7, 2009.

Botargues, P. 2012. The airbus TCAS alert prevention (TCAP). *Safety First—The Airbus Safety Magazine,* January 13, 2012.

Cazes, F., Chabert, M., Mailhes, C., Michel, P., Goupil, P., Dayre, R., and le Berre, H. 2012. Flight control system improvement based on a software sensor derived from partial least squares algorithm. *Eighth IFAC symposium on Fault-Detection, Supervision and Safety of Technical Processes (SAFEPROCESS),* Mexico City, Mexico, August 2012.

Chatrenet, D. 1989. Simulateurs A320 d'Aérospatiale: leur contribution à la conception, au développement et à la certification. *INFAUTOM 89,* Toulouse, France.

DO178A. 1985. *Software Considerations in Airborne Systems and Equipment Certification.* RTCA and European Organization for Civil Aviation Electronics (EUROCAE). ARINC: Annapolis, MD.

DO178B. 1992. *Software Considerations in Airborne Systems and Equipment Certification.* RTCA and European Organization for Civil Aviation Electronics (EUROCAE). Revision C has been published in 2012. ARINC: Annapolis, MD.

DO254. 1995. *Design Assurance Guidance for Complex Electronic Hardware Used in Airborne Systems.* RTCA and by European Organization for Civil Aviation Electronics (EUROCAE). EUROCA: Malakoff, France.

FAR/CS 25. *Airworthiness Standards: Transport Category Airplanes.* Part 25 of "Code of Federal Regulations, Title 14, Aeronautics and Space," for the Federal Aviation Administration, and "Airworthiness Joint Aviation Requirements—Large aeroplane" for the Joint Aviation Authorities.

Favre, C. 1993. Fly-by-wire for commercial aircraft—The Airbus experience. *International Journal of Control,* special issue on *Aircraft Flight Control,* 59(1), 139–157, January 1994, Taylor & Francis, Florence, KY.

Jacob, A., Lignée, R., and Villaumé, F. 2009. The runway overrun prevention system. *Safety First—The Airbus Safety Magazine,* July 8, 2009.

Kubica, F., Livet, T., Le Tron, X., and Bucharles, A. 1995. Parameter-robust flight control system for a flexible aircraft. *Control Eng. Prac.,* 3(9), 1209–1215.

Landi, A. and Nicholson, M. 2011. ARP4754A/ED-79A guidelines for development of civil aircraft and systems—Enhancements, novelties and key topics. *SAE 2011 AeroTech Congress & Exhibition,* Toulouse, France, October 18–21, 2011.

Lelaie, C. 2012. A380: Development of the flight controls. *Safety First—The Airbus Safety Magazine,* January 13, 2012, July 14, 2012, January 15, 2013.

Polchi, J.F. 2005. "Développement système. Un exemple avec l'outil SCADE: les commandes de vol Airbus". *AFIS/Journée outils de l'ingénierie système,* Toulouse, France, 2005.

Souyris, J., Delmas, D., and Duprat, S. 2011. Airbus: Formal verification in avionics. *Static Analysis of Software—The Abstract Interpretation*. John Wiley & Sons: Hoboken, NJ.

Todeschi, M. 2007. A380 flight control actuation—Lessons learned on EHAs design. *Recent Advances in Aerospace Actuation Systems and Components Conference*, Toulouse, France, June 2007.

Todeschi, M. 2010. Airbus—EMA for flight controls actuation system—Perspectives. *Recent Advances in Aerospace Actuation Systems and Components Conference*, Toulouse, France, May 2010.

Traverse, P. 2008. System safety in a few nutshells. *Fourth European Congress ERTS Embedded Real Time Software*, Toulouse, France, January 2008.

Traverse, P., Bezard, C., Camus, J.M., Lacaze, I., Leberre, H., Ringeard, P., and Souyris, J. 2010. Dependable avionics architectures: Examples of a fly-by-wire system. *Safety of Computer Architectures*. John Wiley & Sons, 2010. John Wiley & Sons: Hoboken, NJ.

Traverse, P., Brière, D., and Frayssignes, J.J. 1994. Architecture des commande de vol électriques Airbus, reconfiguration automatique et information équipage. *INFAUTOM 94*, Toulouse, France.

Villaumé, F. 2009. Brake-to-vacate system—The smart automatic braking system for enhanced surface operations. *FAST—Airbus Technical Magazine*, July 44, 2009.

III

Avionics Development: Tools, Techniques, and Methods

This final section is all about tools, techniques, and methods used to implement avionics functions. Topics include information on data buses, avionics architectures, modeling of avionics and some additional details on navigation algorithms. This section is presented in groups of three topical areas. The first group of chapters looks at hardware reliability and architectures used in modern avionics. The chapter on hardware reliability draws together a number of concepts first introduced in Section I of the handbook to show how avionics reliability is demonstrated. This is followed by two chapters on interconnect techniques and another two chapters on avionics architectures. Chapter 37 shows one way of accomplishing control functions within these architectures. Taken together, these six chapters provide an overall picture of some of the fundamental building blocks using in putting together a new avionics platform while ensuring it will meet the necessary reliability and availability requirements for safety-critical avionics.

The section then shifts to look at the approaches currently being used to accomplish the development of that avionics. As avionics systems get more and more complex, modeling tools have been employed to simulate various system components to assure interactions between components and observe system behavior. The next chapter discusses formal methods which are mathematical representations of systems used to prove or disprove the correctness of system requirements and design. Chapter 44 is but one example of complexity of avionics algorithms; this chapter goes into details of the types of considerations that are necessary for navigation system integration and tracking.

32

Electronic Hardware Reliability

P.V. Varde*
*University of Maryland
and
Bhabha Atomic
Research Centre*

Nikhil Vichare
Dell Computers

Ping Zhao
Apple Inc.

Diganta Das
University of Maryland

Michael G. Pecht
University of Maryland

32.1 Introduction

Reliability of avionic systems is crucial to management and control of aircraft and space systems. Failure of avionics can lead to a range of consequences from less critical, like disruption in schedule or aircraft operations to accident conditions. Major systems of avionics are: (1) flight computer, (2) data network, (3) guidance, navigation and control, (4) communication and tracking, and (5) electrical power. The electronics technology makes for significant part of the avionics. Hence, reliability assurance of electronic components plays an important role in design and operation of aviation systems.

Following are the major characteristics/requirements of avionics from the reliability considerations:

1. The avionics are complex systems exposed to unique set of life cycle environmental and operating conditions.
2. Apart from electronic hardware, the software and human factors have significant impact on its successful operation of avionics.
3. The stringent reliability requirements due to mission criticality and safety considerations.
4. Higher expected life requirements >15 years or more.

---

* Senior scientist at BARC and visiting professor at University of Maryland.

5. The technological advances in digital systems require that the reliability modeling approach address the issue in a manner such that reduces uncertainties in the final reliability estimates.
6. Unlike traditional systems where the reliability modeling deals with physical entities like components and systems, in avionics the concept of "service threads" forms essential features of reliability modeling.
7. Modeling for distributed nature of design and operations in aviation systems form inherent part of reliability assessment.
8. Redundancy, diversity and fault tolerant features are inherent to aviation systems design and management.
9. Apart from reliability; availability, maintainability and risk also form important design and operational aspects.
10. Common cause failure is an important issue that need special attention for modeling redundant systems or service threads.
11. Application of risk-informed assessment and management activities often forms part of design and operations procedures.

Apart from this, the considerations like (1) avionics development requires arduous part selection and management process as most of the commercial parts are not made from are not made for avionics life time requirements, (2) implementation through design for reliability to minimize the life cycle cost of the asset, and (3) quality assurance plans to assure consistent reliability of parts from sub-tier supplies for building a complex system-of-systems forms; integral part of acquisition process.

Reliability is defined as the probability that an item can perform its intended function for a specified interval under stated conditions. Reliability has two connotations deterministic and probabilistic that is closely interrelated. The deterministic aspects are generally addressed though design for reliability (DfR) framework while probabilistic aspects are usually evaluated through statistical theory or more precisely probabilistic approach. This chapter deals with both aspects. Formulation of requirements selection, parts selection, field trials/testing to assure quality, failure mode, mechanism, and effects analysis, root cause analysis to understand various failure mechanisms etc, to ensure reliable products/ components are discussed in Section 32.3. The probabilistic assessment of reliability deals with quantification of component reliability based on component failure data and mission time for given set of environmental conditions.

Keeping in view aviation requirements, this chapter covers, apart from reliability and availability, aspects related to "maintainability." In practice, these aspects become very important and can affect component reliability. The deterministic aspects of reliability are covered in Section 32.3 whiles the probabilistic aspects of reliability, availability, and maintainability; have been covered in the section following that. There is a growing interest to shift the approach to management of complex systems from reliability based to risk-based or risk-informed; therefore risk assessment forms an important aspect in support of decision. Hence, a brief discussion on risk assessment has been included. Also included is the use of reliability information in Section 32.6 to manage electronic components in avionics before Section 32.7.

Before initiating discussing on various aspects of reliability as applicable to aviation systems, it is important to discuss major features of aviation taxonomy [4].

The taxonomy is represented using hierarchical objects having "systems of systems" on top while (1) information system, (2) Remote/distributed system, and (3) sub-systems form the sub-nodes.

The information system in turn is comprised of automation system, data system and weather system, while display system forms part of the automation system. These systems have high criticality rating and availability requirements compared to the other two systems. The design of these systems is based on fault tolerant approach, redundancy, automatic fault detection and recovery mechanisms.

The surveillance system, navigation system and RCAG (remote communication between air and ground) forms sub-part of the remote distribution system. These systems are not as critical as information systems and failures in the system can degrade the performance of the system, but do not result in total loss of functions.

The support systems can be divided into two major sub-systems, viz, mission support system and Infrastructure systems. The activities like airspace management, Emergency and alerting and Infrastructure management form sub-activities of mission support systems, while power subsystems, telecom subsystems and HVAC (heating, ventilation, and air conditioning) subsystems form part of infrastructure systems. Even though often specific targets are set for achieving reliability, availability, and maintainability, the requirements are optimized not always by any specific predefined criteria but determined by what is achievable and life cycle cost considerations.

Terminologies, such as Service Thread, National Airspace System Architecture Safety Critical Category either have been explained in the chapter or reproduced in definitions section. The terminology given in MIL-STD-721C [1] is used in this chapter. The readers may refer to the definitions of various terms used in this document at the end of this chapter, before the reference section.

32.2 Background

Traditional avionics during the early period were based on analog technology. The incorporation of digital systems into avionics dated back to early 1960s when computers were employed in flight management systems. Around the same time reliability prediction techniques were developed. These reliability techniques, like MIL-HDBK-217F [2], were employed extensively. However, it was realized that alternative methods were in need to address issues which are mainly related to, (1) increasing complexity of components and systems, (2) limitation of statistical methods in general and exponential modeling approach in particular, (3) limitation for modeling of aging related failures which requires considerations of change in material properties, loading conditions, environmental stresses for improved prediction of reliability. The traditional documents like MIL-HDBK-781D [3] for reliability testing were adequate when the typical MTBF requirements were of the order of 1000 h. However, as the reliability of components/systems increased to the order of 1,000,000 h, the need of advanced methods for reliability assessment was felt. In the traditional approaches, statement of reliability often described in terms of MTBF. This approach had some inherent limitations, like (1) even for non repairable components the reliability was referred in terms of MTBF instead of the correct term, i.e., MTTF, (2) these MTBF statements were based on average failure rate and in the absence of spread or "standard deviation" of the data set, it was a challenge to interpret reliability of the component, and (3) unless the definition of failure criteria is provided with these estimates, it is difficult to compare the results coming from two or more sources.

Apart from this, the other factors that needed improvements in reliability modeling approaches were, (1) inherent increase in system reliability by improving the design and manufacturing techniques, like redundancy, diversity, fault tolerant features, etc; (2) need for capturing the variability by testing and maintenance; (3) increasing use of software components; (4) some management related issues like employment of Commercial off-the shelf components; and (5) issues of inventory storage affecting the reliability.

The new framework called RMA (reliability, maintainability, and availability) came into existence and addressed many of the limitations of the traditional reliability evaluation approach in aviation. In this framework a concept of recovery time was also included to address issues related to "risk" and availability [4].

32.3 Design for Reliability

This section covers the deterministic aspects of reliability. Design for reliability approach deals with designing and developing techniques, methods, material and process that makes the product/services inherently reliable. Major elements of design for reliability are formulation of reliability requirements and specifications, material selection, accelerated testing to understand failure modes and mechanisms within expected service life, testing for specified loads per standard requirements and commissioning tests. Beyond these, it is necessary to develop essential surveillance and test requirement, when the component or system is put into service. In order to achieve product reliability over the complete life cycle demands, the DfR approach requires vast engineering/management commitment and literal enforcement. These tasks impact electronic hardware reliability through the selection of materials, structural geometries and design tolerances, manufacturing processes and tolerances, assembly techniques, shipping and handling methods, operational conditions, and maintenance and maintainability guidelines [5]. The tasks are as follows:

1. *Define realistic product requirements and constraints for certain useful lifetime based on target markets.* This includes defining the functionality, physical attributes, performance, life cycle environment, useful life (with warranties), life cycle cost, testing and qualification methods, schedules, and end-of-life requirements for the product. The manufacturer and the customer must mutually define the product requirements in light of both the customer's needs and the manufacturer's capability to meet those needs.

2. *Define the product life cycle environment* by specifying all expected phases for a product from manufacture to end of life, including assembly, storage, handling, shipping, and operating and non-operating conditions of the product. Identify significant life cycle loads (individual and combination) for each phase and quantify the loads (typical range and variability) a product is expected to experience.

3. *Select the parts required for the product* by using a well-defined assessment procedure to ensure that reliability issues are addressed before and during design. The process enables maximizing the profit and minimizing the time to profit from the product, provides product differentiation, effectively uses the global supply chain, and ensures effective assessment, mitigation, and management of life cycle risks in using the part.

4. *Use a systematic methodology to identify all potential failure modes, mechanisms, and sites by which the product can be expected to fail.* Develop an understanding of the relationships between product requirements and the physical characteristics of the product (and their variation in the production process), the interactions of product materials with loads (stresses at application conditions), and their influence on product failure susceptibility with respect to the use conditions. This involves finding the failure mechanisms and the reliability models to quantitatively evaluate failure susceptibility. Failure mechanisms need to be prioritized to determine the environmental and operational loads that need to be accounted for in the design or to be controlled during operation.

5. *Design the product based on knowledge of the expected life cycle conditions* by using the selected materials and processes. Set appropriate specification limits, considering various stress limits and margins, derating, redundancy, and protective architecture.

6. *Qualify the product design, manufacturing, and assembly processes.* Qualification tests should be conducted to verify the reliability of the product in the expected life cycle conditions. Qualification tests should provide an understanding of the influence of process variations on product reliability. The goal of this step is to provide a physics-of-failure basis for design decisions, with an assessment of all possible failure mechanisms for the product.

7. *Identify, measure, and optimize the key processes in manufacturing and assembly required to make the part.* Monitor and control the manufacturing and assembly processes addressed in the design to reduce defects. Develop screens and tests to assess statistical process control of manufacturing and assembly steps.

8. *Manage the life cycle usage of the product* using closed loop management procedures. This includes inspection, testing, and maintenance procedures.

32.3.1 Product Requirements and Constraints

A product's requirements and constraints are defined in terms of customer demands and the company's core competencies, culture, and goals. There are various reasons to justify the creation of a product, such as to fill a perceived market need, to open new markets, to remain competitive in a key market, to maintain market share and customer confidence, to fill a need of specific strategic customers, to demonstrate experience with a new technology or methodology, to improve maintainability of an existing product, and to reduce the life cycle costs of an existing product.

From the perspective of requirements and constraints, products can be classified into three types: multi-customer products, single-customer products, and custom products with variability of the targeted customer population. To make reliable products, there should be a joint effort between the supplier and the customer. According to IEEE 1332 [6], there are three reliability objectives to achieve in the stage of defining products' requirements and constraints:

- The supplier, working with the customer, shall determine and understand the customer's requirements and product needs, so that a comprehensive design specification can be generated.
- The supplier shall structure and follow a series of engineering activities so that the resulting product satisfies the customer's requirements and product needs with regard to product reliability.
- The supplier shall include activities that assure the customer that the reliability requirements and product needs have been satisfied.

If the product is for direct sale to end users, marketing usually takes the lead in defining the product's requirements and constraints through interaction with the customer's marketplace, examination of product sales figures, and analysis of the competition. Alternatively, if the product is a subsystem that fits within a larger product, the requirements and constraints are determined by the customer's product into which the subsystem fits. The product definition process involves multiple influences and considerations. Figure 32.1 shows a diagram of constraints in the product definition process.

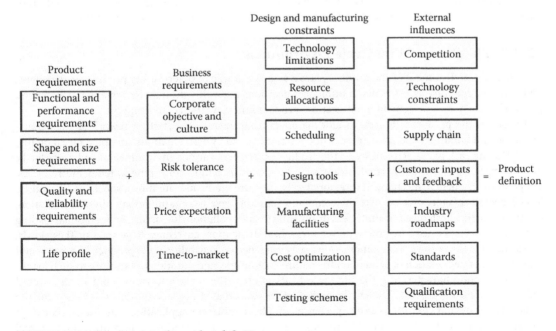

FIGURE 32.1 Constraints in the product definition process.

Two prevalent risks in requirements and constraints definition are the inclusion of irrelevant requirements and the omission of relevant requirements. The inclusion of irrelevant requirements can involve unnecessary design and testing time as well as money. Irrelevant or erroneous requirements result from two sources: requirements created by persons who do not understand the constraints and opportunities implicit in the product definition, and inclusion of requirements for historical reasons. The omission of critical requirements may cause the product to be non-functional or may significantly reduce the effectiveness and the expected market size of the product.

The initial requirements are formulated into a requirements document, where they are prioritized. The requirements document should be approved by engineers, management, customers, and other involved parties. Usually, the specific people involved in the approval will vary with the organization and the product. For example, for human-safety-critical products, legal representatives should attend the approval to identify legal considerations with respect to the implementation of the parts selection and management team's directives. The results of capturing product requirements and constraints allow the design team to choose parts and develop products that conform to company objectives.

Once a set of requirements has been completed, the product engineering function creates a response to the requirements in the form of a specification, which states the requirements that must be met; the schedule for meeting the requirements; the identification of those who will perform the work; and the identification of potential risks. Differences in the requirements document and the preliminary specification become the topic of trade-off analyses.

Once product requirements are defined and the design process begins, there should be constant comparison between the product's requirements and the actual product design. As the product's design becomes increasingly detailed, it becomes increasingly more important to track the product's characteristics (size, weight, performance, functionality, reliability, and cost) in relation to the original product requirements. The rationale for making changes should be documented. The completeness with which a requirement tracking is performed can significantly reduce future product redesign costs. Planned redesigns or design refreshes through technology monitoring and use of technology roadmaps ensure that a company is able to market new products or redesigned versions of old products in a timely, effective manner to retain their customer base and ensure continued profits.

32.3.2 The Product Life Cycle Environment

The life cycle environment of the product plays a significant role in determining product requirements and making reliability assessments. It influences decisions regarding product design and development, parts selection and management, qualification, product safety, warranty, and support.

The phases in a product's life cycle include manufacturing and assembly, testing, rework, storage, transportation and handling, operation (modes of operation, on–off cycles), repair and maintenance, and disposal. During each phase of its life cycle, the product experiences various environmental and usage loads. The life cycle loads can be thermal (steady-state temperature, temperature ranges, temperature cycles, temperature gradients), mechanical (pressure levels, pressure gradients, vibrations, shock loads, acoustic levels), chemical (aggressive or inert environments, humidity levels, contamination), physical (radiation, electromagnetic interference, altitude), environmental (ozone, pollution, fuel spills), or operational loading conditions (stresses caused by power, power surge, heat dissipation, current, voltage spikes). These loads, either individually or in various combinations, may influence the reliability of the product. The extent and rate of product degradation depend upon the nature, magnitude, and duration of exposure to such loads.

Defining and characterizing the life cycle loads is often the most uncertain input in the overall reliability planning process. The challenge is further exacerbated because products designed for the same environmental conditions can experience completely different application conditions, including the application length, the number of applications, the product utilization or non-utilization profile, and maintenance or servicing conditions. For example, typically all desktop computers are designed for office environments. However, the operational profile of each unit may be completely different depending on user behavior. Some users may shut down the computer every time after it is used; others may shut down only once at the end of the day; still others may keep their computers powered on all the time. Thus, the temperature profile experienced by each product, and hence its degradation due to thermal loads, would be different. Four methods used to estimate the life cycle loads on a product are discussed in the following sections.

32.3.2.1 Market Studies and Standards

Market surveys and standards generated independently by agencies* provide a coarse estimate of the actual environmental loads expected in the field. The environmental profiles available from these sources are typically classified per industry type, such as military, consumer, telecommunications, automotive, and commercial avionics. Often the data includes worst case and example use conditions. The data available are derived most often from a similar kind of environment to provide an estimate of the actual environmental loads that the targeted equipment will experience. Care should be taken while using this data as an absolute estimate of environmental loads. The use of these data sources can become inevitable due to time and cost constraints during the design phase.

32.3.2.2 Field Trial, Service, and Failure Records

Field trial records, which estimate the environmental loads encountered by previous or prototype equipment, are also sometimes used to get estimates on environmental profiles. The data available are for shorter durations and are extrapolated to get an estimate of actual environmental conditions. Service records and failure records usually document the causes of unscheduled maintenance and the nature of failure, possibly due to certain environmental or usage conditions. These data give an idea of the critical conditions but should not be used as a basis for developing the entire life cycle profile. They should be used only to accommodate the extreme or special conditions the equipment might encounter.

32.3.2.3 Similarity Analysis and Competitor Benchmark

Similarity analysis is a technique for estimating environmental loads when sufficient field histories for similar products are available. For example, the environmental loads on various sites of an automobile are studied and documented by SAE. The characteristic differences in design and application for the two products need to be reviewed before using data on existing products for proposed designs. Changes and discrepancies in the conditions of the two products should be critically analyzed to ensure the applicability of available loading data for the new product. For example, electronics inside a washing machine in a commercial laundry are expected to experience a wider distribution of loads and use conditions (due to several users) and higher usage rates compared with a home washing machine. These differences should be considered during similarity analysis.

Once the target market segment is defined and similar products are identified, certain competitor products can be secured for benchmark evaluation. This may include design construction analysis and/or reliability benchmark testing. It can be a useful practice for new product development or for a company to expand into a fresh market domain. This may be applicable to a particular technology or component on the product, or simply the overall system level performance. Caution and care must be put into this process, since every product design is unique in a way to focus on certain utility cases and each company may approach then satisfy it from a different perspective.

32.3.2.4 In Situ Monitoring

Environmental and usage loads experienced by the product in its life cycle can be monitored in situ. This data is often collected using sensors, either mounted externally or integrated with the product and supported by telemetry systems. Devices such as health and usage monitoring systems (HUMS)

---

* For example, IPC SM-785 specifies the use and extreme temperature conditions for electronic products categorized under different industry sectors such as telecommunication, commercial, and military.

are popular in aircraft and helicopters for in situ monitoring of usage and environmental loads. Load distributions should be developed from data obtained by monitoring products used by different customers, ideally from various geographical locations where the product is used. The data should be collected over a sufficient period to provide an accurate estimate of the loads and their variation over time. Designers must ensure that the data is not biased, even if the users are aware of the monitoring process. In situ monitoring has the potential to provide the most accurate account of load history for use in health (condition) assessment and design of future products.

32.3.3 Parts Selection and Management

A parts selection and management methodology helps a company to make risk-informed decisions when purchasing and using electronic parts. The parts selection and management process is usually not carried out by a single individual, but by a multidisciplinary team. The parts management team, as a whole, is responsible for the following [7]:

- Assigning parts selection and management responsibilities to groups within the company
- Establishing communication channels within and outside the company
- Managing information flow within the team and to departments outside the team
- Identifying process and assessment criteria and acceptability levels
- Applying the parts selection and management methodology to the candidate part
- Identifying potential supplier intervention procedures, authorizing such action when required, and ensuring the associated effectiveness
- Monitoring periodically and making improvements to the methodology continuously

The methodology in Figure 32.2 describes this approach to parts selection and management. The overall purpose is to help organizations to maximize profits, provide product differentiation, effectively utilize the global supply chain, and assess, mitigate, and manage the life cycle risks in selecting and using parts. Several of these assessment methods directly impact product reliability, and those steps are described next. Several other steps, such as assembly assessment and life cycle obsolescence assessment, are also

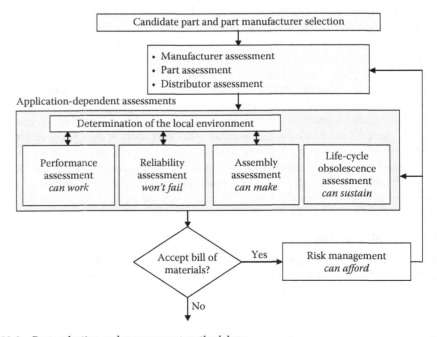

FIGURE 32.2   Parts selection and management methodology.

performed in this process. Although those steps impact overall product performance and profitability, their impact on reliability is not direct.

32.3.3.1 Manufacturer, Part, and Distributor Assessment

The manufacturer assessment step evaluates the part manufacturer's ability to produce parts with consistent quality, and the part assessment step gauges the candidate part's quality and consistency. The distributor assessment evaluates the distributor's ability to provide parts without affecting the initial quality and reliability and to provide certain specific services, such as part problem and change notifications. If the part satisfies the minimum acceptability criteria set by the equipment manufacturer for all three categories, the candidate part then moves to "application-dependent assessments," as shown in Figure 32.2. Appropriate implementation of this step helps improve reliability by reducing the risk of maverick part lots into a product build.

32.3.3.2 Performance Assessment

The goal of performance assessment is to evaluate the part's ability to meet the performance requirements (e.g., functional, mechanical, and electrical) of the product. In general, there are no distinct boundaries for parameters such as mechanical load, voltage, current, temperature, and power dissipation above which immediate failure will occur and below which a part will operate indefinitely [8]. However, there is often a minimum and a maximum limit beyond which the part will not function properly, and in general, those bounds are defined by the recommended operating conditions for the part. It is the responsibility of the parts selection and management team to establish that the electrical, mechanical, or functional performance of the part is suitable for the life cycle conditions of the particular product. If a product must be operated outside the manufacturer-specified operating conditions, then uprating [9] may have to be considered. On the other hand, derating may be utilized to allow better room of reliability by using parts in conditions away from their maximum limits.

32.3.3.3 Reliability Assessment

Reliability assessment results provide information about the ability of a part to meet the required performance specifications in its life cycle application environment for a specified period of time. If the parametric and functional requirements of the system cannot be met within the required local environment, then the local environment may have to be modified, or a different part may have to be used. If part reliability is not ensured through the reliability assessment process, an alternate part or product redesign should be considered.

Reliability assessment is conducted through the use of reliability test data (conducted by the part manufacturer), virtual reliability assessment results, or accelerated test results. If the magnitude and duration of the life cycle conditions are less severe than those of the reliability tests, and if the test sample size and results are acceptable, the part reliability is acceptable. Otherwise, virtual reliability assessment should be considered. Virtual reliability is a simulation-based methodology used to identify the dominant failure mechanisms associated with the part under life cycle conditions to determine the acceleration factor for a given set of accelerated test parameters, and to determine the time-to-failures corresponding to the identified failure mechanisms. If virtual reliability proves insufficient to validate part reliability, accelerated testing should be performed on representative part lots at conditions that represent the application condition to determine the part reliability. This decision process is illustrated in the diagram in Figure 32.3. Several OEMs also implement ongoing reliability tests (ORT) programs to ensure that parts supplied from over time from various manufacturing lots are consistently reliable.

32.3.4 Failure Modes, Mechanisms, and Effects Analysis

Electronic hardware is typically a combination of board, components, and interconnects, all with various failure mechanisms by which they can fail in the life cycle environment. A potential failure mode is

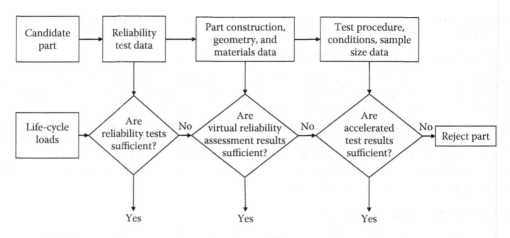

FIGURE 32.3   Virtual reliability simulation.

the manner in which a failure can occur—that is, the ways in which the reviewed item fails to perform its intended design function or performs the function but fails to meet its objectives. Failure modes are closely related to the functional and performance requirements of the product. Failure mechanisms are the processes by which a specific combination of physical, electrical, chemical, and mechanical stresses induce failures. Investigation of the possible failure modes and mechanisms of the product aids in developing failure-free and reliable designs.

Catastrophic failures due to a single occurrence of a stress event when the intrinsic strength of the material is exceeded are termed overstress failures. Failure mechanisms due to monotonic accumulation of incremental damage beyond the endurance of the material are called wearout mechanisms [10]. When the damage exceeds the endurance limit of the component, failure will occur. Unanticipated large stress events can either cause an overstress (catastrophic) failure or shorten life by causing the accumulation of wearout damage. Examples of such stresses are accidental abuse and unexpected exposure to wearout conditions. On the other hand, in well-designed and high-quality hardware, stresses should cause only uniform accumulation of wearout damage; the threshold of damage required to cause eventual failure should not occur within the usage life of the product. Current trends in use of new and unproven materials increase risk.

Electrical performance failures can be caused by individual components with improper electrical parameters, such as resistance, impedance, capacitance, or dielectric properties, or by inadequate shielding from electromagnetic interference (EMI) or ionizing radiation. Protection from ionizing radiation which can cause single event effects (effectively bit flips and logic problems) have been prominent in component design and construction since the trend towards smaller component sizes increases this risk. Some of these effects are only transient but some can leave lasting and permanent effects on the component. Failure modes can manifest as reversible drifts in transient and steady-state responses, such as delay time, rise time, attenuation, signal-to-noise ratio, and crosstalk. Electrical failures common in electronic hardware include overstress mechanisms due to electrical overstress (EOS) and electrostatic discharge (ESD); examples of such failures in semiconductor components include dielectric breakdown, junction breakdown, hot electron injection, surface and bulk trapping, surface breakdown, and wearout mechanisms such as electromigration and stress-driven diffusive voiding.

Thermal performance failures can arise due to incorrect design of thermal paths in an electronic assembly. This includes incorrect conductivity and surface emissivity of individual components, as well as ill-designed convective and conductive paths for heat transfer. Thermal overstress failures are a result of heating a component beyond critical temperatures such as the glass-transition temperature, melting point, fictile point, or flash point. Some examples of thermal wearout failures are aging due to depolymerization, intermetallic growth, and inter-diffusion. Failures due to inadequate thermal design may be manifested as components running too hot or too cold and causing operational parameters to drift beyond specifications, although the degradation is often reversible upon cooling. Such failures can be caused either by direct thermal loads or by electrical resistive loads, which in turn generate excessive localized thermal stresses. Adequate design checks require proper analysis for thermal stress and should include conductive, convective, and radiative heat paths.

Mechanical performance failures include those that may compromise the product performance without necessarily causing any irreversible material damage, such as abnormal elastic deformation in response to mechanical static loads, abnormal transient response (such as natural frequency or damping) to dynamic loads, and abnormal time-dependent reversible (anelastic) response, as well as failures that cause material damage, such as buckling, brittle or ductile fracture, interfacial separation, fatigue crack initiation and propagation, creep, and creep rupture. To take one example, excessive elastic deformations in slender structures in electronic packages can sometimes constitute functional failure due to overstress loads, such as excessive flexing of interconnection wires, package lids, or flex circuits in electronic devices, causing shorting or excessive crosstalk. However, when the load is removed, the deformations (and consequent functional abnormalities) disappear completely without any permanent damage.

Radiation failures—principally caused by uranium and thorium contaminants and secondary cosmic rays—can cause wearout, aging, embrittlement of materials, and overstress soft errors in electronic hardware, such as logic chips. Chemical failures occur in adverse chemical environments that result in corrosion, oxidation, or ionic surface dendritic growth.

There may also be interactions between different types of stresses. For example, metal migration may be accelerated in the presence of chemical contaminants and composition gradients, and thermal loads can accelerate a failure mechanism due to a thermal expansion mismatch.

The design team must be aware of the possible failure mechanisms if they are to design hardware capable of withstanding loads without failing. Failure mechanisms and their related physical models are also important for planning tests and screens to audit nominal design and manufacturing specifications, as well as the level of defects introduced by excessive variability in manufacturing and material parameters.

Failure modes, mechanisms, and effects analysis (FMMEA) is a systematic methodology to identify potential failure mechanisms and models for all potential failures modes and to prioritize failure mechanisms. FMMEA enhances the value of traditional methods such as FMEA (failure modes and effects analysis) and FMECA (failure modes effects and criticality analysis) by identifying high-priority failure mechanisms to create an action plan to mitigate their effects. The knowledge about the cause and consequences of mechanisms found through FMMEA helps in several efficient and cost-effective ways.

FMMEA is based on understanding the relationships between product requirements and the physical characteristics of the product (and their variation in the production process), the interactions of product materials with loads (stresses at application conditions), and their influence on product failure susceptibility with respect to the use conditions. This involves finding the failure mechanisms and the reliability models to quantitatively evaluate failure susceptibility. The FMMEA methodology is shown in Figure 32.4.

32.3.4.1 System Definition, Elements, and Functions

The FMMEA process begins by defining the system to be analyzed. A system is a composite of subsystems or levels that are integrated to achieve a specific objective. The system is divided into various subsystems and continues to the lowest possible level—the component or element. In a printed circuit board system, a location breakdown would include the package, plated through-hole (PTH), metallization, and the board itself.

32.3.4.2 Potential Failure Modes

For the elements that have been identified, all possible failure modes for each given element are listed. For example, in a solder joint the potential failure modes are open or intermittent change in resistance, which can hamper its functioning as an interconnect. In cases where information on possible failure

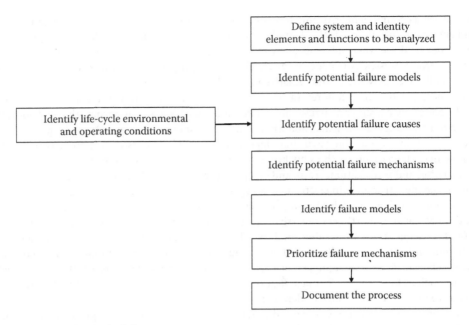

FIGURE 32.4   FMMEA methodology.

modes is not available, potential failure modes may be identified using numerical stress analysis, accelerated tests to failure (e.g., HALT*), past experience, and engineering judgment [11]. A potential failure mode at one level may be the cause of a potential failure mode in a higher level system or subsystem or be the effect of one in a lower level component.

32.3.4.3 Potential Failure Causes

A failure cause is defined as the circumstances during design, manufacture, or use that lead to a failure mode [11]. For each failure mode, all possible ways a failure can result are listed. Failure causes are identified by finding the basic reason that may lead to a failure during design, manufacturing, storage, transportation, or use. Knowledge of potential failure causes can help identify the failure mechanisms driving the failure modes for a given element. For example, in an automotive under-hood environment, the solder joint failure modes such as open and intermittent change in resistance can potentially be caused due to temperature cycling, random vibration, and shock impact.

32.3.4.4 Potential Failure Mechanisms

Failure mechanisms frequently occurring in electronics were discussed earlier in this section. Studies on electronic material failure mechanisms and the application of physics-based damage models to the design of reliable electronic products comprising all relevant wearout and overstress failures in electronics are available in literature [12,13].

32.3.4.5 Failure Models

Failure models use appropriate stress and damage analysis methods to evaluate susceptibility to failure. Failure susceptibility is evaluated by assessing the time-to-failure or likelihood of a failure for a given geometry, material construction, and environmental and operational condition set. Based on JEDEC standards, JEP122B (Failure Mechanisms and Models for Semiconductor Device) and JEP 148 (Reliability Qualification of Semiconductor Devices Based on Physics of Failure Risk and Opportunity Assessment),

---

* Highly accelerated life testing.

TABLE 32.1 Failure Mechanisms, Relevant Loads, and Models for Electronics

Failure Mechanism	Failure Sites	Relevant Loads	Sample Model
Fatigue	Die attach, wirebond/ TAB, solder leads, bond pads, traces, vias/PTHs, interfaces	ΔT, T_{mean}, dT/dt, dwell time, ΔH, ΔV	Nonlinear power; law (Coffin–Manson)
Corrosion	Metallizations	M, ΔV, T	Eyring (Howard)
Electromigration	Metallizations	T, J	Eyring (Black)
Conductive filament formation	Between metallizations	M, ΛV	Power law (Rudra)
Stress-driven diffusion voiding	Metal traces	s, T	Eyring (Okabayashi)
Time-dependent dielectric breakdown	Dielectric layers	V, T	Arrhenius (Fowler–Nordheim)

Δ, cyclic range; V, voltage; T, temperature; s, stress.
Λ, gradient; M, moisture; J, current density; H, humidity.

Table 32.1 provides a list of common failure mechanisms in electronics and associated models. Failure models may be limited by the availability and accuracy of models for quantifying the time to failure of the system. They may also be limited by the ability to combine the results of multiple failure models for a single failure site and the ability to combine results of the same model for multiple stress conditions [11]. If no failure models are available, the appropriate parameter(s) to monitor can be selected based on an empirical model developed from prior field failure data or models derived from accelerated testing.

32.3.4.6 Failure Mechanism Prioritization

Environmental and operating conditions are used for initial prioritization of all potential failure mechanisms. If the stress levels generated by certain operational and environmental conditions are nonexistent or negligible, the failure mechanisms that are exclusively dependent on those environmental and operating conditions are assigned a "low" risk level and are eliminated from further consideration. For all other failure mechanisms, the failure susceptibility is evaluated by conducting a stress analysis to determine if failure is precipitated under the given environmental and operating conditions. To provide a qualitative measure of the failure effect, each failure mechanism is assigned a severity rating. The failure effect is assessed first at the level being analyzed, then the next higher level, the subsystem level, and so on to the system level. The final step involves prioritizing the failure mechanisms into three risk levels: high, medium, and low.

32.3.4.7 Documentation

The FMMEA process facilitates data organization, distribution, and analysis for all the steps. In addition, FMMEA documentation also includes the actions considered and taken based on the FMMEA. For products already developed and manufactured, root-cause analysis may be conducted for the failures that occur during development, testing, and use, and corrective actions can be taken to avoid or to reduce the impacts of those failures. The history and lessons learned, contained within the documentation, provide a framework for future product FMMEA. It is also possible to maintain and update an FMMEA after the corrective actions to generate a new list of high-priority failure mechanisms.

32.3.5 Design Techniques

Once the parts, materials, processes, and stress conditions are identified, the objective is to design a product using parts and materials that have been sufficiently characterized in terms of performance over time when subjected to the manufacturing and application profile conditions. A reliable and cost-effective product can be designed only through a methodical design approach using physics-of-failure analysis, testing, and root-cause analysis.

Design guidelines based on physics-of-failure models can also be used to develop tests, screens, and derating factors. Tests based on physics-of-failure models can be designed to measure specific quantities, to detect the presence of unexpected flaws, and to detect manufacturing or maintenance problems. Screens can be designed to precipitate failures in the weak population while not cutting into the design life of the normal population. Derating or safety factors can be determined to lower the stresses for the dominant failure mechanisms.

32.3.5.1 Protective Architectures

In designs where safety is an issue, it is generally desirable to incorporate some means for preventing a part, structure, or interconnection from failing or from causing further damage when it fails. Fuses and circuit breakers are examples of elements used in electronic products to sense excessive current drain and to disconnect power from the concerned part. Fuses within circuits safeguard parts against voltage transients or excessive power dissipation and protect power supplies from shorted parts. As another example, thermostats can be used to sense critical-temperature limiting conditions and to shut down the product or a part of the system until the temperature returns to normal. In some products, self-checking circuitry can also be incorporated to sense abnormal conditions and to make adjustments to restore normal conditions or to activate switching means to compensate for the malfunction [8].

In some instances, it may be desirable to permit partial operation of the product after a part failure rather than permitting total product failure. By the same reasoning, degraded performance of a product after failure of a part is often preferable to complete stoppage. An example is the shutting down of a failed circuit whose function is to provide precise trimming adjustment within a deadband of another control product; acceptable performance may thus be achieved, perhaps under emergency conditions, with the deadband control product alone [8].

Sometimes, the physical removal of a part from a product can harm or cause failure in another part by removing load, drive, bias, or control. In such cases, the first part should be equipped with some form of interlock mechanism to shut down or otherwise protect the second part. The ultimate design, in addition to its ability to function after a failure, should be capable of sensing and adjusting for parametric drifts to avert failures.

In the use of protective techniques, the basic procedure is to take some form of action after an initial failure or malfunction to prevent additional or secondary failures. By reducing the number of failures, techniques such as enhancing product reliability can be considered, although they also affect availability and product effectiveness. Equally important considerations are the impacts of maintenance, repair, and part replacement. For example, if a fuse protecting a circuit is replaced, the following questions need to be answered: What is the impact when the product is reenergized? What protective architectures are appropriate for postrepair operations? What maintenance guidance must be documented and followed when fail-safe protective architectures have or have not been included?

32.3.5.2 Stress Margins

A properly designed product should be capable of operating satisfactorily with parts that drift or change with time and with changes in operating conditions such as temperature, humidity, pressure, and altitude as long as the parameters remain within their rated tolerances. Figure 32.5 provides a schematic of stress levels and margins for a product. The specification or tolerance value is given by the manufacturer to limit the conditions of customer use. The design margin is the value a product is designed to survive, and the operating margin is the expected value for a recoverable failure of a distribution of a product. The expected value for permanent (overstress) failure of a distribution of a product is called the destruct margin.

To guard against out-of-tolerance failures, the design team must consider the combined effects of tolerances on parts to be used in manufacturing, of subsequent changes due to the range of expected

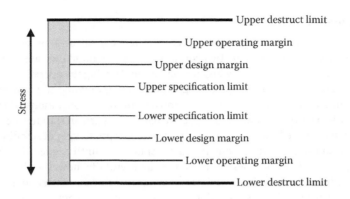

FIGURE 32.5   Stress limits and margins.

environmental conditions, of drifts due to aging over the period specified in the reliability requirement, and of tolerances in parts used in future repair or maintenance functions. Parts and structures should be designed to operate satisfactorily at the extremes of the parameter ranges, and allowable ranges must be included in the procurement or re-procurement specifications.

Statistical analysis and worst-case analysis are methods of dealing with part and structural parameter variations. In statistical analysis, a functional relationship is established between the output characteristics of the structure and the parameters of one or more of its parts. In worst-case analysis, the effect that a part has on product output is evaluated on the basis of end-of-life performance values or out-of-specification replacement parts.

To ensure that the parts used in a system remain within the stated margins, derating can be used. Derating is the practice of limiting thermal, electrical, and mechanical stresses to levels below the manufacturer's specified ratings to improve reliability. Derating allows added protection from system anomalies unforeseen by the designer (e.g., transient loads, electrical surge). For example, manufacturers of electronic hardware often specify limits for supply voltage, output current, power dissipation, junction temperature, and frequency. The equipment design team may decide to select an alternative component or make a design change that ensures that the operational condition for a particular parameter, such as temperature, is always below the rated level. These lower stresses are expected to extend useful operating life where the failure mechanisms under consideration are of wearout type. This practice is also expected to provide a safer operating condition by furnishing a margin of safety when the failure mechanisms are of the overstress type.

The term "derating" suggests a two-step process; first a "rated" stress value is determined from a part manufacturer's data-book, and then some reduced value is assigned. The margin of safety supposed to be provided by derating is the difference between the maximum allowable actual applied stress and the demonstrated limits of the part capabilities. The part capabilities as given by manufacturer specifications are taken as the demonstrated limits. The propensity of the system design team often inclines toward using conservative stresses at the expense of overall productivity. There are reasons to believe that the part manufacturers already provide safety margin when choosing the operating limits. When these values are derated by the users, effectively a second level of safety margin is added.

In order to be effective, derating criteria must target the right stress parameters to address modeling of the relevant failure mechanisms. Once the failure models for the critical failure mechanisms have been identified using the FMMEA process, the impact of derating on the effective reliability of the part for a given load can be determined. Instead of derating the stress rating values provided by the device manufacturers, the goal should be to determine the safe operating envelope for each part and subsystem and then operate within that envelope.

32.3.5.3 Redundancy

Redundancy exists when one or more of the parts of a system can fail and the system can still function with the parts that remain operational. Two common types of redundancy are active and standby. In active redundancy, all the parts are energized and operational during the operation of a system. In active redundancy, the parts will consume life at the same rate as the individual components.

In standby redundancy, some parts are not contributing to the operations of the system and are switched on only when there are failures in the active parts. The parts in standby ideally should last longer than the parts in active redundancies. There are three conceptual types of standby redundancy: cold, warm, and hot. In cold standby, the secondary part is shut down until needed. This lowers the amount of time that the part is active and does not consume any useful life; however, the transient stresses on the part during switching may be high. This transient stress can cause faster consumption of life during switching. In warm standby, the secondary part is usually active but is idling or unloaded. In hot standby, the secondary part forms an active parallel system. The life of the hot standby part is consumed at the same rate as active parts.

A design team often finds that redundancy is (1) the quickest way to improve product reliability if there is insufficient time to explore alternatives, or if the part is already designed; (2) the cheapest solution, if the cost of redundancy is economical in comparison with the cost of redesign; and/or (3) the only solution, if the reliability requirement is beyond the state of the art. However, it is often difficult to realize the benefits of redundancy (both active and standby) due to other reasons. Usually, these reasons include common mode failures, load sharing, and switching and standby failures.

Common mode failures are caused by phenomena that create dependencies between two or more redundant parts, which cause them to fail simultaneously. Common mode failures can be caused by many factors—for example, common electric connections, shared environmental stresses, and common maintenance problems. In system reliability analysis, common mode failures have the same effect as putting in an additional part in series with the parallel redundant configuration.

Load-sharing failures occur when the failure of one part increases the stress level of the other parts. This increased stress level can affect the life of the active parts. For redundant engines, motors, pumps, structures, and many other systems and devices in active parallel setup, the failure of one part may increase the load on the other parts and decrease their times to failure (or increase their hazard rates).

Several common assumptions are generally made regarding the switching and sensing of standby systems. We assume that switching is in one direction only, that switching devices respond only when directed to switch by the monitor, and that switching devices do not fail if not energized. Regarding standby, the general assumption is that standby nonoperating units cannot fail if not energized. When any of these idealizations are not met, switching and standby failures occur. Monitor or sensing failure includes both dynamic (failure to switch when active path fails) and static (switching when not required) failures.

Besides these limitations, the design team may find that the following disadvantages outweigh the benefits of redundancy implementation:

- Costly redundant sensors and switching devices make it too expensive.
- Limitations on size and weight are exceeded.
- It exceeds power limitations, particularly in active redundancy.
- It requires sensing and switching circuitry so complex as to offset the reliability advantage of redundancy.

32.3.5.4 Integrated Diagnostics and Prognostics

Design guidelines and techniques should involve strategies to assess the reliability of the product in its life cycle environment. A product's health is the extent of deviation or degradation from its expected normal (in terms of physical and performance) operating condition [14]. Knowledge of a product's health can be used for the detection and isolation of faults or failures (diagnostics) and for prediction of impending failure based on current conditions (prognostics). Thus, by determining the advent of failure based on actual life cycle conditions, procedures can be developed to mitigate, manage, and maintain the product.

Diagnostics and prognostics can be integrated into a product by (1) installing built-in fuses and canary structures that will fail faster than the actual product when subjected to life cycle conditions [15]; (2) sensing parameters that are precursors to failure, such as defects or performance degradation [16]; (3) sensing the life cycle environmental and operational loads that influence the system's health, and processing the measured data to estimate the life consumed [17,18]. The life cycle data measured by such integrated monitors can be extremely useful in future product design and end-of-life decisions [19].

An example of integrated prognostics in electronic products is the self-monitoring analysis and reporting technology (SMART) currently employed in select computing equipment for hard disk drives (HDDs) [20]. HDD operating parameters, including flying height of the head, error counts, variations in spin time, temperature, and data transfer rates, are monitored to provide advance warning of failures. This is achieved through an interface between the computer's start-up program (BIOS) and the hard disk drive.

32.3.6 Qualification and Accelerated Testing

Qualification includes all activities that ensure that the nominal design and manufacturing specifications will meet or exceed the desired reliability targets. Qualification validates the ability of the nominal design and manufacturing specifications of the product to meet the customer's expectations and assesses the probability of survival of the product over its complete life cycle. The purpose of qualification is to define the acceptable range of variability for all critical product parameters affected by design and manufacturing, such as geometric dimensions, material properties, and operating environmental limits. Product attributes that are outside the acceptable ranges are termed defects, since they have the potential to compromise product reliability [21].

Qualification tests should be performed only during initial product development and immediately after any design or manufacturing changes in an existing product. A well-designed qualification procedure provides economic savings and quick turnaround during development of new products or products subject to manufacturing and process changes.

Investigating failure mechanisms and assessing the reliability of products where longevity is required may be a challenge, since a very long test period under actual operating conditions is necessary to obtain sufficient data to determine actual failure characteristics. The results from FMMEA should be used to guide this process. One approach to the problem of obtaining meaningful qualification data for high-reliability devices in shorter time periods is using methods such as virtual qualification and accelerated testing to achieve test-time reduction.

32.3.6.1 Virtual Qualification

Virtual qualification is a process that requires significantly less time and money than accelerated testing to qualify a part for its life cycle environment. This simulation-based methodology is used to identify and rank the dominant failure mechanisms associated with the part under life cycle loads, to determine the acceleration factor for a given set of accelerated test parameters, and to determine the time-to-failure corresponding to the identified failure mechanisms.

Each failure model is comprised of a stress analysis model and a damage assessment model. The output is a ranking of different failure mechanisms based on the time to failure. The stress model captures the product architecture, while the damage model depends on a material's response to the applied stress. This process is therefore applicable to existing as well as new products. Virtual qualification can be used to optimize the product design in such a way that the minimum time to failure of any part of the product is greater than its desired life. Although the data obtained from virtual qualification cannot fully replace those obtained from physical tests, they can increase the efficiency of physical tests by indicating the potential failure modes and mechanisms that can be expected.

Ideally, a virtual qualification process will involve identification of quality suppliers, quality parts, physics-of-failure qualification, and a risk assessment and mitigation program. The process allows qualification to be readily incorporated into the design phase of product development, since it allows design, manufacturing, and testing to be conducted promptly and cost-effectively. It also allows consumers to qualify off-the-shelf components for use in specific environments without extensive physical tests. Since virtual qualification reduces emphasis on examining a physical sample, it is imperative that the manufacturing technology and quality assurance capability of the manufacturer be taken into account. If the data on which the virtual qualification is performed are inaccurate or unreliable, all results are suspect. In addition, if a reduced quantity of physical tests is performed in the interest of simply verifying virtual results, the operator needs to be confident that the group of parts selected is sufficient to represent the product. Further, it should be remembered that the accuracy of the results using virtual qualification depends on the accuracy of the inputs to the process—that is, the accuracy of the life cycle loads, the choice of the failure models used, the choice of the analysis domain (for example, 2D, pseudo-3D, full 3D), the constants in the failure model, the material properties, and so on. Hence, to obtain a reliable prediction, the variability in the inputs should be specified using distribution functions, and the validity of the failure models should be tested by conducting accelerated tests.

32.3.6.2 Accelerated Testing

Accelerated testing is based on the concept that a product will exhibit the same failure mechanism and mode in a short time under high-stress conditions as it would exhibit in a longer time under actual life cycle stress conditions. The purpose is to decrease the total time and cost required to obtain reliability information about the product under study. Usually it should be possible to quantitatively extrapolate from the accelerated environment to the usage environment with some reasonable degree of assurance.

Accelerated tests can be roughly divided into two categories: qualitative tests and quantitative tests. Qualitative tests, in the form of overstressing the products to obtain failure, are perhaps the oldest form of reliability testing. Those tests typically arise as a result of a single load condition, such as shock, temperature extremes, and electrical overstress. Qualitative tests usually yield failure mode information but do not reveal failure mechanism and time-to-failure information. Quantitative tests target wearout failure mechanisms, in which failures occur as a result of cumulative load conditions and time to failure is the major outcome of quantitative accelerated tests.

The easiest and most common form of accelerated life testing is continuous-use acceleration. For example, a washing machine is used for 10 h/week in average. If it is operated without stop, the acceleration factor would be 16.8. However, this method is not applicable for high-usage products. Under such circumstances, accelerated testing is conducted to measure the performance of the test product at loads or stresses that are more severe than would normally be encountered in order to enhance the damage accumulation rate within a reduced time period. The goal of such testing is to accelerate time-dependent failure mechanisms and the damage accumulation rate to reduce the time to failure. Based on the data from those accelerated tests, life in normal use conditions can be extrapolated.

Accelerated testing begins by identifying all the possible overstress and wearout failure mechanisms. The load parameter that directly causes the time-dependent failure is selected as the acceleration parameter and is commonly called the accelerated load. Common accelerated loads include the following:

- Thermal loads, such as temperature, temperature cycling, and rates of temperature change
- Chemical loads, such as humidity, corrosives, acid, and salt
- Electrical loads, such as voltage or power
- Mechanical loads, such as vibration, mechanical load cycles, strain cycles, and shock and impulses

The accelerated environment may include a combination of these loads. Interpretation of results for combined loads requires a quantitative understanding of their relative interactions and the contribution of each load to the overall damage.

Failure due to a particular mechanism can be induced by several acceleration parameters. For example, corrosion can be accelerated by both temperature and humidity, and creep can be accelerated by both mechanical stress and temperature. Furthermore, a single accelerated stress can induce failure by several wearout mechanisms simultaneously. For example, temperature can accelerate wearout damage accumulation not only by electromigration, but also by corrosion, creep, and so on. Failure mechanisms that dominate under usual operating conditions may lose their dominance as the stress is elevated. Conversely, failure mechanisms that are dormant under normal-use conditions may contribute to device failure under accelerated conditions. Thus, accelerated tests require careful planning if they are to represent the actual usage environments and operating conditions without introducing extraneous failure mechanisms or non-representative physical or material behavior. The degree of stress acceleration is usually controlled by an acceleration factor, defined as the ratio of the life of the product under normal use conditions to that under the accelerated condition. The acceleration factor should be tailored to the hardware in question and can be estimated from an acceleration transform (that is, a functional relationship between the accelerated stress and the life cycle stress) in terms of all the hardware parameters.

Once the failure mechanisms are identified, it is necessary to select the appropriate acceleration load; determine the test procedures and the stress levels; determine the test method, such as constant stress acceleration or step-stress acceleration; perform the tests; and interpret the test data, which includes extrapolating the accelerated test results to normal operating conditions. The test results provide failure information for improving the hardware through design or process changes.

32.3.7 Manufacturing Issues

Manufacturing and assembly processes can significantly impact the quality and reliability of hardware. Improper assembly and manufacturing techniques can introduce defects, flaws, and residual stresses that act as potential failure sites or stress raisers later in the life of the product. The effect of manufacturing variability on time to failure is depicted in Figure 32.6. A shift in the mean or an increase in the standard deviation of key geometric parameters during manufacturing can result in early failure due to a decrease in strength of the product. If these defects and stresses can be identified, the design analyst can proactively account for them during the design and development phase.

Auditing the merits of the manufacturing process involves two crucial steps. First, qualification procedures are required, as in design qualification, to ensure that manufacturing specifications do not compromise the long-term reliability of the hardware. Second, lot-to-lot screening is required to ensure that the variability of all manufacturing-related parameters is within specified tolerances [21,22]. In other words, screening ensures the quality of the product by identifying latent defects before they reach the field.

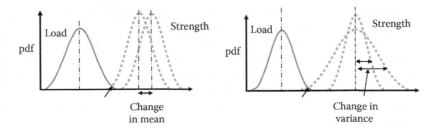

FIGURE 32.6 Influence of quality on failure probability.

32.3.7.1 Process Qualification

Like design qualification, process qualification should be conducted at the prototype development phase. The intent at this step is to ensure that the nominal manufacturing specifications and tolerances produce acceptable reliability in the product. The process needs requalification when process parameters, materials, manufacturing specifications, or human factors change.

Process qualification tests can be the same set of accelerated wearout tests used in design qualification. As in design qualification, overstress tests may be used to qualify a product for anticipated field overstress loads. Overstress tests may also be used to ensure that manufacturing processes do not degrade the intrinsic material strength of the hardware beyond a specified limit. However, such tests should supplement, not replace, the accelerated wearout test program, unless explicit physics-based correlations are available between overstress test results and wearout field-failure data.

32.3.7.2 Manufacturability

The control and rectification of manufacturing defects has typically been the concern of production and process-control engineers, not the design team. In the spirit and context of concurrent product development, however, hardware design teams must understand material limits, available processes, and manufacturing process capabilities to select materials and construct architectures that promote producibility and reduce the occurrence of defects, increasing yield and quality. Therefore, no specification is complete without a clear discussion of manufacturing defects and acceptability limits. The reliability engineer must have clear definitions of the threshold for acceptable quality and of what constitutes nonconformance. Nonconformance that compromises hardware performance and reliability is considered a defect, and failure mechanism models provide a convenient vehicle for developing such criteria. It is important for the reliability analyst to understand which deviations from specifications can compromise performance or reliability and which are benign and can be accepted.

A defect is any outcome of a process (manufacturing or assembly) that impairs or has the potential to impair the functionality of the product at any time. The defect may arise during a single process or may be the result of a sequence of processes. The yield of a process is the fraction of products that are acceptable for use in a subsequent manufacturing sequence or product life cycle. The cumulative yield of the process is approximately determined by multiplying the individual yields of each of the individual process steps. The source of defects is not always apparent, because defects resulting from a process can go undetected until the product reaches some downstream point in the process sequence, especially if screening is not employed.

It is often possible to simplify the manufacturing and assembly processes to reduce the probability of workmanship defects. As processes become more sophisticated, however, process monitoring and control are necessary to ensure a defect-free product. The bounds that specify whether the process is within tolerance limits, often referred to as the process window, are defined in terms of the independent variables to be controlled within the process and the effects of the process on the product or the dependent product variables. The goal is to understand the effect of each process variable on each product parameter to formulate control limits for the process—that is, the points on the variable scale where the defect rate begins to have a potential for causing failure. In defining the process window, the upper and lower limits of each process variable beyond which it will produce defects must be determined. Manufacturing processes must be contained in the process window by defect testing, analysis of the causes of defects, and elimination of defects by process control, such as by closed-loop corrective action systems. The establishment of an effective feedback path to report process-related defect data is critical. Once this is done and the process window is determined, the process window itself becomes a feedback system for the process operator.

Several process parameters may interact to produce a different defect than would have resulted from an individual parameter acting independently. This complex case may require that the interaction of various process parameters be evaluated in a matrix of experiments. In some cases, a defect cannot be detected until late in the process sequence. Thus, a defect can cause rejection, rework, or failure of the product after considerable value has been added to it. These cost items due to defects can reduce a return on investment by adding to hidden factory costs. All critical processes require special attention for defect elimination by process control.

32.3.7.3 Process Verification Testing

Process verification testing, often called screening, involves 100% auditing of all manufactured products to detect or precipitate defects. The aim of this step is to preempt potential quality problems before they reach the field. Thus, screening aids in reducing warranty returns and increases customer goodwill. In principle, screening should not be required for a well-controlled process; however, it is often used as a safety net.

Some products exhibit a multimodal probability density function for failures, as depicted in Figure 32.7, with peaks during the early period of their service life due to the use of faulty materials, poorly controlled manufacturing and assembly technologies, or mishandling. This type of early-life failure is often called infant mortality. Properly applied screening techniques can successfully detect or precipitate these failures, eliminating or reducing their occurrence in field use. Screening should only be considered for use during the early stages of production, if at all, and only when products are expected to exhibit infant mortality field failures. Screening will be ineffective and costly if there is only one main peak in the failure probability density function. Further, failures arising due to unanticipated events such as acts of God (lightning, earthquakes) may be impossible to screen cost-effectively.

Since screening is done on a 100% basis, it is important to develop screens that do not harm good components. The best screens, therefore, are nondestructive evaluation techniques, such as microscopic visual exams, x-rays, acoustic scans, nuclear magnetic resonance (NMR), electronic paramagnetic resonance (EPR), and so on. Stress screening involves the application of stresses, possibly above the rated operational limits. If stress screens are unavoidable, overstress tests are preferred over accelerated wearout tests, since the latter are more likely to consume some useful life of good components. If damage to good components is unavoidable during stress screening, then quantitative estimates of the screening damage, based on failure mechanism models, must be developed to allow the design team to account for this loss of usable life. The appropriate stress levels for screening must be tailored to the specific hardware. As in qualification testing, quantitative models of failure mechanisms can aid in determining screen parameters.

A stress screen need not necessarily simulate the field environment or even use the same failure mechanism as the one likely to be triggered by this defect in field conditions. Instead, a screen should exploit the most convenient and effective failure mechanism to stimulate the defects that can show up in

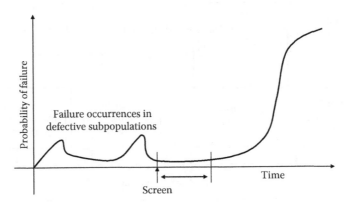

FIGURE 32.7 Candidate for screening due to wearout failure.

the field as infant mortality. Obviously, this requires an awareness of the possible defects that may occur in the hardware and extensive familiarity with the associated failure mechanisms.

Unlike qualification testing, the effectiveness of screens is maximized when screening is conducted immediately after the operation believed to be responsible for introducing the defect. Qualification testing is conducted preferably on the finished product or as close to the final operation as possible; on the other hand, screening only at the final stage, when all operations have been completed, is less effective, since failure analysis, defect diagnostics, and troubleshooting are difficult and impair corrective actions. Further, if a defect is introduced early in the manufacturing process, subsequent value added through new materials and processes is wasted, which additionally burdens operating costs and reduces productivity.

Admittedly, there are also several disadvantages to such an approach. The cost of screening at every manufacturing station may be prohibitive, especially for small batch jobs. Further, components will experience repeated screening loads as they pass through several manufacturing steps, which increases the risk of accumulating wearout damage in good components due to screening stresses. To arrive at a screening matrix that addresses as many defects and failure mechanisms as feasible with each screen test an optimum situation must be sought through analysis of cost-effectiveness, risk, and the criticality of the defects. All defects must be traced back to the root cause of the variability.

Any commitment to stress screening must include the necessary funding and staff to determine the root cause and appropriate corrective actions for all failed units. The type of stress screening chosen should be derived from the design, manufacturing, and quality teams. Although a stress screen may be necessary during the early stages of production, stress screening carries substantial penalties in capital, operating expense, and cycle time, and its benefits diminish as a product approaches maturity. If almost all of the products fail in a properly designed screen test, the design is probably faulty. If many products fail, a revision of the manufacturing process is required. If the number of failures in a screen is small, the processes are likely to be within tolerances, and the observed faults may be beyond the resources of the design and production process.

32.3.8 Closed-Loop Monitoring

Product reliability needs to be ensured using a closed-loop process that provides feedback to design and manufacturing in each stage of the product life cycle, including after the product is shipped and is used in its application environment. Data obtained from field failures, maintenance, inspection and testing, or health (condition) and usage monitoring methods can be used to perform timely maintenance for sustaining the product and preventing catastrophic failures. Figure 32.8 depicts the closed-loop process for managing the reliability of a product over the complete life cycle.

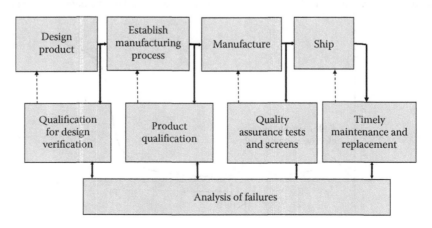

FIGURE 32.8   Reliability management using a closed-loop process.

The objective of closed-loop monitoring is to analyze the failures occurring in testing and field conditions to identify the root cause of failure. The root cause is the most basic casual factor or factors that, if corrected or removed, will prevent recurrence of the situation. The purpose of determining the root cause or causes is to fix the problem at its most basic source so it does not occur again, even in other products, as opposed to merely fixing a failure symptom.

Root-cause analysis is a methodology designed to help (1) describe *what* happened during a particular occurrence; (2) determine *how* it happened; and (3) understand *why* it happened. Only when we are able to determine why an event or failure occurred will we be able to determine corrective measures over time. Root-cause analysis is different than troubleshooting, which is generally employed to eliminate a symptom in a given product, as opposed to finding a solution to the root cause to prevent this and other products from failing.

Correctly identified root causes during design and manufacturing, followed by appropriate actions to fix the design and processes, result in fewer field returns, major cost savings, and customer goodwill. Root-cause analysis of field failures can be more challenging if the conditions during and prior to failure are unknown. The lessons learned from each failure analysis need to be documented, and appropriate actions need to be taken to update the design, manufacturing process, and maintenance actions or schedules.

32.4 RMA

In aviation parlance, it is common to refer the reliability, maintainability and availability, jointly as RMA. While reliability and maintainability are essentially design characteristics which can be tracked during operational phase using real time performance data (failure rate, repair rate and mission time, etc.), availability is operational characteristics of the system. The maintainability aspects of the component and systems are also one of the major concerns of design engineers. The quantification of system maintainability is carried out during operational phase by considering the repair rate (synonymous to failure rate) and actual time for repair The designs of aviation systems employ the concept of redundancy, automatic fault detection and recovery to achieve the required reliability and availability targets for critical services. Here, the objective is to make the "recovery" of the services either with no or minor interruptions to ensure higher reliability and availability.

In practice, it has been seen that apart from quantification of reliability other measures, like mean time to failure, mean time between failure, availability or even the statement of failure rates itself is used as an indicator of the reliability. For example, in aviation industry the MTBF and availability estimates are often used to characterize reliability. Here, it may be noted that MTBF alone may not be adequate to provide the holistic picture on reliability of the component. Without considering the mission time (for non-repairable components) or test interval (for repairable components) the reliability for a specific component cannot be estimated. For example, for same MTTF, for two components the reliability will be different if the mission time is different. Further, for a small increase in recovery time the MTBF can vary exponentially from an initial acceptably MTBF to very low and unacceptable MTBF which may not meet system performance criteria.

Often the term mean time to failure (MTTF) and mean time between failure (MTBF) are used interchangeably in commercial domain. It should be understood that the term MTTF is used for non repairable components while MTBF is used for repairable components. Hence, while estimating reliability for non repairable component, mission time is used while for repairable component repair time should be used. Similarly for standby components which are subjected to periodic testing test interval should be used to estimate availability.

The statistical analysis of data forms basis for modeling reliability, maintainability and availability. This includes, field/experimental data collection, selection of applicable distribution, data analysis employing statistical probability distribution and finally assessment of RMA aspects. The essential feature of probabilistic modeling has been explained in Section 32.4.1, while Sections 32.4.2 and 32.4.3 deal with maintainability and availability.

32.4.1 Reliability Modeling

Even when the components are designed considering the best practices in design and tested before being put into regular operation, the aspects related to random and wear out failure requires probabilistic treatment to have improved understanding of life cycle reliability.

The life cycle reliability of components and products can be characterized using a bath tub curve. The operational life span of the components can be divided into three regions, viz, Infant mortality period, the useful period, and aging period. This can be better appreciated by imagining a test environment where a population of component was subjected to test to collect data on failure times of each component in the population. This information was used to estimate the "instantaneous failure rate" or "hazard rate" of the component. It is expected that during the early phase of the test the failure rate would be high and then as the time pass the failure rate shows a decreasing trend. This phase of the test is characterized by decreasing trend and called as infant mortality period. The early higher failure rate could be due to, design faults, manufacturing defects, or material related issues or it could also be due to lack of learning on the part of test crews or operation or maintenance crews. However, as the initial issues are settled, the failure rate enters the next phase called "useful period" or "useful phase" of the components. This phase is characterized by random failure and failure rate or hazard rate remains generally constant. It is expected that this is the longest operational phase of the components often in years. For electronic components it could range from 1 years to more than 5 years. The end of useful period is characterized by increasing failure rate when the component enters the third phase, i.e., aging period. The challenge lie here is in identifying the start of this phase as this information is crucial in determining the life of the component.

When we see the same bath tub curve in the context of field environment, it can be seen that during initial phase of the component, the failures are relatively frequent and more efforts are required by maintenance and operational crew to identify and correct the failures. As the component enters the constant failure rate region, the scheduled or condition based maintenance activities can address the reliability issues to meet operational targets. However, the entry of component in aging phase is characterized by increase in failure rates, irrespective of having the maintenance schedule in place. In fact during this phase even increasing maintenance efforts may not be adequate to arrest the increasing failure rate trend (Figure 32.9).

Even though the bath tub representation is widely used in reliability engineering to model failure rates of component, there are some specific filed observations and literature that do not agree with the life cycle

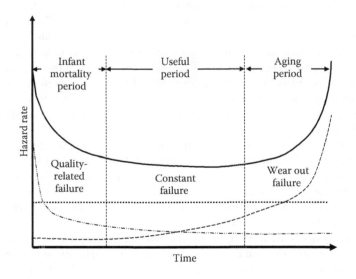

FIGURE 32.9   Bathtub curve.

phenomenon represented by bath tub curve. The Maintenance Steering Group (MSG) data on aviation system shows that only 4% of the filed failure data follows bath tube curve. The major part of the data 68% do not exhibit the aging failure while 14% data shows constant failure rate trend and remaining data follows varying trend which are not in line with bath tub representations [23]. This observation is important and it underlines the need of understanding the entire, i.e., cradle to death cycle of the component in the specific aspects of life cycle reliability. For example, if the product goes through a screening or qualification testing process before entering the field uses then the infant mortality period of the bathtub curve will not be seen as prominently as it appears in bath tub curve. The reason is that screening or qualification testing will eliminate all the design, manufacturing defects related failures. Similarly, the periodic test and maintenance or condition based activities or reliability centered maintenance (RCM) program may ensure the failure rate trend to remain constant as depicted in useful life period. In certain field situation the components are replaced well before aging symptoms are seen. Hence, the aging period of the bath curve for these components may not be as significant as it is seen in the bath tub curve.

Even if the failure rate is constant, the field conditions may require some modification either in design or operating/maintenance procedure that may reduce the failure rate. Hence, it can be argued that the bath tub curve is a general or ideal representation of the life cycle reliability and field conditions have direct bearing on the failure rates of the component.

Assume that the life test experiments is being performed on a population having N components. After some time t the test is terminated. Suppose number of component survived at time $t = N_S(t)$; and number of components failed at time $t = N_F(t)$ then by the definition of reliability

$$R(t) = \frac{N_S(t)}{N} \tag{32.1}$$

Also failure probability $F(t)$ at time t can be given by

$$F(t) = \frac{N_F}{N} \tag{32.2}$$

If we add number of component survived and failed at any instant of time t we get as follows:

$$N_S(t) + N_F(t) = N$$

Hence, when we add Equations 32.1 and 32.2 we get

$$R(t) + F(t) = 1$$

Or

$$R(t) = 1 - F(t)$$

Conversely,

$$F(t) = 1 - R(t)$$

The above, presents the relationship between reliability and failure probability. $F(t)$ is also called cumulative distribution function. For further details, readers can refer any book on Reliability Engineering.

There are two important observations related to the interpretation of "failure." First, it is not possible to predict the exact epoch of time to failure. It suggests that this aspect is basically random in nature.

The second aspect is that a qualitative notion of reliability is not adequate to address issues related to reliability in engineering management of components, and systems in general and complex systems like aviation systems in particular. Hence, it becomes necessary to quantify not only reliability, but availability, maintainability, and risk. The probabilistic approach enables quantification RMA. Another advantage of probabilistic approach is that it enables quantification of uncertainty in the estimates which helps organizations to assess design and safety margins in support of decision-making.

Reliability is defined as the ability of a component, product, or service to perform for a specified period of time under specified conditions. We note here that the term "ability" is a qualitative linguistic expression which can be related to any deterministic parameters related to performance function. When dealing in random nature of performance associated with the component the term ability is replaced by the term called "probability" and enable quantification of reliability by using failure rate and mission time parameters (exponential distribution). Hence, the more general and accepted definition of reliability is as follows:

Reliability is the probability that a component, system, or service thread will perform satisfactorily for a specified period of time under given conditions and expressed mathematically as follows:

$$R(t) = \text{Prob}(T \geq t \mid c_1, c_2, c_3, \ldots)$$

where
    T is a random variable represents actual time to failure of the component
    t is the mission time (for non repairable components)
    $c_1, c_2, c_3, \ldots$ represents the operational and environmental conditions, like temperature, humidity, electrical stresses, etc.

Considering the case of constant failure rate, in which exponential distribution can be used for modeling, the reliability is defined as

$$R(t) = \exp(-\lambda t)$$

where
    λ is the failure rate (per unit time)
    t is mission time (unit time)

For the case of exponential distribution where failure rate is considered as constant (useful life in bath tub curve) the inverse of failure rate is called as MTTF often in aviation parlance it is interpreted as MTBF. Hence,

$$\text{MTTF} = \frac{1}{\lambda}$$

The above, was presentation of introductory aspects of reliability, where exponential distribution was assumed as applicable distribution and direct estimates of failure rate was used to estimate the reliability. In reality, it is required to analyze the data, select a suitable probability distribution, and assess the failure rate for estimating the reliability. Following sections deal with these aspects of reliability modeling.

32.4.1.1 Failure Rate

Estimation of failure rate requires data collection and assessment of applicable distribution to these data. The accuracy of prediction to a large extent depends on adequacy or availability of data, selection of the appropriate failure model to represent and evaluate the data. We have noted that definition of failure and associated concepts forms fundamental component level reliability modeling. It was discussed that failure rates and mission requirements are generally defined in the unit of time. For example, the unit of failure rate has been given as failures per hour (or per days or per year depending on the type of data being modeled). Some time failure rates are also defined in number of failures per cycle. For standby components failure rates are defined as failure per demand. For example the failure rate of an emergency power supply source like a diesel generator that remains standby till a demand is generated, has unit of number of failure per demand. In some specific cases like for aircrafts the failure could be defined as number of failures per landing or take-off. Sometimes it is required that failures are defined in terms of "failure per million-hr-flights" which addresses both the time as well as number of flights for assessing the failure rate. Failure rates should be evaluated keeping in view what type of component being modeled. These models are mission model, standby model, repair model, test model, and demand model. For example, in mission related modeling, mission model will be used where the input will be mission time and λ will represent mission failure rate; for a standby type of component primarily test model where t will represent test interval and λ will be assigned standby failure rate; for a continuously operating or process related component repair model will be used where input will be time to repair and operating failure rate for parameter λ; for a component which is actuated on demand, demand model will be used where demand failure probability represented by parameter Q forms the only input. In certain situation more than one model is required. For example, if a service thread is demanded into the service, it is also required that it should provide continuous service. Here demand as well as mission model both will be used to have assessment of service thread failure probability. The design and operation document provide the information on particular category of the component.

The main input for assessing the failure rates is (if the situation demands assessment in time domain) (1) accumulated operating time for the component, (2) mission time t which depends on the system deterministic requirements, (3) definition of failure. Using these inputs the reliability of the components can be estimated. However, it may be noted that this procedure assumes exponential distribution as the applicable model and availability of the performance data from an operating system (Table 32.2).

32.4.1.2 Probability Distributions

Often these two aspects need to be examined. Before we go to examine these aspects, let us discuss various failure models (or failure distributions) which are fundamental to probabilistic modeling. The failure data can be divided into two categories viz., discrete and continuous data. Discrete data, like number of failure in given time, like failure of a switch or relay to actuate on demand. The second category is continuous data, i.e., number of failure of a continuous operating system like failure of a logic card, or failure of a stepper motor in a specified interval of time. Accordingly there are two types of failure distributions/models, continuous and discrete distribution. The most commonly used distributions which are used for electronic components for life and reliability predictions will be discussed in this section, like exponential distribution, normal distribution, lognormal distribution, and Weibull distribution. The "probability density functions" (pdf) parameter of a distribution characterizes the nature and trend of data and is fundamental to understanding of reliability. Hence, in following section some selected failure rate models will be characterized using "pdf."

32.4.1.3 Hardware Component Reliability Data

The individual piece of physical engineering entity which is considered to be making the building block of a system is called a component. For example when we consider an electronic board as sub-system then the components like microchips, capacitors, resistors, and diodes, etc are refereed as component that makes up the logic board. The integration of board into one module can forms a channel. Further, the redundant channels (say two or three) connected to various distributed devices may make up a system. However, there could be an instance when the analyst considers the electronic card as component or even a commercial of the shelf (Cots) product may also be treated as a component.

Hence, it is the analysts view as to what constitutes a component. In the section component means a lowest level electronic device for which failure rate assessment to be carried out. There are three approaches in assessing component reliability.

TABLE 32.2　Probability Distribution Function (pdf), $f(t)$; Reliability, $R(t)$; Cumulative Distribution Function $F(t)$ and Mean Time to Failure, MTTF for Exponential, Weibull, Normal, and Lognormal Distributions and Its Applications

Distribution	Probability Density Function (pdf) $f(t)$	Reliability $R(t)$	Unreliability or Cumulative Distribution Function $F(t)$	Mean Time to Failure MTTF	Applications
Exponential	$\lambda\exp(-\lambda t)$	$\exp(-\lambda t)$	$1-\exp(-\lambda t)$	$1/\lambda$	Life and reliability prediction considering constant failure rate
Weibull	$\dfrac{\beta}{\alpha}\left(\dfrac{t}{\alpha}\right)^{\beta-1}\exp\left[-\left(\dfrac{t}{\alpha}\right)^{\beta}\right]$	$\exp\left[-\left(\dfrac{t}{\alpha}\right)^{\beta}\right]$	$1-\exp\left[-\left(\dfrac{t}{\alpha}\right)^{\beta}\right]$	$\alpha\Gamma\left(\dfrac{1+\beta}{\beta}\right)$	Life prediction of many wear/ degradation related failures
Normal	$\dfrac{1}{\sqrt{2\pi}\sigma}\exp\left[-\dfrac{1}{2}\left(\dfrac{t-\mu}{\sigma}\right)^{2}\right]$	$1-\displaystyle\int_{t}^{\infty}\dfrac{1}{\sqrt{2\pi}\sigma}\exp\left[-\dfrac{1}{2}\left(\dfrac{\tau-\mu}{\sigma}\right)\right]d\tau$	$\displaystyle\int_{t}^{\infty}\dfrac{1}{\sqrt{2\pi}\sigma}\exp\left[-\dfrac{1}{2}\left(\dfrac{\tau-\mu}{\sigma}\right)\right]d\tau$	μ	Design and manufacturing, life prediction of high stressed components
Lognormal	$\dfrac{1}{\sqrt{2\pi}\sigma_{t}t}\exp\left[-\dfrac{1}{2\sigma_{t}^{2}}(\ln t-\mu_{t})^{2}\right]$	$1-\displaystyle\int_{0}^{t}\dfrac{1}{\sqrt{2\pi}\sigma_{t}\,\theta}\exp\left[-\dfrac{1}{2\sigma_{t}^{2}}(\ln\theta-\mu_{t})^{2}\right]d\theta$	$\displaystyle\int_{0}^{t}\dfrac{1}{\sqrt{2\pi}\sigma_{t}\,\theta}\exp\left[-\dfrac{1}{2\sigma_{t}^{2}}(\ln\theta-\mu_{t})^{2}\right]d\theta$	$\exp\left[\left(\mu_{t}+\dfrac{1}{2}\sigma_{t}\right)\right]$	Data coming from different sources, distribution of defects, life prediction of some components

λ, failure rate; t, time; μ, mean value; σ, standard deviation; β, shape parameter; α, scale parameter; θ, parameter of lognormal distribution.

32.4.1.3.1 Handbook Approach

This approach has been developed based on the life test studies carried out on electronic components under predetermined and simulated operational and environmental stresses in lab environment. The intensity of operational stresses like voltage, current, junction temperature and environment stresses like temperature, humidity, dust load; vibration is simulated to accelerate the failure mechanisms. This information so obtained is extrapolated to use conditions to understand the dominant failure mechanism and life or reliability of the component for use conditions. The data and models are documented handbooks along with the modifying factors.

For an analyst, interested in using handbook approach this knowledge base is available in the form of base/reference failure rates for the specified component, along with modifying factors for quality of the component, technology like CMOS, TTL, etc; construction features like type of package for an integrated circuit, number of pins etc. Similarly, there are modifying factors for operational stresses like junction temperature, humidity, vibration level, etc.

Even though this approach is widely used for estimating reliability of electronic components, there are some concerns expressed in literature on application of handbook approach [24]. The handbook approaches are also available as commercial software tools which provides an analysis platform through an easy to use graphic user interface.

32.4.1.3.2 Operating Experience

Many organizations and institutions keep record of system reliability data. Even if the records are not available, when the reliability studies are performed the history of operations and maintenance records are collected and information on accumulated operating experience, number of failure events, test and maintenance practices, failure modes, etc is elicited. It may be noted that the system specific data forms the best source of input data for reliability estimation. Here the advantage is that even similar types of component operating in the system provide reasonably good data on accumulated operating experience. For example the operating experience on a given type of microchip having a total population of, let us say 200, in the system with operating experience of even 5 years, provide with a total accumulated experience of $5 \times 200 = 1000$ years. Suppose that in these 5 years eight failures have been observed then the average failure rate works out to be 9.2×10^{-7}/h. This operating experience in use condition which carries all the system operating characteristics for future modeling including the stresses seen during the course of its operation provide a vital input for future use or extrapolation in case the design changes from the original. If the component design has not changed much from original specifications this approach works better than other approaches.

The argument against this approach is that the design of electronic system changes so rapidly that by the time new design is built the data on old component may not be applicable. Hence, these issues should be reviewed while selecting an approach for reliability estimation.

32.4.1.3.3 Accelerated Life Tests and Physics-of Failure Approach

Accelerated life test approach is employed extensively for predicting the life and reliability of components and systems. This approach is particularly useful for new design or new generation of components, like a new design of connector or field programmable gate arrays (FPGAs), etc. The handbook approach and system specific approach may not work for new design of components or simply put new components. Even for generic components like connectors, designed using advanced technique and material for fabrication and coating of the contacts, spring back mechanisms, it is required that accelerated life test approach be used to characterize the life and reliability of the component. The life test models as discussed earlier are used to estimate the life of the component.

Physics-of-failure (PoF) approach is essentially a science based approach to understand various failure mechanism or degradation mechanism and there by provide an effective approach to formulate models towards predicting life and reliability of the components. This approach deals with the

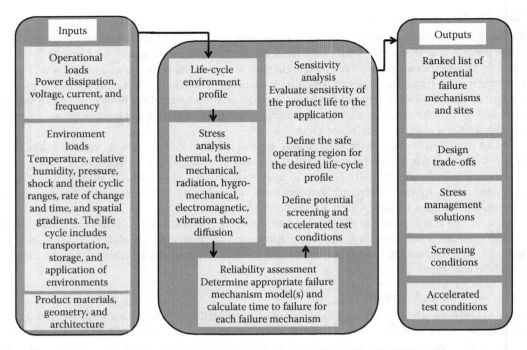

FIGURE 32.10   CALCE physics of failure based approach. (From Pecht, M., Physics of failure for complex systems, *DARPA/NIST Workshop*, Arlington, VA, http://www.isd.mel.nist.gov/meso_micro/Pecht.PDF, March 11–12, 1999.)

application of first principle models to understand the various failure mechanisms and thereby predict the remaining useful life and reliability of the components. The predictions in this approach are based on the component characteristic, like material properties, geometrical attributes and activation energy for applicable degradation processes for given operational and environmental stressors. Accelerated life testing is central to the physics-of failure approach. As PoF enable identification of dominant failure modes, mechanism and mechanism, and there by precursors for monitoring the health of the component. FMMEA form the corner stone of this approach to identify and prioritize the applicable degradation mechanisms (Figure 32.10).

The PoF model can be expressed in general form as:

$$t_{50} = f(x_1, x_2, x_3 \ldots)$$

where

t_{50} is the median life
x_i are the parameter of the model

The commonly known PoF model for life prediction is Arrhenius model, expressed as follows:

$$t_{50} = A \exp\left[\frac{E_a}{kT}\right]$$

where

A is a process constant
E_a is the activation energy of the process in eV (electron volt)
k is the Boltzmann's constant = 8.617×10^{-5} eV/K
T is temperature in Kelvin

The current generation of advanced microchips like the FPGAs are complex and built using millions of transistors and further complex layout and higher densities of interconnects. The net effect is, even though these chips are built with extremely higher capability that includes speed, the reliability of these components remains a question and area of further research. The traditional approach like handbook approach may not provide satisfactory answer. Here, the PoF approach, where FMMEA and Life testing forms integral to formulation of models that predicts the life of the components.

32.4.1.4 System Reliability Modeling

System reliability modeling deals with building a reliability model of the system from its basic constituent components or sub-systems. The overall system configuration and interrelationship of components forms the basic information for system reliability modeling. A system is a composite of equipment and skills, and techniques capable of performing or supporting an operational role or both [26]. In system reliability modeling, it is therefore important to consider, apart from hardware components, human factors, software components to arrive at system reliability estimate. The selection of component failure models (e.g., standby, continuously operating, repairable/non repairable, etc.) and applicable distribution (exponential, normal, log normal, etc.) forms part of procedure in system reliability modeling. Apart from this, the analysis of common cause failure, for redundant trains in a system is crucial to estimating system reliability.

32.4.1.4.1 Reliability Block Diagram

In this technique the system configuration is represented graphically. There are two basic entities that make reliability block diagram (RBD), "blocks" and "line" interconnections. The block may represent, a system, sub-system, equipment, or basic component. It depends at what level the RBD is being developed. For example, if an electronic card is being modeled, microchip, capacitor, resister can be represented by blocks. If a module is being modeled, the logic card, actuation relay and switches, power supply can be the basic component. If a system is being modeled, the sub-system-1, sub-system-2,… can be the basic entity to form a system. Functional relationship between two blocks is represented by line or arrow connections. Each reliability block diagram has one input and one output represented by incoming and outgoing arrow, respectively. Also, each block can have two states viz, Success and Fail. In the same manner, only one mode of failure can be considered in each simulation.

Figure 32.11 shows the series and parallel configurations. Figure 32.11a shows reliability block diagram for a service thread where the two components are connected series. For the success of the service thread each component should function correctly. Interruption of flow between input and output, i.e., failure of any one component is "loss of the thread" or "failure of thread."

Figure 32.11b shows the parallel configuration comprising of two components in the thread.

As can be seen in series system all components should be functional for system's success. Any one component (1 or 2 in Figure 32.11a) failure results in interruption in connection between input (I) and output (O). Most of the component connected on the logic board area treated to be connected in series, unless there is specific provision for backup.

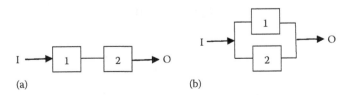

FIGURE 32.11 RBD showing (a) series and (b) parallel configurations.

Reliability of series system = Reliability of component 1 and reliability of component 2, then the reliability of the configuration shown in Figure 32.11a can be given as:

$$R_S(t) = r_1(t) \times r_2(t)$$

where
$R_S(t)$ is system reliability
$r_1(t)$ and $r_2(t)$ is the reliability of component 1 and 2
t is the mission time

If reliability of component 1 and 2 is assumed to follow exponential distribution

$$R_S(t) = \exp(-\lambda_1 t) \times \exp(-\lambda_2(t))$$

where λ_1 and λ_2 are the failure rate of component 1 and 2

$$R_S(t) = \exp[-(\lambda_1 + \lambda_2)t]$$

If $\lambda_1 + \lambda_2 = \lambda_T =$ Sum of failure rate of the component 1 and 2 in the system.

$$R_S(t) = \exp(-\lambda_T)t$$

Hence, in series system the sum total of failure rate of all the components represents total system failure rate. Hence, it is convenient to add failure rate of all the components and assess the system reliability as above.

In parallel configuration as shown in Figure 32.11b, failure of any one component, say component 1 or 2 at a time does not result into interruption between input (I) and output (O) as alternate path is available for successful operation of the system. This can be further clarified by one example. Let us assume there is one primary service thread that is backed up by another service thread that can perform function of primary service thread. Such configuration can be represented by parallel configuration. Please note that, here both the service threads are functioning in parallel and no switching action is required (unlike standby configurations). Hence, in parallel configuration, system keeps operating till last component fails.

If $R_S(t)$ and $Fs(t)$ represent the reliability and failure probability of system S for a mission T then, the system reliability for the parallel configuration can be given modeled as follow:
We know that

$$R(t) + F(t) = 1$$

Which means at any epoch of time t the component would either be working "successfully" or "fail." Hence the sum total of these two probabilities has to be unity.

$$R_S(t) = 1 - Fs(t)$$

or

$$F(t) = 1 - R(t)$$

$Fs(t)$ is system failure probability of parallel configuration
Let $q_1(t)$ and $q_2(t)$ is the failure probability for component 1 and 2, respectively.

Failure probability of parallel configuration can be given as

$$F(t) = q_1(t) \times q_2(t)$$ [21]

If $r_1(t)$ and $r_2(t)$ is reliability for component 1 and 2, then

$$F(t) = [(1 - r_1(t)) \times (1 - r_2(t))]$$

$$R_S(t) = 1 - [(1 - r_1(t)) \times (1 - r_2(t))]$$

Assume that the data follows exponential distribution

$$R_S(t) = 1 - \left[\left(1 - \exp(-\lambda_1 t)\right) \times \left[(1 - \exp(-\lambda_2 t))\right] \right]$$

$$R_S(t) = \exp(-\lambda_1 t) + \exp(-\lambda_2 t) - \exp\left[(\lambda_1 + \lambda_2)t\right]$$

If it assumed that $\lambda = \lambda_1 = \lambda_2$, that means if reliability of component 1 and 2 expressed by say r_1 and r_2, respectively, then $r_1 = r_2 = r$

$$R_S(t) = 2\exp(-\lambda t) - \exp(-2\lambda t)$$

or

$$R_S(t) = 2r - r^2$$

Hence reliability of parallel system having two components, as shown in Figure 32.11b is defined as the sum of reliability of individual components minus product of reliability of individual components.

32.4.1.4.2 Event Tree

The design of aviation system should ensure that a single failure or initiating event should not result in accidental condition. Hence, provision exists in the system to keep the operation with reduced redundancy at first level or safely terminate the operation such that it has consequences in terms of loss in efficiency. If the system degrades further, there should be provision to recover with certain repair or maintenance action and mitigate the safety consequences. Accidents become a reality only when redundant provisions fail and no recovery is possible. Event tree analysis is an effective approach to model event progression phenomenon.

Event tree analysis is a graphical approach. This analysis involves assessment of the system for given sets of undesired initiating events. Of course one initiating event is mapped at a time using event tree approach. Event tree development process involves propagation of the scenario based on system/facility response for a given initiating event. For example, in the event of an initiating event involving power failure, how the on-site redundant provisions like diesel generators, un-interrupted power supply system, batteries, etc., along with human action as recovery provisions, respond to recover/mitigate the consequences. The applicable safety systems, human actions and recovery aspects form the header in the event tree. While the initiating event forms the first header event while the safety systems follow in the functional chronological order. This approach is inductive in nature and often used in assessment of defence-in-depth or adequacy of redundant provision as part of risk analysis in complex systems. Most of the safety critical service threads in aviation systems are have fault tolerant features by incorporating provision of independency, redundancy, diversity and fail safe designs along with provision for recovery.

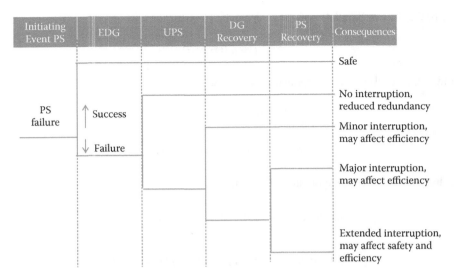

FIGURE 32.12 Example event tree for power supply failure in ATC center. *Note:* PS, power supply; EDG, emergency diesel generator(s), and UPS, uninterrupted power supply.

Ensuring uninterrupted power supply to electronics that comprises aviation equipments at air traffic control is crucial to aviation reliability. Let us assume a scenario where initiating event involves main power supply failure (from the grid) in an air traffic control center. This event is considered as anticipated transient in the design of the ATC centre. Accordingly, provision exists for auto starting of captive diesel generator sets. Apart from this, uninterrupted power system (UPS) has been provided for emergency services. After a power supply interruptions the UPS provides uninterrupted supply to emergency equipments/loads in the ATC as the batteries in UPS provide DC which is converted to AC. The UPS can cater power supply to the emergency loads for 3 h. The above provisions make sure that during normal power supply failure; adequate redundant provisions exist to deal with emergencies. However, even though deterministically these provisions look adequate, it can be further validated by employing event tree analysis to have an assurance that probability of total power supply failure to emergency equipments is reasonably low. Figure 32.12 shows the Event Tree for Power Supply Failure at ATC.

32.4.1.4.3 *Fault Tree*

Fault tree analysis is a deductive, graphical representation for a well defined undesired (top) event. As we have seen in the event tree, the header events are nothing but system safety provision or systems that requires it's evaluation often using fault tree or Markov analysis approach. Hence, the output of faults tree is forms input to event tree headers. The top event in a fault tree is an undesired event. This event is defined using well defined failure criteria. The failure criteria is defined in terms of non fulfilling of a process system parameters, like actuation of a safety service, strength of the communication signal, under-voltage, etc, duration for which the condition not met such that safety function is not fulfilled, minimum number of devices that lead to system failure (like two-out-of-three failure of redundant control/communication channels) etc. There are many input goes into determining the failure criteria.

Fault tree analysis is carried out at two levels, qualitative and quantitative. Table 32.3 shows the basic symbols and their meaning and functions, as used in fault tree analysis. Let us take an example of a service thread in aviation to understand the process of developing a fault tree and what insights can be obtained from a fault tree analysis. In the aviation taxonomy information systems have higher priority compared to remote/distributed system or support system. We assume that there is information system to meet a particular safety function. This service thread comprised of a primary thread, a backup thread and manual procedure which is like a mitigation measure. Figure 32.13 shows the configuration of the service thread.

TABLE 32.3   Fault Tree Symbols

Symbol	Nomenclature	Operation
Top	Top "Undesired" event' of the fault tree.	It is a terminal event has only inputs (more than one) and no output. This event carries the result of the analysis.
	Intermediate event	It carries the nomenclature of the process/event. It has more than one input (output of gate logic) and one output that become input for the next higher level gate.
	Basic event	This is most fundamental or end component/human error that form building block of the fault tree. Basic events forms bottom most entity.
	"OR" gate	Output of this gate is summation of all the inputs. If A & B are input to this gate than output will be (A + B).
	"AND" gate	Output of this gate is product of all the inputs. Ex. If A & B forms input than A*B, i.e., product of A & B is the result of this gate.
k	Combination *n* of events	Failure of "*k*" number of components out of *n* take the system to failed state. Also expressed as *k*-out-of-*n*: failure > system failure. Examples are two-out-of-three voting logic for modeling control system failure.
	House event or external event	An event which is expected to occur.
	Exclusive "OR" gate	Output occurs only if exactly one of the inputs occur.
	Priority "AND" gate	Output fault occurs if all the input fault occurs in a "defined sequence."
	Conditional event	Applicability of specific condition or constraints that apply to a given logic gate.
	Undeveloped event	These events are indicated on the fault tree to indicate that no analysis has been done beyond this point, either information was not available or the analysis does not require going beyond this point.
x	Transfer-in (from page x)	Output of the fault tree from page x is transferred-in to the current location of the higher level fault tree.
x	Transfer-out (to page x)	Output of the current fault tree is transferred-out as input to the fault tree on page x.

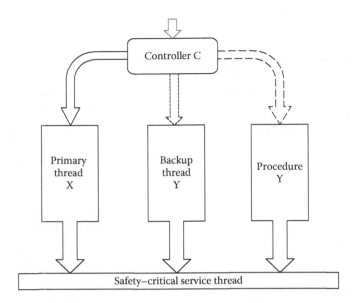

FIGURE 32.13   Representation of a hypothetical service thread.

We make certain assumptions as follows: All the three threads are designed to be independent, hence, the possibility of some common cause events which affect all the three service threads at one point of time is ruled out. However, due to certain aspects like location and sharing of cooling system, etc the possibility of two redundant processors in the same thread cannot be ruled out. Other devices like storage devices in the same thread are not affected by any common cause phenomenon. Each service thread comprised of sub-components. For example, the service thread, X and Y comprised of computer based systems. Hence, for smooth functioning of a service thread, it is required that each sub-component functions properly. For the sake of this example it assumed that the administrative procedure restore the system to the healthy operational state. At this stage it is assumed that transfer of control switch that shifts operation of the system from X thread to Y or vice-versa, works perfectly. The fault tree for the above configuration is developed as shown in Figure 32.14.

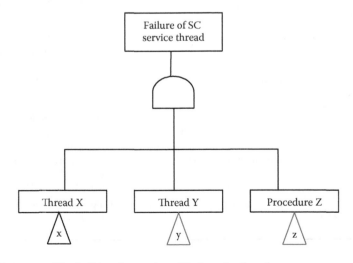

FIGURE 32.14   First stage of the fault tree for service critical service thread.

It can be seen that "Thread X," "Thread Y," and "Procedure Z" have been represented as intermediate events. This is because these events have not reach to basic component level and need to be developed further down.

Further, we have the information that (1) thread X and Y are identical and each comprised of a computer based system which has processor, SM power supply, DD-RAM, mother-board, and input and output cards. (2) Both the computer based systems are powered by the same source of power, (3) They are located in the separate room, but cooled by a common ventilation system. Hence, it becomes important that thread X and Y are identical have potential to fail due to certain common cause failure.

The system procedure involves restoration of the service thread by (1) set of human action and (2) manipulation of set of hardware switches and controls (here we assume that two switches need to be operated correctly to implement the procedure). Hence, implementation of procedure also requires functioning of some hardware in a correct manner.

Keeping above in view the fault tree single thread (X/Y) can be developed as follows (Figure 32.15):

An identical fault tree is developed for thread Y also. The only changes that need to be carried out are replace the component labeled as X with their counterpart component in thread Y. The fault tree drawn for thread X and Y are connecting to the intermediate event Thread X and Thread Y in Figure 32.13 and become part of main fault tree.

The fault tree for "procedure Z" is developed for the manual recovery error and correct operation of two switches is shown in the fault tree in Figure 32.16.

The result of the fault tree analysis carried out using the PSA software ISOGRAPH [27], can be summarized briefly. The qualitative analysis of the fault tree generates list of minimal cut sets. The list of the cut sets has been shown in Table 32.5. It can be seen that there are no first order and second cut sets. Most of the cut sets are third order cut sets. What it shows, it says that the system is robust, as it requires three components to fail simultaneously to result into failure of complete safety critical thread. Further, we can assert this finding by quantization of the fault tree by attaching failure probability to each basic component and human actions (thread Z). The data table (Table 32.4) shows the failure probabilities

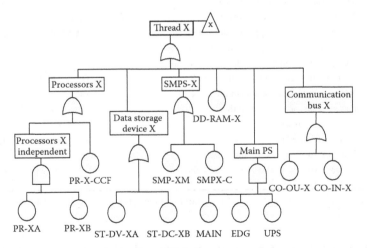

FIGURE 32.15 Fault tree for thread X. Nomenclature: PR-XA, processor A in thread X; PR-XB, processor A in thread B; PR-X-CCF, common cause failure of both the processor in thread X; ST-DV-XA, data storage device A in thread X; ST-DV-XB, data storage device B in thread X; SMP-XM, switch mode power supply module in thread X; SMP-XC, switch mode power supply connector in thread X; MAIN, main power supply to both the thread X and Y; emergency diesel generator (EDG) power supply to both the thread X and Y; UPS, uninterrupted power supply to both the thread X and Y; DD-RAM-X, DD-random access memory in thread X; CO-OU-X, output communication bus in thread X; CO-IN-X, input communication bus to thread X. *Note*: The fault tree for thread Y is similar, except that it will have all component nomenclature identifier Y in place of X.

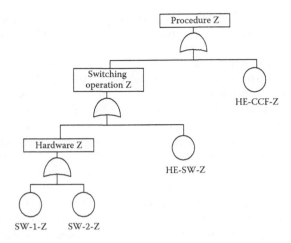

FIGURE 32.16   Fault tree for thread procedure Z. Nomenclature: SW-1-Z, switch-1 in thread Z; SW-2-Z, switch-2 in thread Z; HE-SW-Z, human error in operating the switch; HECCF, human error leading to common cause failure.

TABLE 32.4   Input Data for the Fault Tree

No.	Component Nomenclature	Component Code	Failure Probability
1	Processor A in thread X	PR-XA	2.0e-3
2	Processor B in thread X	PR-XB	2.0e-3
3	Common cause failure of both the processor in thread X	PR-X-CCF	2.0e-4
4	Data storage device A in thread X	ST-DV-XA	5.0e-4
5	Data storage device B in thread X	ST-DV-XB	5.0e-4
6	Switch mode power supply module in thread X	SMP-XM	3.0e-3
7	Switch mode power supply connector in thread X	SMP-XC	3.0e-3
8	Main power supply to both the thread X and Y	MAIN	1.0e-6
9	DD-random access memory in thread X	DD-RAM-X	1.0e-4
10	Output communication bus in thread X	CO-OU-X	7.0e-4
11	Input communication bus to thread X	CO-IN-X	9.0e-4
12	Switch-1 in thread Z	SW-1-Z	1.0e-3
13	Switch-1 in thread Z	SW-1-Z	1.0e-3
14	Human error in operating the switch	HE-SW-Z	5.0e-3
15	Human error induced common cause failure in execution of procedure Z	HE-CCH-Z	6.0e-3

Note: The components in thread Y have same nomenclature, except that it has identifier "Y" in place of "X."

assigned to each component/human actions. The result of the simulation shows that the failure probability of the top event, i.e., "Failure of SC Service Thread" is 4.16×10^{-7}. This result further confirms that the service thread can be considered to be highly reliable. These cut sets can be further prioritized based on the failure probabilities given Table 32.5.

Further, the importance analysis carried out using Fussell–Vessely importance model. The bar chart in Figure 32.18, in the following section, shows the importance measure of components. It can be seen that switch mode power supplies (SMP-YM and SMPX-M) carries highest importance in each thread X and Y, respectively. Further, human error in execution of the emergency procedure (HE-CCF-Z) and

TABLE 32.5 Fault Tree Analysis Results: Cut Set Probability (Only Top Unavailability)

No.	Cut Set	Unavailability
1	SMPX-M. SMP-YM. HE CCF-Z	5.40e-8
2	SMPX-M. SMP-YM. HE-SW-Z	4.50e-8
3	SMPX-M. CO-IN-Y. HE CCF-Z	1.62e-8
4	CD-IN-X. SMP-YM. HE CCF-Z	1.62e-8
5	CD-IN-X. SMP-YM. HE-SW-Z	1.35e-8
6	SMPX-M. CO-IN-Y. HE-SW-Z	1.35e-8
7	SMPX-M. CD-OU-Y. HE CCF-Z	1.26e-8
8	CD-OU-X. SMP-YM. HE CCF-Z	1.26e-8
9	CD-OU-X. SMP-YM. HE-SW-Z	1.05e-8
10	SMPX-M. CD-OU-Y. HE-SW-Z	1.05e-8
11	SMPX-M. SMP-YM. SW-1-2	9.00e-9

actuation of switches (HE-SW-Z) have been shown to be next level of importance. Figure 32.18 showing importance of the component can be referred for details. For details on Fault Tree analyses refer *NASA Handbook* [28].

The above information is crucial in risk analysis. Aspects related to risk analysis will be discussed in the respective section in this chapter.

32.4.2 Maintainability

Maintainability is the probability that a failed service/system/component will be repaired/restored to service in a given period of time when maintenance is performed by trained and qualified persons as per the prescribed procedure. The expression for maintainability is as follows:

$$M(t) = \exp(-\mu t)$$

where
  μ is the repair rate, i.e., no of repair per unit of time
  t is the repair time

The MTTR is also defined as

$$\text{MTTR} = \frac{1}{\mu}$$

MTTR is also referred as recovery time.

It is often argued that maintainability is an operational aspect of the computer based or digital systems. However, the fact is that maintainability is basically design aspect and has bearing on various surveillance programs that includes self test features, manual testing and calibration/periodic services, repair/replacement that will be performed during the operational phase of the systems. The term maintainability also refers to inherent characteristics of the component/thread/system that which provideability to calibrate for drifts or service, repair, or replace a failed component. Hence, designer of the digital systems need to make a decision regarding level of automation, dependence on human factors keeping in view the mission requirements. As discussed during previous sections, the safety threads should be designed in such a way that all the computer systems and its communication link poised for operation such that it is availability to perform the postulated function, when demand arises.

This requires automatic reconfigurable features along with periodic testing facilities/enablers. The automatic reconfigurable feature will recover any failure during the mission, while the testing and condition monitoring features will ensure higher maintainability and availability. The major objective for designing the control should be that during emergency situation the cognitive load on the operator is as minimum as achievable and information related to malfunctions/failure is made available through the man–machine interface such that it helps operator to recover the situation in an effective manner.

Maintenance is one of the major features to ensure implementation of effective maintainability program. Maintenance is an important aspect for ensuring higher reliability and availability of components during the operational phase of components. The proactive maintenance which includes preventive maintenance or predictive maintenance forms major approaches for maintenance. The preventive maintenance programs are based on schedules based on engineering judgment or vendor specifications while predictive maintenance involves condition based maintenance, reliability centered maintenance approach. Even though these approaches are time tested and have served well, there is room for improvement since one approach-only is not sure about the remaining useful life of the part, and that's why most of the approaches are based on very conservative criteria. One of the recommended approaches is risk-based approach of in-service inspection testing and maintenance. In this approach the results of risk analysis, in the form of quantitative estimates of availability, importance measure are used to prioritize the maintenance activities. These approaches also provide optimum test/service intervals for the maintenance activities. Interested readers can refer literature on risk-based approach to in-service inspection, testing, and maintenance.

Apart from this, the prognostics and health management (PHM) is a new paradigm which, showing promising applications. In fact in the present form, this approach is primarily being developed for electronic components. The prediction part in this approach provides answer on the "when to initiate the PM action" and the health management part of this approach provides best course of maintenance actions based on the results of the root cause analysis that forms part of PHM approach [29].

32.4.3 Availability

Availability is defined as the probability that a component will be available when demanded. As mentioned earlier, availability when it is associated with design aspects is refereed as inherent availability while in operational domain it is referred as steady state or transient availability. At this stage we will define availability as "fraction of time the component available to meet a demand or function." Hence in terms of uptime and down time parameters the operational availability can be defined as

$$\text{Availability} = \frac{\text{Uptime}}{\text{Uptime} + \text{Downtime}}$$

Steady state availability, in terms of MTTF and MTTR can be expressed as

$$\text{Availability (steady state)} = \frac{\text{MTTF}}{\text{MTTF} + \text{MTTR}}$$

The concept of "system" automatically calls for treatment of availability and maintainability, as repair and recovery are inherent when the system is put into operation. So when we discuss availability of a system, we know that repair as a mode of recovery form, hence the trade-off between reliability and availability becomes a criteria for decision making. For electronic systems in general and safety critical systems or even efficiency critical systems in particular, the overall availability of the system is improved by designing the systems in such a way that there is a provision for built-in test to facilitate automatic recovery by identification and isolation of the fault. In this way, repair constitutes an alternate form of recovery. This philosophy ensures higher availability as automatic reconfiguration which involves switching control from faulty redundant thread to the healthy thread(s) often with minor transient/ interruption or without an observable interruption. In such situation if the repair can be carried out in off-line mode without affecting the system availability, then there is a significant improvement in system reliability. The other method of improving system availability is by scheduling periodic surveillance programs. In this approach the redundant system is tested by taking out only one thread out of service and keeping the service available with other redundant threads. For example, take the case of three redundant threads (having two-out-of-three voting logic) and catering to monitoring and communication. This configuration allows taking one thread out for service when the service is available with other two threads following one-out-of-two voting logic, of course for short period when the test/maintenance is being performed. Periodic testing also increases system availability, even when the whole system is taken out of service, of course only after ensuring that there are no or minimal operational or safety consequences. The objective of periodic testing is to look for passive or latent faults. By removing these faults the net system availability increases as the time consumed in testing is normally contributed to much less unavailability than the gain in availability by reducing the probability of failure.

One of the example of automatic detection and isolation of fault in electronic system, particularly for standby safety service thread could be deployment of Fine Impulse Test (FIT) facility [30]. Identification of latent or passive fault in a standby thread poses a challenge. This is particularly true when the logic design requires to be, energized on demand. To overcome this limitation, the FIT diagnostic modules are employed for monitoring of the thread. The FIT sends short pulses through the electronic redundant channels at very short Intervals of let us say 2 s and confirms health of the complete thread/channels by comparing if the pre-designated output pattern with the actual output are the same. The data on any mismatch in the pattern is used to detect the fault card/logic. The pulse duration is kept so small that it is not enough to actuate any end actuating device like electromagnetic device. For example an electromagnetic relay actuation requires pulse duration of more than 30 ms to actuate it while the pulse generated by FIT is of much shorter duration. This scheme of testing a logic channel by employing FIT facility ensures higher availability of the redundant threads and finally services.

In the following section we will discuss the application of Markov approach where repair has been assumed as recovery mechanism.

Aviation system are designed employing fault tolerant design with redundant architectures. We consider a case with K-out-of-N voting redundancy which ensures high availability during operations. Here, repair forms a major recovery mechanism. It may be noted that recovery is not an operational characteristic but a design provision that ensures failure free operation of the system even if part of the system is being repaired. Often the fault diagnosis and recovery is performed automatically using reconfigurable logic at processor of logic chip level.

Markov models provide an effective approach to modeling of complex configurations. In Markov models a system can be modeled in more than two states. The basic assumption in Markov process is that the present state of the system depends only on the previous immediately preceding state. The second assumption is that the transition is governed by exponential distribution. The state of the system is represented by circle and the transition by arrow.

Consider a safety critical service comprised of two-out-of-three voting logic. The service is said to be normal when all the three threads operate healthy. The system configuration is such that if one thread fails then repair can be performed to bring back the system to normal state, i.e., all the three threads available. When the second thread fails (when the first failed thread could not be brought back to available state) the system goes to failed state.

The data collected shows that the failure rate λ/unit time and repair rate μ/unit time follows exponential distribution. The objective here is to formulate a Markov model to assess the "availability" or conversely speaking the "unavailability" in this example we assume that there is only one repair facility available. Also we assume that failure rates of the three threads have been arrived by averaging out the failure rate of the three threads and represented by λ.

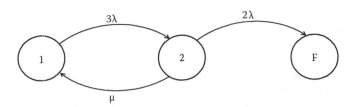

FIGURE 32.17 Markov diagram for two-out-of-three(F) service thread.

If "1" indicates the thread is available and "0" indicates thread as failed, then the system configuration for various states is shown in the bracket

The system has three states as follow:

1. State 1: All three services are available (111)
2. State 2: One service failed, two are available (110), (101), (011)
3. State F: Two services failed that take the system to the failed state (100), (010), (001)

The Markov graph representation for this system is as follows (Figure 32.17):

Suppose the probability of system in state 1, 2, and F at time t is given as $P_1(t)$, $P_2(t)$, and $P_F(t)$, respectively. Then the differential equation for the above systems can be written as

$$\frac{dP_1(t)}{dt} = -3\lambda P_1(t) + \mu P_2(t)$$

$$\frac{dP_2(t)}{dt} = -(2\lambda + \mu)P_2(t) + 3\lambda P_1(t)$$

$$\frac{dP_F(t)}{dt} = 2\lambda P_2(t)$$

The state "F" is absorbing state. This consideration is true for a safety system as failure of a safety system leads to failure consequences and recovery is not applicable to this state.

On solving these differential equation probability of system in state 1 will provide model for $A(t)$, i.e., availability as a function of time in terms of failure and repair rate. Similarly, expression for unavailability can be obtained as probability of system in state F or 1—probability of system in state 1.

Often it is required to model systems for condition where assessment of probability of detection of failure, probability of automatic recovery (by reconfiguration and not by repair) and finally the probability of system going to the failed state needs be evaluated. Markov approach provides an effective framework to model these types of problems. Readers advised to refer *NASA Handbook* [31] on the subject for details of Markov modeling.

32.5 Risk Assessment

Risk assessment in the context of digitals systems in avionics can be related to assessment frequency of failure of avionics systems and evaluation of associated consequences. The probabilistic risk assessment procedure is extensively used for aviation systems [32]. In this procedure the event tree approach is used for propagation of the scenario by inducing system response or human actions or recovery factors to an initiating event. This modeling is carried out by employing event tree analysis approach. The combination of frequency of initiating event and failure of safety functions are represented as accident sequences. For example, failure of (main power source) × (failure of emergency diesel) × (failure of UPS) is a typical accident sequence where the failure of main power source is an initiating event while the other emergency Diesel Generator and UPS are two safety functions. The failure probability of the safety systems which appear as header event in the event tree is performed using fault tree approach. In fault tree basic component failure probabilities form the input while the output of the fault tree is statement of the failure probability of the safety thread and a list of minimal cut sets. The fault tree and event tree approach has been discussed in the previous section of this chapter.

The process of risk assessment should start right away when the requirements specifications are being drafted though conceptual design of the system until the stage when system is put into operations. Even though designers may consider various scenario under which the system may have to perform and meet the mission objectives. It is a well known fact that there always a factor called "residual risk" that need to be assessed once the system is deployed into operation. Even if the postulated scenario has very low frequency of occurrence (rare event category), if the consequences are significant then the studies and subsequent provision should demonstrated that failure of the existing, say automatic provision, will not compromise safety. The provision could be manual actions or some operational procedures.

The definition of the consequences requires considerations of domain specific decision input/ variables. For example, for aviation systems, consequences may cover a wide range of scenarios, involving, right from loss of efficiency to minor delays or loss in business to then injuries or number of fatalities.

When the risk assessment approach is integrated into the design development process or operational activities of the system for identification of weaker aspects, prioritization and configuration optimization then these processes are called "risk management." The objective of fault tolerance risk management framework is to assess "residual risk" and provides suitable design or operational rules to avert or mitigate this risk. The importance and sensitivity analysis plays key role in risk management.

32.5.1 Importance and Sensitivity Analysis

The probabilistic approach to design and operations of complex systems, like aviation systems, provides with two parameters called (1) "importance measure" of the component or service and (2) "sensitivity index" of the assumptions or data or human actions. The basic factors that determine the importance of a component are reliability of component and its location in the system. For example, single computer system catering to a service thread will have higher importance than two identical computer systems operating in parallel to cater the same service. This is true provided the reliability of computer in both the configurations is same. In another case, it is quite possible that a single computer with reliability higher than the net reliability of two computers having very low reliability operating in parallel. In this case the importance of the two computers even though in redundant mode may work out to be higher than the single computer in the first case (Figure 32.18).

Hence, importance and sensitivity analysis form an integral part of the risk assessment. For example, a communication safety critical service comprises of three parallel service threads, viz. "primary," "backup," and "procedure" thread; often the question arises as to what is the contribution of the third alternative, i.e., "procedure thread." This becomes a question for design optimization of the service. Here, the importance analysis will inform the designers about the criticality of the contribution of each thread towards achieving the overall unavailability criteria for the service. After a detailed analysis, recommendation for increasing the redundancy of computational devices or automation of the procedure could form part of the design review. There are well established models for evaluating the importance measures of the component. The most commonly used models are (1) Birnbaum, (2) Criticality, (3) Fussell–Vesely importance measures. When the output of the analysis is in the form of minimal cut sets, then it is appropriate to use two important measures, viz, risk-reduction worth and risk-achievement worth importance. There are many more models available in the literature. Figure 32.16 shows the importance analysis performed using fault tree approach in the example safety critical service thread in the previous section.

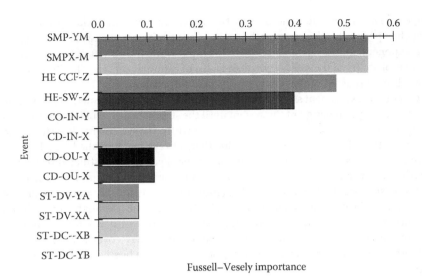

FIGURE 32.18   Importance measure for components.

The sensitivity and importance analysis forms part of the risk-assessment or risk management methodology. This analysis helps identifying and prioritizing the component/services for risk management actions, as discussed in the previous section.

32.5.2 Risk-Informed Approach in Support of Decision-Making

Even though the traditional approach to design, operation, and regulation of aviation are deterministic the probabilistic methods are playing complementary role as part of risk-informed approach. The major steps of risk informed approach are (1) identification of alternatives, (2) risk analysis of alternatives, and (3) risk-informed alternative selection [33]. Furthermore, the risk analysis of alternative involves assessment of safety risk, technical risk, cost-risk, and schedule risk. When seen in the context of regulation, oversight and monitoring forms one of the elements of risk-informed regulation. The reasons why risk-informed approach finding wider acceptability is that (1) it provides flexibility in decision making compared to prescriptive and conservative deterministic approach, (2) decisions are supported by quantified safety indicators and criteria instead of qualitative criteria, (3) it provides best estimate approach to analysis and decisions, (4) this approach allows integration of human factors into the hardware model of the system, and (5) it provides improved framework to characterize uncertainty associated with alternative options. The available literature is testimony to the enhanced role of risk-informed approach in decision-making. The risk-informed approach is different from risk-based approach [33,34]. In risk-based approach the results of risk analysis forms the basis for decisions while in risk informed approach experts deliberations along with engineering arguments and risk-Insights dealt in an integrated framework in support of decision-makings.

32.6 Electronic Component Management Plan per IEC TS 62239

To produce a product, there is usually a complex supply chain of companies that are involved directly and indirectly in producing parts (materials) for the final product. Thus, to produce a reliable product, it is necessary to select parts that have sufficient quality and are capable of delivering the expected performance for the targeted life cycle conditions. IEC TS 62239 is the part 1 of the process management for avionics—management plan, named "Preparation and maintenance of an electronic components management plan." This technical specification has been prepared by IEC Technical Committee 107: Process management for avionics and was released in 2012 [35]. This document provides guidelines to avionics systems manufacturers on selecting and managing electronic components. This section discusses elements of coverage of several topics within that standard and within that context the elements of action that are performed by best in class companies in avionics.

32.6.1 Uprating

Uprating is a process to assess the capability of an electronic part to meet the functional and performance requirements of an application in which the part is used outside the manufacturers' specified operating conditions. The IEC standard requires that components be selected within temperature ranges in excess or matching the temperature range required in the application. However, it accepts the reality that sometime parts are not available to meet the requirements.

If components are used outside the temperature ranges specified by the component manufacturer, then the avionics manufacturer is required demonstrate the control of this process. Recommendations and guidelines on how to perform this uprating are provided in IEC/TR 62240 and use of that document is allowable for the manufacturers. Documents equivalent to the technical report 62240 are also allowable [36]. There are similar GEIA standards that also cover the methods of uprating.

Uprating is carried out after the part, the part manufacturer, and the distributors have been assessed [7,37,38] based on datasheets, application notes, and any other published data. The International Electrotechnical Commission and the Electronics Industry Association [39] accept these methods as industry best practices. Publications for the U.S. Department of Defense acknowledge these methods as effective and rigorous [40]. These three methods are parameter conformance, parameter recharacterization, and stress balancing. These three methods are developed at CALCE at the University of Maryland with support from the CALCE Consortium Members including major avionics manufacturers.

Parameter conformance is a process of uprating in which the part is tested to assess if its functionality and electrical parameters meet the manufacturer's recommended operating conditions over the target temperature range. Electrical testing is performed with the semiconductor manufacturer-specified test setups to assess compliance within the semiconductor manufacturer-specified parameter limits. The tests are of "go/no-go" type, and are generally performed at the upper and lower ends of the target application conditions. A margin may be added to the test, either in a range wider than the target application conditions tighter electrical parameter limits for the test. The electrical parameter specifications in the datasheet are not modified by this method.

Parameter re-characterization is a process of uprating in which the part functionality is assessed and the electrical parameters are characterized over the target application conditions, leading to a possible re-specification of the manufacturer-specified datasheet parameter limits. The parameter re-characterization method of uprating seeks to mimic the part manufacturer's characterization process. The electrical parameter limits of parts rated for multiple temperature ranges are often obtained using the concept of parameter re-characterization [41,42]. Electrical testing is followed by data analysis and margin estimation.

Stress balancing is a process of thermal uprating in which at least one of the part's electrical parameters is kept below its maximum allowable limit to reduce heat generation, thereby allowing operation at a higher ambient temperature than that specified by the semiconductor part manufacturer [43]. The process assesses the possibility that the application may not need to use the full performance capability of the device, and that a power versus operating temperature trade-off for the part may be possible.

32.6.2 HALT

HALT (highly accelerated life testing), typically combined vibration and thermal cycling testing is often performed by avionics manufacturers but not at component level. As a result, the IEC TS 62239 does not cover use of HALT for components. Overstress failures typically arise as a result of a single load (stress) condition. Examples of load conditions that can cause overstress failures are shock, temperature extremes, and electrical overstress. Overstress failure mechanisms include fracture, latch-up, EOS, and ESD. Generally "overstress" tests are performed on small sample sizes and the specimens are subjected to a harsh environment, (i.e., severe levels of stress). If the specimen fails, the destruct limits and the associated failure modes are known but the test data cannot usually be extrapolated to use conditions. Hobbs, widely regarded as the pioneer of HALT stated

> A test in which stresses applied to the product are well beyond normal shipping, storage and in-use levels. From the results of HALT, we cannot determine the field reliability. We make the product a lot better, but do not know how much better.

32.6.3 Lead Free

The IEC standard covers lead free issues for the management of lead-free termination finish and soldering of components and requires that the avionics manufacturers develop a control plan in accordance with IEC/PAS 62647-1 (GEIA-STD-0005-1). Most avionics applications are exempt from lead free requirements. As a response to the EU's Restriction of Hazardous Substances (RoHS), the industry has predominantly settled on tin–silver–copper as the replacement for conventional tin–lead solder. To comply with RoHS, part manufacturers have sought lead-free finishes to replace the traditionally used tin–lead finishes. The finish selection is important in providing corrosion resistance, good solderability, and durable solder joints.

Due to its low cost and compatibility with existing solders, the pure tin and tin-rich lead-free alloys have been adopted by a significant portion of electronic part manufacturers. For the avionics community, the adoption of pure tin and tin-rich lead-free finishes has raised a reliability issue pertaining to the formation of conductive whiskers. Completely avoiding pure tin and tin-rich lead-free finished parts is preferred for high-demand, high-reliability applications like avionics.

To avoid pure tin finish or lead free solder balls, component refinishing methods like solder dipping and BGA reballing are often adopted following the relevant GEIA standards. Most companies also perform length analysis between contacts to determine the risk of shorting by whiskers.

32.6.4 Avoidance of Counterfeiting

The IEC TS defines counterfeit as practice of producing products which are imitations or are fake goods or services and, therefore, infringes the intellectual property rights of the original manufacturer. From a component management point of view, a broader definition of counterfeit electronic part is given by Sood et al. [44] as one whose identity has been deliberately misrepresented. The identity of an electronic part includes: manufacturer, part number, date and lot code, reliability level, inspection/testing, and documentation.

The absence of pedigree verification tools in the electronics part supply chain and the availability of cheap tools and parts to create counterfeits make the counterfeiting of electronic parts a relatively low risk operation for counterfeiters, while the cost of inspection/testing procedures make it harder for part users to detect counterfeits. The effects of a counterfeit part and a sub-standard part may be similar on a finished end-product, but there are two distinctions between the impacts of these two types of parts. First, the liability for the inclusion of a counterfeit part can be wholly on the organization that procures the counterfeit part, since the source of an original part is difficult to confirm. The judicial system and law enforcement may not be able to offer assistance in identifying the original source of the part in order to attribute liability. In addition to liability, there is limited or nonexistent root cause failure analysis support from part manufacturers in case of a counterfeit part that is purchased out of the authorized supply chain. There are three ways of generating counterfeit parts: Relabeling, Illegal manufacturing, and Scrap salvaging. The attributes of the counterfeit parts are different in each case and can be used as a guide for identifying counterfeit parts.

The IEC TS calls out IEC/TS 62668 [45] as guidelines to assist with avoiding counterfeit, fraudulent, and recycled components. The avionics manufactures are required to have a counterfeit, fraudulent, and recycled component risk mitigation plan based on industry documents like SAE AS5553 [46] and/or GIFAS/5052, 2008 [47]. In addition, as per supplier management, the TS requires that a distributor shall have a process for counterfeit electronic parts avoidance if the distributor supplies parts other than from a manufacturer franchise. Also, when a distributor accepts returned stock from customers, these parts will have to be considered in the counterfeit/fraudulent management risk plan of the distributor.

32.6.5 Long Duration Storage

The IEC TC requires that at the component level, the selected plastic parts all need to pass HAST tests as per JESD22-A110 level tests. It also suggests that parts and wafers be stored under dry nitrogen.

The assessment process for long term storage by the avionics manufacturers include part reliability assessment for possible stresses under typical storage conditions such as humidity and corrosive gases, and possible ESD. Generally consideration is given toward possible loss of part solderability, corrosion of terminations and possible damage due to uncontrolled humidity exposure.

In one recent study [48] long term storage reliability of antifuse field programmable gate arrays was evaluated. Aerospace applications typically involve long operating life and there can be a need to store un-programmed antifuse FPGA parts for long periods and program them when necessary to support the system. Field programmable gate arrays can be programmed and used to upgrade the functionality of existing electronics; thus preventing them from becoming obsolete. The results of this study showed that if the strategy of stocking parts is employed to counter antifuse FPGA obsolescence, and if the parts are stored under best practice long term storage conditions (controlled temperature, humidity, ESD packaging), then no failures will be encountered beyond what is specified by the manufacturers.

32.6.6 Obsolescence and Replacement Considerations

The IEC TS covers component obsolescence issues under component availability and associated risk assessment process. The avionics manufacturers are required to have a process to identify risks associated with availability of the component, and methods to mitigate those risks. These risks can include low volume manufacturers, allocation risks, financial stability of manufacturers, single source manufacturers and such items.

The avionics manufactures are required to rate components considered to be at risk using metrics that that account for their susceptibility to technology change and obsolescence. This risk rating covers several issues besides component obsolescence including component changes, lack of qualification data and uprated components.

Regarding component obsolescence, the avionics manufacturer is required to prepare an obsolescence management plan addressing pro-active measures for component obsolescence, component obsolescence awareness, and reaction to component obsolescence. The coverage of the plan should be from design through support of the product in field. The plan is required to document the processes used to resolve obsolete component occurrences for assurance of continued production and support. The processes used to react to component obsolescence occurrences include bridge stock or lifetime buys, identification of alternative sources, equipment re-design and other activities that occur after an obsolescence event occurs.

The avionics manufacturers are required to maintain a periodic report of all existing and impending obsolescence issues, even if the obsolescence issue only has impact beyond the contract period. For example, even if the avionics manufacturers possesses adequate part inventory to meet contractual delivery obligations, but there is a known issue with future procurement of parts, that should be included in report.

Best practices in obsolescence management and the latest practices are covered in publications that cover issues related to design refresh planning [49], forecasting part procurement lifetimes [50], total part cost ownership models [51] and other tools. Summary of many of these tools and tactics can be found in this book on strategies [52].

32.6.7 Radiation Effects

The IEC TS provides requirements regarding selections related to bother atmospheric SEU radiation and total dose radiation effects for space use. At the component level, the following general requirements are laid out regarding part selection and use: selection of components not subject to radiation effects, testing of components from a specific die code for radiation tolerance and then performing one time buy of that that date code part, selection of components not using Boron 10 and derating for voltage.

32.7 Summary

In aviation systems, the approach to performance assurance involves, addressing issues not only related to reliability but also availability and maintainability. It is important to understand that along with deterministic aspects of RMA, the probabilistic modeling and corresponding decision variables/ criteria play an important role in management of life cycle of the component, service, and systems.

High product reliability can only be assured through robust product designs, capable processes that are known to be within tolerances, and qualified components and materials from vendors whose processes are also capable and within tolerances. Quantitative understanding and modeling of all relevant failure mechanisms provide a convenient vehicle for formulating effective design, process, and test specifications and tolerances.

The physics-of-failure approach is not only a tool to provide better and more effective designs, but it also helps develop cost-effective approaches for minimizing the development and sustaining cost of the product. Processes should be implemented for defining more realistic performance requirements and environmental conditions, identifying and characterizing key material properties, developing new product architectures and technologies, developing more realistic and effective accelerated stress tests to evaluate reliability and audit quality, enhancing manufacturing-for-reliability through mechanistic process modeling, and characterization allowing proactive process optimization, increasing first-pass yields, and reducing hidden factory costs associated with inspection, rework, and scrap.

When utilized early in the concept stage of a product's development, reliability aids optimization of design configuration that ensures higher availability while keeping risk to low are acceptable. In the design stage of product development, reliability analysis involves methods to enhance performance over time through the selection of materials, design of structures, design tolerances, manufacturing processes and tolerances, assembly techniques, shipping and handling methods, and maintenance and maintainability guidelines. The use of physics-of-failure concepts coupled with mechanistic and probabilistic techniques are often required to understand the potential problems and trade-offs and to take corrective actions. The use of factors of safety and worst-case studies as part of the analysis is useful in determining stress screening and burn-in procedures, reliability growth, maintenance modifications, field testing procedures, and various logistics requirements.

The role of risk assessment has been growing in the aviation systems, as the tools like probabilistic risk assessment provide an effective platform to create integrated model of the system and thereby an effective strategy for not only design optimization but also identification and prioritization of safety issues in support of operation and maintenance.

The condition monitoring and reliability centered maintenance approaches form integral parts of the maintenance management approach in aviation and space systems. However, there is need to improve the prediction capability in maintenance management to realize higher safety and availability of aviation systems. Towards this, extensive developments are being reported in application of prognostics and health management approach to complex systems in general and aviation systems in particular.

The challenges that the digital system modeling community facing is assessment of reliability of new components, like field programmable gate arrays which are built using new technologies and could suffer from new failure modes. Uncertainty characterization in data and model is another challenge. To address these issues it is required that PoF models are developed for new and unproven technologies. Organizational aspects are crucial to the success of any reliability programmed. Hence, it is required that reliability/risk program should be integrated with the not only design and operations procedures, but also with acquisitions and disposal.

Definitions

Accelerated testing: Tests conducted at stress levels that are more severe than the normal operating levels, in order to enhance the damage accumulation rate within a reduced time period.

Common cause failure: One shared cause among the redundant trains that causes failure of the two or more identical components at the same time or in a very short interval of time.

Criticality: A relative measure of the frequency and consequence of failure.

Damage: The extent of a product's degradation or deviation from a defect-free state.

Derating: The practice of subjecting parts to lower electrical or mechanical stresses than they can withstand to increase the life expectancy of the part.

Failure mechanism: A process (such as creep, fatigue, or wear) through which a defect nucleates and grows as a function of stresses (such as thermal, mechanical, electromagnetic, or chemical loadings) ultimately resulting in the degradation or failure of a product.

Failure mode: Any physically observable change caused by a failure mechanism.

Integrity: A measure of the appropriateness of the tests conducted by the manufacturer and the part's ability to survive those tests.

Overstress failures: Catastrophic sudden failures due to a single occurrence of a stress event that exceeds the intrinsic strength of a material.

Product performance: The ability of a product to perform as required according to specifications.

Qualification: All activities that ensure that the nominal design and manufacturing specifications will meet or exceed the reliability goals.

Quality: A measure of a part's ability to meet the workmanship criteria of the manufacturer.

Reliability: The ability of a product to perform as intended (i.e., without failure and within specified performance limits) for a specified time, in its life cycle application environment.

Service threads: Service threads are strings of system that support one or more of the National Airspace System architecture capabilities. These service threads represent specific data paths (e.g., radar surveillance data) to controllers or pilots. The threads are defined in terms of narratives and reliability block diagrams depicting the system that comprise them.

Service thread loss severity category (STLSC): Each service thread is assigned one of the three service thread loss severity categories based on the severity of impact that loss of thread could have on the safe and efficient operation and control of the aircraft. The service thread loss severity categories are: (1) safety critical: The service thread loss in this category would present an acceptable safety hazard during transition to reduce capacity operation, (2) efficiency critical: The service thread loss in this category could be accommodated by reducing capacity without compromising safety, but the resulting impact has the potential for safety wide impact on National Airspace System efficiency of operations, and (3) essential: The service thread loss in this category could be accommodated by reducing capacity without comprising safety, with only localized impact on National Airspace System efficiency.

Target operational availability: The desired operational availability associated with a given National Airspace System Service/Capability Criticality.

Wear out failures: Failures due to accumulation of incremental damage, occurring when the accumulated damage exceeds the material endurance limit.

References

1. American Military Standard, Definition of terms for reliability and maintainability, MIL-STD-721C, June 12, 1981.
2. American Military Standard, Reliability prediction of electronic equipments, MIL-HDBK-217F, December 1991.
3. American Military Standard, Reliability testing for engineering development, qualification and production, MIL-HDBK-781D, October 18, 1986.
4. Federal Aviation Administration, *Reliability, Maintainability and Availability (RMA) Handbook*, FAA-HDBK-006A, January 7, 2008, FAA, Washington, DC.
5. Pecht, M., *Integrated Circuit, Hybrid, and Multichip Module Package Design Guidelines—A Focus on Reliability*, John Wiley & Sons, New York, 1994.
6. IEEE Standard 1332, IEEE standard reliability program for the development and production of electronic systems and equipment, 1998.
7. Jackson, M., Mathur, A., Pecht, M., and Kendall, R., Part manufacturer assessment process, *Quality and Reliability Engineering International*, 15, 457–468, 1999.
8. Sage, A.P. and Rouse, W.B., *Handbook of Systems Engineering and Management*, John Wiley & Sons, New York, 1999.
9. Humphrey, D., Condra, L., Pendse, N., Das, D., Wilkinson, C., and Pecht, M., An avionics guide to uprating of electronic parts, *IEEE Transactions on Components and Packaging Technologies*, 23, 595–599, 2000.
10. Tullmin, M. and Roberge, P.R., Corrosion of metallic materials, *IEEE Transactions on Reliability*, 44, 271–278, 1995.
11. IEEE 1413.1, Guide for selecting and using reliability predictions based on IEEE 1413, IEEE Standard, February 2003.
12. JEP 122B, Failure mechanisms and models for semiconductor devices, JEDEC Standard, August 2003.
13. JEP 148, Reliability qualification of semiconductor devices based on physics of failure risk and opportunity assessment, April 2004.
14. Vichare, N., Eveloy, V., Rodgers, P., and Pecht, M., In-situ temperature measurement of a notebook computer—A case study in health and usage monitoring of electronics, *IEEE Transactions on Device and Materials Reliability*, 4(4), 658–663, 2004.
15. Mishra, S. and Pecht, M., In-situ sensors for product reliability monitoring, *Proceedings of SPIE*, 4755, 10–19, 2002.
16. Pecht, M., Dube, M., Natishan, M., and Knowles, I., An evaluation of built-in test, *IEEE Transactions on Aerospace and Electronic Systems*, 37(1), 266–272.
17. Ramakrishnan, A. and Pecht, M., A life consumption monitoring methodology for electronic systems, *IEEE Transactions on Components and Packaging Technologies*, 26(3), 625–634.
18. Mishra, S., Pecht, M., Smith, T., McNee, I., and Harris, R., Remaining life prediction of electronic products using life consumption monitoring approach, *Proceedings of the European Microelectronics Packaging and Interconnection Symposium*, Cracow, Poland, June 16–18, 2002, pp. 136–142.
19. Vichare, N., Rodgers, P., Azarian, M.H., and Pecht, M., Application of health monitoring to product take-back decisions, *Joint International Congress and Exhibition—Electronics Goes Green 2004+*, Berlin, Germany, September 6–8, 2004.
20. Self-Monitoring Analysis and Reporting Technology (SMART), PC Guide, http://www.pcguide.com/ref/hdd/perf/qual/featuresSMART-c.html, Last accessed on August 22, 2004.
21. Pecht, M., Dasgupta, A., Evans, J.W., and Evans, J.Y., *Quality Conformance and Qualification of Microelectronic Packages and Interconnects*, John Wiley & Sons, New York, 1994.
22. Upadhyayula, K. and Dasgupta, A., Guidelines for physics-of-failure based accelerated stress testing, *Proceedings of the Annual Reliability and Maintainability Symposium*, New York, 1998, pp. 345–357.

23. Roush, M.L. and Webb, M.W., *Applied Reliability Engineering—Life Models for Repairable Items*, Vol. I, 5th edn., RIAC & University of Maryland, College Park, MD, 2006.

24. White, M. and Bernstein, J.B., Microelectronics reliability: Physics-of-failure based modeling and life time evaluation, National Aeronautical Space Administration, Washington, DC, 2008. http://nepp.nasa.gov. Accessed on November 14, 2012.

25. Pecht, M., Physics of failure for complex systems, *DARPA/NIST Workshop*, Arlington, VA, March 11–12, 1999. http://www.isd.mel.nist.gov/meso_micro/Pecht.PDF. Accessed on January 10, 2013.

26. U.S. Department of Defense, *Military Handbook: Electronic Reliability Design Handbook*, MIL-HDBK-338B, Department of Defense, Washington, DC, October 1998.

27. ISOGRAPH, Fault tree analysis software—Fault tree + version 10, M/S ISOGRAPH, Warrington, U.K.

28. NASA, *Fault Tree Handbook with Aerospace Applications*, Version 1.1, Prepared by NASA Office of Safety and Mission Assurance, NASA Headquarters, Washington, DC, August 2, 2002.

29. Pecht, M.G., *Prognostics and Health Management of Electronics*, Wiley, Hoboken, NJ, 2008.

30. Misra, M.K., Menon, S.P., Sridhar, N., and Sambasivan, S.I., Design of hardwired fine impulse test system for safety logic system of fast breeder test reactor, http://www.vecc.gov.in/~sacet09/downloads/FINAL%20PDF/D12_MISHRA_EID_IGCAR%20_F_.pdf. Accessed on August 7, 2012.

31. Ricky, W.B. and Sally, C.J., Techniques for modeling the reliability of fault-tolerant systems with the Markov State-Space Approach, NASA Reference Publication 1348, Langley Research Center, Hampton, VA, September 1995.

32. Stamatelatos, M. and Dezfuli, H., *Probabilistic Risk Assessment Procedures Guide for NASA Managers and Practitioners*, 2nd edn., NASA/SP-2011-3421, NASA Headquarters, Washington, DC, December 2011.

33. NASA Handbook, *Risk-Informed Decision Making*, Handbook NASA/SP-2010-576, Version 1.0, Office of Safety and Mission Assurance, NASA Headquarters, Washington, DC, April 2010.

34. Stamatelatos, M., Implementation of risk-informed decision making at NASA, *NASA Risk Management Conference 2005*, RMC-VI, Orlando, FL. http://www.rmc.nasa.gov/presentations/Stamatelatos_Implement_Risk_Informed_Decision_Making_at_NASA.pdf. Accessed on March 15, 2013.

35. International Electrotechnical Commission, Process management for avionics—Management plan, Preparation and maintenance of an electronic components management plan, IEC TS 62239, 2012.

36. International Electrotechnical Commission, Process management for avionics—Electronic components capability in operation—Part 1: Temperature uprating, IEC/TR 62240, 2013.

37. Maniwa, R.T. and Jain, M., Focus report: High speed benchmark, *Integrated System Design*, Website: www.isdmag.com, March 1996.

38. Jackson, M., Sandborn, P., Pecht, M., Hemens-Davis, C., and Audette, P., A risk informed methodology for parts selection and management, *Quality and Reliability Engineering International*, 15, 261–271, 1999.

39. International Electrotechnical Commission, Use of semiconductor devices outside manufacturers' specified temperature ranges, IEC/PAS 62240, 1, 2001-04 edn., (Also being developed as GEIA 4900), 2001.

40. Lasance, C.J.M., Accurate temperature prediction in consumer electronics: A must but still a myth, *Cooling of Electronics Systems*, Kakaç, S., Yüncü, H., and Hijikata, K., eds., Kluwer Academic Publishers, Dordrecht, the Netherlands, pp. 825–858, 1993.

41. Pecht, M., Why the traditional reliability prediction models do not work—Is there an alternative? *Electronics Cooling*, 2(1), 10–12, January 1996.

42. Pendsé, N. and Pecht, M., Parameter re-characterization case study: Electrical performance comparison of the military and commercial versions of all octal buffer, *Future Circuits International*, 6, 63–67 (Technology Publishing Ltd., London, U.K.), 2000.

43. McCluskey, P.F., *High Temperature Electronics*, CRC Press, Boca Raton, FL, 1996.
44. Sood, B., Das, D., and Pecht, M., Screening for counterfeit electronic parts, *Journal of Materials Science, Materials in Electronics*, 22, 1511–1522, 2011.
45. International Electrotechnical Commission, Process management for avionics—Counterfeit prevention—Part 2: Managing electronic components from non-franchised source, IEC/TS 62668, 2013.
46. SAE, Counterfeit electronic parts: Avoidance, detection, mitigation, and disposition, SAE AS5553, 2009.
47. GIFAS, Guide for managing electronic component sourcing through non franchised distributors. Preventing fraud and counterfeiting, GIFAS/5052, 2008.
48. Patil, N., Das, D., Scanff, E., and Pecht, M., Long term storage reliability of antifuse field programmable gate arrays, *Microelectronics Reliability*, 53(12), December 2013, 2052–2056, http://dx.doi.org/10.1016/j.microrel.2013.06.016, Accepted June 2013.
49. Zheng, L., Terpenny, J., Sandborn, P., and Nelson III, R., Design refresh planning models for managing obsolescence, *Proceedings of the ASME 2012 International Design Engineering Technical Conferences & Computers and Information in Engineering Conference*, Chicago, IL, August 12–15, 2012.
50. Sandborn, P., Prabhakar, V., and Ahmad, O., Forecasting technology procurement lifetimes for use in managing DMSMS obsolescence, *Microelectronics Reliability*, 51, 392–399, 2011.
51. Sandborn, P. and Prabhakar, V., An electronic part total ownership cost model, *Proceedings DMSMS Conference*, Palm Springs, CA, September 24, 2008.
52. Bartels, B., Ermel, U., Sandborn, P., and Pecht, M.G., *Strategies to the Prediction, Mitigation, and Management of Product Obsolescence*, Wiley, New York, 2012.

33

MIL-STD-1553B Digital Time Division Command/Response Multiplex Data Bus

Chris de Long
Honeywell Aerospace

33.1 Introduction

MIL-STD-1553 is a standard that defines the electrical and protocol characteristics for a data bus. AS-15531 is the Society of Automotive Engineers (SAE) commercial equivalent to the military standard. A data bus is similar to what the personal computer and office automation industry have dubbed a "local area network (LAN)." In avionics, a data bus is used to provide a medium for the exchange of data and information among various systems and subsystems. MIL-STD-1553 has been deployed on fixed- and rotary-wing aircraft, unmanned aircraft, surface and subsurface ships, ground vehicles, spacecraft and satellites, and the International Space Station. It is undoubtedly one of the most successful military standards developed.

33.1.1 Background

In the 1950s and 1960s, avionics were simple stand-alone systems. Navigation, communications, flight controls, and displays were analog systems. Often, these systems were composed of multiple boxes interconnected to form a single system. The interconnections between the various boxes were accomplished with point-to-point wiring. The signals mainly consisted of analog voltages, synchro–resolver signals, and relay/switch contacts. The location of these boxes within the aircraft was a function of operator need, available space, and the aircraft weight and balance constraints. As more and more systems were added, the cockpits became crowded due to the number of controls and displays, and the overall weight of the aircraft increased.

By the late 1960s and early 1970s, it was necessary to share information between various systems to reduce the number of black boxes required by each system. A single sensor providing heading and rate information could provide those data to the navigation system, the weapon system, the flight control system, and the pilot's display system (see Figure 33.1a). However, the avionics technology was still basically analog, and while sharing sensors did produce a reduction in the overall number of black boxes, the interconnecting signals became a "rat's nest" of wires and connectors. Moreover, functions or systems that were added later became an integration nightmare as additional connections of a particular signal could have potential system impacts, plus since the system used point-to-point wiring, the system that was the source of the signal typically had to be modified to provide the additional hardware needed to output to the newly added subsystem. As such, intersystem connections were kept to the bare minimums.

By the late 1970s, with the advent of digital technology, digital computers had made their way into avionics systems and subsystems. They offered increased computational capability and easy growth, compared to their analog predecessors. However, the data signals—the inputs and outputs from the sending and receiving systems—were still mainly analog in nature, which led to the configuration of a small number of centralized computers being interfaced to the other systems and subsystems via complex and expensive analog-to-digital and digital-to-analog converters.

As time and technology progressed, the avionics systems became more digital. And with the advent of the microprocessor, things really took off. A benefit of this digital application was the reduction in the number of analog signals and hence the need for their conversion. Greater sharing of information could be provided by transferring data between users in digital form. An additional side benefit was that

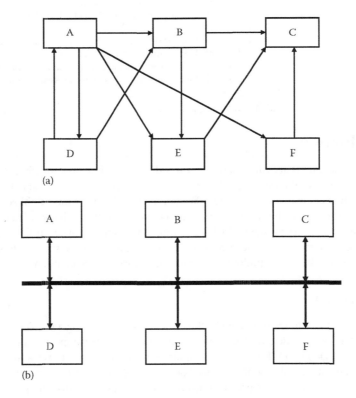

(a)

(b)

FIGURE 33.1 Systems configurations. (a) System using discrete and analog interfaces and (b) system using digital data bus network interfaces.

digital data could be transferred bidirectionally, whereas analog data were transferred unidirectionally. Serial rather than parallel transmission of the data was used to reduce the number of interconnections within the aircraft and the receiver/driver circuitry required with the black boxes. But this alone was not enough. A data transmission medium that would allow all systems and subsystems to share a single and common set of wires was needed (see Figure 33.1b). By sharing the use of this interconnect, the various subsystems could send data among themselves and to other systems and subsystems, one at a time, and in a defined sequence using a methodology called "time division multiplex." Enter the 1553 data bus.

33.1.2 History and Applications

MIL-STD-1553 (USAF) was released in August of 1973. The first user of the standard was the F-16. Further changes and improvements were made and a triservice version, MIL-STD-1553A, was released in 1975. The first user of the "A" version of the standard was again the air force's F-16 and the army's new attack helicopter, the AH-64A Apache. With some "real-world" experience, it was soon realized that further definitions and additional capabilities were needed. The latest version of the standard, 1553B, was released in 1978.

Today, the 1553 standard is still at the "B" level; however, its application has been affected through a series of notices. In 1980, the air force introduced Notice 1. Intended only for air force applications, Notice 1 restricted the use of many of the options within the standard. While the air force felt this was needed to obtain a common set of avionic systems, many in the industry felt that Notice 1 was too restrictive and limited the capabilities in the application of the standard. Released in 1986, the triservice Notice 2 (which supersedes Notice 1) places tighter definitions upon the options within the standard. While not restricting an option's use, it tightly defines how an option will be used if implemented. Notice 2, in an effort to obtain a common set of operational characteristics, also places a minimum set of requirements upon the design of the black box. Notice 2 also removed all references to "aircraft" or "airborne" so as not to limit its applications. The military standard was converted to its commercial equivalent as SAE AS 15531, as part of the government's effort to increase the use of commercial products.

Other notices have since been released. Notice 3 released in 1993 simply states that the standard has been reviewed and is still valid for use. Notice 4 released in 1996 changed the name of the document from a "military standard" to an "interface standard." Notice 5 released in 2006 and Notice 6 released in 2007 were an effort by the air force to increase the data from 1 megabit per second (mbps) up to 200 Mbps. This was to be accomplished by providing an additional orthogonal frequency division multiplex signal "on top of" the existing 1553 signal while utilizing the same cables and couplers. These notices and the addition of the higher-rate signaling were not immediately accepted by the industry, and the air force received lots of comments regarding the operation, use, and testing of this addition. Notice 7, released in 2008, cancelled Notices 5 and 6. So the last "meaningful" notice to 1553B is Notice 2.

Since its inception, MIL-STD-1553 has found numerous applications including direct incorporation as part of MIL-STD-1760 and NATO's STANAG 3910. NATO also published STANAG 3838 AVS, a version of MIL-STD-1553B. The United Kingdom took a similar approach with their publication of Def Stan 0018 (Part 2).

33.2 Standard

MIL-STD-1553B defines the term time division multiplexing (TDM) as "the transmission of information from several signal sources through one communications system with different signal samples staggered in time to form a composite pulse train." For our example in Figure 33.1b, this means that data can be transferred between multiple avionic units over a single transmission media, with the communications among the different avionic boxes taking place at different moments in time, hence time division. Table 33.1 is a summary of the 1553 data bus characteristics. However, before defining how the data are transferred, it is necessary to understand the data bus hardware.

TABLE 33.1   Summary of the 1553 Data Bus Characteristics

Data rate	1 MHz
Word length	20 bits
Data bits per word	16 bits
Message length	Maximum of 32 DWs
Transmission technique	Half duplex
Operation	Asynchronous
Encoding	Manchester II biphase
Protocol	Command–response
Bus control	Single or multiple
Message formats	Controller to terminal (BC-RT)
	Terminal to controller (RT–BC)
	Terminal to terminal (RT–RT)
	Broadcast
	System control
Number of RTs	Maximum of 31
Terminal types	RT
	BC
	BM
Transmission media	Twisted shielded pair cable
Coupling	Transformer or direct

33.2.1 Hardware Elements

The 1553 standard defines certain aspects regarding the design of the data bus system and the black boxes to which the data bus is connected. The standard defines four hardware elements: transmission media, remote terminals (RTs), bus controllers (BCs), and bus monitors (BMs), each of which is detailed as follows.

33.2.1.1 Transmission Media

The transmission media, or data bus, is defined as a twisted shielded pair transmission line consisting of the main bus and a number of stubs. There is one stub for each terminal (system) connected to the bus. The main data bus is terminated at each end with a resistance equal to the cable's characteristic impedance. This termination makes the data bus behave electrically like an infinite transmission line. Stubs, which are added to the main bus to connect the terminals, provide "local" loads and produce an impedance mismatch where added. This mismatch, if not properly controlled, produces electrical reflections and degrades the performance of the main bus. Therefore, the characteristics of both the main bus and the stubs are specified within the standard. Table 33.2 is a summary of the transmission media characteristics.

TABLE 33.2   Summary of Transmission Media Characteristics

Cable type	Twisted shielded pair
Capacitance	30.0 pF/ft max—wire to wire
Characteristic impedance	70.0–85.0 Ω at 1 MHz
Cable attenuation	1.5 dbm/100 ft at 1 MHz
Cable twists	4 twists per foot maximum
Shield coverage	90% minimum
Cable termination	Cable impedance (62%)
Direct coupled stub length	Maximum of 1 ft
Transformer coupled stub length	Maximum of 20 ft

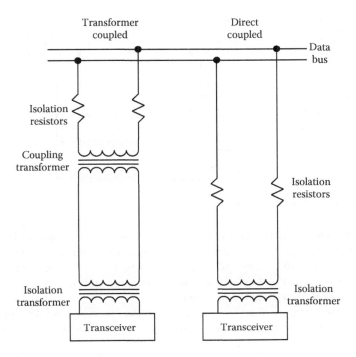

FIGURE 33.2 Terminal connection methods.

The standard specifies two stub methods: direct and transformer coupled. This refers to the method in which a terminal is connected to the main bus. Figure 33.2 shows the two methods, the primary difference between the two being that the transformer coupled method utilizes an isolation transformer for connecting the stub cable to the main bus cable. In both methods, two isolation resistors are placed in series with the bus. In the direct coupled method, the resistors are typically located within the terminal, whereas in the transformer coupled method, the resistors are typically located with the coupling transformer in boxes called data bus couplers. A variety of couplers are available, providing single or multiple stub connections. Bus couplers are available either as boxes using connectors or as "in-line" couplers wherein the stubs are spliced into the main bus cable.

Another difference between the two coupling methods is the length of the stub. For the direct coupled method, the stub length is limited to a maximum of 1 ft. For the transformer coupled method, the stub can be up to a maximum length of 20 ft. Therefore, for direct coupled systems, the data bus must be routed in close proximity to each of the terminals, whereas for a transformer coupled system, the data bus may be up to 20 ft away from each terminal.

33.2.1.2 Remote Terminal

An RT is defined within the standard as "All terminals not operating as the bus controller or as a bus monitor." Therefore, if it is not a controller, monitor, or the main bus or stub, it must be an RT—sort of a "catch all" clause. Basically, the RT is the electronics necessary to transfer data between the data bus and the subsystem. So what is a subsystem? For 1553 applications, the subsystem is the sender or user of the data being transferred.

In the earlier days of 1553, RTs were used mainly to convert analog and discrete data to/from a data format compatible with the data bus. The subsystems were still the sensor that provided the data and computer that used the data. As more and more digital avionics became available, the trend became to embed the RT into the sensor and computer. Today, it is common for the subsystem to contain an embedded RT. Figure 33.3 shows the different levels of RTs possible.

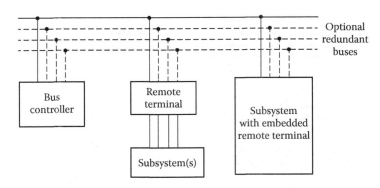

FIGURE 33.3   Simple multiplex architecture.

An RT typically consists of a transceiver, an encoder/decoder, a protocol controller, a buffer or memory, and a subsystem interface. In a modern black box containing a computer or processor, the subsystem interface may consist of the buffers and logic necessary to interface to the computer's address, data, and control buses. For dual redundant systems, two transceivers and two encoders/decoders would be required to meet the requirements of the standard.

Figure 33.4 is a block diagram of an RT and its connection to a subsystem. In short, the RT consists of all the electronics necessary to transfer data between the data bus and the user or originator of the data being transferred.

An RT is more than just a data formatter. It must be capable of receiving and decoding commands from the BC and respond accordingly. It must also be capable of buffering a message-worth of data, detecting transmission errors and performing validation tests upon the data, and reporting the status of the message transfer. An RT must be capable of performing a few of the bus management commands (referred to as mode commands [MCs]), and for dual redundant applications, it must be capable of listening to and decoding commands on both buses at the same time.

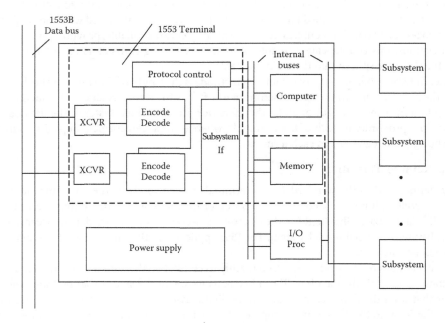

FIGURE 33.4   Terminal definition.

An RT must strictly follow the protocol as defined by the standard. It can only respond to commands received from the BC (i.e., it only speaks when spoken to). When it receives a valid command, it must respond within a defined amount of time. If a message does not meet the validity requirements defined, then the RT must invalidate the entire message and discard the data (not allow the data to be used by the subsystem). In addition to reporting status to the BC, most RTs today are also capable of providing some level of status information to the subsystem regarding the data received.

33.2.1.3 Bus Controller

The BC is responsible for directing the flow of data on the bus. While several terminals may be capable of performing as the BC, only one BC is allowed to be active at any one time. The BC is the only device allowed to issue commands onto the data bus. The commands may be for the transfer of data or the control and management of the bus (referred to as MCs).

Typically, the BC is a function that is contained within some other computer, such as a mission computer, a display processor, or a fire control computer. The complexity of the electronics associated with the BC is a function of the subsystem interface (the interface to the computer), the amount of error management and processing to be performed, and the architecture of the BC. There are three types of BC architectures: a word controller, a message controller, and a frame controller.

A *word controller* is the oldest and simplest type. Few word controllers are built today and they are only mentioned herein for completeness. For a word controller, the terminal electronics transfers one word at a time to the subsystem. Message buffering and validation must be performed by the subsystem.

Message controllers output a single message at a time, interfacing with the computer only at the end of the message or perhaps when an error occurs. Some message controllers are capable of performing minor error processing, such as retransmitting once on the alternate data bus, before interrupting the computer. The computer will inform the interface electronics of where the message exists in memory and provide a control word. For each message, the control word typically informs the BC electronics of the message type (e.g., a remote terminal to bus controller [RT–BC] or remote terminal to remote terminal [RT–RT] command), which bus to use to transfer the message, where to read or write the data words (DWs) in memory, and what to do if an error occurs. The control words are a function of the hardware design of the electronics and are not standardized among BCs.

A *frame controller* is the latest concept in BCs. A frame controller is capable of processing multiple messages in a sequence defined by the computer. The frame controller is typically capable of error processing as defined by the message control word. Frame controllers are used to "off-load" the computer as much as possible, interrupting only at the end of a series of messages or when an error it cannot handle is detected.

There is no requirement within the standard as to the internal workings of a BC, only that it issue commands onto the bus.

33.2.1.4 Bus Monitor

A BM is just that—a terminal that listens to (monitors) the exchange of information on the data bus. The standard strictly defines what BMs may be used for, stating that the information obtained by a BM be used "for off-line applications (e.g., flight test recording, maintenance recording or mission analysis) or to provide a back-up bus controller sufficient information to take over as the bus controller." Monitors may collect all the data from the bus or may collect selected data.

The reason for restricting its use is that while a monitor may collect data, it deviates from the command–response protocol of the standard in that a monitor is a passive device that does not transmit a status word (SW) and therefore cannot report on the status of the information transferred. Therefore, BMs fall into two categories: a recorder for testing or as a terminal functioning as a backup BC.

In collecting data, a monitor must perform the same message validation functions as the RT and, if an error is detected, inform the subsystem of the error (the subsystem may still record the data, but the error should be noted). For monitors that function as recorders for testing, the subsystem is typically a recording device or a telemetry transmitter. For monitors that function as backup BCs, the subsystem is the computer.

TABLE 33.3   Terminal Electrical Characteristics

Requirement	Transformer Coupled	Direct Coupled	Condition
Input Characteristics			
Input level	0.86–14.0 V	1.2–20.0 V	p–p, l–l
No response	0.0–0.2 V	0.0–0.28 V	p–p, l–l
Zero crossing stability	±150.0 ns	±150.0 ns	
Rise/fall times	0 ns	0 ns	Sine wave
Noise rejection	140.0 mV WGN[a]	200.0 mV WGN	BER[b] 1 per 10^7
Common mode rejection	±10.0 V peak	±10.0 V peak	line-gnd, dc-2.0 MHz
Input impedance	1000 Ω	2000 Ω	75 kHz–1 MHz
Output Characteristics			
Output level	18.0–27.0 V	6.0–9.0 V	p–p, l–l
Zero crossing stability	25.0 ns	25.0 ns	
Rise/fall times	100–300 ns	100–300 ns	10%–90%
Maximum distortion	±900.0 mV	±300.0 mV	peak, l–l
Maximum output noise	14.0 mV	5.0 mV	rms, l–l
Maximum residual voltage	±250.0 mV	±90.0 mV	peak, l–l

     [a] WGN, white Gaussian noise.
     [b] BER, bit error rate.

Today, it is common that BMs also contain an RT. When the monitor receives a command addressed to its terminal address (TA), it responds as an RT. For all other commands, it functions as a monitor. The RT portion could be used to provide feedback to the BC of the monitor's status, such as the amount of memory or time left, or to reprogram a selective monitor as to what messages to capture.

33.2.1.5 Terminal Hardware

The electronic hardware among an RT, BC, and BM does not differ much. Both the RT and BC (and BM if it is also an RT) must have the transmitters/receivers and encoders/decoders to format and transfer data. The requirements on the transceivers and the encoders/decoders do not vary between the hardware elements. Table 33.3 lists the electrical characteristics of the terminals.

All three elements have some level of subsystem interface and data buffering. The primary difference lies in the protocol control logic, and often, this is just a different series of microcoded instructions. For this reason, it is common to find 1553 hardware circuitry that is capable of functioning as all three devices.

There is an abundance of "off-the-shelf" components available today from which to design a terminal. These vary from discrete transceivers, encoders/decoders, and protocol logic devices to a single dual redundant hybrid containing everything but the transformers to design code that can be embedded into gate arrays.

33.3 Protocol

The rules under which the transfers occur are referred to as "protocol." The control, data flow, status reporting, and management of the bus are provided by three word types.

33.3.1 Word Types

Three distinct word types are defined by the standard. These are command words (CWs), DWs, and SWs. Each word type has a unique format yet all three maintain a common structure. Each word is 20 bits

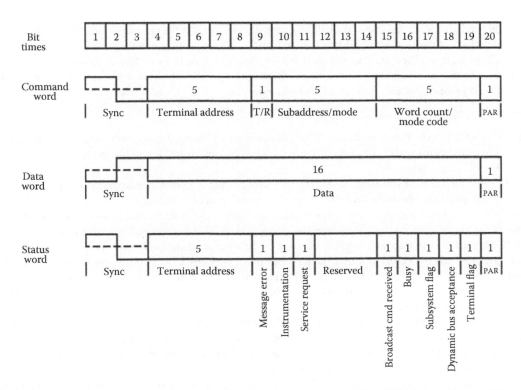

FIGURE 33.5   Word formats. *Note:* PAR, parity bit; Sync, synchronization bit, and T/R, transmit/receive bit.

in length. The first three bits are used as a synchronization field, thereby allowing the decode clock to resync at the beginning of each new word. The following 16 bits are the information field and differ among the three word types. The last bit is the parity bit. Parity is based on odd parity for the single word. The three word types are shown in Figure 33.5.

Bit encoding for all words is based on biphase Manchester II format. The Manchester II format provides a self-clocking waveform in which the bit sequence is independent. The positive and negative voltage levels of the Manchester waveform are dc balanced (same amount of positive signal as there is negative signal) and as such are well suited for transformer coupling. A transition of the signal occurs at the center of the bit time. A logic "0" is a signal that transitions from a negative level to a positive level. A logic "1" is a signal that transitions from a positive level to a negative level.

The terminal's hardware is responsible for the Manchester encoding and decoding of the word types. The interface that the subsystem sees is the 16-bit information field of all words. The sync and parity fields are not provided directly. However, for received messages, the decoder hardware provides a signal to the protocol logic as to the sync type the word was and as to whether parity was valid or not. For transmitted messages, there is an input to the encoder as to what sync type to place at the beginning of the word, and parity is automatically calculated by the encoder.

33.3.1.1 Sync Fields

The first three bit times of all word types are called the sync field. The sync waveform is in itself an invalid Manchester waveform as the transition only occurs at the middle of the second bit time. The use of this distinct pattern allows the decoder to resync at the beginning of each word received and maintain the overall stability of the transmissions.

Two distinct sync patterns are used: the command/status sync and the data sync. The command/status sync has a positive voltage level for the first one and a half bit times, then transitions to a negative voltage level for the second one and a half bit times. The data sync is the opposite—has a negative voltage level for the first one and a half bit times, then transitions to a positive voltage level for the second one and a half bit times. The sync patterns are shown in Figure 33.5. As the sync field is the same for both the CW and the SW, the user merely needs to follow the protocol to determine which is which—only a BC issues a command sync and only an RT issues status sync.

33.3.1.2 Command Word

The CW specifies the function that an RT(s) is to perform. This word is only transmitted by the active BC. The word begins with a command sync in the first three bit times. The following 16-bit information field is defined in Figure 33.5.

The five-bit TA field (bit times 4–8) states to which unique RT the command is intended (no two terminals may have the same address). Note that an address of 00000 is a valid address and that an address of 11111 is reserved for use as the broadcast address. Also note that there is no requirement that the BC be assigned an address; therefore, the maximum number of terminals the data bus can support is 31. Notice 2 to the standard requires that the TA be wire programmable externally to the black box (i.e., an external connector) and that the RT electronics perform a parity test upon the wired TA. The notice basically states that an open circuit on an address line is detected as a logic "1," that connecting an address line to ground is detected as a logic "0," and that odd parity will be used in testing the parity of the wired address field.

The next bit (bit time 9) is the transmit/receive (T/R) bit. This defines the direction of information flow and is always from the point of view of the RT. A transmit command (logic 1) indicates that the RT is to transmit data, while a receive command (logic 0) indicates that the RT is going to receive data. The only exceptions to this rule are associated with MCs.

The following five bits (bit times 10–14) are the subaddress (SA)/MC bits. A logic 00000 or 11111 within this field shall be decoded to indicate that the command is a mode code command. All other logic combinations of this field are used to direct the data to different functions within the subsystem. An example might be that 00001 is position data, 00010 is rate data, 10010 is altitude information, and 10011 is self-test data. The use of the SAs is left to the designer; however, Notice 2 suggests the use of SA 30 for data wraparound.

The next five bit positions (bit times 15–19) define the word count (WC) or mode code to be performed. If the SA/mode code field was 00000 or 11111, then this field defines the mode code to be performed. If not a mode code, then this field defines the number of DWs to be either received or transmitted depending on the T/R bit. A WC field of 00000 is decoded as 32 DWs.

The last bit (bit time 20) is the word parity bit. Only odd parity shall be used.

The best analogy of the CW would be to compare it to a post office. The TA is the address of the post office itself. The SA is the PO box within the post office (directing information to different users/functions), and the WC is the number of letters to be input to or removed from the PO box depending on the T/R bit.

33.3.1.3 Data Word

The DW contains the actual information that is being transferred within a message. DWs can be transmitted by either an RT (transmit command) or a BC (receive command). The first three bit times contain a data sync. This sync pattern is the opposite of that used for command and SWs and therefore is unique to the DW type.

The following 16 bits of information are left to the designer to define. The only standard requirements are that the most significant bit (MSB) of the data be transmitted first and that unused bits be set to a logic 0. While the standard provides no guidance as to their use, Section 80 of MIL-HDBK-1553A and SAE AS-15532 provides guidance and lists the formats (i.e., bit patterns and resolutions) of the most commonly used DWs.

The last bit (bit time 20) is the word parity bit. Only odd parity shall be used.

33.3.1.4 Status Word

The SW is only transmitted by an RT in response to a valid message. The SW is used to convey to the BC whether a message was properly received or the state of the RT (i.e., service request and busy). The SW is defined in Figure 33.5. Since the SW conveys information to the BC, there are two views as to the meaning of each bit—what the setting of the bit means to an RT and what the setting of the bit means to a BC. Each field of the SW, and its potential meanings, is examined in the following.

33.3.1.4.1 Resetting the Status Word

The SW, with the exception of the remote TA, is cleared after receipt of a valid CW. The two exceptions to this rule are if the CW received is a transmit SW mode code and if the CW received is a transmit last CW mode code. Conditions that set the individual bits of the word may occur at any time. If after clearing the SW the conditions for setting the bits still exist, then the bits shall be set again.

Upon detection of an error in the data being received, the message error (ME) bit is set and the transmission of the SW is suppressed. The transmission of the SW is also suppressed upon receipt of a broadcast message. For an illegal message (i.e., an illegal CW), the ME bit is set and the SW is transmitted.

33.3.1.4.2 Status Word Bits

TA. The first five bits (bit times 4–8) of the information field are the TA. These five bits should match the corresponding field within the CW that the terminal received. The RT sets these bits to the address to which it has been programmed. The BC should examine these bits to insure that the terminal responding with its SW was indeed the terminal to which the CW was addressed. In the case of an RT–RT message, the receiving terminal should compare the address of the second CW with that of the received SW. While not required by the standard, it is a good design practice to insure that the data received are from a valid source.

ME. The next bit (bit time 9) is the ME bit. This bit is set to a logic "1" by the RT upon detection of an error in the message or upon detection of an invalid message (i.e., illegal command) to the terminal. The error may occur in any of the DWs within the message. When the terminal detects an error and sets this bit, none of the data received within the message shall be used. If an error is detected within a message and the ME bit is set, the RT must suppress the transmission of the SW (see Section 33.3.1.4.1). If the terminal detected an illegal command, the ME bit is set and the SW is transmitted. All RTs must implement the ME bit in the SW.

Instrumentation. The instrumentation bit (bit time 10) is provided to differentiate between a CW and a SW (remember, they both have the same sync pattern). The instrumentation bit in the SW is always set to logic "0." If used, the corresponding bit in the CW is set to a logic "1." This bit in the CW is the MSB of the SA field and therefore would limit the SAs used to 10000–11110, hence reducing the number of SAs available from 30 to 15. The instrumentation bit is also the reason why there are two mode code identifiers (00000 and 11111), the latter required when the instrumentation bit is used.

Service request. The service request bit (bit time 11) is such that the RT can inform the BC that it needs to be serviced. This bit is set to a logic "1" by the subsystem to indicate that servicing is needed. This bit is typically used when the BC is "polling" terminals to determine if they require processing. The BC upon receiving this bit set to a logic "1" typically does one of the following. It can take a predetermined action such as issuing a series of messages, or it can request further data from the RT as to its needs. The latter can be accomplished by requesting the terminal to transmit data from a defined SA or by using the transit vector word mode code.

Reserved. Bit times 12–14 are reserved for future growth of the standard and must be set to a logic "0." The BC should declare a message in error if the RT responds with any of these bits set in its SW.

Broadcast command received. The broadcast command received bit (bit time 15) indicates that the RT received a valid broadcast command (e.g., a CW with the TA set to 11111). Upon receipt of a valid broadcast command, the RT sets this bit to logic "1" and suppresses the transmission of its SWs. The BC may issue a transmit SW or transmit last CW mode code to determine if the terminal received the message properly.

Busy. The busy bit (bit time 16) is provided as a feedback to the BC when the RT is unable to move data between the RT electronics and the subsystem in compliance to a command from the BC.

In the earlier days of 1553, the busy bit was required because many of the subsystem interfaces (analogs, synchros, etc.) were much slower compared to the speed of the multiplex data bus. Some terminals were not able to move the data fast enough. So instead of potentially losing data, a terminal was able to set the busy bit, indicating to the BC it could not handle new data at that time and for the BC to try again later. As new systems have been developed, the need for the use of busy has been reduced. However, there are systems that still need and have a valid use for the busy bit. Examples of these are radios, where the BC issues a command to the radio to tune to a certain frequency. It may take the radio several milliseconds to accomplish this, and while it is tuning, it may set the busy bit to inform the BC that it is doing as it was told.

When a terminal is busy, it does not need to respond to commands in the "normal" way. For receive commands, the terminal collects the data, but does not have to pass the data to the subsystem. For transmit commands, the terminal transmits its SW only. Therefore, while a terminal is busy, the data it supplies to the rest of the system are not available. This can have an overall effect upon the flow of data within the system and may increase the data latency within time-critical systems (e.g., flight controls).

Some terminals used the busy bit to overcome design problems, setting the busy bit whenever needed. Notice 2 to the standard "strongly discourages" the use of the busy bit. However, as shown in the example earlier, there are valid needs for its use. Therefore, if used, Notice 2 now requires that the busy bit may only be set as the result of a particular command received from the BC and not due to an internal periodic or processing function. By following this requirement, the BC, with prior knowledge of the RT's characteristics, can determine what will cause a terminal to go busy and minimize the effects on data latency throughout the system.

Subsystem flag. The subsystem flag bit (bit time 17) is used to provide "health" data regarding the subsystems to which the RT is connected. Multiple subsystems may logically "OR" their bits together to form a composite health indicator. This single bit is only to serve as an indicator to the BC and user of the data that a fault or failure exists. Further information regarding the nature of the failure must be obtained in some other fashion. Typically, an SA is reserved for built-in-test (BIT) information, with one or two words devoted to subsystem status data.

Dynamic bus control acceptance. The dynamic bus control acceptance bit (bit time 18) is used to inform the BC that the RT has received the dynamic bus control mode code and has accepted control of the bus. For the RT, the setting of this bit is controlled by the subsystem and is based upon passing some level of BIT (i.e., a processor passing its power-up and continuous background tests).

The RT upon transmitting its SW becomes the BC. The BC, upon receipt of the SW from the RT with this bit set, ceases to function as the BC and may become an RT or BM.

Terminal flag. The terminal flag bit (bit time 19) is used to inform the BC of a fault or failure within the RT circuitry (only the RT). A logic "1" shall indicate a fault condition. This bit is used solely to inform the BC of a fault or failure. Further information regarding the nature of the failure must be obtained in some other fashion. Typically, an SA is reserved for BIT information, or the BC may issue a transmit BIT word mode code.

Parity. The last bit (bit time 20) is the word parity bit. Only odd parity shall be used.

33.3.2 Message Formats, Validation, and Timing

The primary purpose of the data bus is to provide a common medium for the exchange of data between systems. The exchange of data is based upon message transmissions. The standard defines 10 types of message transmission formats. All of these formats are based upon the three word types just defined. The 10 message formats are shown in Figures 33.6 and 33.7. The message formats have been divided into two groups. These are referred to within the standard as the "information transfer formats" (Figure 33.6) and the "broadcast information transfer formats" (Figure 33.7).

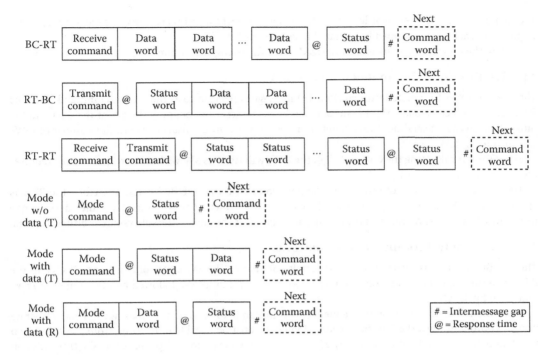

FIGURE 33.6   Information transfer formats.

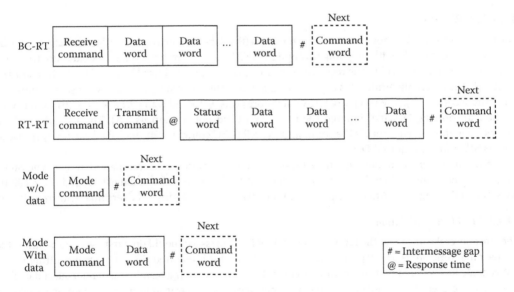

FIGURE 33.7   Broadcast information transfer formats.

The information transfer formats are based upon the command/response philosophy that all error-free transmissions received by an RT be followed by the transmission of an SW from the terminal to the BC. This handshaking principle validates the receipt of the message by the RT.

Broadcast messages are transmitted to multiple RTs at the same time. As such, the terminals suppress the transmission of their SWs (not doing so would have multiple boxes trying to talk at the same time and thereby "jam" the bus). In order for the BC to determine if a terminal received the message, a polling sequence to each terminal must be initiated to collect the SWs.

Each of the message formats is summarized in the following subsections.

33.3.2.1 Bus Controller to Remote Terminal

The bus controller to remote terminal (BC-RT) message is referred to as the receive command since the RT is going to receive data. The BC outputs a CW to the terminal defining the SA of the data and the number of DWs it is sending. Immediately (without any gap in the transmission), the number of DWs specified in the CW is sent.

The RT upon validating the CW and all of the DWs will issue its SW within the response time requirements (maximum of 12 μs).

The RT must be capable of processing the next command that the BC issues. Therefore, the RT has approximately 56 μs (SW response time 12 μs, plus SW transmit time 20 μs, plus intermessage gap minimum 4 μs, plus CW transmit time 20 μs), to either pass the data to the subsystem or buffer the data.

33.3.2.2 Remote Terminal to Bus Controller

The RT–BC message is referred to as a transmit command. The BC issues only a transmit CW to the RT. The terminal, upon validation of the CW, will first transmit its SW followed by the number of DWs requested by the CW.

Since the RT does not know the sequence of commands to be sent and does not normally operate upon a command until the CW has been validated, it must be capable of fetching from the subsystem the data required within approximately 28 μs (the SW response time 12 μs, plus the SW transmission time 20 μs, minus some amount of time for message validation and transmission delays through the encoder and transceiver).

33.3.2.3 RT–RT

The RT–RT command is provided to allow a terminal (the data source) to transfer data directly to another terminal (the data sink) without going through the BC. The BC may, however, collect and use the data.

The BC first issues a CW to the receiving terminal immediately followed by a CW to the transmitting terminal. The receiving terminal is expecting data, but instead of data after the CW, it sees a command sync (the second CW). The receiving terminal ignores this word and waits for a word with a data sync.

The transmitting terminal ignored the first CW (it did not contain its TA). The second word was addressed to it, so it will process the command as an RT–BC command as described earlier by transmitting its SW and the required DWs.

The receiving terminal, having ignored the second CW, again sees a command (status) sync on the next word and waits further. The next word (the first DW sent) now has a data sync, and the receiving RT starts collecting data. After receipt of all of the DWs (and validating), the terminal transmits its SW.

33.3.2.3.1 RT–RT Validation

There are several things that the receiving RT of an RT–RT message should do. First, Notice 2 requires that the terminal time out is 54–60 μs after receipt of the CW. This is required since if the transmitting RT did not validate its CW (and no transmission occurred), then the receiving terminal will not collect data from some new message. This could occur if the next message is either a transmit or receive message, where the terminal ignores all words with a command/status sync and would start collecting DWs beginning with the first data sync. If the same number of DWs were being transferred in the follow-on message and the terminal did not test the command/SW contents, then the potential exists for the terminal to collect erroneous data.

The other function that the receiving terminal should do, but is not required by the standard, is to capture the second CW and the first transmitted DW. The terminal could compare the TA fields of both words to insure that the terminal doing the transmitting was the one commanded to transmit. This would allow the terminal to provide a level of protection for its data and subsystem.

33.3.2.4 Mode Command Formats

Three MC formats are provided for. This allows for MCs with no DWs and for the MCs with one DW (either transmitted or received). The status/data sequencing is as described for the BC–RT or RT–BC messages except that the data WC is either one or zero. Mode codes and their use are described later.

33.3.2.5 Broadcast Information Transfer Formats

The broadcast information transfer formats, as shown in Figure 33.7, are identical to the nonbroadcast formats described earlier with the following two exceptions. First, the BC issues commands to TA 31 (11111), which is reserved for this function. Second, the RTs receiving the messages (those that implement the broadcast option) suppress the transmission of their SW.

The broadcast option can be used with the message formats in which the RT receives data. Obviously, multiple terminals cannot transmit data at the same time, so the RT–BC transfer format and the transmit mode code with data format cannot be used. The broadcast RT–RT allows the BC to instruct all RTs to receive and then instructs one terminal to transmit, thereby allowing a single subsystem to transfer its data directly to multiple users.

Notice 2 allows the BC to only use broadcast commands with mode codes (see broadcast mode codes). RTs are allowed to implement this option for all broadcast message formats. The notice further states that the terminal must differentiate the SAs between broadcast and nonbroadcast messages (see SA utilization).

33.3.2.6 Command and Message Validation

The RT must validate the CW and all DWs received as part of the message. The criteria for a valid CW are as follows: the word begins with a valid command sync, valid TA (matches the assigned address of the terminal or the broadcast address if implemented), all bits are in a valid Manchester code, there are 16 information field bits, and there is a valid parity bit (odd). The criteria for a DW are the same except a valid data sync is required and the TA field is not tested. If a CW fails to meet the criteria, the command is ignored. After the command has been validated, and a DW fails to meet the criteria, then the terminal shall set the ME bit in the SW and suppress the transmission of the SW. Any single error within a message shall invalidate the entire message and the data shall not be used.

33.3.2.7 Illegal Commands

The standard allows RTs the option of monitoring for illegal commands. An illegal command is one that meets the valid criteria for a CW, but is a command (message) that is not implemented by the terminal. An example is if a terminal is designed to only output 04 DWs to SA 01 and a CW was received by the terminal that requested it to transmit 06 DWs from SA 01, then this command, while still a valid command, could be considered by the terminal as illegal. The standard only states that the BC shall not issue illegal or invalid commands.

The standard provides the terminal designer with two options. First, the terminal can respond to all commands as usual (this is referred to as "responding in form"). The data received are typically placed in a series of memory locations that are not accessible by the subsystem or application programs. This is typically referred to as the "bit bucket." All invalid commands are placed into the same bit bucket. For invalid transmit commands, the data transmitted are read from the bit bucket. Remember, the BC is not supposed to send these invalid commands.

The second option is for the terminal to monitor for illegal commands. For most terminal designs, this is as simple as a lookup table with the T/R bit, SA, and WC fields supplying the address and the output being a single bit that indicates if the command is valid or not. If a terminal implements illegal command detection and an illegal command is received, the terminal sets the ME bit in the SW and responds with the SW.

33.3.2.8 Terminal Response Time

The standard states that an RT, upon validation of a transmit CW or a receive message (CW and all DWs), shall transmit its SW to the BC. The response time is the amount of time the terminal has to transmit its SW. To allow for accurate measurements, the time frame is measured from the mid-crossing of the parity bit of the CW to the mid-crossing of the sync field of the SW. The minimum time is 4.0 μs; the maximum time is 12.0 μs. However, the actual amount of "dead time" on the bus is 2–10 μs since half of the parity bit and sync waveforms are being transmitted during the measured time frame.

The standard also specifies that the BC must wait a minimum of 14.0 μs for an SW response before determining that a terminal has failed to respond. In applications where long data buses are used or where other special conditions exist, it may be necessary to extend this time to 20.0 μs or greater.

33.3.2.9 Intermessage Gap

The BC must provide for a minimum of 4.0 μs between messages. Again, this time frame is measured from the mid-crossing of the parity bit of the last DW or the SW and the mid-crossing of the sync field of the next CW. The actual amount of "dead time" on the bus is 2 μs since half of the parity bit and sync waveforms are being transmitted during the measured time frame.

The amount of time required by the BC to issue the next command is a function of the controller type (e.g., word, message, or frame). The gap typically associated with word controllers is between 40 and 100 μs. Message controllers typically can issue commands with a gap of 10–30 μs. But frame controllers are capable of issuing commands at the 4 μs rate and often must require a time delay to slow them down.

33.3.2.10 Superseding Commands

An RT must always be capable of receiving a new command. This may occur while operating on a command on bus A and after the minimum intermessage gap, a new command appears, or if operating on bus A and a new command appears on bus B. This is referred to as a superseding command. A second valid command (the new command) shall cause the terminal to stop operating on the first command and start on the second. For dual redundant applications, this requirement implies that all terminals must, as a minimum, have two receivers, two decoders, and two sets of CW validation logic.

33.3.3 Mode Codes

Mode codes are defined by the standard to provide the BC with data bus management and error handling/recovery capability. The mode codes are divided into two groups: with and without DWs. The DWs that are associated with the mode codes (and only one word per mode code is allowed) contain information pertinent to the control of the bus and do not generally contain information required by the subsystem (the exception may be synchronize with DW mode code). The mode codes are defined by bit times 15–19 of the CW. The MSB (bit 15) can be used to differentiate between the two mode code groups. When a DW is associated with the mode code, the T/R bit determines if the DW is transmitted or received by the RT. The mode codes are listed in Table 33.4.

33.3.3.1 Mode Code Identifier

The mode code identifier is contained in bits 10–14 of the CW. When this field is either 00000 or 11111, then the contents of bits 15–19 of the CW are to be decoded as a mode code. Two mode code identifiers are provided such that the system can utilize the instrumentation bit if desired. The two mode code identifiers shall not convey different information.

TABLE 33.4 Mode Code

T/R	Mode Code	Function	Data Word	Broadcast
1	00000	Dynamic bus control	No	No
1	00001	Synchronize	No	Yes
1	00010	Transmit SW	No	No
1	00011	Initiate self-test	No	Yes
1	00100	Transmitter shutdown	No	Yes
1	00101	Override transmitter shutdown	No	Yes
1	00110	Inhibit terminal flag bit	No	Yes
1	00111	Override inhibit terminal flag bit	No	Yes
1	01000	Reset	No	Yes
1	01001	Reserved	No	TBD
1	•	•	No	•
1	•	•	No	•
1	01111	Reserved	No	TBD
1	10000	Transmit vector word	Yes	No
0	10001	Synchronize	Yes	Yes
1	10010	Transmit last CW	Yes	No
1	10011	Transmit BIT word	Yes	No
0	10100	Selected transmitter shutdown	Yes	Yes
0	10101	Override selected transmitter shutdown	Yes	Yes
1/0	10110	Reserved	Yes	TBD
	•	•	Yes	•
	•	•	Yes	•
1/0	11111	Reserved	Yes	TBD

33.3.3.2 Mode Code Functions

The following defines the functionality of each of the mode codes:

Dynamic bus control. The dynamic bus control mode code is used to provide for the passing of the control of the data bus between terminals, thus providing a "round robin" type of control. Using this methodology, each terminal is responsible for collecting the data it needs from all the other terminals. When it is done collecting, it passes control to the next terminal in line (based on some predefined sequence). This allows the application program (the end user of the data) to collect the data when it needs it, always insuring that the data collected are from the latest source sample and have not been sitting around in a buffer waiting to be used.

Notices 1 and 2 to the standard forbid the use of dynamic bus control for air force applications. This is due to the problems and concerns of what may occur when a terminal, which has passed the control, is unable to perform or does not properly forward the control to the next terminal, thereby forcing the condition of no terminal being in control and having to reestablish control by some terminals. The potential amount of time required to reestablish control could have disastrous effects upon the system (i.e., especially a flight control system).

An RT that is capable of performing as the bus control should be capable of setting the dynamic bus control acceptance bit in the terminal's SW to logic "1" when it receives the mode code command. Typically, the logic associated with the setting of this bit is based on the subsystem's (e.g., computer's) ability to pass some level of confidence test. If the confidence test passes, then the bit is set and the SW is transmitted when the terminal receives the MC, thereby saying that it will assume the role of a BC.

The BC can only issue the dynamic bus control MC to one RT at a time. The command obviously is only issued to terminals that are capable of performing as a BC. Upon transmitting the command, the BC must check the terminal's SW to determine if the dynamic bus control acceptance bit is set. If set, the BC ceases to function as the controller and becomes either an RT or a BM. If the bit in the SW is not set, the RT that was issued the command is not capable of becoming the BC; the current controller must either remain as the BC or attempt to pass the control to some other terminal.

Synchronize. The synchronize mode code is used to establish some form of timing between two or more terminals. This mode code does not use a DW; therefore, the receipt of this command by a terminal must cause some predefined event to occur. Some examples of this event may be the clearing, incrementing, or presetting of a counter, the toggling of an output signal, or the calling of some software routine. Typically, this command is used to time correlate a function such as the sampling of navigation data (i.e., present position and rates) for flight controls or targeting/fire control systems. Other uses have been for the BC to "sync" the backup controllers (or monitors) to the beginning of a major/minor frame processing.

When an RT receives the synchronize MC, it should perform its predefined function. For a BC, the issuance of the command is all that is needed. The terminal's SW only indicates that the message was received, not that the "sync" function was performed.

Transmit SW. This is one of the two commands that does not cause the RT to reset or clear its SW. Upon receipt of this command, the RT transmits the SW that was associated with the previous message, not the SW of the mode code message.

The BC uses this command for control and error management of the data bus. If the RT had detected an error in the message and suppressed its SW, then the BC can issue this command to the RT to determine if indeed the nonresponse was due to an error. As this command does not clear the SW from the previous message, a detected error by the RT in a previous message would be indicated by having the ME bit set in the SW.

The BC also uses this command when "polling." If a terminal does not have periodic messages, the RT can indicate when it needs communications by setting the service request bit in the SW. The BC, by requesting the terminal to transmit only its SW, can determine if the terminal is in need of servicing and can subsequently issue the necessary commands. This "polling" methodology has the potential of reducing the amount of bus traffic by eliminating the transmission of unnecessary words.

Another use of this command is when broadcast message formats are used. As all of the RTs will suppress their SWs, "polling" each terminal for its SW would reveal whether the terminal received the message by having its broadcast command received bit set.

Initiate self-test. This command, when received by the RT, shall cause the RT to enter into its self-test. This command is normally used as a ground-based maintenance function, as part of the system power-on tests, or in flight as part of a fault recovery routine. Note that this test is only for the RT, not the subsystem.

In earlier applications, some RTs, upon receipt of this command, would enter self-test and go "offline" for long periods of time. Notice 2, in an effort to control the amount of time that a terminal could be "offline," limited the test time to 100.0 μs following the transmission of the SW by the RT.

While a terminal is performing its self-test, it may respond to a valid command in the following ways: (1) no response on either bus ("offline"), (2) transmit only the SW with the busy bit set, or (3) normal response. The RT may, upon receipt of a valid command received after this mode code, terminate its self-test. As a subsequent command could abort the self-test, the BC, after issuing this command, should suspend transmissions to the terminal for the specified amount of time (either a time specified for the RT or the maximum time of 100.0 μs).

Transmitter shutdown. This command is used by the BC in the management of the bus. In the event that a terminal's transmitter continuously transmits, this command provides for a mechanism to turn the transmitter off. This command is for dual redundant standby applications only.

Upon receipt of this command, the RT shuts down (i.e., turns off) the transmitter associated with the opposite data bus. That is to say if a terminal's transmitter is babbling on the A bus, the BC would send this command to the terminal on the B bus (a command on the A bus would not be received by the terminal).

Override transmitter shutdown. This command is the complement of the previous one in that it provides a mechanism to turn on a transmitter that had previously been turned off. When the RT receives this command, it shall set its control logic such that the transmitter associated with the opposite bus be allowed to transmit when a valid command is received on the opposite bus. The only other command that can enable the transmitter is the reset RT MC.

Inhibit terminal flag. This command provides for the control of the terminal flag bit in a terminal's SW. The terminal flag bit indicates that there is an error within the RT hardware and that the data being transmitted or the data received may be in error. However, the fault within the terminal may not have any effect upon the quality of the data, and the BC may elect to continue with the transmissions knowing a fault exists.

The RT receiving this command shall set its terminal flag bit to logic "0" regardless of the true state of this signal. The standard does not state that the BIT that controls this bit be halted, but only the results be negated to "0."

Override inhibit terminal flag. This command is the complement of the previous one in that it provides a mechanism to turn on the reporting of the terminal flag bit. When the RT receives this command, it shall set its control logic such that the terminal flag bit is properly reported based upon the results of the terminal's BIT functions. The only other command that can enable the response of the terminal flag bit is the reset RT MC.

Reset RT. This command, when received by the RT, shall cause the terminal electronics to reset to its power-up state. This means that if a transmitter had been disabled or the terminal flag bit inhibited, these functions would be reset as if the terminal had just powered up. Again, remember that the reset applies only to the RT electronics and not to the equipment in which the terminal resides.

Notice 2 restricts the amount of time that a RT can take to reset its electronics. After transmission of its SW, the RT shall reset within 5.0 μs. While a terminal is resetting, it may respond to a valid command in the following ways: (1) no response on either bus ("offline"), (2) transmit only the SW with the busy bit set, or (3) normal response. The RT may, upon receipt of a valid command received after this mode code, terminate its reset function. As a subsequent command could abort the reset, the BC, after issuing this command, should suspend transmissions to the terminal for the specified amount of time (either a time specified for the RT or the maximum time of 5.0 μs).

Transmit vector word. This command shall cause the RT to transmit a DW referred to as the vector word. The vector word shall identify to the BC service request information relating to the message needs of the RT. While not required, this mode code is often tied to the service request bit in the SW. As indicated, the contents of the DW inform the BC of messages that need to be sent.

The BC also uses this command when "polling." Though typically used in conjunction with the service request bit in the SW, wherein the BC requests only the SW (transmit SW mode code) and upon seeing the service request bit set would then issue the transmit vector word mode code, the BC can always ask for the vector word (always getting the SW anyway) and reduce the amount of time required to respond to the terminal's request.

Synchronize with DW. The purpose of this synchronize command is the same as the synchronize without DW, except this mode code provides a DW to provide additional information to the RT. The contents of the DW are left to the imagination of the user. Examples from "real-world" applications have used this word to provide the RT with a counter or clock value, to provide a backup controller with a frame identification number (minor frame or cycle number), and to provide a terminal with a new base address pointer used in extending the SA capability.

Transmit last CW. This is one of the two commands that does not cause the RT to reset or clear its SW. Upon receipt of this command, the RT transmits the SW that was associated with the previous message and the last CW (valid) that it received.

The BC uses this command for control and error management of the data bus. When an RT is not responding properly, then the BC can determine the last valid command the terminal received and can reissue subsequent messages as required.

Transmit BIT word. This MC is used to provide detail with regard to the BIT status of the RT. Its contents shall provide information regarding the RT only (remember the definition) and not the subsystem.

While most applications associate this command with the initiate self-test mode code, the standard requires no such association. Typical uses are to issue the initiate self-test mode code, allow the required amount of time for the terminal to complete its tests, and then issue the transmit BIT word mode code to collect the results of the test. Other applications have updated the BIT word on a periodic rate based on the results of a continuous background test (e.g., as a data wraparound test performed with every data transmission). This word can then be transmitted to the BC, upon request, without having to initiate the test and then wait for the test to be completed. The contents of the DW are left to the terminal designer.

Selected transmitter shutdown. Like the transmitter shutdown mode code, this mode code is used to turn off a babbling transmitter. The difference between the two mode codes is that this mode code has a DW associated with it. The contents of the DW specify which data bus (transmitter) to shutdown. This command is used in systems that provide more than dual redundancy.

Override selected transmitter shutdown. This command is the complement of the previous one in that it provides a mechanism to turn on a transmitter that had previously been turned off. When the RT receives this command, the DW specifies which data bus (transmitter) shall set its control logic such that the transmitter associated with that bus be allowed to transmit when a valid command is received on that bus. The only other command that can enable the selected transmitter is the reset RT MC.

Reserved mode codes. As can be seen from Table 33.4, there are several bit combinations that are set aside as reserved. It was the intent of the standard that these be reserved for future growth. It should also be noticed from the table that certain bit combinations are not listed. The standard allows the RT to respond to these reserved and "undefined" mode codes in the following manner: set the ME bit and respond (see Section 33.3.2.7) or respond in form. The designer of terminal hardware or a multiplex system is forbidden to use the reserved mode codes for any purpose.

33.3.3.3 Required Mode Codes

Notice 2 to the standard requires that all RTs implement the following four mode codes: transmit SW, transmitter shutdown, override transmitter shutdown, and reset RT. This requirement was levied so as to provide the multiplex system designer and the BC with a minimum set of commands for managing the multiplex system. Note that the requirement presented earlier was placed on the RT. Notice 2 also requires that a BC be capable of implementing all of the mode codes; however, for air force applications, the dynamic bus control mode code shall never be used.

33.3.3.4 Broadcast Mode Codes

Notice 2 to the standard allows the broadcast of mode codes (see Table 33.4). The use of the broadcast option can be of great assistance in the areas of terminal synchronization. Ground maintenance and troubleshooting can take advantage of broadcast reset RT or initiate self, but these two commands can have disastrous effects if used while in flight. The designer must provide checks to insure that commands such as these are not issued by the BC or operated upon by an RT when certain conditions exists (e.g., in flight).

33.4 Systems-Level Issues

The standard provides very little guidance in how it is applied. Lessons learned from "real-world" applications have led to design guides, application notes, and handbooks that provide guidance. This section will attempt to answer some of the systems-level questions and identify implied requirements that, while not specifically called out in the standard, are required nonetheless.

33.4.1 Subaddress Utilization

The standard provides no guidance on how to use the SAs. The assignment of SAs and their functions (the data content) is left to the user. Most designers automatically start assigning SAs with 01 and count upwards. If the instrumentation bit is going to be used, then the SAs must start at 16.

The standard also requires that normal SAs be separated from broadcast SAs. If the broadcast option is implemented, then an additional memory block is required to receive broadcast commands.

33.4.1.1 Extended Subaddressing

The number of SAs that a terminal has is limited to 60 (30 transmit and 30 receive). Therefore, the number of unique DWs available to a terminal is 1920 (60 × 32). For earlier applications, where data being transferred were analog sensor data and switch settings, this was more than sufficient. However, in today's applications, in which digital computers are exchanging data or for a video sensor passing digitized video data, the number of words is too limited.

Most terminal designs establish a block of memory for use by the 1553 interface circuitry. This block contains an address start pointer and then the memory is offset by the SA number and the WC number to arrive at a particular memory address.

A methodology of extending the range of the SAs has been successfully utilized. This method either uses a dedicated SA and DW or makes use of the synchronize with DW mode code. The DW associated with either of these contains an address pointer that is used to reestablish the starting address of the memory block. The changing of the blocks is controlled by the BC and can be done based on numerous functions. Examples are operational modes, wherein one block is used for startup messages, a different block for takeoff and landing, a different block for navigation and cruise, a different block for mission functions (i.e., attack or evade modes), and a different block for maintenance functions.

Another example is that the changing of the start address could also be associated with minor frame cycles. Minor frames could have a separate memory block for each frame. The BC could synchronize frames and change memory pointers at the beginning of each new minor frame.

For computers exchanging large amounts of data (e.g., GPS almanac tables) or for computers that receive program loads via the data bus at power-up, the BC could set the pointers at the beginning of a message block, send thirty 32-word messages, move the memory pointer to the last location in the RT memory that received data, and then send the next block of thirty 32-word messages, continuing this cycle until the memory is loaded. The use is left to the designer.

33.4.2 Data Wraparound

Notice 2 to the standard does require that the terminal is able to perform a data wraparound and SA 30 is suggested for this function. Data wraparound provides the BC with a methodology of testing the data bus from its internal circuitry, through the bus media, to the terminal's internal circuitry. This is done by the BC sending the RT a message block and then commanding the terminal to send it back. The BC can then compare the sent data with that received to determine the state of the data link. There are no special requirements upon the bit patterns of the data being transferred.

The only design requirements are placed upon the RT. These are that the terminal, for the data wraparound function, be capable of sending the number of DWs equal to the largest number of DWs sent for any transmit command. This means that if a terminal maximum data transmission is only four DWs, it need only provide for four DWs in its data wraparound function.

The other requirement is that the RT need only hold the data until the next message. The normal sequence is for the BC to send the data, and then in the next message, it asks for it back. If another message is received by the RT before the BC requests the data, the terminal can discard the data from the wraparound message and operate on the new command.

33.4.3 Data Buffering

The standard specifies that any error within a message shall invalidate the entire message. This implies that the RT must store the data within a message buffer until the last DW has been received and validated before allowing the subsystem access to the data. To insure that the subsystem always has the last message of valid data received to work with would require the RT to, as a minimum, double buffer the received data.

There are several methods to accomplish this in hardware. One method is for the terminal electronics to contain a first in, first out (FIFO) memory that stores the data as they are received. Upon validation of the last DW, the terminal's subsystem interface logic will move the contents of the FIFO into memory accessible by the subsystem. If an error occurred during the message, the FIFO is reset.

A second method establishes two memory blocks for each message in common memory. The subsystem is directed to read from one block (block A) while the terminal electronics writes to the other (block B). Upon receipt of a valid message, the terminal will switch pointers, indicating that the subsystem is to read from the new memory block (block B) while the terminal will now write to block B. If an error occurs within the message, the memory blocks are not switched.

Some of the "off-the-shelf" components available provide for data buffering. Most provide for double buffering, while some provided for multilevels of buffering.

33.4.4 Variable Message Blocks

RTs should be able to transmit any subset of any message. This means that if a terminal has a transmit message at SA 04 of 30 DWs, it should be capable of transmitting any number of those DWs (01–30) if so commanded by the BC. The order in which the subset is transmitted should be the same as if the entire message is being transmitted, that being the contents of DW 01 is the same regardless of the WC.

Terminals that implement illegal command detection should not consider subsets of a message as illegal. That is to say, if in our example earlier a command is received for 10 DWs, this should not be illegal. But, if a command is received for 32 DWs, this would be considered as an illegal command.

33.4.5 Sample Consistency

When transmitting data, the RT needs to ensure that each message transmitted is of the same sample set and contains mutually consistent data. Multiple words used to transfer multiple precision parameters or functionally related data must be of the same sampling.

If a terminal is transmitting pitch, roll, and yaw rates, and while transmitting the subsystem updates these data in memory, but this occurs after pitch and roll had been read by the terminal's electronics, then the yaw rate transmitted would be of a different sample set. Having data from different sample rates could have undesirable effects on the user of the data.

This implies that the terminal must provide some level of buffering (the reverse of what was described earlier) or some level of control logic to block the subsystem from updating data while being read by the RT.

33.4.6 Data Validation

The standard tightly defines the criteria for the validation of a message. All words must meet certain checks (i.e., valid sync, Manchester encoding, number of bits, and odd parity) for each word and each message to be valid. But what about the contents of the DW? MIL-STD-1553 provides the checks to insure the quality of the data transmission from terminal to terminal, sort of a "data in equals data out," but is not responsible for the validation tests of the data itself. This is not the responsibility of the 1553 terminal electronics, but of the subsystem. If bad data are sent, then "garbage in equals garbage out." But the standard does not prevent the user from providing additional levels of protection. The same techniques used in digital computer interfaces (i.e., disk drives and serial interfaces) can be applied to 1553. These techniques include checksums, cyclic redundancy check (CRC) words, and error detection/correction codes. Section 80 of MIL-HDBK-1553A that covers DW formats even offers some examples of these techniques.

But what about using the simple indicators embedded within the standard? Each RT provides an SW—indicating not only the health of the RT's electronics but also that of the subsystem. However, in most designs, the SW is kept within the terminal electronics and not passed to the subsystems. In some "off-the-shelf" components, the SW is not even available to be sent to the subsystem. But two bits from the SW should be made available to the subsystem and the user of the data for further determination as to the validity of the data. These are the subsystem flag and the terminal flag bits.

33.4.7 Major and Minor Frame Timing

The standard specifies the composition of the words (command, data, and status) and the messages (information formats and broadcast formats). It provides a series of management messages (mode codes), but it does not provide any guidance on how to apply these within a system. This is left to the imagination of the user.

RTs, based upon the contents of their data, will typically state how often data are collected and the fastest rate they should be outputted. For input data, the terminal will often state how often it needs certain data to either perform its job or maintain a certain level of accuracy. The rates are referred to as the transmission and update rates. It is the system designer's job to examine the data needs of all of the systems and determine when data are transferred from whom to whom. These data are subdivided into periodic messages, those that must be transferred at some fixed rate, and aperiodic messages, those that are typically either event driven (i.e., the operator pushes a button) or data driven (i.e., a value is now within range).

A major frame is defined such that all periodic messages are transferred at least once. This is therefore defined by the message with the slowest transmission rate. Typical major frame rates used in today's applications vary from 40 to 640 μs. There are some systems that have major frame rates in the 1 to 5 s range, but these are the exceptions, not the norm. Minor frames are then established to meet the requirements of the higher update rate messages.

The sequence of messages within a minor frame is again left undefined. There are two methodologies that are predominately used. In the first method, the BC starts the frame with the transmission of all of the periodic messages (transmit and receive) to be transferred in that minor frame. At the end of the periodic messages, either the BC is finished (resulting in dead bus time—no transmissions) until the beginning of the next frame or the BC can use this time to transfer aperiodic messages, error handling messages, or transfer data to the backup BC(s).

In the second method (typically used in a centralized processing architecture), the BC issues all periodic and aperiodic transmit messages (collects the data), processes the data (possibly using dead time during this processing), and then issues all the receive messages (outputting the results of the processing). Both methods have been used successfully.

33.4.8 Error Processing

The amount and level of error processing is typically left to the systems designer but may be driven by the performance requirements of the system. Error processing is typically only afforded to critical messages, wherein the noncritical messages just await the next normal transmission cycle. If a data bus is 60% loaded and each message received an error, the error processing would exceed 100% of available time and thereby cause problems within the system.

Error processing is again a function of the level of sophistication of the BC. Some controllers (typically message or frame controllers) can automatically perform some degree of error processing. This usually is limited to a retransmission of the message either once on the same bus or once on the opposite bus. Should the retried message also fail, the BC software is informed of the problem. The message may then be retried at the end of the normal message list for the minor frame.

If the error still persists, then it may be necessary to stop communicating with the terminal, especially if the BC is spending a large amount of time performing error processing. Some systems will try to communicate with a terminal for a predefined number of times on each bus. After this, all messages to the terminal are removed from the minor frame lists and substituted with a single transmit SW mode code.

An analysis should be performed on the critical messages to determine the effects upon the system if they are not transmitted or the effects of data latency if they are delayed to the end of the frame.

33.4.9 Data Levels

As MIL-STD-1553 has expanded into more and more avionic architectures, system architecture questions have been raised concerning the location and design assurance level of the BC. In general, the BC for a 1553 bus should be located in the highest-criticality system attached to a particular bus. This helps ensure that software and hardware implementing the 1553 functionality have been properly assured per any design assurance requirements that may be applicable to the development (e.g., DO-178B or DO-254) and can exert the necessary control to prevent lower-criticality RTs from interfering with higher-criticality transmissions. See the corresponding chapters in this book for more information on these topics.

33.5 Testing

The testing of a MIL-STD-1553 terminal or system is not a trivial task. There are a large number of options available to the designer including message formats, MCs, SW bits, and coupling methodology. In addition, history has shown that different component manufacturers and designers have made different interpretations regarding the standard, thereby introducing products that implement the same function quite differently.

For years, the air force provided for the testing of MIL-STD-1553 terminals and components. Today, this testing is the responsibility of the industry. The SAE, in conjunction with the government, has developed a series of test plans for all 1553 elements. These test plans are listed in Table 33.5.

TABLE 33.5   SAE 1553 Test Plans

AS-4111	Remote Terminal Validation Test Plan
AS-4112	Remote Terminal Production Test Plan
AS-4113	Bus Controller Validation Test Plan
AS-4114	Bus Controller Production Test Plan
AS-4115	Data Bus System Test Plan
AS-4116	Bus Monitor Test Plan
AS-4117	Bus Components Test Plan

Further Reading

In addition to the SAE test plans listed in Table 33.5, there are other documents that can provide a great deal of insight and assistance in designing with MIL-STD-1553:

- MIL-STD-1553B Digital Time Division Command/Response Multiplex Data Bus
- MIL-HDBK-1553A Multiplex Applications Handbook
- SAE AS-15531 Digital Time Division Command/Response Multiplex Data Bus
- SAE AS-15532 Standard Data Word Formats
- SAE AS-12 Multiplex Systems Integration Handbook
- SAE AS-19 MIL-STD-1553 Protocol Reorganized
- DDC 1553 Designers Guide
- UTMC 1553 Handbook

And lastly, there is the SAE 1553 Users Group. This is a collection of industry and military experts in 1553 who provide an open forum for information exchange and provide guidance and interpretations/clarifications with regard to the standard. This group meets twice a year as part of the SAE Avionics Systems Division conferences.

34

ARINC 429
Digital Information
Transfer System

Paul J. Prisaznuk
ARINC

34.1 Introduction

ARINC 429 is the most widely used data transfer medium used in aviation. It first entered service in the early 1980s as the age of digital avionics introduced. Due to its low cost and simplistic form, ARINC 429 continues to be used in applications requiring low to moderate data rates.

An ARINC 429 interface consists of a single transmitter—or source—connected to up to 20 receivers—or sinks—using a twisted pair wire. Data flows in one direction only, that is, simplex communication. Bi-directional data transfer is accomplished using two ARINC 429 buses. ARINC 429 can send data at a maximum rate of 100 kilobits per second (kbps). The ARINC 429 physical definition and protocols are defined by ARINC Specification 429: Digital Information Transfer System.

34.2 Data Bus Basics

Over the years, ARINC has published a number of data transfer standards for use within the airline industry. ARINC Specification 429 is an example of one of the many documents prepared by the Airlines Electronic Engineering Committee (AEEC) and published by ARINC. The data bus standard documents share a common thread; they describe data communication methods used on commercial transport airplanes, and also used In general aviation, regional, and military airplanes.

Data buses are used to transfer a stream of data to avionics line replaceable units or LRUs. Data buses may be configured in a linear point-to-point topology or used in a more sophisticated topology, commonly known as star topology or network topology. Each topology has benefits in contributing to the overall system design requirements. Each LRU may contain multiple transmitters and receivers communicating on different buses. This simple architecture, almost point-to-point wiring, provides a highly reliable transfer of data.

34.3 ARINC 429

34.3.1 General

ARINC Specification 429, Digital Information Transfer System was first published in July 1977. It has seen service in virtually every aircraft built in the past three decades. Although ARINC 429 is a vehicle for data transfer, it does not fit the normal definition of a data bus. While most data buses provide bi-directional transfer over a single set of wires, ARINC 429 requires two buses to achieve bi-directional data flow.

34.3.2 History

In the early 1970s, the airlines recognized the potential advantage of implementation of digital avionics equipment. The most compelling reason was the significant reduction of weight and power. Digital avionics had already been implemented to a certain degree on airplanes existing at that time. However, there were three new transport airplanes on the horizon: the Airbus A310 and the Boeing B-757 and B-767. The airlines, along with the airframe manufacturers and avionic suppliers, developed an entirely new suite of ARINC 700-series avionics using microprocessors and digital technology. ARINC 429 has been used as the standard for virtually all ARINC 700-series avionics systems. ARINC 429 has been highly-reliable and it continues to be used today in the most modern airplanes.

34.3.3 Design Fundamentals

ARINC 429 uses a single transmitter connected to up to 20 data receivers via a twisted shielded pair of wires. The shields of the wires are grounded at both ends and at any breaks along the installation of the cable.

34.3.3.1 Modulation

Return-to-zero (RZ) modulation is used. The voltage levels are used for this modulation scheme.

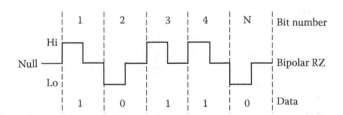

34.3.3.2 Voltage Levels

The differential output voltages across the transmitter output terminal with no load is described in the following table:

	HI(V)	NULL(V)	LO(V)
Line A to line B	+10 ± 1.0	0 ± 0.5	−10 ± 1.0
Line A to ground	5 ± 0.5	0 ± 0.25	−5 ± 0.5
Line B to ground	−5 ± 0.5	0 ± 0.25	5 ± 0.5

The differential voltage seen by the receiver will depend on wire length, loads, stubs, and so on. With no noise present on the signal lines, the nominal voltages at the receiver terminals (A and B) would be the following:

- HI +7.25 V to + 11 V
- NULL + 0.5 V to –0.5 V
- LO –7.25 V to –11 V

In practical installations impacted by noise, the following voltage ranges are typical across the receiver input (A and B):

- HI + 6.5 V to + 13 V
- NULL + 2.5 V to –2.5 V
- LO –6.5 V to –13 V

Line (A or B) to ground voltages are not defined. Receivers are expected to withstand, without damage, steady-state voltages of 30 Vac applied across terminals A and B, or 30 Vdc applied between terminal A or B and the ground.

34.3.3.3 Transmitter Output Impedance

The transmitter output impedance is 70–80 Ω (nominal 75 Ω) and is divided equally between lines A and B for all logic states and transitions between those states.

34.3.3.4 Receiver Input Impedance

The typical receiver input characteristics are as follows:

- Differential Input Resistance R_I = 12,000 Ω minimum
- Differential Input Capacitance C_I = 50 pF maximum
- Resistance to Ground R_H and $R_G \geq$ 12,000 Ω
- Capacitance to Ground C_H and $C_G \leq$ 50 pF

The total receiver input resistance, including the effects of R_I, R_H, and R_G in parallel, is 8000 Ω minimum, that is, 400 Ω minimum for 20 receivers. A maximum of 20 receivers is specified for any one transmitter. See below for the circuit standards.

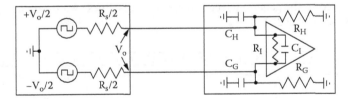

34.3.3.5 Cable Impedance

ARINC 429 wiring may vary between 20 and 26 AWG depending on desired physical integrity of the cable and weight limitations. The characteristic impedance will typically range from 60 to 80 Ω. The transmitter output impedance was chosen at 75 Ω nominal to match this range.

34.3.3.6 Fault Tolerance

ARINC 429 is intended to be robust and components take into account aircraft power variation. Therefore, it should not be susceptible to faults, damage or erratic operation when those variations occur. The ranges of those variations are provided in the following sections.

34.3.3.6.1 *Transmitter External Fault Voltage*

Transmitter failures caused by external fault voltages will not typically cause other transmitters or other circuitry in the unit to function outside of their specification limits or to fail.

34.3.3.6.2 Transmitter External Fault Load Tolerance

Transmitters should indefinitely withstand without sustaining damage a short circuit applied (a) across terminals A and B, (b) from terminal A to ground, (c) from terminal B to ground, or (d) b and c simultaneously.

34.3.3.6.3 Receiver Fault Isolation

Each receiver incorporates isolation provisions to ensure that the occurrence of any reasonably probable internal line replaceable unit (LRU) or bus receiver failure does not cause any input bus to operate outside its specification limits (both under-voltage and over-voltage).

34.3.3.6.4 Transmitter Fault Isolation

Each transmitter incorporates isolation provisions to ensure that it does not undermine any reasonably probable equipment fault condition providing an output voltage in excess of (a) 30 Vac between terminal A and B, (b) +29 Vdc between A and ground, or (c) +29 Vdc between B and ground.

34.3.3.7 Logic-Related Elements

34.3.3.7.1 Numeric Data

ARINC 429 accommodates numeric data encoded in two digital languages: BNR expressed in twos complement fractional notation and binary coded decimal (BCD) per the numerical subset of International Standards Organization (ISO) Alphabet No. 5. An information item encoded in both BCD and BNR is assigned unique labels for both BCD and BNR (see Section 34.4.3).

34.3.3.7.2 Discrete Data

In addition to handling numeric data as specified above, ARINC 429 is also capable of accommodating discrete items of information either in the unused pad bits of data words or, when necessary, in dedicated words.

The rule in the assignment of bits in discrete numeric data words is to start with the least significant bit (LSB) of the word and to continue toward the most significant bit available in the word. There are two types of discrete words: general purpose discrete words and dedicated discrete words. Seven labels (270–276) are assigned to the general purpose discrete words. These words are assigned in ascending label order starting with label 270.

32	31 30	29 28 27 26 25 24 23 22 21 20 19 18 17 16 15 14 13 12 11	10 9	8 7 6 5 4 3 2 1
P	SSM	Data ⟶　　　⟵ — Pad　　　　　⟵ — Discretes	SDI	Label
		MSB　　　　　　　　　　　　　　　　　　LSB		

Generalized BCD word format

P	SSM	BCD CH #1	BCD CH #2	BCD CH #3	BCD CH #4	BCD CH #5	SDI	8 7 6 5 4 3 2 1
		4 2 1	8 4 2 1	8 4 2 1	8 4 2 1	8 4 2 1		
0	0 0	0 1 0	0 1 0 1	0 1 1 1	1 0 0 0	0 1 1 0	0 0	1 0 0 0 0 0 0 1
	Example	2	5	7	8	6	DME distance	

BCD word format example (no discretes)

32	31 30 29	28 27 26 25 24 23 22 21 20 19 18 17 16 15 14 13 12 11	10 9	8 7 6 5 4 3 2 1
P	SSM	Data ⟶　　　⟵ — Pad　　　　　⟵ — Discretes	SDI	Label
		MSB　　　　　　　　　　　　　　　　　　LSB		

Generalized BCD word format

34.3.3.7.3 General Purpose Maintenance Data

The general purpose maintenance words are assigned labels in sequential order, as are the labels for the general purpose discrete words. The lowest octal label assigned to the maintenance words is used when only one maintenance word is transmitted. When more than one word is transmitted, the lowest octal label is used first, and the other labels are used sequentially until the message has been completed. The general purpose maintenance words may contain discrete, BCD, or BNR numeric data. They do not contain ISO Alphabet No. 5 messages. The general purpose maintenance words are formatted according to the layouts of the corresponding BCD, BNR, and discrete data words shown in the word formats above.

34.4 Message and Word Formatting

34.4.1 Direction of Information Flow

The information output of a system element is transmitted from a designated port to the receiving ports of other system elements. In no case does information flow into a port designated for transmission. A separate data bus is used for each direction when data are required to flow both ways between two system elements.

34.4.2 Information Element

The basic information element is a digital word containing 32 bits. There are five application groups for such words: BNR data, BCD data, discrete data, general maintenance data and acknowledgment, ISO Alphabet No. 5, and maintenance (ISO Alphabet No. 5) data. The relevant data-handling rules are set forth in Section 34.4.6. When less than the full data field is needed to accommodate the information conveyed in a word in the desired manner, the unused bit positions are filled with binary zeros or, in the case of BNR and BCD numeric data, valid data bits. If valid data bits are used, the resolution may exceed the accepted standard for an application.

34.4.3 Information Identifier

A six-character label identifies the type of information contained in an ARINC 429 word. The first three characters are octal characters coded in binary in the first eight bits of the word. The eight bits will identify the information contained within BNR and BCD numeric data words (e.g., DME distance, static air temperature, etc.).

The last three characters of the label are hexadecimal characters used to identify ARINC 429 bus sources. Each triplet of hexadecimal characters identifies a system element with one or more ARINC 429 ports. Each three-character code (and LRU) may have up to 255 eight-bit labels assigned to it. The code is used administratively to retain distinction between unlike parameters having like label assignments.

Octal label 377 has been assigned for the purpose of electrically identifying the system element. The code appears in the three least significant digits of the 377 word in a BCD word format. Although data encoding is based on the BCD word format, the sign/status matrix (SSM) encoding is per the discrete word criteria to provide enhanced failure warning. The transmission of the equipment identifier word on a bus will permit receivers attached to the bus to recognize the source of the information. Since the transmission of the equipment identifier word is optional, receivers should not depend on that word for correct operation.

34.4.4 Source/Destination Identifier

Bit numbers 9 and 10 of numeric data words are used for a data source and destination identification function. They are not available for this function in alphanumeric (ISO Alphabet No. 5) data words or when the resolution needed for numeric (BNR/BCD) data necessitates their use for valid data. The source and destination identifier function may be applied when specific words need to be directed to a specific system of a multisystem installation or when the source system of a multisystem installation needs to be

TABLE 34.1   SDI Encoding

Bit No.		Installation
10	9	See text
0	0	—
0	1	1
1	0	2
1	1	3

Note: In certain specialized applications of the SDI function, the all-call capability may be forfeited so that code 00 is available as an installation no. 4 identifier.

recognizable from the word content. When it is used, the source equipment encodes its aircraft installation number in bits 9 and 10 as shown in Table 34.1. Sink equipment will recognize words containing its own installation number code and words containing code 00, the all-call code.

Equipment will fall into the categories of source only, sink only, or both source and sink. Use of the SDI bits by equipment functioning only as a source or only as a sink is described above. Both the source and sink texts above are applicable to equipment functioning as both a source and a sink. Such equipment will recognize the SDI bits on the inputs and also encode the SDI bits, as applicable, on the outputs. DME, VOR, ILS, and other sensors are examples of source and sink equipment generally considered to be only source equipment. These are actually sinks for their own control panels. Many other types of equipment are also misconstrued as source only or sink only. If a unit has an ARINC 429 input port and an ARINC 429 output port, it is both a source and sink. With the increase of equipment consolidation (e.g., centralized control panels), the correct use of the SDI bits cannot be overstressed.

When the SDI function is not used, binary zeros or valid data should be transmitted in bits 9 and 10.

34.4.5 Sign/Status Matrix

This section describes the coding of the SSM field. In all cases, the SSM field uses bits 30 and 31; for BNR data words, the SSM field also includes bit 29. The SSM field is used to report hardware condition (fault/normal), operational mode (functional test), or validity of data word content (verified/no computed data). The following definitions apply:

34.4.5.1 Invalid Data

This is defined as any data generated by a source system whose fundamental characteristic is the inability to convey reliable information for the proper performance of a user system. There are two categories of invalid data: (1) no computed data and (2) failure warning.

34.4.5.2 No Computed Data

This is a particular case of data invalidity in which the source system is unable to compute reliable data for reasons other than system failure. This inability to compute reliable data is caused exclusively by a definite set of events or conditions whose boundaries are uniquely defined in the system characteristic.

34.4.5.3 Failure Warning

This is a particular case of data invalidity in which the system monitors have detected one or more failures. These failures are uniquely characterized by boundaries defined in the system characteristic.

Displays are normally flagged invalid during a failure warning condition. When a no computed data condition exists, the source system indicates that its outputs are invalid by setting the SSM of the affected words to the no computed data code, as defined in the sections that follow. The system indicators may or may not be flagged depending on system requirements.

While the unit is in the functional test mode, all output data words generated within the unit are coded with functional test. Pass-through data words are those words received by the unit and retransmitted without alteration.

When the SSM code is used to transmit status and more than one reportable condition exists, the condition with the highest priority is encoded in bits number 30 and 31. The order of condition priorities is shown in the following table:

Failure warning	Priority 1
No computed data	Priority 2
Functional test	Priority 3
Normal operation	Priority 4

Each data word type has its own unique utilization of the SSM field. These various formats are described in the following sections.

34.4.5.4 BCD Numeric

When a failure is detected within a system, which may cause one or more of the words normally output by that system to be unreliable, the system stops transmitting the affected words on the data bus. Some avionics systems are capable of detecting a fault condition that results in less than normal accuracy. In these systems, when a fault of this nature is detected, for instance, partial sensor loss which results in degraded accuracy, each unreliable BCD digit is encoded 1111 when transmitted on the data bus. For avionics equipment having a display, the 1111 code should, when received, be recognized as representing an inaccurate digit and a dash or equivalent symbol is normally displayed in place of the inaccurate digit.

The sign (plus/minus, north/south, etc.) of BCD numeric data is encoded in bits 30 and 31 of the word as shown in Table 34.2. Bits 30 and 31 of BCD numeric data words are zero where no sign is needed.

The no computed data code is annunciated in the affected BCD numeric data words when a source system is unable to compute reliable data for reasons other than system failure. When the functional test code appears in bits 30 and 31 of an instruction input data word, it is interpreted as a command to perform a functional test.

34.4.5.5 BNR Numeric Data Words

The status of the transmitter hardware is encoded in the status matrix field (bit numbers 30 and 31) of BNR numeric data words as shown in Table 34.3.

A source system annunciates any detected failure that causes one or more of the words normally output by that system to be unreliable by setting bit numbers 30 and 31 in the affected words to the failure warning code defined in Table 34.3. Words containing this code continue to be supplied to the data bus during the failure condition.

The no computed data code is annunciated in the affected BNR numeric data words when a source system is unable to compute reliable data for reasons other than system failure.

TABLE 34.2 BCD Numeric Sign/Status Matrix

Bit No.		
31	30	Function
0	0	Plus, north, east, right, to, above
0	1	No computed data
1	0	Functional test
1	1	Minus, south, west, left, from, below

TABLE 34.3 Status Matrix

Bit No.		
31	30	Function
0	0	Failure warning
0	1	No computed data
1	0	Functional test
1	1	Normal operation

TABLE 34.4 Status Matrix

Bit No. 29	Function
0	Plus, north, east, right, to, above
1	Minus, south, west, left, from, below

TABLE 34.5 Accuracy Status

Bit No. 11	Function
0	Nominal accuracy
1	Degraded accuracy

When the functional test code appears as a system output, it is interpreted as advice that the data in the word result from the execution of a functional test.

If, during the execution of a functional test, a source system detects a failure that causes one or more of the words normally output by that system to be unreliable, it changes the states of bits 30 and 31 in the affected words such that the functional test annunciation is replaced with a failure warning annunciation.

The sign (plus, minus, north, south, etc.) of BNR numeric data words are encoded in the sign matrix field (bit 29) as shown in Table 34.4. Bit 29 is zero when no sign is needed.

Some avionics systems are capable of detecting a fault condition that results in less than normal accuracy. In these systems, when a fault of this nature is detected, for instance, partial sensor loss, which results in degraded accuracy, the equipment will continue to report normal for the SSM while indicating the degraded performance by coding bit 11 as shown in Table 34.5.

This implies that degraded accuracy can be coded only in BNR words not exceeding 17 bits of data.

34.4.5.6 Discrete Data Words

A source system annunciates any detected failure that could cause one or more of the words normally output by that system to be unreliable. Three methods are defined. The first method is to set bits 30 and 31 in the affected words to the failure warning code defined in Table 34.6. Words containing the failure warning code continue to be supplied to the data bus during the failure condition. When using the second method, the equipment may stop transmitting the affected word or words on the data bus.

TABLE 34.6 Discrete Data Words

Bit No.		
31	30	Function
0	0	Verified data, normal operation
0	1	No computed data
1	0	Functional test
1	1	Failure warning

This method is used when the display or utilization of the discrete data by a system is undesirable. The third method applies to data words, which are defined such that they contain failure information within the data field. For these applications, the associated ARINC Standard specifies the proper SSM reporting. Designers are urged not to mix operational and built-in test equipment (BITE) data in the same word.

The no computed data code is annunciated in the affected discrete data words when a source system is unable to compute reliable data for reasons other than system failure. When the functional test code appears as a system output, it is interpreted as advice that the data in the discrete data word contents are the result of the execution of a functional test.

34.4.6 Data Standards

Data units, ranges, resolutions, refresh rates, number of significant bits, and pad bits for the items of information to be transferred by ARINC 429 are administered by the AEEC staff and incorporated in ARINC Specification 429.

ARINC Specification 429 calls for numeric data to be encoded in BCD and binary, the latter using two's complement fractional notation. In this notation, the most significant bit of the data field represents half of the maximum value chosen for the parameter being defined. Successive bits represent the increments of a binary fraction series. Negative numbers are encoded as the two's complements of positive value and the negative sign is annunciated in the SSM.

In establishing a given parameter's binary data standards, the unit's maximum value and resolution are first determined in that order. The LSB of the word is then given a value equal to the resolution increment, and the number of significant bits is chosen such that the maximum value of the fractional binary series just exceeds the maximum value of the parameter (i.e., it equals the next whole binary number greater than the maximum parameter value less one LSB). For example, to transfer altitude in units of feet over a range of zero to 100,000 feet with a resolution of 1 ft, the number of significant bits is 17 and the maximum value of the fractional binary series is 131,071 (i.e., 131,072 − 1).

Note that because accuracy is a quality of the measurement process and not the data transfer process, it plays no part in the selection of word characteristics. Obviously, the resolution provided in the data word should equal or exceed the accuracy not to degrade it.

For the binary representation of angular data in ARINC 429, degrees are divided by the unit of data transfer and ±1 (semi-circle) as the range for two's complement fractional notation encoding, ignoring, for the moment, the subtraction of the LSB. Thus, the angular range 0–359.XXX degrees is encoded as 0 through ±179.XXX degrees; the value of the most significant bit is one half semi-circle.

This may be illustrated as follows. Consider encoding the angular range 0–360 in one degree increments. Per the general encoding rules above, the positive semi-circle will cover the range 0–179 (one LSB less than full range). All bits are set to zero for 0, all bits are set to one for 179, and the SSM will indicate the positive sign. The negative semi-circle will cover the range 180–359. All the bits are set to zero for 180. The codes for angles between 181 and 359 degrees are determined by taking the two's complements of the fractional binary series for the result of subtracting each value from 360. Thus, the code for 181 is the two's complement of the code for 179. Throughout the negative semi-circle, which includes 180, the SSM contains the negative sign.

34.5 Timing-Related Elements

The data transfer parameters related to the timing aspects of the signal circuit are as described in the following sections.

34.5.1 Bit Rate

34.5.1.1 ARINC 429 High-Speed Operation

The bit rate for high-speed operation of the system is 100 kilobits per second (kbps) ±1%.

TABLE 34.7   ARINC 429 Data Rates and Timing

Parameter	High-Speed Operation	Low-Speed Operation
Bit rate	100 kbps ± 1%	12–14.5 kbps
Time Y	10 μs ± 2.5%	Z^a μs ± 2.5%
Time X	5 μs ± 5%	Y/2 ± 5%
Pulse rise time	1.5 ± 0.5 μs	10 ± 5 μs
Pulse fall time	1.5 ± 0.5 μs	10 ± 5 μs

[a] Z = 1/R where R = bit rate selected from 12 to 14.5 kbps range.

34.5.1.2 ARINC 429 Low-Speed Operation

The bit rate for low-speed operation of the system is within the range 12.0–14.5 kbps. The selected rate is maintained within 1% (Table 34.7).

34.5.2 Information Rates

Data works are transferred at prescribed rates. The minimum and maximum transmit rates for each item of information is specified within ARINC Specification 429. Data words with the same label ID, but with different SDI codes are treated as unique items of information. Each and every unique item of information is transmitted once during an interval bounded in length by the specified minimum and maximum values. Stated another way, a data word having the same label and four different SDI codes will appear on the bus four times (once for each SDI code) during that time interval.

Discrete bits contained within data words are transferred at the bit rate and repeated at the update rate of the primary data. Words dedicated to discrete data are repeated continuously at specified rates.

34.5.3 Clocking Method

Clocking is inherent in the data transmission. The identification of the bit interval is related to the initiation of either a HI or LO state from a previous NULL state in a bipolar RZ code.

34.5.4 Word Synchronization

The digital word should be synchronized by reference to a gap of four bit times minimum between the periods of word transmissions. The beginning of the first transmitted bit following this gap signifies the beginning of the new word.

34.5.5 Timing Tolerances

The waveform timing tolerances are shown below.

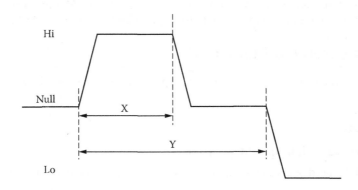

Note: Pulse rise and fall times are measured between the 10% and 90% voltage amplitude points on the leading and trailing edges of the pulse and include time skew between the transmitter output voltages A-to-ground and B-to-ground.

34.6 Communications Protocols

34.6.1 Development of File Data Transfer

AEEC adopted ARINC Specification 429 in July 1977. However, in October 1989, AEEC updated the file data transfer procedure with a more comprehensive definition to support the transfer of both bit-oriented and character-oriented data. The new protocol became known as the ARINC 429 Williamsburg protocol also known as ARINC 429W.

34.6.1.1 File Data Transfer Techniques

This file data transfer techniques specification describes a system in which an LRU may generate binary extended length messages on demand. Data is sent in the form of link data units (LDUs) organized in 8-bit octets. System address labels (SALs) are used to identify the recipient.

34.6.1.2 Data Transfer

The same principles of the physical layer implementation apply to file data transfer. Any avionics system element having information to transmit does so from a designated output port over a single twisted and shielded pair of wires to all other system elements having need of that information. Unlike the simple broadcast protocol that can deliver data to multiple recipients in a single transmission, the file transfer technique can be used only for point-to-point message delivery.

34.6.1.3 Broadcast Data

The broadcast transmission technique described above can be used concurrently with file data transfer.

34.6.1.4 Transmission Order

The most significant octet of the file and LSB of each octet should be transmitted first. The label is transmitted ahead of the data in each case. It may be noted that the label field is encoded in reverse order, that is, the LSB of the word is the most significant bit of the label. This reversed label is a legacy from past systems in which the octal coding of the label field was, apparently, of no significance.

34.6.1.5 Bit-Oriented Protocol Determination

An LRU will require internal logic to determine whether to use character-oriented protocol or bit-oriented protocol, and which protocol to use when prior knowledge is not available.

34.6.2 Bit-Oriented Communications Protocol

This section describes the bit-oriented ARINC 429 Williamsburg protocol and message exchange procedures for file data transfer between units desiring to exchange bit-oriented data assembled in data files. The bit-oriented protocol is designed to accommodate data transfer between sending and receiving units in a form compatible with the Open Systems Interconnect (OSI) model developed by the ISO. This document directs itself to an implementation of the link layer; however, an overview of the first four layers (physical, link, network, and transport) is provided.

Mixing bit-oriented file transfer data words with conventional data words, which contain label codes, is permitted. If the sink should receive a conventional data word during the process of accepting a bit-oriented file transfer message, the sink should accept the conventional data word and resume processing of the incoming file transfer message.

The data file and associated protocol control information are encoded into 32-bit words and transmitted over the physical interface. At the link layer, data are transferred using a transparent bit-oriented data file transfer protocol designed to permit the units involved to send and receive information in multiple word frames. It is structured to allow the transmission of any binary data organized into a data file composed of octets.

34.6.2.1 Physical Medium

The physical interface is described above.

34.6.2.2 Physical Layer

The physical layer provides the functions necessary to activate, maintain, and release the physical link that will carry the bit stream of the communication. The interfacing units use the electrical interface, voltage, and timing described earlier. Data words will contain 32 bits; bits 1–8 will contain the SAL and bit 32 is the parity bit, set to odd parity.

34.6.2.3 Link Layer

The link layer is responsible for transferring information from one logical network entity to another and for enunciating any errors encountered during transmission. The link layer provides a highly reliable virtual channel and some flow control mechanisms.

34.6.2.4 Network Layer

The network layer performs a number of functions to ensure that data packets are properly routed between any two terminals. The network layer expects the link layer to supply data from correctly received frames. The network layer provides for the decoding of information up to the packet level to determine where the message should be transferred. To obtain interoperability, this basic process must be reproduced using the same set of rules throughout all the communications networks on-board the aircraft and on the ground. The bit-oriented data link protocol was designed to operate in a bit-oriented network layer environment. Specifically, ISO 8208 would typically be selected for the subnetwork layer protocol for air and ground subnetworks.

34.6.2.5 Transport Layer

The transport layer controls the transportation of data between a source end-system to a destination end-system. It provides network independent data delivery between these processing end-systems. It is the highest order function involved in moving data between systems and it relieves higher layers from any concern specifically with the transportation of information between them.

34.6.2.6 Link Data Units

A LDU contains binary encoded octets. The octets may be set to any possible binary value. The LDU may represent raw data, character data, bit-oriented messages, character-oriented messages, or any string of bits desired. The only restriction is that the bits must be organized into full 8-bit octets. The interpretation of those bits is not a part of the link layer protocol. The LDUs are assembled to make up a data file.

LDUs consist of a set of contiguous 32-bit data words, each containing the SAL of the sink. The initial data word of each LDU is a start of transmission (SOT). The data described above are contained within the data words that follow. The LDU is concluded with an end of transmission (EOT) data word. No data file should exceed 255 LDUs.

Within the context of this document, LDUs correspond to frames and files correspond to packets.

34.6.2.7 Link Data Unit Size and Word Count

The LDU may vary in size from 3 to 255 ARINC 429 words including the SOT and EOT words. When a LDU is organized for transmission, the total number of ARINC 429 words to be sent is calculated. This is called word count. It is the sum of the SOT word, the data words in the LDU, and the EOT word. In order to obtain maximum system efficiency, the data is typically encoded into the minimum number of LDUs.

The word count field is 8 bits in length. Thus, the maximum number of ARINC 429 words that can be counted in this field is 255. The word count field appears in the request to send (RTS) and clear to send (CTS) data words. The number of LDUs needed to transfer a specific data file will depend upon the method used to encode the data words.

34.6.2.8 System Address Labels

LDUs are sent point-to-point, even though other systems may be connected and listening to the output of a transmitting system. In order to identify the intended recipient of a transmission, the label field (bits 1–8) is used to carry a SAL. Each on-board system is assigned a SAL. When a system sends an LDU to another system, the sending system (the source) addresses each ARINC 429 word to the receiving system (the sink) by setting the label field to the SAL of the sink. When a system receives any data containing its SAL that is not sent through the established conventions of this protocol the data are ignored.

In the data transparent protocol data files are identified by content rather than by an ARINC 429 label. Thus, the label field loses the function of parameter identification available in broadcast communications.

34.6.2.9 Bit Rate and Word Timing

Data transfer may be set to either high-speed or low-speed. The source introduces a gap between the end of each ARINC 429 word transmitted and the beginning of the next. The gap should be at least four bit times, and the sink should be capable of receiving the LDU with the minimum word gap of four bit times between words. The source should not exceed a maximum average of 64 bit times between data words of an LDU.

The maximum average word gap is intended to compel the source to transmit successive data words of an LDU without excessive delay. This provision prevents a source that is transmitting a short message from using the full available LDU transfer time. The primary value of this provision is realized when assessing a maximum LDU transfer time for short fixed-length LDUs, such as for automatic dependence surveillance (ADS).

If an ARINC 429W source device were to synchronously transmit long-length or full LDUs over a single data bus to several sink devices, the source may not be able to transmit the data words for a given LDU at a rate fast enough to satisfy this requirement because of other bus activity. In aircraft operation, given the asynchronous burst mode nature of ARINC 429 Williamsburg LDU transmissions, it is unlikely that an ARINC 429W source device would synchronously begin sending a long-length or full LDU to more than two ARINC 429W sink devices. A failure to meet this requirement will result in either a successful LDU transfer or an LDU retransmission due to an LDU transfer time-out.

34.6.2.10 Word Type

The Word Type field occupies bits 31–29 in all bit-oriented LDU words and is used to identify the function of each ARINC 429 data word used by the bit-oriented communication protocol.

34.6.2.11 Protocol Words

The protocol words are identified with a Word Type field of 100 and are used to control the file transfer process.

The Protocol Identifier field occupies bits 28–25 of the protocol word and identifies the type of protocol word being transmitted. Protocol words with an invalid protocol identifier field are ignored.

Some protocol words contain a Destination Code. The Destination Code field (bits 24–17) indicates the final destination of the LDU. If the LDU is intended for the use of the system receiving the message, the destination code may be set to null (00). However, if the LDU is a message intended to be passed on to another on-board system, the Destination Code will indicate the system to which the message is to be passed. The Destination Codes are assigned according to the applications involved and are used in the Destination Code field to indicate the address of the final destination of the LDU.

In an OSI environment, the link layer protocol is not responsible for validating the destination code. It is the responsibility of the higher-level entities to detect invalid destination codes and to initiate error logging and recovery. Within the pre-OSI environment, the Destination Code provides network layer information. In the OSI environment, this field may contain the same information for routing purposes between OSI and non-OSI systems.

Some protocol words contain a Word Count field. The Word Count field (bits 16–9) reflects the number of ARINC 429 words to be transmitted in the subsequent LDU. The maximum word count value is 255 ARINC 429 words and the minimum is three ARINC 429 words. A LDU with the minimum word count value of three words would contain a SOT word, a data word, and an EOT word. A LDU with the maximum word count value of 255 ARINC 429 words would contain a SOT word, 253 data words, and an EOT word.

34.7 Applications

34.7.1 Initial Implementation

ARINC 429 was first used in the early 1980s on the Airbus A310 and Boeing B-757 and B-767 airplanes. In some cases, approximately 150 separate buses are used to interconnect LRUs comprising computers, radios, displays, controls, and sensors on these airplanes.

34.7.2 Evolution of Controls

The introduction of ARINC 429 proved to be a major step toward reduction of wires. Until that time, radio equipment was controlled using an approach called two-out-of-five tuning. Each digit of the desired radio frequency was encoded on each set of five wires. Multiple digits dictated the need for multiple sets of wires for each radio receiver. With ARINC 429 a tuning unit needs only one bus to tune multiple radios of the same type. An entire set of radios and navigation receivers could be tuned with a few control panels, using approximately the same number of wires previously required to tune a single radio.

As space in the flight deck became scarce, the need to reduce the number of control panels became critical. The industry recognized that a single control panel, properly configured, could replace most of the existing control panels. The Multi-purpose Control/Display Unit (MCDU), which came from the industry effort, was derived essentially from the need to provide control and display for the Flight Management Computer (FMC). Since then, the MCDU has become the primary data entry device for controlling the FMC and other equipment.

A special protocol was developed for ARINC 429 to provide the capability of addressing different units connected to a single ARINC 429 bus from the MCDU. The protocol employed two-way communications using two pairs of wires between the controlling unit and the controlled device. An addressing scheme provided for selective communications between the controlling unit and any one of the controlled units. Only one output bus from the controller is required to communicate addresses and commands to the receiving units. With the basic ARINC 429 design, up to 20 controlled units could be connected to the output of the controller. Each of the controlled units is addressed by an assigned SAL. This is further defined in ARINC Characteristic 739.

34.7.3 Longevity of ARINC 429

New airplane designs in the twenty-first century continue to employ the ARINC 429 bus for routine data transfer. The relative simplicity and integrity of the bus, as well as the ease of certification, are characteristics that contribute to the continued selection of the ARINC 429 bus. New ARINC Standards are expected to emerge to fill the industry need for additional data volume and increased data rates. Additional information regarding ARINC Standards may be found at http://www.aviation-ia.com/aeec.

35

RTCA DO-297/EUROCAE ED-124 Integrated Modular Avionics (IMA) Design Guidance and Certification Considerations

Cary R. Spitzer
AvioniCon

Leanna Rierson
Digital Safety Consulting

35.1 Introduction

RTCA document DO-297, *Integrated Modular Avionics (IMA) Design Guidance and Certification Considerations,* is one of several documents that are key to the approval of avionics and ultimately the certification of the host aircraft. These documents include DO-160 [1], DO-178 [2,3], DO-254 [4], SAE ARP 4754 [5,6], and ARP 4761 [7], as well as many others, such as technical standard orders (TSOs). DO-297, like DO-178 and DO-254, provides guidance, not regulations, to be followed in the development of avionics. Consultation by the avionics developers and aircraft manufacturers with the certification authorities determines which portions of the document will be used as part of the aircraft certification basis. All of the listed RTCA documents (DO designation) become sanctioned by the Federal Aviation Administration (FAA) for certification use through the issuance of advisory circulars (ACs). The FAA has recognized DO-297 in AC 20-170, "Integrated Modular Avionics Development, Verification, Integration, and Approval Using RTCA/DO-297 and Technical Standard Order-C153." The FAA states in AC 20-170 section 1-4.c, "We find that the objectives, processes, and activities in RTCA/DO-297, plus the additional guidance material contained in this AC, constitute an acceptable means of compliance for the development, integration, verification, and installation approval of IMA systems" [8]. As can been seen from the title of AC 20-170, DO-297 is closely related to TSO-C153, "Integrated Modular Avionics Hardware Elements" [9]. TSO-C153 was developed prior to DO-297 and is often used to gain approval of hardware elements (such as circuit cards and cabinets) that are part of the IMA platform.

The European Organisation for Civil Aviation Equipment (EUROCAE) equivalent of DO-297 is ED-124. Any reference in this chapter to DO-297 also infers ED-124. At the time of writing, the European Aviation Safety Agency (EASA) has not yet recognized ED-124 by advisory material; however, it is referenced on a project-by-project basis using certification review items. EASA is in the process of developing their version of TSO-C153 and AC 20-170.

The need for DO-297 is derived from the emergence of pioneering IMA architectures on the Airbus A-380 and Boeing B-787. Although DO-297 was too late to be part of the certification basis for these aircraft, the lessons learned from them did guide, in part, the content of DO-297.

According to DO-297, IMA is a "shared set of flexible, reusable, and interoperable hardware and software resources that, when integrated, form a platform that provides services, designed and verified to a defined set of safety and performance requirements, to host applications performing aircraft functions" [10]. The document provides guidance for IMA developers, integrators, applicants, and those involved in the approval and continued airworthiness of IMA systems. It provides specific guidance for the assurance of IMA systems as differentiated from traditional federated avionics.

The key contribution of DO-297 is guidance on the objectives, processes, and activities related to the development and integration of IMA modules, applications, and systems to *incrementally accumulate* design assurance toward the installation and approval of an IMA system on an approved aviation product. Incremental acceptance of individual items of the IMA platform (including the core software) and hosted applications enables the reduction of follow-on certification efforts without compromising system safety.

35.2 Key Terms

The following terms are used throughout DO-297 and are important to understanding and properly applying the guidance [10]:

- *Application software*: The part of an application implemented through software. It may be allocated to one or more partitions.
- *Approval*: The act or instance of giving formal or official acknowledgment of compliance with regulations.
- *Component*: A self-contained hardware or software part, database, or combination thereof that may be configuration controlled.
- *Integral process*: A process that assists the system, software, or hardware development processes and other integral processes and, therefore, remains active throughout the life cycle. The integral processes are the verification process, the quality assurance process, the configuration management process, and the certification liaison process.*
- *Module*: A component or collection of components that may be accepted by themselves or in the context of an IMA system. A module may also comprise other modules. A module may be software, hardware, or a combination of hardware and software, which provides resources to the IMA system hosted applications.

---

* Although the DO-297 definition of *integral process* doesn't explain it, DO-297 chapter 5 also identifies the following as integral processes: safety assessment, system development assurance, and requirements validation.

- *Partition*: An allocation of resources whose properties are guaranteed and protected by the platform from adverse interaction or influences from outside the partition.
- *Partitioning*: An architectural technique to provide the necessary separation and independence of functions or applications to ensure that only intended coupling occurs.

35.3 Outline

DO-297 contains six chapters:

- Chapter 1 is an overview of the document.
- Chapter 2 introduces the concept of IMA and highlights the system, hardware, and software characteristics.
- Chapter 3 provides a description of IMA-specific development and integration guidelines.
- Chapter 4 is the core of the document and provides guidelines for the acceptance of IMA and describes the relationship to the approval of the installed IMA system.
- Chapter 5 describes integral (umbrella) processes for IMA development.
- Chapter 6 gives guidelines for continued airworthiness of IMA systems.

Figure 35.1 illustrates the relationships between the chapters. The following four annexes are also included:

- Annex A has six tables that summarize the objectives to be satisfied. These objectives are based on the six tasks described in Chapter 4.
- Annex B is the glossary.
- Annex C is the list of acronyms and abbreviations used in the document.
- Annex D contains IMA examples.

35.4 Step-by-Step IMA Development and Approval

Six tasks define the incremental acceptance of IMA systems in the certification process:

- Task 1: Module acceptance
- Task 2: Application software or hardware acceptance
- Task 3: IMA system acceptance
- Task 4: Aircraft integration of IMA system, including validation and verification (V&V)
- Task 5: Change of modules or applications
- Task 6: Reuse of modules or applications

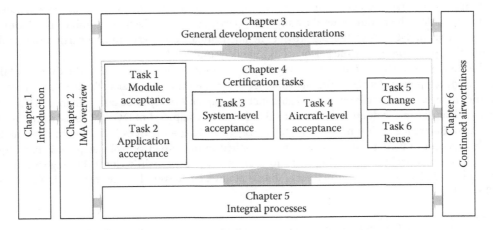

FIGURE 35.1 Chapters and their relationships. (Courtesy of RTCA, Washington, DC.)

TABLE 35.1 Task Index to Description and Objectives

Task	Reference/Objectives
Task 1: Module acceptance	Section 4.2/Table A-1
Task 2: Application acceptance	Section 4.3/Table A-2
Task 3: IMA system acceptance	Section 4.4/Table A-3
Task 4: Aircraft integration of IMA system—including V&V	Section 4.5/Table A-4
Task 5: Change of modules or applications	Section 4.6/Table A-5
Task 6: Reuse of modules or applications	Section 4.7/Table A-6

DO-297 provides detailed, comprehensive guidance on each of these tasks. Table 35.1 lists each task along with the section of DO-297 that presents the details on how to accomplish the task. Each task has multiple objectives spelled out in tables in Annex A.

There are several traditional aspects of avionics development that take on added significance for IMA. Foremost among these are definition of the requirements, liaison among the stakeholders, and robust partitioning. Closely related to these items is the need for very close liaison with the certification authority.

Defining the IMA requirements is a significant challenge in that a very long-term perspective is required. In recognition of the growing role of avionics in modern aircraft operations, coupled with the cost of new or upgraded avionics, aircraft customers demand a large amount of growth in the IMA platform capabilities. The customer may expect the newly delivered IMA to use perhaps only about half of its installed processing power, input/output, and memory. Typically, the IMA cabinet also will have growth slots for additional modules. It is recommended to be very liberal in establishing the spare capabilities, especially the growth slots in the IMA cabinet.

In contrast to traditional avionics, the IMA platform and its functions may come from many stakeholders. It is the role of the integrator (typically the aircraft manufacturer) to ensure these stakeholders communicate with the integrator and with each other to ensure that all requirements are identified, validated, accommodated, and verified. Open, effective communication between the integrator and the application providers is especially important.

Partitioning is perhaps the toughest design challenge in developing an IMA system. Because of the multiple functions hosted on a typical IMA platform, the extra demands placed on the partitioning techniques cause them to be referred to as *robust partitioning*. In the most basic, lowest-level properties of robust partitioning, no application function in one partition can (1) access memory of any another partition in an adverse manner, (2) affect the timing of any another partition in an adverse manner, or (3) adversely affect the resources used by any other partition. Proof of achieving robust partitioning is especially difficult. DO-297 section 3.5 provides detailed guidance for performing a partitioning analysis to ensure the partitioning scheme performs its intended function.*

Table 35.2 is from DO-297 and shows the 10 objectives to be accomplished in the IMA module or platform development. Other tables in Annex A cover the other five tasks. Table 35.2 includes the reference to the text that spells out general guidance in Chapter 3 and specific task guidance in Chapter 4. Specific remarks on the documentation are also found in Chapter 4 of DO-297 as listed in the "Life Cycle Data Reference" column. As with DO-178B/C, DO-254, and ARP4754A, the control category sets the amount of configuration control to be applied to the data: Control Category 1 data require the application of problem reporting and a formal release process, while Control Category 2 data do not. Table 35.3 identifies the minimal set of data that must be delivered to the certification authorities for Tasks 1–4.

---

* See Chapter 21 of *Developing Safety-Critical Software* by Leanna Rierson for more information on partitioning [12]; additionally, chapters 20 (real-time operating systems) and 22 (configuration data) provide information closely related to IMA.

TABLE 35.2 IMA Module/Platform Development Process (Task 1) Objectives

ID	Objective	Doc Ref	Life Cycle Data Description	Life Cycle Data Reference	Control Category
1	Module/platform development and acceptance life cycle and associated processes are planned and implemented consistently with the guidance of DO-160, DO-178, DO-254, and this document.	4.2.1a 3.1.1a	Module/platform acceptance plan	4.2.3	CC1
2	Module/platform requirements specifications are defined, traceable, and verifiable.	4.2.1b 3.1.1b	Module/platform requirements specifications Traceability Data	4.2.4 4.2.5	CC1 CC2
3	Module/platform design is documented and addresses the IMA unique failure modes, safety analysis, and functionality.	3.1.1c,d 4.2.1b,c 5.1	Module/platform design data Module/platform failure analyses and safety analyses	4.2.4 4.2.12b	CC1 CC1
4	Verification and development tools are assessed and qualified, as needed.	4.2.1i 3.4 5.2.3	Module/platform tool qualification data	4.2.12c	CC2 or CC1[a]
5	Partitioning ensures that the behavior of any hosted application is prevented from adversely affecting the behavior of any other application or function.	3.5 3.1.1c,d 4.2.1c,d 5.1, 5.3, 5.4	Partitioning analysis data	4.2.4j	CC1
6	Compliance with module requirements, resource requirements, etc., is demonstrated.	3.1.1d,e 4.2.1c 4.2.1d 4.2.1e	Module/platform V&V data	4.2.5	CC2
7	Ensure module users have the information needed to integrate and interface the module.	4.2.1g,k,l 3.4	Module/platform acceptance data sheet	4.2.10	CC1
8	Platform integration is complete.	4.2.1h 3.1.1d 5.3, 5.4	Platform integration, verification, and validation data	4.2.5	CC2
9	Health monitoring and fault management functions of the IMA platform are provided and documented for use by the hosted applications and the IMA system.	4.2.1c 3.6 5.1.5.5 5.1.5.6	Platform requirements specification	4.2.4f	CC1
10	Quality assurance, configuration management, integration, validation, verification, and certification liaison for the module/platform are implemented and completed.	4.2.1f,j,k 5.3–5.7	Module/platform QA records Module/platform CM records Module/platform V&V data Module/platform acceptance accomplishment summary Module/platform configuration index Module/platform problem reports	4.2.6 4.2.8 4.2.5 4.2.9 4.2.7 4.2.11	CC2 CC2 CC2 CC1 CC1 CC2

Source: Courtesy of RTCA, Washington, DC.

[a] Control category for tool qualification data is defined in DO-178/ED-12 (Ref. [2]) or DO-254/ED-80 (Ref. [4]).

TABLE 35.3 Life Cycle Data to be Submitted to Certification Authority

Life Cycle Data Item	Ref.	Task 1	Task 2	Task 3	Task 4
Module acceptance plan (MAP)	4.2.3	X			
Module configuration index(es) (MCI)	4.2.7	X			
Module acceptance accomplishment summary (MAAS)	4.2.9	X			
Module acceptance data sheet	4.2.10	X			
Plan(s) for hardware aspects of certification (PHAC)	4.2.12.a 4.3.2	X	X		
Plan(s) for software aspects of certification (PSAC)	4.2.12.a 4.3.2	X	X		
Software configuration indices (SCIs)	4.2.12.a 4.3.2	X	X		
Hardware configuration indices (HCIs) (i.e., top-level drawings)	4.2.12.a 4.3.2	X	X		
Software accomplishment summary(ies) (SAS)	4.2.12.a 4.3.2	X	X		
Hardware accomplishment summary(ies) (HAS)	4.2.12.a 4.3.2	X	X		
Safety assessment analysis/report(s)	4.2.12.b			X	X
Hosted application acceptance data sheet	4.3.2		X		
IMA certification plan (system and aircraft level)	4.4.3 4.5.3			X	X
IMA V&V plan (system and aircraft level)	4.4.4 4.5.4			X	X
IMA configuration index (system and aircraft level)	4.4.5 4.5.5			X	X
IMA accomplishment summary (system and aircraft level)	4.4.6 4.5.6			X	X
EQT plan	DO-160	X	X	X	X
EQT reports	DO-160	X	X	X	X

Source: Courtesy of RTCA, Washington, DC.

35.5 Recommendations for IMA Certification

This section provides seven recommendations for aircraft manufactures to consider when developing, integrating, and certifying IMA systems using DO-297. There is some intentional overlap between the interrelated recommendations in order to provide a complete view [11].

35.5.1 Recommendation #1: Plan Ahead

Failure to plan is essentially equivalent to planning to fail. The following planning practices are recommended for IMA systems:

- Ensure that all stakeholders develop and submit their plans (e.g., safety plans, platform plans, software plans, airborne electronic hardware plans) as early as possible.
- Review suppliers' plans early in the program to ensure that they agree with the aircraft-level plan(s).
- Hold technical meetings with the certification authorities as early as possible to explain the plans and implement feedback.
- Submit plans to certification authorities as early as possible and seek approval.

- Follow approved plans and communicate any deviations or updates to the certification authorities as needed.
- Develop a realistic schedule, communicate it, and work to it.
- Develop the process so that technical issues can be proactively addressed.
- Identify how requirements and problem reporting will be managed between the multiple stakeholders.

In order to document the plans, the aircraft developer, in conjunction with the IMA system supplier, should develop an aircraft-level IMA certification plan.* The certification plan provides the following information:

- A description of the IMA system and its intended installation.
- Identification of stakeholder roles and responsibilities (including all suppliers, their subcontractors, and any offshore subsidiaries) and an explanation of how communication will occur between the stakeholders.
- A summary of the means of compliance to the regulations.
- A list of life cycle data to be generated with identification of what data will be submitted to the certification authority; if multiple certification authorities are involved, an explanation of what data are going to each authority.
- An explanation of how each DO-297 objective will be satisfied and who will satisfy it.
- An explanation of how the guidance of AC 20-170, particularly the guidance that goes above and beyond DO-297, will be satisfied.
- If using TSO-C153 in conjunction with DO-297, an explanation of the approach in order to explain which parts of the guidance material are being utilized; this will typically involve an explanation of how each objective of DO-297 will be satisfied, as well as how guidance of TSO-C153 and AC 20-170 will be met.
- An explanation of how integration, partitioning, health management, data configuration, resource allocation, part marking, tool usage, and testing will be managed and executed.
- An explanation of how any issue papers (or equivalent) from the certification authorities will be addressed.
- Inclusion of information listed in DO-297 Sections 4.4.3 and 4.5.3.

35.5.2 Recommendation #2: Identify Roles and Responsibilities

Developing and certifying an IMA system generally involves numerous stakeholders. For example, a typical program involves the aircraft manufacturer, IMA platform provider, operating system supplier, software subcontractors, and numerous hosted application teams or suppliers. The roles and responsibilities for all suppliers should be clearly identified in writing early in the program. Roles and responsibilities typically include the following for each stakeholder:

- Tasks to be performed
- Data to be generated
- Criteria for completion
- Schedule for data review and submittal
- Planned review process
- Submittal and approval process

---

* In many projects, a system-level IMA certification plan is also developed.

Any proprietary data issues that exist between stakeholders should be identified and resolved as early as possible. It is also important to clearly define who is performing environmental qualification tests, since these are often shared among stakeholders.

A summary of each stakeholder's roles and responsibilities should be included in the aircraft-level IMA certification plan; however, each stakeholder should detail their roles and responsibilities in their own plans (e.g., system plans, software plans, hardware plans, environmental qualification test [EQT] plans).

35.5.3 Recommendation #3: Define Stakeholder Communication Methods

In addition to defining the roles and responsibilities, it is important to define the methods of communication between the stakeholders. The methods should support timely communication of essential data and issues. Communications include technical meetings, e-mails, websites, data deliveries, and problem reports. As problems arise among stakeholders, they should be communicated and worked proactively.

The communication between the platform supplier and the hosted application developers is particularly critical. This communication typically includes an IMA platform user's manual, ongoing technical exchanges, design materials and data sheets, and technical support (perhaps via a hotline and/or website).

It is also important to identify what data items are considered certification *artifacts* and which are not (e.g., a platform user's guide may not be considered a certification artifact; however, the platform data sheet is). All certification artifacts (i.e., life cycle data) should be appropriately controlled in order to satisfy the regulations.

35.5.4 Recommendation #4: Define Supplier Management Approach

Aircraft developers should define how aircraft-level requirements will be developed, managed, flowed down, validated, and verified. Each developer should do the same for their suppliers in order to ensure a seamless implementation and to demonstrate compliance with the aircraft requirements. It should be noted that commercial off-the-shelf (COTS) components developed without the higher-level requirements in mind can be difficult to show traceability to the aircraft requirements and may require extra effort. For example, the traceability between the platform requirements and the COTS operating system may need to be developed to ensure all platform requirements allocated to the operating system are satisfied and that any additional functionality in the operating system is properly deactivated.

Additionally, supplier management should address how problem reports from multiple suppliers and subcontractors will be fed up, evaluated, assessed for safety impact, classified, maintained, etc. Again, this may be difficult for COTS components, since their problem reports are often not available.

Each stakeholder that uses suppliers, subcontractors, or offshore subsidiaries should clearly identify how they will manage and oversee these entities. As a minimum, plans should address how technical, certification, and quality assurance oversight will be carried out.

35.5.5 Recommendation #5: Determine Integration Approaches

Integration of the multiple components is a key factor in successful IMA certification. Integration typically occurs in the context of the V&V effort. The following integration and V&V activities should be considered:

- Determine the overall integration approach (top-down requirements flow and bottom-up integration is typical).
- Strive to mature the IMA platform before integrating hosted applications.

- If a test platform (or emulator) is used for software testing, coordinate the details with the certification authorities, IMA platform provider, and hosted applications as early as possible.
- Define what equipment will be needed for hosted applications to perform their tests. In particular, consider if the actual IMA platform hardware and software will be used or if it will be some representative subset or emulation. If the actual IMA platform hardware and software is not used for hosted applications testing, perform the necessary analysis (or qualification) to demonstrate that equivalent test results are obtained.

35.5.6 Recommendation #6: Perform Ongoing Certification Activities

Early and frequent communication with the certification authorities is essential to successful certification of IMA systems. Each stakeholder should employ certification liaison personnel to help identify and address certification requirements and to communicate with the certification authorities throughout the project. Here are some suggestions for coordinating with the authorities:

- Coordinate plans as early as possible and throughout the project. Both verbal (meetings, phone calls, etc.) and written (document submittals, memos, e-mails, presentations, etc.) communication are important.
- Communicate any changes to the plans or overall strategy to the certification authorities.
- When multiple certification offices are involved, coordinate with all of the appropriate certification offices. It is also important to keep records of the communication and agreements in case personnel or circumstances change. As certification authorities often emphasize, "if it isn't written down, it didn't happen."
- Communicate new and novel concepts to promote early identification of issues (e.g., issue papers and certification review items).
- Communicate responses to issue papers and certification review items as soon as possible after issues are identified.
- Schedule technical meetings and status meetings with the certification authorities throughout the program. Typically, monthly or quarterly meetings are needed for IMA systems.
- Seek certification authorities' input on technical issues as they arise. Be sure to contact and involve the appropriate technical specialists.
- Respond promptly to actions and issues raised by the certification authorities.
- Obtain written agreements with certification authorities (typically through meeting minutes, memos, or approved plans).

35.5.7 Recommendation #7: Address Challenges as They Arise

Certifying an IMA system leads to a number of technical and programmatic challenges. These should be considered early and addressed proactively throughout the project. Here are some of the more challenging technical issues faced when developing and certifying IMA systems:

- Determining and verifying the allocation of resources
- Defining how configuration of all components will be managed
- Designing the part marking scheme (especially when using electronic part marking) to satisfy regulations and support maintenance
- Identifying the approach to be taken for field-loadable software and possibly field-loadable hardware
- Establishing partitioning requirements, identifying partitioning vulnerabilities, and showing mitigation for partitioning vulnerabilities
- Tracking problem reports for all components and analyzing them for impact on functionality, safety, operation, performance, and regulatory compliance

- Establishing the integration and testing scheme for all stakeholders (including number of labs, equipment need for each lab, lab schedules)
- Identifying the certification approach for all stakeholders and obtaining agreement with certification authorities

References

1. RTCA DO-160, Environmental conditions and test procedures for airborne equipment (Washington, DC: RTCA, Inc., various dates).
2. RTCA DO-178B, Software considerations in airborne systems and equipment certification (Washington, DC: RTCA, Inc., December 1992).
3. RTCA DO-178C, Software considerations in airborne systems and equipment certification (Washington, DC: RTCA, Inc., December 2011).
4. RTCA DO-254, Design assurance guidance for airborne electronic hardware (Washington, DC: RTCA, Inc., April 2000).
5. SAE ARP4754, Certification considerations for highly-integrated or complex aircraft systems (Warrendale, PA: SAE Aerospace, November 1996).
6. SAE ARP4754A, Guidelines for development of civil aircraft and systems (Warrendale, PA: SAE Aerospace, December 2011).
7. SAE ARP4761, Guidelines and methods for conducting the safety assessment process on civil airborne systems and equipment (Warrendale, PA: SAE Aerospace, December 1996).
8. Federal Aviation Administration, Integrated modular avionics development, verification, integration, and approval using RTCA/DO-297 and Technical Standard Order-C153, Advisory Circular 20-170 (Washington, DC: FAA, October 2010).
9. Federal Aviation Administration, TSO-C153, Integrated modular avionics hardware elements (Washington, DC: FAA, May 2010).
10. RTCA DO-297, Integrated modular avionics (IMA) development guidance and certification considerations (Washington, DC: RTCA, Inc., November 2005).
11. L. Rierson, Best practices for certifying IMA systems in civil aircraft, *IEEE Digital Avionics Systems Conference*, St. Paul, MN, 2008, pp. 1–8.
12. L. Rierson, *Developing Safety-Critical Software: A Practical Guide for Aviation Software and DO-178C Compliance* (Boca Raton, FL: CRC Press, 2012).

36

ARINC Specification 653, Avionics Application Software Standard Interface

Paul J. Prisaznuk
ARINC

36.1 Introduction

This chapter acquaints the reader with the real-time operating system (RTOS) and the associated avionics application software standard widely used in digital avionics systems. It addresses the following questions: Why use an avionics RTOS? What makes an avionics RTOS different from general computer RTOS? What are the basic concepts? And how do they work?

36.2 Why Use an Avionics Operating System?

The early years of the jet age readily produced aircraft requiring a three-man flight crew. In those days, it was the navigator who provided the captain and first officer with many aircraft performance and prediction computations. Times have changed. The introduction of microelectronics and the related software applications have dramatically improved the ability of avionics systems to introduce automation, improve flight crew situational awareness, and create operational efficiencies—all with ever-increasing margins of safety. Today, the two-man flight crew configuration and pilot work load management are made possible by rapidly evolving hardware and software technology.

For many years, the benefits of reduced weight, volume, and power afforded by microelectronics have been very attractive to the avionics development community. While avionics continue to grow in terms of functions and capability, the physical mass and volume needed to implement these functions continue to get smaller. This trend has continued along a predictable path. Software development efforts have enabled avionics to be developed with ever-increasing levels of sophistication. While the industry has taken these trends for granted, it has recognized that complex software must be managed properly, and software parts must be developed as well-defined modules.

ARINC Specification 653 was written to define software modules at an abstract level and, in particular, define the interface boundary between the application software modules and the underlying computer operating system, also called the RTOS.

The RTOS creates a uniform environment that allows software modules or applications to execute and provides a standard set of system services to schedule software execution, map memory areas, and, in some cases, partition memory accesses, perform input/output (I/O) operations, and handle errors or faults. An RTOS further refines the environment to support temporal characteristics critical to the proper reliable operation of avionics application software.

In the earliest generation of digital avionics, each unit relied on a microprocessor with unique instruction set and unique RTOS. This was costly to develop and maintain. A survey of digital avionics in the early 1990s revealed what we know today—all digital avionics have a common set of components that generally include the following hardware resources: the computer processing unit, memory, I/O, and power supply. In the same time period, it was recognized that the cost of avionics was no longer driven by hardware costs. Rather, the cost of avionics is attributed largely to software development costs, which is estimated to be up to 85% of the overall development cost of digital avionics systems. The need for a standardized RTOS became clear.

36.3 Why Develop an Operating System Interface?

A standardized RTOS interface for avionics is viewed to be necessary and desirable for two primary reasons. First, it establishes a known interface boundary for avionics software development, thus allowing independence of the avionics software applications and the underlying RTOS. This enables concurrent development of both RTOS and application software. Second, the standard RTOS interface allows the RTOS and the underlying hardware platform to evolve independent of the software applications. Together, this enables cost-effective upgrades over the life of the airplane.

The air transport industry has developed a standard RTOS interface definition, which is specified as the interface boundary between the avionics software applications and the RTOS proper. The standardization effort was initiated by the Airlines Electronic Engineering Committee (AEEC) and involved many interested parties representing major stakeholders in the industry. Software specialists convened and identified the specific needs for avionics equipment:

- Safety critical (as defined by Federal Aviation Regulations [FAR] Part 25.1309)
- Real time (responses must occur within a prescribed time period)
- Deterministic (the results must be predictable and repeatable)

Determinism is the ability to produce a predictable outcome generally based on the preceding operations. The outcome occurs in a specified period of time with some degree of repeatability (RTCA DO-297/EUROCAE ED-124).

The emergence of integrated modular avionics (IMA) was another drive. The IMA concept encourages the integration of many software functions. Mechanisms are necessary to manage the communication flows within the system, between applications and the data on which they operate, and provide control routines to be performed as a result of system-level incidents that affect the aircraft operation. Clear interface specifications are necessary between both the elements within the software architecture and at the interface between the software and the other physical elements.

Software functionality within the IMA equipment is provided by software applications that rely on the available resources provided by the avionics platform. The use of application software offers greater flexibility to meet user and customer needs, including functional enhancement. As the size and complexity of embedded software systems increase, modern software engineering techniques (e.g., object-oriented analysis and design, functional decomposition with structured design, top-down design, bottom-up design) are desired to aid the development process and to improve productivity. There are many advantages to this approach. For example, software can be defined, developed, managed, and maintained as modular components.

Avionics software is qualified to the appropriate level of RTCA DO-178/EUROCAE ED-12, a process that requires rigor and attention to detail. The software must be demonstrated to comply with the appropriate government standards set to assure safe operation. Therefore, the RTOS is designed to be simple and deterministic in its operational states.

36.4 Overall System Architecture

The ARINC 653 software interface specification was developed for IMA systems and for federated equipment that may contain more than one partitioned function. ARINC 653 services are defined in a way that enables the most critical of avionics functions, commonly known as RTCA DO-178 Level A applications, to be installed in a system that can be certifiable by the FAA, EASA, and other regulatory authorities.

An example of software architecture is shown in Figure 36.1. The primary components of the system are application software that performs the avionics functions and core software that provides a standard and common environment for software applications, which are further divided into the following functions:

- The RTOS manages logical responses to applications and demands. It allocates processing time, communication channels, and memory resources. This function maps application requests to system-level logical mechanisms and provides the uniform logical interface to the applications.
- The health monitor (HM) function initiates error recovery or reconfiguration strategies that perform a specific recovery action (see Chapter 22).
- The hardware interface system (HIS) manages the physical hardware resources on behalf of the RTOS. The HIS maps the logical requests made by the RTOS onto the particular physical configuration of the core hardware.

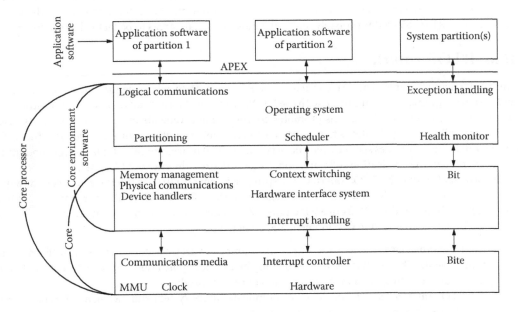

FIGURE 36.1 System architecture.

The partition is similar to that of a multitasking application within a general-purpose computer. Each partition consists of one or more concurrently executing processes sharing access to processor resources based upon the requirements of the application. All processes are uniquely identifiable, having attributes that affect scheduling, synchronization, and overall execution.

To enable software portability, communication between partitions is independent of the location of both the source and destination partition. An application sending a message to, or receiving a message from, another application will not contain explicit information regarding the location of its own host partition or that of its communication partner. The information required to enable a message to be correctly routed from the source to a destination is contained in configuration tables that are developed and maintained by the system integrator, not the individual application developer. The system integrator configures the environment to ensure the correct routing of messages between partitions hosted on an avionics platform.

36.5 Software Modularity

With the basic IMA architecture in mind, it is possible to develop software in well-defined modules. The process of structurally decomposing avionics software into well-partitioned units yields greater benefits than simply breaking the system down into manageable units. This includes grouping elements that are more likely to change and those that are dependent on specific temporal characteristics. Moreover, those elements susceptible to change can be further divided into those subject to functional enhancement and those that are implementation specific to the physical environment in which they run. This enables both functional enhancement of the software and its porting to other IMA platforms.

Multiple software applications may be hosted on an IMA platform. These applications, which can originate from different avionics sources, must be integrated into the selected implementation of the core hardware. It is necessary to create "brick walls" between applications to ensure reliable software partitioning, especially when these applications have different levels of software criticality. Detailed specifications are necessary to allow these applications to be integrated not only with each other but also with the IMA platform. Software modules can be made portable and reusable as a result of a standardized RTOS. Though ARINC 653 standardizes the interface between software applications and the operating system, it does not standardize the data content of messages sent and received by partitions. This is viewed to be a large contributor to software costs in IMA systems. Data content is clearly outside the scope of ARINC 653 but is probably one of the most important areas to be tackled in the future.

36.6 RTOS Interface

Among the many features of an IMA system is the RTOS interface, which defines the environment for application programs to communicate with the RTOS proper. An open standard allows for healthy competition among RTOS suppliers. This promotes continual improvements in the RTOS products made available yet allows for applications to be ported between implementations with relative ease. The RTOS interface definition provides services and consists primarily of procedure calls. It places rigid bounds on the RTOS proper, the application software, and, to some degree, the entire IMA system.

The principal goal of the RTOS interface is to provide a general-purpose interface between the application software and the RTOS itself. Standardization of the RTOS interface allows the use of application code, hardware, and RTOS from a wide range of suppliers, encouraging competition and allowing reduction in the costs of system development and ownership:

1. The RTOS interface provides the minimum number of services consistent with fulfilling IMA requirements. The interface will consequently be easy for application software developers to efficiently use, assisting them in the production of reliable products.
2. The RTOS interface is extendable to accommodate future system enhancements. It retains compatibility with previous versions of the software.

3. The RTOS interface satisfies the common real-time requirements for high-order programming languages such as Ada and C. Applications use the underlying services provided by the RTOS interface in accordance with their certification and validation criticality level.

4. The RTOS interface decouples application software from the actual processor architecture. Therefore, hardware changes will be transparent to the application software. The RTOS interface allows application software to access the executive services but insulates the application software from architectural dependency.

5. The RTOS interface specification is language-independent. This is a necessary condition for the support of application software written in different high-level languages or application software written in the same language but using different compilers. The fulfillment of this requirement allows flexibility in the selection of the following resources: compilers, development tools, and development platforms.

The RTOS interface places layering and partitioning demands on the RTOS to allow growth and functional enhancement. RTOS requests, particularly communication requests, are made in one of several modes. The application can request that the function be performed and be suspended awaiting completion of that request, or it can continue running and either poll the status of the request or merely be alerted that the transaction is complete. It is anticipated that the interface is split into groups or areas (e.g., memory management, data I/O) and that each area be further subdivided.

The processing of a communication request within the RTOS consists of setting up the appropriate I/O format and passing the message to the HIS. To ensure temporal determinism, each message definition includes a maximum and minimum time of response. Certain message types also include a time-out specification, which could be set up by the application. Hence, all communication (including requests, commands, responses, data I/O) between the applications and the RTOS is rigorously defined by this interface.

Partitioning between applications and between applications and the RTOS is controlled by standard hardware mechanisms. It can therefore be shown that if an application can be proved to correctly meet the RTOS interface, then it will correctly interface with the RTOS. Given a rigorous definition of the RTOS, it is possible to integrate and test applications on a general-purpose computer that simulates the RTOS interface. It is also possible to host a variety of applications without the need to modify the RTOS. Standardized RTOS message definitions allow a growth path for future enhancement of the interface. As long as any future RTOS definition is a superset of earlier definitions, no conflict will arise as a result of any enhancement.

36.7 Software Applications

Application software performs a specific avionics function. It is specified, developed, and verified to the level of criticality appropriate to its function. The latest version of RTCA DO-178/EUROCAE ED-12 will apply to the development of software applications. Within an IMA platform, the method and the level of redundancy, fault detection, fault isolation, and reconfiguration is transparent to the application software.

The application software is responsible for the redundancy management of a specific function and for signal selection and failure monitoring of inputs from external sensors or other systems. Modular software design enables the implementation of software partitions to isolate avionics functions within a common hardware environment. Application software can be developed independent of the underlying hardware. However, some applications will require dedicated I/O, such as a pitot-static probe.

Software modules may be developed by independent sources and integrated into the IMA platform. Therefore, it is necessary for software within a partition to be completely isolated from other partitions so that one partition cannot cause another adverse affects. To ensure partition integrity, partition load images are statically and separately built.

Process scheduling attributes are used by the partition code to change the execution or state of a process within that partition. In this way, the integrity of one application is not compromised by the behavior of another, whether the other application is deemed more or less critical. All communications between applications are performed through the RTOS, whose mechanisms ensure that there is no violation of the interface and that no application either monopolizes a resource or leaves another permanently suspended.

I/O handling is one significant area that is very specific to the aircraft configuration of sensors in traditional application development. In the interest of portability and reuse, the partitioning of the application software architecture should identify the aircraft-specific I/O software and partition it from the functional and algorithmic elements of the application. Such sensor I/O conditioning is logically defined as a separate function associated with the sensor.

Most applications require functional data at specific rates. Much of the design of applications that ties them specifically to a particular aircraft system is the sensor handling. The removal of this function from the application increases portability and, furthermore, concentrates the sensor handling into a single area, allowing sensor data with specific characteristics to be generally available to more than one application. Changes to sensor characteristics are confined to sensor data managers, thereby increasing portability and reuse of application software. Application software is invoked by the scheduling component of the RTOS in a deterministic manner.

36.8 RTOS Proper

The main role of the RTOS is to ensure functional integrity in scheduling and dispatching application programs. It should be possible to prove that the level of temporal determinism (i.e., specific behavior at a specific time) is unaffected by the addition of other applications to the IMA platform. The RTOS ensures partition isolation, allocates processing time to the partitions, and dispatches processes for execution.

One method of achieving this is to implement a method of time slicing and to split applications into "strict tempo" and "scheduled" groups. Each strict tempo application is assured a specific amount of processing time in each time slice so that it can perform a certain number of defined algorithms in each time slice. If an application attempts to overrun its time slot, then it is timed out by the RTOS. The scheduling provides a sufficient amount of time to each time slice.

The RTOS is capable of recognizing both periodic and aperiodic processes and scheduling and dispatching all processes. It provides health status information and fault data for each partition. Since the RTOS will need to perform with a high degree of integrity for critical applications, it will have certification criteria that are commensurate with the collection of functions. It should, therefore, be as simple as possible.

The RTOS also manages the allocation of logical and physical resources on behalf of the applications. It is responsible for the management of memory and communications and receives interrupts associated with power failure and hardware error, relaying these incidents to the HM function, which in turn directs the necessary actions to enable recovery or otherwise. It should also channel application-specific software interrupts or events to the appropriate application to a defined timescale. As manager of all physical resources in this multiprogramming environment, the RTOS monitors requests for resources and controls access for those resources that cannot be used concurrently by more than one application. It has access to all memory, interrupts, and hardware resources.

The RTOS software should manage the redundancy within the IMA platform and be capable of reporting faults and executing subsequent actions. In its allocation and management of resource requests and communication interfaces from and to applications, it has limited access to the applications' memory. The RTOS monitors the hardware responsible for the integrity of the software partitions and communicates with the HM function about software and hardware integrity failures. The health monitor function is specific to the IMA platform and to the specific aircraft installation; therefore, this software is partitioned into an RTOS HM and a recovery strategy table. The latter requires configuration definition and recovery strategies embedded within it. The RTOS contains an error-reporting capability that can be accessed by applications. If an application detects a fault in its operation, it is able to report this to the RTOS, which in turn invokes the health monitoring function. It is the role of the application to inquire about health status and any reconfiguration that might have been performed. The RTOS interface enables interoperability to be achieved among application software products developed by different teams.

There are combinations of applications that have varying overall complexity. Simple applications use only the basic features of a full RTOS.

36.9 Health Monitor Software Function

The HM function resides with the RTOS and interfaces with a recovery strategy table defined by the aircraft designer or system integrator. The HM is responsible for monitoring hardware and software faults within the avionics platform and functional faults within the RTOS. Health monitoring responsibilities are divided among software applications and the RTOS. The high-level strategy is for applications to perform their own monitoring and advise the RTOS of their status. On a practical level, not every application-level fault can be self-monitored. Therefore, the RTOS has the ability to detect rogue applications and, when needed, shut down applications that may cause faults to propagate through the IMA system.

Faults detected within the RTOS include application violations, communication failures with remote devices, and faults detected by applications and reported back to the RTOS. The recovery table of faults is used to specify the action to be taken in response to the particular fault. This action would be initiated by the HM and could include the termination of an application and the initiation of an alternative application together with an appropriate level of reporting. The recovery action is largely dependent on the design of the IMA system. The HM should determine the need for action and initiate the recovery process. It is worth pointing out that HM tables are configured at design time and then loaded into the system using methods specific to the platform. Therefore, health monitoring decisions are configured, not programmed.

The HIS manages the physical hardware resources on behalf of the RTOS. The HIS maps the logical requests made by the RTOS onto the particular physical configuration of the core hardware. Hardware faults detected by the built-in test (BIT) may include memory and processor faults and faults with the I/O interface. Faults are reported to the central maintenance system, but some faults need to be reported and acted upon at a higher level. Therefore, an interface between the HM and built-in test equipment (BITE) functions is recommended.

36.10 Summary

Avionics software requirements place performance and integrity demands on the IMA platform, including the core software and RTOS. The core software supports one or more avionics applications and allows independent execution of those applications. Partitioning and functional isolation are the key to IMA. A partition can be thought of as a software program running in a single application environment. For systems that require large software applications, the concept of multiple partitions providing a single application should be recognized. The RTOS interface provides a set of services to application software for control of scheduling, communication, and status information of its internal processing elements. Together, these principles enable millions of lines of avionics software code to be readily developed, approved, and updated as necessary to meet operational needs of an aircraft over its entire lifetime.

The emergence of the ARINC 653-compliant RTOS has enabled the operation of multiple federated systems on a single computer. This was a major step forward over the strictly federated approach where each avionics function resided in a dedicated unit. However, many software applications remain largely federated. The next step is the development of standards that support the development of integrated applications that provide time and space partitioning in a system of processing cores, or even a system of modules. The next generation will move from a single-processor module to modules that contain many processors. Soon it is likely that individual modules will likely host more processors that were attached to the entire aircraft network in the early IMA systems. The requirement for multiple processing modules to ensure reliability, availability, and safety will continue to be paramount.

Bibliography

ARINC Report 651, Design Guidance for Integrated Modular Avionics. Published by ARINC, www.aviation-ia.com/aeec. Accessed on July 7, 2014.

ARINC Specification 653, Avionics Application Software Standard Interface. Published in multiple parts by ARINC, www.aviation-ia.com/aeec. Accessed on July 7, 2014.

    Part 0—Overview of ARINC 653
    Part 1—Required services
    Part 2—Extended services
    Part 3—Conformity test specification
    Part 4—Subset services
    Part 5—Core software required capabilities

RTCA DO-178/EUROCAE ED-12, Software Considerations in Airborne Systems and Equipment Certification. Published by RTCA, Washington, DC, www.rtca.com; Published by EUROCAE, Paris, France, www.eurocae.org. Accessed March 2014.

RTCA DO-248B/ED-94-B, Final Report for Clarification of DO-178B, Software Considerations in Airborne Systems and Equipment Certification. Published by RTCA, Washington, DC, www.rtca.org; Published by EUROCAE, Paris, France, www.eurocae.org. Accessed March 2014.

RTCA DO-255/EUROCAE ED-96, Requirements Specification for an Avionics Computer Resource (ACR). Published by RTCA, Washington, DC, www.rtca.org; Published by EUROCAE, Paris, France, www.eurocae.org. Accessed March 2014.

RTCA DO-297/EUROCAE ED-124, Integrated Modular Avionics (IMA) Guidance and Certification Considerations, Published by RTCA, Washington, DC, www.rtca.org; Published by EUROCAE, Paris, France, www.eurocae.org. Accessed March 2014.

37

Time-Triggered Protocol

Mirko Jakovljevic
TTTech Computertechnik AG

37.1 Introduction

Time-triggered protocol (TTP) is a high-speed, masterless, multicast, dual-channel 5 Mbit/s (bus) 20 Mbit/s (star) field bus communication protocol for safety-critical embedded applications in the transportation industry. TTP communication controllers provide built-in health monitoring, system synchronization, and redundancy services for straightforward development of fault-tolerant embedded systems. A network designed with TTP guarantees manageable modular system design, simplified application development with minimized integration effort, strictly deterministic communication, and new levels of safety at lower total life-cycle costs. TTP enables the development of physically distributed, but fully integrated, time-triggered architectures (TTAs) for modern avionics or aerospace control systems. Hence, the development of distributed safety-critical, hard real-time computing and networking for smart control systems (see Figure 37.1) in *more electric* or *all electric* aircraft is fully supported.

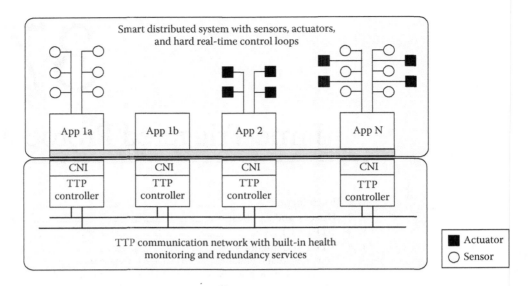

FIGURE 37.1   Distributed TTP-based system with LRUs or LRMs connected by the dual-channel communication network.

37.2 History and Applications

TTP has been developed over the last 25 years as a fully distributed and strictly deterministic safety-critical computing and networking platform. TTP (SAE AS6003) is an open industry standard available on the SAE website [1]. TTP is designed to provide levels of safety required for the aerospace applications as well as to support its wider cross-industry use.

TTP is deployed in a variety of aerospace applications such as full authority digital engine control systems for Lockheed Martin F-16 and Aermacchi M-346, cabin pressure control systems in Airbus A380, and electric and environmental control systems in Boeing 787 Dreamliner and new regional aircraft, and it is used in flight control systems for Embraer Legacy and Bombardier CSeries. This makes TTP a key contender for safety-critical subsystems in aircraft under development. TTP is also targeted for onboard space control applications and is applied in development and series projects in other industries such as automotive, off-highway, and railway.

TTP is suited for a range of applications in which the following objectives must be accomplished:

- Configurable, table-driven integrated modular platforms with built-in health monitoring and fault localization suitable for *all electric* aircraft functions
- Optimization of aircraft systems through replacement of centralized functions with smaller physically distributed but tightly integrated line-replaceable units (LRUs) and corresponding sensors and actuators found in *more electric* and *all electric* aircraft
- Simplified design of complex distributed systems with straightforward integration of fault-tolerant hard real-time subsystems delivered by different suppliers or teams

37.3 TTP and TTA

In event-triggered systems, the system environment drives temporal behavior. The system reacts to external demand for computing or communication resources when requested. Delays, hazards, and jitter are the most likely consequences if the system is not able to satisfy demand for resources. This may lead to rare system failures that cannot be easily analyzed. The systems based on the TTA paradigm anticipate resource demand in advance and provide just-in-time access to resources. Instead of being driven by events, TTA is guided by the progression of time with an exact resource use plan tailored to avoid adverse operating conditions. Time-triggered systems make optimal use of resources beyond the saturation point, where an event-driven system spends its useful time working on the resource-sharing conflict resolution or recovery operations. As a core component of the TTA, TTP uses the time-triggered approach to system communication. TTA needs a common time base to provide precise temporal specification of communication interfaces with static distribution of available communication bandwidth to the nodes. Hence, the communication conflict is ruled out by the protocol design.

The deterministic and predictable behavior of distributed fault-tolerant systems is not only dependent on the availability of resources and precise temporal interfaces. Every subsystem should have the same notion of operational status of all other subsystems to carry out a correct action compatible with actions of other subsystems. Furthermore, asymmetric failure conditions among subsystems can create casual ambiguity about failure root causes or inconsistent notion of system status. As a result, health diagnosis, failure localization, and fault-tolerance mechanisms can be difficult to maintain, and the system can become unstable, which can impede the development, integration, and maintenance of complex systems. Manageable and scalable safety-critical systems should completely separate the communication system from the fault-tolerant behavior and application software functionality. Robust partitioning must prevent any latent failure propagation path.

During the creation of TTP, all these issues were addressed to reduce system complexity, simplify integration, and enable the development of a fully deterministic, fault-tolerant, hard real-time platform for distributed computing and networking. This chapter begins with fundamentals of TTP communication and basic LRU structure and continues with a more detailed presentation of available safety, redundancy, and communication services. It also provides a concise description of application development for leveraging key strengths of TTA. Finally, this chapter discusses the reasons for straightforward integration, composability, interoperability, and scalability of TTP-based systems and gives a future outlook for this communication technology.

37.4 TTP Fundamentals

37.4.1 Time-Triggered Communication Principles

TTP implements a time division multiple access (TDMA) scheme that avoids collisions on the bus; therefore, communication is organized into TDMA rounds of the same length. A TDMA round is divided into slots with flexible length. Each LRU in the communication system has one slot—its sending slot—and sends frames in every round (Figure 37.2).

The cluster cycle is a recurring sequence of TDMA rounds. In different rounds, different messages can be transmitted in the frames, but in each cluster cycle, the complete set of state messages is repeated. The messages are broadcast to every node on the bus at predefined times with known latency, thus guaranteeing hard real-time arrival of the message at the target LRU. Safety-critical features of the TTP are implemented at higher abstraction levels and do not impose limitations on the physical layer. Therefore, the physical layer is not a part of the protocol specification, and different well-established media, transformer coupling, and coding schemes are supported by available TTP controllers [2].

A TTP cluster consists of a maximum of 64 LRUs or line-replaceable modules (LRMs) with a TTP controller. Typically, all relevant data in distributed control loops are sent periodically, once in a TDMA round. This means that the update frequency can be as short as 1 ms or shorter, depending on the message length and number of LRUs in the TTP cluster. Future generations of TTP controllers can offer much higher communication speed range and faster control loops, using standard physical layers based on EIA-485, 100BASE-T, and MIL-1553.

A TTP communication system autonomously establishes a fault-tolerant global time reference and coordinates all communication activities based on the globally known message schedule specified at the

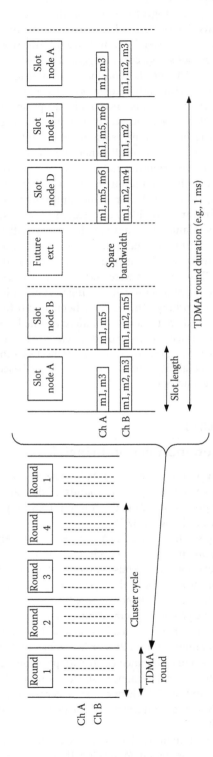

FIGURE 37.2 TTP communication cluster cycle with scheduled slots and rounds. Different sets of messages can be sent in different rounds in the same slot.

design time. It requires all communication participants to comply with an exactly specified and rigidly enforced temporal communication schedule that serves as a strict communication interface definition.

TTP-based systems can be designed to tolerate any single LRU fault. As a result of redundant communication channels, TTP tolerates a single channel failure. A correct receiver will consider the sender correct if the sender is correct on at least one of its replicated network interfaces (provided the respective channel is correct). The fault assumptions in TTP cover communication faults such as incorrect timing, inconsistent communication, and state notion differences in different controllers caused by asymmetric (Byzantine) transmission errors.

Incorrect outputs from a single data source (logic faults) caused by the host software or external I/O as an out-of-range sensor data or single-event upset (SEU) are tolerated by voting from replicated data sources supported by the fault-tolerant communication (FT-COM) layer and TTP's precise notion of time and state. In special cases, other application-specific diagnostics or specific architectural or design measures such as redundant topologies, error-correcting memories, and fault-tolerant host and software design can be used to meet the safety requirement of an application.

37.4.2 Data Frame Format

The frame size allocated to a node can vary from 2 to 240 bytes in length, each frame usually carrying application messages (see Figure 37.3). Four-bit frame header data are sent before the data frame that is protected by a 24-bit cyclic redundancy checksum (CRC). The first bit in the frame header shows whether or not the frame explicitly carries protocol state information. Status information can be transmitted implicitly as a part of the CRC or explicitly within the message data field. The other three bits represent the code of the requested mode change. Due to the static definition of the TDMA schedule frame and known frame arrival time, message and target identifiers are not required.

37.4.3 Line-Replaceable Unit with TTP

A TTP cluster consists of a set of LRUs or LRMs connected by a dual replicated communication channel. Each LRU contains hardware components such as a TTP communication controller and a host controller with software modules (application, FT-COM, and real-time operating system [RTOS]) running on the host (see Figure 37.4).

The TTP field bus connects different nodes and transmits data frames. Frames contain application data and state messages directed to different distributed applications running on a node or subsystem. An application consists of application tasks executable on one or several hosts. Depending on the application software and I/O hardware, the host can acquire and log data from sensors, control the actuators, provide data-processing capabilities for a distributed control loop, or perform a combination of these tasks.

37.4.4 Configurable Table-Driven Communication System

The configuration of message schedule and system networking architecture, as well as other safety-related properties, rely on the message descriptor list (MEDL). The MEDL is defined at design time and used by the TTP controller to autonomously communicate with its host and other TTP cluster nodes (LRUs).

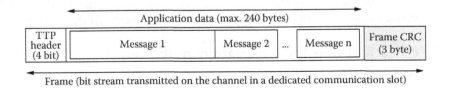

FIGURE 37.3 TTP frame format with low communication overhead (N-frame).

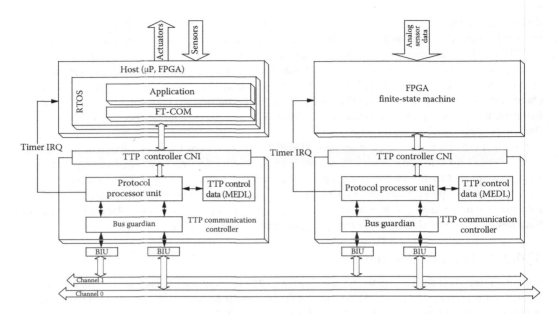

FIGURE 37.4 Electronic modules (LRU or LRM) consisting of the host processor (left) or FPGA state machine (right, e.g., smart sensor configuration), TTP controller, and bus guardian.

Overall temporal behavior of the communication system is governed by the information stored in the MEDL. System design tools are used to configure and verify the configuration of all LRUs in a system. The configuration of communication system and host application can be simply uploaded over the TTP network. If some communication bandwidth and slots are reserved for the system upgrades (e.g., new subsystems and LRUs), changes in the MEDL are not needed if such upgrades are later implemented.

MEDL stores information such as clock setup data for global timing and communication rate; communication schedules with slots, rounds, and cycles; transmission delays taking into account distances between nodes; bus guardian parameters; startup parameters; and various service and identification parameters. The configuration data also contain CRCs used for continuous scrubbing and self-checking of all configuration data structures.

37.4.5 Global Time Base

The decentralized clock synchronization in TTP provides all nodes with an equivalent time base without the use of any dedicated external time source. Accurate global time is essential for the design of deterministic fault-tolerant distributed systems. It supports the coordinated operation of redundant systems and prevents a common-mode clocking failure by built-in and formally proven fault-tolerant mechanisms [3].

Distributed LRUs in a TTP communication system develop a masterless fault-tolerant global network time reference. The globally known schedule stored in the MEDL prescribes expected message arrival time for every LRU in the system. The difference between the expected and the actual message arrival time is used to correct the local time for every correctly received frame. If an LRU is authorized for the time synchronization, the measured time of the arrival of its frame will be used in the synchronization process.

37.4.6 TTP-Based LRU as Fault Containment Region

37.4.6.1 Communication Network Interface

The communication network interface (CNI) of a TTP controller is a dual-port memory used for data exchange with other hosts. Control signals are not required for the communication with other LRUs in

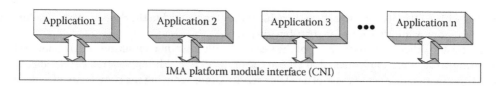

FIGURE 37.5 The CNI enables robust partitioning and independent operation of communication system. The CNI operates as a conflict-free shared memory for data exchange between distributed applications.

the network; therefore, control signal errors cannot propagate to the network. As the communication system decides when to transfer data, the application or host cannot influence the timing of the message transmission stored in the CNI. The communication subsystem reads message data from the TTP controller's CNI at the predefined fetch instant and delivers it to the CNIs of all receiving LRUs of the cluster at the known delivery instant, overwriting the previous version of the frame.

The CNI separates the behavior of the application software from the operation of the communication network and makes them mutually independent. It is a *temporal firewall* that prevents the error propagation that may influence temporal properties of the communication system. Thus, the CNI creates a well-defined fault containment region; faulty host hardware or application software will never be able to influence the operation of the TTP communication network or cause data collisions. In fact, CNI behaves as a shared conflict-free data exchange interface. It encapsulates and hides all functional properties that are not relevant for the system-level operation while making globally relevant data available throughout the network (Figure 37.5). The CNI integrates all applications functionality in the network as defined at design time. It represents a clean interface between application subsystems and provides robust partitioning between the application software and the TTP communication network.

37.4.6.2 Bus Guardian

The fault containment of an LRU is maintained by the controller's CNI, but additional protection mechanisms should ensure that the failure of the TTP controller will not propagate to the network. The bus guardian is an independent unit that protects the TTP network from timing failures of the controller. Any observable effect of a faulty node that impacts temporal behavior (e.g., physical faults, syntactic timing errors) of the prescribed communication interface will be masked by the guardians and will not propagate any further. An internal bus guardian has its own oscillator and is not dependent on controller timing. Depending on the system safety requirements, an external bus guardian can be also added. The coupling of bus guardians determines the topology of the system.

37.4.7 Real-Time Health Monitoring and Redundancy Management

The system health monitoring and continuous error prevention using redundancy management are integral parts of the TTP communication network, and they are robustly separated from the application software functionality. Any LRU connected to the system is continuously monitored. Adherence to the interface specification is strictly enforced in real time without any additional measures in the application software. A noncompliant or faulty component is not allowed to interfere with the continued operation of the system. The failures of intelligent sensors and data-processing LRUs that do not result in temporally changed behavior at the application level may simply cause the delivery of false data (e.g., SEUs, sensor failure). Such problems are tolerated by active redundancy in real time and voting.

In some safety-critical systems such as flight control computer (FCC) software, the application software represents only one-third of the total lines of the source code. At the same time, the portion of code written to support system redundancy management and fault detection monitoring exceeds 55% of the total code (e.g., Boeing 757 FCC software [4]). This part of the software code can be surprisingly complex to develop and verify, thus adding disproportional costs for system development and integration.

The TTA, with TTP in its core, significantly reduces the effort linked to the development of redundancy and health monitoring algorithms in distributed systems.

In contrast to other field buses, many of the essential fault-tolerant computing and distributed health monitoring services are implemented in TTP at the protocol level. The application development can be focused on functional issues and use of available services, thus reducing the complexity of software and the development effort for safety-critical distributed systems.

37.5 TTP Communication Protocol Layering

The interaction of different communication services, fault-tolerance management, and software applications can be easily represented by using a conceptual reference architecture in the form of a straight-forward layered model (Figure 37.6). The most important aspect of layering in TTP is the separation of the communication system, the redundancy and fault-tolerance management, and the application development. They are separated by clean data-sharing interfaces configured at design time—host CNI and TTP controller CNI. The underlying layers govern their temporally deterministic behavior, without any control signals propagating from the application to the network. Interlayer interfaces such as CNI significantly contribute to the partitioning and fault containment of a single LRU.

37.5.1 Physical Layer

TTP does not prescribe a specific bit encoding or the physical media used. The TTP bus can consist of two independent physical channels that may be based on different physical layers. TTP may work using a shared broadcast medium in bus, a star, or mixed bus–star topologies, which are determined by the location and capacity of the bus guardian devices in the system. The boundaries of the propagation delay must be known and given in the timing schedule stored in MEDL. The choice of a suitable physical layer depends on the application constraints regarding transmission speed, physical environment, and distance among nodes.

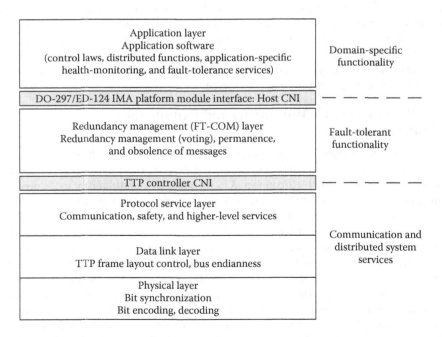

FIGURE 37.6   TTP layers as a conceptual reference architecture.

37.5.2 Data Link Layer

The data link layer provides the capability to exchange data frames among the LRUs. This layer defines the format of TTP frames with header, status, data, and CRC information. Frames are transmitted on both channels in parallel for safety-critical messages, but if the system design permits relaxed replication of less critical messages, different messages can be sent on both channels. An LRU's status information (i.e., its own perception of the system status) for maintaining system consistency, such as global time, current slot and round position, current mode, pending mode change, and membership vector, is stored in the controller state and broadcast to the network for status and data transmission agreements.

During the synchronized operation, TTP understands three types of data frames:

- *Initialization (I-)frames* are used for the integration of nodes and contain explicit controller state information but do not carry any data. Unused communication slots are filled with these types of frames. These frames are sent when no application data transfer is required in the system.
- *Normal (N-)frames* are used for clock synchronization on correctly transmitted frames and low-overhead data transfer. They carry data and implicit controller state information hidden in the CRC.
- *Extended (X-)frames* are used for integration, data transfer, and high-speed processing of explicit controller state information. An X-frame is, in fact, an I-frame with application data transfer and clock synchronization capability.

During the startup, TTP controller uses another type of frame similar to I-frame:

- *Cold-start frames* contain the global time of the sender and its position in the message schedule (round and slot). At that time, other status data are unknown (e.g., membership) and not a part of the frame.

The frame CRC is calculated by using a configuration table identifier as a seed value common to all MEDLs in the network and uses different parts of its binary value for each channel, thus preventing crossed-out channel connections or communication of LRUs with incompatible configurations.

37.5.2.1 Bus Access Control

Normally, every LRU has its own slot in every round, but different LRUs can use multiplexed slots in different rounds if defined in the MEDL.

37.5.2.2 Physical Distances

The maximal distance between nodes exceeds 130 m, which is beneficial for a larger airplane and is not dependent on protocol features such as collision avoidance (e.g., controller area network [CAN] CSMA/collision avoidance [CA]). The maximum distance that can be achieved in a specific TTP architecture depends on the underlying physical layer, coding schemes, shielding, electromagnetic interference (EMI) environment, and driving electronics.

37.5.2.3 Bus Latency, Jitter, and Efficiency

Delays are not introduced by protocol services such as clock synchronization, membership, acknowledgment, or message ordering. The share of control data in the message is reduced to bare minimum. No additional latency is imposed on the system in case of transmission errors, as no retries are attempted and there are no collision checks or resolutions. The predefined communication latency of TTP networks is very low, and the bandwidth can be almost fully utilized.

The message format is kept simple. Theoretically, it provides up to the 98% message data transfer efficiency rate. The remaining 2% includes the bandwidth used for sending CRC and frame header bits. Typically, 50%–80% of the available bandwidth is used for message data transfer in avionics applications with existing TTP controllers (i.e., AS8202NF). New TTP controllers with faster internal protocol processing may provide much higher data efficiencies close to the theoretical limit of 98% as a result of significantly reduced interframe gap (IFG). The overhead caused by the IFG is dependent on the protocol processing speed and the communication speed.

Minimal latency jitter is very important in fast control loops and by-wire applications. Depending on the oscillator and physical layer, the latency jitter can be reduced to the submicrosecond range. Typical values are between 1 and 10 μs.

37.5.3 Protocol Layer

The protocol service layer establishes the protocol operation with communication and safety services and other higher-level services. Communication services cover redundant data transmission, startup and reintegration of nodes, fault-tolerant clock synchronization, and distributed acknowledgments. They simply establish the communication in a distributed time-triggered real-time system. Furthermore, communication services contain all functions required for a temporal firewall and thus completely decouple the communication system from the host.

Safety services include membership for network-wide status notion, clique avoidance for sustaining consistent communication in cases beyond fault assumption, life sign algorithms for the detection of host failures, and network protection against the timing failures of the TTP controller in the form of independent bus guardians. Safety services guarantee fail-silent behavior of a faulty node in the time domain and prevent inconsistent interactions with the existence of different LRU cliques with different controller states. This maintains the consistency of communication and prevents the distribution of faults within the cluster.

Safety and communication services guarantee consistent communication and fail-operational behavior of the TTP network by establishing an LRU as a fault containment region. Therefore, faults cannot propagate from one LRU to the rest of the TTP network. Communication services tolerate single communication faults and detect faulty nodes, while safety services prevent distribution of faults within the cluster. Membership service and distributed acknowledgment from the protocol layer provide distributed real-time health monitoring, diagnosis, and localization of faults in the system without any increase in the bus data bandwidth or delays in communication. Higher-level services contain real-time mode changes of the network schedule, external clock synchronization, and reconfiguration of the node.

37.5.3.1 Controller State as Local Notion of System State

An internal and locally calculated TTP controller status (C-state) represents a notion of the network status from the LRU's perspective based on continuous monitoring of all frame transmissions in the network. Only LRUs with agreed C-state can synchronize with the network and participate in distributed computing and networking. A C-state disagreement between an LRU and the majority of the other LRUs means that the LRU is faulty or it simply has an inconsistent status notion. The C-state contains the global time of the corresponding slot transmission, round slot position, cluster mode data, and membership information for all LRUs in the network. The agreement on C-state is possible among LRUs because of common knowledge of slot transmissions, round positions, and other network architecture design data stored in configuration tables (MEDL).

37.5.3.2 Communication Services

37.5.3.2.1 Redundant Data Transmission

TTP utilizes redundant data transmission on both channels or on the same channel at different times to suppress transient errors. Redundant units can send the same messages at different times in their own slot. The receivers decide which messages will be selected as correct from the set of redundant messages. This is done in the FT-COM layer using different replica-determinate agreement (RDA) algorithms such as majority vote, fault-tolerant average, and high/low wins, among others.

Communication on two channels is protected by the 24-bit CRC unique to the respective TTP channel. In the case of permanent failure on one channel, the second channel still makes data transmission possible. The safety-critical data are always replicated and transmitted on both channels.

Maintenance errors such as false wiring or use of incompatible LRUs are detected by the TTP because the integration into the communication system is not possible for LRUs with incorrect network wiring. If an LRU's channel A and B connectors were mistakenly connected to the network's B and A channels, respectively, this LRU would not be able to integrate into the network communication. Additional user-specific fault detection services can be defined at the application level.

37.5.3.2.2 *Cluster Startup and (Re)Integration*

Reintegration of individual components or the restart of the whole system is self-stabilizing services that support transition from an unsynchronized into a synchronized state. At power-on or reset, all nodes in a cluster try to become mutually synchronized. After power-up and initialization of the host and TTP controller, the TTP controller listens to all frames with explicit C-state for the duration of the listen timeout (Figure 37.7). The listen timeout is unique for each LRU in the network (Figure 37.8). The reception of such frames starts integration, and LRUs start to send data. Other nodes are added to the membership as the LRU recognizes their activity. This can be done only if the host is alive and ready to send data and the controller has the permission in his MEDL to perform a cold start. The number of allowed cold starts after an activation of the TTP controller is limited by a design parameter to prevent an indefinite number of startups.

A controller expelled from the system communication and membership may want to restart and try to reintegrate on I-frames or X-frames, which contain C-state information to initialize membership and the controller's clock. The following round cycle is closely observed and used to check if the settings acquired from the C-state have resulted in behavior consistent with other participants in the TTP network. After a positive decision, the reintegrating LRU starts communication on the bus.

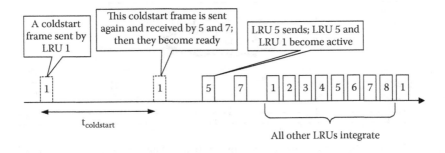

FIGURE 37.7 TTP communication startup.

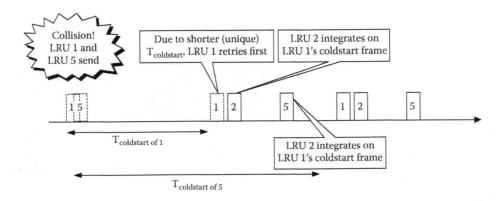

FIGURE 37.8 Startup collision between LRUs 5 and 1 leads to delayed submission.

37.5.3.2.3 Distributed Acknowledgment

In order to provide the capability to determine the status of every unit in the network and the agreement on frame transmissions in the network, TTP provides a frame acknowledgment mechanism distributed over all functioning LRUs in the network.

The distributed acknowledgment provides the sending LRU with a confirmation of successful (or failed) data transmission from other correctly working LRUs in the network. A receiving (successor) LRU considers a sending LRU *alive* if at least one of the replicated frames sent on either channel has arrived correctly at the receiver. The correctness of the frame is confirmed by a CRC check. An LRU simply listens to frame transmissions of its successors and compares its own status perception, stored in the membership information submitted in the frame, with their notion of the system status. The sending node expects either a single positive acknowledgment or two mutually reaffirming negative acknowledgments from fault-free LRUs (see Figures 37.9 and 37.10). By listening to the judgment of its successors, a sending LRU determines whether its transmission was successful or unsuccessful.

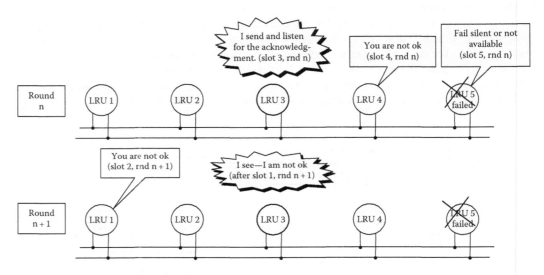

FIGURE 37.9 Update of the membership vector from the stream of mutual distributed acknowledgments for a TTP network consisting of five nodes. First two correctly working successors can provide negative acknowledgment to the LRU 3. The same applies for all other sending LRUs in the system. The failure of one or more LRUs does not invalidate the distributed acknowledgment service.

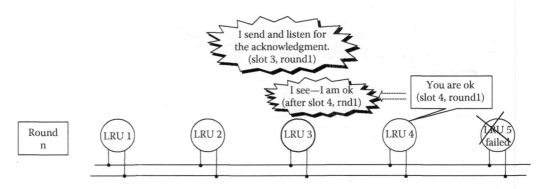

FIGURE 37.10 Positive acknowledgment for LRU 3. Only one positive acknowledgment from two operating successors is required to reassure that LRU 3 works correctly.

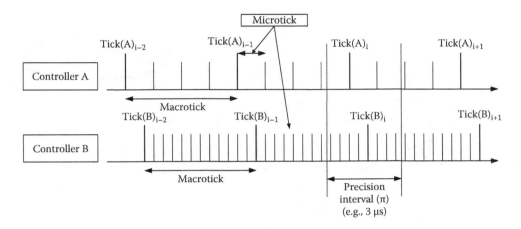

FIGURE 37.11 Sparse time base (macroticks), microticks, and precision interval.

37.5.3.2.4 Fault-Tolerant Clock Synchronization

The model of clock synchronization in TTP is based on a sparse time base called macroticks, a common time resolution in the TTP network. Every macrotick consists of microticks; the ratio of macroticks to microticks may be different for different TTP controllers depending on their oscillator. Nevertheless, the configuration data in the MEDL consider these local differences so that the sparse time base is consistent throughout the network. The precision interval is determined to enable correction of the clock. This interval is smaller than one macrotick. All nodes within the network should fit the precision interval and global time base accuracy.

The synchronization means that the common understanding of time within the precision interval $\pm\pi/2$ will be continuously adjusted and maintained (see Figure 37.11). This is done by using a fault-tolerant synchronization algorithm at the end of every round. Every TTP controller measures the difference in microticks between the slot arrival time and the expected arrival time stored in the MEDL. The difference is determined by using a fault-tolerant distributed synchronization algorithm, which takes the last four measured values, rejects the maximal and minimal ones, and averages the remaining two. As a result, the correction term for the local time in microticks is calculated. Abrupt (at once) or smooth (over several slots) adjustment can be selected for the correction of time synchronization. The communication controllers periodically check whether their macrotick is within the system-wide precision interval and will raise an error if it is not. At least four LRUs are required for the fault-tolerant masterless time base to tolerate a single asymmetric (Byzantine) timing failure.

The network can be configured to hold a specific number of nodes with spares to contribute to the clock calculation and others excluded from the global time base calculation. This can have an impact on total system cost reduction, because the maximum accuracy in the system may be governed only by a smaller number of LRUs with very accurate internal clocking.

37.5.3.3 Safety Services

37.5.3.3.1 Membership

The update of the membership vector from the continuous stream of mutual network-wide distributed acknowledgments shows which units participate in the communication and deliver correct data frames. Only correctly sending LRUs can participate in TTP communication. The membership service informs all nodes in the TTP communication network about the operational state of each LRU in the network within a latency of one TDMA round, thus simplifying the localization of error sources. With each frame submission (broadcast), LRUs provide their own perception of the network status, which can be

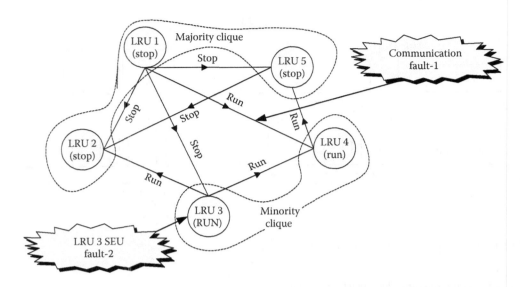

FIGURE 37.12 Development of cliques is resolved by LRU 3 and LRU 4 restart and reintegration. It can be also eliminated by other application-specific recovery operation defined at the development time.

used by other senders to acknowledge the sender's data transmission. The membership information can be either provided explicitly in the frame or hidden in the checksum at the end of each frame.

37.5.3.3.2 Clique Avoidance

If the single-fault assumption is violated (see Figure 37.12), two or more cliques with a different notion of the network status may emerge. Cliques can only occur in multifault scenarios or with asymmetric transmission faults. For example, in a system with the two possible states, RUN and STOP, two faults occur in parallel. LRU 4 misunderstands LRU 1 as a result of a communication fault and believes that the system is in the RUN state. LRU 3, as a result of internal SEU, has a false understanding of the transmission from LRU 1 and firmly believes that everybody else should be in the RUN state. All other LRUs are convinced that the system state is STOP; therefore, there are two groups (cliques) with a different notion of system status. The clique avoidance algorithm solves this kind of problem.

The consistency of the membership vector is continuously agreed upon among all nodes in the system, and conflicts are resolved within two TDMA rounds. Clique avoidance signals to the application whether it is in the majority clique or not, based on the comparison of membership vectors transmitted in the CRC of every node. Before the node sends a frame, it checks if it is in a majority clique. If it is part of the minority clique, the node signals to the application to decide how to proceed with the operation—by conducting a complete restart and new reintegration or by executing other activities appropriate to the level of safety required by the system.

37.5.3.3.3 Protection against Timing Failures

A bus guardian enforces the temporal fail silence of an LRU in the case of arbitrary faults such as "babbling idiot" or "slightly-off-specification" (SOS). A bus guardian error is immediately reported to the controller and host software. An internal bus guardian in a TTP controller has an independent copy of the communication schedule, and its clock is separate from the TTP controller. The internal bus guardian is synchronized by a start-of-round TTP controller signal.

In TTP, mixed topologies may be defined by combining bus and star topologies. Other system architecture decisions, such as parallel use of two or more double-channel TTP buses, can further improve the safety of the systems.

37.5.3.4 Higher-Level Protocol Services

37.5.3.4.1 Mode Changes

In order to provide consistent online change in the message schedule timing for all LRUs, mode changes are possible at the end of every cluster cycle. All LRUs will be able to switch at the same time to a different schedule. This permits different operating modes with different communication behavior. If several mode changes are requested within the round, the latest will be used. Any node can prevent the mode change. The current mode request is given in the header of every frame.

37.5.3.4.2 External Clock Synchronization

TTP supports external clock synchronization to provide synchronization among different TTP clusters or provide synchronization with external time sources. By adding drift correction values to the network global time, the differences between global TTP network time base and external time can be equalized.

37.5.3.4.3 Reconfiguration of LRU

This service is part of the communication controller and permits download of the MEDL data and protocol software data independently of the application. It also provides a link for the download client capable of downloading data and new versions of the software to the host.

37.5.4 Fault-Tolerant Layer

In a TTP-based system, the fault-tolerance mechanisms are implemented in a dedicated FT-COM layer. The number of replicas, voting, and reintegration remain completely transparent to the application software. FT-COM reduces redundant messages from replicated applications from other subsystems (e.g., LRUs, LRMs) to a single, agreed-upon value before they are presented to the local application but also determines the formatting and endianness of the message data. An application that uses such a fault-tolerance communication layer can thus operate with the same functionality and without any modifications in a fault-tolerant system or a standard system.

With a FT-COM layer, TTP implements all services needed to handle complex fault-tolerant distributed software systems effectively and efficiently. For the application developer, it means much simpler and faster application and control algorithm development because there are no mutual interactions between functions and they can be developed independently.

37.5.4.1 Fault-Tolerant Units and Redundant LRUs

Safety-critical, real-time systems impose high demands on the reliability and availability of the system. Active redundancy by replication and voting are the fastest fault-tolerance techniques because there is no dead time of the service. The realization of active replication demands a number of mechanisms, such as replica coordination, voting, group membership, internal state alignment, and reintegration of nodes after a transient failure. All of these mechanisms are available in TTP and are separated from the application functionality. The failures of intelligent sensors, actuators, and data-processing LRUs are tolerated by hardware replication in real time by the TTP (including the FT-COM layer). TTP does not set upper limits for the number of redundant LRUs (Figure 37.13) in a fault-tolerant unit (FTU) that behaves functionally and temporally as a single LRU.

37.5.4.2 Clique Detection, Interactive Consistency, and Replica Determinism

The integrity of network communication and fault detection beyond the single-fault hypothesis is maintained by the clique detection mechanism. Clique detection helps prevent subsystems from having an incorrect notion of system status. This mechanism is also of the utmost importance for interactive consistency and replica determinism.

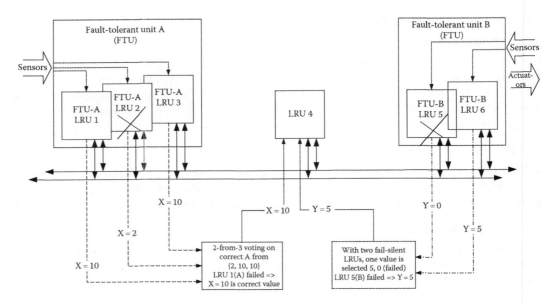

FIGURE 37.13   FTUs improve availability and safety. Active replicated LRUs provide the needed functionality if one LRU in the FTU fails. LRU 4 reads data from FTU-A (voting from triple-redundant LRUs tolerates an incorrect output from one of the three LRUs, 1, 2, and 3) and FTU-B (LRUs 5 and 6 are designed as fail-silent; in this figure, LRU 5 has failed, and the output is taken from LRU 6).

Interactive consistency requires that any transmission results in an identical (atomic) reception at all receivers. With distributed acknowledgment, membership, and clique detection services, interactive consistency guarantees that all nodes receive the same sequence of messages. In basic terms, replica determinism ensures that inputs from one data source received at the same time will be bit identical to results received at a later instant from all other working replicas. Based on the global time base and interactive consistency, the same results can be delivered only if all replicas pass through the same system states at the same time. This is possible due to the time-triggered design of LRU application and system tasks synchronized to the global time. Therefore, any event-driven behavior (e.g., interrupts) in replica deterministic FTUs and safety-critical systems in general should be avoided if possible.

Replica determinism guarantees consistent computations and transfer of functionality in the case of transient or permanent failure of one replica in the FTU. Otherwise, the switching from one replica to another can lead to a serious error and upset the controlled system. The lack of those mechanisms can severely increase the design effort of safety-critical distributed systems and add disproportionate costs to the system design.

37.5.5 Application Layer

The application layer contains application tasks running on the LRU's host microcontroller. In distributed applications, tasks are executed on several LRUs and communicate over a TTP communication network to accomplish the desired functionality. A strictly deterministic system behavior can be realized if both the application and communication systems have fully predictable temporal behavior synchronized to global system time. In Figure 37.14, the context of application tasks within the LRU and in relation to the FT-COM layer and RTOS is presented. The application software consists of established manageable software tasks with a task start and deadline during the design.

Tasks communicate over local messages within a single LRU and over global messages between tasks in different LRUs. Global messages (state variables) are transmitted through the network. The message structure, content, and time of the transmission and access are defined design. Therefore, the message

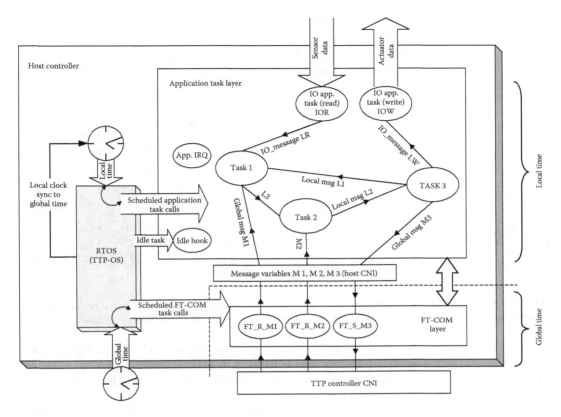

FIGURE 37.14 Application and system software (RTOS) with statically scheduled tasks and polling access to sensors and actuators.

communication and all other processing steps are always conflict-free and do not impose any delays in system behavior. Messages in TTP-based systems are transferred just-in-time.

Figure 37.15 shows how this works for every LRU in the system. Global messages (state variables) M1, M2, and M3 are handled as normal variables by application tasks. M1 is taken from two different slots on both channels of the TTP bus and unified into one value in an F1 task in the fault-tolerant layer. Depending on design, averaging or voting on all four correct values can be done in the task F1. The message M2 is taken from one TTP channel and prepared for use by task T2. The message M3 is a result of processing done by task T3 and sent in slot 5 on both TTP channels. All other messages (MX, MY) transmitted through the network are invisible to application tasks as defined at design time. An idle task is executed if no immediate application action is required. This task can be used for self-diagnosis operations or even for running another operating system (e.g., RTAI-Linux). FT-COM tasks and application tasks are executed in an order that permits timely transfer of messages from the communication channels to the processing applications tasks and vice versa. Obviously, the scheduling of the system communication and running of applications on a distributed system is the key issue in the design of hard real-time safety-critical systems with TTP. The FT-COM configuration is generated from network and communication specifications by the design tools. The developer just has to write application code on top of it.

37.5.5.1 Time-Triggered Software Behavior

In order to keep tasks simple and free from event-driven behavior, any temporally indeterministic constructs (e.g., semaphores, blocking read/writes, or variable task execution timing) should be avoided. Furthermore, event-driven functions are not needed for the operation of TTA.

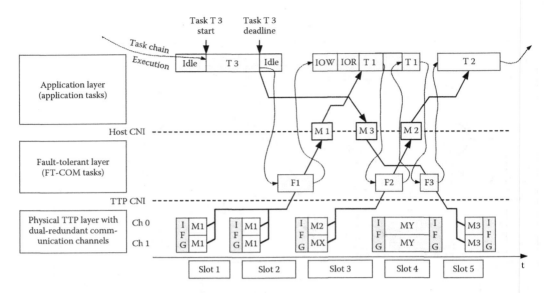

FIGURE 37.15 Statically scheduled application tasks in relation to communication message schedule with conflict-free data transfer. The temporal behavior of an LRU and the system is defined at design time.

Cyclic sensor and actuator polling scheduled at design time can be accomplished through dedicated lightweight tasks with temporally fixed task execution (e.g., IOW and IOR from Figure 37.14). Therefore, this guarantees the temporal determinism of application execution and very low system latency and jitter in distributed high-speed control loops.

Event-driven software design principles (e.g., interrupts) may be used carefully only if surplus processing capacity is reserved to prevent violation of deadlines for all combinations of interrupts. TTP can also accommodate software applications that are event driven and do not require fully deterministic system behavior but can profit from deterministic communication, built-in health monitoring, fault containment, system synchronization, and straightforward integration.

37.5.5.2 Error Handling

Even in the case of host failure, the robust separation of the communication system from application software over the CNI guarantees that temporal behavior of the TTP-based system does not change and the failure does not spill over to the network. The most severe failure mode of host computers is thus to supply either wrong data as a result of bit flips or sensor failures or no data at all at the required points in time. The host computer has strict requirements of when to output what data. Its local TTP controller and all other controllers on the network monitor its temporal behavior.

All other LRUs in the network will immediately recognize its incompliance or failure based on controller-driven membership and acknowledgment services. LRUs sending the correct amount of data at the correct time with correct frame CRC, but with faulty message content as a result of host (or sensor) failure, cannot be detected by these mechanisms. It is up to the FT-COM layer or application (for fine-grained application-specific services) to deal with this type of fault. Usually triple modular redundancy (TMR) or lane-switching logic is employed to deal with these faults.

Temporally incorrect task behavior that exceeds the given time budget is captured using error-handling tasks. In this case, the TTP controller does not send any data from the failed host. The LRU will have a chance to recover and integrate in one of the following communication rounds after completion of recovery operations. In the case of repetitive failures, the LRU may withdraw permanently from the network communication based on the TTP controller decision as prescribed at design time in the TTP controller's configuration tables.

37.5.5.3 Distributed Application Software Development and Interactive Consistency

Inconsistent communication and different notions of system state are immediately recognized and reported to the application. Failure of one LRU in an FTU will be immediately recognized by safety and communication services and tolerated by voting in the FT-COM layer.

Application developers usually expect interactive consistency in distributed systems and write their software or develop hardware assuming that every node has consistent input data. Without such distributed services based on a global time base, distributed acknowledgment, membership, and clique avoidance, the design of distributed applications becomes a painstaking process. The interactive consistency has to be implemented by the user at the application level, which, in turn, requires deep understanding of specific distributed computing issues. The implementation of interactive consistency at the application level contains potential pitfalls and risks that significantly complicate application software structure, certification processes, and system integration. The requirement for consistency may be sometimes neglected or not well understood during the planning of complex fault-tolerant control systems, but all required support is already built into the TTP communication protocol.

37.5.5.4 Application Development with TTP

The developer can design the system at a higher level of abstraction, simulate physical and control systems using *MATLAB® [5], formally verify its design [6], manually write the application (task) code, or use code generation tools from simulated models [5,6]. As the redundancy and communication layer is completely separated from the application layer, the software application developer can focus solely on functions, sensor data acquisition, control laws, and data formats.

A major part of the development for TTP-based system development is driven by the design of configuration data (e.g., communication, timing and scheduling, fault tolerance, message formatting). This significantly reduces the development effort for the design of distributed safety-critical systems with built-in redundancy and health monitoring. Using built-in TTP services can significantly reduce the design effort for complex distributed safety-critical systems.

37.6 System Integration with TTP

37.6.1 Interoperability

Because the timely delivery of messages and the accurate moment of their submission cannot be easily estimated in all cases, subsystems do not have perfect knowledge of what can be expected from other subsystems. Therefore, the interoperability between subsystems organized to jointly accomplish a function is at risk, which presents a major roadblock to straightforward integration of complex systems. TTP implements global control of network traffic management with clearly defined temporal interfaces and eliminates interoperability issues.

37.6.2 Composability

One of the key factors in guaranteeing temporal composability is the robust separation of temporal control structure from logical control structure, so a subsystem or LRU can be validated in isolation [7].

---

* MATLAB® is a registered trademark of The MathWorks, Inc. For product information, please contact:

The MathWorks, Inc.
3 Apple Hill Drive
Natick, MA 01760-2098 USA
Tel: 508-647-7000
Fax: 508-647-7001
E-mail: info@mathworks.com
Web: www.mathworks.com

In TTP, the temporal behavior is precisely defined by the design of the communication schedule and the separation of communication system and application software.

An architecture is temporally composable if the system integration does not invalidate the independently established temporal behavior of subsystems or LRUs. This also implies that the separate testing of a stand-alone LRU, as well as its testing in the system, delivers the same temporal behavior and verification results.

In TTP, none of the subsystems designed to work independently in the same network will cause unintended interaction with other subsystems. All subsystems participate in distributed fault-tolerant time synchronization, distributed acknowledgment, membership, and clique detection and jointly support interactive consistency. The fault of one LRU does not influence the network-wide services and network data transport. The applications running on subsystems and LRUs will be able to share data as defined at the design time (see Figures 37.16 and 37.17).

The communication schedule relies on a global time base. A master-free, fault-tolerant time synchronization algorithm is used to prevent any common-mode timing failures. The "babbling idiot" and "SOS" errors caused by an erroneous LRU transceiver or oscillators are captured by the bus guardian, which permits submission of frames only in a dedicated slot. The temporal composability would be much harder to achieve without stable interfaces designed to robustly separate a system into fault containment regions (i.e., LRUs or FTUs), which prevent the fault propagation to other subsystems.

The logical composability of the fault-tolerant system is supported by the robust separation of fault-tolerant functions from application software and their inclusion in the TTP. Interactive consistency and replica determinism are important properties of a composable fault-tolerant system in a temporal and logical domain (see Figure 37.18).

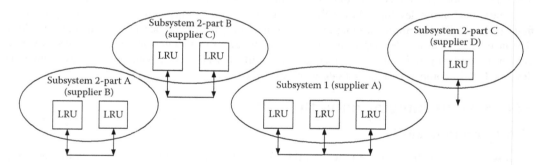

FIGURE 37.16 TTP-based subsystems can be tested separately by different suppliers without composability and interoperability challenges that multiply the integration effort for complex systems.

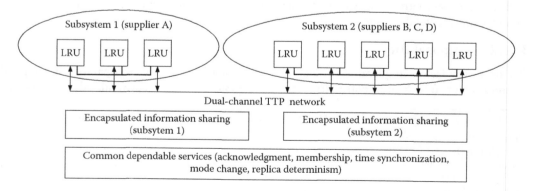

FIGURE 37.17 TTP-based distributed network management services established at the subsystem level work smoothly after system integration.

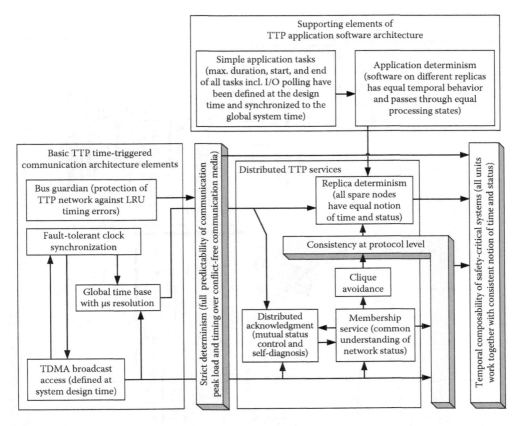

FIGURE 37.18 Origin of temporal composability of a safety-critical TTP-based system. Clique detection, distributed acknowledgment, and membership provide additional support for logical (functional) composability of fault-tolerant systems.

37.6.3 Integration and Cooperative Development

Ultimate responsibility for functional composability and integration lies with suppliers and their cooperative development processes. TTP provides all interfaces (state message formats and timing schedules) needed for accurate and unambiguous common specification and represents the baseline in a joint development effort. Independent communication with predefined temporal behavior supports separate development of the network architecture (integrator), subsystems (supplier), and LRUs (teams). This causes an unparalleled reduction of integration effort.

The physical components of distributed TTP-based systems can be verified separately with the simulated behavior of the remaining system (e.g., other subsystems to accomplish required functionality, aircraft network, flight dynamics) to provide high predictability of system-level behavior as a result of composability and interoperability of TTP systems (Figure 37.19).

37.7 Modularity and Scalability

Scalable platforms are based on a number of understandable generic elements (distributed services, software/hardware modules, and components) with clean interfaces that support the development of complex systems and scalable architectures. Scalable architectures are open to changes and additional capacity upgrades without exponential increase in development effort.

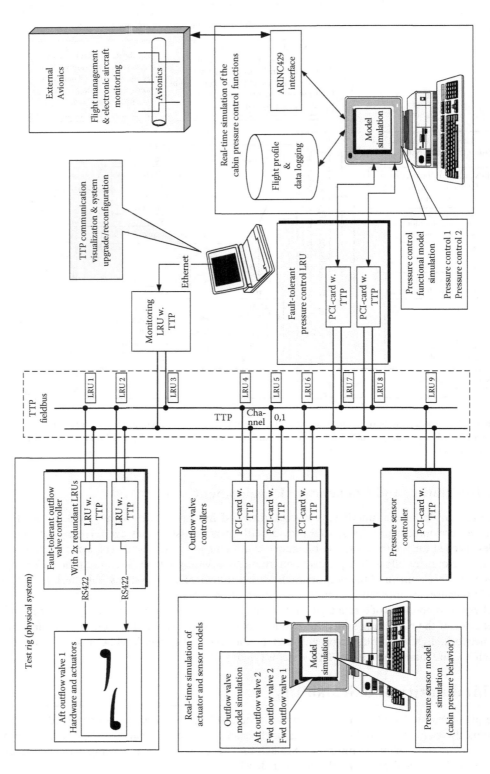

FIGURE 37.19 The development and verification of TTP-based SETTA cabin pressure control system (CPCS) demonstrator architecture using a blend of physical subsystems and simulated system behavior. (Adapted from www.vmars.tuwien.ac.at/projects/setta/index4.htm.)

In general, integration of different subsystems by different suppliers and modular reuse of designed components with sustainable airworthiness throughout several programs represent major challenges, technically as well as regulatory. TTA enables scalable and manageable development of safety-critical fault-tolerant systems and reduces system complexity. Even the increasing system complexity and a rising number of components and system size do not affect the ability to understand the behavior of TTP-based systems. Horizontal layering (abstraction) and vertical layering (partitioning) help to reduce the perceived complexity of large systems [9]. Both principles have been used in TTA platforms.

37.7.1 TTP and DO-297/ED-124 IMA Guidelines

The TTP communication data bus represents an integrated modular avionics (IMA) module and resource in compliance with the terminology and guidelines of DO-297/ED-124. TTP-based platforms incorporate an advanced fault-tolerant distributed IMA concept and also provide full support for forthcoming distributed safety-critical, real-time applications. According to DO-297/ED-124, "IMA is described as a shared set of flexible, reusable, and interoperable hardware and software resources that, when integrated, form a platform that provides services, designed and verified to a defined set of requirements, to host applications performing aircraft functions."

A TTP-based platform supports the IMA concept in compliance with DO-297/ED-124 [10] but offers some unique features for the development of distributed applications with built-in safety and fault-tolerant features.

In order to satisfy DO-297/ED-124 IMA considerations, TTP meets the following requirements:

- Shares resources with multiple applications (i.e., the TTP double-channel bus and CNI are shared by an application [a distributed application can consist of one or more tasks executed on different hosts in the network])
- Provides autonomously robust partitioning of shared resources (with spatial and temporal partitioning)
- Allows only hosted applications to interact with the platform and all others through well-defined interfaces (i.e., application software can interact only by using a host CNI on top of the FT-COM, which interacts directly with CNI and communication hardware)
- Allows configuration of resources to support reuse and modular certification of the platform (i.e., downloadable table-driven FT-COM and MEDL configuration)

The capability to protect shared resources is part of the TTP data bus design based on the use of TDMA bus access and bus guardians, which prevent access at arbitrary times. The TTP-based IMA platform provides distributed fault management and health monitoring using distributed fault-tolerant services, acknowledgment, and membership.

37.7.2 Platform Components and Modules

TTP-based DO-297 IMA platform consists of the following components and modules (Figure 37.20):

- An RTOS (TTP-OS) represents a reusable software module running on the host processor and uses the services of a TTP communication network module. RTOS integrates both the host component and TTP network module. Together with FT-COM running local TTP-OS tasks, it provides the application program interface (API) IMA functionality.
- The TTP controller module contains the TTP controller chip and the protocol software executed by the chip. The protocol software and RTOS are developed to support DO-178B Level A; the TTP controllers are developed to support DO-254 certification.

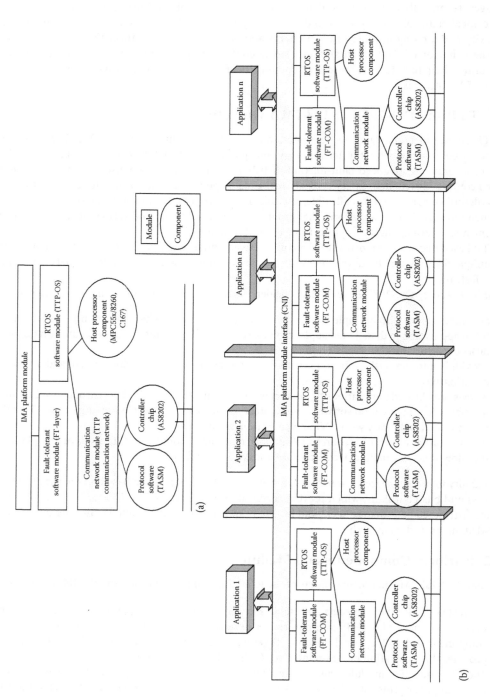

FIGURE 37.20 (a) Single TTP-based DO-297 IMA module. (b) Distributed IMA platform. All modules and applications are spatially and temporally partitioned.

37.7.3 Application Reuse

Robust partitioning at the communication level is guaranteed by the CNI, fault-tolerant time base, static access schedule (defined by the configuration defined in the MEDL), and bus guardian. At the application level, tasks are scheduled during the design to provide required functionality—this guarantees temporal partitioning at the node level, as well as spatial partitioning at the system level. The host resources for every task are predefined at design time, and the application fulfills the constraints set by the resource availability. An application can be designed independently of other applications and unintended interaction with other applications is avoided. Therefore, the application represents a reusable component that is independently modifiable.

37.8 Summary

37.8.1 Aircraft Architecture and TTP

The TTP works well in a serial backplane as well as for distributed field bus applications (see Figure 37.21). TTP-based systems can be physically distributed and separated similarly to federated systems, but the level of integration is comparable to IMA. In many cases, critical avionics systems require real-time reaction times and full synchronization of all subsystems with exchange of all essential data within milliseconds over large distances. The TTP has been developed from scratch for deterministic, safety-critical applications. Reduced weight and improved dispatchability, as well as lower total life-cycle costs, are major reasons for the use of innovative distributed architectures in modern civil aircraft. TTP seamlessly supports those objectives.

TTP enables weight optimization at the system level by placing the fault-tolerant electronics controls in the vicinity of sensors and actuators, which reduces the wiring weight and provides additional options for the replacement of hydraulic or mechanical systems with electrical or electrohydrostatic systems.

Dispatchability and maintainability are supported by longer and controllable maintenance intervals and simplified diagnostics. TTP has built-in health monitoring and redundancy management with

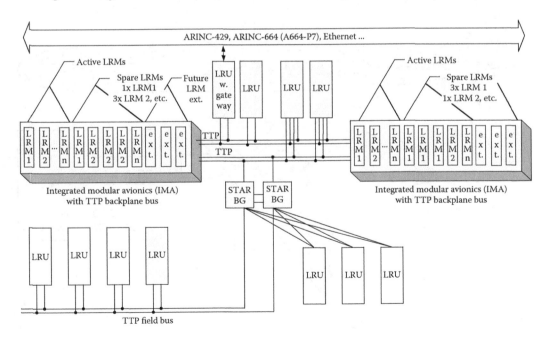

FIGURE 37.21 Integrated and federated architectures in different topologies tailored for specific needs can be created with TTP.

accurate fault localization that prevents casual ambiguity about the error sources. This supports the design of integrated vehicle health management (IVHM) as a system engineering discipline.

A modern airplane may have more complex flight dynamics as a result of more flexible composite materials. Theoretically, the engine and flight controls may be tightly integrated even on the same bus. With TTP, those functions are partitioned and remain separately certifiable. Another example is distributed power control in *more electric* aircraft, which ensures synchronized work of all systems with strictly deterministic FT-COM.

New relationships between suppliers and integrators include risk sharing and integration of complete subsystems. The suppliers will have more freedom to choose communication systems with lower life-cycle costs and straightforward integration that are appropriate for their application and system-level optimization (Figure 37.22). This, in turn, will boost the use of airworthy, modularly certifiable commercial off-the-shelf (COTS) components.

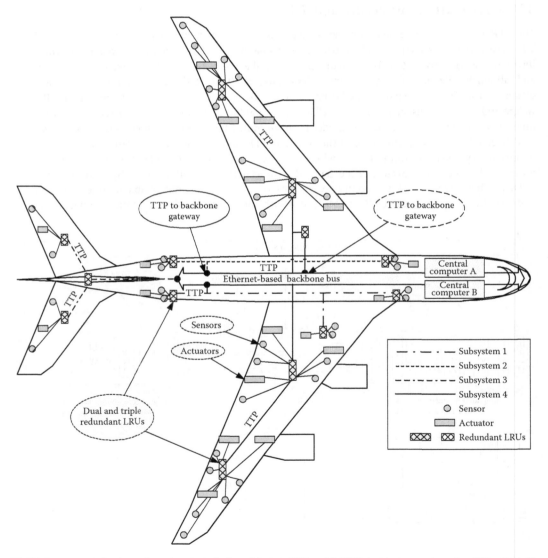

FIGURE 37.22 Subsystems based on strictly deterministic TTAs with TTP support new aircraft architectures for distributed but tightly integrated systems.

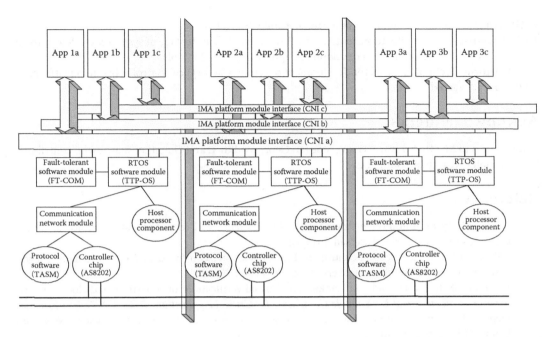

FIGURE 37.23 Distributed modular applications executed on distributed LRMs with full temporal and spatial partitioning at the module and system level.

37.8.2 Future Outlook

The speed of TTP is limited only by the specific implementation of controller technology and the physical layer in use. Therefore, higher speeds on two or more channels in different topologies can be expected in the future. Future generations of TTP controllers can also be extended to provide scalable levels of communication safety and extended support for avionics architecture development. A group of completely independent distributed software applications (Figure 37.23) at different criticality levels can be executed on a fault-tolerant system with completely deterministic behavior, mutually synchronized with microsecond resolution. Hence, TTP-based systems will establish the networked system as a fully partitioned, certifiable, embedded, fault-tolerant computer with reusable applications at different criticality levels.

Definitions

Deterministic: A system whose time evolution and behavior can be exactly predicted.

Event Driven: A system triggered by events at arbitrary instants.

Fault-Tolerant Region (FCR): A set of components that is considered to fail as an atomic unit and in a statistically independent way with respect to other FCRs.

Fault-Tolerant Unit (FTU): A set of replica-determinate nodes.

Jitter: Oscillations in the latency of data transmission from sensors to computing units.

Latency: Delay between message transmission and reception.

LRM: Line-replaceable module.

LRU: Line-replaceable unit.

Membership: A vector containing a common notion of system status and health information.

TDMA: Time division multiple access.

Time-Triggered: A system triggered by the progression of time at predefined instants.

TMR: Triple modular redundancy, supported naturally by TTP.
TTA: Time-triggered architecture; contains TTP and offers full advantage of system-level determinism including the application development approach with TTP tools.
TTA Platform: Technology used for development of reusable aerospace control subsystem platforms.
TTP: Time-triggered protocol.

Acknowledgments

Many thanks to Martin Schwarz, Georg Stoeger, Georg Walkner, and Guenther Bauer, who contributed to this chapter with their comments.

References

1. SAE (Society of Automotive Engineers). SAE AS6003 "TTP communication protocol." http://standards.sae.org/as6003. Accessed on April 30, 2014.
2. TTTech. TTP communication controllers. http://www.tttech.com/products/aerospace/flight-rugged-hardware/components/ttp-controller/. Accessed on April 30, 2014.
3. Pfeifer, H., Schwier, D., and von Henke, F.W. Formal verification for time-triggered clock synchronization, *Proceedings of Dependable Computing for Critical Applications 7 (DCCA 7)*, San Jose, CA, Dependable Computing and Fault-Tolerant Systems Series, IEEE Computer Society, New York, January 1999, pp. 207–226.
4. Spitzer, C.R. Digital avionics systems, Tutorial, *24th DASC Conference*, Washington, DC, October 31, 2005.
5. Mathworks. http://www.mathworks.com/model-based-design/. Accessed on July 14, 2014.
6. Esterel Technologies. http://www.esterel-technologies.com/products/scade-system/. Accessed on July 14, 2014.
7. Kopetz, H. *Real Time Systems: Design Principles for Distributed Embedded Applications*, Kluwer Academic Publishers, Norwell, MA, 1997.
8. SETTA, SETTA downloads. www.vmars.tuwien.ac.at/projects/setta/index4.htm.
9. Maier, R., Bauer, G., Stöger, G., and Poledna, S. Time triggered architecture: A consistent computing platform, *IEEE Micro*, 22(4): 35–45, 2002.
10. DO-297/ED-124. Integrated modular certification avionics (IMA) design guidelines and certification considerations, RTCA and European Organization for Civil Aviation Electronics (EUROCAE), 2005.

38

Digital Avionics Modeling and Simulation

Jack Strauss
Xcelsi Group

Joseph Lyvers
Xcelsi Group

Terry Venema
Xcelsi Group

Andrew Shupe

38.1 Introduction

In order to realize unprecedented but operationally essential levels of avionics system performance, reliability, supportability, and affordability, commercial industry and the military will draw on advanced technologies, methods, and development techniques in virtually every area of aircraft design, development, and fabrication. Federated and integrated avionics architectures, hybrid systems architectures, and special-purpose systems such as flight control systems, engine control systems, navigation systems, reconnaissance collection systems, electronic combat systems, weapons delivery systems, and communications systems all share certain characteristics that are germane to digital systems modeling and simulation. All of these classes of avionics systems are increasing in complexity of function and design and are making increased use of digital computer resources. Given these advancements, commercial and military designers of new avionics systems and of upgrades to existing systems must understand, incorporate, and make use of state-of-the-art methods, disciplines, and practices of digital systems design. This chapter presents fundamentals, best practices, and examples of digital avionics systems modeling and simulation.

38.2 Underlying Principles

The following comparison illuminates the conundrum of modeling and simulation. The results of most mathematical processes are either correct or incorrect, but modeling and simulation has a third possibility. The process can yield results that are correct but irrelevant (Strauss, 1994). Given this, perhaps, startling but true realization of the potential results of modeling and simulation, it is important to understand the different perspectives that give rise to the motivation for modeling and simulation, the trade space for the development effort to include the users and systems requirements, and the technical underpinnings of the practice.

38.2.1 Historic Perspective

The past 45 years of aviation has seen extraordinary innovation in all aspects of complex systems design and manufacturing technology. Digital computing resources have been employed in all functional areas of avionics, including communication, navigation, flight controls, propulsion control, and all areas of military weapon systems. As analog, mechanical, and electrical systems have been replaced or enhanced with digital electronics, there has been an increased demand for new, highly reliable, and secure digital computing techniques and for high-performance digital computing resources.

Special-purpose data, signal, and display processors were commonly implemented in the late 1960s and early 1970s (Swangim et al., 1989). Special-purpose devices gave way to programmable data, signal, and display processors in the early to mid-1980s. These devices were programmed at a low level; assembly language programming was common. The late 1980s and early 1990s have seen commercial and military avionics adopting the use of high-performance general-purpose computing devices programmed in high-order languages. The USAF F-22 fighter, for example, incorporates general-purpose, commercially available microprocessors programmed in Ada. The F-22 has an integrated avionics architecture with an operational flight program consisting of nearly 1 million lines of Ada code and onboard computing power on the order of 20 billion operations per second of signal processing and 500 million instructions per second (MIPS) of data processing. Additionally, there is increasing use of commercial off-the-shelf (COTS) products such as processor boards, storage devices, graphics displays, and fiber-optic communications networks for many military and commercial avionics applications. The F-35 Joint Strike Fighter uses commercial processors programmed predominantly in the C language. Current program data count nearly 6 million lines of code in onboard systems with another 9 million lines of code for off-board systems.

COTS product designers and avionics systems developers have made it a standard engineering practice to model commercial computer and communications products and digital avionics products and systems at all levels of design abstraction. As the complexity of electronics design dramatically has increased, so too has modeling and simulation technology in both functional complexity and implementation. Complex computer-aided design (CAD) software can be several hundred thousand source lines of code without taking into account the extensive libraries that may be purchased with these systems. These software products require advanced engineering computing resources with sophisticated file and storage structures and data management schemes. Workstations with multiple GHz processors with several terabyte disk drives, connected with high-speed local area networks (LANs), are common. Additionally, special-purpose hardware environments, used to enhance and accelerate simulation and modeling, have increased in performance and complexity to supercomputing levels. Hardware emulators and rapid prototype equipment can reach near real-time system performance. At this point in history, the complexity and performance of modeling and simulation technology is every bit equal to the digital avionics products they are used to develop.

38.2.2 Economic Perspective

For commercial systems and product designers, time to market is a critical product development factor that has significant impact on the economic viability of any given product release. The first product to market generally recoups all of the nonrecurring engineering and development costs and commonly captures as much as half the total market. This is why technologies aimed at decreasing time to market remain important to all commercial developers. Early analysis shows the amount of

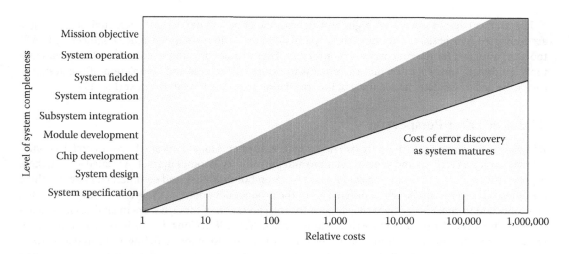

FIGURE 38.1 Relative cost of error discovery as design progresses.

development time saved as a result of digital system modeling and simulation (Donnelly, 1992). At a lower level, cost-sensitive design has two significant issues: the learning curve and packaging.

The learning curve is best described as an increase in productivity over time. For device manufacturing, this can be measured by a change (increase) in yield or the percentage of manufactured devices that survive testing. Whether it is a chip, a board, or a system, given sufficient volume, designs that have twice the yield will generally have half the cost (Hennessy and Patterson, 2011). The design reuse inherent in digital modeling and simulation directly enhances the learning curve. Packaging at the device, board, or system level has cost implications related to fundamental design correctness and system partitioning. A case study of performance modeling within specific system partitions is presented later in this text.

For military systems designers, many of the same issues affecting commercial systems designers apply, especially as more commercial technologies are being incorporated as implementation components. Additionally, as will be shown (in Section 38.2.3), mission objectives and operational requirements are the correct point of entry for modern, top-down design paradigms. However, once a development effort has been initiated, the relative cost of error discovery (shown notionally in Figure 38.1) is more expensive at each successive level of system completeness (Portelli et al., 1989). Thus, a low-price solution that does not fully meet system requirements can turn into a high-cost problem later in the development cycle.

38.2.3 Design Perspective

Bell and Newell divided computer hardware into five levels of abstraction (Bell and Newell, 1973): processors–memory–switches (system), instruction set (algorithm), register transfer, logic, and circuit. Hardware systems at any level can also be described by their behavior, structure, and physical implementation. Walker and Thomas combined these views and proposed a model of design representation and synthesis (Walker and Thomas, 1985) that includes the architecture level (system level), algorithmic level, functional block level (register transfer level), logic level, and circuit level. Each of these levels is defined in terms of their behavioral domain, structural domain, and physical domain. Behavioral design describes the behavior or function of the hardware using such notations as behavioral languages, register transfer languages, and logic equations. Structural design specifies functional blocks and components such as registers, gates, flip-flops, and their interconnections. Physical design describes the implementation of a design idea in a hardware platform, such as the floor plan of a circuit board and layout of transistors, standard cells, and macrocells in an integrated circuit (IC) chip.

Hierarchical design starts with high-level algorithmic or behavioral design. These high-level designs are converted (synthesis) to circuits in the physical domain. Various computer-aided engineering (CAE) tools are available for design entry and conversion. Digital modeling and simulation technologies and tools are directly incorporated into the process to assure correctness and completion of the design at each level and to validate the accuracy of the translation from one level to the next (Liu, 1992).

38.2.4 Market Perspective

It is generally assumed that there are two major market segments for avionics—the commercial avionics market (targeting airlines and general aviation) and the military avionics market. As stated earlier, there is great similarity in the technological forces at work on both military and commercial systems. There are, however, several fundamental differences in the product development cycle and business base that are important to consider because they impact the interpretation of the cost–benefit analysis for modeling and simulation. These differences are summarized in Table 38.1. Consider the impact of commercial versus military production volumes on a capital investment decision for modeling and simulation technology. The relative priority of this criterion is likely to differ for commercial products as compared to military systems.

38.2.5 Requirements in the Trade Space

Technology (commercial or military) without application, tactical, or doctrinal context is merely engineering curiosity. The development of an avionics suite or the implementation of an upgrade to an existing suite requires the careful balance of many intricate constraints. More than any other type of development, avionics has the most intricate interrelationships of constraints. The complex set of issues associated with volume, weight, power, cooling, capability, growth, reliability, and cost create some of the most complex engineering trades of any systems development effort. The risks associated with the introduction of new technologies and the development of enabling technologies create mitigation plans that have the characteristics of complete parallel developments. It is little wonder that avionics systems are becoming the most expensive portion of aircraft development.

To fully exploit the dollars available for avionics development, it is necessary to invest a significant effort in an intimate understanding of the system requirements. Without knowledge of what the pilot needs to accomplish the mission, and how each portion of the system contributes to the satisfaction of that need, it is impossible to generate appropriate trades and understand the impacts on the engineering process. Indeed, often the mission needs are vague and performance requirements are not specified. This critical feature of the development is further complicated by the fact that pilots often are unaware or do not have the technical background to articulate the detailed technical features of the system or requirements set and cannot specifically identify critical technical system parameters and requirements.

TABLE 38.1 Market Factors Comparison for Commercial and Military Market Segments

Criterion	Commercial	Military
Financial basis	Market	Budget
Development focus	Product	Capability
Production volume	Medium–high	Low
System complexity	Medium–high	High
System design cycle	Short	Medium–long
Life cycle	Short–medium	Long–very long
Contractual concerns	Warranty, liability	Reliability, mortality

The avionics suite is a tool used by the pilot to accomplish a task. The avionics are an extension of his senses and his capabilities. They provide orientation, perception, and function while he is attempting to complete an endlessly variable set of tasks. With this in mind, the first and most important step in the design and development of an avionics package is the development of the requirements. Modeling and simulation is well suited for this task.

38.2.6 Technical Underpinnings of the Practice

Allen defines modeling as the discipline for making predictions of system performance and capacity planning (Allen, 1994). He further categorizes techniques in terms of rules of thumb, back-of-the-envelope calculations, statistical forecasting, analytical queuing theory modeling, simulation modeling, and benchmarking. When applied intelligently, all methods have utility. Each has specific cost and schedule impacts. For nontrivial modeling and simulation, the areas of analytical queuing theory modeling, simulation modeling, and benchmarking have the greatest information content, while rules of thumb often hold the most wisdom. For quantitative estimates, at the system component level, back-of-the-envelope calculations are appropriate.

Analytical queuing theory seeks to represent the system algorithmically. The fundamental algorithm set is the M/M/1 queuing system (Klienrock, 1975), which is an open system (implying an infinite stream of incoming customers or work) with one server (that which handles customers or does the work) that provides exponentially distributed service. Mathematically, the probability that the provided service will require not more than t time units is given by

$$P[S \leq t] = 1 - e^{-t/s}$$

where S is the average service time.

For the M/M/1 queuing system, the interarrival time, that is, the time between successive arrivals, also has an exponential distribution. Mathematically, this implies that if τ describes the interarrival time, then

$$P[\tau \leq t] = 1 - e^{-\lambda \tau}$$

where λ is the average arrival rate.

Therefore, with the two parameters, the average arrival rate λ and the average service time S, we completely define the M/M/1 model.

Simulation modeling is defined as driving the model of a system with suitable inputs and observing the corresponding outputs (Bratley et al., 1987). The basic steps include construction of the model, development of the inputs to the model, measurement of the interactions within the model, formation of the statistics of the measured transactions, analysis of the statistics, and validation of the model. Simulation has the advantage of allowing more detailed modeling and analysis than analytical queuing theory but has the disadvantage of requiring more resources in terms of development effort and computer resources to run.

Benchmarking is the process of running specialized test software on actual hardware. It has the advantage of great fidelity in that the execution time associated with a specific benchmark is not an approximation, but the actual execution time. This process is extremely useful for comparing the results of running the same benchmark on several different machines. There are two primary disadvantages: it requires actual hardware, which is difficult if the hardware is developmental, and, unless your application is the benchmark, it may not represent accurately the workload of your specific system in operation.

38.2.7 Summary Comments

Historically, there have been dramatic increases in the complexity and performance of digital avionics systems. This trend is not likely to change. This increase in complexity has required many new tools and processes to be developed in support of avionics product design, development, and manufacture.

The commercial and military avionics marketplaces differ in significant ways. Decisions made quantitatively must be interpreted in accordance with each marketplace. However, both commercial and military markets have economic forces that have driven the need for shorter development cycles. A shorter development cycle generally stresses the capabilities of design technology; thus, both markets have similar digital system design process challenges.

The most general design cycle proceeds from a concept through a design phase to a prototype test and integration phase, ending finally in release to production. That release date (be it a commercial product introduction or a military system deployment) generally does occur earlier than originally planned. Designs often take longer than scheduled due to complexity and coordination issues. The time between prototype release to fabrication and release to production, which should be spent in test and debug, gets squeezed. Ideally, it is desirable to lengthen the test and debug phase without compromising either design time or the production release date. Modeling and simulation helps this by allowing the testing of designs before they are fabricated, when changes are easier, faster, and cheaper to make.

The design process for any avionics development must begin with the development of the requirements. A full understanding of the requirements set is necessary because of the inherent constraints it places on the development and the clarity it provides as to the system's intended use.

There are several techniques available for the analysis and prediction of a system's performance. These techniques include Allen's modeling techniques (Section 38.2.6): rules of thumb, back-of-the-envelope calculations, statistical forecasting, analytical queuing theory modeling, simulation modeling, and benchmarking. Each technique has advantages and disadvantages with respect to fidelity and cost of resources to perform. When taken together and applied appropriately, they form the rigor base for the best practices in digital avionics systems modeling and simulation.

38.3 Best Practices

In order for the results of any modeling and simulation effort to be effective, attention must be paid to the conundrum of modeling and simulation, that any model may return correct but irrelevant results. To guard against this dilemma, manage the modeling and simulation effort as a complete and disciplined technical project with a thorough understanding of the customer's requirements, the physical environment in which those operate, and the flow down through system implementation with a completely traceable pedigree. Significant progress is being made toward the generation of automated tool sets enabling development of fully compliant traceable requirements. The first step in crafting an understandable, predictable system is ensuring the proper requirements are being implemented correctly. Market forces and system engineering practices are encouraging the use and maturation of appropriate tool sets that will enable avionics system designers to accomplish these goals. The following sections will summarize the best practices in requirements generation and system modeling and simulation.

38.3.1 Requirements Engineering

Establishing the broad parameters of the system performance is critical to the design phase beginning with the generation of requirements. The process of developing requirements begins with a stepwise refinement of the system culminating in the articulation of quantifiable mission effectiveness metrics that flow down to the implementation subsystems. Attention paid to the early modeling of the system will often yield unexpected results that feed back to the highest level of requirement understanding.

Refining and narrowing the engineering trade space with high-level understanding gained by require-ment modeling processes prevents costly missteps and programmatic embarrassments, failures, or even termination.

Begin with the end in mind. The first step is the generation of final (end-state) user requirements. The constraints of cost, schedule, and technical considerations need not be considered initially so as to allow the formation of a maximally effective system solution to implement what the user truly desires. Simulation of the initial system then allows for a benchmark to be used in the subsequent evaluation of necessary compromises, which are driven by programmatic constraints. Predicted performance metrics then reflect the deviations from the maximum represented by the benchmark. Likewise, if the original baseline was suboptimal, performance under certain perturbations of system functionality would likely spawn improvements relative to the benchmark. Conditional use exercises of the system model—which normally occur in the successive refinement and rationalization of requirements with the usual cost, schedule, and technical constraints—assist in the formation of early, deep understanding of the system performance.

Once the initial requirements set has been established, whether or not it is acknowledged to be opti-mal or suboptimal, the associated system performance predictions may be evaluated within the trade space available to the designer. The effects of the system requirements may be evaluated by a design-of-experiments approach to the system mission scenarios that are used to stimulate the rudimentary system model. A thorough understanding of the impacts of the requirements on customer constraints is critical to a balanced system design.

A key step is the system model and simulation used in assessing the utility of the various trade-off design approaches during the process of refining the requirements needs to be validated so as to be a reliable indicator of system performance. Many models have been thoroughly tested to be trusted as reliable. As the expense of avionics systems escalates, the value and trust placed in the models escalates; for this reason, the validation and verification of modeling and simulation continues to be a subject of much interest within government channels. Tremendous efforts continue to be made in validation of the models. Early in the program phase, attention must be paid to the validation of any modeling activity because the fidelity and accuracy of the model will be used as a representation of the problem set. Hence, the wrong model produces wrong inferences of system utility. The Department of Defense has published standards and implementing instructions just because of this importance. See MIL STD 3022, for example. Accreditation of the modeling activity in and of itself has taken on a significantly more important role in the validation and verification of model themselves, thus indicating how impor-tant that department considers the value in getting the simulations correct.

Digital avionics simulation is an integral part of the avionics design of the modern cockpit. Whereas once it was possible to present crude non-real-time or elementary functioning controls and display systems to aid in the evaluation of human interaction with the system, it is no longer an option to treat the system as anything other than an extension of the human interacting with his environment. The crucial nature of the data presented to the pilot must be understood in the context of the reaction time-dependent decision response expected. Simulation prior to physically rendering the system is the acceptable way to build confidence in the end-to-end effectiveness and performance of the entire closed-loop system, including the human response.

For this reason, extensive man-in-the-loop (MITL) simulations have been built to evaluate proposed systems. Whether the scope of the MITL simulations is limited to partial or full-scale mission simula-tions, extensive use of the resultant observations may be proffered as corroboration of the validity of the requirement set. Without validation derived from MITL, the likelihood that the requirements will be correct but irrelevant only increases. Validation of the requirement set then is the starting point toward a balanced design. Confidence in the utility of the resultant system must be established prior to development.

If, however, certain aspects of the proposed system are amenable to an abstraction of human inter-action, then the MITL may be itself stimulated with an appropriately faithful synthetic stimulus.

Understanding the limitations of the simulated MITL results a priori then tempers the extrapolation to the expected validation of requirement set. Research continues in the performance modeling of pilots in extremely high-demand task environments. Refined understanding will result in a more thorough system model of human interaction with complex systems, which will enable a higher level of trust in the results derived from simulated MITL models.

Finally, requirements validation engenders confidence in the end-to-end system performance prior to commitment to development and fielding of new systems. The MITL simulation folds together the abstracted levels of system performance in the chain from receiver sensor sensitivity to effective stimulation and response of the human operator. The questions modeling and simulation may answer prior to commitment of scarce resources are twofold. First, does the proposed system perform as expected—was it built right? Second, does the system satisfy the capability the user needs—was it the right thing to build? Analysis of requirements then is the fundamental initiative in answering those questions.

38.3.2 TDSS

Top-down system simulation (TDSS) begins the process of refining the design with the aim of reducing the risk that the system will not perform properly. If the system can be designed with confidence that what is being built will perform correctly, with all of the subsystems integrating seamlessly and the interfaces mating properly, then the overall risk of false starts is minimized. Obvious benefits include eliminating hardware and software redesign tasks with cost and schedule implications, preventing false starts in integration, and promoting early resolution of requirements ambiguities. Management visibility into the design and development process of complex systems is heightened as well. The costs and resources incurred in committing to the implementation of a total or partial TDSS will often reap benefits far outweighing any short-term negative (incorrectly perceived) programmatic impact.

Early design practices were a stepwise refinement of specifications, beginning with the topmost requirement specification and followed by subsequent partitioning of the system into component parts until the lowest level of decomposition was achieved. Interfaces were captured; agreements were struck. Individual subsystem design processes within the respective disciplines refined the approaches, often resulting in being characterized as "stove-piped." Not until the system integration phase were differences in assumptions and interpretations discovered, with resolution often recovered only at the expense of the schedule. Modeling and simulation of the system offers a way to avoid these pitfalls.

The key to using TDSS as a risk reduction tool is to validate the design requirements, the functional decomposition into subsystems, and the data sets and their associated timing relationships. TDSS provides a visualization and virtual realization of the system prior to committing hardware and software resources to implement a potentially fundamentally flawed design, which avoids expensive reworking of the system to accommodate and resolve problems discovered late in the system integration phase.

An example of this methodology is the USAF Advanced Tactical Fighter (ATF) Demonstration and Validation (DEM-VAL) development effort. The interoperability of designs of five critical interfaces was tested through simulation. Of the five interfaces involved, testing revealed over 200 problems, both with the designs and with the specifications on which they were based. These problems would have resulted in many iterations of hardware fabrication during the integration phase, and several of them would probably not have been detected until the system was fielded. The air force concluded that the application of a TDSS methodology to the DEM-VAL program alone resulted in a savings of approximately $200 million, 25 times the cost of the initiative itself.

38.3.3 TDSS Plan

At the outset of a design program, all the relevant parties must commit to a TDSS effort with clearly articulated goals and timeframes for achieving those goals, and each of the stakeholders must agree to the allocation of resources to accomplish the goals. TDSS is indeed a virtual implementation performed

in parallel with and ahead of the desired system. To be effective, it must precede the design by an appropriate time phase. TDSS performed exactly in phase or slightly behind the real system will not allow
problems to be resolved without cost and schedule implications because the results would be available
too late to influence the build.

To be an effective representation, TDSS needs to be planned with enough detail and granularity to
mirror the final system. Detailed planning of the implementation steps and coordination of the interfaces between the segments of the system are critical to the validity of the simulation. The assignment
of responsibilities to the various integrated product teams, along with performance targets, will ensure
that the same assumptions, design ambiguity resolutions, and approaches are used by the teams in the
implementation of the final system. For example, the system architecture team defines the subsystem
partitioning and functionality to implement the requirements of the system. Subsequently, it hands off to
the subsystem "A" development team a self-consistent requirements specification with defined interfaces
and timing constraints. The subsystem "A" development team, in turn, develops its own domain-specific
model that then produces outputs corresponding to the input stimuli it receives from other subsystems.
The net result is a virtual system that simulates the performance of the ultimate system in all important
domains. Problems with interfaces and timing are resolved early, prior to commitment to the actual
hardware and software realizations of the system.

Given the importance of TDSS to the success of the system development, it is paramount that the
design team approaches the implementation of the TDSS itself with the same rigor it uses with the system design planning. If it is to be a faithful representation of the system, the TDSS cannot be done ad
hoc. Adequate resources with sufficient scope and visibility must be committed to gain a valid realization of the requirements.

Exit criteria for each phase in the TDSS development must exist and be followed rigorously. Rushing
through the simulation in an effort to "get through it" will nullify the utility of performing the simulation in the first place. The simulation needs be matured enough to produce acceptable results prior to
committing hardware designs to fabrication or software to the implementation phase. At each phase in
the system design, whether preliminary or critical design, suitable reviews of the results of the simulation must be conducted and approved as entrance criteria to the next step. The consequences of slighting
the results of the simulation are simply postponing problem discovery to a more expensive recovery
stage later. Attention to virtual design assurance that the TDSS brings will conserve the resources, the
schedule, and the budget.

Planning the TDSS includes defining the data sets and formats passed between each of the teams
involved in the system design. Prior to initiating the design, all parties must agree to the adequacy and
accuracy of the data sets. To minimize unique definitions and usage, individual data sets ought to be
common to all related teams sharing common domains, thus maximizing utility and consistent interpretations of the data. Reducing opportunities to interpret data differently will minimize errors driven by
the translation process. Examples of errors are legion; one is the common definition and usage of "units."

Lastly, successively refining the model is akin to the successive reduction of the design from the
abstract to the physical realization of hardware and software. It is critical to keep focused on the intent
of the simulated system as a virtual representation of the real system as it transitions from concept
through to integration and ultimately fielding.

38.3.4 TDSS Process

The process for TDSS follows the structure of the normal system development cycle with decomposition
of the system functionality proceeding from higher levels of abstraction successively to lower levels of
implementation. The development of TDSS begins with the system-level definition modeling and simulation. Appropriate validation of the system model is performed to assure a thorough and complete
representation of the system. Requirements are traced from the specification to the proposed implementation to assure completeness, rigor, and compliance. Shortfalls are shored up; overdesign is eliminated;

assumptions are clarified. Early in the program, the parties agree that the model faithfully represents the intended, specified system. Successive refinement of detail then takes place in the modeling domain, as it would in the real system. Each lower level of design detail is modeled with domain-specific knowledge again to represent the performance of each component. Performance encompasses all the relevant information pertaining to the component itself: data transformation, accuracy, precision, timeliness of resultants, and adherence to other constraints, as well as statistical measures of allowable performance variance. Figure 38.2 illustrates the typical system model development flow. At each review, appropriate scrutiny is applied to the development of the model. At system requirements review, the planning for developing the model is reviewed to ascertain whether the proposed model is adequately complex to represent the system. At system design review, the maturity of the model is reviewed to assess whether there is sufficient detail to provide the benchmark for the system so as to be a useful tool in allowing critiques of alternative design approaches that various potential subcontractors and suppliers may propose.

Detailed design follows the modeled system partitioning into subsystem, function, modules, and components. Machine-executable representations of the system are developed to refine the system further. The system architects review and approve the detailed models and verify that they indeed perform the expected partitioned functionality. Ensuring the correctness of the mid- and low-level models enables the system to be accurately and validly integrated in a synthetic sense prior to actually building the real hardware and software. Participating in the reviews, building the models, testing the modeled behavior, comparing results, and solving the discrepancies between differing team interpretations all contribute to the collective building of understanding within the entire implementation team. Confidence in the resulting system increases as does trust in the validity of the model.

Finally, the models are built to reflect the lowest level of hardware and software functionality. Designs are tested against the models, and deviations are repaired so as to yield the same results as the models predict. Remembering that the models are the standard by which the hardware and software is measured ensures that, once implemented, the lowest levels of the system perform as expected, and successively higher levels will likewise integrate properly. The TDSS models must be subjected to configuration control with the same rigor as the system hardware and software. Not doing so will allow the modeling effort to diverge.

By utilizing TDSS in flowing the system model to the lowest level, complete with executable models with common data sets implemented and understood, repeatable mechanizations will resolve the ambiguous interpretations arising between teams. By far, the majority of system integration problems are due to those differing interpretations of ambiguous specifications. Giving the system designers the opportunity to perform multiple tests on the simulated system modeled top to bottom in virtual mode is a powerful tool. The system architects may then predict with confidence how the system will behave in a variety of conditions before the system is built. Demonstrating the system via modeling and simulation is the key to predictable and controlled system design. Discovering, understanding, solving, and testing the model interface differences goes a long way toward reducing the final system integration errors where the costs and schedule disruption associated with fixing problems is very high. These two results—confidence in the correctness of the system and verifiable integrated system behavior—justify the expense and resources required to create and maintain the TDSS.

38.4 Performance Characterization for an Airborne Receiver Upgrade: Case Study

The following case study describes a practical application of modeling and simulation used to characterize and predict system-level performance implications of architecture modifications to an existing airborne receiver system. Figure 38.3 describes the initial system concept (ISC) for an airborne communications system upgrade, with the proposed modifications highlighted in gray. The overall system design utilizes COTS where possible and is minimally integrated with the existing aircraft systems to

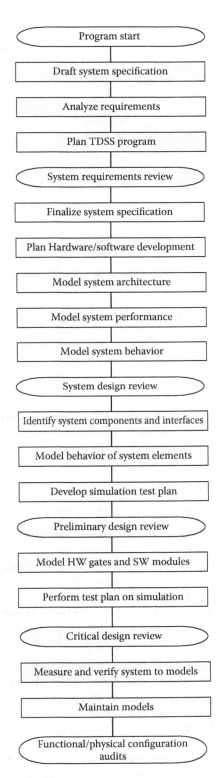

FIGURE 38.2   TDSS development flow.

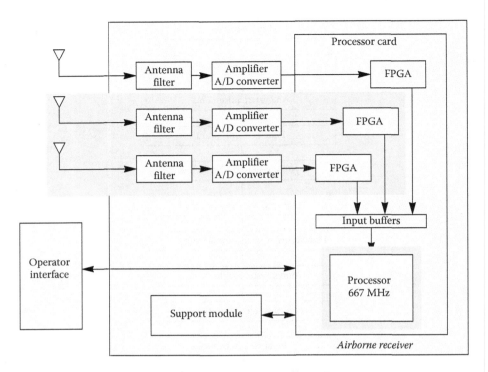

FIGURE 38.3   ISC for system upgrade.

reduce impacting overall program schedule and cost. The system will be placed in an environment with high levels of electromagnetic interference (EMI), as well as radio-frequency (RF) jamming, which combine to degrade the overall signal-to-noise ratio (S/N).

38.4.1 System Description

The existing architecture includes the following functional components:

- Single antenna input consisting of two loop antennas combined to form an omnidirectional pattern in the azimuthal direction
- Analog front end that filters, amplifies, and digitizes the antenna signal
- Processor card containing the main processor that provides overall system power-up and initialization, system control, and waveform processing
- Digital front-end field-programmable gate array (FPGA) (contained within the processor card) that provides mixer/tuner, filtering, and signal processing functionality
- Operator interface consisting of a screen display and keypad

Given this existing architecture, the system designer must first determine whether the system as is can satisfy the maximum processor utilization requirement for worst-case (stressed) EMI and RF jamming environments. To allow for future waveform growth, this maximum processor utilization requirement has been set at 40%. Independent laboratory benchmark testing data provide recommended utilization levels for high-reliability operations under benign, normal, and worst-case environments and are shown in Table 38.2.

Significant S/N improvements can be been achieved in stressed environments using multiple antenna channels to support polarization diversity processing of received signals. Using signal processing

TABLE 38.2 Processor Utilization for Various Environmental Conditions

Environment	Description	Processor Utilization (%)
Benign—High S/N	Minimal levels of EMI, no RF jamming	Min: 10 Max: 25
Normal—Mid S/N	Average conditions, which encompasses both benign and stressed conditions	Min: 10 Max: 30
Stressed—Low S/N	Maximum levels of EMI and RF jamming	Min: 15 Max: 30

algorithms, the proposed receiver upgrade will be able to spatially isolate and mitigate the effects of one or two interfering signals, depending whether two or three independent antenna input signals are used.

The proposed upgrade will require an additional antenna, antenna filter assembly, amplifier and A/D converter assembly, and FPGA for each added channel. Each added channel will have similar processor utilization characteristics as for the existing one-channel receiver described in Table 38.2. The additional channel data will be sent to the existing input buffers, adding to the data that are processed by the main processor. It is estimated that the existing processor will consume an additional 5% and 10% processor capacity for handling the parallel processing of two and three channels, respectively. An upgraded processor with twice the capacity of the existing processor is proposed in the ISC to provide additional processing capability.

38.4.2 Model Development

Modeling, simulation, and demonstration of the existing and proposed receiver design is accomplished using the computational software program Mathematica® from Wolfram Research, Inc. The model is partitioned into three sections: initialization, which sets the underlying parameters of the model; controls, which provides real-time model manipulation; and output, which presents model diagrams, graphs, and statistics. The model controls and output for a three-channel, upgraded processor design under a stressed environment are shown in Figure 38.4.

38.4.2.1 Initialization

Based on the processor utilization data from the existing design, three data arrays are constructed based on the uniform random distribution, representing processor utilization under benign, normal, and stressed environments. Depending on the number of additional channels used for each model scenario, the random data arrays are combined with the expected operating system overhead and parallel processing overhead to form a new random distribution, from which the desired statistics can be calculated. The following assumptions are made regarding the upgraded design:

- Independent antenna channels experience the same RF environment, that is, if a benign scenario is chosen with three channels, all three channels will each be given a uniform distribution based on the benign parameters.
- The RF environment utilization and operating system overhead utilization values for the upgraded processor are half of the utilization values for the existing processor.
- The three-channel design has twice the parallel processing overhead as compared to the two-channel design (one-channel design has no parallel processing overhead).
- Existing input buffers have sufficient capacity to handle additional channels.

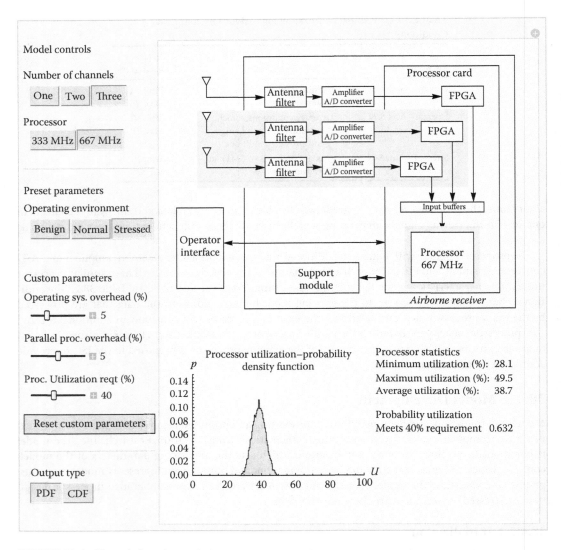

FIGURE 38.4   Upgraded receiver system.

38.4.2.2 Controls

Nearly all model parameters are made customizable in real time for the model operator. A description of each control is provided in the following:

- Number of channels: determines whether one-, two-, or three-antenna channels are modeled
- Processor: determines whether the existing (333 MHz) or upgraded (667 MHz) processor is modeled
- Operating environment: determines which RF environment is modeled (benign, normal, or stressed)
- Operating system overhead: sets the processing consumed by the operating system, relative to the existing processor
- Parallel processing overhead: sets the amount of processing consumed by the addition of antenna channels, relative to the existing processor

- Processor utilization requirement: sets the maximum permissible processor utilization percentage requirement
- Reset custom parameters: sets operating system overhead, parallel processing overhead, and processing requirement variables to model defaults of 5%, 5%, and 40%, respectively
- Output type: selects the probability density function or cumulative distribution function as the primary output type

38.4.2.3 Output

The model output displays a functional block diagram representing the receiver with the selected number of antenna channels and existing or upgraded processor. The default output shows a functional block diagram of the existing one-channel receiver design. The second output is a probability density function or cumulative distribution function of the processor utilization percentage, depending on which output type is selected. Finally, statistics for the processor utilization for the options chosen are displayed, including the probability that the processor utilization of the receiver design selected will fall under the maximum permissible processor utilization requirement. The Mathematica environment updates the model outputs dynamically after any control value is modified, allowing the modeling and simulation results to be presented as part of an interactive presentation, as opposed to a static display of preset results.

38.4.3 Model Results

The modeling and simulation results present several options for the system designer to consider when deciding how many channels and which processor to include in the upgraded design. A summary for both the existing and upgraded processor design with one to three channels under stressed conditions is shown in Table 38.3.

From the summary data, it becomes apparent that the existing processor is not suitable for a two- or three-channel design. This shows that if additional antenna channels are desired, then the upgraded processor must be incorporated in the design. Further analysis reveals that one option, relaxing the processor utilization requirement to 50%, allows the three-channel design with the upgraded processor to meet the requirement, as shown in Figure 38.5.

TABLE 38.3 Model Output Statistics, Stressed Environment

Antenna Channels	Processor Utilization Statistics: Existing Processor		Processor Utilization Statistics: Upgraded Processor	
One	Minimum	10%	Minimum	7.5%
	Maximum	30%	Maximum	15%
	Average	22.5%	Average	11.3%
	Probability utilization	100%	Probability utilization	100%
	Meets 40% requirement		Meets 40% requirement	
Two	Minimum	30.1%	Minimum	17.5%
	Maximum	59.9%	Maximum	32.4%
	Average	44.9%	Average	25%
	Probability utilization	22.1%	Probability utilization	100%
	Meets 40% requirement		Meets 40% requirement	
Three	Minimum	46%	Minimum	28.1%
	Maximum	88.8%	Maximum	49.5%
	Average	67.6%	Average	38.7%
	Probability utilization	0%	Probability utilization	63.2%
	Meets 40% requirement		Meets 40% requirement	

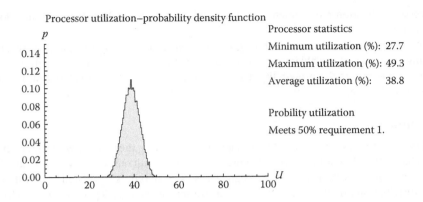

FIGURE 38.5   Option 1: Three-channel, upgraded processor utilization PDF.

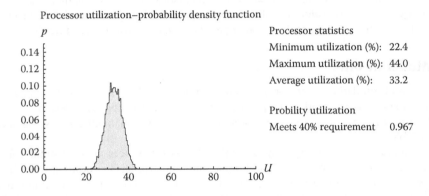

FIGURE 38.6   Option 2: Three-channel, upgraded processor utilization PDF.

As a second option, if the operating system overhead and parallel processing overhead can be reduced to 2% and 3%, respectively, then the 40% processor loading requirement can be met by the three-channel, upgraded processor design with 96.7% probability, as shown in Figure 38.6.

38.4.4 Case Study Summary

The modeling and simulation effort that formed the basis for this case study was performed as part of the requirements development process for an airborne receiver upgrade program. The results guided the tailoring of system-level requirements to include options for a three-channel design utilizing the upgraded processor. By characterizing processor performance early in the requirements phase, the ISC for the receiver could be matured, which avoided costly system characterization of multiple processor and antenna channel options later in the development phase of the program.

This case study further demonstrates the advantages of modeling and simulation when starting with a known design with benchmarked data, especially when a development program is schedule-driven. Rather than starting the design phase with an unknown design and immature requirements, modeling and simulation efforts can be used to refine requirements and guide the focus from ISCs to more mature preferred system concepts and ultimately the preliminary design.

38.5 Research Issues and Summary

Not every aspect of a real system can be modeled completely in every tool (nor is it desired), but the known characteristics of the system that relate to performance form a quality model for performance estimation and capacity planning. Clearly, market forces will continue to drive the quantity, quality, completeness, and rate of change of system engineering environments. As a practical matter, better integration and traceability between requirements automation systems and system/software engineering environments are called for. Improved traceability between proposed system concepts, their driving requirements, the resultant technical configuration(s), and their cost and schedule impacts will be a powerful result of this integration.

Program-level commitment to the requirement for integration of the trade space across technical, program, financial, and market boundaries is likely to continue and remain incomplete. To paraphrase Brooks, there is more common ground than the technical community will admit, and there are differences that most managers do not understand (Brooks, 1995). Automation alone will not reduce the complexities of this effort until there is a common, multidisciplinary, quantitative definition of cost versus price in system performance trades. As to the automation issue, the size, content, and format issues of current and legacy technical, financial, programmatic data bases will continue to grow and diverge until the stewards of college curricula produce graduates who solve more problems than they create (Parnas, 1989). These new practitioners must develop future information systems and engineering environments that encompass the disciplines, languages, and methods critical to improving the practice.

Definitions

CAD: Computer-automated design.
CAE: Computer-aided engineering.
CDF: Cumulative distribution function.
COTS: Commercial off the shelf (products).
EMI: Electromagnetic interference.
FPGA: Field-programmable gate array.
ISC: Initial system concept.
LAN: Local area network.
MIPS: Millions of instructions per second.
MITL: Man in the loop.
PDF: Probability density function.
RF: Radio frequency.
S/N: Signal-to-noise ratio.
TDSS: Top-down system simulation.

Further Reading

Arnold Allen's text, *Computer Performance Analysis with Mathematica®* (Academic Press, 1994), is an excellent and wonderfully readable introduction to computing systems performance modeling. It is out of print; if you can find a copy, buy it.

Hennessy and Patterson's fifth edition of their textbook *Computer Architecture: A Quantitative Approach* (Morgan Kaufmann, 2011) remains the definitive text in undergraduate and graduate-level computer engineering. The text is replete with "Case Studies and Examples" sections that form boundaries around which modeling and simulation is useful.

For performance modeling and capacity planning, the following organizations provide information through periodicals and specialized publications:

- The Computer Measurement Group (CMG) (www.cmg.org)
- Association for Computing Machinery (www.acm.org)

For CAE, the following websites provide information on vendors of major tools and environments:

- www.opnet.com
- www.cadence.com
- www.wolfram.com
- www.mentor.com
- www-cdr.stanford.edu/SHARE/DesignNet.html

References

Allen, A.O., *Computer Performance Analysis with Mathematica®*, Academic Press, New York, 1994.

Bell, C.G. and Newell, A., *Computer Structures: Readings and Examples*, McGraw-Hill, New York, 1973.

Bratley, P., Fox, B., and Schrage, L., *A Guide to Simulation*, 2nd edn., Springer-Verlag, New York, 1987.

Brooks Jr., F.P., *The Mythical Man Month*, Addison-Wesley, Reading, MA, 1995.

Donnelly, C.F., Evaluating the IOBIDS specification using gate-level system simulation, in *Proceedings of IEEE National Aerospace Electronics Conference*, Dayton, OH, May 18–22, 1992, p. 748.

Hennessy, J.L. and Patterson, D.A., *Computer Architecture: A Quantitative Approach*, 5th edn., Morgan Kaufmann, San Francisco, CA, 2011.

Kleinrock, L., *Theory, Queueing Systems*, vol. 1, John Wiley & Sons, New York, 1975.

Liu, H.-H., Software issues in hardware development, in *Computer Engineering Handbook*, McGraw-Hill, New York, 1992, Chapter 2, pp. 2.1–2.13.

Parnas, D.L., Education for computing professionals, Technical. Report. 89-247, March, 1989, ISSN 0836-0227.

Portelli, W., Oseth, T., and Strauss, J.L., Demonstration of avionics module exchangeability via simulation (DAMES) program overview, in *Proceedings of IEEE National Aerospace Electronics Conference*, Dayton, OH, May 22–26, 1989, p. 660.

Strauss, J.L., The third possibility, in *Modeling and Simulation of Embedded Systems, Proceedings of Embedded Computing Inst.*, La Jolla, CA, July 18–19, 1994, p. 160. Society for Computer Simulation, San Diego CA.

Swangim, J., Strauss, J.L. et al., Challenges of tomorrow—The future of secure avionics, in *Proceedings of IEEE National Aerospace Electronics Conference*, Dayton, OH, May 22–26, 1989, p. 580.

Walker, R.A. and Thomas, D.E., A model of design representation and synthesis, in *Proceedings of the 22nd Design Automation Conference*, Las Vegas, NV, 1985, pp. 453–459.

39

Model-Based Development with AADL

Julien Delange
*Carnegie Mellon Software
Engineering Institute*

Bruce Lewis
*U.S. Army Aviation
and Missile Research
Development and
Engineering Center*

39.1 Introduction

Aviation systems, a safety-critical complex subset of cyber-physical systems (CPSs), require a rigorous development process that includes validated requirements, careful design and implementation of functions, and certification/qualification before its release. While these systems have strong development practices and requirements, the number of collocated and interacting functions has increased significantly, driving the cost and risk of integration up exponentially. Increased functionality has also driven the use of integrated modular avionics (IMA), a much more complex style of architecture. Thus, developing new systems or updating existing ones is very hard to achieve. The statistics show that 70% (at least) [NIST02] of the issues are found during or after integration. It strongly points to the conclusion that the primary issues driving costs relate to architectural issues. We fail to find these issues in the requirements and design phases, discovering only 3.5% [NIST02] of the errors, during the time that we establish the architecture and elaborate it. We can also fail to build the architecture we design. As a consequence, on the newest aircraft, we have paid over 70% of the system development cost on software, most of that in software rework. In fact, software rework is now a, and in some cases the, major cost of total aircraft system development.

To overcome these software/system architectural issues, one idea is to abstract system concerns using a formal architecture description language. Such a language would be designed to support quantitative analysis, software system composition, analysis of execution semantics, and automated system integration. In that context, model-based engineering technologies can provide a high-level description of the system along with tools to support the development process in a predictable, precise way. Depending on the technology and the specification language of the system (level of abstraction, description of characteristics and specific aspects) and its toolset, approaches support

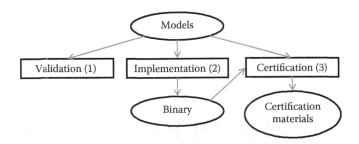

FIGURE 39.1 Support of development process using models.

different aspects of the development process. An ideal model-based engineering approach would support all aspects of the development platform (as illustrated in Figure 39.1):

1. *Validation*: Models are analyzed and processed to check system requirements, correctness, and the feasibility of the system. This validation can start at the requirements phase against conceptual architectures through a virtual integration process. During the development phases, the architecture is validated incrementally as the design is elaborated. During this time, suppliers provide models of their subsystems/components, and architectural constraints are incrementally verified as the system is composed.

2. *Implementation*: To a precise specification to maintain analysis predictions, code can be automatically produced using code generators that transform architectural models and components into an executable code that runs on the execution environment (processors plus operating systems, buses, memory, etc.). All configuration, deployment, and integration aspects are automated so that engineers do not have to hand integrate components in the execution environment.

3. *Certification and verification*: Generated binaries and models are processed in order to check that the produced system enforces system requirements and also assist production of certification materials. Depending on models' details and requirements description, several materials could be automated, such as the traceability of systems requirements from documentation to code and automation of integration and unit tests. A new field of research is the formal analysis and composition of the system using a formal architecture description language.

Among existing model-based engineering approaches, the Architecture Analysis and Design Language [AADL] has been demonstrated in several projects to support these different aspects of the development of avionics systems, either design, validation, implementation, or recent research in certification/verification. In particular, the technology was successfully demonstrated in the context of the System Architecture Virtual Integration (SAVI) [AADL-WWW] project that incorporates a predictive, analytical virtual integration of the software and system using AADL as a means to discover integration problems both early and incrementally throughout the lifecycle.

The following chapter presents this technology and its related ecosystem for developing avionics architecture. First, we present the core language and its principles. Then, we explain its tailoring for representing avionics architecture. In particular, the AADL has a dedicated additional document (the ARINC653 annex [AADL-ARINC653]) to capture IMA architecture with respect to their requirements (such as time and space isolation). We also present tools that process AADL, check specific avionics system architectural requirements, and automate system implementation production by generating the code. Finally, we enumerate several new challenging research areas in this context, explaining how AADL models will enable support of avionics system certification needs. Those interested in the SAVI process, analysis methods, and demonstrations can find documentation on the official AADL website [AADL-WWW].

39.2 Brief Overview of AADL

The AADL is a standard published by the Society of Automotive Engineers (SAE). It specifies a modeling language for defining software and hardware of a system architecture using a textual or a graphical notation. The core language distinguishes three components categories:

- Software components
  - *Process* component represents a memory space to store program code/data and provides an execution environment for application threads.
  - *Thread* and *thread group* components represent concurrent entities that execute code (characterized with a stack, instruction pointer, etc.).
  - *Subprogram* and *subprogram group* components model the code in terms of programming objects. They reference the source code (indicating the programming language, procedure/function names, etc.) to be invoked by the architecture.
  - *Data* component may specify a type or a shared variable.
- Hardware and deployment components
  - *Processor* and *virtual processor* components model both the physical/virtual hardware (processor architecture such as x86) and software execution environment on each processor/virtual processor (OS such as Linux).
  - *Bus* and *virtual bus* components represent either the physical bus (such as Ethernet or MIL-1553) that connects two or more distributed nodes or its virtual separation into different layers (such as channel allocation of a bus, layering according to a specific quality-of-service policy).
  - *Device* components represent hardware devices (sensor, actuator, keyboard, display, camera, etc.) connected to a runtime platform. Devices can be extended into systems if one needs to later model the internals, for instance, if application software needs to be loaded onto the component.
- Hybrid components
  - *System* components represent either the root component of a model (system as a whole) or a node of a distributed architecture (a subsystem).
  - *Abstract* components represent a generic component whose type shall be refined into a specific component category before implementation. This is useful when modeling a function that can be realized either in software or in hardware. In that case, a high-level component models its interface and behavior using an *abstract* component, and its implementation refines it into an appropriate type (e.g., a *process* when implementing the functionality with software or a *device* when using a hardware approach). This approach avoids modeling the same high-level requirements twice and ensures a consistency between the two realizations of the component.

Components are defined with separated views, the *type* (external specification) and *implementation* (internal blueprint):

- Component *type* defines a component category, its name, properties, and its interfaces. The component interface specifies communication mechanisms (send or receive an event and/or data) and its dependencies (e.g., if a *process* requires access to a specific *bus*).
- Component *implementation* extends the type and includes subcomponent declarations (each an instance of a defined component), internal connections, subcomponent properties, and modes of operation. Subcomponent instances also have a type and implementation that define them, resulting in a compositional hierarchy. For example, a *process* (space that stores code and data) contains a *thread* calling one (or more) *subprogram*(s). The language defines legality rules for component composition so that a model represents a feasible architecture (e.g., a *subprogram* cannot contain a *processor*). The AADL also checks for architectural completeness but allows incrementally defined models and analysis. For instance, the architecture in Figure 39.2 would require a bus to connect the CPU to the RAM when fully specified.

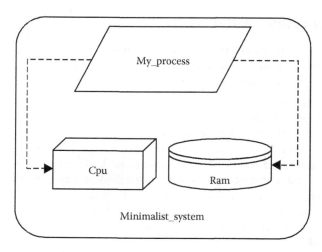

FIGURE 39.2   AADL graphical model representing a minimalist system.

A complete and legal AADL model is a hierarchical assembly of components, the topmost being a *system* that contains at least one *process* (modeling the software) associated to one *processor* (which provides the execution environment). Figure 39.2 shows such a model using the graphic representation of the AADL, while Listing 39.1 shows its textual representation.

System requirements and characteristics are captured by associating *properties* to components. The standard includes a set of predefined properties representing the most used characteristics when representing architecture, but it can be refined and extended by the user, as described in the following. Note in Figure 39.2 the process bindings to the CPU and the memory and the textual properties in the textual specification.

While the core language may suit the needs of most engineers, AADL provides two extension mechanisms for defining domain-specific language capabilities:

1. *User-defined properties* extend the potential set of characteristics associated to each component. In the same manner as the predefined set of properties, users can define their own property name, usage restrictions, and values to be added. User-defined properties are often used to support specific analysis tools. Property values with relevant architectural data are extracted and provided to the analysis tool.

2. *The annex extension mechanism* can be used to associate other languages to AADL components and specify other system aspects. Several annexes have already been standardized, such as the behavior annex (components' behavior in terms of state machine), the error model annex (error faults and propagation), or the ARINC653 annex (modeling of IMA architecture). The next section focuses on the latter, illustrating the support of avionics architectures and in particular the ARINC653 architecture for time and space partitioning.

There is a large set of tools supporting the AADL. Among them, the Open Source AADL Tool Environment (OSATE) provides a modeling environment on top of the Eclipse modeling environment. While its development is led by the Software Engineering Institute (SEI), many other industrial and academic partners contribute to the toolset base. OSATE has been evaluated on a number of projects, such as the SAVI project (see AADL wiki [AADL-WWW] for reports, papers, demonstrations, and presentations). SAVI [SAVI] includes organizations from the avionics industry such as Airbus, Boeing, Embraer, Goodrich, Honeywell, Rockwell Collins, and agencies such as the FAA, NASA, and Army. Another popular AADL tool is Ocarina, developed by French research institutes. Ocarina [OCARINA] is a command-line-based tool for model processing and code generation. It has been evaluated in different

```
processor i5
end i5;

processor implementation i5.four_cores
end i5.four_cores;

memory ram
end ram;

memory implementation ram.i
end ram.i;

process application
end application;

process implementation application.impl
end application.impl;

system minimalist_system
end minimalist_system;

system implementation minimalist_system.i
subcomponents
    cpu        : processor i5.four_cores;
    ram        : memory ram.i;
    my_process : process application.impl;
properties
    Actual_Processor_Binding => (reference (cpu))
                                    applies to my_process;
    Actual_Memory_Binding => (reference (ram))
                                    applies to my_process;
end minimalist_system.i;
```

LISTING 39.1 AADL textual model representing a minimalist system.

projects and is currently integrated in TASTE, a model-based engineering platform supported by the European Space Agency. A commercial AADL toolset with editor, documentation, and analysis support is STOOD. A documented list of tools is available on the AADL wiki [AADL-WWW].

39.3 Modeling Avionics Architecture with AADL

The AADL committee standardized modeling patterns for representing partitioned avionics architecture in a document called the ARINC653 annex [AADL-ARINC653]. It consists of a set of rules that define AADL component assemblies to capture avionics architecture and requirements. It also enriches the language with new properties dedicated to the description of IMA system supporting time and space isolation provided by ARINC653 OS [ARINC653].

The following sections illustrate the use of the annex through a partitioned architecture example composed of two communicating partitions:

1. The first partition (data_acquisition) retrieves data from hardware sensors (such as speed/temperature). Each sensor is controlled by one task (ARINC653 process corresponding to an AADL *thread*) within this partition. Acquired data are sent to the second partition using interpartition communication.
2. The second partition (statistics) receives sensor data from the first partition and performs computations. It is composed of two tasks: one that receives and filters the raw data from the sensors (e.g., removing inconsistent or out of bounds values) and the other that calculates some statistics and measurements.

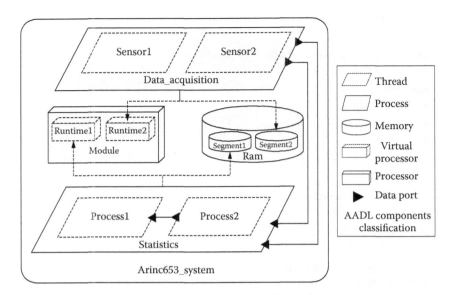

FIGURE 39.3   AADL model of our ARINC653 architecture case study with description of AADL graphical notation.

The following paragraphs detail the application of AADL ARINC653 modeling patterns on this example, demonstrating how to build an AADL model from scratch. We provide both the textual (in the attached annex to this chapter) and the associated graphical AADL model (depicted in Figure 39.3).

IMA architectures isolate system applications into partitions. Partitions are described as AADL *process* components. The AADL core language defines this component as a memory space that contains application code and data, and thus, it fits with the description of an ARINC653 partition. Within partitions, applications are executed by one or more ARINC653 processes that share a partition's resources. ARINC653 processes are similar to the concept of a task or thread, and thus, we use the AADL *thread* component to represent the ARINC653 process.

The underlying OS on a physical processor isolates partitions in terms of time by guaranteeing a predefined timeslot for their execution. Within each of these timeslots, there is a runtime to execute the AADL *thread* components (ARINC653 processes) within that partition, executing conceptually on a virtual processor. To represent the separated runtime environments, the ARINC653 annex requires separating (or building on top of) the main *processor* component (modeling the hardware architecture and the OS) into several *virtual processor* components, each one representing the runtime of each partition. In addition, each partition (AADL *process* component) is associated with a dedicated runtime (AADL *virtual processor* component) that defines partition runtime requirements (represented by a dashed arrow associating *process* and *virtual processor* components in Figure 39.3). *Actual_Processor_Binding* of the virtual processors to the physical processor is shown graphically with containment.

In addition, to model the separation of partition runtimes, three dedicated properties are specified to capture the time isolation policy:

1. *Partitions_Slots*: Enumerates the list of slots for executing partitions
2. *Slots_Allocation*: Describes the allocation of slots for each partition
3. *Module_Major_Frame*: Specifies the major frame of the module (the pace at which the scheduling slots are repeated)

These three properties are specified on the *processor* component, since it ensures the enforcement of the timeslot-based isolation policy.

ARINC653 systems also support space isolation. Partitions are confined in separated memory segments so that a partition has its own space to store its code and data and cannot get access into another partition's memory. To capture this requirement in AADL models, we separate the main AADL *memory* component (the one that models the hardware memory) into logical memory segments using *memory* subcomponents. Each memory subcomponent represents a memory segment. Then, for each subcomponent, we define the type of objects it contains (code and/or data), along with its requirements (word size, word count, etc.—represented by the *Byte_Count* property in the textual model). Each partition (AADL *process*) is associated with at least one memory segment (AADL *memory* subcomponent). In our case study (see Figure 39.3), the hardware memory component (RAM) is separated into two segments (segment1 and segment2), each partition being associated with one of them (dashed arrow in the graphical version and *Actual_Memory_Binding* property in the textual one).

ARINC653 processes and partitions communicate using either intrapartition (ARINC653 blackboard, buffer, semaphore, or event) or interpartition (ARINC653 queuing and sampling ports) communication services. The former connects ARINC653 processes (represented by AADL *thread* components) within a partition, while the latter connects ARINC653 partitions (AADL *process* components). Both communication levels are represented using the appropriate AADL *ports* (*event port, data port*, or *event data port*) connected between components.

The following modeling rules map ARINC653 interpartition services to AADL models:

1. An ARINC653 queuing port is specified using an AADL *event data port* connected between two AADL *process* components.
2. An ARINC653 sampling port is represented using an AADL *data port* connected between two AADL *process* components.

The following modeling patterns map ARINC653 intrapartition services to AADL models:

1. An ARINC653 blackboard is specified using an AADL *data port* connected between several AADL *thread* components.
2. An ARINC653 buffer is specified using an AADL *event data port* connected between several AADL *thread* components.
3. An ARINC653 event is specified using an AADL *event port* connected between several AADL *thread* components.
4. An ARINC653 semaphore is specified using a shared AADL *data* component between several *thread* components (access to a shared resource with a specific protection mechanism—specified with an AADL property—to avoid concurrent accesses).

Our example (see Figure 39.3) uses both interpartition and intrapartition services:

• Two ARINC653 sampling ports (AADL *data ports*) connected between both partitions (AADL *process* components)
• One blackboard (AADL *data ports*) connected between the two ARINC653 processes (AADL *thread* components) of the statistics partition

The ARINC653 standard also defines a health monitoring service that associates a recovery policy for each fault that may occur at the different execution levels (module, partition, or process). For example, the system designer can specify to restart a task if its application code triggers a divide by zero exception or also to restart a partition if one of its tasks issues an illegal request (such as trying to access a memory area it does not own). This service is translated in AADL using specific properties that associate faults and recovery strategies for each level of the layered architecture (module, partition, and process). The annex includes the faults specified in the ARINC653 standard but can be extended according to user needs or OS specification.

```
processor implementation arinc_module.two_partitions
subcomponents
        runtime1 : virtual processor partition.i;
        runtime2 : virtual processor partition.i;
properties
        ARINC653::Partition_Slots = > (10 ms, 20 ms);
        ARINC653::Slots_Allocation = > (reference (runtime2), reference
(runtime1));
        ARINC653::Module_Major_Frame = > 30 ms;
        ARINC653::HM_Errors = > (Module_Config, Module_Init,
Module_Scheduling);
        ARINC653::HM_Module_Recovery_Actions = > (Ignore, Reset, Reset);
end arinc_module.two_partitions;
```

LISTING 39.2 Specification of the ARINC653 module with two partitions.

We specify a health monitoring policy for the module of the case study. As AADL properties are not shown in graphical models and only in their textual representation, the textual definition of the ARINC653 module (AADL *processor*) is illustrated in Listing 39.2. According to the *HM_Errors* and *HM_Module_Recovery_Actions* properties, three potential faults can be detected:

1. *Module_Config* corresponding to a configuration error. In this example, the error is ignored.
2. *Module_Init* is triggered when an erroneous operation is detected when starting the module. When detected, the module is restarted (reset).
3. *Module_Scheduling* is triggered when an error is raised during partition scheduling. When detected, the module is restarted (reset).

Using the same approach, users can specify the health monitoring policy for ARINC653 partitions and processes by adding properties to AADL *virtual processor* (runtime associated with a partition) and AADL *thread* (ARINC653 process) components.

39.4 Validation of Avionics Architecture Requirements

Models provide an abstract representation of the system and a good communication mechanism across system stakeholders. In addition, thanks to their nonambiguous notation, analysis tools process them to check requirements enforcement. This provides support to engineers during the design process, enabling checking for potential issues before starting implementation efforts. In addition, by detecting issues earlier, propagation across the development processes is avoided, reducing reengineering efforts and development costs and keeping delivery time under control.

In the context of avionics architecture, several requirements must be inspected and validated, either consistency requirements with respect to the modeling patterns or system-level characteristics, such as resources dimensioning and performance analysis. In the following, we choose two analysis methods in order to explain the validation process and the use of models using the following approaches:

1. *Model-consistency validation* checks model compliance against ARINC653-specific patterns. It guarantees that models can be processed by tools that support the ARINC653 AADL annex.
2. *Performance analysis* provides metrics (such as latency) with respect to runtime properties and deployment concerns (such as scheduling and use of shared resources).

Model-consistency validation consists in analyzing models and ensuring enforcement of ARINC653 modeling patterns. Supporting tools process the system architecture and perform the following operations:

1. *Check component aggregation* to guarantee that the model enforces modeling rules defined in the ARINC653 annex. For example, this would consist in validating that an AADL *processor* (ARINC653 module) contains AADL *virtual processor* components, each one representing a

runtime supporting an ARINC653 partition. In our example, this would be validated by checking that the main AADL *processor* component (module) is separated into two AADL *virtual processor* components (runtime1 and runtime2).

2. *Check property definition correctness* with regard to system requirements. For example, an analysis tool checks that each AADL *processor* (ARINC653 module) schedules each AADL *virtual processor* (ARINC653 partition), ensuring that each partition is executed at least once in each major frame. This is validated by processing the textual model. The AADL processor component must reference each AADL *virtual processor* component in the *ARINC653::Slots_Allocation* property.

Performance analysis is done by inspecting the component definition, configuration, and requirements. Several analyses and validations could be done, and the more the better. In ARINC653 systems, evaluation of latency between connected partitions could be a major concern because it varies according to the execution environment and system configuration (such as partition scheduling policy). In that context, we designed a dedicated tool for evaluating latency between two partitions. It analyzes the system under the worst-case scenario, when tasks output data at the end of their period and interpartition communication ports are flushed when the major frame is reached. As a consequence, depending on the partition scheduling, this policy may have a significant impact on data flows involving several connections with different components, even located on the same processor. Figure 39.4 illustrates this aspect; the data sent by the first partition (data_acquisition) are available to the second one (statistics) only after the major frame.

Using our example, our tool evaluates the latency between the data_acquisition completion and the start of the statistics partition and gives a result of 30 ms, the minimal latency at this stage of the development, ignoring OS overheads. Indeed, assuming that interpartition communications are realized at the major frame, data sent by the data_acquisition partition are not available to other partitions until the start of the next major frame. As a result, data would be available to the statistics partition at the next period. This is illustrated in the scheduling diagram in Figure 39.4: the data_acquisition partition sends data at the completion of its period. The data are available only at the occurrence of the start of the next major frame and can be processed by the statistics partition only when it is being executed again. This specific scheduling concern may have a major impact on system performance and the underlying functional behavior. Moreover, this latency may vary according to the target OS and its scheduling policy when the OS is considered. For these reasons, evaluation of system performance before starting system implementation is an important assessment that would show potential issues before investing in code development efforts.

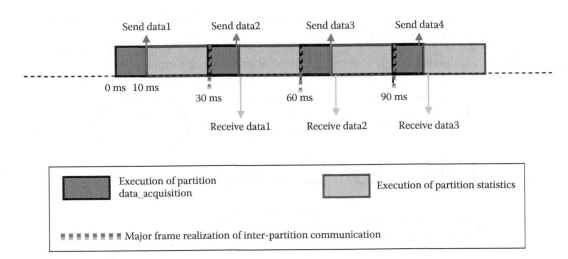

FIGURE 39.4 **(See color insert.)** Scheduling of partitions.

39.5 Automatic Implementation from Models

In order to automate system development and produce implementation that matches stakeholder's requirements, we produce the system from AADL models. Using the same consistent notation among all development steps avoids traditional pitfalls when reusing different requirements specifications and improves system reliability and robustness. Generating the implementation is done during a process that transforms the model into code amenable to being integrated with the execution platform (OS, device drivers, etc.). To do so, the modeling language semantics should provide the necessary information to produce the final application without having to make assumptions about system/component communication and execution requirements. In our context, the semantics of the AADL describes the system architecture with its configuration and deployment concerns; it is an appropriate language to produce the code for executing system functions. In particular, code generated from models would consist in the following:

- *Creation of system threads* with respect to their characteristics (period, deadline, priority, etc.) for supporting the execution of system application code.
- *Instantiation of shared resources* (such as variables or memory area) according to their specific properties (locking mechanism, priority ceiling).
- *Activation and configuration* of system devices (network interfaces) according to their connection with other components (link with other nodes on a bus, setting IP address, etc.).
- *Establishment of connections across system applications* (communication channels) and automatic dimensioning of distribution concerns (size of an input or output interfaces such as the ARINC653 queuing ports or buffers).
- *Management of system state* such as initialization, finalization, fault tolerant error handling, and mode changes across system state and within components. The AADL formally defines system state and state changes that can be used by generators and simulators.

The generated code is dedicated to support execution of system application code, referenced in the model as AADL `subprogram` components called by *thread* components. These *subprogram* components specify properties that define their implementation characteristics (source language, source code, etc.). Source code components are provided by the user and referenced within the AADL model, along with a supporting OS (compliant with the ARINC653 standard for avionics systems) in order to produce a complete implementation, as shown in Figure 39.5.

Producing the implementation from models relies on the following workflow:

1. The *code generation process* produces the architecture code with the characteristics from (potentially previously validated) AADL models.
2. A *compilation process* integrates the generated code with the functional code on top of an ARINC653-compliant dedicated OS (in the case of avionics architectures).

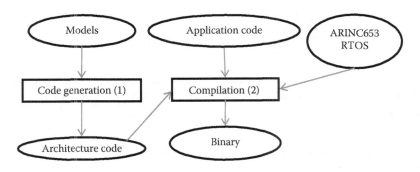

FIGURE 39.5   Automatic production of system implementation.

In order to create the system implementation code, the generator translates the source language (AADL models) into implementation code (C or Ada) using predefined code patterns. Each pattern produces a code block parameterized according to the AADL component's type and properties. For example, for each AADL `thread` component, the generator outputs code that creates an ARINC653 process. Then, the instantiated ARINC653 process is configured according to the AADL component properties (period, deadline, task type—periodic, sporadic, etc.). Also, when processing a connection between two components, the generator outputs code that configures a communication channel according to the model requirements (size of the data being exchanged, output rate, etc.).

However, as depicted in Figure 39.5, producing the application uses a specific OS so that the generation process must be tailored for the OS's needs. Since each OS has its own build directives, it is necessary to adapt some generated artifacts (such as Makefile) in order to ensure a smooth integration of generated code. At the end of the process, the user gets the binary, ready to be deployed on the target architecture. The target architecture specified can be multiprocessor and integrate multiple styles of architecture, such as IMA and federated.

Compared to existing development methods, this process improves system production on several aspects and criteria:

1. *Avoidance of bugs*: By producing code from predefined and established code patterns from models, this process avoids mistakes related to manual code production.
2. *Enforcement of system requirements*: By a direct translation between specifications (AADL models) and implementation code (C/Ada).
3. *Enablement of requirements traceability*: By automating the production of the implementation from models, it is possible to annotate the code and add information (e.g., in terms of comments) to trace and justify each line of source code from the specification to the implementation.
4. *Compliance with coding guidelines*: Standards such as DO178-C require that implementation code enforces specific coding rules and may restrict the syntax and/or semantics of the target language. By implementing these rules in code generation patterns, produced code will be compliant automatically to these standards.
5. *Performance improvements and memory optimization*: By deriving the implementation code from the AADL architecture specification and automating resource instantiation according to system needs, we avoid any useless system resources. This improves the code coverage (removal of useless code) and memory overhead (resource creation limited to system needs), two requirements of code that target high-reliability CPSs.

This process is implemented in Ocarina, an AADL framework that provides code generation facilities, and has been evaluated and demonstrated through several projects. In particular, Ocarina is able to produce ARINC653-compliant code, and many experiments have been done to automatically integrate the generated code with POK [POK], an open-source OS supporting ARINC653 services. Moreover, this code generation functionality can be adapted to any commercial OS; changes are mostly limited to the integration of generated artifacts into the target build system.

39.6 Other Model-Related Research Interests

Models are an important asset for developing CPSs. Interest in model-based engineering technologies will continue to increase in the incoming years. Architecture models will be a central artifact in the production process, either for developing new systems or for updating existing ones. While current research outcomes show the value of models for the specification, validation, and implementation of architecture, models are not currently used at their full potential. In particular,

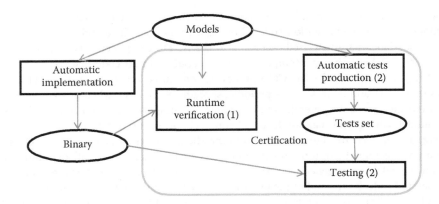

FIGURE 39.6   Use of models for supporting system certification.

they can be a valuable asset for supporting certification efforts. In particular, they could be leveraged for testing the system automatically, as depicted in Figure 39.6:

1. Implementation is automatically tested according to the specification: the system is executed and performance metrics are compared against system specifications and requirements (e.g., checking that the execution time for a task is compliant with its description).
2. Certification tests are automatically generated from system architecture specification and executed against the produced application.
3. Assumptions and assurance case obligations are referenced in the architecture and tested or validated in some form.

The following paragraphs present research topics that would demonstrate additional value of models in the area of system certification.

Aviation systems have strong requirements in terms of performance and resources consumption. Some service must not be delivered or performed as quickly as possible but rather just on time. On one hand, computing and communication resources must be analyzed during design to guarantee their feasibility. On the other hand, their enforcement at execution must be also checked. Architecture models would be a valuable asset for automating this verification process (depicted as the first test case in Figure 39.6): by processing architecture models and extracting their characteristics, it is possible to enumerate the list of constraints (timing, performance, etc.) a system would meet. Then, from this list, performance characteristics that would be monitored at runtime could be established to test the system and check constraint enforcement.

In addition to this verification of compliance between implementation and specification, the system must be stress-tested to show is behavioral correctness when operating in hostile environment. Until now, these tests are mostly written and performed manually, which is still costly and prone to error. On the other hand, as for code generation, models could be processed to automate test production (depicted as the second test case in Figure 39.6) and automated to execute them on the produced application. In particular, architecture specifications reveal the potential strengths and weaknesses of a system and thus would help to design tests that will raise potential errors and/or failures.

While reducing development costs and improving system robustness, automating these verification efforts and tests has other benefits. In particular, it improves customer confidence in produced applications. It also constitutes a valuable support for production of certification materials: stress-testing the system and/or checking execution behavior against specification is required by most standards. Thus, having a precise predictive model for the design would enable stronger verification against analysis results of the implementation through the automation of test generation and documentation, along with traceability to system requirements.

39.7 Conclusion

Recent research efforts have demonstrated the relevance of predictive model-based engineering technologies for developing CPSs and avionics architectures. These models become a valuable asset when starting or updating a project, either for communication, validation, or implementation purposes. In our context, the AADL is of special interest: it provides the ability to use the same notation for carrying all development aspects—design/specification, validation, and implementation. Ongoing research topics plan to use this notation to support certification efforts as well and thus would cover the complete development process. Using the same artifacts for performing and automating all development steps would avoid usual traps and pitfalls of translating system specifications and requirements from one notation to another and manual development efforts.

In the context of avionics systems, the AADL and its ARINC653 annex supply a valuable capability by providing analyzable avionics-specific modeling patterns for specifying IMA architecture. Its associated analysis framework includes functions for validating architecture correctness, evaluating performance (such as latency), and automating the production of the implementation. Validation and automatic implementation of avionics architecture has been already demonstrated [SAVI] [RAMSES], and these validation and code generation capabilities have been implemented in software engineering tools [ASSERT].

These modeling techniques can be applied on other aspects on the development process. In particular, as software complexity continues to grow, certification and testing activities are demanding more attention and thus require more effort. An analytical, predictive model-based approach extended into the testing and certification domain could provide a valuable asset, including greater automation and targeted architecture-related analysis. Using the same specification for multiple development aspects such as design, incremental verification, implementation, and then testing and certification would strengthen the whole development process and make it more consistent.

Annexes

Textual AADL Model for the ARINC653 Case Study

```
thread thr_sensor
features
        value : out data port dummy;
end thr_sensor;
thread implementation thr_sensor.i
end thr_sensor.i;
process pr_data_acq
features
        sensor1out : out data port dummy;
        sensor2out : out data port dummy;
end pr_data_acq;
process implementation pr_data_acq.i
subcomponents
        sensor1 : thread thr_sensor.i;
        sensor2 : thread thr_sensor.i;
connections
        port sensor1.value -> sensor1out;
        port sensor2.value -> sensor2out;
end pr_data_acq.i;
thread thr_filter
features
        sensor1in : in data port dummy;
        sensor2in : in data port dummy;
        filteredout : out data port dummy;
```

```
end thr_filter;
thread implementation thr_filter.i
end thr_filter.i;
thread thr_collector
features
        filteredin : in data port dummy;
end thr_collector;
thread implementation thr_collector.i
end thr_collector.i;
process pr_stats
features
        sensor1in : in data port dummy;
        sensor2in : in data port dummy;
end pr_stats;
process implementation pr_stats.i
subcomponents
        process1 : thread thr_filter.i;
        process2 : thread thr_collector.i;
connections
        port process1.filteredout -> process2.filteredin;
        port sensor1in -> process1.sensor1in;
        port sensor2in -> process1.sensor2in;
end pr_stats.i;
memory segment
properties
        ARINC653::Memory_Type = > (Data_Memory, Code_Memory);
end segment;
memory implementation segment.i
end segment.i;
memory ram
end ram;
memory implementation ram.i
subcomponents
        segment1 : memory segment.i;
        segment2 : memory segment.i;
end ram.i;
virtual processor partition
end partition;
virtual processor implementation partition.i
end partition.i;
processor arinc_module
end arinc_module;
processor implementation arinc_module.two_partitions
subcomponents
        runtime1 : virtual processor partition.i;
        runtime2 : virtual processor partition.i;
properties
        ARINC653::Partition_Slots = > (10 ms, 20 ms);
        ARINC653::Slots_Allocation = > (reference (runtime1), reference
(runtime2));
        ARINC653::Module_Major_Frame = > 30 ms;
end arinc_module.two_partitions;
system arinc653_system
end arinc653_system;
system implementation arinc653_system.i
```

```
subcomponents
      module : processor arinc_module.two_partitions;
      ram : memory ram.i;
      data_acquisition : process pr_data_acq.i;
      statistics : process pr_stats.i;
properties
      Actual_Processor_Binding = > (reference (module.runtime1)) applies
to statistics;
      Actual_Memory_Binding = > (reference (ram.segment1)) applies
to statistics;
      Actual_Processor_Binding = > (reference (module.runtime2)) applies
to data_acquisition;
      Actual_Memory_Binding = > (reference (ram.segment2)) applies to
data_acquisition;
end arinc653_system.i;
```

Disclaimer

The views expressed herein are those of the author and do not reflect the official policy or position of the Department of the Army, Department of Defense, or the U.S. government.

Reference herein to any specific commercial, private or public products, process, or service by trade name, trademark, manufacturer, or otherwise, does not constitute or imply its endorsement, recommendation, or favoring by the U.S. government.

Copyright

References

[ARINC653] Avionics application software standard interface: ARINC specification 653P1-3. Aeronautical Radio, Inc., Annapolis, MD, November 15, 2010.

[NIST02] The economic impacts of inadequate infrastructure for software testing, NIST Planning report 02-3. National Institute for Standards and Technology, Washington, DC, 2002. http://www.nist.gov/director/planning/upload/report02-3.pdf.

[AADL] Architecture Analysis and Design Language (AADL). http://standards.sae.org/as5506b/, 2013.

[AADL-WWW] Architecture Analysis and Design Language (AADL) official website, http://www.aadl.info, 2013.

[AADL-ARINC653] ARINC653 annex for the AADL. http://standards.sae.org/as5506/2/, 2013.

[SAVI] System Architecture Virtual Integration (SAVI). http://www.avsi.aero/research/current_projects.HTML.

[POK] Partitioned Operating Kernel. http://pok.tuxfamily.org.

[OCARINA] J. Delange, L. Pautet, and F. Kordon. Code generation strategies for partitioned systems. In *29th IEEE Real-Time Systems Symposium (RTSS'08)*, Barcelona, Spain, Work in Progress, IEEE Computer Society, Washington, DC, December 2008.

[RAMSES] F. Cadoret, E. Borde, S. Gardoll, and L. Pautet. Design patterns for rule-based refinement of safety critical embedded systems models. In *International Conference on Engineering of Complex Computer Systems (ICECCS'12)*, Paris, France, 2012.

[ASSERT] J. Hugues, L. Pautet, B. Zalila, P. Dissaux, and M. Perrotin. Using AADL to build critical real-time systems: Experiments in the IST-ASSERT Project. In *Fourth European Congress ERTS*, Toulouse, France, January 2008.

40

Mathworks Approach to MBD

Bill Potter
The MathWorks, Inc.

Pieter Mosterman
The MathWorks, Inc.

Tom Erkkinen
The MathWorks, Inc.

40.1 Introduction

Miniaturization of electronics has been on a tear over the past decades with the number of transistors per area doubling about every 18 months. This has unlocked an unparalleled world of opportunity in terms of implementing the most sophisticated behaviors with a future of functionality that is yet to be imagined. Given this almost limitless behavioral flexibility of computation, technical systems have come to rely on software in the role of universal system integration glue as the main enabler of system functionality, simply because of the extreme flexibility in behavior that can be designed in software. Moreover, beyond being an enabler, software has become a driver of innovation as feature differentiation of modern products such as aircraft, automobiles, and consumer electronics relies to a large extent on pure information technology functionality by supporting features that would not be attainable with any technology other than computation.

However, while software provides unprecedented behavioral flexibility, the development of large code bases is also fraught with complexity. Many software engineering methods have been put forward over the years, but these mostly isolate the software as "the system" and fail to address the complexity of embedding software in a physical environment. Most notable is the lack of time and resource utilization as well as first-class support for concurrency in common software engineering paradigms. In contrast to software engineering paradigms, the embedded systems community has tackled the software producibility challenge by model-based design where the level of abstraction of computational functionality is raised from software to models. Consequently, engineers solve their design problems in the problem space (e.g., difference equations), rather than forcing them to think in terms of the solution space (e.g., function call graphs). From a design perspective, engineers are thus empowered to explore their design space in an intuitive manner without having to understand and be concerned with a broad range of software implementation details.

Even though raising the level of abstraction is effective at tackling the productivity gap (i.e., the number of transistors per area goes up more quickly than the number of transistors that can be efficiently designed for), from an implementation perspective, the system still must be realized in software at some point. Model-based design derives much of its value from automatic code generation where embedded software can be automatically generated from a high-level model for a variety of different embedded targets. Whereas automatic code generation technology bridges the gap between design and implementation, for large-scale software projects, the resulting software must still be integrated with (1) the software of other subsystems, (2) the software of the run-time platform, and (3) the target hardware. For safety-critical systems such as an aircraft, it is critical to demonstrate the integrity of the software development.

The Federal Aviation Authority (FAA) recently adopted the certification standard DO-178C "Software considerations in airborne systems and equipment certification" [1] and the corresponding DO-331 "Model-based development and verification supplement to DO-178C and DO-278A" [2] that support model-based design to enable efficient certification. In particular, the various artifacts that are developed during design in order to arrive at an implementation enable incremental evidence building toward full certification. This chapter describes the high-level architecture of the development and verification tool chain used for model-based design in a DO-178C/DO-331 process. There are two types of tools used in the process, development tools and verification tools, where formal methods have enabled bridging the gap by providing tools that automatically generate tests from development tools for use in verification tools.

The chapter is structured to first provide an overview and introduction to the development tools in model-based design. Next, tools that enable verification of the developed artifacts are presented. After this, it is discussed how such verification tools can be applied to the developed models and source code. It is then discussed how tests can be automatically generated from models. Following this, the development and verification of the object code are treated. Finally, conclusions and a summary of model-based design for efficient DO-178C development are presented.

40.2 Description of Development Tools

This section introduces the development tools used in the example model-based design tool chain and discusses their dependencies.

40.2.1 Model Development

*Simulink® is the platform environment for simulation and model-based design. It provides an interactive graphical environment and a customizable set of block libraries to model mathematical algorithms. Simulink supports the description of both discrete- and continuous-time models in a graphical block-diagram language. Simulink models can further include state-transition diagrams, flowcharts, or truth tables provided by Stateflow®. Stateflow supports three types of state-transition diagrams: classic, Mealy, or Moore machines. Mealy and Moore state machines are based on formal semantics.

A Simulink/Stateflow model consists of interconnected function blocks. The function blocks describe the algorithmic or logical components of the model. Connections between the blocks represent the control and data communicated between the blocks. Components within the block-diagram hierarchy can

---

* Simulink® is a registered trademark of The MathWorks, Inc. For product information, please contact:

The MathWorks, Inc.
3 Apple Hill Drive
Natick, MA 01760-2098 USA
Tel: 508-647-7000
Fax: 508-647-7001
E-mail: info@mathworks.com
Web: www.mathworks.com

aggregate other components or be elemental. An important aspect of the Simulink environment is that it provides extensive capabilities for simulation that attributes an executable semantics to Simulink models.

Simulink and Stateflow are separate tools used for the development of models. Simulink may be used without Stateflow, but when Stateflow is used, Simulink is also required.

40.2.2 Source Code Development

The code generators *MATLAB Coder®, Simulink Coder, and Embedded Coder™ facilitate source code generation from Simulink diagrams and Stateflow charts.

Simulink Coder facilitates C and C++ code generation from Simulink diagrams and Stateflow charts. The generated source code can be used for real-time and non-real-time applications, including simulation acceleration, rapid prototyping, and hardware in-the-loop (HIL) testing. Embedded Coder adds that capability to target embedded real-time systems that require efficient and traceable code. It extends Simulink Coder by providing configuration options and advanced optimizations to generate code for on-target rapid prototyping boards and microprocessors used in production. Embedded Coder furthermore improves code efficiency and facilitates integration with legacy code.

MATLAB Coder, Simulink Coder, and Embedded Coder are separate tools used for the development of a source code. MATLAB Coder is a prerequisite for Simulink Coder and Embedded Coder. In the following sections of this chapter, references to Embedded Coder are intended to include Simulink Coder and MATLAB Coder as the entire code generation tool set.

40.2.3 Executable Object Code Development

Third-party compiler tools are used to translate the source code into executable object code (EOC).

40.2.4 Requirements Management

A requirements management interface allows users to link Simulink and Stateflow elements within models to requirements documents and to create links between model elements. Document types supported by the interface are Microsoft Word documents, Microsoft Excel spreadsheets, IBM Rational DOORS databases, MuPAD notebooks, Simulink DocBlocks, text documents, hypertext markup language (HTML) files, portable document format (PDF) files, and uniform resource locator (URL).

The requirements management interface is provided as part of Simulink Verification and Validation.

The left-hand side of Figure 40.1 illustrates the model and source code development processes including requirements management. The EOC development process is depicted in the middle row of Figure 40.3.

40.3 Description of Verification Tools

This section introduces the verification tools and report generators used in the example model-based design tool chain and discusses their dependencies.

---

* MATLAB® is a registered trademark of The MathWorks, Inc. For product information, please contact:

The MathWorks, Inc.
3 Apple Hill Drive
Natick, MA 01760-2098 USA
Tel: 508-647-7000
Fax: 508-647-7001
E-mail: info@mathworks.com
Web: www.mathworks.com

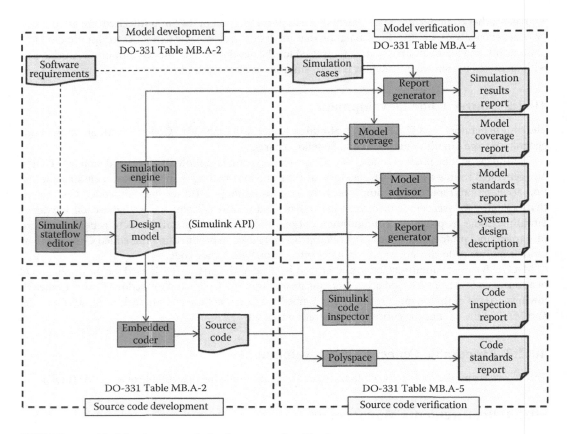

FIGURE 40.1   Model and source code development and verification.

40.3.1 Report Generation

Report generators provide components for reporting on Simulink and Stateflow models.

Reporting components for Simulink and Stateflow models are provided by Simulink Report Generator. These components can interrogate a model using the Simulink application programming interface (API) to read data from the model loaded in memory. The System Design Description is created from this tool and includes requirements traceability links that have been inserted into the model using the requirements management interface. The report generator also can run simulations in batch mode, set up test conditions, and generate simulation reports that include plots or analysis of data. The Simulink Report Generator uses functionality provided by MATLAB Report Generator.

The MATLAB and Simulink Report Generators are two separate tools, with the MATLAB Report Generator being a prerequisite for the Simulink Report Generator.

40.3.2 Model Verification

Model Advisor is a static analysis tool for Simulink and Stateflow models. Model Advisor performs static checks on the model that can be used for verification of modeling standards as well as aspects of accuracy and consistency of the models.

The Model Advisor checks are provided in several different products: Simulink, Embedded Coder, Simulink Code Inspector, Simulink Verification and Validation, and Simulink Control Design. The Model Advisor uses the Simulink API to read data from the model.

The model coverage tool analyzes structural coverage achieved during model simulations. The tool instruments a Simulink and Stateflow model loaded into memory prior to simulation and evaluates the coverage criteria as the simulation progresses. Model coverage also has the capability to merge multiple simulation runs into a combined coverage report. Model coverage also provides a measurement of the cyclomatic complexity of the model.

The model coverage capability is provided as part of Simulink Verification and Validation. The user can run simulations with coverage enabled and disabled to insure there has been no effect on the behavior of the model due to the instrumentation.

Simulink Design Verifier is a tool with three capabilities: design error detection, property proving, and test case generation. Simulink Design Verifier contains formal analysis engines that operate on an internal representation derived from but in a different form than the Simulink model. Design error detection can find specific design errors in the model, such as divide-by-zero or numeric overflows, using formal methods. Property proving, which also uses formal methods, can prove properties that are defined by the user in conjunction with assumptions that are also defined by the user. Property proving may assist in the validation of system safety properties as addressed in ARP4754A [3] and ARP4761 [4]. Simulink Design Verifier can automatically generate test cases based on the model that can be used to verify if the EOC complies with the model. The basis for the test cases can be a combination of user-defined constraints, model coverage criteria for blocks in the model, and user-defined test objectives. In order to verify the code using the generated test cases, the test cases must be run on the model in order to produce expected results for the code. The completeness of those test cases may be assessed using the model coverage tool and the expected results may be assessed via review of the results from the simulation.

40.3.3 Source Code Verification

Simulink Code Inspector [5] is a separate tool that can be used to verify C source code developed from Embedded Coder. This tool interrogates the model using the Simulink API to read data from the model. The model is converted into a different intermediate representation, in the form of an abstract syntax tree, for use in the code inspection process. The Simulink Code Inspector also uses the generated C code files as input and parses these into a different intermediate representation that can be compared to the model's intermediate representation. The tool outputs an inspection report and a traceability report for the source code. Polyspace is a tool that has two capabilities: coding standards checking (e.g., MISRA AC AGC [6]) and run-time error detection. The main input to Polyspace is the source code; however, it can optionally read range specification data from the model using the Simulink API. When using the Polyspace Model Link SL product, it can trace defects found in the source code back to the source blocks in the model. Polyspace also supports any C code, whether it is automatically generated or manually developed. For run-time error detection, Polyspace uses abstract interpretation in its formal methods engine.

40.3.4 Executable Object Code Verification

The EOC developed from the models may be verified using processor-in-the-loop (PIL) capability provided with Simulink and Embedded Coder. The PIL mode allows a target CPU to be connected to the Simulink in order to provide an efficient test environment for the EOC.

Third-party code coverage tools are used to analyze structural coverage of the source and/or object code while the EOC is being executed in the PIL mode.

The right-hand side of Figure 40.1 illustrates the model and source code verification processes. The EOC testing process is depicted in the upper part of Figure 40.3. The next sections provide a description and discussion of these processes.

40.4 Model and Source Code Development and Verification Processes

When models are used in a DO-178C [1] process, then DO-331 [2] becomes an applicable supplement for that project. In a workflow where code is generated from the Simulink and Stateflow models using Embedded Coder, the models are considered to be design models as defined in DO-331. This represents the most common use case for the models and also requires that there are higher-level requirements from which the model is developed. See Figure 40.1—Model and code development and verification. It is also possible to use Simulink and Stateflow to model requirements, thus representing a specification model as defined in DO-331, but that is not a typical usage of the tools.

In Figure 40.1, the software requirements are typically in the form of a document or data within a requirements management tool such as IBM Rational DOORS. These requirements represent the high-level software requirements as defined in DO-178C, section 5.1. Verification of these requirements is performed using reviews and analyses as defined in section 6.3.1 of DO-178C.

The design model in Figure 40.1 represents the low-level software requirements as defined in DO-331, section MB.5.2. The modeling elements can be traced to the high-level requirements using the requirements management interface. The verification of the models may be performed using a combination of reviews, analysis, and simulation, as defined in DO-331 section MB.6.3.2 and MB.6.3.3. DO-331 introduces simulation as a verification method and also model coverage analysis as an activity for the verification of models.

As shown in Figure 40.1, simulation cases are developed from the software requirements and executed on the design model. Simulation result reports and model coverage reports are generated during this activity and reviewed as part of the verification activities. These two reports address the objectives associated with compliance to high-level requirements, accuracy and consistency, verifiability, and algorithm aspects. The model standards report, generated using the Model Advisor, addresses the conformance to standards objective. The System Design Description document, generated by the Simulink Report Generator, provides the design description data defined in DO-331, section MB.11.10, and some of the trace data defined in DO-331 MB.11.21. Review of the trace data in this document addresses the objective for traceability to the high-level requirements.

The source code (see DO-178C section 5.3) is developed from the design model using Embedded Coder. Embedded Coder can generate ANSI C code that is platform independent but also has the optional capability to generate code that uses custom target features. Embedded Coder only supports discrete-time fixed-step solvers in Simulink models and also constrains which types of Simulink blocks may be used. By adhering to certain modeling constraints, enforced by the Model Advisor, the generated code can comply with the MISRA AC AGC coding standard [6]. The generated source code can also include comments that trace the code back to the model, and also to the software requirements document.

As shown in Figure 40.1, the source code can be verified using a combination of Simulink Code Inspector and Polyspace. The Simulink Code Inspector provides a Code Inspection Report that addresses the objectives associated with compliance to low-level requirements, compliance with software architecture, verifiability, traceability, and portions of accuracy and consistency. Polyspace can be used to address the objective of compliance to coding standards, specifically to the MISRA AC AGC rules. Polyspace also provides a run-time error report that includes some aspects of accuracy and consistency for the source code. Credit can also be taken for Polyspace run-time error detection to address robustness of EOC, and that is discussed later.

40.5 Test Case Development Processes

The DO-178C standard calls out three types of testing, all of which are based on the software requirements: hardware/software integration tests, software integration tests, and low-level tests. For DO-178C, test cases should include normal-range test cases and robustness test cases.

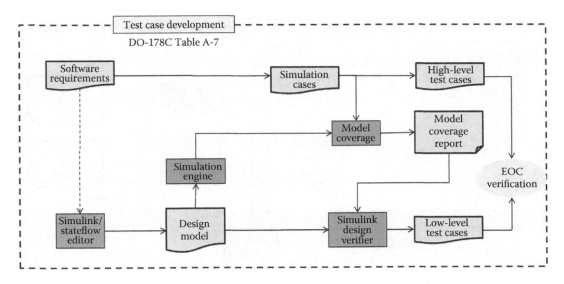

FIGURE 40.2   Test case development.

For the EOC developed from models, the high-level test cases and expected results can be the same as the simulation cases and expected results (see Figure 40.2). These are developed from the high-level requirements document and should also include robustness cases. These test cases can be executed using PIL capability in conjunction with the Simulink environment used as a test harness, or on a completely separate software test harness.

The low-level test cases and expected results are based on the models, which represent the low-level requirements. Simulink Design Verifier may be used to develop these test cases (see Figure 40.2). Simulink Design Verifier uses the model as its primary input and also has the capability to input model coverage data. DO-178C (see note in section 6.4) calls out that if it can be shown that high-level tests cover low-level requirements, then those low-level requirements do not need to be covered by specific low-level tests. Model coverage can be used as evidence that high-level tests cover low-level requirements, in particular for logical decisions within the models, but also for lookup table data and signal range data within the models. Simulink Design Verifier can then be used to generate tests for the remaining low-level requirements that are not covered by high-level testing, for example, derived requirements within the model. The user can also insert signal constraints and user-defined test objectives within the models or in model test harnesses to complete the testing. The use of test objectives on the inputs to a model to insert test data beyond normal ranges is a good way to verify robustness, for example.

The hardware/software integration cases and the software integration cases (see Figure 40.3) are typically developed manually based on the high-level software requirements. These test cases are executed on the final target in an environment independent of the modeling environment. The final target would include an RTOS or scheduler and the device drivers that interface to the target hardware.

40.6 Object Code Development and Verification Processes

Figure 40.3 shows the EOC development and verification activities, including the use of the PIL mode and target integration testing. These activities are downstream of the model and source code development and verification activities. The compiler or IDE are third-party tools that are not provided by MathWorks. Errors injected by the compiler are detectable by the testing process. The code coverage tool is also provided by a third party, rather than MathWorks, and this tool is normally qualified for the project.

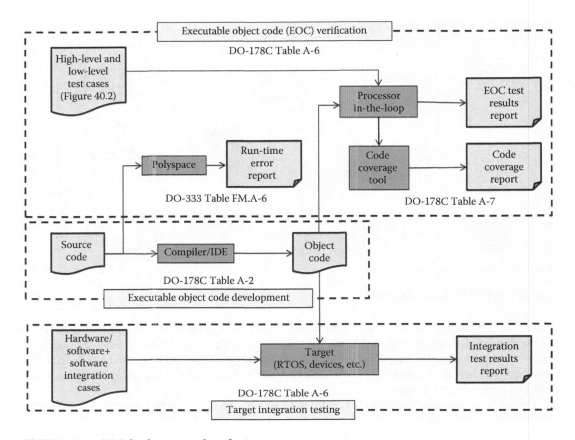

FIGURE 40.3   EOC development and verification.

The testing done in PIL mode is using an alternate test environment, as described in the note in section 6.4.1 of DO-178C. This environment can be used to test software components individually, which allows for finer control of test inputs and monitoring of component outputs. It is typical to obtain structural coverage results in a more controlled test environment like PIL mode. PIL mode testing addresses the objectives for compliance of the EOC to high- and low-level requirements, as well as robustness with respect to high- and low-level requirements. Of course this does not eliminate the need for software–software and software–hardware integration testing of these components on the final target. Target integration testing addresses the compatibility of the EOC with the target computer.

Also shown in Figure 40.3 is the use of Polyspace analysis to eliminate certain robustness test cases on the object code. An example of this would be replacement of out-of-bounds data testing by the range analysis capability in Polyspace. This type of credit is explained in DO-333, the formal methods supplement.

40.7 Summary and Conclusions

This chapter presented development and verification processes using model-based design and a corresponding tool chain to enable the efficient development and certification of aerospace software according to the DO-178C and DO-331 standards. The presented approach has been developed and refined in collaboration with leading aerospace companies.

The DO-178C standards and its supplements lay out additional objectives not discussed in this chapter, for example, the qualification of software verification tools. This activity can be supported by dedicated tool qualification kits provided by the respective tool vendors, such as the DO Qualification Kit that covers a variety of the model and code verification tools discussed in the preceding sections.

References

1. RTCA DO-178C, Software considerations in airborne systems and equipment certification, December 2011. RTCA, Inc.: Washington, DC.
2. RTCA DO-331, Model-based development and verification supplement to DO-178C and DO-278A, December 2011. RTCA, Inc.: Washington, DC.
3. Aerospace Recommended Practice ARP4754A, Guidelines for development of civil aircraft and systems, SAE International, December 2010.
4. Aerospace Recommended Practice ARP4761, Guidelines and methods for conducting the safety assessment process on civil airborne systems and equipment, SAE International, December 1996.
5. M. Conrad, M. Englehart, T. Erkkinen, X. Lin, A. R. Nirakh, B. Potter, J. Shankar, P. Szpak, J. Yan, and J. Clark, Automating code reviews with Simulink Code Inspector, *Proceedings of VIII Workshop Model-Based Development of Embedded Systems (MBEES 2012)*, Dagstuhl, Germany, February 2012, pp. 31–36.
6. MISRA AC AGC, Guidelines for the Application of MISRA-C:2004 in the Context of Automatic Code Generation, MIRA Ltd., November 2007.

41

Esterel SCADE
Approach to MBD

Jean-Louis Camus
Esterel Technologies

41.1 Introduction

SCADE Suite® is a model-based development environment dedicated to critical embedded software. With native integration of the Scade language and its formal notation, SCADE Suite is the integrated design environment for critical applications spanning model-based design, simulation, verification, qualifiable/certified code generation, and interoperability with other development tools and platforms, including requirements traceability. SCADE Suite has been developed specifically to address mission and safety-critical embedded applications. The SCADE Suite KCG code generator is certified/qualified according to the following international safety standards including DO-178B qualified up to level A and DO-178C ready for civilian and military aeronautics. This chapter is intended as an introduction to model-based development and verification (MBDV) process using tool sets that have been used in designing, developing, and simulating an avionics system.

41.2 SCADE Overview

41.2.1 SCADE Product Overview

The name "SCADE" stands for "safety-critical application development environment." SCADE stands for the toolchain; "Scade" stands for the underlying modelling language. This chapter addresses model-based software development and verification with SCADE.

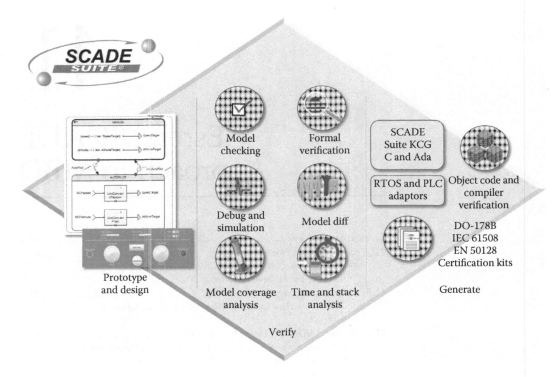

FIGURE 41.1 The SCADE integrated development environment.

Esterel SCADE is a product line that includes the following products (Figure 41.1):

- SCADE Suite for the development of embedded reactive software application
- SCADE Display for the development of embedded display graphics part
- SCADE System for system design
- SCADE LifeCycle for the application lifecycle management of these applications

For the sake of simplicity, this paper only addresses the SCADE Suite part.

41.2.2 SCADE Product Origin

From its early industrial stages, SCADE Suite was developed in conjunction with companies developing safety-critical software. SCADE Suite was used on an industrial basis for the development of safety-critical software, such as flight control software (Airbus, Eurocopter), nuclear power plant safety systems (Rolls-Royce Civil Nuclear, formerly Schneider Electric), and railway switching systems (Ansaldo STS, formerly CSEE Transport). These companies used to develop software with in-house toolchains. They contributed to the development of a third-party solution, in order to share a standard environment with their suppliers and partners.

41.2.3 SCADE Position in the Embedded Software

SCADE addresses the applicative part of software, including graphics and associated logic, as illustrated in Figure 41.2. This is usually the most complex and changeable aspect of software.

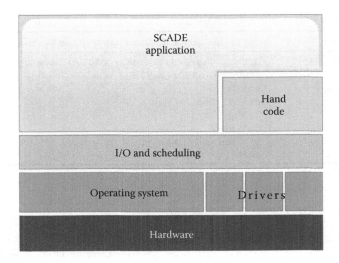

FIGURE 41.2 SCADE addresses the applicative part of software.

41.3 Scade Language

41.3.1 SCADE Data Flow Language Elements

The Scade language properties play a key role for tools and for users, due to its fitness for reactive systems and to its integrity principles.

The Scade language is a bridge between control engineering and software engineering. This reduces errors and cost in the interface between system lifecycle and software lifecycle. It is a so-called synchronous language (Benveniste et al., 2003). This usually means software computing cyclically its output, based on a "sample-and-hold" approach.

SCADE Suite supports a free combination of two notations that are familiar to control engineers:

- Block diagrams (also called data flow diagrams) to specify the control algorithms (control laws, filters)
- State machines to specify modes and transitions in an application (e.g., taking off, landing)

The basic building block in Scade is called an operator. An operator is either a predefined operator (e.g., +, delay) or a user-defined operator that decomposes itself using other operators. It is thus possible to build complex applications in a structured way.

The Scade language allows both a textual notation such as "x = y + z" and a graphical notation. The type of notation has no impact regarding the semantics and the code generation (Figure 41.3).

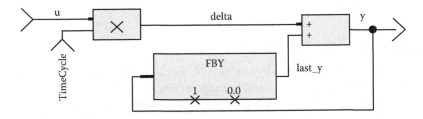

FIGURE 41.3 Integrator in graphical Scade notation.

TABLE 41.1   Elements of a Scade Operator

Component	Textual Notation for an Integrator Operator
Formal interface	`node IntegrFwd(U: real; hidden TimeCycle: real)` ` returns (Y: real);`

Local variables
 declarations
	`var`
	`delta: real;`
	`last_y: real;`

Equations	`delta = u * TimeCycle;`
	`y = delta + last_y;`
	`last_y = fby(y, 1, 0.0);`

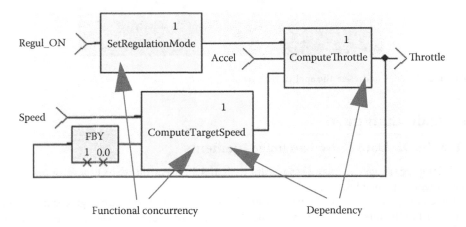

FIGURE 41.4   Concurrency and functional dependencies.

The language includes predefined primitive data types (Boolean, integer, real, char, enumerated) with related primitive operators such as Boolean/arithmetic operators. The language allows building structured data types (arrays, struct) that can be composed with no limit (Table 41.1).

The Scade language provides a simple and clean expression of concurrency and functional dependency at the functional level, as illustrated by Figure 41.4:

- Operators SetRegulationMode and ComputeTargetSpeed are functionally parallel.
- ComputeThrottle functionally depends on an output of ComputeTargetSpeed. SCADE Suite KCG code generator takes this into account: it generates code that executes ComputeTargetSpeed before ComputeThrottle.

41.3.2 Scade Advanced Construct

In addition to its data flow historic foundations, SCADE supports advanced constructs. Structured state machines support parallelism, hierarchy, and preemption (Figure 41.5).

The language also supports safe iterative processing. This allows the development of a wide range of iterative computations such as matrix computation, Kalman filtering, and management of sets of reactive graphical user interface elements.

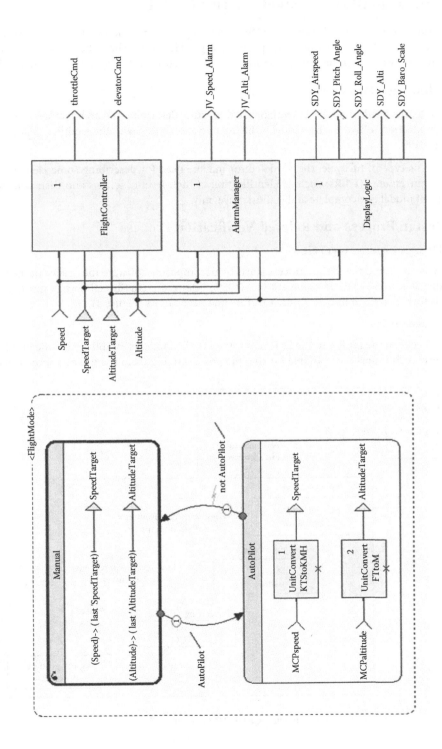

FIGURE 41.5 State machine combined with data flow.

41.4 Typical MBDV Lifecycle with SCADE

There are various possible lifecycles with SCADE, depending on the position of the models in the lifecycle and on techniques. In this document, we present a typical lifecycle, where the Scade model is essentially used to specify a part of the application software architecture and low-level requirements (LLR).

41.4.1 Lifecycle Overview

In this lifecycle, the Scade model is essentially used for fully describing the architecture and LLR of a part of the application software. This model is the input to autocoding with the qualified code generator KCG (Figure 41.6).

Note: With this type of lifecycle, the Scade editor may be used for describing some elements of the high-level requirements (HLRs) such as identification of high-level functions and their connections. But this type of model is incomplete and is illustrative only.

41.4.2 Design Process and Related Verification Process

41.4.2.1 Design Process Overview

Using the HLRs as input, the design process starts by decomposing software into major parts. Some of these are allocated to SCADE development (40%–95%), the rest to traditional development. Then the Scade part is further refined into Scade architecture and Scade LLRs (Figure 41.7).

41.4.2.2 Traceability

Traceability between the HLR and the LLRs contained in the model is established while building the model. Several techniques are available for that purpose, ranging from the use of a spreadsheet tool

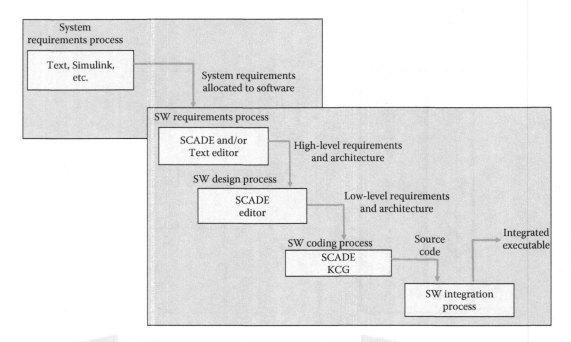

FIGURE 41.6 Typical lifecycle overview.

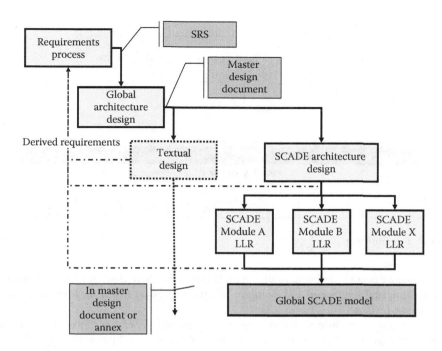

FIGURE 41.7 Design process overview.

or model annotations up to dedicated tools such as the SCADE Requirements Management Gateway and/or DOORS. This traceability is verified during model verification.

41.4.2.3 Design Integrity

There are two possible approaches for achieving software integrity:

1. Use a general-purpose permissive language/notation and attempt to impose/verify rules that constrain the use of that language. For instance, the C language at source code level or the UML at model level usually need a lot of coding/design rules.
2. Design and use a language for specific classes of applications with appropriate built-in properties. These are sometimes called "application-oriented languages."

The Scade language has been designed from its origin for the development of safety-critical software and is based on integrity principles. These are formalized in the language reference manual and verified by the qualified semantic checker.

The Scade language is deterministic, and this is reflected in code generated from the model: for a given input sequence, there will always be the same output sequence, determined by the language reference manual, whatever the version of the code generator and the code generation options.

The language is strongly typed, in the sense that each data flow has an explicit type and that type consistency in Scade models is verified by the SCADE Suite tools. There is no language construct with a risk of side effect: a module body cannot write to any other data than its formal output, which is itself explicitly connected to other modules in its calling modules. There is no risk related to undefined data or data overriding: any data that are consumed in a cycle have to be produced once and only once in that cycle.

Scade is modular: the behavior of an operator does not vary from one context to another. Thus, verification of the contents of a module remains valid for any use of that module.

Scade makes it possible to deal properly with issues of timing and causality. Causality means that if datum x depends on datum y, then y has to be available before the computation of x starts. A recursive

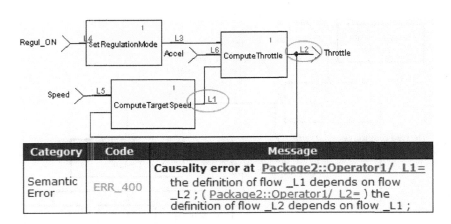

FIGURE 41.8   Causality loop.

data circuit poses a causality problem, as shown on Figure 41.8, where the "Throttle" output depends on itself via the ComputeTargetSpeed and ComputeThrottle operators. Inserting a unit delay in the loop resolves the causality issue.

As a summary, a large amount of design integrity is systematically ensured by the inherent properties of the Scade language, which prevents a large number of risks either by not offering risky constructs (e.g., pointer handling) or by qualified verification of language integrity rules (e.g., typing, causality).

41.4.2.4 Model Review and Simulation

As a complement to verification ensured by syntax and semantic checks, model review, analysis, and simulation are used for verification of the following:

- A-4.1: Compliance of the model to its HLR (simulation is well suited to verification of dynamic aspects).
- A-4.5 and A-4.12: Conformance to modeling rules that the user has added (if any) to the pre-defined (imposed) rules of the language reference manual.
- A-4.6: Traceability to the model's HLR.
- A-4.7: Algorithms accuracy (simulation is well suited to dynamic aspects).
- A-4.8: Compatibility of architecture with HLR.
- A-4.9: Consistency of interface with non-Scade software.
- A-4.3 and A-4.10: Preliminary assessment of amount of resources needed for compatibility with target computer. This is ultimately verified during HW/SW integration testing.
- Software partitioning (A-4.13) analysis is performed in a traditional way.

Simulation cases shall be based on HLR (Figure 41.9).

41.4.3 Coding Process and Related Verification Process

41.4.3.1 Coding and Integration Process

The coding process combines two threads (Figure 41.10): qualified autocoding with KCG and manual coding. Both codes are compiled and integrated for building the final software.

KCG generates structured code suited to safety-critical embedded software, for instance:

- Portability (target and OS independent).
- Readability and traceability with respect to the design.
- It is based on static memory allocation and contains no pointer arithmetic.
- It contains no recursion and has bounded loops only and bounded execution time.

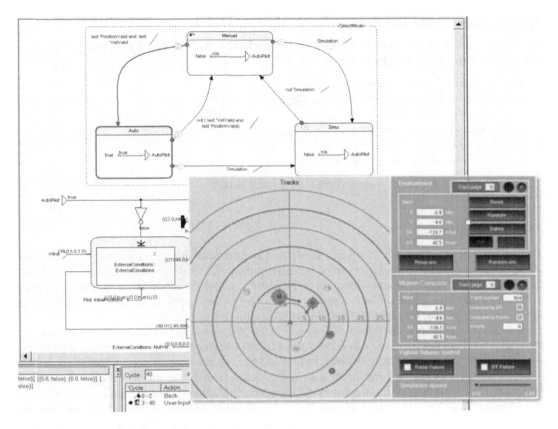

FIGURE 41.9 Requirements-based model simulation.

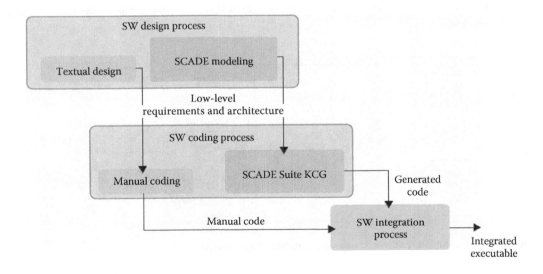

FIGURE 41.10 Coding process overview.

41.4.3.2 Source Code Verification

The qualification of KCG at TQL-1 (development tool for level A in DO-178B) saves human review of the source code with respect to objectives of table A-5:

A-5.1 Source code complies with LLR.
A-5.2 Source code complies with software architecture.
A-5.3 Source code is verifiable.
A-5.4 Source code conforms to standards.
A-5.5 Source code is traceable to LLR.
A-5.6 Source code is accurate and consistent.

Verification of output of integration process is done traditionally.

41.4.4 Testing

There is an efficient approach consisting in sharing requirements-based verification cases between model simulation and testing. The qualified test environment (QTE) allows running these shared verification cases both for model simulation, for target testing and for the model coverage analysis (Figure 41.11). In all cases, the actual software response is compared to the expected response, which is included in the verification cases that are developed and verified with independence.

HLR-based tests are first developed. Wherever possible, they are also used to cover nonderived LLRs. Dedicated low-level tests need to be developed for derived LLRs (if any), since by definition they are not traceable (their accidental coverage by HLR tests would have no value).

LLR coverage analysis and resolution are used in order to detect insufficient testing, inadequate HLRs, and dead/deactivated/unintended model elements. This analysis is supported by SCADE model test coverage (MTC). Scade model coverage criteria are designed in order to support accurate analysis of data flow and control flow coverage of the LLRs contained in the model. This analysis addresses both fine-grain coverage of primitive constructs (e.g., MC/DC, time operators) and activation of libraries functionality (e.g., watchdog firing, filter reset).

In addition to the primary purpose of MTC, which is LLR coverage analysis, there is a secondary benefit: KCG preserves coverage from model to code and MTC analyzes structural code coverage with traceback to the code . This allows the user to concentrate in analyzing coverage at the model level rather than at code level.

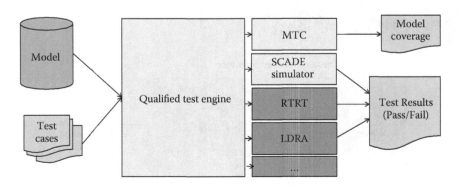

FIGURE 41.11   Unified requirements-based simulation and testing.

41.5 SCADE Aeronautics Applications

SCADE has been first qualified as level A development tool in 1999 for the autopilot of the Eurocopter EC 135. Since then, it has been deployed on more than 100 different airborne equipments including but not limited to:

- Autopilot and flight control systems
- Braking and landing gear systems
- Cockpit display systems
- FADEC
- Flight warning systems
- Fuel management systems
- Navigation, guidance, and inertial units
- Electrical load management

41.6 Summary and Conclusions

Experience has shown that MBDV is an efficient means to achieve high integrity at reduced cost when combining the following:

- First of all, a real understanding of the principles of DO-178B/C with an appropriate application to MBDV. This was first done in the frame of DO-178B and is now transposed into DO-331 (MBDV supplement to DO-178C and DO-278A). We would like to stress that a model is no more no less than other development lifecycle data: there is no mystery; there shall be no less rigor.
- Use of an application-oriented language with a formal basis.
- Qualified model semantic check.
- Qualified code generation.
- Unified simulation and testing with model coverage analysis.

Glossary

Acronym	Definition
FADEC	Full authority digital engine controller
HLR	High-level requirement
HW	Hardware
LLR	Low-level requirement
KCG	SCADE qualifiable code generator
MBDV	Model-based development and verification
MTC	Model test coverage
QTE	Qualified test environment
Scade	Scade language
SCADE	Safety-critical application development environment
SW	Software
UML	Unified modeling language

References

[DO-178C] Software Considerations in Airborne Systems and Equipment Certification, RTCA DO-178C, RTCA Inc, 2012.

[DO-248C] Final report for clarification of DO-178C Software Considerations in Airborne Systems and Equipment Certification, RTCA DO-248C, RTCA Inc, 2012.

[DO-330] Software Tool Qualification Considerations, RTCA DO-330, RTCA Inc, 2012.

[Benveniste et al., 2003] A. Benveniste, P. Caspi, S.A. Edwards, N. Halbwachs, P.L. Guernic, R. de Simone, The synchronous languages 12 years later, *Proceedings of the IEEE,* 91(1): 64–83, January 2003.

[Halbwachs, 1991] N. Halbwachs, P. Caspi, P. Raymond et Daniel Pilaud, The synchronous dataflow programming language Lustre, *Proceedings of the IEEE,* 79(9):1305–1320, September 1991.

42

Model Checking

Tingting Hu
*Institute of Electronics,
Computer and
Telecommunication
Engineering
and
Polytechnic University
of Turin*

Ivan Cibrario
Bertolotti
*Institute of Electronics,
Computer and
Telecommunication
Engineering*

42.1 Introduction

Model checking [5] is a widespread formal verification technique aimed at evaluating the properties of a complex system. In recent years, the general approach of model checking has been applied to a broad spectrum of application areas, ranging from hardware components design to software and system engineering. To limit the scope of the discussion, in the following, we will focus exclusively on the verification of concurrent, distributed software systems. In this domain, one of the most prominent tools is the SPIN model checker [1,10], and this chapter is focused on it.

Like virtually all other model checkers, SPIN requires and works on a model of the system under analysis and a property to be verified, both specified by means of a suitable formal language. Informally speaking, SPIN aims at proving that the property of interest holds by systematically exploring every possible *state* the system under analysis can reach during its execution, starting from its initial or starting state. Collectively, these states are called the *state space* of the system. Each individual state is represented by a *state vector*, which contains all the information to fully characterize that state, for example, state variables.

The system transitions from one state to another by means of elementary *computation steps*. The possible computation steps originating from a given state are specified by the model. Therefore, a path from the starting state up to a final state, in which no more transitions are possible, represents one of the possible finite *computations* the system can perform. Infinite computations are also possible, if the path contains a cycle. Unless the model specifies a strictly sequential system, many different computations are possible, leading to different results.

In its original form, and unlike other model checking tools, SPIN does not include *time* as a native concept. When this concept is needed, it must be modeled explicitly, for instance, by introducing additional state variables to keep track of time. A notable example of model checker that directly supports the concept of time, by means of the timed automata formalism [4], is UPPAAL [3,7]. SPIN itself has been extended to support discrete time, too, leading to DTSPIN [8].

Similarly, Spin is only able to assess whether or not a certain state can be reached in the computation. Other model checkers, for instance, Prism [2,15], allow the users to specify a probability for a transition to occur, and hence, they are also able to calculate the *probability* of the system reaching a certain state. In both cases, the price to be paid to have a richer model is often a loss of performance in the verification.

The goal of this chapter is to give an overview of how Spin works and a short example of its application. It is structured as follows: Section 42.2 briefly describes the Spin input languages, for instance Promela [9,6]. Then, Section 42.3 provides more details on the verification flow and gives some hints to improve performance. Lastly, Section 42.4 contains a practical example on how it is possible to model and verify a couple of interesting properties of a simple communication protocol. The last section contains a small number of references that can be useful as a starting point for further reading.

42.2 Promela Modeling Language

This section provides a general idea about the structure of the Promela modeling language, whose syntax is similar to the C programming language [13]. The description will be given informally for the most part and should be taken as a starting point to thoroughly learn the language. Additional information can be found in more specialized textbooks, for instance, [9,6].

42.2.1 Processes

The concept of concurrency is expressed by means of *processes* in Promela. For example, in a distributed system, a node can be represented as a set of processes, modeling functions of interest performed by that node. Processes can execute concurrently and, hence, interleave with each other. For this reason, processes are the main component in a Promela model. They contain all the executable statements of a model and can be declared by means of the `proctype` keyword:

```
proctype P(<param>) {
  <body>
}
```

Previously, P is the name of the process, and `<param>` is a list of parameters, which is optional, while `<body>` is a sequence of statements constituting the body of the process. Here and in the following examples, the notation `<body>` (with angular brackets) is used as a placeholder for parts of code that are not shown. Unlike in the C language, statements within a sequence are *separated* (but not terminated) by a semicolon. Processes declared as in the previous text must be instantiated explicitly, zero or more times, by means of the `run` operator, that is to say, the `run` keyword followed by the name of a process and its instance parameters, if appropriate. An alternative way of process declaration makes use of the `active` keyword:

```
active [<n>] proctype P() {
  <body>
}
```

With the `active` keyword, the process will be instantiated implicitly at verification start-up. However, in this case, the process cannot have any parameters. By specifying a positive integer value `<n>`, the same process can be instantiated `<n>` times, but all instances will be identical. If `[<n>]` is omitted, the number of instances of P defaults to one. If necessary, a special process called `init` can also be declared as follows:

```
init {
  <body>
}
```

The `init` process, if declared, is always the first process to be instantiated. It is typically used to instantiate other processes and initialize global variables in a more complex way than simply giving them a fixed, predefined initial value.

42.2.2 Variables and Data Types

Variables are the most basic elements in the PROMELA language and they are typed. Table 42.1 lists all basic data types supported in PROMELA. Data type representation on a given architecture depends on the underlying C compiler, as it will be better explained in Section 42.3.1. Table 42.1 gives the range of numeric data types on a typical 32-bit architecture. The main factor affecting the choice of which data type to use for a certain numeric variable is a compromise between its range requirements and the size of the state vector. It should also be noted that some data types, which are common in other programming languages, are not available in PROMELA, such as characters, strings, pointers, and floating-point numbers [9].

In terms of scope, variables can be classified into *local* variables and *global* variables, depending on where they are declared with respect to processes. If a variable is declared within a process, then it is local to that process and it is *private*, that is, it cannot be referenced outside that process. On the other hand, if a variable is declared out of all processes, then it is global and it is *shared* among them. Variables can be declared as follows:

```
byte x = 4
```

This statement declares a variable named x, of type `byte`, and initializes it to 4. In PROMELA, all numeric variables are initialized to 0 by default if no initial value is given explicitly.

Moreover, PROMELA also provides support for arrays. Array elements are of the same type, and they can be referred to by a numeric index, which indicates its position within the array. Similar to the C language, the first element is at index zero. Arrays can be declared as follows:

```
int a[5] = 8
```

In this example, `int` is the data type of the array elements, a is the array name, and the number within the square brackets indicates the total number of elements in this array. Only one-dimensional arrays are directly supported in PROMELA, but it is possible to declare multidimensional data structures using a `typedef` declaration, to be discussed later. If an initial value is specified, like 8 in the example, it is assigned to all array elements.

The channel is a quite useful data type for modeling concurrent and distributed systems, where nodes and communication networks can be modeled as concurrent processes and channels over which processes exchange messages, respectively. Channels can be declared as follows:

```
chan ch = [<capacity>] of {<type>,…, <type>}
```

TABLE 42.1 Summary of Basic PROMELA Data Types with Their Size and Range on a Typical 32-Bit Architecture

Name	Size (Bits)	Range
bit	1	0, 1
bool	1	`false, true`
byte	8	0, …, 255
short	16	$-32768, …, 32767$
int	32	$-2^{31}, …, 2^{31} - 1$
unsigned	$n \leq 32$	$0, …, 2^n - 1$
chan	n/a	n/a

In this declaration, ch is the name of a channel, and <capacity> is either zero, representing a *rendezvous* unbuffered channel, or greater than zero, which means that the channel is *buffered*. In turn, this affects the behavior of the send and receive operations on the channel itself, as it will be better described in Section 42.2.5. The sequence of data type names <type>,..., <type> defines the message structure.

It is also worth noting that the channel capacity must be taken into account when assessing the state vector size. Using a large capacity without good reasons can impair the verification process.

Generally in PROMELA, channels are declared as global variables. In this way, any process can send/receive messages to/from any channel. So, somewhat contrary to their name, channels are not constrained to be point to point. Local channels are possible [6], too, but their usage is beyond the scope of this chapter.

In order to make the code more readable, symbolic names can be given to values by means of an mtype declaration. For instance,

```
mtype = {mon, tue, wed, thu, fri, sat, sun};
mtype date = fri
```

Internally, values of this type are represented as positive values of type byte. It should be noted that only one set of names of this type can be defined for the whole program. If symbolic names are given in multiple mtype declarations, they will all be part of the same set. The mtype type was originally introduced to use symbolic names instead of numbers to represent different types of messages.

Compound data types can be defined by using the typedef keyword, followed by the name of the new data type and by a list of components. For instance,

```
typedef message {
  mtype msg;
  <type> var_1;
  <type> var_2
}
```

declares a compound data type called message with three fields: msg, var_1, and var_2. Compound data types are especially useful, for instance, to represent the structure of a message. Individual fields within a compound data type can be referenced by the well-known C-language "dot notation." For instance, if x is a variable of type message, x.var_1 refers to field var_1 of x.

42.2.3 Expressions, Statements, and Their Execution

The way a PROMELA expression is constructed is, for the most part, the same as in the C language and most C-language operators are supported. They are listed in Table 42.2 in decreasing precedence order; operators with the same precedence are listed together. The most important difference relates to the fact that it must be possible to repeatedly evaluate a PROMELA expression to determine whether or not a process is executable. Expressions are therefore constrained to be *side-effect-free*, that is, their evaluation must not modify the system state in any way.

For this reason, unlike in C, PROMELA assignments are not expressions (because they change the value of the variable on their left side), and the postfix increment and decrement operators (++ and --) can be used only in a stand-alone assignment, like a++, and not in an expression. All other aspects of the expression syntax are the same as in C, except for the conditional expression operator (in which the ? operator is replaced by ->).

Both individual expressions and assignments are valid PROMELA *statements*. Besides these, there are five more *control statements* in PROMELA, to model how the computation flow proceeds in a process, namely, sequence, selection, repetition, jump, and the *unless* clause. The sequence has already been discussed in Section 42.2.1. Among the others, only the most basic ones (i.e., selection and repetition) will be described here for conciseness.

TABLE 42.2 Summary of PROMELA Operators in Decreasing Precedence Order

Operators	Assoc.	Meaning
() [] ·	L	Parentheses, array indexing, field selection
! ~	R	Negation, bitwise complement
* / %	L	Multiplication, division, modulus
+ −	L	Addition, subtraction
<< >>	L	Left/right bitwise shift
< <= >= >	L	Relational
== !=	L	Equality/inequality
&	L	Bitwise and
^	L	Bitwise exclusive or
|	L	Bitwise or
&&	L	Logical and
||	L	Logical or
-> :	R	Conditional expression

Before discussing control statements in detail, it is necessary to briefly introduce two key concepts in PROMELA, that is, the *executability* of statements and *atomicity* of execution. These concepts are extremely important because they are tied to how *passive wait* and *synchronization*—two crucial aspects of any concurrent system—are formalized in a PROMELA model.

The basic definition of executability is straightforward. Any PROMELA statement may or may not be executable in a given state. Whenever a process encounters a nonexecutable statement during its execution, it blocks (passively waits) without proceeding further, until the statement becomes executable again in a future state of the computation. The executability rules that apply to a given statement depend on the kind of statement. The rules for some kinds of statement, like expressions and assignments, are fairly simple. Namely, an expression is executable if its value is not *false*, whereas assignments are always executable. The executability rules for the other, more complex, kinds of statement will be given along with their description.

Expressions can profitably be used to synchronize multiple processes. For example, the following fragment of code (unfortunately not completely correct, as we will see later) is an attempt to force multiple processes to enter a critical region in a mutually exclusive way:

```
bool lock = false;

!lock; lock = true;
  <critical region>
lock = false
```

The reasoning behind the code is that the value of the global variable `lock` tells whether or not any process is within the critical region at any given instant. At the beginning, no processes are in the critical region, and hence, its initial value is `false`. Thus, the expression `!lock` will not be executable when `lock` is `true`, that is, it will block any process trying to enter the critical region while another process is already inside it.

When a process successfully goes beyond that expression—meaning that it is allowed to enter the critical region—it will execute the assignment `lock = true` (i.e., always executable) to prevent other processes from doing the same. The `lock` variable is reset to `false` when the process is about to exit from the critical region. This assignment will make the expression `!lock` executable and, hence, allow another process formerly blocked on it to proceed and enter the critical region.

As said before, this solution to the mutual exclusion problem is not fully correct, for a reason related to the *atomicity* of execution. In PROMELA, unless specified otherwise, only expressions and assignments are executed as an indivisible unit, that is, *atomically*. On the contrary, any sequence of statements, like the one shown before the critical region in the example, is *not* indivisible and its execution by a process can *interleave* with other processes' activities.

In our example, it is therefore possible that a process *A* evaluates !lock and proceeds beyond it, because lock is false at the moment. Then, before *A* sets lock = true in the next statement, it is possible that another process *B* evaluates !lock, too. As a result, both *A* and *B* are allowed to enter the critical region together, leading to a *race condition*, a well-known pitfall in concurrent programming.

A straightforward solution to this issue, often available at the instruction set level of modern microprocessors, is to execute both the test expression and the subsequent assignment atomically. This can be modeled in PROMELA by means of the following construct:

```
atomic {!lock; lock = true}
```

In PROMELA, an atomic sequence is executable if any only if its first statement is executable. When a process begins executing an atomic sequence, it does not interleave with any other process until either the sequence ends or a nonexecutable statement is encountered within it. In the second case, the process blocks until the offending statement becomes executable again. At that point, execution is resumed within the atomic sequence and without interleaving, as before.

It is important to remark that introducing an atomic sequence in the model is reasonable if and only if it is guaranteed that the modeled system works in exactly the same way. Otherwise, verification results may be incorrect.

42.2.4 Control Statements

The concept of *sequence* of statements has already been introduced in Section 42.2.1 and models unconditional sequential execution. On the other hand, the *selection* statement models the choice among a number of execution alternatives. Its general syntax is as follows:

```
if
:: <guard_1> -> <sequence_1>
...
:: <guard_n> -> <sequence_n>
fi
```

Overall, the selection statement is executable if at least one guard expression among <guard_1>, ..., <guard_n> is true. In this case, execution proceeds with the execution of the sequence of statements that follows a true guard. The special guard else is true when no other guards in the same statement are true. When more than one guard is true, a *nondeterministic choice* exists among the corresponding sequences of statements and *all* the possible execution paths will be considered during verification.

For instance, when both i and j are 1, the following fragment of code specifies a nondeterministic choice between the two assignments. As a result, k can be set to *either 1 or 2*, and verification will consider both possibilities:

```
if
:: i == 1 -> k = 1
:: j == 1 -> k = 2
fi
```

This concept is particularly important, because nondeterminism appears in all sorts of systems of practical interest. For instance, even in a very simple distributed system, a node may have to react to a number of external events. A suitable way to model this behavior is, therefore, a selection statement in which the guards check whether or not a certain event has occurred and the related sequences of statements contain the reactions.

It is of course quite possible that several events occur roughly together, and hence, the exact order in which they are handled cannot be uniquely determined in advance. In all these cases, it is crucial that the verification process considers *all* possible orders, instead of just one or some, to get reliable results.

The syntax of the *repetition* statement is quite similar to the selection statements, with the keywords if and fi replaced by do and od, respectively:

```
do
:: <guard_1> -> <sequence_1>
...
:: <guard_n> -> <sequence_n>
od
```

Execution proceeds in the same way for the most part, with the important difference that, after a certain sequence of statements has been executed, execution goes back to the beginning of the repetition statement itself, leading to a loop. Within the repetition statement, the keyword break can be used to abandon the loop and execute the next statement.

42.2.5 Channel Operations

Statements and functions associated with communication channels, mentioned in Section 42.2.2, can roughly be divided into two groups, that is, send/receive and channel status operations. Although all kinds of send and receive statements available in PROMELA perform message transmission and reception through a communication channel as their primary goal, they differ in a few very important aspects. The most basic form of the send (!) and receive (?) statements is as follows:

```
ch ! <exp_1>, ..., <exp_n>
ch ? <var_1>, ..., <var_n>
```

The send statement evaluates the expressions <exp_1>, ..., <exp_n> and sends their values as a message through channel ch. The corresponding receive statement receives a message from channel ch and assigns the values found in it to the sequence of variables <var_1>, ..., <var_n>. In both cases, the number of expressions and variables, as well as their type, must match the channel declaration. In a receive statement, the special variable _ (underscore) means that the corresponding value shall be discarded.

If ch is a *rendezvous* channel—that is, it has a capacity of [0]—the send statement becomes executable when another process reaches the receive statement and vice versa. At this point, data transfer between the two processes takes place atomically. On the other hand, when ch is *buffered* and has a capacity for [k] messages, with $k > 0$, the send statement is executable as long as the channel holds less than k messages, that is, it is not full. Symmetrically, the receive statement is executable if the channel holds at least one message, that is, it is not empty.

Other, more powerful variants of the send and receive statements exist. They are able to store and retrieve messages into and from a buffered channel in an order that is not necessarily first in, first out (FIFO). In addition, a nondestructive receive statement, which does not remove the message from the channel, is available. Several expressions check the number of messages stored in a buffered channel and can be used as guards to execute a certain operation only if, for instance, the channel is not empty. Last, a *polling receive* expression has a syntax similar to a receive statement. It is true if and only if the corresponding receive statement could be executed without blocking. As the previous ones, this expression is mostly used as a guard.

42.2.6 Property Specification

When the model is ready, properties expressing requirements on the behavior of the system should be specified. After that, the model checker can be used to check if the properties hold on the model. Simple correctness properties, for instance, mutual exclusion, can be specified by means of an assertion of the PROMELA language:

assert(<expression>)

Assertions can be placed anywhere between two statements in the model. The <expression> is evaluated when the assertion is encountered during state space exploration and an assertion failure (i.e., <expression> being false) leads to a verification error. Since an assertion is checked only in a specific location of a specific process, it lacks the ability to evaluate global properties regardless of what processes are executing, for example, absence of deadlocks. Complex properties like those can be specified by means of a higher-level formalism, namely, linear temporal logic (LTL) [19] formulae.

In its simplest form, an LTL formula can be built from the following elements (SPIN syntax shown in parentheses; refer to [16] for a thorough treatment of this topic):

- *Atomic propositions.* Any Boolean PROMELA expression is a valid atomic proposition in LTL.
- *Propositional calculus operators*, such as *not* (!), *and* (&&), *or* (||), *implication* (->), and *equivalence* (<->). Atomic propositions can be combined by means of propositional calculus operators to get propositional calculus formulae.
- Besides, *temporal operators*, such as *always* ([]) and *eventually* (<>), can be applied to the previous two elements to specify temporal features.

LTL formulae can be automatically translated to PROMELA code before verification and can become part of the model being verified as a *never claim*. As an example, the following LTL formula specifies that, along a computation, it will always be true that if F is true, then it implies that sooner or later G will eventually be true:

[] (F -> <> G)

The example highlights an important distinction between atomic propositions and can propositional calculus formulae, on the one hand, and temporal operators on the other. Both atomic propositions and propositional calculus formulae can be evaluated by looking only at a single state in the computation. Instead, the evaluation of temporal operators involves looking at future states in the computation. In the previous example, whenever there is a state within which F is true, it is necessary to examine the following states to see whether there is a state within which G is true.

42.3 SPIN Usage Notes

42.3.1 SPIN Verification Flow

Starting from a model, SPIN can perform two different activities: *simulate* the system behavior in various ways and *verify* whether or not a property holds by means of state space exploration. Simulation is especially useful when developing and debugging a model but will not be discussed in detail here, whereas verification is usually the main, ultimate users' goal. The basic SPIN verification flow is enumerated in the following and is shown in Figure 42.1:

- The main input of SPIN consists of the model and the property to be verified, both shown as light gray boxes in Figure 42.1. Both the model and simple properties, like assertions, are specified directly in PROMELA.
- More complex properties are specified in LTL, and as explained in Section 42.2.6, they must be translated into a PROMELA never claim before use. The translation is done by SPIN itself, when invoked with either the -f or the -F command-line option.

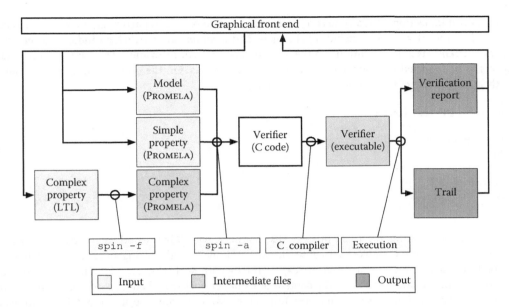

FIGURE 42.1 Simplified view of the SPIN verification flow.

- When both the model and the property to be verified are available, SPIN generates the *verifier*, that is, a C-language program that will carry out the verification. In order to do this, it must be invoked with the command-line option -a.
- The verifier source code is compiled into an executable program by means of the native C compiler of the computer on which the verification will run. Traditionally, the verifier is called pan.
- The verifier is then run to produce a verification report and, possibly, a *trail*. Both outputs are shown as dark gray boxes in Figure 42.1. The verification report summarizes the outcome of the verification and the trail contains the full details about counterexamples, if some were found.

Most often, SPIN is not invoked directly by the user, but through a graphical front end. Besides offering a more user-friendly interface to the tool, front ends also provide other useful functions. In particular, they are able to postprocess SPIN textual output and present it in a more convenient form. A typical example is the generation of a detailed message sequence chart (MSC) [14] derived from the SPIN trail. The MSC is very useful to truly understand a counterexample, because it lists in an intuitive, graphical form the computation steps and message exchanges leading to the violation of the property being verified. Several different front ends exist, with different levels of sophistication and supporting a variety of operating systems. More details can be found in [1].

In addition, SPIN itself is able to perform a certain amount of trail postprocessing, when invoked with the -t option. Of special interest is a simulation mode, in which the simulator is forced to follow the trail leading to a counterexample. This feature, used in conjunction with the graphical display of variables and communication channels contents offered by most front ends, often sheds more light on the counterexample itself.

42.3.2 Performance Hints

Time and memory resources to be used during verification are always a critical issue, common to all model checkers. This is because verification is assumed to be done over the whole state space and it is quite easy to get state space explosion. Although it is impossible to fully describe how SPIN works internally in a little space—see [9] for an authoritative reference on this topic—some high-level information is often useful to use it more effectively.

As defined in Section 42.1, a state is represented as a state vector. In particular, it consists of values of global and local variables, including channels, and *location counters*, which indicate where each process currently is in its execution. On the one hand, nondeterminism and interleaving are important aspects to be modeled for concurrent and distributed systems. On the other hand, they can significantly affect the size of the state space. For example,

```
chan c = [N] of {byte};

active proctype fill() {
  do
  :: c ! 1
  :: c ! 0
  od
}
```

In the previous listing, the `fill` process fills channel c with N byte-sized values, chosen in a nondeterministic way, and then blocks because the channel buffer is full. During an exhaustive state space exploration, the number of states x grows exponentially with N, that is, $x \simeq 2^{N+1}$. This is because, just considering the channel variable c, the content of its buffer goes from being completely empty (at the beginning of verification) to being completely full (at the end), and at any step, it can contain any sequence of values, from all 0s to all 1s. In order to perform an exhaustive search, it is necessary to consider and check the property of interest in all these states.

It is not the case that the whole state space is built all at once, and then verification is carried out on it. Instead, state vectors are calculated on the fly and stored into a hash table. Namely, when storing a state vector, a hash function is applied to calculate an index that indicates the position where the state vector should be stored within the table. If there is already a state vector stored in that position, and it is not the same as the current one, a hash conflict occurs. In this case, the new state vector is stored elsewhere and linked to the existing state vector through a linked list. By intuition, as the number of hash conflicts increases, storing a new state becomes less efficient, because the linked lists grow. As a consequence, more time is required to linearly scan the lists, in order to check whether or not a certain state vector is already in the table.

Verification is the process to check whether a given property holds in each state. As a consequence, if a state has already been checked, there is no need to check it again. Let us assume that the verification is currently considering a certain state, and it is about to execute a statement. After execution, the values of some variables in the model may change. This leads to another state, namely, a state transition happens. What SPIN does next is to look into the hash table. If the new state has not been stored yet, then it checks whether the property holds on that state. If it does, SPIN stores the state vector into the hash table, otherwise SPIN just found a counterexample, that is, a point of the state space in which the property does not hold.

If the state has already been stored in the table, it is a proof that the property holds on that state. There is no need to store it again and the program can move to the next step. Overall, during verification, SPIN keeps storing state vectors into the hash table and looking up newly created state vectors to see whether or not the corresponding states have already been visited before. This process may be highly time and memory consuming, depending on a lot of different aspects. For instance, if the quality of the hash function is not good enough, it is possible to have long linked lists somewhere in the hash table, whereas other parts of the table are still empty.

When trying to improve the performance of SPIN, the goal is to achieve the best trade-off between speed and memory requirements. More specifically, speed depends on many factors, such as how large the hash table is, how effective its management algorithms are, and how efficient SPIN is in updating and searching it. However, speed improvements in most cases have some impact on memory requirements.

This topic can be addressed in different ways and at different levels, starting from the model itself down to tune how some Spin algorithms work internally:

- *Writing efficient models.* More efficient models can be developed by carefully considering which aspects of a model have a bigger effect on state vector size and on the number of states. For instance, using as few concurrent processes as possible, enclosing sequences of statements into atomic sequences when possible, and introducing nondeterministic constructs only when necessary considerably reduce the number of distinct states to be considered during verification.

  Similarly, in order to reduce the state vector size, it is useful to keep the amount of channel buffer as small as possible and declare numeric variables with the narrowest size compatible with their range requirements.

  The approaches mentioned earlier will improve the performance in terms of both memory and time, because less space will be required to store states and state comparison will be faster, too. On the other hand—as discussed in Section 42.2.3 for atomic sequences—special care is necessary to make sure that the simplified model is still a faithful representation of the real system or, at least, the side effects of the simplification are well understood.

- *Allocating more memory for the hash table.* As discussed earlier, the time spent walking through the linked list associated with a hash table entry is significantly higher than the time spent on calculating its index. If more memory is allocated for the hash table, the probability to have hash conflicts will usually be reduced and the verification time will improve. However, in this case, we need to sacrifice memory for efficiency.

  Another important point to consider is that, in any case, the hash table must fit in the *physical* memory available on the machine Spin is running on. Although most operating systems do provide an amount of *virtual* memory that is much bigger than the physical memory, the overhead associated with disk input/output operations due to virtual memory paging would certainly degrade Spin's performance more than hash conflicts. This is especially true because paging algorithms rely on memory access predictability (or locality) to achieve good performance, whereas hash table accesses are practically pseudorandom and, hence, very hard to predict.

- *Compressing state vectors.* State vectors of several hundred bytes are quite common. Instead of storing them as they are, it is possible to compress them. In this case, the memory they require will be smaller. However, more time will be spent on compression whenever it is necessary to store a state vector. This option therefore represents a trade-off between memory and time requirements.

- *Partial order reduction.* The details of partial order reduction are quite complex and fully described in [12]. Informally speaking, it may happen that, starting from a certain state, it can be proved that the execution order of several computation steps does not affect the final state, and the different execution orders cannot be distinguished in any way. This may happen, for instance, when two concurrent processes are working independently on their own local variables.

  In this case, instead of considering all the possible execution orders and generating the corresponding intermediate states, it is enough to follow just one execution order. In some cases, this is a quite effective method since it can reduce the size of the state space sharply.

- *Bitstate hashing and hashing compact.* Instead of allocating more memory to accommodate a large hash table, it is also possible to reduce its memory requirements.

  In bitstate hashing, a state is identified by its index in the hash table. As a result, a single bit is sufficient to indicate whether a state has already been visited or not. However, two different states may correspond to the same index, due to hash conflict. If the "visited" bit is already set for a certain state, when coming to verify whether the property of interest holds on a different, but conflicting, state, the result is positive, even if this may not be the case in reality. In other words, some parts of the state space may not be searched and it is possible to miss some counterexamples. However, if a counterexample is found, it does represent a true error.

In hashing compact, instead of storing a large hash table, the indexes of visited states in that table are stored in another, much smaller, hash table. It has the same issue as bitstate hashing, because two different states may have the same index in the large hash table and thus collide.

Both methods are quite effective to reduce the memory requirements of state storage. However, they are *lossy* and entail a certain probability of having a *false positive* in the verification results, that is, a property may be considered true although some counterexamples do exist. The false-positive probability can be estimated and often brought down to an acceptable level by tuning some of the algorithm parameters [11].

42.4 Example

In this example, it will be shown how to model and check the correctness of a simple communication protocol. The goal of the protocol—proposed and thoroughly discussed in [18]—is to enable a set of agents to pass to each other a private value, by means of a communication network, and reach an agreement on the value each agent has got. The protocol must still function correctly even if some of the agents are faulty and may send fake messages.

The specific protocol being discussed was originally developed to address a few important design issues in the software implemented fault tolerance (SIFT) fault-tolerant computer for aircraft control [20]. Many of its offspring are still very relevant and widespread in avionics and other fault-tolerant systems.

42.4.1 Interactive Consistency Protocol

The protocol and the proof of its correctness presented in [18] are based on the following assumptions: a set of n agents is considered, and it is known that *at most m* of them are faulty. Agents communicate by means of point-to-point messages sent through a perfect network, which never drops, alters, or duplicates messages.

Faulty agents are allowed to send messages with whatever content they wish, but it is assumed that the recipient can always identify the true sender of a message. In other words, faulty agents cannot lie about their identity on the network. It is also assumed that faulty agents will nevertheless send all the messages required by the protocol. Otherwise, a time-out mechanism on the receiving side would detect their failure.

Under these hypotheses, the protocol guarantees that, for any $n \geq 3m + 1$, each nonfaulty agent will be able to compute an *interactive consistency* vector of n elements, that is, a vector containing one element for each agent in the system. The interactive consistency vectors computed by the nonfaulty agents have the following two properties:

1. They are exactly the same.
2. The vector element that corresponds to a nonfaulty agent A holds the private value of A.

It should be noted that the element that corresponds to a faulty agent F may not correspond to the private value of F. However, all the vectors computed by nonfaulty agents will still have *the same* (albeit arbitrary) *value* in that element. In turn, this enables the nonfaulty agents to have a consistent, shared view of each agent's value, including the faulty ones.

For the sake of simplicity, only the simplest case of $m = 1$, giving a minimum value of $n = 4$, will be analyzed in detail. In this case, the protocol performs two rounds of message exchanges, in which each agent sends $n - 1 = 3$ messages. Namely,

1. Each agent sends to the others a message containing its own private value
2. Each agent sends to the others a message containing all the values it received during the first round

After the second round, each agent A_i will therefore have 3 *reports* about the private value of agent A_j, $i \neq j$: one from A_j itself (received during the first round) and two more from the other two agents present in the system (received during the second round). Agent A_i will then construct its interactive consistency vector V_{A_i} based on those reports. In particular,

- The element of V_{A_i} corresponding to A_i itself is set to the private value of A_i
- The element of V_{A_i} corresponding to another agent A_j is set to the value contained in the majority of the reports about A_j received by A_i, if such a value exists. Else, it is set to a reserved or default value, called `nil` in the following

Although the protocol works with private values of any kind, to further simplify modeling, it will be assumed that these values are a simple Boolean flag. This restriction does not change the protocol structure in any way.

42.4.2 Modeling the Protocol

The very first modeling step is the definition of the data types required by the protocol agents. In this case, they are as follows:

```
typedef buffer_t {
  bool agent[NA]
}

typedef report_t {
  byte i; /* Number of reports */
  bool value[NB] /* Their values */
}

#define nil 2

typedef result_t {
  byte agent[NA]
}

typedef results_t {
  result_t opinion[NA]
}
```

Namely, `buffer_t` represents the message buffer used in the second round for message exchanges, `report_t` represents the reports about the private value of a single agent collected during protocol execution, and `results_t` represents the overall results of the protocol, that is, the opinion each agent has about the private value of every other agent. The compound data type `report_t` has two fields: The array `value` holds the reports themselves, while `i` counts how many reports have already been received and holds the index of the first free element of `value`. Unlike in the other data type definitions, the elements of `results_t` are of type `byte` instead of `bool` because they can assume the value `nil`.

In the model, `NA` is a macro that represents the total number of agents in the system, 4 in this case, and `NB` is set to `NA-1`. To facilitate property verification, as it will be discussed in Section 42.4.3, the results of the protocol are held in the global variable `results` declared as follows:

```
results_t results;
```

Figure 42.2 summarizes the main data structures used by the interactive consistency protocol agents. It shows that each agent holds its private value in the local variable `bool  my_value`, collects the reports about other agents' values in the local array `report_t reports[NA]`, and stores its interactive consistency vector into row `id` of global variable `results_t  results`. A local variable

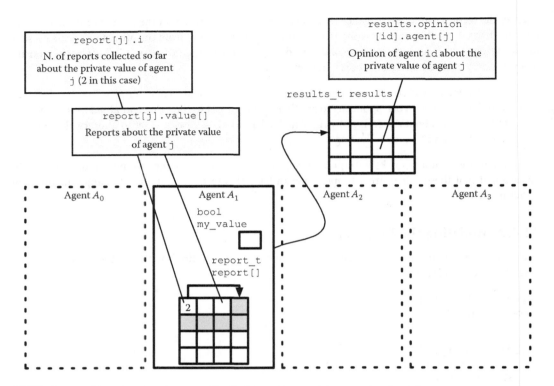

FIGURE 42.2 Main data structures used by the interactive consistency protocol agents.

`buffer_t buffer` also exists, but it is used only as temporary storage during the second round of message exchanges and it is not shown for clarity.

Then, the communication channels among agents are declared as follows:

```
chan c1[NA] = [NB] of {byte, bool}
chan c2[NA] = [NB] of {byte, buffer_t}
```

Two distinct arrays of NA input channels are used: c1 for the first round and c2 for the second. This equates to assuming that no confusion can arise between messages pertaining to different rounds. In each group, each input channel is uniquely associated with an agent. The channel buffers are big enough to store all the messages expected in one round, that is, NB messages. The first message field, of type byte, is always filled with the agent identity, in both rounds. On the contrary, as explained earlier, the structure and contents of the second field depend on the round.

The behavior of each agent is specified by means of the following proctype definition:

```
proctype P(byte id) {
  bool my_value;
  report_t report[NA];

  buffer_t buffer = false;
  byte i, j;
  choose_my_value(my_value);

  first_round();
  second_round();
  compute_result()
}
```

In the definition, argument `id` is a value that uniquely identifies each agent upon instantiation and can assume any integer value in the range from 0 to 3 included. As shown in Figure 42.2, the local variable `my_value` represents the private value of agent `id`, and `report` is an array holding the reports received by agent `id` about the private values held by the other agents. Element `[id]` of this vector corresponds to the agent itself and is unused. The other local variables are used as temporary storage during execution.

For clarity, the definition makes use of the *inline* construct, a convenient way to give a symbolic name to a sequence of statements. When the name of the sequence is found elsewhere, it is replaced by the corresponding sequence of statements, like in a macro expansion. Inline sequences may have formal parameters. In this case, they are replaced by the corresponding actual parameters during the expansion, by means of a simple textual substitution. Formal parameters are therefore not typed, and syntax and semantics checks are performed only after expansion.

For instance, the inline sequence `choose_my_value` is defined as follows:

```
inline choose_my_value(v)
{
  if
  :: v = false
  :: v = true
  fi
}
```

When its body is expanded in P, the formal parameter `v` is replaced by `my_value`. The effect of the code is therefore to set `my_value` to either `false` or `true` in a nondeterministic way and, hence, model how each agent chooses its private value.

The inline sequence `first_round` performs the first round of message exchanges and is defined as follows:

```
inline first_round()
{
  byte source, dest;
  bool val;

  dest = 0; skip_id(id, dest);
  do
  :: (dest<NA) ->
      {
        c1[dest] ! id, my_value;
        dest++; skip_id(id, dest)
      }
  :: else break
  od;

  i = 0;
  do
  :: (i<NB) ->
      {
        c1[id] ? source, val;

        add_report(source, val);
        i++
      }
  :: else break
  od
}
```

The first repetition (do) construct sends the private value my_value of agent id to every other agent. The agent identifier currently being handled is held in variable dest. The skip_id inlined sequence (whose expansion is not shown for conciseness) increments its second argument by one if it is equal to the first, and hence, it prevents agents from sending messages to themselves.

The second repetition construct waits for the NB messages that agent id expects to receive during the first round, by means of a blocking receive statement. Each message contains a report about the private value of another agent. The inlined sequence add_report handles it, by appending the new value val to the set of reports collected about agent source.

No deadlock may come from putting the send and receive phases of the first round in strict sequence, because the channel buffers can hold all the messages they are expected to contain during the round without blocking the sender. The same technique is used in the second round as well.

The second round of message exchanges is modeled similarly. The added complexity comes from the fact that, in the second round, each message contains more than one report. Each agent gathers the reports it will send out into the variable buffer, of type buffer_t, taking them from its report variable. Since exactly one report about the value of each agent was collected during the first round, and they were stored in the first free element of each report_t data structure (i.e., element [0]), the report concerning agent i is retrieved from report[i].value[0]. The element [id] of buffer_t is not actually used and defaults to false, because agent id does not send reports about its own value in the second round.

The initialization of buffer has been enclosed in a d_step (or *deterministic step*), a more restrictive form of atomic sequence, in which nondeterminism is forbidden and blocking statements are allowed only at the very beginning of the sequence. Provided these restrictions are acceptable in the model, verification handles the whole sequence in a very efficient way, that is, as a single deterministic execution step. This is the case here because, from the protocol point of view, the buffer setup is an activity internal to each agent that cannot possibly affect any other agent. The send loop that follows is very similar to the previous one:

```
inline second_round()
{
    d_step {
        i=0;
        do
        :: (i<NA) ->
            {
                buffer.agent[i] =
                    ((i==id) -> false : report[i].value[0]);
                i++
            }
        :: else break
        od
    }

    dest=0; skip_id(id, dest);
    do
    :: (dest<NA) ->
        {
            c2[dest] ! id, buffer;
            dest++; skip_id(id, dest)
        }
    :: else break
    od;

    atomic {
        i=0;
        do
```

```
        :: (i<NB) ->
            {
                c2[id] ? source, buffer;

                j=0;
                do
                :: (j<NA) ->
                    {
                        if
                        :: (j==source || j==id) -> skip
                        :: else   add_report(j, buffer.agent[j])
                        fi;

                        j++
                    }
                :: else break
                od;

                i++
            }
        :: else break
    od
    }
}
```

The receive phase of the second round is instead more complex, because each of the NB messages received by an agent contains NA-2 reports to be processed. It is therefore made of two nested repetition constructs. The outer one is executed NB times (using `i` as an index), and its body handles one single incoming message. The inner one is executed NA times (using `j` as an index) and processes message contents by means of the `add_report` construct discussed previously. Elements `[source]` and `[id]` of the incoming message are discarded, because agent `id` does not collect reports about a certain agent `source` from the message sent from `source` itself; moreover, agent `id` does not collect reports about itself.

The receive phase as a whole is defined as an `atomic` sequence of statements. Therefore, during verification, it will be assumed that, after an agent resumes execution after receiving a message, its execution will continue uninterrupted—that is, without interleaving with other agents—until it blocks again, waiting for the next one. As explained in Section 42.3.2, this assumption speeds up verification, because it reduces state space size, without affecting its correctness. In fact, all operations executed by an agent upon receiving a message are internal to that agent and cannot affect the others in any way. A `d_step` cannot be used instead, because it cannot contain any blocking statements except the first one.

The very last phase of each agent's activity is to generate its own interactive consistency vector from the reports it got. This is modeled by the following code:

```
inline compute_result() {
  d_step {
    i=0;
    do
    :: (i<NA) ->
        {
            if
            :: (i==id) ->
                results.opinion[id].agent[id] = my_value
            :: else ->
                majority_report(
                    results.opinion[id].agent[i], report[i])
```

```
        fi;
        i++
    }
    :: else break
    od
  }
}
```

Previously, the inlined sequence `majority_report` stores into its first argument the value found in the majority of the elements of its second argument. Its code is not shown here, due to lack of space. Some efficient algorithms to do this are presented, for instance, in [17]. All the interactive consistency vectors are stored in the global `results` data structure for analysis.

42.4.3 Analysis Results

The primary goal of the analysis is to check whether or not the protocol is able to ensure that the properties of the interactive consistency vector listed in Section 42.4.1 are satisfied, even if up to one agent may be faulty. Furthermore, we will look for *counterexamples*, that is, scenarios in which some of the properties are *not* satisfied, when two or more agents are faulty. The second aspect of the analysis is important, too, because it may give useful hints on why a protocol or system is not working as it should and on how to improve it.

In order to model faulty agents, an alternate agent model, called `faulty_P`, has been defined. According to the model, a faulty agent completely ignores the messages it receives and sends fake messages—whose contents are chosen nondeterministically—to the other agents. At the same time, the model reflects the restrictions on faulty behavior set forth in Section 42.4.1, which were taken for granted when the protocol was designed.

The properties of interactive consistency vectors have been modeled as LTL formulas. For instance, when agent A_0 is faulty, the first property can be written as

```
<> [] p
```

where p is defined as

```
(results.opinion[1].agent[0] == results.opinion[2].agent[0] &&
 results.opinion[1].agent[1] == results.opinion[2].agent[1] &&
 results.opinion[1].agent[2] == results.opinion[2].agent[2] &&
 results.opinion[1].agent[3] == results.opinion[2].agent[3] &&
 results.opinion[2].agent[0] == results.opinion[3].agent[0] &&
 results.opinion[2].agent[1] == results.opinion[3].agent[1] &&
 results.opinion[2].agent[2] == results.opinion[3].agent[2] &&
 results.opinion[2].agent[3] == results.opinion[3].agent[3])
```

This LTL property is true if and only if in every computation, *eventually* (<>) p is *always* ([]) true. In turn, p is true if and only if the interactive consistency vectors of agents A_1 to A_3 (the good ones) are identical. Very informally speaking, this means that sooner or later in their execution, the good agents will compute identical interactive consistency vectors and they will be identical from then on, regardless of what the faulty agents may do. The second property can be modeled in a similar way and is not shown for conciseness. Other properties of interest can be specified in a similar way. For instance, it is possible to verify that all agents conclude protocol execution, by having them set a termination flag at the end and checking that in every computation those flags are *eventually* set.

Quite unsurprisingly, since the protocol being analyzed is known to be correct in these scenarios, SPIN confirms that both properties are satisfied when four agents, of which at most one can be faulty, are instantiated and their behavior is checked against the properties. On the other hand, instantiating two faulty agents (A_0 and A_1) along with two good ones (A_2 and A_3) leads SPIN to find several interesting counterexamples. Two of them will be shortly commented here.

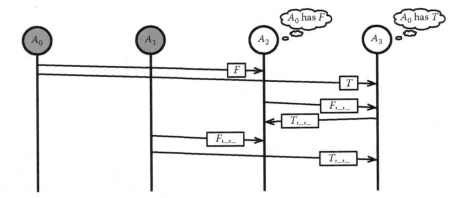

FIGURE 42.3 First counterexample: A_2 and A_3 have different views of A_0's private value, violating property 1 of interactive consistency (irrelevant messages and message fields not shown for clarity).

The first counterexample is illustrated by the simplified MSC shown in Figure 42.3 and derived from the SPIN output *trail*, which lists the computation steps and message exchanges leading to the violation of the property being verified. The messages irrelevant to illustrate how the counterexample works have been removed for clarity. In the counterexample, the two faulty agents coalize and lead the two good agents to disagree on the private value of faulty agent A_0, thus violating property 1 of interactive consistency:

- In the first round, A_0 sends to A_2 and A_3 contrasting reports about its own private value.
- The anomaly would be detected during the second round, if A_0 were the only faulty agent in the system, by means of the reports about A_0's value sent by the other agents.
- In the counterexample, however, the other faulty agent A_1 sends to A_2 and A_3 contrasting reports on the value it got from A_0 in the first round.

Although A_2 and A_3 send to each other correct reports about A_0's value, this is no longer enough to ensure correctness. Namely, A_2 receives two reports that A_0's value is `false` (F in the figure), from A_0 itself and from A_1. The third, minority report from A_2 says that the value is `true` (T), but A_2 concludes anyway that A_0's value is `false`. On the other hand, A_3 receives two reports that A_0's value is `true` (from A_0 itself and from A_1) and one conflicting report (from A_2), coming to the conclusion that A_0's value is indeed `true`.

The outcome of the second counterexample, shown in Figure 42.4, is even worse because it leads good agent A_3 to a false belief about the private value of the other good agent, A_2. This violates the second

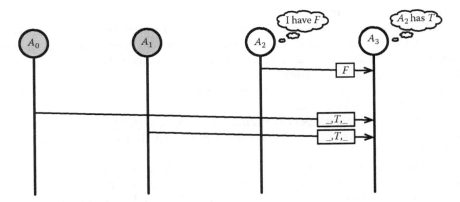

FIGURE 42.4 Second counterexample: A_3 has an incorrect view of A_2's private value, violating both properties of interactive consistency (irrelevant messages and message fields not shown for clarity).

property of interactive consistency and, incidentally, also the first one—because the interactive consistency vectors of A_2 and A_3 are not identical. Namely,

- In the first round, A_2 sends to A_3 a correct report about its own private value, which is `false`
- In the second round, A_0 and A_1 lead A_3 into believing that A_2's private value is `true` instead, by sending it two false reports about A_2's value

References

1. Unix group of the Computing Science Research Centre of Bell Labs, On-the-fly, LTL model checking with Spin. Berkeley Heights, NJ, Online. Available at http://spinroot.com/. Accessed on March 2014.
2. D. Parker, G. Norman and M. Kwiatkowska. PRISM—probabilistic symbolic model checker. Online. Available at http://www.prismmodelchecker.org/. Accessed on March 2014.
3. Department of Information Technology, Uppsala University, Sineden and Dept. of Computer Science, Aalborg University, Denmark. UPPAAL home. Online. Available at http://uppaal.org/. Accessed on March 2014.
4. R. Alur and D.L. Dill. A theory of timed automata. *Theoretical Computer Science*, 126(2):183–235, 1994.
5. C. Baier and J.-P. Katoen. *Principles of Model Checking*. The MIT Press, Cambridge, MA, 2008.
6. M. Ben-Ari. *Principles of the Spin Model Checker*. Springer-Verlag, London, U.K., 2008.
7. J. Bengtsson, K. Larsen, F. Larsson, P. Pettersson, and W. Yi. Uppaal—A tool suite for automatic verification of real-time systems. In *Hybrid Systems III, LNCS 1066*, pp. 232–243. Springer-Verlag, Berlin, Germany, 1995.
8. D. Bošnački and D. Dams. Discrete-time Promela and Spin. In *Formal Techniques in Real-Time and Fault-Tolerant Systems*, volume 1486 of *Lecture Notes in Computer Science*, pp. 307–310. Springer, Berlin, Germany, 1998.
9. G. Holzmann. *The Spin Model Checker: Primer and Reference Manual*. Pearson Education, Boston, MA, 2003.
10. G.J. Holzmann. The model checker SPIN. *IEEE Transactions on Software Engineering*, 23(5):279–295, 1997.
11. G.J. Holzmann. An analysis of bitstate hashing. *Formal Methods in System Design*, 13(3):289–307, November 1998.
12. G.J. Holzmann and D. Peled. An improvement in formal verification. In *Proceedings of the Seventh IFIP WG6.1 International Conference on Formal Description Techniques*, pp. 197–211, Berne, Switzerland, 1994.
13. *International Standard ISO/IEC 9899, Programming Languages—C*. International Organization for Standardization, Geneva, Switzerland, 2011.
14. ITU-T. *Recommendation Z.120—Message Sequence Chart (MSC)*, International Telecommunication Union, Geneva, Switzerland, 2004.
15. M. Kwiatkowska, G. Norman, and D. Parker. PRISM: Probabilistic symbolic model checker. In P. Kemper, ed., *Proceedings of Tools Session of Aachen 2001 International Multiconference on Measurement, Modelling and Evaluation of Computer-Communication Systems*, pp. 7–12, September 2001. VDE Verlag, Berlin, Germany.
16. Z. Manna and A. Pnueli. *The Temporal Logic of Reactive and Concurrent Systems—Specification*. Springer-Verlag, New York, 1992.
17. J. Misra and D. Gries. Finding repeated elements. *Science of Computer Programming*, 2(2):143–152, 1982.
18. M. Pease, R. Shostak, and L. Lamport. Reaching agreement in the presence of faults. *Journal of the ACM*, 27(2):228–234, 1980.
19. A. Pnueli. The temporal logic of programs. In *Proceedings of the 18th Annual Symposium on Foundations of Computer Science*, pp. 46–57, November 1977. IEEE Computer Society, Washington, DC.
20. J.H. Wensley, L. Lamport, J. Goldberg, M.W. Green, K.N. Levitt, P.M. Melliar-Smith, R.E. Shostak, and C.B. Weinstock. SIFT: Design and analysis of a fault-tolerant computer for aircraft control. *Proceedings of the IEEE*, 66(10):1240–1255, October 1978.

43

Formal Methods

Ben Di Vito
Langley Research Center

43.1 Introduction

Formal methods provide powerful tools and techniques for modeling and analyzing software designs and implementations. They make use of mathematics and formal logic to achieve high levels of assurance that software components have important behavioral properties or will not raise runtime errors. For safety-critical avionics, formal methods can help achieve higher degrees of software dependability than would be possible using only testing, simulation, and other nonformal techniques.

Since the publication of the first edition of the Digital Avionics Handbook, the field of formal methods has experienced substantial growth. This growth has occurred in core technical capabilities, in the maturity and variety of software tools, and in adoption by engineers across several industries. From 2012 onward, increased recognition and acceptance by certification authorities in civil aviation is expected to lead to higher adoption of formal methods by avionics developers.

In the previous handbook edition, Sally Johnson and Ricky Butler provided a chapter on formal methods that emphasized the techniques of writing formal specifications. Although the example application from that chapter has been retained, this edition examines different aspects of formal methods technology. Formal modeling and specification are still discussed, but more attention will be focused on techniques for analysis and verification. In addition, current regulatory guidance is factored in, owing to recent favorable developments in that area.

No prior familiarity with formal methods is assumed. Nevertheless, familiarity with engineering mathematics or computer science would be quite helpful. Monin and Hinchey [14] have provided a valuable survey of the mathematical and logical background used in this field. Due to space limitations, specification examples will be presented with limited explanation so that the exposition can focus on verification methods. The example application is of modest complexity so the reader should be able to grasp the essence of the formal models without a full understanding of the language features.

43.2 Formal Methods Landscape

While the technology of formal methods has advanced steadily, more recent developments of note concern avionics certification guidance. Opportunities to use the results of formal analysis in certification are on the rise. In addition to the obvious importance for digital avionics developers, the expected increase in applications could motivate technology developers to enhance their tools to accommodate avionics-oriented users.

43.2.1 Certification after RTCA DO-333

Formal methods have been available for use by digital avionics developers as an aid in creating dependable designs and implementations. Their use to support certification, however, has been limited to special cases where an "alternative means of compliance" would be permitted after sufficient justification has been provided. This deliberately modest endorsement in DO-178B reflected the technological immaturity of formal methods in the early 1990s. Consequently, formal methods have yet to assume any significant role in the certification process. Recent events promise to change the outlook.

In January 2012, RTCA and European Organization For Civil Aviation Equipment (EUROCAE) released several documents, products of RTCA Special Committee 205 (SC-205) and EUROCAE Working Group 71 (WG-71), that are key to the future of certification of avionics software. Foremost among these are DO-178C (ED-12C), Software Considerations in Airborne Systems and Equipment Certification, and DO-278A (ED-109A), Software Integrity Assurance Considerations for Communication, Navigation, Surveillance, and Air Traffic Management (CNA/ATM) Systems. As the long-awaited successors to DO-178B and DO-278, these documents provide updated guidance for developers of airborne and ground-based software.

SC-205 and WG-71 also created four supplements to accompany DO-178C and DO-278A. One of these, Formal Methods Supplement to DO-178C and DO-278A, has been published as RTCA document DO-333 [21]. The corresponding EUROCAE document is ED-216. For the first time, avionics developers will have extensive guidance on applying formal methods technology to produce evidence that certification authorities can accept for certification credit. This will provide a pathway for the gradual introduction of formal methods into the development life cycle and the eventual certification of aircraft and engines whose software airworthiness is assured (in part) by formal methods.

Much of DO-333 is concerned with how formal methods can be used to satisfy relevant certification objectives in DO-178C and DO-278A. DO-333 neither endorses nor restricts the types of formal methods that might be suitable for generating certification evidence. It does, however, stipulate that an analysis method can only be regarded as formal if its determination that a property holds is *sound*. A sound analysis is deemed to be one that never asserts a property to hold when it does not. The converse, asserting a property fails to hold when it does, is acceptable, albeit undesirable. An outcome such as this is regarded as merely a "false alarm."

43.2.2 Overview of Analysis Techniques

In the 1990–2010 time frame, formal methods experienced substantial growth by almost every measure. This includes the variety of modeling and analysis techniques, the number of researchers and practitioners, the number and maturity of software tools, and the record of successful applications in project settings. In one industry, namely, commercial digital electronics, formal techniques and specialized tools to support them have been adopted as standard practices in the design workflow.

In this chapter, we describe only a small subset of current analysis techniques. We will first sketch the three broad categories of formal methods cited in DO-333: (1) deductive (theorem-proving) techniques, (2) model checking, and (3) abstract interpretation. Later, we will choose a representative tool from the first two categories and focus on their capabilities, illustrated via examples.

Most formal methods need to have formal models or specifications to work from. Normally, these are written by users. In some cases, models and other formalizations are extracted directly from source code in an automated fashion. The techniques have different approaches to modeling, but they all share an underlying reliance on formal logic(s) and a formalization of some mathematical domains. Where they exhibit greater differences is in the analysis and verification techniques that form the heart of their methodologies.

Key characteristics of the DO-333 analysis categories mentioned earlier can be summarized as follows:

- *Deductive techniques* establish the truth of explicitly stated properties using a theorem-proving system. Properties are typically expressed in the same notation as the formal models. Theorem provers can be either automatic or interactive. In both cases, users need to understand the underlying logical formalism to construct proofs effectively.
- *Model checking* makes use of efficient exploration of large state spaces to verify properties of formal models. Most models are based on state transition systems of various kinds. An important tool feature is the ability to generate counterexamples or other diagnostic information in response to failed verification attempts.
- *Abstract interpretation* relies on formalized semantics of programming languages to analyze properties over abstract domains. The technique is normally realized through sound algorithms that examine source code to ascertain whether various predefined properties hold. Tools usually take the form of static analyzers for mainstream programming languages.

Not all analysis tools and techniques can be classified to fit these categories neatly. Some involve the hybrid application of multiple techniques. Others provide core deduction capabilities that are not seen as end user tools but more as components to be incorporated in other tool suites. An example is the category of satisfiability modulo theories (SMT) solvers. These tools accept logical formulas and apply disparate deduction algorithms knitted into a framework called SMT. SMT-based techniques are making rapid progress and are appearing as key components in several verification tools.

Another distinction in analysis tools is whether they act on formal models or act directly on program source code. This latter category is enjoying greater success and renewed interest. Significant tools in both the model checking and abstract interpretation categories are capable of analyzing source code. Moreover, deductive techniques originated from the study of "program verification" in the 1970s, which advocated program proving as an analog to unit testing. Although we will not have an opportunity to illustrate these techniques in this chapter, this category is likely to expand in future years.

43.3 Example Application

We will demonstrate two different formal techniques using a highly simplified but representative avionics subsystem. The following mode control panel (MCP) is based loosely on features found in early Boeing 737 autopilots. While newer aircrafts use more sophisticated flight management systems, the core elements of modes, displays, and controls are still relevant. Informal software requirements are presented in the succeeding text in a form similar to what software developers often encounter in practice (minus the "shall" verbiage). Readers might wish to skim over some requirement details and return to them later when examining the formal models.

43.3.1 Requirements for the Mode Control Panel

The following specification for the MCP focuses on a limited aircraft function and its pilot interface. Other interfaces and commands that would be needed in a real design are omitted for brevity's sake:

1. The MCP contains four buttons for selecting modes and three displays for dialing in or displaying values, as shown in Figure 43.1a. The system supports the following four modes: (a) attitude control wheel steering (`att_cws`), (b) flight path angle select (`fpa_sel`), (c) altitude engage

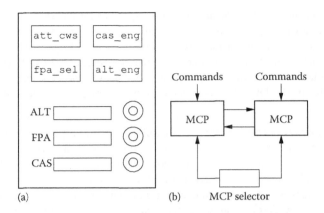

FIGURE 43.1 MCP: (a) pilot interface and (b) dual-panel configuration.

(alt_eng), and (d) calibrated air speed (cas_eng). Exactly one of modes (a)–(c) should be engaged at all times. Mode (d), cas_eng, can be engaged at the same time as any of the others. The pilot engages a mode by pressing the corresponding button on the panel. Engaging any of the modes (a)–(c) disengages the other two.

2. Three displays lie on the panel: altitude (ALT), flight path angle (FPA), and calibrated air speed (CAS). The displays usually show the current values for these quantities. The pilot can, however, enter a new value by dialing the knob next to the display to choose a target or "preselected" value for the aircraft to attain. For example, to climb to 25,000 feet, the pilot dials 25,000 into the ALT display and then presses the alt_eng button. Once the target is reached or the mode is disengaged, the display reverts to showing the "current" value.

3. If the pilot dials in an ALT that is more than 1200 feet above the current ALT and then presses the alt_eng button, the ALT mode will not directly engage. Instead, alt_eng mode changes to "armed" and fpa_sel mode becomes engaged. The pilot must then dial in an FPA for the flight control system to follow until the aircraft attains the desired ALT. fpa_sel mode will remain engaged until the aircraft is within 1200 feet of the desired ALT, after which the alt_eng mode engages.

4. The CAS and the FPA values need not be preselected before the corresponding modes are engaged— the current values displayed will be used. The pilot can dial in a different target value after the mode is engaged. Conversely, an ALT must be preselected before pressing the alt_eng button; otherwise, the command is ignored.

5. The cas_eng and fpa_sel buttons toggle on and off every time they are pressed. In contrast, if either the att_cws or alt_eng button is pressed while the corresponding mode is already engaged, the button is ignored.

6. Whenever a mode other than cas_eng is engaged, all other preselected displays should be returned to their current values.

7. If the pilot dials in a new ALT while the alt_eng button is already engaged or armed, then the alt_eng mode is disengaged and the att_cws mode is engaged. If the alt_eng mode was armed, then the fpa_sel should be disengaged as well.

43.3.2 Requirements for Coordinating Two Panels

The MCP example can be extended by postulating a dual-panel arrangement to support two pilots. The following specification stipulates a minimal capability along these lines:

1. There will be two MCPs, one for each pilot, as shown in Figure 43.1b. Only one panel can be active at a time. Only commands issued on the active side will have the effects described in Section 43.3.1. Commands issued on the passive side will be ignored.

2. Communication channels will exist for the two panels to exchange information. After each pilot command is processed, the active panel will forward any resulting state changes to the other panel. The passive panel will accept these updates and keep its state information current. It will also update its three displays (ALT, FPA, and CAS) so that both pilots will be able to see the current values.

3. A switch or other control mechanism available to the crew will be provided to choose when the passive side should become active. When the switch is operated to initiate a changeover, a deactivate signal will be sent to the currently active panel. For a brief instant during the changeover, it is possible that neither panel will be active.

4. When the currently active panel receives the deactivate signal, it first completes any command processing that might be underway and then disables its command processing functions. Next, it notifies the other panel that it should now assume the role of the active panel. On receiving this notification, the passive panel immediately enables its command processing functions.

43.4 Deductive Methods

Historically, deductive techniques were the original formal methods, dating back to the late 1960s. The other types of formal methods were introduced later in response to theoretical advances that made them possible. Theorem proving has existed as an organized academic discipline since the 1950s. In the early days, theorem proving was mostly of interest to artificial intelligence researchers. Its application to formal methods arose in the early 1970s as the use of mathematical logic to reason about program behavior began to gain traction with researchers.

Today, there are relatively few formal verification tool suites organized around major theorem-proving systems. Two of the more significant tools in the United States are prototype verification system (PVS) [17,18] (described in the next section) and a computational logic for applicative common lisp (ACL2) [11]. In Europe, the Coq [10], Isabelle [19], and higher order logic (HOL) [7] systems are popular. Most make use of a higher-order logic. ACL2 uses a more restrictive logic; in return, it offers fully automated proofs. The other systems opt for greater theoretical power at the expense of requiring users to carry out proofs interactively. They are often said to represent a "heavyweight" style of formal methods.

A number of smaller research efforts that offer more automated verification using weaker logics are currently under study. Many of these efforts are built around SMT solvers.

43.4.1 PVS Verification System

PVS was introduced by SRI International in 1992. It features an interactive theorem prover for a typed form of classical higher-order logic. The specification language, also called PVS, is based on a syntax similar to modern programming languages. The PVS user interface is implemented using the Emacs text editor as a front end to the proof system. Through the Emacs interface, users manage and edit files, submit analysis and verification commands, and carry out all other system interactions including interactive proving. PVS is available from SRI International as open-source software.

A user group at NASA Langley Research Center has compiled a substantial collection of PVS libraries [15] covering various mathematical domains as well as several theories specific to computing. Other PVS users have constructed libraries and made them available to the user community. Some have been incorporated into the NASA Langley collection.

43.4.2 Formally Specifying the Mode Control Panel

In the previous handbook edition, Johnson and Butler illustrated the process of developing formal specifications using the MCP example. In this chapter, our focus is to describe several analysis techniques, which precludes a detailed discussion of the modeling task. Readers wishing to understand the reasoning behind the PVS model can consult the previous handbook chapter. We use the same example here with only minor modifications.

The MCP requirements described in Section 43.3.1 have been modeled in PVS as shown in Section 43.A.1. In PVS, the basic modular unit is called a *theory*. Three theories are included: defs, tran, and panel. The first theory (lines 1–24) introduces types needed to describe the interface, commands, and internal state of an MCP. The second theory (lines 26–115) formalizes the state transition function, which captures high-level design details, namely, how the MCP responds to commands and other inputs. The final theory (lines 117–160) formalizes some of the requirements expressed informally in Section 43.3.1. To accomplish this, it introduces a constraint on valid initial states, a *trace* notion used to express invariant properties, and two invariants that are to be proved about the model.

In theory tran, the function nextstate (lines 101–113) plays the key role in formalizing the transition relation for the state machine model: $S_{n+1} = $ nextstate(S_n, e). The preceding definitions in that theory handle the specifics for each type of input event. States belong to a record type (lines 11–18). Each subordinate transition function maps from current state st into a new state using IF expressions and WITH expressions. The expression "st WITH [f := v]" yields a state value identical to st, except on field f where it takes the value v.

Unfortunately, space limitations prevent a more thorough presentation of the PVS model. We encourage the reader to try to grasp the general outline of the model without being overly concerned with its details.

43.4.3 Formally Verifying the Mode Control Panel

Verification activities take several possible forms depending on the verifier's goals. Proving that a design or implementation meets its specification is one such goal. In the case of the MCP model, the tran theory captures design details that can be proved to satisfy the requirements expressed in theory panel.

Several deductive aids have been placed in theory panel to facilitate the proof of invariant properties. The trace type (lines 134–137) defines the set of all state sequences that can result from repeated application of the nextstate function. Each trace can contain only those states reachable from a valid initial state. Traces are formalized here as infinite sequences. Since any finite trace will be a prefix of an infinite trace, this formulation is general enough to serve all cases of interest.

A parameterized definition of the invariant concept appears in `is_invariant` (line 139). Given a predicate P on states, `is_invariant` expresses succinctly that *P* is an invariant if and only if it holds on every state in every trace. This definition is used to express conjectures (lines 156 and 158) that will be submitted to the theorem prover. Such conjectures could be proved by induction using the definitions for trace and `is_invariant`. Simpler proofs are obtained, though, by appealing to the lemma `invariant_cases` (lines 141–144), which makes the induction cases explicit. Higher-order logic facilitates the expression and proof of such deductive utilities.

Although seldom encountered in most engineering disciplines, mathematical induction is a commonly used proof technique in deductive formal methods. Its most basic form gives a method for proving that a proposition $P(n)$ holds for all natural numbers n. To proceed, first prove the base case, $P(0)$. Then show that $P(k)$ implies $P(k + 1)$ for any natural number k. This latter case is generally known as the *induction step* and $P(k)$ is known as the *induction hypothesis*. Variants of this basic scheme are used to handle different numerical ranges or different data types. Theorem provers typically provide built-in commands for carrying out induction proofs.

In theory tran, the predicate `mode_rqmt` (lines 146–152) captures how the modes are related to one another. These relationships stem from requirements (1) and (3). For example, the requirement that exactly one of the modes `att_cws`, `fpa_sel`, and `alt_eng` be engaged at all times is expressed as follows. First, specify that at least one of the modes is engaged:

```
att_cws(st) = engaged OR fpa_sel(st) = engaged OR alt_eng(st) = engaged
```

Then, specify that at most one mode is engaged:

```
(alt_eng(st)/= engaged OR fpa_sel(st)/= engaged) AND
(att_cws(st) = engaged IMPLIES
 alt_eng(st)/= engaged AND fpa_sel(st)/= engaged)
```

The conjunction of these two conditions yields the desired result.

A claim that the formal requirement on modes is an invariant of the MCP model appears in the theorem on line 156. It is worth noting that all the conditions in `mode_rqmt` need to be present to achieve a provable invariant. If one were to break these into separate invariant conditions and attempt to prove them one by one, the attempt would fail. This is a typical situation when proving invariants. Being individually true does not mean they are provable in isolation. To succeed, an induction proof often, but not always, requires an induction hypothesis broad enough to cover the interdependent relationships among state components.

A second invariant, `armed_rqmt` (line 154), expresses an additional condition from requirement (3) concerning the *armed* mode. A claim of its invariance appears on line 158. `armed_rqmt` is a fairly simple relationship that expresses a necessary condition for entering the *armed* mode. Thus, it can be proved in isolation, in contrast to the invariant described in the previous paragraph.

Using the PVS interactive prover involves submitting commands to advance the proof toward its conclusion incrementally. Each step causes the proof state to be altered in a sound manner consistent with the underlying logic. The prover displays the new goal that results after each proof step. Some steps can cause the proof to branch into two or more paths. PVS proofs are therefore inherently tree structured.

To illustrate interactive proving, we show the proof of theorem `armed_invariant`:

```
armed_invariant:
|- - - - - - -
{1} is_invariant(armed_rqmt)
```

Initially, the prover displays the conjecture to be proved. The user proceeds by entering commands (inference rules) in the form of parenthesized expressions. The next command requests the prover to import and apply the lemma `invariant_cases`:

```
Rule? (use "invariant_cases")
Using lemma invariant_cases,
this simplifies to:
armed_invariant:
{-1} (FORALL st:
   (is_initial(st) IMPLIES armed_rqmt(st)) AND
         (armed_rqmt(st) IMPLIES
                  (FORALL (e: events): armed_rqmt(nextstate(st, e)))))
         IMPLIES is_invariant(armed_rqmt)
   |- - - - -
   [1] is_invariant(armed_rqmt)
```

The current goal of a proof takes a particular form known as a *sequent*. A sequent appears as two numbered lists of formulas (Boolean expressions) separated by a symbol called the *turnstile*: |— — —. The meaning is that the conjunction of the antecedent formulas (those above the turnstile) implies the disjunction of the consequent formulas (those below the turnstile).

Next, the user requests the prover to apply some basic simplification.

```
Rule? (ground)
Applying propositional simplification and decision procedures,

this simplifies to:
armed_invariant:
|- - - - -
{1} FORALL st:
      (is_initial(st) IMPLIES armed_rqmt(st)) AND
               (armed_rqmt(st) IMPLIES
                  (FORALL (e: events): armed_rqmt(nextstate(st, e))))
    [2] is_invariant(armed_rqmt)
```

At this point, consequent formula 2 is no longer needed and it would be best to remove it from the sequent before proceeding:

```
Rule? (hide 2)
Hiding formulas: 2,
this simplifies to:
armed_invariant:
   |- - - - -
   [1] FORALL st:
          (is_initial(st) IMPLIES armed_rqmt(st)) AND
                  (armed_rqmt(st) IMPLIES
                     (FORALL   (e: events): armed_rqmt(nextstate(st, e))))
```

Completing the proof only requires one additional command. It is, however, quite a powerful command that carries out a large variety of deductive heuristics, some of which require much computation. In this case, the invocation of the (grind) command does lead to many individual actions, whose details have been omitted:

```
Rule? (grind)
is_initial rewrites is_initial(st) to…
  <many lines of output deleted>
  Trying repeated skolemization, instantiation, and if-lifting,
  Q.E.D.
  Run time = 0.48 secs.
  Real time = 11.65 secs.
```

Needing only four commands, the proof completed successfully, establishing that predicate `armed_rqmt` is indeed an invariant property of all possible state sequences. This result holds for sequences of all lengths, up to and including infinitely long sequences. The proof of theorem `mode_invariant` is conducted in exactly the same manner.

By way of comparison, the proof of support lemma `invariant_cases` involves a few more steps, although it is not much longer. Only the completed "proof script" is shown in the following; intermediate sequents resulting from the proof steps are omitted:

```
(""
  (skosimp*)
  (expand "is_invariant")
  (skolem!)
  (induct "i")
  (("1" (inst?) (flatten) (hide-2) (typepred "T!1") (ground))
   ("2" (skosimp*) (inst?) (typepred "T!1") (grind))))
```

This proof contains a few of the prover's more "technical" commands. If theorems `mode_invariant` and `armed_invariant` were proved directly without using lemma `invariant_cases`, they would require most of these same proof steps. Hence, using the lemma is advantageous because proof steps subsequent to invoking the lemma are less challenging.

43.5 Model Checking

The field of model checking dates to a pair of seminal publications from 1981 [2,20]. Since then, it has grown dramatically to become the most frequently used type of formal method. This appeal can be attributed to several factors, including a more gentle learning curve, more highly automated analysis techniques, and an ability to provide diagnostic information such as counterexamples. The latter trait has also enabled some model checkers to serve as advanced debugging tools.

Shared by most model checkers is the ability to search large state spaces during the analysis process. Moderately different approaches, though, are used by the various tools. For example, SPIN [8,9], one of the most widely used verifiers, belongs to the category of explicit-state model checkers. These tools generate and "visit" each reachable state within the user's model, simultaneously testing whether the states or paths violate the desired properties.

Exploring a vast set of states obviously requires substantial memory and processor resources. Model checking tools try to guard against the "state explosion" problem that comes with increasing model size. Many language and tool features are designed to keep state sizes manageable. Users must be aware of how language features contribute to state growth so they can write models that minimize the problem. It is not hard to create models whose analysis is infeasible. Clarke et al. [3] present basic model checking algorithms and describe techniques for containing state growth.

Symbolic model checkers form a second category, making use of special techniques such as binary decision diagrams (BDDs) to represent state information and transition functions in a more indirect manner. BDDs encode large Boolean functions using a data structure based on directed acyclic graphs, leading to much smaller structures than alternatives such as binary trees. This allows larger state spaces to be explored, although the states are not literally visited individually. A key drawback is a limit on the complexity of transition functions. Explicit-state model checkers, by comparison, can work with models having more complicated transition functions. Nevertheless, symbolic model checkers have enjoyed much success in verifying digital hardware designs. This tool class is typified by symbolic model verifier (SMV) [1] and its successor NuSMV [16].

Other model checking approaches have been pursued by researchers. Symbolic Analysis Laboratory (SAL) supports multiple model checking algorithms within a common language and modeling framework. SAL, and other tools such as Kind, makes use of SMT solving within their algorithms. Another concept is that of "software model checking," where the checker analyzes software source code directly. Java PathFinder (JPF) is an example of this category, as is the Berkeley lazy abstraction software verification tool (BLAST) checker for C programs. Beyond these are a host of other approaches such as real-time and probabilistic model checking.

43.5.1 SPIN Model Checker

We use SPIN in this chapter to explore the coordination aspects of the MCP example and illustrate the use of analysis techniques different from that of PVS. SPIN is more of a loosely coupled collection of tools than PVS. There is the core SPIN analyzer that runs as a command-line tool. Then there is a graphical front end (called XSPIN) to provide a user-friendly way to invoke the analyzer and examine its output. A variety of third-party utilities also have been created to improve various aspects of SPIN modeling and verification.

In a companion chapter in this edition of the handbook, Hu and Bertolotti introduce the concept of model checking and present the use of SPIN in some detail. We refer the reader to that chapter for background information on the SPIN tool and especially its modeling language, process metalanguage (PROMELA).

43.5.2 Modeling the Two-Panel Subsystem

A second set of requirements for the example application was presented in Section 43.3.2. These requirements concern the behavior of a dual-panel configuration, as depicted in Figure 43.1b, where one panel is provided for each pilot. Coordination between the two panels is necessary to maintain consistency of operation and avoid ambiguous states that might result from unconstrained operation. Most importantly, we wish to ensure that only one panel is active at a time.

SPIN was designed to model and analyze the interactions of concurrent processes and communication protocols. A message-passing style of communication is supported by the PROMELA modeling language. It is well suited, therefore, to analyze coordination in the two-panel configuration specified in Section 43.3.2. Section 43.A.2 lists the model for the two-panel subsystem expressed in the PROMELA language.

The primary language features at the heart of the model are processes and channels for interconnecting them. Other items include type and variable declarations. The central process type of interest in the model is named MCP (lines 11–41). It has the following general structure:

```
proctype MCP (bit side) /* Mode-control panel process */
{
        byte panel_state,c,s; bit b;
        do
        :: guard-1 → action-1
        :: guard-2 → action-2
        :: guard-3 → action-3
        :: guard-4 → action-4
        od
}
```

This declaration introduces the process type MCP, which will be instantiated twice, once for each side of the dual-panel MCP. The parameter side of the process declaration will assume the values 0 and 1 so each process knows which side it is operating on. The do... od structure is a do-loop whose body has been presented in skeleton form. Details within its body (lines 17–39) represent the actions performed by the MCP process in response to received commands and other inputs.

Each alternative of the do-loop has the form G → S, where G is known as a *guard* or *guard statement*. Guard G is either a Boolean expression or a channel operation (sending or receiving a message). If a guard expression is true or its channel operation is enabled, the statement sequence that follows is executable. When multiple guarded sequences are eligible, one is selected for execution nondeterministically. All four guards within the do-loop of process MCP are channel receive operations. Channel operations are written chan?v to receive into variable v and chan!e to send value e. The statements after each guard specify how the process responds to the corresponding type of incoming message.

A local variable panel_state (line 13) provides a stub for the state information represented in the PVS model. Because the SPIN model is only concerned with the panel interactions and not the details of internal MCP behavior, an abstraction of the MCP state is used here. SPIN, like other model checkers, favors concrete data types in contrast to the more abstract approach used by PVS and similar systems. In this case, the type byte is used to represent the internal MCP state.

Several pairs (arrays) of channels are provided to connect to the MCP processes (lines 5–9). Each is indexed by a "side" ID (0 or 1) to refer to a single channel.

```
chan MCP_commands[2] = [1] of {byte};        /* Represents buttons, dials */
chan switch_over[2] = [1] of {bit};          /* Signal to activate other MCP */
chan to_MCP[2] = [1] of {mtype, byte};       /* Pipes between MCPs */
```

Each channel provides inputs to the MCP processes, corresponding to the arrows in Figure 43.1b. Channel MCP_commands[i] represents inputs resulting from pilot commands (button presses and dial rotations) as well as sensor inputs such as ALT-related events. Channel `switch_over[i]` carries the signal used to indicate when the crew has requested activation of the passive MCP. Channel `to_MCP[i]` allows the two MCP processes to communicate with one another to exchange state information and notify the other when it has been requested to become the active side.

Two additional processes are provided to serve as sources of inputs for the MCP processes:

```
proctype crew_commands () {/* Simulates commands from pilots */
    do
    :: MCP_commands[0]!0 → skip
    :: MCP_commands[1]!0 → skip
    od
    }
```

The crew _ commands process (lines 50–55) emits "messages" to the command inputs of the MCP processes in a nondeterministic manner. These messages simulate the crew commands and ALT notifications that were included in the PVS model. No attempt is made to model any details; only the occurrence of command inputs is of interest in this model.

Likewise, the switcher process (lines 43–48) emits activate/deactivate signals to both MCP processes in a nondeterministic manner:

```
proctype switcher () {/* Generates switch-over inputs */
    do
    :: switch_over[0]!0 → skip
    :: switch_over[1]!0 → skip
    od
    }
```

In practice, activation signals would occur infrequently. Nonetheless, to achieve a thorough verification, generating all possible interleaved MCP process inputs is desirable. Moreover, explicitly providing generator processes can be useful during model debugging and simulation.

Finally, a special process called init (lines 62–67) is a standard part of SPIN models. Its primary purpose is starting the other processes in the model and passing any parameters that might be needed:

```
init {atomic {side_active[0] = true;
              run MCP(0); run MCP(1);
              run crew_commands();
              run switcher();
              run monitor()
}}
```

Two instances of MCP are started with the IDs 0 and 1. The two generator processes are also started as well as a monitor process (described in the next section).

43.5.3 Analyzing the Two-Panel Subsystem

SPIN provides a variety of tools for analysis and verification. Included are simulation features that allow users to perform random simulations, guided simulations from previously recorded event traces ("trails" in SPIN terminology), and interactive simulations, where the user selects the next nondeterministic choice. Output information includes detailed trace information and message sequence charts, which are graphical renditions of the communication events in a simulation showing chronological relationships.

Verification applies the state exploration algorithm to search for any of several types of errors. One type is the failure of user-supplied assertions. PROMELA allows the introduction of assertions throughout a model. During verification (and simulation) runs, the assertions are tested and any that evaluate to false are flagged as errors.

Another class of detectable errors is *deadlock*. SPIN will report an error type of "invalid final state" to indicate that the model was found to have a final state in which at least one process is not at its expected termination point. Usually, such a condition indicates a deadlock, that is, an unintended state from which no transition is possible. Deadlocks typically happen when multiple processes are waiting for each other, leaving no opportunity to proceed without receiving input from the other process(es).

Invariants can be verified in SPIN using several techniques. One way is to add a process to the model that includes statements of the form assert(*P*). Assertions are always executable, so they would be interleaved with all other actions in the model, causing them to be repeatedly checked during analysis. If the condition *P* ever evaluates to false, SPIN recognizes an assertion violation and reports it as such.

A variant of this approach uses a guarded statement of the form atomic {!*P* → assert(*P*)}. As before, the condition !*P* is repeatedly checked during analysis. If it ever evaluates to true, a violation of *P* has been found. Now executable, the assert statement triggers SPIN's reaction to an assertion violation. This variant is a standard SPIN idiom that achieves the same effect as the simpler assert(P) but executes more efficiently.

This approach was used to verify an invariant of the MCP model by introducing the monitor process (lines 57–60):

```
proctype monitor () {/* Checks on invariant assertions */
    atomic { (side_active[0] && side_active[1]) →
          assert(!(side_active[0] && side_active[1]))}
}
```

Expressed in the monitor invariant earlier is the condition that at most one side will be active at a time. The array variable side _ active (line 3) records whether each side thinks it is the active panel. Ruled out by the invariant is the situation where both sides think they are active, an obviously undesirable circumstance:

Several other techniques exist for expressing invariants in SPIN models. One makes use of a *never claim*, a special SPIN construct whose analysis is based on Büchi automata, theoretical structures having suitable semantics for this task. Another approach is to write formulas in linear temporal logic (LTL), which SPIN first translates to *never claims*. LTL can be used to express additional properties besides invariants.

Figure 43.2 shows the output reported by the SPIN verifier after being presented with the two-MCP model. The report includes information on which kinds of errors were checked during the search. In this case, assertion violations and invalid end states were selected. Also displayed are counts of states visited, the maximum extent of the depth-first search, and various statistics on memory usage.

Listed at the end of the report is the final verdict: "no errors found." For this model, the absence of errors means (1) the invariant on side_active held true in every state, and (2) no deadlock states were found in the model.

It would be possible to verify additional invariants or LTL formulas that express desired properties. An example would be an invariant that the panel state is always the same on both sides, except during the middle of processing actions that update these values. Another possibility would be the property that after side *k* receives a signal to deactivate, side 1–*k* eventually becomes active. Expressing this property requires a temporal logic (LTL) formula.

```
(Spin Version 6.2.3 -- 24 October 2012)
+ Partial Order Reduction

Full statespace search for:
never claim      - (not selected)
assertion violations +
cycle checks     - (disabled by -DSAFETY)
invalid end states +

State-vector 72 byte, depth reached 9773, errors: 0
        51373 states, stored
        106910 states, matched
        158283 transitions (= stored+matched)
            5 atomic steps
hash conflicts:        542 (resolved)

Stats on memory usage (in Megabytes):
        4.115 equivalent memory usage for states (stored*(State-vector + overhead))
        3.413 actual memory usage for states (compression: 82.93%)
              state-vector as stored = 58 byte + 12 byte overhead
       64.000 memory used for hash table (-w24)
        0.343 memory used for DFS stack (-m10000)
       67.664 total actual memory usage

pan: elapsed time 0.11 seconds
No errors found -- did you verify all claims?
```

FIGURE 43.2 Verification output from running SPIN on the model.

It is worth noting that the PVS model in Section 43.A.1 could be expressed and analyzed using SPIN or several other model checkers. In fact, Lüttgen and Carreño [12] examined three model checkers to determine how well they can analyze designs for possible sources of mode confusion, a problem domain similar to that explored in this chapter. On the other hand, if the MCP model of Section 43.A.1 was more numerical in nature, or it formalized nontrivial data structures, then deductive verification would likely be a more appropriate choice than model checking.

43.6 Abstract Interpretation

Abstract interpretation was introduced in 1977 [4]. Based on a broad array of mathematical underpinnings, this method is primarily used to develop static analyzers that directly examine source code. Rather than verifying user-supplied models or properties, these static analyzers look for code defects that can result in a variety of runtime errors. The analyzers are intended for use by software development teams even though the theories and algorithms have a deep mathematical basis.

Despite ongoing research to expand the range of abstract interpretation tools, they have already been incorporated into two commercial products. PolySpace was developed in France and later purchased by MathWorks to incorporate into their product line [13]. PolySpace can analyze software written in C, C++, and Ada. A second-generation tool known as analyseur statique de logiciels temps-réel embarqués (real-time embedded software static analyzer, ASTRÉE) [5] has been used to analyze large amounts of software (over 400 KLOC) for Airbus aircraft. It is now distributed by a commercial tool vendor.

A noncommercial effort that also produced a successful system was the C Global Surveyor [22], developed at the NASA Ames Research Center and used to verify array-bound compliance. This tool was applied to the flight software of the NASA Mars Exploration Rover mission (over 550 KLOC). In addition, it analyzed software for missions Deep Space 1 (280 KLOC) and Mars Pathfinder (140 KLOC).

Unfortunately, the development of tools based on abstract interpretation is sufficiently costly that few offerings are available. Outside of the products mentioned, the only other implementations are research prototypes. Few tools of any significance are available in open-source form. Also, abstract interpretation can be computationally expensive; implementations occasionally require fine tuning by experts. In spite of these limitations, the track record of successful applications to aerospace software makes abstract interpretation a viable technique for avionics projects.

43.6.1 General Characteristics

Several static analysis tools have been introduced in recent years that have become popular with software developers. These include products from companies such as Coverity and Klocwork. While undoubtedly useful at bug finding, these tools apply heuristics that can lead to both false-positives and false-negatives. Results that harbor false-negatives, that is, undetected defects, are problematic for dependable software.

In contrast, analysis by abstract interpretation is *sound*—no false-negatives are possible. If an analyzer fails to find a particular class of defect in a section of code, then the code is guaranteed to be free of such defects. For this reason, the analysis constitutes a strong form of verification.

False-positives, however, are still possible with abstract interpretation. Tool developers try to improve their precision and reduce their false-positive rate. Greater precision typically requires more computation, so there are practical limits to achieving higher-quality results. False-positive results often stem from inadequate information about the runtime environment in which the software will run. Careful examination by software engineers can determine when the runtime errors in question cannot occur because the conditions that enable them cannot arise.

At the heart of abstract interpretation is a notion of *abstract domains* and a disciplined form of discrete approximation. An abstract domain A includes operations and properties forming an abstraction of a concrete domain C. Although less precise than C's semantics, A and its associated approximation algorithms yield an abstract semantics sufficient to analyze specific program properties. Moreover, computation of the abstract semantics is generally tractable.

Several abstract domains, such as intervals and convex polyhedra, are commonly used. By mapping between concrete domains and abstract domains, the methods achieve a type of "overapproximation" that allows useful inferences to be drawn about possible values of program variables. For instance, by using intervals to constrain the value of an integer variable in a loop, the algorithms can infer the range of possible values after n iterations. In the process, they derive a loop invariant that constrains the space of variable values.

For a more detailed illustration, consider the following code fragment:

```
x = 0;
for (i = 0; i < 10; i++) {
    if (f(i, x) > 20)     x + = 2;
    else                  x + = 1;
}
```

A possible analysis approach would use the 2D space formed by the values of variables i and x. Since variable x is incremented by 1 or 2 on each iteration, typical algorithms could deduce the invariant $i \leq x \leq 2i$. Using a convex hull approach, the value of variable i at loop exit can be found by a narrowing process, and the range of variable x can be narrowed to the interval [10,20]. In turn, this range could be used to show that other operations (e.g., array accesses) are within their bounds. While these results are obvious for such a trivial example, the method works for much more complicated cases where manual analysis is far from practical.

Besides the mathematical theories mentioned so far, abstract interpretation relies on fully accurate representations of programming language semantics. Given that most languages have areas of semantic uncertainty (e.g., ambiguities and undefined behaviors), analyzers need to accommodate these semantic difficulties. Any features that can affect program execution are potentially involved in the analysis process. In fact, some types of analysis that compilers have historically performed during code generation can be considered instances of abstract interpretation.

43.6.2 Properties Verifiable Using Abstract Interpretation

Software defects that lead to several kinds of runtime errors can be detected by abstraction interpretation. Verifiers can determine when code is defect-free with respect to many of these categories:

- Out-of-bound array accesses and buffer overflows
- Read accesses to uninitialized data
- Pointer dereferencing problems (null pointers, out-of-bound accesses)
- Invalid arithmetic operations (division by zero, square root of negative numbers, others)
- Illegal-type conversions
- Integer and floating point overflow and underflow
- Some cases of nontermination of loops
- Concurrent accesses to shared data
- Various cases in the C language standard that are earmarked as having undefined behavior
- Violations of certain user-defined runtime properties

Research in abstract interpretation is aimed at exploring new abstract domains and increasing precision without incurring too large a computational cost. A team at NASA Ames, for example, is developing the Inference Kernel for Open Static Analyzers (IKOS). Rather than leading to a self-contained application, IKOS will create a C++ library designed to facilitate the development of sound static analyzers based on abstract interpretation. When combined with other software packages, developers will eventually be able to use IKOS to create custom analyzers having the characteristics needed for a particular application domain.

Microsoft Research has pursued another approach with their cccheck tool [6]. Also called Clousot, cccheck uses abstract interpretation to perform static analysis of program contracts in a language-independent manner. It has been added to the Code Contracts framework for design-by-contract programming. Preconditions, postconditions, and object invariants are the types of contracts supported. cccheck aims to decide if a program violates its contracts using analysis only.

43.7 Summary

In this chapter, we have presented a brief sampler of formal methods tools and techniques. Although the example used (MCP) is quite simple compared to avionics systems of realistic complexity, it nevertheless suffices to illustrate elementary use of formal languages and their tools. Readers are encouraged to explore the websites of PVS, SPIN, and other tools to learn more about their capabilities. In the end, there is no substitute for actually trying to carry out small modeling and verification projects. Such exploratory ventures are advisable before choosing to embark on a serious verification effort.

Most of the methods require significant investments of time to achieve proficiency. Lesser levels of expertise, however, are often sufficient for determining whether a method is appropriate for one's current needs. In any case, the growing sophistication of formal methods will lead to higher rates of adoption in safety-critical avionics. As the engineering workforce becomes more conversant with formal methods, their use in mainstream projects is likely to increase.

As should be apparent, none of the three categories of formal methods discussed, nor any of the individual tools or methodologies, constitutes a complete analysis approach. In real-world usage, developers are likely to incorporate several tools and techniques to achieve their verification objectives. While few of the tools are designed to facilitate such heterogeneous usage, there is a growing awareness by formal methods tool developers that interoperability should receive more serious attention.

Recently completed by Rockwell Collins is a set of case studies intended as an aid for practitioners and certifiers who expect to work with formal methods certification evidence in accordance with DO-333. These case studies were recently published as a NASA Contractor Report [23].

43.A Appendix

43.A.1 Listing of PVS Specification

```
1    defs: THEORY
2    BEGIN
3
4    md_status: TYPE = {off, armed, engaged}
5    off_eng: TYPE = {md: md_status | md = off OR md = engaged}
6
7    disp_status: TYPE = {pre_selected, current}
8
9    altitude_vals: TYPE = {away, near_pre_selected, at_pre_selected}
10
11   state: TYPE = [# att_cws: off_eng,
12                    cas_eng: off_eng,
13                    fpa_sel: off_eng,
14                    alt_eng: md_status,
15                    alt_disp: disp_status,
16                    fpa_disp: disp_status,
17                    cas_disp: disp_status,
18                    altitude: altitude_vals #]
19
20   events: TYPE = {press_att_cws, press_cas_eng, press_fpa_sel,
21                    press_alt_eng, input_alt, input_fpa, input_cas,
22                    alt_reached, alt_gets_near, fpa_reached}
23
24   END defs
25
26   tran: THEORY
27   BEGIN
28
29   IMPORTING defs
30
31   event: VAR events
32   st: VAR state
33
34   tran_att_cws(st): state =
35      IF att_cws(st) = off THEN
36      st WITH [att_cws := engaged, fpa_sel := off, alt_eng := off,
37      alt_disp := current, fpa_disp := current]
38   ELSE    st%%IGNORE

39   ENDIF
40
41   tran_cas_eng(st): state =
42      IF cas_eng(st) = off THEN
43      st WITH [cas_eng := engaged]
44   ELSE
45      st WITH [cas_eng := off, cas_disp := current]
46   ENDIF
47
48   tran_fpa_sel(st): state =
49     IF fpa_sel(st) = off THEN
50       st WITH [fpa_sel := engaged, att_cws := off,
51               alt_eng := off, alt_disp := current]
```

```
52    ELSE
53         st WITH [fpa_sel := off, fpa_disp := current, att_cws := engaged,
54                 alt_eng := off, alt_disp := current]
55    ENDIF
56
57    tran_alt_eng(st): state =
58    IF alt_eng(st) = off AND alt_disp(st) = pre_selected THEN
59    IF altitude(st)/= away THEN %% ENG
60       st WITH [att_cws := off, fpa_sel := off, alt_eng := engaged,
61               fpa_disp := current]
62    ELSE %% ARM
63       st WITH [att_cws := off, fpa_sel := engaged, alt_eng := armed]
64    ENDIF
65    ELSE
66       st%% IGNORE request
67    ENDIF
68
69    tran_input_alt(st): state =
70      IF alt_eng(st) = off THEN
71        st WITH [alt_disp := pre_selected]
72    ELSE
73        st WITH [alt_eng := off, alt_disp := pre_selected,
74                att_cws := engaged, fpa_sel := off, fpa_disp := current]
75    ENDIF
76
77    tran_input_fpa(st): state =
78         IF fpa_sel(st) = off THEN st WITH [fpa_disp := pre_selected] ELSE st ENDIF
79
80    tran_input_cas(st): state =
81         IF cas_eng(st) = off THEN st WITH [cas_disp := pre_selected] ELSE st ENDIF
82
83    tran_alt_gets_near(st): state =
84         IF alt_eng(st) = armed THEN
85             st WITH [altitude := near_pre_selected, alt_eng := engaged,
86                  fpa_sel := off, fpa_disp := current]
87    ELSE
88             st WITH [altitude := near_pre_selected]
89    ENDIF
90
91    tran_alt_reached(st): state =
92         IF alt_eng(st) = armed THEN
93             st WITH [altitude := at_pre_selected, alt_disp := current,
94                  alt_eng := engaged, fpa_sel := off, fpa_disp := current]
95    ELSE
96             st WITH [altitude := at_pre_selected, alt_disp := current]
97    ENDIF
98
99    tran_fpa_reached(st): state = st WITH [fpa_disp := current]
100
101   nextstate(st, event): state =
102     CASES event OF
103       press_att_cws: tran_att_cws(st),
104       press_alt_eng: tran_alt_eng(st),
105       press_fpa_sel: tran_fpa_sel(st),
106       press_cas_eng: tran_cas_eng(st),
```

```
107          input_alt: tran_input_alt(st),
108          input_fpa: tran_input_fpa(st),
109          input_cas: tran_input_cas(st),
110          alt_reached: tran_alt_reached(st),
111          fpa_reached: tran_fpa_reached(st),
112          alt_gets_near: tran_alt_gets_near(st)
113     ENDCASES
114
115     END tran
116
117     panel: THEORY
118     BEGIN
119
120     IMPORTING tran
121
122     event: VAR events
123     st: VAR state
124     P: VAR pred[state]
125
126     is_initial(st): bool = att_cws(st) = engaged
127                            AND cas_eng(st) = off
128                            AND fpa_sel(st) = off
129                            AND alt_eng(st) = off
130                            AND alt_disp(st) = current
131                            AND fpa_disp(st) = current
132                            AND cas_disp(st) = current
133
134     trace: TYPE = {s: sequence[state] |
135                     is_initial(s(0)) AND
136                     FORALL (i: nat):
137                         EXISTS (e: events): s(i + 1) = nextstate(s(i), e)}
138
139     is_invariant(P): bool = FORALL (T: trace): FORALL (i: nat): P(T(i))
140
141     invariant_cases: LEMMA
142      (FORALL st: (is_initial(st) IMPLIES P(st)) AND
143          (P(st) IMPLIES FORALL (e: events): P(nextstate(st, e))))
144     IMPLIES is_invariant(P)
145
146     mode_rqmt(st): bool =
147        (att_cws(st) = engaged OR fpa_sel(st) = engaged
148          OR alt_eng(st) = engaged) AND
149        (alt_eng(st)/= engaged OR fpa_sel(st)/= engaged) AND
150        (att_cws(st) = engaged IMPLIES
151        alt_eng(st)/= engaged AND fpa_sel(st)/= engaged) AND
152        (alt_eng(st) = armed IMPLIES fpa_sel(st) = engaged)
153
154     armed_rqmt(st): bool = (alt_eng(st) = armed IMPLIES altitude(st) = away)
155
156     mode_invariant: THEOREM is_invariant(mode_rqmt)
157
158     armed_invariant: THEOREM is_invariant(armed_rqmt)
159
160     END panel
```

43.A.2 Listing of SPIN Model

```
1     mtype = {pstate, activate}
2
3     bool side_active[2] = false;
4
5     chan MCP_commands[2] = [1] of {byte};/* Represents buttons, dials */
6
7     chan switch_over[2] = [1] of {bit};/* Signal to activate other MCP */
8
9     chan to_MCP[2] = [1] of {mtype, byte};/* Pipes between MCPs */
10
11    proctype MCP (bit side)/* Mode-control panel process */
12    {
13      byte panel_state,c,s;
14      bit b;
15
16      do
17      :: switch_over[side]?b →/* Need to activate other side */
18         if
19      :: side_active[side] →
20            side_active[side] = false;
21            to_MCP[1-side]!activate,panel_state
22      :: else → skip
23         fi
24      :: MCP_commands[side]?c →
25         /* Following represents a panel state transition */
26         panel_state = 1
27      :: to_MCP[side]?activate,s →/* Other MCP is passing the baton */
28         if
29         :: !side_active[side] →
30             panel_state = s
31             side_active[side] = true
32         :: else → skip
33         fi
34      :: to_MCP[side]?pstate,s →/* Received a state update from other MCP */
35         If
36         :: !side_active[side] →
37            panel_state = s
38         :: else → skip
39         fi
40      od
41    }
42
43    proctype switcher () {/* Generates switch-over inputs */
44        do
45        :: switch_over[0]!0 → skip
46        :: switch_over[1]!0 → skip od
47        od
48    }
49
50    proctype crew_commands () {/* Simulates commands from pilots */
51        do
52        :: MCP_commands[0]!0
```

```
53            :: MCP_commands[1]!0 od
54            od
55       }
56
57       proctype monitor () {/* Checks on invariant assertions */
58            atomic {(side_active[0] && side_active[1]) ->
59                assert(!(side_active[0] && side_active[1]))}
60       }
61
62       init {atomic {side_active[0] = true;
63                      run MCP(0); run MCP(1);
64                      run crew_commands();
65                      run switcher();
66                      run monitor()
67       }}
```

References

1. J.R. Burch, E.M. Clarke, K.L. McMillan, D.L. Dill, and L.J. Hwang. Symbolic model checking: 10^{20} states and beyond. *Information and Commutation*, 98, 142–170, 1992.

2. E.M. Clarke and A. Emerson. Synthesis of synchronization skeletons for branching time temporal logic. In *Logic of Programs Workshop*, volume 131 of *Lecture Notes in Computer Science*, Springer-Verlag, Yorktown Heights, NY, 1981.

3. E.M. Clarke, O. Grumberg, and D. Peled. *Model Checking*. MIT Press, Cambridge, MA, 1999.

4. P. Cousot and R. Cousot. Abstract interpretation: A unified lattice model for static analysis of programs by construction or approximation of fixpoints. In *Fourth Symposium on Principles of Programming Languages*, pp. 238–353, 1977.

5. P. Cousot, R. Cousot, J. Feret, L. Mauborgne, A. Mine, D. Monniaux, and X. Rival. The ASTREE analyzer. In *European Symposium on Programming (ESOP'05)*, volume 3444 of *Lecture Notes in Computer Science*, pp. 21–30, Springer, Heidelberg, Germany, 2005.

6. M. Fähndrich and F. Logozzo. Static contract checking with abstract interpretation. In *Formal Verification of Object-Oriented Software FoVeOOS*, pp. 10–30, Springer-Verlag, Paris, France, 2010.

7. M.J.C. Gordon and T.F. Melham. *Introduction to HOL: A Theorem Proving Environment for Higher-Order Logic*, Cambridge University Press, Cambridge, England, 1993.

8. G. Holzmann. The model checker SPIN. *IEEE Transactions on Software Engineering*, 23(5): 279–295, 1997.

9. G. Holzmann. *The SPIN Model Checker: Primer and Reference Manual*. Addison-Wesley, Pearson Education, Boston, 2004.

10. INRIA. *The Coq Proof Assistant Reference Manual*. http://coq.inria.fr/documentation.

11. M. Kaufmann, P. Manolios, and J.S. Moore. *Computer-Aided Reasoning: An Approach*. Kluwer Academic Press, Dordrecht, Netherlands, 2000.

12. G. Lüttgen and V. Carreño. Analyzing mode confusion via model checking. Technical Report NASA/CR-1999-209332 (ICASE Report No. 99–18), NASA Langley Research Center, May 1999.

13. The MathWorks. PolySpace verifier. http://www.mathworks.com/products/polyspace.

14. J.F. Monin and M.G. Hinchey. *Understanding Formal Methods*. Springer, NY, 2003.

15. NASA Langley PVS library collection. Theories and proofs available at http://shemesh.larc.nasa.gov/fm/ftp/larc/PVS-library/pvslib.html.

16. NuSMV symbolic model checker. http://nusmv.fbk.eu/NuSMV.

17. S. Owre, J. Rushby, and N. Shankar. PVS: A prototype verification system. In *11th International Conference on Automated Déduction (CADE)*, volume 607 of *Lecture Notes in Artificial Intelligence*, pp. 748–752, Saratoga, NY, 1992.

18. S. Owre, J. Rushby, N. Shankar, and F. von Henke. Formal verification for fault-tolerant architectures: Prolegomena to the design of PVS. *IEEE Transactions on Software Engineering*, 21(2): 107–125, 1995.

19. L.C. Paulson. *Isabelle: A Generic Theorem Prover*, volume 828 of LNCS, Springer, Heidelberg, Germany, 1994.

20. J.P. Queille and J. Sifakis. Specification and verification of concurrent systems in Cesar. In *Fifth International Symposium on Programming*, volume 137 of *Lecture Notes in Computer Science*, pp. 195–220. Springer-Verlag, Heidelberg, Germany, 1981.

21. RTCA, Inc. Washington, DC. *DO-333, Formal Methods Supplement to DO-178C and DO–278A*, December 13, 2011. http://www.rtca.org.

22. A. Venet and G.P. Brat. Precise and efficient static array bound checking for large embedded C programs. In *International Conference on Programming Language Design and Implementation (PLDI)*, pp. 231–242, 2004.

23. Darren Cofer and Steven Miller. Formal Methods Case Studies for DO-333. Technical Report NASA/CR-2014-218244, NASA Langley Research Center, April 2014. Available at http://ntrs.nasa.gov.

44

Navigation and Tracking

James Farrell
VIGIL, Inc.

Maarten Uijt
de Haag
Ohio University

44.1 Introduction

The task of navigation ("nav") interacts with multiple avionics functions. To clarify the focus here, this chapter will not discuss tight formations, guidance, steering, minimization of fuel/noise/pollution, or managing time of arrival. The accent instead is on determining position and velocity (plus, where applicable, other variables: acceleration, verticality, heading)—with maximum accuracy reachable from whatever combination of sensor outputs are available at all times. Position can be expressed as a vector displacement from a designated point or in terms of latitude/longitude/altitude above mean sea level, above the geoid—or both. Velocity can be expressed in a locally level coordinate frame with various choices for an azimuth reference (e.g., geodetic north, Universal Transverse Mercator [UTM] grid north, runway centerline, wander azimuth with or without Earth sidereal rate torquing). In principle, any set of axes could be used—such as an Earth-centered, Earth-fixed (ECEF) frame for defining position by a Cartesian vector and velocity in Cartesian coordinates or in terms of ground speed, flight path angle, and ground track angle—in either case, it is advisable to use accepted conventions.

Realization of near-optimal accuracy with any configuration under changing conditions is now routinely achievable. The method uses a means of dead reckoning (DR)—preferably an inertial navigation system (INS)—which can provide essentially continuous position, velocity, and attitude in three dimensions by performing a running accumulation from derivative data. Whenever a full or partial fix is available from a nav sensor, a discrete update is performed on the entire set of variables representing the state of the nav system; the amount of reset for each state variable is determined by a weighting computation based on modern estimation. In this way, "initial" conditions applicable to the DR device are in effect reinitialized, as the "zero" time is advanced (and thus kept current), with each update. Computer-directed operation easily accommodates conditions that may arise in practice (incomplete fixes, inconsistent data rates, intermittent availability, changing measurement geometry, varying accuracies) while providing complete flexibility for backup with graceful degradation. The approach inherently combines short-term accuracy of the DR data with the navaids' long-term accuracy. A commonly cited example of synergy offered by the scheme is a tightly coupled GPS/INS wherein the inertial information provides short-term aiding that vastly improves responsiveness of narrowband code and/or carrier tracking, while GPS information counteracts long-term accumulation of INS error.

The goal of navigation has long progressed far beyond mere determination of geographic location. Efforts to obtain double and triple "mileage" from inertial instruments, by integrating nav with sensor stabilization and flight control, are over a decade old. Older yet are additional tasks such as target designation, precision pointing, tracking, antenna stabilization, and imaging sensor stabilization (and therefore transfer alignment). Digital beamforming (DBF) for array antennas (including graceful degradation to recover when some elements fail) needs repetitive data for instantaneous relative position of those elements; on deformable structures, this can require multiple low-cost transfer-aligned inertial measuring units (IMUs) and/or fitting of spatial data to an aeroelastic model. The multiplicity of demands underlines the importance of integrating the operations; the rest of this chapter describes how integration should be done.

44.2 Fundamentals

To accomplish the goals just described, the best available balance is obtained between old and new information—avoiding both extremes of undue clinging to old data and jumping to conclusions at each latest input. What provides this balance is a modern estimation algorithm that accepts each data fragment as it appears from a nav sensor, immediately weighing it in accordance with its ability to shed light on every variable to be estimated. That ability is determined by accounting for all factors that influence how much or how little the data can reveal about each of those variables; those factors include

- Instantaneous geometry (e.g., distance along a skewed line carries implications about more than one coordinate direction)
- Timing of each measurement (e.g., distance measurements separated by known time intervals carry implications about velocity as well as position)
- Data accuracy, compared with accuracy of estimates existing before measurement

Only when all these factors are taken into account, accuracy and flexibility as well as versatility are maximized. To approach the ramifications, gradually consider a helicopter hovering at constant altitude, which is to be determined on the basis of repeated altimeter observations. After setting the initial *a posteriori* estimate to the first measurement $\hat{Y}_1$, an *a priori* estimate $\hat{x}_2^{(-)}$ is predicted for the second measurement and that estimate is refined by a second observation:

$$\hat{x}_2^{(-)} = \hat{x}_1^{(+)}; \quad \hat{x}_2^{(+)} = \hat{x}_2^{(-)} + \frac{1}{2}z_2; \quad z_2 \triangleq \hat{Y}_2 - \hat{x}_2^{(-)} \tag{44.1}$$

and a third observation

$$\hat{x}_3^{(-)} = \hat{x}_2^{(+)}; \quad \hat{x}_3^{(+)} = \hat{x}_3^{(-)} + \frac{1}{3}z_3; \quad z_3 \triangleq \hat{Y}_3 - \hat{x}_3^{(-)} \tag{44.2}$$

and then the fourth observation

$$\hat{x}_4^{(-)} = \hat{x}_3^{(+)}; \quad \hat{x}_4^{(+)} = \hat{x}_4^{(-)} + \frac{1}{4}z_4; \quad z_4 \triangleq \hat{Y}_4 - \hat{x}_4^{(-)} \tag{44.3}$$

which now clarifies the general expression for the *m*th observation

$$\hat{x}_m^{(-)} = \hat{x}_{m-1}^{(+)}; \quad \hat{x}_m^{(+)} = \hat{x}_m^{(-)} + \frac{1}{m}z_m; \quad z_m \triangleq \hat{Y}_m - \hat{x}_m^{(-)} \tag{44.4}$$

which can be rewritten as

$$\hat{x}_m^{(+)} = \frac{m-1}{m}\hat{x}_m^{(-)} + \frac{1}{m}\hat{Y}_m, \quad m > 0 \tag{44.5}$$

Substitution of $m = 1$ into this equation produces the previously mentioned condition that the first *a posteriori* estimate is equal to the first measurement; substitution of $m = 2$, combined with that condition, yields a second *a posteriori* estimate equal to the average of the first two measurements. Continuation with $m = 3, 4, \ldots$ yields the general result that, after m measurements, estimated altitude is simply the average of all measurements.

This establishes an equivalence between the *recursive* estimation formulation expressed in (44.1) through (44.5) and the *block* estimate that would have resulted from averaging all data together in one step. Since that average is widely known to be optimum when all observations are equally accurate statistically, the recursion shown here must then be optimum under that condition. For measurement errors that are sequentially independent random samples with zero mean and variance R, it is well known that mean squared estimation error $P_m^{(+)}$ after averaging m measurements is just R/m. That is the variance of the *a posteriori* estimate (just *after* inclusion of the last observation); for the *a priori* estimate, the variance $P_m^{(-)}$ is $R/(m-1)$. It is instructive to express the last equation earlier as a blended sum of old and new data, weighted by factors

$$\frac{R}{P_m^{(-)} + R} \equiv \frac{R/P_m^{(-)}}{1 + R/P_m^{(-)}} = \frac{m-1}{m} \tag{44.6}$$

and

$$\frac{P_m^{(-)}}{P_m^{(-)} + R} = \frac{1}{m} \tag{44.7}$$

respectively; weights depend on variances, giving primary influence to information having lower mean squared error. This concept, signified by the left-hand sides of the last two equations, is extendible to more general conditions than the restrictive (uniform variance) case considered thus far; we are now prepared to address more challenging tasks.

As a first extension, let the sequence of altimeter measurements provide repetitive refinements of estimates for both altitude x_1 and vertical velocity x_2. The general expression for the mth observation now takes a more inclusive form:

$$\hat{\mathbf{x}}_m^{(-)} = \Phi_m \hat{\mathbf{x}}_{m-1}^{(+)}; \quad \hat{\mathbf{x}}_m^{(+)} = \hat{\mathbf{x}}_m^{(-)} + \mathbf{W}_m z_m, \quad z_m \triangleq \hat{Y}_m - \hat{x}_{1,m}^{(-)}, \quad \mathbf{x}_m \triangleq \begin{bmatrix} x_{m,1} \\ x_{m,2} \end{bmatrix} \tag{44.8}$$

so that the method accommodates estimation of multiple unknowns, wherein the status of a system is expressed in terms of a state vector ("state") $\mathbf{x}$, in this case a 2×1 vector containing two *state variables* ("states"); superscripts and subscripts continue to have the same meaning as in the introductory example, but for these states, the conventions $m,1$ and $m,2$ are used for altitude and vertical velocity, respectively, at time t_m. For this dynamic case, the *a priori* estimate at time t_m is not simply the previous *a posteriori* estimate; that previous state must be premultiplied by the *transition matrix*:

$$\Phi_m = \begin{bmatrix} 1 & t_m - t_{m-1} \\ 0 & 1 \end{bmatrix} \tag{44.9}$$

which performs a time extrapolation. Unlike the static situation, elapsed time now matters since imperfectly perceived velocity enlarges altitude uncertainty between observations—and position measurements separated by known time intervals carry implicit velocity information (thus enabling vector estimates to be obtained from scalar data in this case). Weighting applied to each measurement is influenced by three factors:

- A sensitivity matrix $\mathbf{H}_m$ whose (i, j) element is the partial derivative of the ith component of the mth measured data vector to the jth state variable. In this scalar measurement case $\mathbf{H}_m$ is a 1×2 matrix $[1\ 0]$ for all values of m.
- A covariance matrix $\mathbf{P}_m$ of error in state estimate at time t_m (the ith diagonal element = mean squared error in estimating the ith state variable and, off the diagonal, $P_{ij} = P_{ji} = \sqrt{P_{ii}P_{jj}} \times$ (correlation coefficient between ith and jth state variable uncertainty)).
- A covariance matrix $\mathbf{R}_m$ of measurement errors at time t_m (in this scalar measurement case, $\mathbf{R}_m$ is a 1×1 *matrix*—i.e., a scalar variance R_m).

Although formation of $\mathbf{H}_m$ and $\mathbf{R}_m$ follows directly from their definitions, $\mathbf{P}_m$ changes with time (e.g., recall the effect of velocity error on position error) and with measurement events (because estimation errors fall when information is added). In this "continuous–discrete" approach, uncertainty is decremented at the discrete measurement events:

$$\mathbf{P}_m^{(+)} = \mathbf{P}_m^{(-)} - \mathbf{W}_m\mathbf{H}_m\,\mathbf{P}_m^{(-)} \tag{44.10}$$

and, between events, dynamic behavior follows a continuous model of the form

$$\dot{\mathbf{P}} = \mathbf{A}\mathbf{P} + \mathbf{P}\mathbf{A}^T + \mathbf{E} \tag{44.11}$$

where $\mathbf{E}$ acts as a forcing function to maintain positive definiteness of $\mathbf{P}$ (thereby providing stability and effectively controlling the remembrance duration—the "data window" denoted herein by T—for the estimator) while $\mathbf{A}$ defines dynamic behavior of the state to be estimated ($\dot{\mathbf{x}} = \mathbf{A}\mathbf{x}$ and $\dot{\Phi} = \mathbf{A}\Phi$). In the example at hand,

$$\mathbf{A} = \begin{bmatrix} 0 & 1 \\ 0 & 0 \end{bmatrix}; \quad \begin{bmatrix} \dot{x}_1 \\ \dot{x}_2 \end{bmatrix} = \mathbf{A} \begin{bmatrix} x_1 \\ x_2 \end{bmatrix} \tag{44.12}$$

Given $\mathbf{H}_m$, $\mathbf{R}_m$, and $\mathbf{P}_m^{(-)}$, the optimal (Kalman) weighting matrix is

$$\mathbf{W}_m = \mathbf{P}_m^{(-)} \mathbf{H}_m^T \left(\mathbf{H}_m\,\mathbf{P}_m^{(-)}\,\mathbf{H}_m^T + \mathbf{R}_m \right)^{-1} \tag{44.13}$$

which for a scalar measurement produces a vector $\mathbf{W}_m$ as the previous inversion simplifies to division by a scalar (which becomes the variance R_m added to P_{11} in this example):

$$\mathbf{W}_m = \mathbf{P}_m^{(-)} \mathbf{H}_m^T \left(\mathbf{H}_m\,\mathbf{P}_m^{(-)}\,\mathbf{H}_m^T + R_m \right)^{-1} \tag{44.14}$$

The preceding (hovering helicopter) example is now recognized as a special case of this vertical nav formulation. To progress further, horizontal navigation addresses another matter, that is, location uncertainty in more than one direction—with measurements affected by more than one of the unknowns (e.g., lines of position [LOPs] skewed off a cardinal direction such as north or east; Figure 44.1). In the classic "compass-and-dividers" approach, DR would be used to plot a running accumulation of position increments until the advent of a fix from two intersecting straight or curved LOPs; position would then

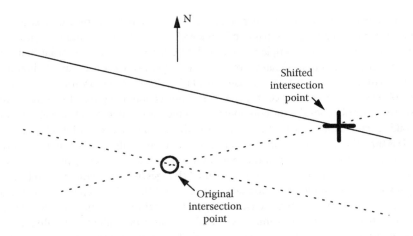

FIGURE 44.1   Nonorthogonal LOPs.

be reinitialized at that fixed position, from whence DR would continue until the next fix. For integrated nav, we fundamentally alter that procedure as follows:

- In the reinitialization, data imperfections are taken into account. As already discussed, Kalman weighting (Equations 44.13 and 44.14) is based on accuracy of the DR extrapolation as well as the variance of each measurement *and* its sensitivity to each state variable. An optimal balance is provided between old and new information—and the optimality inherently applies to updating of every state variable (e.g., to velocity estimates as well as position, even when only position is observed directly).
- Fixes can be incomplete. In this example, one of the intersecting LOPs may be lost. An optimal update is still provided by the partial fix data, weighted by $\mathbf{W}_m$ of (44.14).

Implications of these two alterations can be exemplified by Figure 44.1, depicting a pair of LOPs representing partial fixes, not necessarily synchronous. Each scalar measurement allows the entire state vector to be optimally updated with weighting from (44.14) in the relation

$$\hat{\mathbf{x}}_m^{(+)} = \hat{\mathbf{x}}_m^{(-)} + \mathbf{W}_m z_m \tag{44.15}$$

where z_m is the predicted residual formed by subtracting the predicted measurement from the value observed at time t_m and acceptance-tested to edit out wild data points:

$$z_m = y_m + \epsilon = Y_m - \hat{Y}_m^{(-)} + \epsilon = \hat{Y}_m - \hat{Y}_m^{(-)}; \quad \hat{Y}_m^{(-)} = Y\left(\hat{\mathbf{x}}_m^{(-)}\right) \tag{44.16}$$

The measurement function $Y(\mathbf{x})$ is typically a simple analytical expression (such as that for distance from a designated point, the difference between distances from two specified station locations, GPS pseudorange or carrier-phase difference). Its partial derivative with respect to each position state is obtained by simple calculus; other components of $\mathbf{H}_m$ (e.g., sensitivity to velocity states) are zero—in which case, updating of those states occurs due to dynamics, from off-diagonal elements of $\mathbf{P}$ in the product $\mathbf{P}_m^{(-)}\mathbf{H}_m^T \cdot R_m$—whether constant or varying (e.g., with signal strength)—is treated as a known quantity; if not accurately known, a conservative upper bound can be used. The same is true for the covariance matrix $\mathbf{P}_0$ of error in state estimate at the time of initiating the estimation process—after which, the changes are tracked by (44.10) at each measurement event and by (44.11) between measurements—thus $\mathbf{P}$ is always available for Equations 44.13 and 44.14.

It is crucial to note that the updates are *not* obtained in the form of newly measured coordinates, as they would have been for the classical "compass-and-dividers" approach. Just as old navigators knew how to use partial information, a properly implemented modern estimator would not forfeit that capability. The example just shown provides best updates, even with no dependably precise way of obtaining a point of intersection when motion occurs between measurements. Furthermore, even with a valid intersection from synchronized observations, the north coordinate of the intersection in Figure 44.1 would be more credible than the east. To show this, consider the consequence of a measurement error effectively raising the dashed LOP to the solid curve as shown; the north coordinate of the new intersection point "+" exceeds that of point "O"—but by less than the east–west coordinate shift. Unequal sensitivity to different directions is automatically taken into account via H_m—just as the dynamics of P will automatically provide velocity updating without explicitly forming velocity in terms of sequential changes in measurements—and just as individual values of R_m inherently account for measurement accuracy variations. Theoretically then, usage of Kalman weighting unburdens the designer while ensuring optimum performance; no other weighting could provide lower mean squared error in the estimated value of any state. Practically, the fulfillment of this promise is realized by observing additional guidelines, some of which apply across the board (e.g., usage of algorithms that preserve numerical stability) while others are application dependent. Now that a highly versatile foundation has been defined for general usage, the way is prepared for describing some specific applications. The versatility just mentioned is exhibited in the examples that follow. Attention is purposely drawn to the standard process cycle; models of dynamics and measurements are sufficient to define the operation.

44.3 Applications

Various operations will now be described, using the unified form to represent the state dynamics* with repetitive instantaneous refresh via discrete or discretized observations (fixes, whether full or partial). Finite space necessitates some limitations in scope here. First, all updates will be from position-dependent measurements (e.g., Doppler can be used as a source of continuous DR data but is not considered herein for the discrete fixes). In addition, all nav reference coordinate frames under consideration will be locally level. In addition to the familiar north–east–down (NED) and east–north–up (ENU) frames, this includes any wander azimuth frame (which deviates from the geographic by only an azimuth rotation about the local vertical). Although these reference frames are not inertial (thus the velocity vector is not exactly the time integral of total acceleration as expressed in a nav frame), known kinematical adjustments will not be described in any depth here. This necessitates restricting the aforementioned data window T to intervals no greater than a tenth of the 84 min Schuler period. The limitation is not very severe, when considering the amount of measured data used by most modern avionics applications within a few minutes duration.

Farrell [1] is cited here for expansion of conditions addressed, INS characterization, broader error modeling, increased analytical development, physical basis for that analysis, and myriad practical "dos and don'ts" for applying estimation in each individual operation.

44.3.1 Position and Velocity along a Line

The vertical nav case shown earlier can be extended to the case of time-varying velocity, with accurately (not necessarily exactly) known vertical acceleration Z_V:

$$\begin{bmatrix} \dot{x}_1 \\ \dot{x}_2 \end{bmatrix} = \begin{bmatrix} 0 & 1 \\ 0 & 0 \end{bmatrix}\begin{bmatrix} x_1 \\ x_2 \end{bmatrix} + \begin{bmatrix} 0 \\ Z_V \end{bmatrix} \tag{44.17}$$

* A word of explanation is in order: For classical physics, the term *dynamics* is reserved for the relation between forces and translational acceleration, or torques and rotational acceleration—while *kinematics* describes the relation between acceleration, velocity, and position. In the estimation field, all continuous time variation of the state is lumped together in the term *dynamics*.

which allows interpretation in various ways. With a positive upward convention (e.g., as in the ENU reference), x_1 can represent altitude above any datum while x_2 is upward velocity; a positive downward convention (NED reference) is also accommodated by simple reinterpretation. In any case, the previous equation correctly characterizes actual vertical position and velocity (with true values for Z_V and all x's) and likewise characterizes *estimated* vertical position and velocity (denoted by circumflexes over Z_V and all x's). Therefore, by subtraction, it also characterizes *uncertainty in* vertical position and velocity (i.e., error in the estimate, with each circumflex replaced by a tilde ~). That explains the role of this expression in two separate operations:

1. Extrapolation of the *a posteriori* estimate (just after inclusion of the last observation) to the time of the next measurement, to obtain an *a priori* estimate of the state vector—which is used to predict the measurement's value. If a transition matrix can readily be formed (e.g., Equation 44.9 in the example at hand), it is sometimes, but not always, used for that extrapolation.
2. Propagation of the covariance matrix from time t_{m-1} to t_m via Equation 44.11 initialized at the *a posteriori* value $\mathbf{P}_{m-1}^{(+)}$ and ending with the a priori value $\mathbf{P}_m^{(-)}$. Again an alternate form using (44.9) is an option.

After these two steps, the cycle at time t_m is completed by forming gain from (44.14), predicted residual from (44.16), update via (44.15), and decrement by (44.10).

The operation just described can be visualized in a generic pictorial representation. Velocity data in a DR accumulation of position increments predicts the value of each measurement. The difference z between the prediction and the observed fix (symbolically shown as a discrete event depicted by the momentary closing of a switch) is weighted by position gain W_{pos} and velocity gain W_{vel} for the update. Corrected values, used for operation thereafter, constitute the basis for further subsequent corrections.

For determination of altitude and vertical velocity, the measurement prediction block in Figure 44.2 is replaced by a direct connection; altimeter fixes are compared versus the repeatedly reinitialized accumulation of products (time increment) × (vertical velocity). In a proper implementation of Figure 44.2, time history of *a posteriori* position tracks the truth; root mean square (RMS) position error remains near $\sqrt{P_{11}}$. At the first measurement, arbitrarily large initial uncertainty falls toward sensor tolerance—and promptly begins rising at a rate dictated by $\sqrt{P_{22}}$. A second measurement produces another descent followed by another climb—now at gentler slope, due to implicit velocity information gained from repeated position observations within a known time interval. With enough fix data, the process approaches a quasi-static condition with $\sqrt{P_{11}}$ maintained at levels near RMS sensor error.

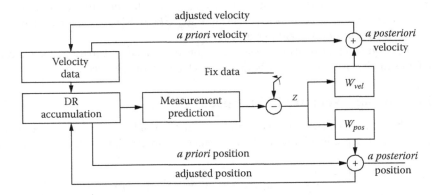

FIGURE 44.2   Position and velocity estimation.

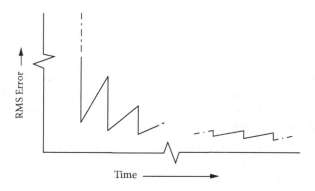

FIGURE 44.3 Time history of accuracy.

Extensive caveats, ramifications, etc., could be raised at this point; some of the more obvious will be mentioned here.

- In analogy with the static example, the *left* side of (44.7)—with $P_{m11}^{(-)}$ substituted for $P_m^{(-)}$—implies high initial weighting followed by lighter weights as measurements accumulate. If fixes are from sensors with varying tolerance, the entire approach remains applicable; only parameter values change. The effect in Figure 44.3 would be a smaller step decrement and less reduction in slope, when RMS fix error is larger.
- Vertical velocity can be an accumulation of products, involving instantaneous vertical acceleration—which comes from data containing an accelerometer offset driven by a randomly varying error (e.g., having spectral density in conformance to **E** of [44.11]). With this offset represented as a third state, another branch would be added to Figure 44.2, and an augmented form of (44.17) could define dynamics in instantaneous altitude, vertical velocity, and vertical acceleration (instead of a constant bias component, extension to exponential correlation is another common alternative):

$$\begin{bmatrix} \dot{x}_1 \\ \dot{x}_2 \\ \dot{x}_3 \end{bmatrix} = \begin{bmatrix} 0 & 1 & 0 \\ 0 & 0 & 1 \\ 0 & 0 & 0 \end{bmatrix} \begin{bmatrix} x_1 \\ x_2 \\ x_3 \end{bmatrix} + \begin{bmatrix} 0 \\ 0 \\ e \end{bmatrix} \tag{44.18}$$

Rather than ramping between fixes, position uncertainty then curves upward faster than the linear rate in Figure 44.3; curvature starts to decrease after the *third* fix. It takes longer to reach quasi-static condition, and closeness of "steady-state" $\sqrt{P_{11}}$ to RMS sensor error depends on measurement scheduling density within a data window.
- (44.12) and Figure 44.2 can also represent position and velocity estimation along another direction, for example, north or east—or both—as developed in the next section.

44.3.2 Position and Velocity in 3D Space

For brevity, only a succinct description is given here. First, consider excursion over a meridian with position x_1 expressed as a product [latitude (*Lat*) increment] × [total radius of curvature (R_M + altitude)]:

$$R_M = \frac{a_E\left(1-e_E^2\right)}{\left[1-e_E^2\sin^2\left(Lat\right)\right]^{3/2}}; \quad a_E = 6,378,137\,m; \quad e_E^2 = \left(2-f\right)f, \quad f = \frac{1}{298.25722} \tag{44.19}$$

so that, for usage of **A** in (44.12), x_2 is associated with north component V_N of velocity. North position fixes could be obtained by observing the altitude angle of Polaris (appropriately corrected for slight deviation off the north pole). To use the formulation for travel in the east direction, the curvature radius is $(R_P + h)$:

$$R_P = \frac{a_E}{\sqrt{1 - e_E^2 \sin^2(Lat)}}; \quad h = \text{altitude} \tag{44.20}$$

and, while the latitude rate is $V_N/(R_M + h)$, the longitude rate is $V_E\sec(Lat)/(R_P + h)$. Even for limited distance excursions within a data window, these spheroidal expressions would be used in kinematic state extrapolation—while our short-term ground rule allows a simplified ("flat-Earth" Cartesian) model to be used as the basis for matrix extrapolation in (44.11). The reason lies with *very* different sensitivities in Equations 44.15 and 44.16. The former is significantly less critical; a change $\delta\mathbf{W}$ would modify the *a posteriori* estimate by only the second-order product $z_m\delta\mathbf{W}_m$. By way of contrast, small variations in an anticipated measurement (from seemingly minor model approximations) can produce an unduly large deviation in the residual—a small difference of large quantities.

Thus, for accuracy of *additive* state *vector* adjustments (such as *velocity* × Δ*time* products in dynamic propagation), Equations 44.19 and 44.20 properly account for path curvature and for changes in direction of the nav axes as the path progresses. At the poles, the well-known singularity in {sec(*Lat*)} of course necessitates a modified expression (e.g., Earth-centered vector).

In applying (44.12) to all three directions, a basic decision must be made at the outset. Where practical, it is desirable for axes to remain separated, which produces three uncoupled 2-state estimators. An example of this form is radar tracking at long range—long enough so that, within a data window duration, the line-of-sight (LOS) direction remains substantially fixed (i.e., nonrotating). If all three axes are monitored at similar data rates and accuracies, experience has shown that even a fully coupled 6-state estimator has position error ellipsoid axes aligned near the sensor's range/azimuth/elevation directions. In that case, little is lost by ignoring coupling across sensor reference axes—hence, the triad of uncoupled 2-state estimators, all in conformance to (44.12). To resolve vectors along cardinal directions at any time, all that is needed is the direction cosine matrix transformation between nav and sensor axes—which is always available.

When the conditions mentioned earlier do not hold, the reasoning needs to be revisited. If LOS direction rotates (which happens at short range) or if all three axes are not monitored at similar data rates, decoupling may or may not be acceptable; in any case, it is suboptimal. If one axis (or a pair of axes) is unmonitored, a fully coupled 6-state estimator can dramatically outperform the uncoupled triad. In that case, although (44.12) represents uncoupled dynamics for each axis, coupling comes from multiple changing projections in measurement sensitivity **H** as the sensor sight-line direction rotates.

Even the coupled formulation has a simple dynamic model in partitioned form; for a relative position vector **R** and velocity **V** driven by perturbing acceleration **e**,

$$\begin{bmatrix} \dot{\mathbf{R}} \\ \dot{\mathbf{V}} \end{bmatrix} = \begin{bmatrix} \mathbf{0} & \mathbf{I} \\ \mathbf{0} & \mathbf{0} \end{bmatrix} \begin{bmatrix} \mathbf{R} \\ \mathbf{V} \end{bmatrix} + \begin{bmatrix} \mathbf{0} \\ \mathbf{e} \end{bmatrix} \tag{44.21}$$

where **0** and **I** are null and identity partitions, respectively. The next section extends these concepts.

44.3.3 Position, Velocity, and Acceleration of a Tracked Object

In this chapter, it has been repeatedly observed that velocity can be inferred from position-dependent measurements separated by known time intervals. In fact, a velocity *history* can be inferred. As a further generalization of methods just shown, the position reference need not be stationary; in the example now to be described, the origin will move with a supersonic jet—carrying a radar and INS. Furthermore, the object whose state is being estimated can be external, with motions that are independent of the platform carrying the sensors that provide all measurements.

For tracking, first, consider the uncoupled case already described, wherein each of three separate estimator channels corresponds to a sensor reference axis direction—and each channel has three kinematically related states, representing that directional component of relative (sensor-to-tracked-object) position, relative velocity, and total (not relative) acceleration of the tracked object.* The expression used to propagate state estimates between measurements in a channel conforms to standard kinematics, that is,

$$
\begin{bmatrix} \hat{x}_{m,1}^{(-)} \\ \hat{x}_{m,2}^{(-)} \\ \hat{x}_{m,3}^{(-)} \end{bmatrix} = \begin{bmatrix} 1 & t_m - t_{m-1} & \frac{1}{2}\left(t_m - t_{m-1}\right)^2 \\ 0 & 1 & t_m - t_{m-1} \\ 0 & 0 & 1 \end{bmatrix} \begin{bmatrix} \hat{x}_{m-1,1}^{(+)} \\ \hat{x}_{m-1,2}^{(+)} \\ \hat{x}_{m-1,3}^{(+)} \end{bmatrix} - \begin{bmatrix} \frac{1}{2}\left(t_m - t_{m-1}\right)q_m \\ q_m \\ 0 \end{bmatrix}
\tag{44.22}
$$

where q_m denotes the component, along the sensor channel direction, of the change in INS velocity during $(t_m - t_{m-1})$. In each channel, $\mathbf{E}$ of (44.11) has only one nonzero value, a spectral density related to data window and measurement error variance σ^2 by

$$
E_{33} = \left(\frac{\left(20\sigma^2 / T^5\right)}{g^2} \right) (\text{g/s})^2 / \text{Hz}
\tag{44.23}
$$

To change this to a fully coupled 9-state formulation, partition the 9×1 state vector into three 3×1 vectors $\mathbf{R}$ for relative position, $\mathbf{V}_r$ for relative velocity, and $\mathbf{Z}_T$ for the tracked object's total acceleration—all expressed in the INS reference coordinate frame. The partitioned state transition matrix is then constructed by replacing each diagonal element in (44.22) by a 3×3 identity matrix $\mathbf{I}_{33}$, each zero by a 3×3 null matrix, and multiplying each above-diagonal element by $\mathbf{I}_{33}$. Consider this transition matrix to propagate covariances as expressed in sensor reference axes, so that parameters applicable to a sensing channel are used in (44.23) for each measurement. Usage of different coordinate frames for states (e.g., geographic in the example used here) and $\mathbf{P}$ (sensor axes) must of course be taken into account in characterizing the estimation process: an orthogonal triad $\mathbf{I}_b \mathbf{J}_b \mathbf{K}_b$ conforms to directions of sensor sight line $\mathbf{I}_b$, its elevation axis $\mathbf{J}_b$ in the normal plane, and the azimuth axis $\mathbf{K}_b = \mathbf{I}_b \times \mathbf{J}_b$ normal to both. The instantaneous direction cosine matrix $\mathbf{T}_{b/A}$ will be known (from the sensor pointing control subsystem) at each measurement time. By combination with the transformation $\mathbf{T}_{A/G}$ from geographic to airframe coordinates (obtained from INS data), the transformation from geographic to sensor coordinates is

$$
\mathbf{T}_{b/G} = \mathbf{T}_{b/A} \mathbf{T}_{A/G}
\tag{44.24}
$$

which is used to resolve position states along $\mathbf{I}_b \mathbf{J}_b \mathbf{K}_b$

$$
\frac{1}{|\mathbf{R}|} \mathbf{T}_{b/G} \mathbf{R} = \begin{bmatrix} 0 \\ p_A \\ -p_E \end{bmatrix}
\tag{44.25}
$$

where p_A and p_E—small fractions of a radian—are departures above and to the right, respectively, of the *a priori* estimated position from the sensor sight line (which due to imperfect control does not look exactly where the tracked object is anticipated at t_m).

---

* Usage of relative acceleration states would have sacrificed detailed knowledge of INS velocity history, characterizing ownship acceleration instead with the random model used for the tracked object. To avoid that unnecessary performance degradation, the dynamic model used here, in contrast to (44.18), has a forcing function with nonzero mean.

For application of (44.16), p_A and p_E are recognized in the role of *a priori* estimated measurements—adjusting the "dot-off-the-crosshairs" azimuth ("AZ") and elevation ("EL") observations—so that a full 3D fix (range, AZ, EL) in this operation would be

$$
\begin{bmatrix} y_R \\ y_{AZ} \\ y_{EL} \end{bmatrix} = \begin{bmatrix} 1 & 0 & 0 \\ 0 & \dfrac{1}{|\mathbf{R}|} & 0 \\ 0 & 0 & \dfrac{1}{|\mathbf{R}|} \end{bmatrix} \mathbf{T}_{b/G}\mathbf{R} - \begin{bmatrix} 0 \\ p_A \\ -p_E \end{bmatrix} \tag{44.26}
$$

Since **R** contains the first three states, its matrix coefficient in (44.26) provides the three nonzero elements of **H**; for example, for scalar position observables, these are comprised of

- The top row of $\mathbf{T}_{b/G}$ for range measurements
- The middle row of $\mathbf{T}_{b/G}$ divided by scalar range for azimuth measurements
- The bottom row of $\mathbf{T}_{b/G}$ divided by scalar range, $\times\ (-1)$, for elevation measurements

By treating scalar range coefficients as well as the direction cosines as known quantities in this approach, *both* the dynamics *and* the observables are essentially linear in the state. This has produced success in nearly all applications within experience of the writers the sole need for extension arising when distances and accuracies of range data were extreme (the cosine of the angle between sensor sight line and range vector could not be set at unity). Other than that case, the top row of $\mathbf{T}_{b/G}$ suffices for relative position states—and also for relative velocity states when credible Doppler measurements are available.

A more thorough discourse would include a host of additional material, including radar and optical sensing considerations, sensor stabilization—with its imperfections isolated from tracking, error budgets, kinematical correction for gradual rotation of the acceleration vector, extension to multiple track files, sensor fusion, myriad disadvantages of alternative tracking estimator formulations, etc. The ramifications are too vast for inclusion here.

44.3.4 Position, Velocity, and Attitude in 3D Space (INS Aiding)

In the preceding section, involving determination of velocity history from position measurement sequences, dynamic velocity variations were expressed in terms of an acceleration vector. For nav (as opposed to tracking of an external object) with high dynamics, the history of velocity is often tied to the angular orientation of an INS. In straight-and-level northbound flight, for example, an unknown tilt ψ_N about the north axis would produce a fictitious ramping in the indicated east velocity V_E; in the short-term, this effect will be indistinguishable from a bias n_{aE} in the indicated lateral component (here, east) of the accelerometer output. More generally, velocity *vector* error will have a rate

$$
\dot{\mathbf{v}} = \psi \times \mathbf{A} + \mathbf{n}_a = -\mathbf{A} \times \psi + \mathbf{n}_a \tag{44.27}
$$

where
   bold symbols (**v**, **n**) contain the geographic components equal to corresponding scalars denoted by italicized quantities (v, n)
   A represents the vector, also expressed in geographic coordinates, of total nongravitational acceleration experienced by the IMU

Combined with the intrinsic kinematical relation between **v** and a position vector error **r**, in a nav frame rotating at $\tilde{\omega}$ rad/s, the 9-state dynamics with a time-invariant misorientation ψ can be expressed via 3×3 matrix partitions:

$$\begin{bmatrix} \dot{\mathbf{r}} \\ \dot{\mathbf{v}} \\ \dot{\psi} \end{bmatrix} = \begin{bmatrix} -\tilde{\omega}\times & \mathbf{I} & 0 \\ 0 & 0 & (-\mathbf{A}\times) \\ 0 & 0 & 0 \end{bmatrix} \begin{bmatrix} \mathbf{r} \\ \mathbf{v} \\ \psi \end{bmatrix} + \begin{bmatrix} 0 \\ \mathbf{n}_a \\ \mathbf{e} \end{bmatrix} \tag{44.28}$$

which lends itself to numerous straightforward interpretations; for brevity, these will simply be listed here:

- For strapdown systems, it is appropriate to replace vectors such as **A** and $\mathbf{n}_a$ by vectors initially expressed in vehicle coordinates and transformed into geographic coordinates—so that parameters and coefficients will appear in the form received.
- Although both $\mathbf{n}_a$ and **e** appear as forcing functions, the latter drives the highest order state and thus exercises dominant control over the data window.
- If $\mathbf{n}_a$ and **e** contain both bias and time-varying random (noisy) components, (44.28) is easily reexpressible in augmented form, wherein the biases can be estimated along with the corrections for estimated position, velocity, and orientation. Especially for accelerometer bias elements, however, observability is often limited; therefore, the usage of augmented formulations should be adopted judiciously. In fact, the number of states should in many cases be *reduced*, as in the next two examples:
    - In the absence of appreciable sustained horizontal acceleration, the azimuth element of misorientation is significantly less observable than the tilt components. In some operations, this suggests replacing (44.28) with an 8-state version obtained by omitting the ninth state—and deleting the last row and column of the matrix.
    - When the last *three* states are omitted—while the last three rows and columns of the matrix are deleted—the result is the fully coupled 3D position and velocity estimator (44.21).

The options just described can be regarded as different modes of the standard cyclic process already described, with operations defined by dynamics and measurement models. Any discrete observation could be used with (44.28) or an alternate form just named, constituting a mode subject to restrictions that were adopted here for brevity (position-dependent observables only, with distances much smaller than Earth radius).

At this point, expressions could be given for measurements as functions of the states and their sensitivities to those state variables: (44.26) provides this for range and angle data; it is now appropriate to discuss GPS information in the context of integrated nav and tracking, while including considerations for usage in operation.

44.4 Operational Developments

This extension to the "Navigation and Tracking" chapter in earlier editions of the *Avionics Handbook* can begin with a brief historical perspective. Much of the methodology needed for nav integration was developed decades ago. Optimal, near-optimal, and suboptimal estimation all arrived before 1960, and inertial navigation existed long before that (though strapdown, with its greater demand for processing capability, came about a decade later). In addition to those two pillars of the operation, much of the requisite theory—though not yet all of the required processing power demanded by the algorithms—is likewise decades old. Even the first attempt [1] to collect applicable theory together with models for all nav sensor data in existence, to cover integrated nav in an all-inclusive modern estimation framework, is over 35 years old at the time of this writing. What is different now is an opportunity to capitalize on all these ingredients, due to continued (and in fact progressive) advances in that processing power—in combination with another landmark occurrence in capability enhancement: satellite navigation.

The explosive growth of navigation applications within the past few decades has been largely attributed to GPS. Never before had there been a nav data source of such high accuracy, reachable from any Earth surface location at any time. Elsewhere in this handbook, the reader is shown how GPS data can be used to

- Solve for 3D position and user clock offset with pseudorange observations received simultaneously from each of four space vehicles (SVs)
- Use local differential GPS corrections that combine, for each individual SV, compensation for propagation effects plus SV clock and ephemeris error
- Compensate via wide-area augmentation that, though not as straightforward as local, is valid for much greater separation distances between the user and reference station
- Use differencing techniques with multiple SVs to counteract user clock offsets while multiple receivers enable compensation of the errors mentioned
- Apply these methods to carrier phase as well as to pseudorange so that, once the cycle count ambiguities are resolved, results can be accurate to within a fraction of the L-band wavelength

The following will describe usage of satellite data in navigation and tracking applications. Examples will begin with discussion of methodology, followed by progression toward results—first simulation, then testing (both van and flight test). All of this material clearly goes significantly beyond current conventional practice; justification is evident from dramatically improved performance with robustness and potential for ease of operation.

44.4.1 Individual GPS Measurements as Observables

Immediately, we make a definite departure from custom here; each scalar GPS observable will call for direct application of Equation 44.15. To emphasize this, results are first summarized for instances of sparse measurement scheduling: initial runs were made long ago with real SV data, taken before activation of selective availability (SA) degradations, collected from a receiver at a known stationary location but spanning intervals of several hours. Even with that duration consumed for the minimum required measurements, accuracies of 1 or 2 m were obtained—not surprising for GPS with good geometry and no SA.

The results just mentioned, while not remarkable, affirm the realization that full fixes are not at all necessary with GPS. They also open the door for drawing dependable conclusions when the same algorithms are driven by simulated data containing errors from random number generators. Section 8.1.1 of [2] describes a subsequent high-speed aircraft simulation with no more than one pseudorange observation every 6 s (and furthermore with some gaps, even in that slow data rate). Since the results are again unremarkable, only a brief synopsis suffices here:

- Position and velocity estimates settled as soon as measurement accumulation was sufficient to produce a nav solution (e.g., two asynchronous measurements from each of three noncoplanar SVs for a vehicle moving in three dimensions, with known clock state, or four SVs with all states initially unknown).
- Initial errors tended to wash out; the estimation accuracies just mentioned were determined by measurement error levels amplified through geometry.
- Velocity errors tended toward levels proportional to the ratio of RMS measurement error to (T), where T here represents either time elapsed since the first measurement on a course leg or the data window, whichever is smaller. The former definition of the denominator produced a transient at the onset and when speed or direction changed.
- Doppler data reduced the transient, and INS velocity aiding minimized or removed it.

- Extreme initial errors interfered with these behavioral patterns somewhat—readily traceable to usage of imprecise direction cosines—but the effects could be countered by reinitialization of estimates with *a posteriori* values and measurement recycling. These results mirror familiar real-world experience (including actual measurement processing by many authors); they are mentioned here to emphasize adequacy of partial fixes at low rates that many operational systems to this day still fail to exploit [3].

Although the approach just described is well known (i.e., in full conformance to the usual Kalman filter updating cycle) and the performance unsurprising, the last comment is significant. There are numerous applications wherein SV sight lines are often obscured by terrain, foliage, buildings, or structure of the vehicle carrying the GPS receiver. In addition, there can be SV outages (whether from planned maintenance or unexpected failures), intermittently strong interference or weak signals, unfavorable multipath geometry in certain SV sight-line directions, etc., and these problems often arise in critical situations.

At the time of this writing, there remain widespread opportunities, accompanied by urgent need, to replace loose (cascaded) configurations by tightly coupled (integrated) configurations. Accentuating the benefit is the bilateral nature of ultratight integration. As tracking loops (code loop and, where activated, carrier-phase track) contribute to the estimator, rapid dynamics maintenance enhances ability to maintain stable loop operation. For a properly integrated GPS/INS, this enhancement occurs even with narrow bandwidth in the presence of rapid change. Loop response need not follow high dynamics—but only error in the perceived dynamics—with ultratight integration.

It is also noted that the results just described are achievable under various conditions and can be scaled over a wide accuracy range. Sensitivity **H** of an individual SV observation contains the SV-to-receiver unit vector; when satellite observations are subtracted, that sensitivity contains a difference of two SV-to-receiver unit vectors. Position measurements may be pseudoranges, differentially corrected pseudoranges, or carrier phase with their ambiguities resolved, with typical RMS accuracies of <10 m, 2 m, or <1 cm, respectively. Attainable performance is then determined by those values and by the span of **H** for each course leg. An analogous situation has long been recognized for other navaids when used with the standard updating procedure presented herein.

44.4.2 Velocity-Related Observables: Sequential Changes in Carrier Phase

All observables considered thus far have been restricted to direct dependence on position, with capability to infer dynamics being traceable to

- Coupling via time variations in the covariance matrix by Equation 44.11
- Aiding (e.g., by inertial data)
- A combination—with IMU updating via position-dependent observables also in coordination with Kalman gain influenced by effects of Equation 44.11

Changes in carrier phase over 1 s intervals offer an extension of that basic procedure. The methodology constitutes a major departure from convention, using a segmented estimator wherein

- Phase-related measurements are for DR only (not for position)
- Feedforward of the resulting streaming velocity is followed by integration, with correction via pseudoranges
- Just as carrier phase is not used for position, pseudoranges are not used for dynamics; estimation of position is conducted separately from dynamics
- The dynamic observables are linearly proportional to time integrals of the lowest-order states

Due primarily to that last item, space limitations preclude self-sufficiency of this chapter. Theory and in-depth analytical basis for all processing, with and without IMU aiding, are fully documented—along with a host of dramatic advantages, supported by extensive graphical results from van and flight test presented in [2]. An example from the latter shows accuracy with a low-cost IMU, illustrated in the table on page 104 of [2] (also appearing near the bottom of a one-page description on the website noted with that reference citation). Those tabulated 1 s phase increment residual magnitudes were all zero or 1 cm for the seven satellites (six across-SV differences) observed at a measurement instant chosen for illustration. Over almost an hour of flight at altitude (i.e., excluding takeoff, when heading uncertainty caused larger lever-arm vector error), cm/s RMS velocity accuracy was obtained. For performance without the IMU, Section 8.1.2 (pp. 154–162 of [6]) shows dm/s RMS velocity errors outside of turn transients.

Accuracies thus verified, important as they are, constitute only one feature among many. This methodology offers other benefits, *absent* from conventional approaches using carrier-phase data. These can best be explained by comparison versus conventional usage of carrier-phase measurements, that is, as a product,

$$(\text{L-band wavelength}) \times (\text{Integer} + \text{Fraction}) \tag{44.29}$$

where the fraction is precisely measured while the integer must be determined.

When that integer is known exactly, the result is of course extremely accurate. *However*, even the most ingenious methods of integer extraction occasionally produce a highly inaccurate result. The outcome can be catastrophic, and there can be an unacceptably long delay before correction is possible. Eliminating that possibility provided strong motivation for using 1 s changes: all phases can be forever ambiguous, that is, integers can remain unknown; they cancel in forming sequential differences. Furthermore, discontinuities can be tolerated; a reappearing signal is instantly acceptable as soon as two successive carrier phases differ by an amount that satisfies single-measurement receiver autonomous integrity monitoring (RAIM) testing. The technique is especially effective with receivers using FFT-based processing, which provides unconditional access to all correlation cells (rather than a limited subset offered by a tracking loop), with no phase distortion (because the FFT phase-versus-frequency characteristic is linear). Another benefit is subtle but highly significant: acceptability of submask carrier-phase changes. Ionospheric and tropospheric timing offsets change very little over a second. Conventional systems are designed to reject measurements from low-elevation satellites. Especially in view of improved geometric spread, retention here prevents unnecessary loss of important information. Demonstration of that occurred in flight data when a satellite dropped to the horizon; submask pseudoranges of course had to be rejected, but the 1 s carrier-phase changes were perfectly acceptable until the satellite was no longer detectable. For a parallel reason, the phase changes are insensitive to ephemeris errors; even with satellite mislocation, *changes* in satellite position are precise.

As if all those advantages weren't enough, the technique also lends itself to easy operation with satellites from other Global Navigation Satellite System (GNSS) constellations (*Galileo*, GLONASS, etc.). That interoperability is now one of the features attracting increased attention. Although the important benefits just noted were realized with usage of only one constellation (i.e., GPS), the advantage is amplified with GNSS; sequential changes in carrier phase are much easier to mix than the phases themselves. Usage is described for navigation in the preceding discussion and for tracking in the next section. In all cases, sequential changes in carrier phase over 1 s offer a host of dramatic performance improvements over convention. It is believed that, despite current absence from operational systems, sheer necessity will drive the industry toward wide acceptance.

44.4.3 Tracking: Relative Velocity and Position Determination

The discussion in Sections 44.4.1 and 44.4.2 can be easily extended to *relative* position and velocity determination between ownship and intruder. Not just by computing GNSS position and velocity estimates at each of the aircraft and exchanging these estimates through some data link, but also by exchanging the GNSS observables themselves [3]. The latter solution is particularly effective for applications that require high-accuracy and high-integrity solutions. In the former method, the intruder track can be formed from the history of position (and velocity) reports using the methods described in Section 44.3.3. In an operational scenario, these position and velocity reports will be transmitted to the ground and other aircraft using automatic dependent surveillance—broadcast (ADS-B) and will include parameters representing the accuracy and integrity performance of transmitted estimates. More details can be found in Chapter 23.

In the measurement-based method, both aircraft participants also transmit raw GNSS measurements via the ADS-B data link; on the receiving end, the incoming observables (e.g., GNSS pseudoranges and carrier-phase measurements) are then combined with the local observables to establish a track of the intruder. Given the raw observables from both ownship and intruder, range single differences, double differences, and sequential differences can be used to accurately derive the separation vector between ownship and traffic as well as the change in separation vector over a time. The latter quantity is directly related to the average velocity over a time epoch. Since in any conflict scenario most errors in the observables are highly correlated spatially and temporally, they will cancel out resulting in a better and more robust estimate of the relative position and velocity. Advantages of this method are increased observability, independence on a datum reference, the availability of an intrinsic duality indicator through the estimator's covariance matrix, a guaranteed incorporation of existing correlations in the estimator, optimal weighting of the measurements, and the ability to perform data screening and other integrity methods such as RAIM. For a more in-depth description of this method, the reader is referred to [4,5]. Using flight test data, it has been shown that such a measurement-based method can achieve meter-level relative position accuracies and mm/s-level relative velocity estimates.

The uncertainty in the current position estimate is formally referred as the estimated position uncertainty (EPU) [6]. Using only the state transition part of the tracking filter (i.e., Equation 44.8), the trajectory can be predicted for the *time horizon* T_i. The uncertainty in this prediction can be estimated as well (i.e., Equation 44.10) and is referred to as the estimated trajectory uncertainty (ETU). Both the EPU and ETU are illustrated in Figure 44.4 for an example conflict scenario. The ETU plays an important role in determining the predicted closest point of approach and loss of separation in a statistical sense, which, in this case, occurs when the two ETU circles intersect. Therefore, the tracking performance is an essential component of aircraft surveillance applications (ASA), in particular conflict detection and resolution (CDR) methods.

The quality of the predicted ownship and traffic positions forward in time and, hence, conflict detection is dependent on (1) the time horizon T_i, (2) the dynamics of both ownship and traffic,

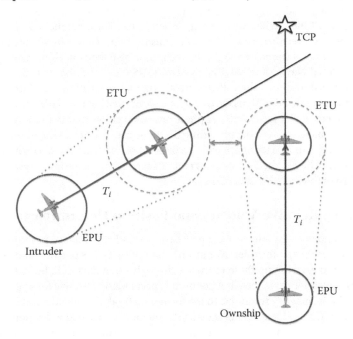

FIGURE 44.4   Estimated position and trajectory uncertainty.

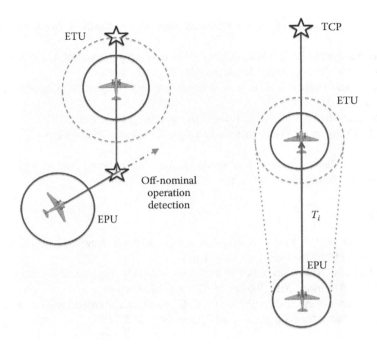

FIGURE 44.5 Detection of off-nominal operation.

and (3) the probabilistic aspects of the prediction process. While the EPU due to the initial position estimate is statistically bounded, the ETU will grow over time due to an error in the velocity estimate, which should be incorporated in the conflict detection algorithm.

Figure 44.5 illustrates how better velocity accuracy could help detect an off-nominal scenario in which the traffic intruder deviates from its nominal flight path [7]: the better the accuracy, the shorter the time-to-detection of this off-nominal behavior. Therefore, accurate knowledge of velocity is very important since it reduces the magnitude of the ETU and enables detection of off-nominal maneuvers of intruder from its nominal flight trajectory.

44.5 Conclusion

Principles and applications of nav system integration and tracking have been described. Within space limitations, usage in operation has also been added to the discussion. Inevitably, restrictions in scope were adopted; those wishing to pursue topics in greater depth may consult the sources that follow.

Further Reading

1. Journal and Conference Proceedings from the Institute of Navigation, Alexandria, VA.
2. Tutorials from Conferences sponsored by the Institute of Navigation (Alexandria, VA) and the *Position Location and Navigation Symposium (PLANS) of the Institute of Electrical and Electronic Engineers* (IEEE).
3. *Transactions of the Institute of Electrical and Electronic Engineers (IEEE) Aerospace and Electronic Systems Society (AES).*

4. Bierman, G. J., *Factorized Methods for Discrete Sequential Estimation*, Academic Press, New York, 1977.
5. Institute of Navigation Redbooks (reprints of selected GPS papers); Vol. 1: 1980, Vol 2: 1984, Vol. 3: 1986, Vol. 4: 1993, Vol. 5: 1998; Alexandria, VA. 703/683-7101.
6. Brown, R. G. and Hwang, P. Y. C., *Introduction to Random Signals and Applied Kalman Filtering*, Wiley, New York, 1996.
7. Kayton, M. and Fried, W. R. (eds.), *Avionics Navigation Systems*, 1997.
8. Farrell, J. L., Collision avoidance by speed change, *Coordinates Magazine*, VIII(9), September 2012, 8–12.
9. Farrell, J. L., Collision avoidance by speed change, *International Journal of Unmanned Systems Engineering (IJUSEng)*, 1(1), 2013, 1–8.

References

1. Farrell, J. L., *Integrated Aircraft Navigation*, Academic Press, New York, 1976. (Now available in paperback only; 800/628-0885 or 410/647-6165.)
2. Farrell, J. L., *GNSS Aided Navigation & Tracking—Inertially Augmented or Autonomous*, American Literary Press, Baltimore, MD, 2007 (http://JamesLFarrell.com).
3. Farrell, J. L., McConkey, E. D., and Stevens, C. G., Send measurements not coordinates, *Navigation (Journal of the Institute of Navigation)*, 46(3), Fall 1999, 203–215.
4. Duan, P. and Uijt de Haag, M., Flight test results of a conflict detection method using ADS-B with raw GNSS measurements, *Proceedings of the ION GNSS*, 2012, Nashville, TN.
5. Duan, P., Uijt de Haag, M., and Farrell, J. L., Transmitting raw GNSS measurements as part of ADS-B: Why, how, and flight test results, *Proceedings of the AIAA/IEEE 31st Digital Avionics Systems Conference (DASC)*, Williamsburg, VA, October 2012.
6. RTCA SC-186, Minimum aviation system performance standards (MASPS) for aircraft surveillance applications, DO-289, Washington, DC, December 9, 2003.
7. Bezawada, R., Duan, P., and Uijt de Haag, M., Hazard tracking with integrity for surveillance applications, *Proceedings of the 30th Digital Avionics Systems Conference (DASC)*, Seattle, WA, October 2011, pp. 8C1-1–8C1-15.

IV

Conclusion

Throughout this handbook, the authors have focused on technologies, approaches, and functions that are all in service or are actively under development to facilitate NextGen capabilities in the National Airspace. To conclude, it seemed appropriate to postulate what might lie ahead. In this concluding chapter, Mark Ballin connects the dots between the current avionics capabilities and one possible set of outcomes that some have called FarGen. It is the editors' hope that the content of this chapter and the handbook overall will help its readers make these capabilities a reality in the years to come.

45

Next Frontier: Sharing the Airspace with Increased Autonomy

Mark G. Ballin
Langley Research Center

45.1 Introduction

The United States National Airspace System is responsible for the safe, secure, and efficient operations of over 50,000 flights per day. It is a large and complex operating infrastructure made up of thousands of interacting participants, hundreds of air traffic control facilities, and thousands of pieces of equipment and software systems in the air and on the ground. Applying new concepts of operation and advanced technology to the system should enable it to accommodate an even higher number of operations than today. System users should be able to achieve higher levels of operating efficiency and schedule reliability while lowering impacts to the environment.

The high complexity of the system makes transition to new modes of operation very challenging. Safety and security are of highest priority. Large capital investments by the Federal Aviation Administration (FAA), airports, airlines, and technology vendors are required for any change. Any new equipment and operating procedures must interoperate with legacy systems and procedures. Requirements for training of human operators to accommodate new technology and procedures are substantial. Changes require the agreement and concurrence of system stakeholders and international civil aviation organizations. Therefore, the FAA values predictability over change, consensus over innovation, adherence to plan over rapid response to change, and operational assurance over rapid deployment of innovation.[1] However, costs to operate and maintain the current system are also very large, and the need for system modernization is recognized. The development of the Next Generation Air Transportation System (NextGen) is being undertaken to increase significantly the capacity, safety, efficiency, and security of air transportation in the United States,[2] and modernization of air traffic management (ATM) is a central focus of that activity. Coordinated activities are also taking place in Europe as part of the Single European Sky Air Traffic Management Research (SESAR) program.

In the early days of aviation, air navigation and traffic management was the sole responsibility of the pilot. Over time, the world has migrated to a traffic management system that is primarily ground-based, driven by a desire for high safety, the need to manage limited resources such as runways, and the availability of ground-based surveillance technology. The direction of migration appears to be slowing or partially reversing. We are now exploring and defining a new frontier in airborne capabilities that will enable the flight deck to assist and assume partial responsibility for ATM. This chapter describes some of the modernization ideas and trends that call for an increased role of the flight deck. It also describes the advanced concepts of operation and the enabling flight deck technologies that may drive these trends in airspace modernization.

45.2 First Step: Utilize Existing Aircraft Capabilities

For the near-term future, there is an emphasis in both NextGen and SESAR to make increased use of existing airborne capabilities. Modern flight management systems (FMSs) are capable of accurately predicting and flying energy-efficient flight paths that meet a required time of arrival (RTA) at a point in the flight plan. The NextGen Avionics Roadmap[3] and the RTCA Task Force 5[4] identify and recommend the use of existing single-RTA compliance capabilities as a means for efficient use of airspace and efficient flight during descent to increase system capacity and reduce environmental impacts of air transportation. The concept of tailored arrivals makes use of existing aircraft trajectory compliance capabilities by having aircraft equipped for data link capability follow efficient fixed-route arrival trajectories, computed and uplinked to the aircraft by the air navigation service provider (ANSP).[5]

The use of existing aircraft capabilities in the current ATM paradigm, while offering great potential for improvements, will require new functions for exchange of information between the air and ground. The functions will be needed for the exchange of trajectories and traffic management constraints and to facilitate access to data that will reduce trajectory prediction uncertainties in airborne and ground-based systems. The use of the existing airborne capabilities will also require new ground-based automation systems and air traffic controller decision support tools to generate accurate and stable traffic management constraints. New procedures for negotiated solutions between airborne and ground-based systems will also be necessary.

Many operational feasibility issues must be resolved to make use of existing aircraft capabilities. Less-capable aircraft must be managed using traditional air traffic control procedures. Controllers may be required to vary their control strategies depending on equipage types, and some of the strategies may require limiting the controller's flexibility and problem-solving degrees of freedom. To illustrate the operational challenges, consider a controlled time-of-arrival study that investigated the use of two modern aircraft flying in trail, predicting and flying an open-loop descent profile to meet a time-based spacing requirement at the bottom of descent.[6] The study found that a small percentage of separation violations would occur for a target spacing of 120 s or less at the bottom-of-descent terminal-entry RTA fix. These violations occur because of differences in aircraft performance envelopes, speed strategies, and guidance strategies. Therefore, to keep spacing small, either some form of active control will be needed to ensure separation or spacing tailored for each individual aircraft pair will be required. Such capabilities must be based on in-depth knowledge of each aircraft's current flight state and each type's performance and its trajectory generation and guidance strategies. Active control may require an air traffic controller to monitor and intervene with limited control options, thereby increasing task complexity and workload. Providing the ANSP knowledge of each aircraft's specific strategies may require complex and costly data management and the real-time exchange of company-proprietary information.

Therefore, as the ANSP continues to be a trajectory manager for every aircraft in the near term, extensive new ANSP capabilities will be required to achieve benefits. The reliance on existing aircraft capabilities may introduce operational feasibility issues that severely limit the realizable benefits.

45.3 Midterm NextGen: Increased Airborne Capabilities to Support Airspace Management

When NextGen was initiated by Joint Planning and Development Office (JPDO) in 2003, it was intended to be transformational, providing revolutionary changes to today's operations to achieve a system that scales to meet a tripling of traffic demand or more. Near-term and midterm NextGen implementations seek to provide capability increases that are deemed achievable by 2025, provided organizational, technological, and political challenges are resolved.[7] Midterm systems to be used in meeting 2025 targeted capabilities are under development, with a goal of completion by 2018. Figure 45.1 illustrates these capabilities at a high level. There are no changes to the roles and responsibilities of pilots and controllers, but they will have new decision support tools supported by new capabilities in surveillance, communication, and trajectory management.

Two new midterm capabilities are the establishment of trajectory-based operations (TBO), requiring planning of, compliance with, and exchange of 4D flight trajectories consisting of the three spatial dimensions and time, and performance-based operations and services (PBO), in which an aircraft's ability to meet specific performance standards is used to increase the capacity and efficiency of the National Airspace System. Future airspace management may require 4D trajectories defined by multiple constraints associated with multiple objectives. Compliance with 4D trajectory clearances may include RTA at several specified waypoints and continuous containment to a planned trajectory.

45.3.1 Midterm NextGen Concepts of Operation

Midterm NextGen concepts such as TBO and PBO require significant additional airborne and ground-based infrastructure to allow the aircraft to exchange data with ground-based systems and take a more active role through limited delegation of trajectory management. Development of a universal method of exchanging and interpreting trajectory information between airborne and ANSP systems will be necessary for the former, and airborne surveillance will be necessary for the latter. Both capabilities will require significant additions to FMS trajectory prediction and guidance functions.

TBO is based on the premise that increased predictability of flight operations is possible by using precise trajectories and sharing plans for these trajectories with the ANSP and other airspace users. Such increased predictability in turn results in increased capacity, efficiency, and ANSP productivity while maintaining safety.[2] TBO is a shift from today's tactical management of aircraft through clearances

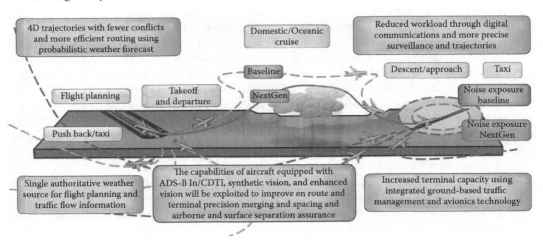

FIGURE 45.1 **(See color insert.)** Targeted NextGen 2025 capabilities for a typical flight. (From Joint Planning and Development Office, Targeted NextGen capabilities for 2025, November 2011.)

by controllers responsible for regions of airspace to a more strategic approach of managing individual aircraft trajectories over longer time horizons. TBO relies on future aircraft systems to generate and execute a 4D trajectory, defined as the centerline of an aircraft path in space and time plus a position uncertainty.

Operators who equip their aircraft to conduct TBO would receive services from the ANSP that allow them to achieve operating benefits. Many unequipped aircraft may also benefit if Class B terminal airspace can be reduced to active arrival and departure corridors, thereby opening access to the remaining airspace to other operators. TBO also has the potential to facilitate precise control of aircraft separation and spacing in congested environments. Combined with improved weather information, it is expected to allow access to more airspace most of the time, thereby facilitating increased capacity and better utilization of limited airspace and airport resources. TBO will also potentially improve aircraft abilities to fly precise noise-sensitive and reduced-emission departure and arrival paths.

PBO will provide the basis for defining procedures and airspace access in NextGen. Under PBO, communication, navigation, and surveillance (CNS) performance would be the basis for operational approval rather than specific equipage, as in today's system. PBO opens opportunities for limited airspace resources to be managed according to real-time demand and available airborne capabilities. The ANSP will provide performance-based services that give operational advantages to aircraft with higher CNS performance.

All of the 4D trajectory-related concepts and procedures identified by NextGen rely on required navigation performance (RNP) area navigation (RNAV) capability. RNP is defined as a statement of the navigation performance accuracy necessary for operation with a defined airspace. In addition, RNP RNAV is defined as an extension of RNP that also includes the containment requirements and area navigation functional and performance standards defined in RTCA DO-236B.[8]

New 4D trajectory-related concepts also require the use of actual navigation performance (ANP), the navigation computed accuracy with associated integrity for the current FMS-generated aircraft position. ANP is a measure of the quality of the FMS navigational position estimate.

Concepts and procedures defined for midterm NextGen include the following[2]:

- *Trajectory-based separation management*: Automation and shared trajectory information are used to manage separation among aircraft, airspace, and hazards such as weather and terrain. If airborne surveillance is available, some aircraft may be temporarily delegated the authority to self-separate by the ANSP. For aircraft not delegated with separation, ANSP automation manages short-term conflict-driven updates to the 4D trajectory. Intent-based conflict detection and resolution is assumed to be necessary. Therefore, a common awareness of the 4D trajectory is required, which requires exchange of the trajectory between the aircraft and the ANSP.
- *Controlled time of arrival (CTA)*: One or more waypoints of a 4D path may be constrained by the ANSP to require the aircraft to arrive at a specific time within a prescribed performance tolerance. Referred to as a CTA, these constraints are equivalent to the aircraft-centered concept of RTA. Multiple CTA constraints may be specified for flow control purposes.
- *Flow corridors*: Large numbers of separation-capable aircraft may be bundled into corridors during times of high demand. Aircraft self-separate within the corridors, while the ANSP maintains separation between the corridor and other aircraft. The concept requires 4D trajectories, CTA, and RNP capabilities for participating aircraft.
- *Airborne merging and spacing (also known as flight deck interval management [FIM])*: Aircraft capable of 4D trajectory management are instructed to achieve and, in some operational concepts, maintain spacing from an ANSP designated lead aircraft. This procedure requires airborne surveillance and a concept of relative 4D trajectory management rather than Earth-referenced or absolute time-based trajectory management.
- *Trajectory-based surface operations*: Procedures proposed for surface operations include self-separation under low-visibility conditions, the scheduling of entry into FAA-controlled surface movement areas, and scheduling of arrival at locations within them, such as active runway crossings.

45.3.2 Midterm NextGen Technology Enablers

Midterm NextGen systems will continue to make maximum use of avionics systems already present on most modern aircraft, such as the FMS and FMS-coupled autoflight systems. In addition, some new systems will be required for operators who choose to equip to take advantage of new services. Data communications between airborne and ground-based system elements will be a key enabler. Additional technologies to support these operations include airborne surveillance and 4D trajectory prediction and guidance.

45.3.2.1 Data Communications

Several midterm concepts of operation rely on data communications between aircraft and the ANSP. Some voice-based communications will be replaced with digital data links for equipped aircraft, which will be used predominantly for issuance of revised ANSP clearances and instructions. Some of these will involve the uplink of FMS-compatible flight plan updates. Anticipated benefits include more efficient operations, reduced greenhouse gases, and reduced operational costs through the use of trajectory-based routing and optimized profile descents; improved controller and crew productivity through faster clearances; and reduced communication errors.[7]

45.3.2.2 Airborne Surveillance

Automatic dependent surveillance—broadcast (ADS-B) has been established as a surveillance infrastructure that will provide substantial benefits to both airspace users and the ANSP in the future. Concepts of operation that increase the operational autonomy of aircraft typically rely on ADS-B surveillance to provide traffic position and state information to airborne systems. Because the surveillance is dependent on navigation systems of the traffic aircraft, ADS-B provides highly accurate position and state, and depending on the development of standards, it may provide FMS-coupled intent information in the future. The FAA has mandated that all aircraft operating in current mode C airspace be equipped with ADS-B transmit capability ("ADS-B Out") by 2020. ADS-B is described in detail in Chapter 23. TCAS (described in Chapter 22) also provides traffic surveillance for the purposes of collision detection and avoidance.

45.3.2.3 4D Trajectory Prediction and Guidance

A primary capability of the FMS is the generation of 4D flight trajectories for a given flight plan. Most systems use aircraft performance data and forecast winds to generate trajectories that are flyable by the airplane. The path definitions are typically a function of a user-defined cost index (CI), which enables the operator to provide a preference for the value of flight time relative to the value of fuel efficiency. Since FMS trajectory predictions normally contain time as an element of the trajectory, much research and development has been performed to control the flight time of the aircraft to achieve a specified time goal.[9–12] Current systems are very accurate, subject to the accuracy of the performance models, forecast winds, and temperature. Since some CTA concepts may involve multiple RTA constraints, future FMS capabilities may include the need to generate a trajectory with multiple time constraints. Several research systems have been developed by FMS manufacturers and research labs. In one research system, the end conditions of an RTA waypoint are used as the initial conditions for the route segments to the next RTA waypoint.[13] Separate CI limits and tolerance values are computed and maintained for each RTA route segment. Multiple RTA systems may also require a constraint management and relaxation capability to provide optimal trajectory solutions for flow-management constraints that may be incompatible. These concepts will probably require some form of RTA tolerance limit, as could be provided by a future specification of a longitudinal (i.e., along track) RNP containment area. A longitudinal RNP value could easily be derived from a specified time tolerance or temporal RNP value.

Three potential methods of time control guidance may be required for NextGen time-based concepts of operation. RTA predictive guidance is designed to achieve RTAs at discrete waypoints using trajectory prediction techniques, similar to the method used by existing FMSs. Continuous time control guidance that steers the aircraft along a precomputed trajectory while continuously maintaining or achieving time accuracy may be required for some concepts. Pair-dependent speed guidance may also be needed, especially for capacity-constrained terminal-arrival operations.

RTA predictive guidance relies on trajectory prediction to generate a flight trajectory that achieves the desired arrival time. The trajectory generator will iterate on possible trajectories until the estimated time of arrival at the RTA waypoint is within a prespecified tolerance. There are a number of methods for accomplishing this iteration. The prototype system of Ref. [13] uses the CI as the independent variable for this iteration because past research has shown this method results in a trajectory that is fuel optimal for the prescribed flight time.[14,15] Once a predicted trajectory that achieves the RTA has been generated, the airplane flies normal FMS roll, pitch, and speed guidance relative to it. Periodically, the trajectory generator will update the estimated arrival time, as well as the maximum and minimum arrival times, from the current aircraft location along the reference trajectory to the RTA waypoint. If the estimated time of arrival is earlier or later than the RTA by more than a set time error tolerance, the FMS will trigger a new trajectory iteration to meet the RTA. The FMS will then provide guidance relative to this new trajectory. This process continues until the aircraft arrives at the RTA waypoint or the RTA is deleted from the flight plan.

Continuous time control guidance provides the capability to manage the flight time of the aircraft along a precomputed 4D reference trajectory. Time errors are based on the current location of the aircraft relative to this computed trajectory rather than on the estimated arrival time at some future RTA waypoint. This guidance method may be necessary for operational concepts that require prenegotiated trajectories to ensure separation. If the trajectory has been computed to achieve an RTA at a specified waypoint, the aircraft achieves this RTA by adjusting speed throughout the flight to track the time profile of the reference trajectory. Within the aircraft performance envelope, the reference trajectory does not need to be recomputed to achieve compliance. However, continuous time control guidance may cause unnecessary fuel consumption because it must compensate for wind forecast errors within the reference trajectory computation. The errors may also cause the aircraft to encounter its performance limits, thereby preventing trajectory compliance. Periodic updates to the reference trajectory improve the fuel efficiency of this technique by incorporating updated wind information and reducing the need for throttle and/or speed brake to track the time. Because the periodic updates modify the reference trajectory, they may not be appropriate for concepts that require a trajectory contract with the ANSP.

Pair-dependent speed guidance may also be a midterm NextGen electronic flight bag- or FMS-based guidance mode. This form of guidance is used in airborne relative spacing concepts. The mode would make use of ADS-B or other surveillance to establish a prescribed spacing relative to another aircraft. The guidance can be used to maintain a time interval or distance or, as described in a following section, to achieve a spacing interval at a future point in the aircraft's trajectory. The use of trajectory prediction as an element of spacing guidance enables the aircraft to follow a lead aircraft while flying a different route.

45.4 Long-Term Concepts of Operation: "FarGen"

Although midterm NextGen should result in substantial improvements, a full resolution of the operational feasibility issues and full attainment of NextGen goals may require concepts of operation that involve a significant change of roles and responsibilities between the human operator and automation and between the ANSP and the flight deck. These concepts will rely on even more advanced airborne and ground-based technologies. Some organizations have dubbed this next evolutionary step in NAS Operations as "FarGen." One proposed approach for a long-term solution is to support and maintain the ANSP role by increasing the use of computer automation and air/ground data exchange. While that approach may lead to improved system performance by leveraging the advantages of automation, it may be a highly complex and costly solution. Another approach is to continue down the path of PBO, increasing airborne capabilities through evolutionary improvements in aircraft avionics and introducing new avionics systems that allow further transition of specified responsibilities from the ANSP to flight crews. The latter approach is explored in the discussion that follows.

The introduction of airborne surveillance concepts such as ADS-B led to a new operational paradigm in which reliance on centralized airspace system modernization would be reduced in favor of decentralized modernization. In 1995, RTCA Task Force 3 defined "free flight" as a safe and efficient flight operating capability under instrument flight rules in which operators have the freedom to select their path and speed in real time, thereby freeing the ANSP to concentrate on traffic flow management.[16] Proponents of free flight believed the current system imposed a substantial handicap on the user community, forced to rely on a large fixed infrastructure that was extremely difficult to modernize due to its complexity and size. They also believed that centralized air traffic control restricted competition-related business decisions. Therefore, free flight had a goal not only to optimize the system but also to open the system up to allow each user to self-optimize.

The free flight movement led to research and development in the United States and Europe based on these principles. Distributed Air/Ground Traffic Management (DAG-TM), investigated by NASA from 1997 to 2004, was a concept that evolved from free flight. DAG-TM was based on the premise that large improvements in system capacity as well as flexibility and efficiency for the airspace user would be enabled through (1) sharing information related to flight intent, traffic, and the airspace environment; (2) collaborative decision making among all involved system participants; and (3) distributing decision authority to the most appropriate decision maker.[17] Flight crews, the ANSP, and airline operational control organizations would interact as both information suppliers and users, thereby enabling collaboration and cooperation in all levels of traffic management decision making. DAG-TM postulated that distributing decision-making authority would be a key enabler in increasing the capacity of National Airspace System by minimizing the occurrence of human workload bottlenecks. It offered the potential of a linearly scalable system that accommodated an increase in demand through a proportional increase in infrastructure and human decision-making capability because each additional aircraft would contribute directly to traffic management. System-wide reliability and safety improvements would also result from the increased redundancy of traffic management capability. It was considered desirable for users to plan and operate according to their preferences as the rule, with ATM deviations occurring only as needed. The DAG-TM research identified and established feasibility of two new long-term flight deck roles: airborne separation assurance and airborne relative spacing.

45.4.1 Airborne Separation Assurance

Airborne separation assurance is a potential future airspace management function that allows flight crews to operate autonomously in constrained en route and unconstrained terminal-arrival environments. Variations of the fundamental concept may also be used in surface operations to provide an additional layer of safety. Using advanced flight deck capabilities, flight crews are responsible for maintaining separation with other aircraft and with restricted airspace. Flight deck-based separation concepts rely on using the best information available for determining whether a loss of separation with another aircraft or an encroachment on hazardous airspace is predicted and whether the flight crew should take action in each particular situation. Trajectory information on traffic aircraft may include only state data, or it may also include varying degrees of intent ranging from target states to FMS-level intent when the autoflight system is engaged.

Primary functions required for airborne separation are surveillance of traffic in proximity, a cockpit display of traffic information (CDTI), and a system of traffic conflict detection and resolution. Surveillance of traffic is typically assumed to be provided by current or future instantiations of ADS-B, although other sources of traffic information may be available and needed for augmentation. CDTI has been assumed necessary to improve the flight crew situational awareness of traffic. Human factors research indicates that flight crew awareness alone is not sufficient and that conflict detection and resolution automation is necessary as flight crew decision aids.[18] Because equitable management and allocation of limited resources is inherently a centralized function, determination for operational constraints for congested airspace and terminals is performed by the ANSP, and airborne separation must be performed while complying with them. Constraints may be applied in the form of controlled times of arrival at metering or destination fixes, a requirement to follow or maintain spacing with another aircraft, or a requirement to avoid regions of airspace.

The ultimate benefit of airborne separation may be that it is critical in decoupling the tactical management of short-term events from the strategic management of limited airspace resources. Management of limited resources is best accomplished through planning, whereas tactical management in the presence of high uncertainty may require more responsiveness in the presence of changing situations. If it is possible to decouple tactical control such as separation assurance from strategic issues such as throughput management, both problems may become simpler and can be solved independently. Because aircraft have greater flexibility and response bandwidth to overcome disturbances than the ANSP can provide, they may be more capable of complying with an assigned schedule. The improved schedule compliance in turn may provide a more stable planning environment, thereby allowing for simpler ANSP-based scheduling algorithms and enabling better management of the limited system resources. Further, flow restrictions are often established not only because of real system limitations (such as higher demand than capacity at a destination) but also to prevent traffic density to exceed acceptable levels in a controller's sector. If aircraft can be fully responsible for separating themselves from all other aircraft (including conventionally managed aircraft), they may add no workload to the sector controller. It may be possible to relieve self-separating aircraft from compliance with some flow restriction initiatives, thereby providing an incentive for airspace users to equip for the needed airborne capability.

45.4.2 Airborne Relative Spacing

Airborne relative spacing, commonly referred to as FIM, involves flight crew responsibility for maintaining a spacing interval with another aircraft or achieving the interval at a specified location. It was originally envisioned as a method for flight crews to contribute actively in maximizing airport arrival throughput when instrument approaches are in use. Early instantiations of FIM are planned for midterm NextGen, but widespread use leading to optimal flight-efficient arrival and departure throughput at multiple-runway airports may be possible over a longer term. The ANSP provides a spacing interval to follow a leading aircraft and issues a clearance for the flight crew to adhere to algorithm-generated speed cues to achieve the spacing. Again, surveillance is assumed to be provided by current or future instantiations of ADS-B. The controller maintains responsibility for separation, so the spacing interval is envisioned to be significantly larger than the minimum separation standard. When an arrival stream contains large numbers of aircraft equipped for FIM, large contiguous streams of self-spacing aircraft are possible, each aircraft spacing on an assigned lead. Spacing algorithms that are based on predicted trajectories of the leading aircraft offer operational advantages over simpler spacing algorithms. Aircraft flying on different routes may be spaced relative to one another, assuming a defined route and speed profile for the lead aircraft is shared with the trailing aircraft. Trajectory-based algorithms enable aircraft to conduct optimal-profile descents while spacing. Trajectory-based spacing may also account for differing final approach speeds and wind environments.

In 2011, the "ADS-B In" Aviation Rulemaking Committee introduced the concept of "defined interval,"[19] which would authorize the controller to assign spacing boundaries dynamically. Dynamic interval spacing may allow aircraft to achieve spacing closer than currently allowed by static separation standards. Airborne spacing with a defined interval may increase system throughput significantly when conditions permit, as when a leading aircraft's wake is blown away from a trailing aircraft's path by a crosswind.

Airborne relative spacing can provide increased conformance with constraints that maximize throughput because flight crews are capable of high precision in managing their own trajectory. Increased precision leads to a reduction in spacing buffers and hence higher throughput. Several factors prevent a human air traffic controller from achieving such precision. Because the required separation based on wake avoidance can vary based on the weight class of each leading and trailing airplane, a controller must determine the specific required spacing for each pair, leading to the use of large spacing buffers for some combinations.[20] Inherent communication time requirements and discretized speed clearances add closed-loop system delays that require a reduction of control gains for stability, thereby reducing precision. Party-line communication and the simultaneous management of several aircraft by the controller cause only one aircraft in the arrival stream to receive a clearance at a time. Airborne relative spacing may also facilitate cost-effective growth for today's underutilized airports that will see increased demand in the future. Because each aircraft brings with it a significant portion of the needed infrastructure and human decision-making capability, minimal ground infrastructure additions will be required for these terminal areas as demand increases. Some ANSP-based preconditioning of the traffic flow is required.

Although airborne spacing research was begun with the goal of leveraging the aircraft's ability to control its trajectory precisely, it was not clear at the start of research whether relative guidance or absolute time guidance would produce the best results. The relative guidance method controls the ownship trajectory relative to the trajectory of another aircraft, while the absolute time method guides the ownship trajectory relative to an ANSP assigned constraint, such as RTA. Relative control is anticipated to be less dependent on accurate ANSP scheduling predictions. In environments with high uncertainty, this robustness would lead to less workload for a human controller and reduced arrival slot prediction precision for an ANSP-based scheduling algorithm. If arrival capacity is underestimated or if some aircraft need to exit the arrival stream, either an absolute time system would create missed arrival slots unless the arrival schedule is periodically recomputed, with a new RTA assigned to each arriving aircraft. To compensate, ANSP scheduling research systems intentionally allocate some delay absorption to terminal airspace by scheduling a terminal entry fix time that is earlier than the capacity-limited runway can accommodate. The scheduling compensation allows them to be better positioned to recover lost slots. Absolute time guidance, whether airborne- or ANSP-controlled, would therefore require aircraft to absorb this delay by flying slower than their nominal speeds, at a cost of reduced flight efficiency. This terminal airspace front-loading may not be necessary if airborne relative spacing is used. Aircraft use their precise flight guidance to compensate for trajectory prediction uncertainty, and their speed adjustments propagate up the arrival stream.

45.4.3 Far-Term NextGen Technology Enablers

These far-term concepts will further increase the roles and responsibilities of flight deck-based systems and crew. Continuing research will be required to create safe, efficient, and cost-effective operations based on the new roles, but the foundational technologies and principles are well understood. Technology enablers are networked communications, conflict detection and resolution, flight optimization, and flight crew decision support. These technologies must be developed into certifiable and affordable airborne systems.

45.4.3.1 Networked Communications

In addition to the data links planned for NextGen, in the future there may be increased reliance on high-bandwidth networked communications to provide information between airborne and ground-based systems. In-flight Internet systems already provide high-bandwidth two-way communication capabilities to passengers. These or separate dedicated systems may be able to provide high-quality winds aloft and convective weather data. They may also be able to receive information from airborne systems or the crew and provide cloud-based flight management services to the cockpit.

45.4.3.2 Conflict Detection, Prevention, and Resolution

A conflict is defined by the International Civil Aviation Organization (ICAO) as a predicted loss of separation between two aircraft. Airborne separation concepts will require flight deck automation that

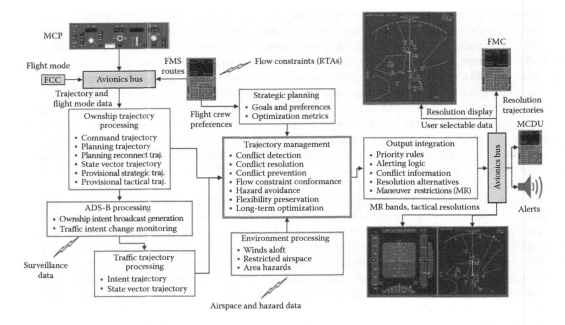

FIGURE 45.2 The NASA AOP research system.

detects traffic conflicts with sufficient time to maneuver and then determine a maneuver strategy to avoid loss of separation. To develop these technologies and explore behaviors of future systems with airborne responsibility for separation, NASA has developed a flight deck research system referred to as the autonomous operations planner (AOP).[21] In the discussion that follows, the elements of separation assurance technologies will be described using examples from AOP. Figure 45.2 is a functional overview of the system. AOP receives ownship state, flight plan, and flight mode information from the aircraft systems. It receives traffic surveillance, ATM constraints, and information about the airspace environment through data links or crew inputs based on voice communication. Using preferences for problem solution provided by the flight crew, AOP performs ownship trajectory management and provides trajectory change advisories to the crew. In some modes, it provides modified trajectories to aircraft systems for execution. AOP relies on the trajectory generation function of the aircraft's FMS. In operation, AOP functions would probably be elements of an advanced FMS and would have access to accurate performance models and real-time aircraft status information.

Automated conflict detection is a critical element for any future automated separation assurance function, which would be used in concert with collision avoidance functions that provide an inner layer of safety. Many airborne conflict detection algorithms have been developed by the ATM research and unmanned aerial systems communities. Depending on the application, conflict detection may be required for cooperative and noncooperative traffic. Cooperative conflict detection relies on traffic information sources such as ADS-B. If future instantiations of ADS-B provide limited intent information, conflict detection may be performed based on both traffic state and a portion of the traffic aircraft's flight plan. This has the potential for increasing detection accuracy over a larger look-ahead time horizon, reducing both false and missed alerts, and providing some predictive capability for potential blunders. Noncooperative conflict detection relies on active surveillance, so traffic intent must be inferred by the automation if it is used.

Some separation assurance concepts also make use of a function that performs conflict detection for potential changes to the ownship's trajectory. Referred to as conflict prevention, this function is used in concepts that prohibit intentional generation of a new conflict by an ownship's maneuver. The conflict prevention function of AOP probes potential trajectory changes such as the FMS MOD route; the reconnect route on transition to an autoflight-coupled mode; and for tactical maneuvering, headings or vertical speeds that would create a new conflict if followed.

Two types of conflict resolution are typically considered. Tactical conflict resolution has been defined as a maneuver that resolves a conflict but does not account for the ownship's flight plan in doing so. Strategic resolution has been defined as a trajectory change that, if executed, includes reconnection with the intended flight plan and recovery of the flight plan's constraints. Both types of resolution have advantages and disadvantages. Tactical systems may be most appropriate for providing a certifiably safe conflict resolution function, and they facilitate a transition to future operations by providing a limited capability at low cost. Strategic systems may be most appropriate for airspace users who desire to optimize their trajectories in airspace environments dominated by severe flow-management and airspace constraints, or in environments prone to high crew workload. These systems may also be required for the integration of airborne separation and ANSP-managed operations in the same region of airspace. In dense and highly constrained traffic environments, strategic conflict resolution functions may need to account for all known constraints while determining the appropriate trajectory for the ownship. Constraints include nearby traffic, special use airspace, weather and other environmental hazards, and arrival time, speed, or altitude constraints imposed by the ANSP for safety or to expedite traffic flow. The sooner a conflict is detected, the greater the possibility of coordinating the resolution with flight planning goals. For conflicts detected with little notice, flight planning goals may be set aside in the interest of safety. Therefore, airborne separation concepts such as NASA's Autonomous Flight Rules[22] use a combination of tactical and strategic detection and resolution.

AOP selects a conflict resolution strategy for each conflict based on the time remaining before the point of closest approach. If several minutes are available, AOP selects strategic resolution. Consideration of the ownship flight plan and all flight plan constraints can reduce unnecessary and excessive maneuvering, thereby promoting overall system stability by minimizing the number of trajectory changes for all aircraft. To further promote system stability, AOP generates resolution maneuvers that do not create a new conflict with known traffic. Strategic algorithms utilize intent information provided by traffic if available, and they may consider several potential paths of each traffic aircraft to determine a maneuver that is robust to traffic trajectory uncertainties and blunders. Strategic resolution algorithms may also employ the concept of flexibility preservation: resolutions avoid airspace regions that are predicted to be congested. If the strategic solution is integrated into the FMS flight plan, the resolution maneuver is accomplished while remaining in FMS guidance mode during the entire event. An important shared benefit of integrating the strategic resolution into the flight plan is that the modified FMS flight plan is available for broadcast to others as the new ownship intent. If less time is available to resolve a conflict, AOP utilizes tactical resolution based on right-of-way rules. Using rules rather than coordination messages is referred to as implicit coordination and is important in reducing requirements for data exchange between aircraft. Tactical resolutions may utilize intent information from traffic if available, but the pilot must manually reconnect with the aircraft's flight plan after executing a resolution maneuver. An aircraft employing a tactical resolution can provide no intent information to traffic without some form of intent inference being used either by the ownship's automation or by the conflict detection systems of traffic aircraft.

45.4.3.3 Flight Optimization

Efficient flight path management and passenger comfort are very important to airspace users. To perform trajectory planning that accounts for these operator flight goals as well as conflicts with traffic, area hazards, aircraft performance limits, and flight constraints imposed by the ANSP, strategic conflict resolution algorithms may employ an iterative search to achieve an optimal resolution trajectory. When a conflict is detected, an optimizing strategic resolution function attempts to provide the best conflict-free solution based upon a user-specified cost function and other user preference inputs. The AOP research system uses such a function. It utilizes the FMS trajectory generation capability to obtain predictions of the ownship trajectory to ensure that the resolution trajectories are within the aircraft's performance envelope. If a conflict is detected, the resolution algorithm repetitively perturbs, evaluates, and selects a set of trajectories. Each of these trajectories is produced from a call to the FMS trajectory generator. A set of predefined perturbation patterns are used by AOP to reduce the search computations. Perturbed trajectories that are flyable by the airplane and resolve the conflict without creating new conflicts are compared against a cost function. Using evolutionary algorithm approaches, the best of these trajectories are used to provide new perturbed trajectories for the next iteration. The iteration is continued until optimality conditions are met.

45.4.3.4 Flight Crew Decision Support

For human-centered flight management concepts, the flight crew is provided final authority in all flight decisions. Therefore, future airborne systems will provide situational awareness information and decision-supporting advisories to the flight crew. Future flight deck capabilities will probably need to provide interactive tools to increase crew situational awareness and, if necessary, to generate alternative solutions to those proposed by the automation. Interfaces will either provide direct flight guidance to the crew, who will have the option to disregard it, or alert the crew of a threat, allowing the crew to either develop a resolution or evaluate a recommended resolution before executing it. The crew will also require the ability to provide inputs to the flight deck automation. The flight deck tools should be designed to enable the crew to execute a recommended action in a way that minimally impacts crew workload. In one long-term vision, humans and intelligent machines can be integrated to form a partnership similar to that of a horse and its rider. This concept is known as the *H-metaphor*.[23] Future vehicles would interact with the flight crew as a semiautonomous agent with limited situational awareness. The aircraft would be provided with self-preservation "instinct" and a limited independent will. Like a horse directed to jump a hurdle it does not believe it can clear, the aircraft would be capable of a set of reactive behaviors in response to environmental or situational stimulus and provide feedback to the pilot in situations it deems dangerous. A horse provides feedback to the rider through nervous behaviors. Haptic control allows the aircraft to provide tactile feedback to the pilot by simulating kinesthesia in the flight controls that the pilot is touching. This enables strong bidirectional communication between the pilot and the aircraft. Within the H-metaphor, researchers investigate the concepts of loose reins, in which more autonomy and responsibility is given to the aircraft side of the partnership, and tight reins, in which the pilot has more responsibility.

45.5 Very Long View: Operational Autonomy

From the broadest perspective, the airspace is a societal resource. All potential airspace uses should be enabled to the extent possible while ensuring the safety and security of those on the ground and in the air. The system that manages the airspace resource must also facilitate efficient flight for those using it, minimize the costs involved, and protect the environment. Scheduled airline operations will almost certainly continue to exist over the long term, but in the future, they may be a small element of a shared airspace. Anticipated new vehicles include autonomous unpiloted air vehicles acting both independently and in coordinated groups, unpiloted cargo carriers, and very large lighter-than-air freight-carrying vehicles. Some vehicles will keep station over fixed locations, and some may be tethered to the ground. Groups of aircraft may fly in formation for flight efficiency. The airspace must also accommodate an anticipated explosion in growth of personal air vehicles and small-scale point-to-point transports. These new uses of the airspace have the potential to increase societal mobility, transport freight at lower cost and with lower environmental impact, improve the study of the Earth's atmosphere and ecosystem, and increase safety and security by improving or drastically lowering the cost of critical services such as firefighting, emergency medical evacuation, search and rescue, border and neighborhood surveillance, and the inspection of our infrastructure. Therefore, the number of vehicles and other systems using the National Airspace in the long term may be orders of magnitude greater than the number we have today. Figure 45.3 is a snapshot of such a future. Personal aircraft routinely share the airspace with each other, robotic vehicles, and future airliners.

FIGURE 45.3 **(See color insert.)** Safe coexistence of many airspace users will be critical in the future.

The large number of vehicles and their proximity to each other will make centrally planned and executed control of each vehicle both impractical and very expensive. Communications bandwidth will be excessive, communications latencies will have direct impacts on system response, and system management issues will be complex. A potential solution is placing maximum reliance on vehicle operational autonomy. In a future scenario of full operational autonomy, each flight can be unique in its purpose and independent in control of its mission and associated flight trajectory. Each vehicle would be responsible for safely sharing the airspace with all other vehicles. Safety functions would be the responsibility of each operator, based upon common operating rules and procedures. Wireless access to dynamic databases would permit avoidance of all operational hazards during flight, using common rule sets or real-time peer-to-peer coordination of actions among aircraft. Fixed ground-based navigation and surveillance systems would be minimal to facilitate a cost-effective scalable and demand-adaptive infrastructure. Advanced aircraft-centric technologies are critical to such a future. Enabling concepts may include network-centric operations (NCO) and decentralized command and control.

45.5.1 Network-Centric Operations

The concept of NCO relies on a widely recognized trend that tremendous capability increases are occurring for communications bandwidth, data exchange reliability, and data currency, security, and customizability. Network-centric concepts make use of these capabilities to supply a very large number of distributed system elements with a very rich level of information. The value of these concepts may be independent of debates regarding which system elements can be automated or human controlled. Benefits to be gained include the ability of each element to pull information from the network and to make its information available to the others. In theory, this leads to a common and detailed situational awareness. For many years, NCO has been popular among management theorists, especially within the business community, and it can be argued that the Internet is a successful example of NCO.

NCO may enable airspace operations to make use of cloud, cluster, and network computing concepts. These concepts are applied in large-scale computing networks to increase the efficiency of applications development and their upgrades and to reduce duplication of development and data storage. They enable the location of use to be separated from the location of processing or storage, thereby allowing for effective and consistent access of the same capability to multiple locations or parties. Access to the information can be either synchronous or asynchronous. For widely distributed applications or data, modifications and upgrades of the capabilities (such as a trajectory model of an aircraft) need to be done only once to a central application or data repository. This may reduce down time of aircraft, the ANSP, and airline operations centers for upgrades of software, models, databases, and other functions. The separation of functions from location of use also allows minimal processing, storage, and software to be resident at the location of use. Because the capabilities must be accessible on demand, a reliable and secure networking capability is critical to the successful use of these network computing concepts for safety-critical applications.

NextGen and ICAO recommend an initial NCO-like capability for multinational airspace, termed System Wide Information Management (SWIM),[24] although available information indicates that SWIM's purpose is different from the general concept of NCO. SWIM advocates see it as a secure broker of information between producers and consumers of data. SWIM would also seek to reduce the number of interfaces between peer-to-peer systems. Each system will send its information to SWIM's central repository, which then translates the data and makes it available to others. Many envision SWIM to enable "virtual facilities"—facilities that do not change in function but no longer need brick-and-mortar infrastructure, such as an airport tower, which can be located anywhere because the information it needs can be available anywhere.

45.5.2 Decentralized Command and Control

In addition to a shared awareness, network-centric communications may in fact enable a completely different approach to operations: highly decentralized decision making. Brynjolfsson and Hitt[25] and many social network theorists have researched the potential for operations transformation enabled by the communications revolution. If NCO is considered to be much more than an increase in situational awareness, an operational organization can change in significant ways. Layered and tightly structured management and control is replaced with highly responsive dispersed control. Adaptation to disturbances is increased. Individual training requirements are increased, but fewer people are required to accomplish a mission. These ideas have been given the name "Power to the Edge" in an influential book published in 2003.[26] Although the book's subject is military operations, the fundamental principles expressed apply to many types of operations, including civil aviation. The book's authors argue that true transformation is caused by distributing decision making to those within an organization who have the timeliest and highest-quality information, which are typically those who interact directly with the operating environment. Central managers control only that which should be kept centralized. For all other decisions, the managers provide general mission goals and allow the units on the edge of the organization to decide autonomously how to accomplish them. This is termed "self-synchronization" in the reference and is achieved at a cost of requiring highly skilled system participants. The use of machine-based decision making may offset some of the costs of having such widespread skills.

Ironically, self-synchronization is actually antithetical to a fundamental goal of NCO, which is to provide all with a shared level of deep awareness. If everyone has the same awareness, then there is no need to decentralize the decision making; currently centralized functions such as separation assurance can remain so. All decisions would continue to be made in a centralized location, with high-bandwidth communications using thin-client systems acting to mimic localized control. The price to be paid for such an architecture is a high reliance on the availability and integrity of communications. In fact, military organizations began employing distributed authority concepts because they are more robust when communications break down. Shared situational awareness is still a critical element, however, because units on the edge do interact with each other and because shared awareness is critical for central functions to be effective.

A substantial increase in system agility may prove to be a large benefit of decentralized command and control of air traffic. Agility is the capability to be flexible and adapt quickly to changing or unanticipated situations and has obvious benefits for a military system. While airspace management generally does not involve defeating a deliberately unpredictable opponent, it does involve many forms of uncertainty management. Increased awareness, capability, and authority on the part of those who interact directly with the disturbances allow the system to respond more quickly. A pilot or onboard machine intelligence can make decisions much faster and with much better local information than a centralized decision system can. Local flight deck control may be especially important as long as human decision makers are in the loop. Experience-based expertise often cannot be translated into a communication message sent to a central system. Therefore, information processing based on decentralized human expertise cannot be performed by a central system, no matter how fast communications become.

A decentralized infrastructure may also facilitate demand-adaptive and self-modernizing system characteristics. Today's ground-based radar provides full coverage of controlled airspace even when there are no aircraft within an installation's surveillance range. A surveillance system that adapts to demand has the potential to be much more affordable. If surveillance is distributed to the edge (in this case, to each aircraft), then surveillance capability becomes great when it is needed—at times when many aircraft are occupying a volume of airspace. It is nonexistent when it is not needed—when there are no aircraft in the airspace. Self-modernization may ultimately prove to be the most valuable attribute of operational autonomy. Centralized control often requires a fixed infrastructure, such as ground-based radar surveillance and large networked computing platforms. The infrastructure is expensive to maintain and difficult to replace with more modern equipment due to issues associated with funding and management of large-scale systems. It is also difficult to retire obsolete infrastructure if a segment of the user community relies on it. A distributed system with autonomous participants has the potential to be largely self-modernizing. As aircraft with the enabling infrastructure and capabilities are decommissioned, they would be replaced with new aircraft that have updated infrastructure and increased capabilities.

References

1. Pyster, A., More disciplined than agile at the federal aviation administration, Presentation by the Deputy Assistant Administrator for Information Services and Deputy Chief Information Officer, Federal Aviation Administration, March 19, 2003.
2. Joint Planning and Development Office, Concept of operations for the next generation air transportation system, Version 2.0, June 13, 2007.
3. Joint Planning and Development Office, NextGen avionics roadmap, Version 1.0, October 24, 2008.
4. RTCA, Task Force 5: NextGen Mid-Term Implementation Task Force Report, Washington, DC, September 2009.
5. Chong, R. S. and Smith, E. C., Using data communications to manage tailored arrivals in the terminal domain: A feasibility study, *Ninth USA/Europe Air Traffic Management Research and Development Seminar*, Berlin, Germany, June 2011.
6. Klooster, J. K. and de Smedt, D., Controlled time-of-arrival spacing analysis, *Ninth USA/Europe Air Traffic Management Research and Development Seminar*, Berlin, Germany, June 2011.
7. Joint Planning and Development Office, Targeted NextGen capabilities for 2025, November 2011.
8. RTCA, *Minimum Aviation System Performance Standards: Required Navigation Performance for Area Navigation*, RTCA DO-236B, October 28, 2003.
9. Lee, H. P. and Leffler, M. F., *Development of the L-1011 Four-Dimensional Flight Management System*, NASA CR 3700, February 1984.
10. DeJonge, M. K., Time controlled navigation and guidance for 737 aircraft, *Aerospace and Electronics Conference*, NAECON, 1988.
11. Jackson, M. R. C., Sharma, V., Haissig, C. M., and Elgersma, M., Airborne technology for distributed air traffic management, Decision and Control, *2005 European Control Conference, CDC-ECC '05, 44th IEEE Conference on Decision and Control*, December 2005.

12. Korn, B. and Kuenz, A., 4D FMS for increasing efficiency of TMA operations, *25th Digital Avionics Systems Conference*, IEEE/AIAA, October 2006.

13. Ballin, M. G., Williams, D. H., Allen, B. D., and Palmer, M. T., Prototype flight management capabilities to explore temporal RNP concepts, *27th Digital Avionics Systems Conference*, IEEE/AIAA, October 2008, Minneapolis, MN.

14. Sorensen, J. A. and Waters, M. H., Airborne method to minimize fuel with fixed time-of-arrival constraints, *Journal of Guidance and Control*, 4, May 1981.

15. Burrows, J. W., Fuel-optimal aircraft trajectories with fixed arrival times, *Journal of Guidance and Control*, 6, January 1983.

16. Final Report of the RTCA Task Force 3: Free Flight Implementation. RTCA, Inc., Washington, DC, October 1995.

17. NASA advanced air transportation technologies project: Concept definition for distributed air/ground traffic management (DAG-TM), Version 1.0, September 1999.

18. Wickens, C. D., Mavor, A. S., Parasuraman, R., and McGee, J. P., eds., *The Future of Air Traffic Control: Human Operators and Automation*, National Academy Press, Washington, DC, 1998.

19. Federal Aviation Administration, Recommendations to define a strategy for incorporating ADS–B in technologies into the national airspace system. A report from the ADS–B in aviation rulemaking committee to the federal aviation administration, September 30, 2011.

20. Ballin, M. G. and Erzberger, H., Potential benefits of terminal airspace traffic automation for arrivals, *Journal of Guidance, Control, and Dynamics*, 21(6), 1998.

21. Karr, D. A., Vivona, R. A., Roscoe, D. A., DePascale, S. M., and Wing, D. J., Autonomous operations planner: A Flexible Platform for Research in Flight-Deck Support for Airborne Self-Separation, AIAA-2012-5417, September 2012.

22. Wing, D. J. and Cotton, W. B., *Autonomous Flight Rules: A Concept for Self-Separation in U.S. Domestic Airspace*, NASA TP-2011-217174, November 2011.

23. Flemisch, F. O., Adams, C. A., Conway, S. R., Goodrich, K. H., Palmer, M. T., and Schutte, P. C., *The H-Metaphor as a Guideline for Vehicle Automation and Interaction*, NASA TM 2003-212672, 2003.

24. Federal Aviation Administration, System Wide Information Management information link: http://www.faa.gov/nextgen/swim. Accessed on May 19, 2014.

25. Brynjolfsson, E. and Hitt, L., Beyond computation: Information technology, organizational transformation and business performance, *Journal of Economic Perspectives*, 14(4), 23–48, 2000.

26. Alberts, D. S. and Hayes, R. E., *Power to the Edge: Command... Control... in the Information Age*, Department of Defense Command and Control Research Program, CCRP Publications, 2003.

Index

S

W